OF THE ELEMENTS

						0
						2
						He
						Helium
	III A	IV A	V A	VI A	VII A	4.00260
	5	6	7	8	9	10
	B	**C**	**N**	**O**	**F**	**Ne**
	Boron	Carbon	Nitrogen	Oxygen	Fluorine	Neon
	10.81	12.011	14.0067	15.9994	18.99840	20.179

I B	II B	13	14	15	16	17	18
		Al	**Si**	**P**	**S**	**Cl**	**Ar**
		Aluminum	Silicon	Phosphorus	Sulfur	Chlorine	Argon
		26.98154	28.086	30.97376	32.06	35.453	39.948

28	29	30	31	32	33	34	35	36
Ni	**Cu**	**Zn**	**Ga**	**Ge**	**As**	**Se**	**Br**	**Kr**
Nickel	Copper	Zinc	Gallium	Germanium	Arsenic	Selenium	Bromine	Krypton
58.71	63.546	65.38	69.72	72.59	74.9216	78.96	79.904	83.80
46	47	48	49	50	51	52	53	54
Pd	**Ag**	**Cd**	**In**	**Sn**	**Sb**	**Te**	**I**	**Xe**
Palladium	Silver	Cadmium	Indium	Tin	Antimony	Tellurium	Iodine	Xenon
106.4	107.868	112.40	114.82	118.69	121.75	127.60	126.9045	131.30
78	79	80	81	82	83	84	85	86
Pt	**Au**	**Hg**	**Tl**	**Pb**	**Bi**	**Po**	**At**	**Rn**
Platinum	Gold	Mercury	Thallium	Lead	Bismuth	Polonium	Astatine	Radon
195.09	196.9665	200.59	204.37	207.2	208.9804	(210)a	(210)a	(222)a

metals ← → nonmetals

63	64	65	66	67	68	69	70	71
Eu	**Gd**	**Tb**	**Dy**	**Ho**	**Er**	**Tm**	**Yb**	**Lu**
Europium	Gadolinium	Terbium	Dysprosium	Holmium	Erbium	Thulium	Ytterbium	Lutetium
151.96	157.25	158.9254	162.50	164.9304	167.26	168.9342	173.04	174.97
95	96	97	98	99	100	101	102	103
Am	**Cm**	**Bk**	**Cf**	**Es**	**Fm**	**Md**	**No**	**Lr**
Americium	Curium	Berkelium	Californium	Einsteinium	Fermium	Mendelevium	Nobelium	Lawrencium
(243)a	(247)a	(249)a	(251)a	(254)a	(253)a	(256)a	(254)a	(257)a

Chemistry for the Health Professions

Charles H. Henrickson   Larry C. Byrd
Western Kentucky University

D. Van Nostrand Company
New York Cincinnati Toronto London Melbourne

To our daughters

Lynn and Jennifer
Gina and Sherry

Cover photo by Dan McCoy—Black Star

D. Van Nostrand Company Regional Offices:
New York   Cincinnati

D. Van Nostrand Company International Offices:
London   Toronto   Melbourne

Copyright © 1980 by Litton Educational Publishing, Inc.

Library of Congress Catalog Card Number: 79–64469
ISBN: 0–442–23258–6

All rights reserved. No part of this work covered by the copyright
hereon may be reproduced or used in any form or by any means—graphic,
electronic, or mechanical, including photocopying, recording, taping,
or information storage and retrieval systems—without written permission
of the publisher. Manufactured in the United States of America.

Published by D. Van Nostrand Company
135 West 50th Street, New York, N.Y. 10020

10 9 8 7 6 5 4 3 2 1

Preface

Chemistry for the Health Professions provides an understanding of chemical principles and an appreciation for their importance in health and health care. The text is for students preparing for careers in the allied health areas, such as nursing, dental hygiene, dietetics, respiratory therapy, and medical laboratory technology. It also provides a chemistry background suitable for students in home economics, physical education, community health, and the liberal arts. No assumptions have been made concerning previous training in the sciences. Each topic is treated as if it is seen by the student for the first time.

One of the most challenging tasks that can be given a chemistry instructor is that of bringing a class of students from the fundamental concepts of matter and energy to the complex nature of biochemical systems in a single course of study. We have written *Chemistry for the Health Professions* to help both students and teachers meet this challenge. This textbook is one that *teaches,* as opposed to one that simply *presents,* the principles of chemistry. The discussions of many chemical principles have been expanded beyond that usually seen in other texts to help students understand them better. Often principles are examined from two or more points of view, and analogy is used wherever possible to aid understanding. An ample number of figures further clarifies discussions. We have extensively applied the principles of chemistry to health care and to the chemical operations of the body so that students can understand better why chemistry is included in their course of study. For example, the discussion of electronic transitions in atoms with the absorption or emission of light is given a practical perspective in that it is related to the method used to measure sodium and potassium levels in blood serum. Furthermore, as chemical principles are re-examined in new contexts, we often redefine the principles and describe how they were used previously. In this way, students can gain an appreciation of the broad significance and utility of what may have seemed a highly specific fact.

Students are asked to learn many new concepts in a beginning chemistry course, and these concepts are often described in language that is not part of their common experience. It is important that students learn the language of chemistry, but care must be taken in an introductory course to ensure that language does not become a barrier to understanding. We have, therefore, attempted to limit chemical terminology to that needed to describe the principles and have avoided unnecessary technical jargon. New terms are printed in bold type and defined as they are introduced. A list of important new terms appears at the end of each chapter. An extensive glossary listing over 500 terms appears at the end of the book.

In those areas that require students to perform calculations or to derive structures, formulas, or names of compounds, numerous examples are presented with detailed solutions that explain how the facts are analyzed and used. Unit analysis is introduced in Chapter 2 and is included in a discussion of the techniques of problem solving. Frequently, students need to be refreshed in the use of mathematics and exponential notation. For this purpose, an appendix describes common math operations and includes sets of questions with answers, so that students can check their skills.

As a further aid to students, an extensive list of questions appears at the end of each chapter, and answers to the starred (*) questions are provided at the end of the book. Other questions with answers immediately following are distributed throughout each chapter in brief "check tests" that allow students to "check" their understanding of topics immediately after they are discussed.

We have structured the text in a format that has proven successful at many institutions, providing a separate but balanced and integrated discussion of general, organic, and biochemistry. The presentation is generally qualitative, though it becomes quantitative when describing stoichiometry, solution concentrations, and the gas laws. We have chosen not to use SI units because of their lack of use in the health professions today. Generally, students find that atomic structure, bonding, acid-base chemistry, and certain areas of biochemistry require a greater effort to master than most other topics. Special care has been taken in these areas to overcome the difficulties students seem to have. Atomic structure is based on the Bohr model of the atom since it is readily grasped by students and also provides a satisfactory model in the discussions of chemical bonding and molecular structure. Nevertheless, an optional discussion of the quantum mechanical model of the atom is also provided for those who desire its inclusion.

A Study Guide for this text has been prepared by John R. Wilson, Shippensburg State College. It includes for each chapter a self-test on terms, questions on the objectives for each chapter, a chapter quiz, and answers to all of the above. Also available is an Instructor's Manual that contains chapter tests and answers to questions and problems not answered in the text.

As the manuscript neared completion, we became increasingly aware of our indebtedness to Harriet Serenkin and Ralph DeSoignie at D. Van Nostrand Company for their help, encouragement, and enthusiasm. We also wish to express our appreciation to those who reviewed the manuscript, especially Barbara Bogner, Montgomery County Community College; Thomas J. Cassen, University of North Carolina at Charlotte; Clyde E. Davis, Humbolt State University; David F. Dever, Macon Junior College; Judith E. Durham, Saint Louis University; and Kenneth J. Hughes and E. B. Williams of the University of Wisconsin at Oshkosh. Their critical comments and suggestions provided invaluable guidance as the manuscript developed into a textbook. But the ultimate success of our work will be measured by the teachers and students who use this text, and we continue to seek your comments and suggestions.

Contents

Chapter 1   **Chemistry and Health—A Natural Alliance**   1

Learning Chemistry

Chapter 2   **Measurements in Chemistry and the Health Sciences**   6

Objectives; The Metric System; Mass and Weight; The Measurement; Problem Solving Using the Factor-Label method; Measurement of Length; Measurement of Volume; Temperature; Time; Measurements Using Combined Units; Terms; Questions.

Chapter 3   **Matter, Molecules, and Atoms**   37

Objectives; Solids, Liquids, and Gases; Properties of Matter; Physical and Chemical Changes; Pure Substances; Elements; Atoms; Compounds; Molecules; Ions; Mixtures; Terms; Questions.

Chapter 4   **A Closer Look at Atoms, Molecules, and Chemical Changes**   60

Objectives; The Atomic Weight of the Elements; Formula Weights of Compounds; The Mole and Avogadro's Number; Percent Composition of Compounds; Balancing Chemical Equations; Calculations Using Balanced Equations; Terms; Questions.

Chapter 5   **Energy—Why Changes Occur**   82

Objectives; Energy; The Different Forms of Energy; Kinetic Energy and Potential Energy; The Measurement of Heat Energy; Heat of Fusion and Vaporication; Energy and Chemical Changes; Energy and Nutrition; Terms; Questions.

Chapter 6   **The Structure of the Atom, and the Periodic Table**   106

Objectives; The Particles That Form the Atom; Atomic Numbers and Mass Numbers; Isotopes; The Arrangement of the Electrons; The Structures of Atoms; The Valence Electrons; The Periodic Table of the Elements; Electron-Dot Symbols for the Elements; Periodic Properties of the Elements; The Modern Picture of the Atom; Terms; Questions.

Chapter 7   The Nucleus—Radioactivity and Radiation   134

Objectives; Radiation and Nuclear Equations; Natural and Artificial Radioactivity; The Biological Effects of Radiation; Half-Life; Detection of Radiation; Units for Measuring Radiation; Radiation Safety; Use of Radiation in Therapy; Use of Radioisotopes in Diagnosis; Terms; Questions.

Chapter 8   Atoms in Combination—Ionic and Covalent Compounds   157

Objectives; The Octet Rule; Ionic Compounds—The Sodium Chloride Case; The Ionic Bond; Ions and the Periodic Table; Predicting Formulas of Ionic Compounds; Naming Ionic Compounds; Covalent Compounds; Multiple Bonds; Polar and Nonpolar Covalent Bonds; The Electronegativity of Atoms; Naming Covalent Compounds; The Coordinate Covalent Bond; Covalent Bonds in Polyatomic Ions; Deriving Lewis Structures; Resonance; Exceptions to the Octet Rule; The Shapes of Molecules; Predicting Molecular Shapes; Comparing Ionic and Covalent Compounds; Terms; Questions.

Chapter 9   Metals and Their Biological Importance   201

Objectives; The Metallic Bond; Alloys; Metals in the Body; The Biologically Important Metals; Terms; Questions.

Chapter 10   Gases, Liquids, Solids, and the Forces between Molecules   213

Objectives; The Kinetic Molecular Theory of Gases; Pressure; Gases and the Gas Laws; The Blood Gases—CO_2 and O_2; Forces between Molecules; Liquids; Solids; Terms; Questions.

Chapter 11   Chemical Changes   248

Objectives; Types of Chemical Reactions; Oxidation-Reduction Reactions; The Rates of Chemical Reactions; How Chemical Reactions Take Place; The Activated Complex; Factors That Influence the Reaction Rate; Reversible Reactions; Chemical Equilibrium; Terms; Questions.

Chapter 12   Water, Solutions, and Colloids   **271**

Objectives; The Water in Our Bodies; Solutions; Water as a
Solvent; Expressing the Concentration of Solutions; Dilution
of Solutions; The Physical Properties of Solutions; Colloids;
Dialysis; The Purification of Water; Terms; Questions.

Chapter 13   Acids, Bases, and Salts   **304**

Objectives; The Arrhenius Theory of Acids and Bases; The
Properties of Acids; Five Common Acids; Strong and Weak
Acids; The Properties of Bases; Some Common
Bases—Strong and Weak; Neutralization Reactions; Salts;
The Ionization of Water; pH—Another Way to Express $H^+_{(aq)}$
Concentration; Measuring pH; pOH; Buffers—Keeping the pH
Constant; The Blood Buffers; Acidosis and Alkalosis; The
Gram Equivalent Weight; Normality—Another Concentration
Term; Acid-Base Titration; Terms; Questions.

Chapter 14   An Introduction to Organic Chemistry   **344**

Objectives; Carbon—The Exceptional Element; The Shapes of
Organic Compounds; Writing Formulas for Organic Com-
pounds; Modeling Organic Compounds; The Three Major
Types of Organic Compounds; The Functional Groups of
Organic Chemistry; Terms; Questions.

Chapter 15   The Saturated Hydrocarbons   **361**

Objectives; The Alkanes; Structural Isomers of the Alkanes;
Naming Alkanes and Their Derivatives; Structural Formulas
from IUPAC Names; Sources of Organic Compounds;
Physical Properties of the Alkanes; The Chemical Properties
of Alkanes; The Cycloalkanes; Terms; Questions.

Chapter 16   Alkenes, Alkynes, and the Aromatic Hydrocarbons   **389**

Objectives; The Alkenes; Naming the Alkenes; *Cis-Trans*
Isomers of Alkenes; Sources of Alkenes; The Physical Proper-
ties of Alkenes; The Chemical Properties of Alkenes; The
Alkynes; The Chemical Properties of Alkynes; The Aromatic
Hydrocarbons; The Chemical Properties of Aromatic
Hydrocarbons; Naming the Derivatives of Benzene; Polycyclic
Aromatic Compounds; Terms; Questions.

Chapter 17   Alcohols, Phenols, and Ethers   419

Objectives; Alcohols; Naming the Alcohols; Physical Proper-
ties of Alcohols; The Preparation of Alcohols; The Chemical
Properties of Alcohols; Some Important Alcohols;
Thiols—Sulfur Analogs of Alcohols; Phenols; Phenols as Ger-
micides; Other Important Phenols; Ethers; Naming Ethers;
The Cyclic Ethers; Properties of Ethers; Ethers as Anesthetics;
Terms; Questions.

Chapter 18   Aldehydes, Ketones, Carboxylic Acids, and Esters   454

Objectives; Aldehydes and Ketones; Naming the Aldehydes;
Naming Ketones; Physical Properties of Aldehydes and
Ketones; The Preparation of Aldehydes and Ketones;
Chemical Properties of Aldehydes and Ketones; Some Impor-
tant Aldehydes and Ketones; The Carboxylic Acids; Naming
the Carboxylic Acids; The Preparation of Carboxylic Acids;
The Physical Properties of Carboxylic Acids; Some Important
Carboxylic Acids; Esters; Naming Esters; Physical Properties
of Esters; Chemical Properties of Esters; The Esters of
Salicylic Acid; Terms; Questions.

Chapter 19   Organic Nitrogen Compounds   497

Objectives; Amines; Amides; Heterocyclic Nitrogen Compounds; Some Nitrogen-Containing Drugs; Terms; Questions.

Chapter 20   Lipids and the Fat-Soluble Vitamins   532

Objectives; Classification of Lipids; The Fatty Acids; Simple
Lipids; Compound Lipids; Steroids; The Fat-Soluble Vitamins;
Digestion and Absorption of Fats; Terms; Questions.

Chapter 21   Carbohydrates   568

Objectives; Photosynthesis; The Classes of Carbohydrates;
Monosaccharides; Chemical Tests for Glucose; Disaccharides; Polysaccharides; Digestion and Absorption of Carbohydrates; Diabetes and the Blood Sugar Level; Optical
Isomerism; Terms; Questions.

Chapter 22   Proteins   606

Objectives; The Composition and Molecular Weights of Proteins; The Function and Classification of Proteins; Amino
Acids; Essential Amino Acids; Optical Activity of Amino

Acids; The Dipolar Nature of Amino Acids; Peptides; Separating Biological Mixtures; A Close-Up Look at Two Important Proteins; The Structure of Proteins; Collagen, Bone, and Teeth; The Properties of Proteins; Tests for Proteins; The Digestion and Absorption of Proteins; Terms; Questions.

Chapter 23   Nucleic Acids   648

Objectives; The Composition of Nucleic Acids; The Sugars in Nucleic Acids—Ribose and Deoxyribose; The Bases in Nucleic Acids—Pyrimidines and Purines; The Nucleosides; The Nucleotides; Combining the Nucleotides—The Primary Structure of Nucleic Acids; The Secondary Structure of Nucleic Acids—DNA and the Double Helix; Heredity, Chromosomes, and DNA; DNA Replication; The Ribonucleic Acids; Protein Synthesis; Mutations; Genetic Diseases; Genetic Engineering; Viruses; Drugs That Inhibit Nucleic Acid and Protein Synthesis; Interferon—The Body's Defense Against Viral Infection; Terms; Questions.

Chapter 24   Enzymes, Vitamins, and Hormones   695

Objectives; The Names of Enzymes and Their Classification; Enzymes as Functioning Units; Vitamins as Cofactors; Enzyme Action; Enzyme Specificity; Factors That Influence Enzyme Action; Enzyme Inhibition; Uses of Enzymes; Hormones; Terms; Questions.

Chapter 25   Cellular Metabolism   736

Objectives; Cellular Energy—The Role of ATP; Substrate Level Phosphorylation; Oxidative Phosphorylation—The Electron Transport System; The Krebs Cycle; Carbohydrate Metabolism; Glucose Metabolism; The Embden-Meyerhof Pathway; Fermentation; The ATP Yield from the Embden-Meyerhof Pathway; Lipid Metabolism; The ATP Yield from Fatty Acid Metabolism; Acetyl Coenzyme A—The Supreme Intermediate; Metabolism Out of Balance—Ketosis; Protein Metabolism; The Urea Cycle; Terms; Questions.

Appendix I   A Review of Mathematics for the Health Sciences   786

A. Fractions; B; Scientific Notation (Exponential Notation); C. Algebraic Operations; D. Practice Problems: Fractions; E. Practice Problems: Scientific Notation; F. Practice Problems: Algebraic Operations.

Appendix II The Cell   C1

Answers to Selected Questions   A1

Glossary   G1

Index   I1

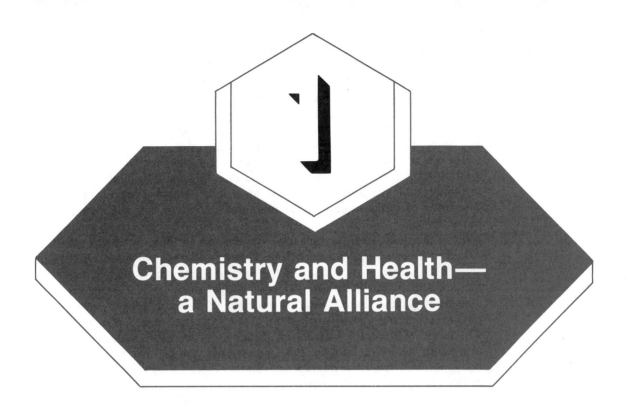

Chemistry and Health—
a Natural Alliance

Perhaps you wish to become a nurse or a dental hygienist, an x-ray technician, therapist, or a medical technician. Careers like these, as well as several others in the health care field, will place you in a rapidly growing and vital part of society. Though you will have many responsibilities, you will be able to function with increased confidence and personal satisfaction if you have a good understanding of the composition and operation of the human body. Much of this can be acquired at the introductory levels of biology and chemistry. At first glance you might think these two subjects are quite different, but in living systems they are closely related. The biological functions of the body can actually be described as chemical functions. For example, the digestion of food in the gastrointestinal tract is really a series of chemical reactions. These reactions reduce food to smaller particles so it can be absorbed and used by the body. Also, the biological functions of eyesight, memory, and the beating of a heart are actually the result of different chemical processes that take place over and over again in our bodies.

Chemistry is going on in and around us everyday. The conversion of iron ore to steel is chemistry, and so is bread baking or paint drying. Devising a simple definition for chemistry might seem difficult since it is involved in so many different things. But if we regard iron ore, bread, paint, and our own bodies as part of the material world, then **chemistry** can be described as that area of endeavor that seeks to understand what makes up the material world and how that material changes form. That part of chemistry devoted to the understanding of our own body chemistry and that of other living organisms is a specialized area called **biochemistry.** As the name implies, biochemistry involves both biology and chemistry, and it is one of the five general areas of chemistry. Let us briefly look at the other four areas.

The second area of chemistry is **organic chemistry,** which primarily deals with substances containing the element carbon. Since many of these substances are important in the body, a section of this text will deal with organic chemistry. The study of metals such as iron, mercury, and silver, and the chemistry of minerals is called **inorganic chemistry.** Many metals are known to play important roles in the body, even though they may be present in very small quantities. The fourth area is called **analytical chemistry.** People in this area develop schemes and carry out procedures to determine the composition of materials. Medical technologists are doing analytical chemistry when they measure the amount of glucose (blood sugar) in a sample of blood serum. The last area is **physical chemistry.** It deals with many of the properties of matter as well as the way matter and energy interact. At times it can be quite mathematical and abstract though it has increased our understanding of biologic processes in the body. We will avoid the mathematics in this text. Though it is convenient to divide chemistry into these five specialty areas, each one overlaps with the others quite a lot, and chemists will often have a good understanding of all the areas.

The birth of modern chemistry can be traced back to the 1700's, a time when people began to carry out carefully controlled experiments to study matter and the way it changes. The results of countless experiments allowed people to formulate many chemical principles (general truths) that summarized what was known about matter and the way it behaved. Experimentation is still going on today with the discovery of new principles and the refinement of old ones. You will become acquainted with many of these principles.

The material world in which we live is a complex mixture of many different chemical substances. The same is true of the human body, and undoubtedly you have heard of some of the parts of this mixture: proteins, fats, carbohydrates, and nucleic acids. Thousands of chemical processes take place within this carefully arranged mixture to maintain life, and each of these processes can be described in terms of chemical principles.

Many chemical processes in the body, like the digestion of food, take place in definite sequences. If one step of a sequence is seriously disrupted, a diseased condition may result which will display a particular set of symptoms. In the last several decades, the chemical basis of disease has been much better understood, and as a result many chemical agents, called drugs, have been developed to combat disease. This knowledge has also allowed the level of various chemical substances in blood, urine, and tissue to be used as powerful diagnostic tools and as a means for the physician to follow the progress of treatment. As our knowledge of the chemistry of the body increases, it will surely be possible to treat disease more effectively in the future and perhaps eliminate certain ones altogether.

A practical example of how knowledge of body chemistry has been applied is a new method that is used to treat alcoholism. The problems associated with an individual who drinks too much alcohol over extended periods of time are well known. Though rehabilitation is possible it is very difficult to begin a successful rehabilitation program if the individual continues to drink on a daily basis. But if the intake of alcohol could be stopped for a month or more, then, with the help of others, the underlying causes of the condition can be soberly assessed and rehabilitation can begin.

In the body, alcohol ends up being converted to carbon dioxide and water through a series of chemical processes. Before it is converted, though, it can circulate in the bloodstream to the brain and other tissues and cause all sorts of behavioral changes, some pleasant, some not. In order to understand the new approach to treat alcoholism, we first need to look, in a simplified way, at the processes that convert alcohol to carbon dioxide and water. Most of the details of the processes are known, but for the sake of clarity let us call them ''process 1,'' ''process 2,'' and ''process 3.''

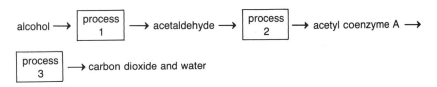

Starting on the left, the alcohol that is drunk enters process 1 and is converted to a different substance called acetaldehyde (ă-set-al'-de-hide). Once the acetaldehyde is formed, it is rapidly changed to another species called acetyl coenzyme A in process 2. The coenzyme is then changed in the third process to carbon dioxide and water. You can see the entire sequence is something like an assembly line. As the alcohol is passed from one work station to another (that is, from one process to another), it is changed each time until it ends up as carbon dioxide and water.

Let us back up now and look at the substance produced in the first process, acetaldehyde. In the body, acetaldehyde causes nausea, vomiting, sweating, confusion, rapid heart beat, low blood pressure, and, as you might suspect, severe discomfort. This condition is known as acetaldehyde syndrome. Most people can have one or two alcoholic drinks and never experience the syndrome since acetaldehyde is consumed by process 2 nearly as fast as it is formed. Because of this, there is not enough acetaldehyde in the body at any one time to bring on the discomforting effects. But, if these discomforting effects could be made to occur in alcoholics, the desire for alcohol would be considerably reduced, perhaps allowing a rehabilitation program to begin. But how can the syndrome be made to occur? Perhaps if a drug could be administered that would slow down process 2, the acetaldehyde would not be removed quickly and it would build up in the body and cause the expected results. Once the chemistry of the second process was understood, a drug was discovered that would slow it down enough to cause the acetaldehyde syndrome each time alcohol was consumed. The drug is called disulfiram, and once administered, it stays in the body for a long enough time to keep the patient "sensitized" to alcohol for several days. The alcoholic then has added reinforcement to avoid alcohol and the chances for rehabilitation are increased. It is interesting to note that orientals commonly experience acetaldehyde syndrome when they drink alcohol. This may be responsible for the reduced incidence of alcoholism among Oriental people.

Though you may not be familiar with the chemical names at this time, you can surely appreciate the significance of understanding the chemical processes in the body, and how a carefully designed drug can alter them to bring about a constructive change.

Learning Chemistry

As you leaf through this text for the first time, you might think an insurmountable volume of chemistry lies ahead. But do not despair. Chemistry, like a musical score, can be learned one measure at a time. It is the kind of subject that can be broken down into easily handled "blocks" of information that build on one another as the course of study develops. Though it is natural to have a few doubts at the beginning, you really will be able to learn a good amount of chemistry in this way.

You should use the various learning aids that are included in each chapter. The introduction to each chapter includes a list of *Chapter Objectives*. They list the facts and skills you should acquire in that chapter. You can use them to review the material after you finish the chapter, and you should resist moving on to the next chapter until you are confident you can meet the objectives. If your instructor assigns only part of the chapter, you only need to be concerned with those objectives that relate to that part. Brief *Check Tests* are scattered throughout each chapter. They contain two or three questions about the topic just discussed. They will help you "check" how you are doing, and since the answers are given, you will know right away if you need to go over a section again.

An important part of chemistry, or any field for that matter, is the terminology that is used. For this reason, a list of important terms is presented at the end of each chapter to alert you to their importance. As a further aid to learning the language of chemistry, each term is defined in the glossary at the end of the text. Often the meanings of these terms have to be memorized. Students often rebel at the idea of rote memorization, but it is a valid learning method.

Finally, the many questions at the end of each chapter will help you review important concepts and give you a chance to use the facts you have learned. To check yourself, you will find the answers to the starred (*) questions at the end of the text.

As your study of chemistry progresses, you should gain an appreciation for the chemical basis of the material world and of the body and its processes. If this happens, you will have a better understanding of the principles underlying many procedures you will encounter as you work in the health care area.

2

Measurements in Chemistry and the Health Sciences

Neither chemists nor health scientists could perform their services without making measurements. A physician usually cannot tell if a patient has high blood pressure by simple observation. Rather, the physician must carefully measure it, and the treatment prescribed will depend on the result of the measurement. A medical technologist must combine carefully measured volumes of serum and chemical reagents to determine the glucose level of a patient suspected of having diabetes. The amount of a drug administered to a patient must be large enough to be effective but not so large as to cause unpleasant side effects, so the physician must take several measurements into account in order to determine the proper dosage. These are just a few of the thousands of examples that could be mentioned in which measurements are required to carry out a task properly.

In this chapter we will study the metric system of measurement since it is widely used in the health care professions and is slowly replacing the English measurement system in the United States.

6

Objectives

By the time you finish this chapter you should be able to do the following.

1. State the meaning of the prefixes used in the metric system.
2. Describe the metric units of length, volume, and mass.
3. Describe the tools used to measure length, volume, and mass.
4. Convert quantities expressed in metric units to English units and vice versa using the factor-label method.
5. Describe the Celsius, Fahrenheit, and Kelvin temperature scales and be able to convert a temperature on one scale to the corresponding temperature on another.
6. Carry out calculations using quantities with units that are combinations of the fundamental units.
*7. Express numbers in scientific notation (also called exponential notation).

The Metric System

The **metric system** of measurement provides units that are used to express mass, volume, and length. The principal advantage of the metric system, besides its wide use throughout the world, is that the various measurement units for mass or volume or length are related by factors of 10. For example, the fundamental unit of length in the metric system is the meter (abbreviated m), a length about three inches longer than a yard. The next shorter unit of length is the decimeter, which is exactly one-tenth the length of the meter, and the next longer unit is the decameter, which is 10 times the length of one meter. The prefixes deci- and deca- indicate units that are one-tenth and 10 times the length of the fundamental unit, the meter.

The reason we use different units of length such as the decimeter and meter is simply for convenience. A moderately short length might be more easily described as being one decimeter long as opposed to being one-tenth of a meter, though both terms represent exactly the same length.

Several prefixes used in the metric system are shown in Table 2.1 along with their abbreviations and the numerical relationship each has to the fundamental unit. If you are unfamiliar with scientific notation (numbers written as 10^3 and 10^{-6}) study the mathematics review in Appendix I.

---

*A discussion of objective 7 is found in Appendix I, A Review of Mathematics for the Health Sciences. You should review those parts of Appendix I that are used in this chapter, especially the sections on scientific notation, algebraic manipulations, and fractions.

TABLE 2.1 Prefixes Used in the Metric System		
Prefix	Abbreviation	Relation to the fundamental unit
Mega-	M	1,000,000 (10^6) times larger
Kilo-	k	1,000 (10^3) times larger
Hecto-	h	100 (10^2) times larger
Deca-	da	10 (10^1) times larger
Fundamental unit	—	—
Deci-	d	0.1 (10^{-1}) of the fundamental
Centi-	c	0.01 (10^{-2}) of the fundamental
Milli-	m	0.001 (10^{-3}) of the fundamental
Micro-	μ*	0.000001 (10^{-6}) of the fundamental
Nano-	n	0.000000001 (10^{-9}) of the fundamental

*μ is the Greek letter mu, pronounced "mew."

The information in Table 2.1 clearly shows how the prefixes are related to each other by factors of 10. Unfortunately, the English system does not have a consistent relationship between units. One yard is three feet and one foot is 12 inches. To express an 8-in. length in terms of feet or yards requires arithmetic that would certainly be more complicated than changes from one unit of length to another in the metric system.

Mass and Weight

The terms mass and weight are often used to mean the same thing, though there is an important difference between them. The quantity of matter contained in an object is described by its **mass,** whereas the gravitational force pulling the body toward the center of the earth is its **weight.** The greater the mass of an object, the greater will be its weight. Also, the greater the mass of the body to which it is attracted, the greater will be its weight. For this reason, an astronaut weighs six times more on the earth than on the moon since the mass of the earth is six times greater than that of the moon (see Figure 2.1). Even though the *weight* of the astronaut would change, the *mass* would be the same on earth and on the moon since the quantity of matter that makes up the astronaut's body would not change.

There is another factor that can also affect weight. The weight of an object decreases as the distance from the center of the earth increases. Consequently, you would weigh slightly less on top of Mt. Everest than you would in Death Valley, though your mass would be the same in both places.

If two objects with the same mass are weighed in the same place, the weights of each will be equal because the effect of gravity on each will be identical.

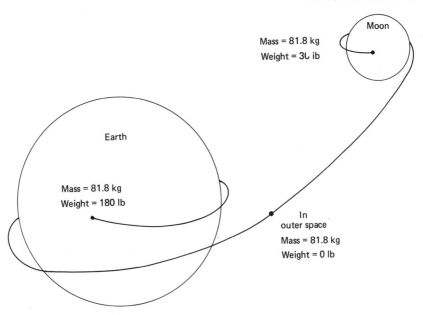

Fig. 2.1. As an astronaut travels from the earth to the moon, his mass remains constant, but his weight changes from 180 lb on earth to zero pounds during weightless space flight, and ends as 30 lb on the moon. Can you explain what causes the change in weight?

This is why we often interchange the terms mass and weight and speak of the weight of something when we actually mean its mass. Since we all live on the same planet at about the same distance from the earth's center, we can do this without causing serious problems, though you should understand the difference between the terms mass and weight.

The Measurement of Mass

The common unit of mass in the metric system is the **gram** (abbreviated g), which is one one-thousandth of the **standard kilogram.** A penny has the mass of about 3 g and a nickel about 5 g (see Figure 2.2). One pound is equal to about 454 grams and one kilogram is equivalent to 2.20 pounds.

Mass is measured using a **balance,** a device which compares the mass of an object with others of known mass. In Figure 2.3, a sample of unknown mass is placed on one pan of the balance, and a number of objects of known mass (commonly called weights) are placed on the opposite pan until the pointer indicates the two pans are in balance. The sum of the known masses will then

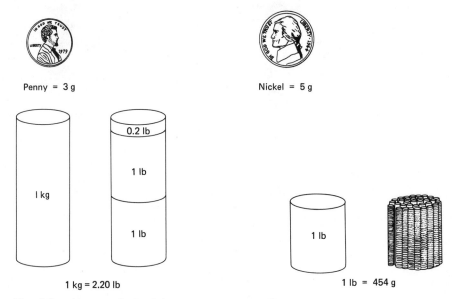

Penny = 3 g

Nickel = 5 g

0.2 lb

1 lb

I kg

1 lb

1 lb

1 kg = 2.20 lb

1 lb = 454 g

Fig. 2.2.   A comparison of the common mass units.

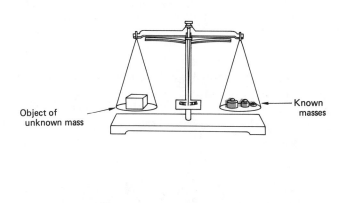

Object of
unknown mass

Known
masses

5.0 g + 1.0 g + 0.2 g = 6.2 g
Unknown mass=6.2 g

Fig. 2.3.   Mass is measured using a balance. The mass of the unknown object is equal to the sum of the known masses added to the opposite pan of the balance to reach the "balance" point.

equal the mass of the sample. Commercial balances operate on the same principle even when only one pan is visible. Some different kinds of balances are shown in Figure 2.4.

Table 2.2 presents the common units of mass in the metric system and the equivalent of each in units of the gram.

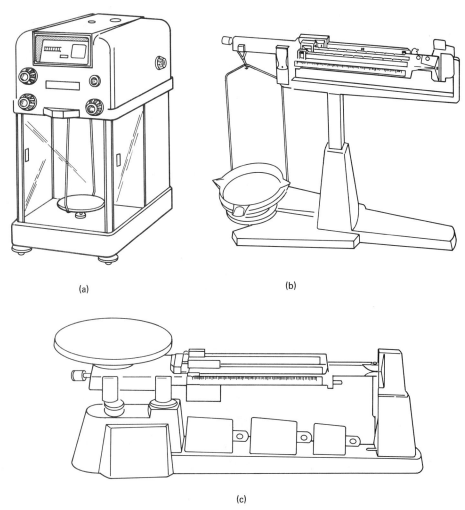

(a)

(b)

(c)

Fig. 2.4. Tools for measuring mass. An analytical balance (a) can determine mass to the 0.0001 g. Multiple beam balances [(b) and (c)] can determine mass to the 0.01 g. Though only one pan is used, these balances work on the same principle as that in Figure 2.3.

Notice in Table 2.2 how each mass unit is related to the others by factors of 10. There are 10 decigrams in one gram and 100 (10 × 10) centigrams in one gram. It follows then, that there would be 10 centigrams in one decigram. You can change from one unit of mass to another by just shifting the decimal point either to the right or left. For example, to change a mass of 15 grams to milligrams, the decimal would be moved three places to the *right* since there are 1000 mg in one gram and the answer in milligrams should be 1000 times larger than that in grams:

$$15 \text{ g} = \quad? \quad \text{mg}$$
$$15.000 \text{ g} = 15000 \text{ mg}$$

The answer in milligrams can also be written in scientific notation:

$$15 \text{ g} = 1.5 \times 10^4 \text{ mg}$$

To convert 15 grams to kilograms, you would need to move the decimal three places to the *left*. There are 1000 grams in one kilogram, so the answer in kilograms should be 1000 times smaller than in grams.

$$15 \text{ g} = \quad? \quad \text{kg}$$
$$015. \text{ g} = .015 \text{ kg}$$
$$= 1.5 \times 10^{-2} \text{ kg}$$

A different approach to these kinds of unit conversion problems that has several advantages is called the factor-label method or dimensional analysis. It uses the units that are attached to numbers to guide you as you make conversions and solve problems. The factor-label method is very useful and it is a great help in organizing your thoughts as you plan a method to solve a problem. A logical approach will give successful results, as shown in Figure 2.5.

TABLE 2.2 Mass Units in the Metric System			
Mass unit	Abbreviation	Mass equivalent in grams	
Kilogram	kg	1 kg = 1000 g	10^{-3} kg = 1 g
Gram	g	1 g = 1 g	
Decigram	dg	1 dg = 0.1 g	10 dg = 1 g
Centigram	cg	1 cg = 0.01 g	100 cg = 1 g
Milligram	mg	1 mg = 0.001 g	1000 mg = 1 g
Microgram	μg	1 μg = 10^{-6} g	10^6 μg = 1 g

Problem Solving Using the Factor-Label Method

The **factor-label** approach to problem solving will increase your awareness of the labels or units that accompany numbers. In this approach, numbers with

Fig. 2.5. A logical approach to a problem will give a successful result. The young lady first carefully studies and identifies the parts of her bicycle and works out a successful plan. The young man does not bother to develop a plan, but rather he jumps right in randomly fitting parts together. As you might expect the result is failure. The logical approach will also bring success when working chemistry problems.

their units are written in a series so that when multiplied all units that are *not* needed in the answer cancel (factor) out. Units that accompany numbers, such as 10 ft or 3 cg, can cancel out in calculations just as the numbers can. The fraction $(3 \times 5)/5$ can be simplified by dividing the numerator (the top of the fraction) and the denominator (the bottom of the fraction) by 5:

$$\frac{(3 \times \cancel{5})}{\cancel{5}} = 3$$

Likewise, the units in (ft × lb)/ft reduce to pounds when both the numerator and the denominator are divided by "foot":

$$\frac{(\cancel{ft} \times lb)}{\cancel{ft}} = lb$$

The following fraction can be greatly simplified by eliminating identical units in the numerator and denominator while the math is being done:

$$\frac{10 \text{ g} \times 6 \text{ m}}{5 \text{ m}} = \frac{60 \text{ g } \cancel{m}}{5 \cancel{m}} = 12 \text{ g}$$

Before detailing the factor-label method, it would be helpful to outline a four-step approach to follow when attempting to solve a problem:

1. *Identify what is being sought.*   Read the problem carefully and write down the units the answer should have.
2. *Identify what information is given.*   Read the problem and list the items given with their correct units.
3. *Outline a method to solve the problem.*   Plan a sequence of steps that will lead from the information given to that which is sought. You may need additional information, such as conversion factors, from tables in the text. Use the example problems that are worked throughout the chapter to guide you along.
4. *Solve the problem.*   Put in the numbers and carry out the arithmetic. Check the answer to see if it is reasonable.

This may seem like a long procedure to follow, but it will help you get started if you are having trouble.

Let us go back to an example seen earlier and solve it step by step using the factor-label method.

$$15 \text{ g} = \underline{\ ?\ } \text{ mg}$$

Step 1. An answer in milligrams is sought.

Step 2. The known value is 15 g.

Step 3. We need to convert a number in grams to one in milligrams. In Table 2.2, under the column entitled "mass equivalent in grams" we find a relationship between these units:

$$1000 \text{ mg} = 1 \text{ g}$$

This relationship is called a **conversion factor,** and it can be rewritten as

$$\frac{1000 \text{ mg}}{1 \text{ g}} = 1 \quad \text{or} \quad \frac{1 \text{ g}}{1000 \text{ mg}} = 1$$

To convert a number in grams to a number in milligrams, it must be multiplied by a conversion factor that will cancel out the gram units so only the milligram unit remains in the numerator. This can be done by multiplying "gram" by "milligram per gram":

$$\text{g} \times \frac{\text{mg}}{\text{g}} = \cancel{\text{g}} \times \frac{\text{mg}}{\cancel{\text{g}}} = \text{mg}$$

This arrangement of the units outlines the method that will solve the problem.

Step 4. Putting the numbers in will give the answer:

$$15 \cancel{\text{g}} \times \frac{1000 \text{ mg}}{\cancel{\text{g}}} = 15000 \text{ mg}$$
$$= 1.5 \times 10^4 \text{ mg}$$

The following examples will utilize the factor-label method to perform conversions.

Example 2.1

A male patient weighs 138 lb. What would his weight be in kilograms?

Step 1. Sought: kilograms

Step 2. Given: 138 lb

Step 3. The conversion factor relating pounds and kilograms is found in the text: 1 kg = 2.20 lb. We can arrange the units in this way to cancel out pounds and give kilograms:

$$\text{lb} \times \frac{\text{kg}}{\text{lb}} = \cancel{\text{lb}} \times \frac{\text{kg}}{\cancel{\text{lb}}} = \text{kg}$$

So the conversion factor should be written with kg in the numerator and lb in the denominator:

$$\frac{1 \text{ kg}}{2.20 \text{ lb}} = 1$$

Step 4. $138 \,\cancel{\text{lb}} \times \dfrac{1 \text{ kg}}{2.20 \,\cancel{\text{lb}}} = 62.7 \text{ kg}$

Example 2.2

How many grams are in 3.5 ounces?

Step 1.   Sought: grams

Step 2.   Given: 3.50 oz

Step 3.   The conversion factors we need are

$$16 \text{ oz} = 1 \text{ lb}$$
$$454 \text{ g} = 1 \text{ lb}$$

We can arrange the units of the given value and the conversion factors in this sequence to make the conversion:

$$\text{oz} \times \frac{\text{lb}}{\text{oz}} \times \frac{\text{g}}{\text{lb}} = \cancel{\text{oz}} \times \frac{\cancel{\text{lb}}}{\cancel{\text{oz}}} \times \frac{\text{g}}{\cancel{\text{lb}}} = \text{g}$$

So the conversion factors would be written in this way:

$$\frac{1 \text{ lb}}{16 \text{ oz}} = 1 \qquad \text{and} \qquad \frac{454 \text{ g}}{1 \text{ lb}} = 1$$

Step 4. $3.50 \,\cancel{\text{oz}} \times \dfrac{1 \,\cancel{\text{lb}}}{16 \,\cancel{\text{oz}}} \times \dfrac{454 \text{ g}}{1 \,\cancel{\text{lb}}} = \dfrac{3.50 \times 454}{16} \text{ g} = 99.3 \text{ g}$

Example 2.3

How many milligrams are in 35 dg?

Step 1.   Sought: milligrams

Step 2.   Given: 35 dg

Step 3.   From Table 2.2

$$10 \text{ dg} = 1 \text{ g}$$
$$1000 \text{ mg} = 1 \text{ g}$$

Both 10 dg and 1000 mg are equal to 1 g. By setting these two quantities equal to one another, another conversion factor can be derived:

$$10 \text{ dg} = 1000 \text{ mg}$$
$$1 \text{ dg} = 100 \text{ mg}$$

(Note: We could have combined the two conversion factors in the previous example to relate ounces to grams, 1 oz = 28.4 g, and perhaps simplified the problem.)

The units can be arranged to eliminate decigrams in this way:

$$\cancel{dg} \times \frac{mg}{\cancel{dg}} = mg$$

and the conversion factor will be written

$$\frac{100 \text{ mg}}{1 \text{ dg}} = 1$$

Step 4. $\quad 35 \ \cancel{dg} \times \dfrac{100 \text{ mg}}{1 \ \cancel{dg}} = 3500 \text{ mg}$

Once you have become familiar with the factor-label method, test your skill on the first check test.

Check Test Number 1

Perform the following conversions using the factor-label method.

a.   200 mg = _____ g

b.   2.50 oz = _____ g

c.   91 dg   = _____ lb

Answers: (a) 0.200 g, (b) 70.9 g, (c) 2.0×10^{-2} lb

Measurement of Length

The standard unit of length in the metric system is the **meter** (abbreviated m), which is 39.37 in., about 3 in. longer than the yard (see Figure 2.6). Larger distances are measured in kilometers (1000 m), and smaller ones in centimeters (1/100 m) or millimeters (1/1000 m). The metric units of length are shown in Table 2.3.

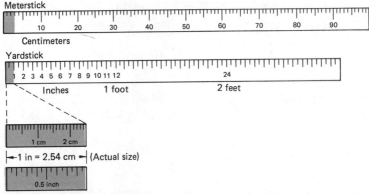

Fig. 2.6.   Tools for measuring length and a comparison of common units of length.

Length units	Abbreviation	Length equivalent in meters	
TABLE 2.3			
Metric Length Units			
Kilometer	km	1 km = 1000 m	0.001 km = 1 m
Meter	m	1 m = 1 m	
Decimeter	dm	1 dm = 0.1 m	10 dm = 1 m
Centimeter	cm	1 cm = 0.01 m	100 cm = 1 m
Millimeter	mm	1 mm = 0.001 m	1000 mm = 1 m
Nanometer	nm	$1 \text{ nm} = 1 \times 10^{-9} \text{ m}$	$10^9 \text{ nm} = 1 \text{ m}$
Angstrom	Å	$1 \text{ Å} = 1 \times 10^{-10} \text{ m}$	$10^{10} \text{ Å} = 1 \text{ m}$

The English and metric units of length are related by the following:

$$1 \text{ in.} = 2.54 \text{ cm}$$
$$1 \text{ yd} = 0.914 \text{ m}$$
$$1 \text{ mile} = 1.61 \text{ km}$$

In most instances the units of length most commonly used are the centimeter and the millimeter. The length of 1 mm is about the thickness of a dime, and a nickel has a diameter of about 2 cm. The angstrom is an extremely short unit of length which is used to express the size of very small particles.

Conversions between different units of length can be done in the same way as conversions between units of mass by using the factor-label method.

Example 2.4

Suppose you have a friend who is 5 ft 7 in. tall. What would be the height expressed in meters or in centimeters?

Step 1. Sought: One answer in meters and a second in centimeters

Step 2. Given: 5 ft 7 in.

Step 3. From Table 2.3

$$2.54 \text{ in.} = 1 \text{ cm} \quad \text{and} \quad 100 \text{ cm} = 1 \text{ m}$$

[handwritten: $Cm = 1 m$]

If the height is converted to inches it can be converted to centimeters:

$$\text{in.} \times \frac{\text{cm}}{\text{in.}} = \text{cm}$$

The height in centimeters can then be converted to meters:

$$\text{cm} \times \frac{\text{m}}{\text{cm}} = \text{m}$$

Step 4. Convert 5 ft to inches and then add 7 in. to it to get the height in inches:

$$5 \text{ ft} \times \frac{12 \text{ in.}}{\text{ft}} = 60 \text{ inches}$$

$$60 \text{ in.} + 7 \text{ in.} = 67 \text{ in.}$$

The height in cm is

$$67 \text{ in.} \times \frac{2.54 \text{ cm}}{1 \text{ in.}} = 170 \text{ cm}$$

Then converting cm to m,

$$170 \text{ cm} \times \frac{1 \text{ m}}{100 \text{ cm}} = 1.7 \text{ m}$$

Example 2.5

A piece of sodium metal is composed of a large number of very small particles (atoms) packed together in a highly organized arrangement. The diameter of one sodium atom in sodium metal is 3.60 Å. What is the diameter in centimeters and in inches?

(Starting with Example 2.5, we will no longer write out the four-step problem-solving sequence as was done. Use the previous examples as a guide to recognize what is given and what is being sought, and then plan the method of solution in your head, making a few notes as you go along.)

$$3.60 \text{ Å} = \underline{\text{ ? }} \text{ cm} = \underline{\text{ ? }} \text{ in.}$$

We can use these conversion factors:

$$\frac{10^{-10} \text{ m}}{1 \text{ Å}} = 1, \frac{100 \text{ cm}}{1 \text{ m}} = 1, \frac{2.54 \text{ cm}}{1 \text{ in.}} = 1$$

Converting angstroms to centimeters:

$$3.60 \text{ Å} \times \frac{10^{-10} \text{ m}}{1 \text{ Å}} \times \frac{100 \text{ cm}}{1 \text{ m}} = 3.60 \times 10^{-8} \text{ cm}$$

angstroms to meters; meters to centimeters

Converting centimeters to inches:

$$3.60 \times 10^{-8} \text{ cm} \times \frac{1 \text{ in.}}{2.54 \text{ cm}} = 1.42 \times 10^{-8} \text{ in.}$$

Check Test Number 2

Perform the following length conversions.

a. 3.50 m = _____ mm

b. 12.0 in. = _____ cm

c. 1.8 Å   = _____ in.

Answers: (a) 3500 mm, (b) 30.5 cm, (c) 7.1×10^{-9} in.

Measurement of Volume

The standard unit of volume in the metric system is the **liter** (abbreviated l), which is a little larger than the quart. The liter is the volume occupied by 1000 g of water at approximately 4°C.*

---

*Water has its maximum density at 3.96°C, and it is at this temperature that 1000 g of water occupies 1 liter. Density will be discussed later in this chapter.

In the laboratory, a more convenient unit of volume is the milliliter, one one-thousandth the volume of the liter. The milliliter is equal to the volume of a cube that is one centimeter long on each side, and so the milliliter is also given the name cubic centimeter (see Figure 2.7). The cubic centimeter is abbreviated cc (read as "see-see"), or cm^3 (read as cubic centimeter).

The volume of a cube is equal to the product of its length, width, and height. In a cube, these lengths would all be equal. For example, if the length, width, and height of a cube were each equal to 10 cm, then its volume would be 1 l.

$$\begin{aligned} \text{volume} &= \text{length} \times \text{width} \times \text{height} \\ &= 10 \text{ cm} \times 10 \text{ cm} \times 10 \text{ cm} \\ &= 1000 \text{ cm}^3 = 1000 \text{ ml} = 1.000 \text{ l} \end{aligned}$$

The most frequently used units of volume in the metric system are listed in Table 2.4.

The following conversion factor relates the English and metric units of volume:

$$946 \text{ ml} = 1 \text{ qt}$$

or

$$1 \text{ l} = 1.06 \text{ qt}$$

There are several commonly used tools to measure volumes of liquids in chemical and medical laboratories. Volumes as large as 3 l can be measured

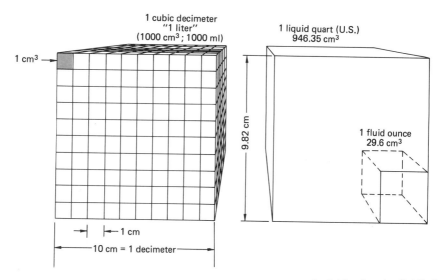

Fig. 2.7. A cube 10 cm on a side contains the volume of 1 l. The liter is slightly larger than the U.S. quart. There are 32 fluid ounces in one quart.

TABLE 2.4 Metric Volume Units			
Volume units	Abbreviation	Volume equivalent in liters	
Liter	l		$1 \, l = 1 \, l$
Deciliter	dl	$1 \, dl = 0.1 \, l$	$10 \, dl = 1 \, l$
Milliliter	ml	$1 \, ml = 0.001 \, l$	$1000 \, ml = 1 \, l$
Cubic centimeter	cc or cm^3	$1 \, cc = 0.001 \, l$	$1000 \, cc = 1 \, l$
Microliter	μl	$1 \, \mu l = 10^{-6} \, l$	$10^6 \, \mu l = 1 \, l$

using glass cylinders that are marked off in units of volume. They are called graduated cylinders, or more often, graduates. When a volumetric flask is filled to the proper point (indicated by an etched line), it will contain the exact volume indicated on its label. Pipets are often used for transferring liquids from one container to another. Pipets are long glass tubes that are graduated to allow them to deliver variable volumes, or they are designed to deliver only one fixed volume. Burets are used to accurately measure out variable volumes of liquids through a stopcock fitted to the end of a graduated glass tube. Measured volumes of liquids, usually medications, are delivered using syringes. These tools for measuring volumes are pictured in Figure 2.8.

Example 2.6

The normal adult's body contains approximately 6.0 l of blood. What is this volume in quarts?

$$1 \text{ qt} = 946 \text{ ml} = 0.946 \text{ l}$$
$$6.0 \, l \times \frac{1 \text{ qt}}{0.946 \, l} = 6.3 \text{ qt}$$

Example 2.7

After fasting for 10 hours, the blood glucose level of a healthy person might be 80 milligrams per deciliter, 80 mg/dl. If the total blood volume of the person is 6.0 l, what is the total number of milligrams of glucose in the blood?

If we determine the number of milligrams of glucose in one liter of blood, we can multiply it by 6.0 l, the total number of liters of blood, to get the total glucose content.

To convert deciliters to liters, we can use the conversion factor found in Table 2.4:

$$\frac{10 \text{ dl}}{1 \text{ l}} = 1$$

The number of milligrams of glucose in one liter would be

$$\frac{80 \text{ mg}}{\cancel{dl}} \times \frac{10 \cancel{dl}}{1} = 800 \frac{\text{mg}}{l}$$

In 6.0 l the amount of glucose would be:

$$800 \frac{\text{mg}}{\cancel{l}} \times 6.0\cancel{l} = 4800 \text{ mg} \quad \text{or} \quad 4.8 \times 10^3 \text{ mg}$$

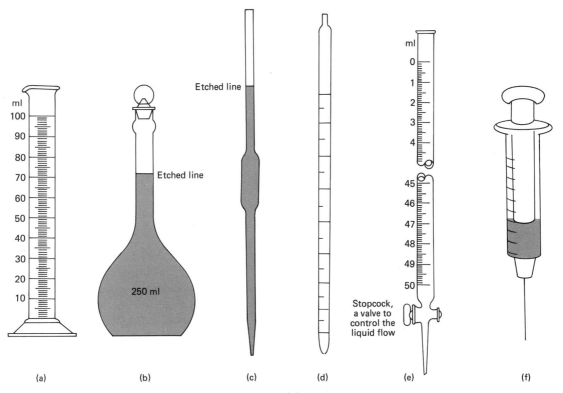

Fig. 2.8. Tools for measuring volume: (a) graduated cylinder, (b) volumetric flask, (c) fixed volume pipet, (d) graduated pipet, (e) buret, and (f) syringe.

Check Test Number 3

Perform the following volume conversions.

a. 5.0 cc = _____ dl

b. 0.150 qt = _____ ml

c. 1.0×10^{-3} l = _____ cc

Answers: (a) 0.50 dl, (b) 142 ml, (c) 1.0 cc

Temperature

The measurement of temperature is of critical importance in the health sciences. The body temperature of a patient or the temperature of an incubation bath in a blood chemistry laboratory must both be known and controlled for obvious reasons.

When we speak of the **temperature** of an object we are stating a measure of its hotness as compared to some other object that is used as a reference. The hotter an object, the higher its temperature will be. When you place a pan of water on a hot stove, heat flows into the water causing its temperature to increase. As more heat is added, the temperature climbs still higher. **Heat** is a form of energy, and its presence in the water is what causes its temperature. Sometimes the terms "heat" and "temperature" are incorrectly used to mean the same thing. The temperature of an object is proportional to the amount of heat contained in it. Temperature is a measure of "hotness," and you should remember the difference between these terms. More will be said about heat and other forms of energy in Chapter 5.

Temperature can be measured using any one of three different temperature scales. Two of the scales are commonly used in the health sciences, the Fahrenheit scale and the Celsius (centigrade) scale. The third is useful in chemistry and is called the Kelvin or absolute temperature scale.

Temperature is measured by use of thermometers. Common thermometers use the expansion and contraction of mercury or colored alcohol in a glass tube to indicate temperature. Because of convenience, electronic thermometers are finding increased use in hospitals since they indicate temperature accurately in a matter of a few seconds (Figure 2.9).

On the Celsius scale, the temperature at which water freezes is assigned a value of zero degrees Celsius, written as 0°C. The superscript (°) is used to symbolize degrees on all temperature scales. The freezing temperature of water on the Fahrenheit scale is thirty-two degrees, 32°F. The temperature at which water boils at sea level is assigned a value of 100°C on the Celsius scale and 212°F on the Fahrenheit scale. You can see there are 100 degrees between the freezing and boiling temperatures of water on the Celsius scale and 180 degrees between these points on the Fahrenheit scale. It follows then, that a 1° temperature change on the Celsius scale would correspond to a change of 1.80° on the Fahrenheit scale. On either temperature scale negative temperatures exist below the 0° mark. Ten degrees below zero on a Celsius scale would be written −10°C.

There are no negative temperatures on the Kelvin or absolute temperature scale, since the zero degree point is assigned to the lowest temperature that is theoretically possible, a point called the absolute zero. (There is no theoretical maximum temperature, by the way.) Zero degrees Kelvin, 0°K, corresponds to about −273°C. Since the size of the Celsius and Kelvin degrees are the same

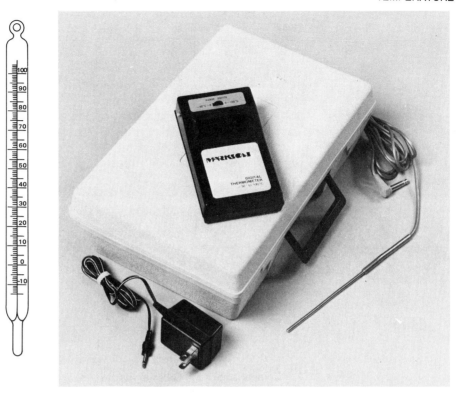

Fig. 2.9. Tools for measuring temperature. (Markson Science, Inc., Del Mar, California)

(both 1.8 times larger than one Fahrenheit degree) 0°C would equal 273°K. The three temperature scales are compared in Figure 2.10.

You can convert a Celsius temperature to a Kelvin temperature knowing that the Kelvin temperature is always 273° larger than the corresponding Celsius temperature:

$$\text{Kelvin temperature} = \text{Celsius temperature} + 273$$
$$°K = °C + 273$$

To convert a temperature in degrees Fahrenheit to the corresponding temperature in degrees Celsius, you must first subtract 32° from the Fahrenheit temperature and then divide the difference by 1.8 as shown in the following formula:

$$\text{Celsius temperature} = \frac{(\text{Fahrenheit temperature} - 32)}{1.8}$$
$$°C = \frac{(°F - 32)}{1.8}$$

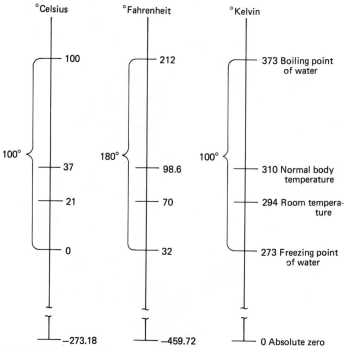

Fig. 2.10.   Comparison of the Celsius, Fahrenheit, and Kelvin temperature scales.

This formula can be rearranged to convert a Celsius temperature to Fahrenheit. In this case you multiply the Celsius temperature by 1.8 and then add 32 to the product:

$$°F = (°C \times 1.8) + 32$$

Example 2.8

A particular autoclave, used to sterilize surgical instruments, operates at a temperature of 255°F. What does this temperature correspond to on the Celsius scale?

The conversion from Fahrenheit to Celsius will be done using the equation relating the two temperature scales:

$$°C = \frac{(°F - 32)}{1.8}$$
$$= \frac{(255 - 32)}{1.8} = \frac{223}{1.8} = 124°C$$

Example 2.9

When negative temperatures are used, be certain you do the calculations correctly.

Dry ice (solid carbon dioxide) has a temperature of $-78°C$. What is the temperature of dry ice in °F? In °K?

$$°F = (°C \times 1.8) + 32$$
$$= (-78 \times 1.8) + 32 = (-140) + 32 = -108°F$$

The temperature of dry ice on the Kelvin scale would be

$$°K = °C + 273°$$
$$°K = -78° + 273° = 195°K$$

Check Test Number 4

Perform the following temperature conversions.

a. $70°F = $ _____ °C

b. $0°F = $ _____ °C $= $ _____ °K

c. $-40°C = $ _____ °F

Answers: (a) 21°C; (b) $-18°C$, 255°K; (c) $-40°F$

Time

The unit of time in the English and metric system is the **second** (abbreviated sec). There are 60 seconds in one minute and 60 minutes in one hour.

Measurements Using Combined Units

There are many quantities we measure that are labeled with units that are combinations of the units we have already discussed. For example, an intravenous solution might be administered to a patient at a rate of 4 ml each minute. The rate of delivery is expressed in terms of a given volume per unit of time. The unit for rate would be a combination of two units and written as 4 ml/min.

When you see labels such as this using a combination of units, they can be read as "four milliliters in one minute" or "four milliliters per minute."

Knowing the rate of delivery, it is not difficult to calculate the amount of time needed to administer a known volume of solution.

Example 2.10

The delivery rate of an intravenous solution is 8.0 ml/min. How much time is required to administer 600 ml of the solution?

The units outline how the problem can be solved:

$$\cancel{ml} \times \frac{min}{\cancel{ml}} = min$$

Entering the numbers, the correct answer is obtained:

$$600 \cancel{ml} \times \frac{1\ min}{8.0\ \cancel{ml}} = 75\ min$$

Combined units also are used to describe the dosage of certain potent drugs. An effective dosage might be stated as 3 mg of drug per kilogram of the patient's body weight, written as 3 mg/kg. The proper dose must then be calculated.

Example 2.11

If a patient weighed 70 kg (154 lb), and the effective dosage rate of a drug was given as 1.5 mg per kilogram of body weight, what would be the correct dosage?

We multiply the dosage rate in mg/kg by the body weight in kg to get the correct dosage. The kg unit divides out, leaving the dose in mg:

$$70 \cancel{kg} \times \frac{1.5\ mg}{1\ \cancel{kg}} = 105\ mg$$

Drugs are often supplied in solution for intramuscular injection, and it is important to be able to calculate the volume of drug solution needed to provide the required dosage. For example, a solution might contain 10 mg of drug per milliliter of solution, written as 10 mg/ml.

Example 2.12

A patient requires a dosage of 25 mg of sedative, and the drug is provided in a solution that contains 15 mg of drug per milliliter. What volume must be injected to provide the correct dosage?

The units of the given quantities can be set up to eliminate the milligram unit and leave the volume unit in the numerator:

$$\text{mg} \times \frac{\text{ml}}{\text{mg}} = \text{ml}$$

Entering the numbers, we can calculate the correct volume of sedative solution to inject:

$$25 \text{ mg} \times \frac{1 \text{ ml}}{15 \text{ mg}} = 1.66 \text{ ml} = 1.7 \text{ ml}$$

Example 2.13

A patient weighs 165 lb and is to receive a single injection of a drug that should be administered at a rate of 2 mg per kilogram of body weight. The drug solution as provided contains 50 mg/ml of the drug. How many milliliters of the solution must be injected?

First, convert the patient's body weight to kilograms:

$$165 \text{ lb} \times \frac{\text{kg}}{2.2 \text{ lb}} = 75 \text{ kg}$$

Second, calculate the number of milligrams of drug needed in the dose:

$$\text{milligrams of drug needed} = 75 \text{ kg} \times \frac{2 \text{ mg}}{\text{kg}} = 150 \text{ mg}$$

Third, calculate the volume of solution required:

$$\text{volume of solution} = 150 \text{ mg} \times \frac{1 \text{ ml}}{50 \text{ mg}} = 3.0 \text{ ml}$$

So, the correct dosage would be given by injecting 3.0 ml of the solution.

Check Test Number 5

The following problems contain quantities that use combinations of units.
a. What volume of a solution of xylocaine that contains 10 mg of xylocaine per milliliter is needed to provide a dose of 35 mg?
b. An I.V. solution flows at a rate of 6 ml/min. How long would be required to deliver 250 ml of solution?

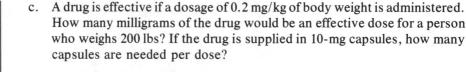

c. A drug is effective if a dosage of 0.2 mg/kg of body weight is administered. How many milligrams of the drug would be an effective dose for a person who weighs 200 lbs? If the drug is supplied in 10-mg capsules, how many capsules are needed per dose?

Answers: (a) 3.5 ml; (b) about 41 minutes; (c) 18 mg, 2 capsules/dose would be sufficient.

Density is a physical property of matter that uses the combined units of mass and volume. **Density** is defined as the mass of an object divided by its volume:

$$\text{density} = \frac{\text{mass}}{\text{volume}} = \frac{M}{V}$$

One gram of water has a volume of exactly 1 ml at approximately 4°C. The density D of water at this temperature, then, would be 1.00 g/ml:

$$D = \frac{1.00\ \text{g}}{1.00\ \text{ml}} = 1.00\ \text{g/ml}$$

At higher temperatures, for example 25°C, the volume occupied by one gram of water is slightly larger than one milliliter, actually 1.003 ml. Consequently, the density of water is slightly less at 25°C (0.997 g/ml) than at 4°C.

The density of silver is 10.5 g/ml. This means that one milliliter of silver would weigh 10.5 times more than one milliliter of water with its density of 1.00 g/ml. A silver ring would sink in water because its density is greater than that of water. On the other hand, a piece of ice will float on water because its density is less, about 0.92 g/ml.

Be sure you understand the difference between mass and density. The mass of water in each of three different beakers would depend on how much water each contains, say 50, 100, and 200 g, but the density of water in each beaker would be identical (see Figure 2.11).

The densities of several common substances are found in Table 2.5. Note the unit of density for gases is stated in grams per liter, while for liquids and solids it is stated in grams per milliliter.

The density of a substance is a useful property that can be used to fix its identity. For example, one could distinguish carbon tetrachloride from benzene (both liquids) by knowing the density of carbon tetrachloride is about 1.6 g/ml and that of benzene is around 0.88 g/ml (see Table 2.5). To determine the density of an unknown liquid, you could weigh a 10-ml volume (V) of the liquid. If it had a mass M of about 16 g, the density would be 1.6 g/ml and the liquid would be carbon tetrachloride:

$$D = \frac{M}{V} = \frac{16\ \text{g}}{10\ \text{ml}} = 1.6\ \text{g/ml}$$

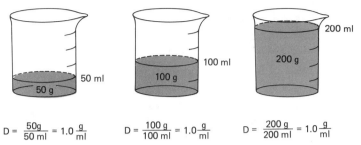

$$D = \frac{50g}{50\ ml} = 1.0\ \frac{g}{ml} \qquad D = \frac{100\ g}{100\ ml} = 1.0\frac{g}{ml} \qquad D = \frac{200\ g}{200\ ml} = 1.0\frac{g}{ml}$$

Fig. 2.11. The density of water is the same no matter what mass of water is studied. Density does not depend on the amount of matter being studied, but mass does.

Another approach to the problem would use the fact that both benzene and carbon tetrachloride are insoluble in water. You could add a few drops of the unknown liquid to a quantity of water, and if it sinks, it has a density greater than one and would be carbon tetrachloride. If it floated on water, then it would be benzene.

TABLE 2.5 Densities of Common Substances					
Solids (g/ml at 20°C)		**Liquids (g/ml at 20°C)**		**Gases (g/l at 0°C)**	
white pine	0.35–0.50	ethyl alcohol	0.789	hydrogen	0.090
ice (0°C)	0.918	methyl alcohol	0.792	helium	0.178
aluminum	2.70	benzene	0.879	ammonia	0.771
silver	10.5	water (4°C)	1.000	nitrogen	1.251
lead	11.3	carbon tetrachloride	1.59	air	1.293
gold	19.3	sulfuric acid	1.84	oxygen	1.429
bone (normal)	1.7–2.0	mercury	13.6	carbon dioxide	1.963

Example 2.14

The density of iron is 7.86 g/ml. What is the mass of a cube of iron that is 4.00 cm on a side?

$$D = \frac{M}{V}$$

The equation can be rearranged to solve for mass:

$$M = D \times V$$

The density is given in the problem, and we can calculate the volume knowing that for a cube, the volume equals length × width × height:

$$V = \text{length} \times \text{width} \times \text{height} = 4.00 \text{ cm} \times 4.00 \text{ cm} \times 4.00 \text{ cm}$$

$$= 64 \text{ cm}^3 = 64 \text{ ml}$$

The mass of iron in the cube is then

$$M = 7.86 \frac{\text{g}}{\text{ml}} \times 64 \text{ ml} = 503 \text{ g}$$

Another method that is primarily used for liquids, and provides the same information as density, is **specific gravity.** Specific gravity is the ratio of the density of one liquid to that of a reference liquid, usually water at 4°C.

$$\text{specific gravity} = \frac{\text{density of liquid}}{\text{density of reference liquid}}$$

For example, the specific gravity (sp. gr.) of ethyl alcohol would be determined as follows:

$$\text{sp. gr.} = \frac{D \text{ ethyl alcohol}}{D \text{ water at 4°C}} = \frac{0.789 \text{ g/ml}}{1.000 \text{ g/ml}} = 0.789$$

Note that specific gravity does not have units. Since the units of density appear both in the numerator and the denominator, they cancel out. The specific gravity of a liquid states how the mass of a given volume of the liquid compares to an identical volume of the reference liquid. Here, 1 l of ethyl alcohol would weigh only about eight-tenths as much as 1 l of water.

Example 2.15

A 20.75-ml volume of blood plasma had a mass of 21.29 g. What is the density of this plasma sample? What would be its specific gravity compared to water at 4°C?

$$D = \frac{M}{V} = \frac{21.29 \text{ g}}{20.75 \text{ ml}} = 1.026 \text{ g/ml}$$

$$\text{sp. gr.} = \frac{D \text{ plasma}}{D \text{ water}} = \frac{1.026 \text{ g/ml}}{1.000 \text{ g/ml}} = 1.026$$

The normal specific gravity range for blood plasma is between 1.024 and 1.028.

Check Test Number 6

a. What volume would be occupied by a 100-g sample of gold? The density of gold is given in Table 2.5.

b. What volume of methyl alcohol would weigh the same as 10.0 ml of carbon tetrachloride? Densities are in Table 2.5.

c. The density of mercury is 13.6 g/ml. What is its specific gravity related to water at 4°C?

Answers: (a) 5.18 ml, (b) 20.1 ml, (c) 13.6

Terms

Several important terms appeared in Chapter 2. You should understand the meaning of each of the following.

the metric prefixes

mass

weight

Celsius temperature scale

Fahrenheit temperature scale

Kelvin temperature scale

meter

liter

temperature

heat

kilogram

scientific notation

density

specific gravity

conversion factor

Questions

Answers to the starred questions appear at the end of the text.

*1. State how the first term in each pair is different from the second.
   a. mass–weight
   b. temperature–heat
   c. density–specific gravity
   d. mass–density

2. Why would a person weigh more in a sailboat at sea level than in an airplane flying at 50,000 feet?

3. What advantages are there in the metric system compared to the English system of measurement?

*4. What unit of length in the metric system would be most appropriate when measuring the following?
   a. distances between cities
   b. the length of a room
   c. the length of an incision

*5. What is the advantage of using scientific notation when recording very large or very small numbers?

6. Water at the temperature of its maximum density is used to relate the metric units of length, volume, and mass. Outline these relationships.

*7. One kilogram of water has a temperature of 20°C, and a second 1-kg sample of water has a temperature of 30°C. Which sample contains the larger amount of heat energy?

*8. How would the numerical value of the specific gravity of methyl alcohol change if the reference liquid was changed from water at 4°C to carbon tetrachloride at 20°C?

9. Prove that 100 cm/1 m = 1. Start with the fact that 1 meter = 100 cm.

Mathematical review: The following four questions (10, 11, 12, and 13) concern topics covered in Appendix I.

10. Express each of the following in scientific notation.

    a. 1805   **d.** 0.0000050
    b. 0.00483   **e.** 1000000
    c. 195.3   **f.** 32

11. Write the following numbers in conventional form.

    *a. 7.3×10^{-2} ***c.** 2.9×10^{8}
    b. 1.94×10^{-3} **d.** 7.608×10^{2}

12. Solve the following problems that involve fractions.

    *a. $3 + \frac{5}{8} =$
    b. $\frac{5}{9} + \frac{8}{3} =$ **e.** $\dfrac{10 \text{ mi}}{50 \text{ mi/hr}} =$
    *c. $\frac{5}{4} - \frac{5}{6} =$
    *d. $\frac{7}{10} \div \frac{1}{2} =$ **f.** $\dfrac{3/5}{4/3} + \frac{4}{5} =$

13. Solve the following algebraic equations for x, that is, the answer should have the form $x =$ (other quantities).

    *a. $5x = 15$
                                    *e. $x - 16 = y + 8$
    b. $\dfrac{x}{y} = ab$
                                    f. $\dfrac{xy}{ab} = c$
    *c. $\dfrac{a}{x} = b$
                                    *g. $a + b = c(x + d)$
    d. $5 - x = a$
                                    h. $\dfrac{a + b}{x} = \dfrac{c + d}{y}$

The following problems refer to topics discussed in Chapter 2.

14. Using the factor-label method, carry out the following conversions.

    *a. 75 mm = _____ cm   ***e.** 3.50 lb = _____ g

    b. 110 mg = _____ g   **f.** 7.5 l = _____ dl

    *c. 24 in. = _____ cm   ***g.** 10.0 gal = _____ l

    d. 10 cm = _____ μm   **h.** 40 mi = _____ km

15. Perform the following temperature conversions.

    *a. 72°F = _____ °C   ***d.** 53°C = _____ °K

    b. 500°F = _____ °C   **e.** 50°F = _____ °K

    *c. −10°C = _____ °F   **f.** −10°F = _____ °C

16. Do the following conversions using the factor-label method.

*a. 15 dg = _____ mg   ***e.** 1.3 Å = _____ cm

b. 5.2 qt = _____ ml   **f.** 3 μl = _____ ml

*c. 7.5 in. = _____ μm   ***g.** 2000 lb = _____ g ✗

d. 0.250 lb = _____ mg   **h.** 1.4×10^{-3} g = _____ lb

*17. A solid has a volume of 100 ml and a mass of 450 g. What is its density?

18. The density of mercury is 13.6 g/ml. What volume would be occupied by 1 kg of mercury?

*19. Which of the following would float on a pool of liquid gallium which has a density of 5.9 g/ml? Refer to Table 2.5.
a. mercury
b. silver
c. aluminum

20. Arrange the following temperatures from the coldest to the hottest.

80°C, 180°F, 345°K

*21. How much time is required to administer 400 ml of a glucose I.V. solution if the rate of flow is 7.5 ml/min?

22. How many milliliters of an injectable drug that contains 2.5 milligrams of drug per milliliter are needed to provide a dose of 8 mg to a patient?

23. A 250-ml volumetric flask was filled to the etched line with diethyl ether, a common anesthetic. The density of diethyl ether is 0.716 g/ml. How many grams of ether are contained in the flask?

*24. If a person smoked 100 cigarettes a week, how many *weeks* of smoking would be needed to inhale one pound of tar? Each cigarette contains 40 mg of tar.

*25. It is possible to do a blood count using just one drop of blood. If there are 20 drops in 1 ml, what would the volume of one drop be in liters?

26. A single aspirin tablet contains five grains of aspirin. If there are 15 grains in one gram, how many aspirin tablets could be made from 9 g of aspirin?

27. Which quantity is larger in each pair listed below?

*a. 1000 cg or 1000 mg

b. 5.2 cm or 2.0 in.

*c. 10 kg or 25 lb

d. −25°C or −20°F

*e. 20 $\dfrac{mg}{min}$ or 20 $\dfrac{mg}{hr}$

f. the volume of 50 g of a substance with a density of 5.3 g/cc

or

the volume of 70 g of a substance with a density of 7.9 g/cc

28. Calculate the volume of a box that is 60 cm long, 0.20 m wide, and 15 mm deep.

*29. A graduated cylinder is filled to the 25.0 ml mark, and a 50.0 g piece of metal is submerged in the water raising the water level to the 29.4 ml mark. Referring to Table 2.5, what is the identity of the metal?

30. If the blood contains 10.5 mg/dl of calcium and the human body normally contains 6 qt of blood, what mass of calcium will be found in the total blood volume?

*31. A standard test tube holds about 30 ml. How many test tubes would be needed to hold a quart of water?

32. A large dose of an antileukemia drug is to be administered to a 190-lb patient by I.V. injection. Assume the recommended dosage is 50 mg per kilogram of body weight, and the drug is supplied as a solution that contains 20 mg per milliliter. The I.V. has a flow rate of 3.0 ml/min. How many milliliters of the solution will be required and how long will it take to give the recommended dose?

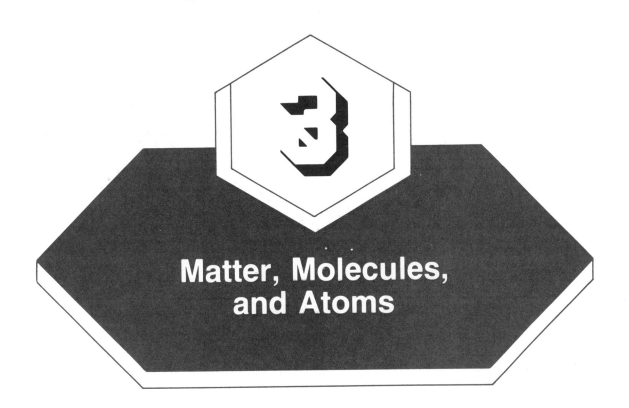

Matter, Molecules, and Atoms

The human body and all that which forms its environment are composed of matter. **Matter** is considered anything that occupies space and has mass. The air we breathe, the food we eat, and the clothes we wear are just different kinds of matter.

For centuries we have tried to understand matter and control the way it changes. Early attempts to do this relied on superstition and magic. Eventually, as society developed, day to day needs in agriculture, metallurgy, and medicine provided the driving force that launched a careful, organized study of matter based on observation and experiment. Since the problem is an immense one, the study is still going on today. Our steady advancement in the health care areas parallels our increased understanding of matter. It would be nearly impossible to understand modern medicine without first understanding matter. Much of what is discussed in this chapter will be examined in greater detail later on in the text, especially the material that deals with atoms, ions, and compounds. It is appropriate to introduce these topics here so you can become familiar with the language of chemistry.

Objectives

By the time you finish this chapter you should be able to do the following.

1. Describe the properties of a solid, liquid, and gas.
2. Describe and give examples of physical and chemical properties.
3. Describe and give examples of physical and chemical changes.
4. Write word equations to describe chemical changes.
5. Define the following terms and give examples of each: substance, element, compound, atom, and molecule.
6. Write the symbols for the more common elements.
7. Describe the properties of metals and nonmetals.
8. Define and give examples of homogeneous and heterogeneous mixtures.

Solids, Liquids, and Gases

There are three common physical forms in which matter is found: solid, liquid, and gas. Each can be described by its own unique set of characteristics.

Solids are rigid and have definite shapes that are not easily deformed. The particles that make up solids are strongly attracted to one another and are packed closely together. Because of this strong attraction, particles in a solid are rigidly held in place by those around it. Consequently, the volume of a solid remains essentially constant even under very high pressures, and it changes ever so slightly as temperature changes.

Liquids do not have definite shapes, since they assume the shape of their containers. Liquids can easily flow from one container to another because the particles that make them up are not attracted to one another as strongly as they are in solids. This allows them to flow over and around each other with relative ease. But since the particles are still close together, the volume of a liquid will change only a small amount when subjected to high pressures. Liquids do expand a little when heated, which leads to their use in thermometers.

Gases do not have definite shapes, and they completely fill the containers in which they are kept. The particles that make up gases are not strongly attracted to each other and so are far apart. Like liquids, gases can flow, but unlike liquids, they can undergo a substantial volume expansion when their temperature is increased or when the pressure exerted on them decreases. If the vessel that contains a gas is not closed, the gas will escape and expand to fill the room, the building, and eventually the atmosphere. So another characteristic of gases is their ability to expand indefinitely.

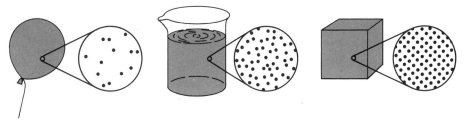

Fig. 3.1. The three states of matter. Each differs in the number and arrangement of particles in a given volume. Solid: particles are close together and packed in an organized arrangement; liquid: particles are close together and in a random arrangement; gas: particles are far apart with no particular arrangement.

Why are some things solid, some liquid, and yet others gaseous at room temperature? A detailed answer will be given in Chapter 10, but at this point, it is sufficient to say that the greater the attraction between the particles that make up a sample of matter, the greater will be the likelihood that it will be a solid at room temperature. If the attraction is weaker, then it may be a liquid, and if much weaker, a gas. (See Figure 3.1.)

Properties of Matter

We are able to distinguish one kind of matter from another because each has its own unique set of properties. **Properties** are those characteristics that allow us to identify different kinds of matter just as body temperature and blood pressure are properties that allow us to identify disease. **Physical properties** are those that can be measured without changing the identity (chemical composition) of the sample. Some physical properties are color, hardness, density, odor, boiling and melting temperatures, and surface luster.

Properties that can be determined only by changing the identity of the sample are called **chemical properties.** Methane is the major component of natural gas. One of the chemical properties of methane is that it burns in air. As it burns, methane combines with oxygen and is changed to carbon dioxide and water. The fact that methane can combine with oxygen was learned only by changing it into other kinds of matter; thus it is a chemical property. A chemical property of iron is that it can combine with oxygen in moist air, forming iron rust. Once the iron is consumed, the properties associated with iron are lost. Each of the materials that form when methane burns and iron rusts has its own set of identifying properties. No other kind of matter known has a set of properties identical to those of carbon dioxide and none has the unique set of either water or iron rust. Because properties change when methane burns and iron rusts, we know that matter changes. Some of the properties of these materials are shown in Table 3.1. Note that each set is different.

TABLE 3.1
Some Properties of Six Kinds of Matter

	Physical state at 25°C	Color	Density	Melting temperature	Reacts with oxygen in air
Methane	gas	colorless	0.72 g/l	−184°C	yes
Water	liquid	colorless	1.00 g/ml	0°C	no
Carbon dioxide	gas	colorless	1.98 g/l	—	no
Iron	solid	silver-white	7.86 g/ml	1535°C	yes
Oxygen	gas	colorless	1.43 g/l	−218°C	no
Iron rust (as Fe_2O_3)	solid	red-brown	5.24 g/ml	1565°C	no

Physical and Chemical Changes

The burning of methane and the rusting of iron are considered **chemical changes**. Each produces matter that is different from that which existed before the change took place. Chemical changes, then, are those that cause the identity of matter to change. Chemical changes are commonly called **chemical reactions,** or simply, **reactions.** Thousands of reactions take place every second in the body ranging from digestion to the development of memory.

Chemists have devised a relatively easy way to express chemical changes by writing them in the form of equations. **Chemical equations,** as they are called, use symbols, formulas, or names to indicate the materials involved in the change. Since symbols and formulas will be discussed later in this chapter, equations will be written here using the names of the materials involved. The reaction of methane with oxygen forming carbon dioxide and water would be written

$$\text{methane + oxygen} \longrightarrow \text{carbon dioxide + water}$$

The plus signs (+) can be read as "and," and the arrow (→) as "reacts to form" or "yields." So the equation could be read as, "methane and oxygen react to form carbon dioxide and water." The materials on the left side of the arrow are called **reactants** and those on the right are called the **products** of the reaction. When calcium carbonate (limestone) is heated to high temperatures, it decomposes to form calcium oxide (lime) and carbon dioxide. The equation that describes this chemical change would be

$$\text{calcium carbonate} \xrightarrow{\;\Delta\;} \text{calcium oxide} + \text{carbon dioxide}$$

The Greek delta (Δ) over the arrow is used to symbolize that heat is required to carry out the change.

As a chemical change takes place, there is neither a gain nor loss in the total mass of matter. This fact of nature is known as the **law of conservation of mass,** and it will serve as the basis for calculations using chemical equations in Chapter 4. If 100 g of calcium carbonate is decomposed, the combined masses of the products, calcium oxide and carbon dioxide, will also equal 100 g.

The reaction of 10 g of methane with 40 g of oxygen produces 27.5 g of carbon dioxide and 22.5 g of water. Here, 50 g of matter changes into 50 g of matter, the only difference being its identity.

Physical changes do not alter the composition or identity of the material being studied. Ice melting to form water is a physical change. The only change is from the solid to the liquid state, water remaining as water. Some other physical changes are boiling water to form water vapor, bending an iron bar, crushing a boulder, and sharpening the blade of a knife.

Check Test Number 1

Classify each of the following as either a physical or chemical property or a physical or chemical change.

a. frying an egg

b. boiling of sulfur at 444°C

c. resistance of gold to corrosion

d. expansion of mercury in a thermometer

e. dissolving alcohol in water

f. transparency of glass

g. burning of coal

h. ability of copper to be rolled into thin sheets

Answers: (a) chemical change, (b) physical property, (c) chemical property, (d) physical change, (e) physical change, (f) physical property, (g) chemical change, (h) physical property

Pure Substances

A **pure substance** is any sample of matter that has identical physical and chemical properties and composition throughout. Pure water is a substance; if you obtained samples of pure water from a dozen different sources, each sample would have the same density, color, boiling temperature, and other properties. Furthermore, each sample would have the identical composition, containing about 8 g of oxygen for each gram of hydrogen. Oxygen and hydrogen are elements, and water is a compound. Elements and compounds represent the two classes of pure substances.

Elements

Elements are pure substances that cannot be decomposed into simpler sub-stances by ordinary chemical changes. At the present time there are 106 known elements; 105 of them are listed in Table 3.2 and inside the front cover of this text. Element 106 is man-made and as yet unnamed. Some common elements that are familiar to you are carbon, oxygen, aluminum, iron, copper, nitrogen, and gold. The elements are the building blocks of matter just as the numerals 0 through 9 are the building blocks for numbers. To the best of our knowledge, the elements that have been found on the earth also comprise the entire universe.

Element	Not found in nature	Solid at 25°C	Liquid at 25°C	Gas at 25°C	Metal	Nonmetal	Metalloid
actinium		X			X		
aluminum		X			X		
americium	X	X			X		
antimony		X					X
argon				X		X	
arsenic		X					X
astatine	X					X	
barium		X			X		
berkelium	X	X			X		
beryllium		X			X		
bismuth		X			X		
boron		X					X
bromine			X			X	
cadmium		X			X		
calcium		X			X		
californium	X	X			X		
carbon		X				X	
cerium		X			X		
cesium		X			X		
chlorine				X		X	
chromium		X			X		
cobalt		X			X		
copper		X			X		
curium	X	X			X		
dysprosium		X			X		

TABLE 3.2
The Known Elements

	TABLE 3.2 The Known Elements (Continued)						
Element	Not found in nature	Solid at 25°C	Liquid at 25°C	Gas at 25°C	Metal	Nonmetal	Metalloid
einsteinium	x	x			x		
erbium		x			x		
europium		x			x		
fermium	x	x			x		
fluorine				x		x	
francium	x				x		
gadolinium		x			x		
gallium		x			x		
germanium		x					x
gold		x			x		
hafnium		x			x		
hahnium	x	x			x		
helium				x		x	
holmium		x			x		
hydrogen				x		x	
indium		x			x		
iodine		x				x	
iridium		x			x		
iron		x			x		
krypton				x		x	
lanthanum		x			x		
lawrencium	x	x			x		
lead		x			x		
lithium		x			x		
lutetium		x			x		
magnesium		x			x		
manganese		x			x		
mendelevium	x	x			x		
mercury			x		x		
molybdenum		x			x		
neodymium		x			x		
neon				x		x	
neptunium	x	x			x		
nickel		x			x		
niobium		x			x		
nitrogen				x		x	
nobelium	x	x			x		

Element	Not found in nature	Solid at 25°C	Liquid at 25°C	Gas at 25°C	Metal	Nonmetal	Metalloid
TABLE 3.2 **The Known Elements (Continued)**							
osmium		X			X		
oxygen				X		X	
palladium		X			X		
phosphorus		X				X	
platinum		X			X		
plutonium	X	X			X		
polonium		X					X
potassium		X			X		
praseodymium		X			X		
promethium	X	X			X		
protactinium		X			X		
radium		X			X		
radon				X		X	
rhenium		X			X		
rhodium		X			X		
rubidium		X			X		
ruthenium		X			X		
rutherfordium	X	X			X		
samarium		X			X		
scandium		X			X		
selenium		X				X	
silicon		X					X
silver		X			X		
sodium		X			X		
strontium		X			X		
sulfur		X				X	
tantalum		X			X		
technetium	X	X			X		
tellurium		X					X
terbium		X			X		
thallium		X			X		
thorium		X			X		
thulium		X			X		
tin		X			X		
titanium		X			X		
tungsten		X			X		
uranium		X			X		

TABLE 3.2
The Known Elements (Continued)

Element	Not found in nature	Solid at 25°C	Liquid at 25°C	Gas at 25°C	Metal	Nonmetal	Metalloid
vanadium		x			x		
xenon				x		x	
ytterbium		x			x		
yttrium		x			x		
zinc		x			x		
zirconium		x			x		

About 85% of the elements can be found in nature, usually combined with other elements in minerals and vegetable matter or in substances like water and carbon dioxide. Copper, silver, gold, and about 20 other elements can be found in highly pure forms. Sixteen elements are not found in nature; they have been produced in generally small amounts in nuclear explosions and nuclear research. These man-made elements are included in Table 3.2.

Nearly 99% of the earth's crust is made up of only eight of the 106 elements. The human body is composed primarily of only six elements. Oxygen is the predominant element in each. (See Table 3.3.)

The names of many of the elements are derived from Latin or Greek terms that usually describe one of their properties. Chlorine comes from the Greek *chloros,* which means "greenish yellow," the color of chlorine gas. Phosphorus comes from the Greek *phosphoros,* meaning "light bearing," for its "glow in the dark" property. Other elements are named after people or places such as einsteinium (for Albert Einstein), curium (for Madame Curie), californium (for the state), and uranium (for the planet Uranus).

TABLE 3.3
Elemental Composition of the Earth's Crust and the
Average Human Body, by Weight

Earth's crust		Human body	
oxygen	46.60%	oxygen	65.0%
silicon	27.72%	carbon	18.0%
aluminum	8.13%	hydrogen	10.0%
iron	5.00%	nitrogen	3.0%
calcium	3.63%	calcium	2.0%
sodium	2.83%	phosphorus	1.0%
potassium	2.59%		
magnesium	2.09%		

Just as symbols and abbreviations are widely used in medical areas to simplify communications, chemists use symbols to represent the names of the elements. Each element has a different symbol made up of one or two letters. If one letter is used, as it is for 13 of the elements, it is written as a capital: oxygen, O; nitrogen, N. If two letters are used, only the first is capitalized: calcium, Ca; aluminum, Al. The symbols are more than just abbreviations since they also stand for certain amounts of the element. The symbol H means not only hydrogen, but one atom of hydrogen. Atoms are discussed in the next section.

Most symbols suggest the name of the element they represent, while others seem unrelated to their English names. The symbols for this latter class of elements are derived from their early names (often Latin), which were widely used in the past.

When you write symbols that use two letters, it is important that only the first letter be capitalized. The symbol for nickel is Ni, but if it were written NI it would indicate a substance formed from the elements nitrogen (N) and iodine

TABLE   3.4
Symbols for Several Common Elements

Element	Symbol*	Element	Symbol*
aluminum	Al	magnesium	Mg
antimony	Sb (Latin, Stibium)	manganese	Mn
argon	Ar	mercury	Hg (Latin, Hydrargyrum)
arsenic	As	neon	Ne
barium	Ba	nickel	Ni
bismuth	Bi	nitrogen	N
boron	B	oxygen	O
bromine	Br	phosphorus	P
cadmium	Cd	potassium	K (Latin, Kalium)
calcium	Ca	silicon	Si
carbon	C	silver	Ag (Latin, Argentum)
chlorine	Cl	sodium	Na (Latin, Natrium)
chromium	Cr	strontium	Sr
cobalt	Co	sulfur	S
copper	Cu (Latin, Cuprum)	tin	Sn (Latin, Stannum)
fluorine	F	titanium	Ti
gold	Au (Latin, Aurum)	tungsten	W (German, Wolfram)
helium	He	uranium	U
hydrogen	H	zinc	Zn
iodine	I		
iron	Fe (Latin, Ferrum)		
lead	Pb (Latin, Plumbum)		
lithium	Li		

*Old names are also given if they are the source of the symbol.

(I). No such substance exists. If CO were written for the element cobalt (Co), it would be taken to mean carbon monoxide, a toxic gas made from carbon (C) and oxygen (O).

The symbols for the most common elements are listed in Table 3.4. A complete list of symbols can be found inside the front cover of this text.

You should begin to learn the symbols of the common elements in Table 3.4 right away, since they will be used in writing formulas of compounds and in equations throughout the remainder of the text.

Elements can be classified into two general categories, **metals** and **nonmetals.** Most of the elements are metals. With the exception of mercury, all the metals are solids at room temperature. They have shiny surfaces, are good conductors of electricity, and are malleable (can be hammered or rolled into sheets) and ductile (can be drawn into a wire). Most metals have high densities. The metals are those elements to the left of the stairstep line beginning at boron (B) in Figure 3.2. Several metals that are important trace elements in the body will be discussed further in Chapter 9.

Nonmetals can exist as solids, liquids, or gases at room temperature. Solid nonmetals, such as sulfur or phosphorus, do not have shiny surfaces and they are brittle and not malleable or ductile. With the exception of carbon (graphite), nonmetals are not conductors of electricity. Most nonmetals have relatively

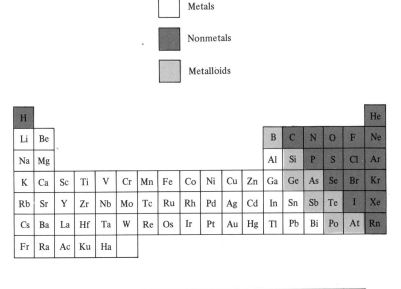

Fig. 3.2. The Periodic Table. Metals are to the left of the stairstep line, and nonmetals are to the right. The metalloids lie along the stairstep line. The periodic table arranges the elements according to their properties.

low densities. The nonmetals are those elements located to the right of the stairstep line in Figure 3.2, and they also include hydrogen. Hydrogen is a unique element. Many of its properties are like those of the nonmetals, yet in chemical changes there are times when it acts more like a metal. But since its physical properties and much of its chemistry are similar to the nonmetals, we will think of hydrogen as a nonmetal.

Except for aluminum, the elements that lie along the stairstep line display properties of both metals and nonmetals to varying degrees. These elements are often called **metalloids** to reflect their intermediate behavior. They are all solids and they may or may not conduct electricity. They all possess the surface luster of metals, yet are brittle.

The display of elements in Figure 3.2 is known as the Periodic Table. A more detailed Periodic Table appears inside the front cover, and it will be referred to several times throughout the text. The development of the Periodic Table is discussed in Chapter 6.

Check Test Number 2

Complete the following chart.

Name	Symbol	Metal or nonmetal
a. sodium	——	——
b. ——	Zn	——
c. ——	P	——
d. bromine	——	——

Answers: (a) Na, metal; (b) zinc, metal; (c) phosphorus, nonmetal; (d) Br, nonmetal

Atoms

The basic structural unit that makes up a sample of an element is the atom. The **atom** is the smallest particle of an element that can exist and still have the chemical properties of that element. A sample of iron is actually a collection of millions upon millions of iron atoms packed together in an orderly way. Atoms are not destroyed in physical or chemical changes. Iron atoms can be forced to move around one another as a piece of iron is bent into a different shape, and they can react chemically with oxygen to form iron rust, but in neither case are the atoms divided into smaller particles. The atoms retain their identity as iron atoms. If we now try to measure properties, we will be measuring those of iron rust and not of iron or oxygen individually.

Atoms are exceedingly small. Some examples of the relative sizes of atoms are shown in Figure 3.3. An iron atom can be viewed as a sphere with a diameter of 2.6×10^{-10} m, and 40 million of these spheres placed side by side would form a line only 1 cm long. Atoms of other elements have diameters from about one-fourth of that of iron (the hydrogen atom) to roughly twice that of iron (the francium atom). The nature of the atom will be discussed in more detail in Chapters 6, 7, and 8.

Compounds

Compounds are pure substances composed of two or more elements combined in a fixed weight ratio and which can be decomposed by ordinary chemical means. Let us look at an example of a substance to see what this definition means.

The most common pure compound is water. In water, oxygen and hydrogen are chemically combined in such a way that for every 1.000 g of hydrogen there

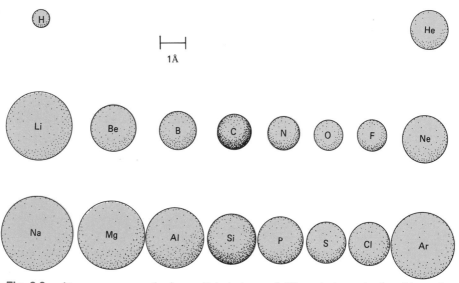

Fig. 3.3. Atoms are exceedingly small, but atoms of different elements do not have the same size.

are 7.936 g of oxygen. Every sample of pure water contains hydrogen and oxygen in this same weight ratio. The fact that a fixed mass ratio of elements combine to form a compound was known about 180 years ago and it is stated as the **law of definite composition**. The law states that all samples of a pure compound always contain the same elements in the same mass ratio. All samples of carbon dioxide contain 2.664 g of oxygen for each 1.000 g of carbon, and all samples of methane contain 2.979 g of carbon for each 1.000 g of hydrogen.

Though there are only 106 elements, there are well over five million known compounds formed from the combinations of these elements. Exactly what happens when two elements combine chemically will be discussed later on in this text, but at this point you can regard a chemical combination as a strong attraction between two or more atoms that holds them together in a definite, organized arrangement.

Molecules

Suppose you could subdivide a drop of water into smaller and smaller portions. You might wonder how long you could continue before you reached the smallest particle that would still be considered water. The answer is that the limit of subdivision would be reached when the particle was a single water molecule. A **molecule** is the smallest part of a compound that retains the chemical properties of that compound. Further subdivision would break the molecule apart into atoms, destroying its identity.

Molecules are made up of two or more atoms chemically joined together forming a single particle. The atoms can be of the same element or different elements. Two atoms of hydrogen and one atom of oxygen combine to form one molecule of water. The composition of water in terms of its elements can be expressed in a **formula.** Formulas are abbreviations for compounds and they tell us the number of atoms of each element that combine to form the molecule or formula unit.

The following rules will guide you as you write formulas for compounds.

1. The formula is composed of the symbols of the elements that form the compound.
2. Small whole numbers are written as subscripts after the symbols to indicate the number of atoms of that element in the formula. If only one atom of an element is in a formula, the subscript 1 is not used since the symbol itself stands for a single atom. The formula of ethyl alcohol, C_2H_6O, shows that there are two atoms of carbon, six atoms of hydrogen, and one atom of oxygen in the formula.
3. If the compound contains a metal and a nonmetal, the symbol for the metal is given first followed by the symbol for the nonmetal. If there are three

or more elements in the formula, and one of them is oxygen, the symbol for oxygen is given last. The symbol for sodium carbonate would be written Na_2CO_3.

From what was said about the water molecule, the formula for water would be written H_2O, and it would tell several things about the compound.

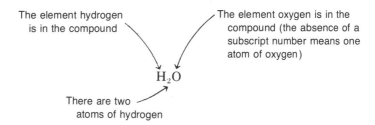

One molecule of carbon dioxide contains two atoms of oxygen and one atom of carbon. Its formula is CO_2. Though formulas show the number and kinds of atoms in a compound, they do not show how the atoms are grouped together in a molecule.

Other compounds which exist as molecules are given in Table 3.5 with their formulas. Notice that each compound contains only nonmetal atoms. Compounds formed by combination of nonmetal elements almost always exist as molecules.

Several elements in their pure state also exist as molecules. In most cases only two atoms combine to form the molecule so they are often called the

TABLE 3.5 The Formulas of Compounds Indicate their Elemental Composition				
Name	Methane	Carbon monoxide	Benzene	Chloroform	Ammonia
Formula	CH_4	CO	C_6H_6	$CHCl_3$	NH_3
Composition of the molecule	1 atom of carbon + 4 atoms of hydrogen	1 atom of carbon + 1 atom of oxygen	6 atoms of carbon + 6 atoms of hydrogen	1 atom of carbon + 1 atom of hydrogen + 3 atoms of chlorine	1 atom of nitrogen + 3 atoms of hydrogen

diatomic elements (di-, "having two"). The diatomic elements are: hydrogen (H_2), nitrogen (N_2), oxygen (O_2), fluorine (F_2), chlorine (Cl_2), bromine (Br_2), and iodine (I_2). Other elements exist as larger molecules. Two of them are phosphorus (P_4) and sulfur (S_8). Be certain you realize the difference between the symbol H_2 (which describes the molecule) and H (which symbolizes an atom of hydrogen).

Coefficients can also be used with symbols and formulas to indicate a certain number of atoms or molecules. A coefficient is a number written in front of the symbol or formula. For example, $3H_2O$ would stand for three molecules of water. The 3 is a coefficient. Five hydrogen atoms would be written as 5H, and 10 methane molecules as $10CH_4$. Do not confuse the coefficient number with the subscript numbers used in formulas, since they do not mean the same thing. To represent a molecule of hydrogen, you would write H_2, but to represent two atoms of hydrogen not joined in a molecule, you would have to write 2H. Coefficients will be very useful to us in later chapters. Table 3.6 lists formulas of several common compounds and gives some of their uses.

TABLE 3.6 Formulas of Several Common Compounds and Their Uses		
Formula	**Name**	**Uses**
HCl	hydrochloric acid	common laboratory acid—present in the human stomach
H_2SO_4	sulfuric acid	common laboratory acid
$C_2H_4O_2$	acetic acid	vinegar, salad dressing—common laboratory acid
NaOH	sodium hydroxide	common laboratory base
$CaCO_3$	calcium carbonate (limestone)	antacid
CaO	calcium oxide (lime)	glass making, metallurgy
NaCl	sodium chloride	table salt, saline solution
$BaSO_4$	barium sulfate	x-ray examination of internal organs
$(NH_4)_2CO_3$	ammonium carbonate	smelling salts
$MgSO_4$	magnesium sulfate	laxative
$AgNO_3$	silver nitrate	antiseptic
$NaHCO_3$	sodium hydrogen carbonate (baking soda)	antacid
ZnO	zinc oxide	calamine lotion
NH_4Cl	ammonium chloride	diuretic, expectorant
SO_2	sulfur dioxide	fumigant, air pollutant
CH_4O	methyl alcohol	antiseptic
C_2H_6O	ethyl alcohol	antiseptic, drinking alcohol
$C_4H_{10}O$	diethyl ether	anesthetic
$C_6H_{12}O_6$	glucose	blood sugar
$C_{12}H_{22}O_{11}$	sucrose	table sugar

Check Test Number 3

a. Write the formula for carbon tetrachloride, a compound formed from one atom of carbon and four atoms of chlorine.
b. How would you symbolize seven molecules of carbon tetrachloride?
c. How would you symbolize 10 molecules of bromine; 10 atoms of bromine? Which would contain the greater number of bromine atoms.

Answers: (a) CCl_4, (b) $7CCl_4$, (c) $10Br_2$, $10Br$. There are 20 atoms in 10 molecules of bromine. Two atoms per molecule times 10 molecules is 20 atoms, more atoms than in the other case.

Ions

Earlier we said that when two nonmetals react, the compound formed usually exists as molecules. But when metals react with nonmetals, very often the compounds formed are solids composed of particles that bear positive and negative electrical charges. Particles that bear an electrical charge are called **ions.** Metals form ions that bear a positive charge, and nonmetals form ions that bear a negative charge. The reason why will be discussed in Chapter 8. In a compound, the positive and negative ions are strongly held together due to the attraction between the opposite charges. Molecules do not exist in compounds that contain ions. Rather they are made up of ions arranged in an orderly way forming solids that are crystalline, that is, they have regular geometric shapes like cubes, needles, and plates. Formulas can readily be written for these kinds of compounds since for every ion with a 1+ charge there will be one with a 1− charge in order to make the overall compound electrically neutral, that is, have no charge. An example of a compound composed of ions (called an ionic compound) is sodium chloride, common table salt. Sodium is in the form of an ion with a 1+ charge and is written as Na^+, and chlorine forms an ion that bears a 1− charge and is written as Cl^-. For each Na^+ in the compound there would be one Cl^-, so the simplest formula that would show the composition of sodium chloride would be NaCl. Though no molecules exist, the formula clearly tells us that there is one sodium species for every chlorine species in the compound. Since molecules do not exist here, we usually refer to each of the species that comprise the formula of the compound as a **formula unit.** One formula unit of sodium chloride would be one Na^+ and one Cl^-. (See Figure 3.4.)

Many different ions are found in nature. Ions that bear negative charges are collectively called **anions,** and those with positive charges are called **cations.** If an ion is formed from a single atom, it is called a **simple ion,** but if it is made up of several atoms joined together it is a **polyatomic ion.** The chloride ion,

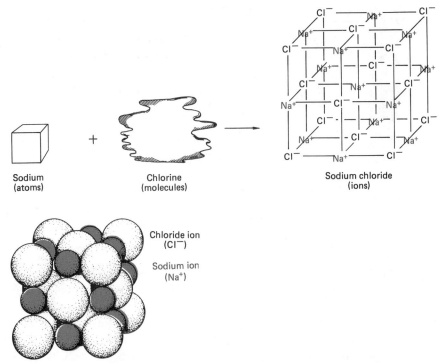

Fig. 3.4. When metals react chemically with nonmetals, the products are usually crystalline solids composed of ions. Sodium, a metal, and chlorine, a nonmetal, form sodium chloride, a crystalline solid. Note the orderly arrangement of sodium ions, Na^+, and chloride ions, Cl^-, in sodium chloride. This orderly arrangement is responsible for the shape of salt crystals.

TABLE   3.7
Some Common Simple and Polyatomic Ions

Simple ions		Polyatomic ions	
H^+	hydrogen ion	OH^-	hydroxide ion
Li^+	lithium ion	NO_3^-	nitrate ion
K^+	potassium ion	CN^-	cyanide ion
Mg^{2+}	magnesium ion	HCO_3^-	hydrogen carbonate ion (bicarbonate ion)
Ca^{2+}	calcium ion		
Al^{3+}	aluminum ion	CO_3^{2-}	carbonate ion
F^-	fluoride ion	SO_4^{2-}	sulfate ion
Cl^-	chloride ion	PO_4^{3-}	phosphate ion
Br^-	bromide ion	$C_2H_3O_2^-$	acetate ion
I^-	iodide ion	NH_4^+	ammonium ion
O^{2-}	oxide ion		

Cl^-, would be a simple anion, and the nitrate ion, NO_3^-, would be a polyatomic anion. Most common polyatomic ions are anions. Several common ions are listed in Table 3.7.

Check Test Number 4

For each pair of elements, indicate whether the compound formed when they react would be composed of ions or molecules.
a. carbon and nitrogen
b. lithium and nitrogen
c. magnesium and chlorine
d. sulfur and fluorine

Answers: (a) molecules, (b) ions, (c) ions, (d) molecules

Mixtures

In nature, matter is most often found as mixtures of substances. A **mixture** is made up of two or more pure substances that do not interact chemically, so they retain their individual identities and properties. The relative amounts of each substance in a mixture can vary over a wide range, since there is no fixed ratio of one substance to another as we find in compounds. Air is a mixture of several substances: oxygen, nitrogen, carbon dioxide, water vapor, and others. The air over a lake has a higher water vapor content than air over a desert, and air would contain more carbon dioxide around a coal burning power plant than over the ocean. Being a mixture, the composition of air varies slightly, but each component retains its identity.

Under the right conditions, the differences in the physical properties of the substances in a mixture will allow them to be separated from one another. Separation techniques that rely on differences in physical properties are called **physical methods** since they will separate the substances without changing their identities (their chemical composition).* For example, sugar dissolved in water is a mixture. By simply allowing the water to evaporate, the sugar would be left behind and obtained pure. Evaporation is a physical change, so separation of water from sugar in this way would not change the identity of either com-

---

*By contrast, a chemical method of separation would result in a change in identity of at least one component of the mixture since chemical reactions would be used. A mixture of iron and sand could be separated by reacting the iron with acid, converting it to a water soluble compound. The sand would not react and it would be separated by filtering the mixture once the iron had reacted.

pound. If you travel to the Great Salt Lake in Utah, you might collect samples of salt mixed with sand in certain places along the shore. One physical means that could separate the components of the mixture would be to dissolve the salt in water, leaving the sand behind. The sand could be removed by filtering the solution through a porous paper held in a funnel. When the water evaporated, the pure salt could be collected. This process is illustrated in Figure 3.5.

Mixtures can be classified as being either **homogeneous** or **heterogeneous**. A homogeneous mixture is one that has no visible boundaries within it. Both air and sugar–water are homogeneous mixtures since one cannot visibly distinguish particles of one kind from those of the others (no visible boundary exists between them). Solutions formed by dissolving one substance in another, such as alcohol in water, are homogeneous mixtures. Alloys such as brass and bronze are solutions of one solid dissolved in another.

A **heterogeneous mixture** is one that has visible boundaries within it. The sand and salt mixture discussed earlier is heterogeneous, since you can actually see individual particles of salt mixed with individual particles of sand. Some other heterogeneous mixtures are granite, human tissue, smoke, and blood. Though

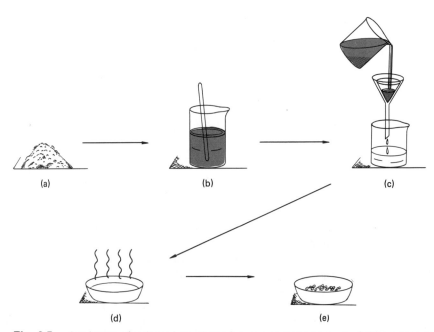

Fig. 3.5. A mixture of salt and sand can be separated because of differences in their physical properties. Salt is soluble in water but sand is not. Separations based on differences in physical properties are called physical methods of separation. (a) Mixture of sand and salt; (b) water is added and the mixture is stirred to dissolve the salt; (c) the salt–water solution is filtered through porous paper to remove the sand that remains in the paper; (d) the salt–water solution is set out to allow the water to evaporate; (e) after evaporation is complete, the pure salt is collected and the separation is complete.

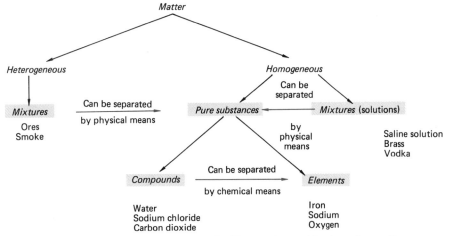

Fig. 3.6. Relationships between terms in Chapter 3 used to describe matter.

blood looks homogeneous to the naked eye, its heterogeneous composition is evident when viewed under a microscope.

Figure 3.6 relates many of the terms used in this chapter to describe matter.

Terms

Several important new terms appeared in Chapter 3. You should understand the meaning of each of the following.

matter	heterogeneous mixture
solid	reactant
liquid	product
gas	substance
property	ion
physical property	element
chemical property	compound
physical change	symbol
chemical change	formula
solution	metal
chemical equation	nonmetal
atom	law of definite composition
molecule	mixture
formula unit	law of conservation of mass
homogeneous	

Questions

Answers to the starred questions appear at the end of the text.

1. Carbon dioxide can exist as a solid, a liquid, or a gas under the proper conditions. Compare the way in which the molecules are arranged in each physical state.

*2. In what way is the strength of the attractive forces between molecules related to the physical state of a compound?

3. Iron can be magnetized. Is this a physical or chemical property of iron? Milk will sour if left out at room temperature. Is this a physical or chemical property of milk?

*4. A small balloon filled with air can be compressed to a smaller volume by squeezing it in your hands, but if filled with water it cannot be. Why?

5. Write word equations for the following chemical changes.
    *a. Sodium metal reacts with water to form sodium hydroxide and hydrogen gas.
    *b. Sucrose (table sugar) decomposes when heated in the absence of air to form carbon and water.
    c. Carbon dioxide and carbon react to form carbon monoxide at 500°C.
    d. Hydrogen and nitrogen react to form ammonia.

6. 5 g of hydrogen completely react with 15 g of carbon to form methane. How many grams of methane are formed?

*7. 18 g of iron react with chlorine to form 52.3 g of iron(III) chloride, a compound of iron and chlorine. How many grams of chlorine are consumed?

*8. In what important way does a compound formed from copper and sulfur differ from a mixture of the two elements?

9. Without looking at the list of elements in the text, give the correct symbol for the following:

hydrogen	iron	oxygen
copper	calcium	sulfur
fluorine	potassium	sodium
carbon	phosphorus	chlorine
nitrogen	silicon	lead
aluminum	neon	silver

10. Describe the difference between the symbols in each pair.
    *a. Si and SI
    b. HOH and HoH
    *c. SO_2 and SO_3
    d. I_2 and I

11. List three ways in which metals differ from nonmetals.

*12. What is the diameter of an iron atom in angstroms? The diameter in meters is given in the text.

13. Write the formula for each of the following compounds from the data given.
    *a. sodium fluoride—1 atom of sodium and 1 atom of fluorine
    b. mercury(II) oxide—1 atom of mercury and 1 atom of oxygen
    *c. aluminum oxide—2 atoms of aluminum and 3 atoms of oxygen

   d. copper sulfate—1 atom of copper, 1 atom of sulfur, and 4 atoms of oxygen

   *e. ethyl alcohol—2 atoms of carbon, 6 atoms of hydrogen, and 1 atom of oxygen

   f. sulfuric acid—2 atoms of hydrogen, 1 atom of sulfur, and 4 atoms of oxygen

14. How many atoms are in one molecule or one formula unit of each of the compounds in question 13?

*15. Describe the kind of compound that often forms when a metal reacts with a nonmetal.

*16. Which of the following are mixtures?

   a. an alloy of gold and platinum

   b. mercury

   c. milk

   d. ammonia

   e. pizza

   f. natural gas

*17. Which of the following is a physical method of separating a mixture?

   a. Separating particles according to size with a sieve.

   b. Decomposing one component of the mixture with heat.

   c. Removing particles of aluminum by reaction with a solution of sodium hydroxide.

   d. Removing a component by melting it and letting it drain from the mixture.

   e. Pouring the mixture into water to separate the wood chips from denser materials.

18. How does a chemical formula differ from a symbol?

4

A Closer Look at Atoms, Molecules, and Chemical Changes

The previous chapter dealt with several facts concerning the composition of matter and how its changes can be described. The discussion was basically **qualitative** in that it described matter in terms that did not require mathematics. The discussion will continue in this chapter, but on a more **quantitative** level that will introduce you to the basics of chemical arithmetic. Unlike Chapter 3, emphasis will be placed on the masses of atoms and molecules. Compounds will be described in terms of the percent by weight of each element contained in the molecule, and chemical equations will be used to calculate "how much" product can be formed from a given amount of reactant.

Not everyone gets excited about doing chemistry problems, but they do serve as an excellent device for learning the facts of chemistry. Study the example problems carefully, and be certain to do the check tests before moving on to the next section since each section uses information presented in those that precede it.

Objectives

By the time you finish this chapter you should be able to do the following.

1. Describe the atomic weight scale.
2. Define these terms: atomic weight, formula weight, gram atomic weight, gram formula weight, and mole.
3. Convert a given number of grams of substance into moles and vice versa.
4. State the value of Avogadro's number and describe its meaning in terms of one mole of an element or compound.
5. Calculate the percent composition by weight of a compound, given its formula.
6. Balance chemical equations and describe what a balanced equation means in terms of moles, molecules, and grams of all substances present.
7. Calculate the number of moles of product formed in a reaction given the balanced equation and the number of moles of reactant.

The Atomic Weight of the Elements

In Chapter 3 you learned that no two elements are alike, since each has a different set of properties which give it a unique identity. One important difference between elements is that atoms of different elements do not have the same mass. For example, a typical hydrogen atom, the lightest element, has a mass of about 1.7×10^{-24} g, while a typical atom of uranium is about 238 times heavier, weighing 4.0×10^{-22} g. One gram of hydrogen would contain the same number of atoms as about 238 g of uranium. Why? Because the uranium atom weighs about 238 times more than a single hydrogen atom, and to obtain an identical number of atoms of each element, the uranium sample would have to weigh 238 times that of the hydrogen sample.

Individual atoms do not weigh very much, and because of this it is not practical to use the actual masses of atoms in laboratory calculations. Instead, a scale of relative weights for the elements has been devised called the **atomic weight scale.** The term "atomic weight" has been in common use for many years and it will be used here, though the more correct term would be "atomic mass."

Some examples of comparative actual and atomic weights of atoms are shown in Figure 4.1.

Before delving into the atomic weight scale, we should examine what is meant

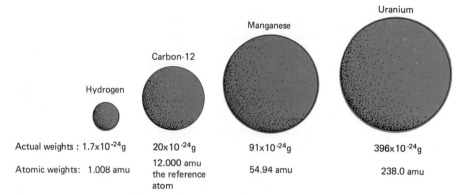

Actual weights : 1.7x10^{-24}g   20x10^{-24}g   91x10^{-24}g   396x10^{-24}g

Atomic weights: 1.008 amu   12.000 amu the reference atom   54.94 amu   238.0 amu

Fig. 4.1.   The actual weights of atoms are very small. For convenience, scientists use the relative weights of atoms instead of actual weights in calculations. The relative weights are called atomic weights, and they state the weight of an atom relative to the weight of the carbon-12 atom. Carbon-12 is assigned an atomic weight of exactly 12.000 atomic mass units.

by "relative weights." Suppose you weighed a penny, a nickel, and a quarter on a balance and found their *actual weights* (masses) to be

penny: 3.01 g
nickel: 4.83 g
quarter: 5.67 g

You could express the weight of each on a relative scale by choosing one coin as a reference, say the penny, and then stating the weights of the other two as being so many times greater than that of the penny. This is done by dividing the mass of each coin by the mass of the reference coin:

	Actual weight	÷	Weight of reference	=	Relative weight
penny:	3.01 g	÷	3.01 g	=	1.00
nickel:	4.83 g	÷	3.01 g	=	1.60
quarter:	5.67 g	÷	3.01 g	=	1.88

The same kind of thing is done with the elements. The mass of the atoms of an element is expressed as a relative weight, or **atomic weight**, using the mass of a particular kind of carbon atom as a reference, the carbon-12 isotope.*

---

*Isotopes will be discussed in Chapter 6. There are three kinds of carbon atoms (isotopes) found in nature. They are chemically identical though each has a slightly different mass. Naturally occurring carbon is a mixture of the three kinds of carbon atoms, and the mixture has an average atomic weight of 12.011 amu, slightly higher than that of the C-12 isotope by itself.

To establish a fixed point on the atomic weight scale, the carbon-12 reference atom is arbitrarily assigned a mass of exactly 12.000 atomic mass units (amu). Since hydrogen atoms weigh about one-twelfth as much as the carbon-12 atom, the atomic weight (AW) of hydrogen is 1.008 amu. Magnesium atoms weigh about twice as much as carbon-12 and so have an atomic weight of 24.305 amu. Krypton atoms are about seven times heavier and so they have an atomic weight of about seven times that of carbon-12, 83.8 amu. To give you an idea of the size of the atomic mass unit, one amu is equal to 1.6×10^{-24} g. When atomic weights are used in calculations, they can often be rounded off to simplify the work. Accurate atomic weights of several common elements along with their approximate values (that you can use to simplify calculations) are given in Table 4.1. The atomic weights of all the elements can be found inside the front cover of this text.

TABLE 4.1
Accurate and Approximate Atomic Weights of Some of the Common Elements
(C-12=12.000 amu)

Element	Accurate atomic wt. (amu)	Approximate atomic wt. (amu)	Element	Accurate atomic wt. (amu)	Approximate atomic wt. (amu)
aluminum	26.9815	27.0	magnesium	24.305	24.3
antimony	121.75	122	manganese	54.938	55.0
arsenic	74.922	75.0	mercury	200.59	201
barium	137.34	137	nickel	58.71	58.7
bismuth	208.980	209	nitrogen	14.0067	14.0
boron	10.81	10.8	oxygen	15.9994	16.0
bromine	79.904	79.9	phosphorus	30.9738	31.0
cadmium	112.40	112	potassium	39.098	39.1
calcium	40.08	40.1	silicon	28.086	28.1
carbon	12.011	12.0	silver	107.868	108
chlorine	35.453	35.5	sodium	22.9898	23.0
chromium	51.996	52.0	strontium	87.62	87.6
cobalt	58.933	59.0	sulfur	32.06	32.0
copper	63.546	63.5	tin	118.69	119
fluorine	18.998	19.0	titanium	47.90	48.0
gold	196.966	197	tungsten	183.85	184
hydrogen	1.008	1.0	uranium	238.03	238
iodine	126.90	127	zinc	65.38	65.4
iron	55.847	55.8			
lead	207.2	207			
lithium	6.941	6.9			

Formula Weights of Compounds

Since the atoms of different elements have different masses, it would be logical to expect molecules and formula units of compounds to have different masses also. Because compounds are composed of atoms each with its own relative weight, the formula weight of a compound will also be a relative weight, relative to that of carbon-12. The **formula weight** (FW) of a compound is equal to the sum of the atomic weights of all the atoms in its formula. Another term, **molecular weight** (MW), is also used to describe the relative weight of a compound. The molecular weight is equal to the sum of the atomic weights of all the atoms in one molecule of a compound. Molecular weight and formula weight are not identical terms, though for a molecular compound they would be numerically equal. Formula weight is the broader term since it is determined from the formulas of compounds, whether they exist as molecules or aggregates of ions. To avoid confusion, we will use formula weight here to describe the relative weight of a compound.

Once you know the formula of a compound, it is an easy task to determine its formula weight. If the compound contains ions, as in NaCl, the atomic weight of each ion is identical to that of the neutral parent atom. So, the atomic weight of Na^+ would be 23.0 amu, the same as that for the sodium atom, and for Cl^- it would be the same as that for the chlorine atom, 35.5 amu.

Example 4.1

What is the formula weight (FW) of water, H_2O?
There are two hydrogen atoms and one oxygen atom in the molecule. Let us use the accurate atomic weights (AW) from Table 4.1.

$$2 \times \text{AW of H} = 2 \times 1.008 \text{ amu} = 2.016 \text{ amu}$$

$$1 \times \text{AW of O} = 1 \times 15.999 \text{ amu} = \underline{15.999 \text{ amu}}$$
$$18.015 \text{ amu}$$

$$\text{FW of } H_2O = 18.015 \text{ amu}$$

If only an approximate formula weight is needed, you can use the approximate atomic weights in Table 4.1 (H=1.0 amu, O=16.0 amu) to get 18.0 amu for the FW of H_2O.

Example 4.2

Calculate the approximate formula weight of calcium nitrate, $Ca(NO_3)_2$. There are two nitrate ions (NO_3^-) and single calcium ion (Ca^{++}) in the formula for a total of two nitrogen atoms, six oxygen atoms, and one calcium atom.

$$1 \times \text{AW Ca} = 1 \times 40.1 \text{ amu} = 40.1 \text{ amu}$$

$$2 \times \text{AW N} = 2 \times 14.0 \text{ amu} = 28.0 \text{ amu}$$

$$6 \times \text{AW O} = 6 \times 16.0 \text{ amu} = \underline{96.0 \text{ amu}}$$
$$164.1 \text{ amu}$$

$$\text{FW Ca(NO}_3)_2 = 164.1 \text{ amu}$$

If you are considering the formula weight of the diatomic elements (H_2, N_2, O_2, F_2, Cl_2, Br_2, and I_2) as they exist in pure form, you need to use twice the atomic weight. The formula weight of nitrogen (N_2) would be two times the atomic weight of nitrogen, $2 \times 14.0 \text{ amu} = 28.0 \text{ amu}$, and for hydrogen ($H_2$), $2 \times 1.0 \text{ amu} = 2.0 \text{ amu}$.

Check Test Number 1

1. Atom X has an atomic weight of 12.5 amu and a mass of 2×10^{-23} g. Atom Z has an atomic weight of 50 amu. What is the actual mass of atom Z in grams?
2. Calculate the approximate formula weights of ethyl alcohol, C_2H_6O, and aspirin, $C_8H_8O_3$.

Answers:
1. Since the atomic weight of Z is four times that of X, its actual mass will be four times greater, 8×10^{-23} g.
2. Ethyl alcohol, 46.0 amu; aspirin, 152.0 amu.

The Mole and Avogadro's Number

Though atomic weights and formula weights are important physical properties of elements and compounds, at first glance they might seem to be of limited use since, in practice, amounts of different substances are usually weighed out in grams and not in atomic mass units. But by expressing the relative weights in units of grams, both can be converted into terms that are useful in the laboratory.

Gram Atomic Weight (GAW)—An amount of an element equal to its atomic weight in grams. One GAW of hydrogen is 1.008 g of that element.

Gram Formula Weight (GFW)—An amount of a compound equal to its formula weight in grams. One GFW of sodium chloride is 58.443 g of NaCl.

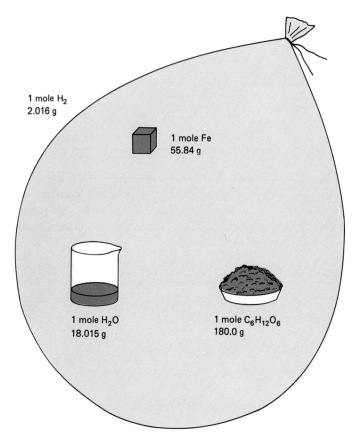

1 mole H$_2$
2.016 g

1 mole Fe
55.84 g

1 mole H$_2$O
18.015 g

1 mole C$_6$H$_{12}$O$_6$
180.0 g

Fig. 4.2.   One mole of H$_2$, Fe, H$_2$O, and C$_6$H$_{12}$O$_6$.

As a matter of convenience, each of the above terms is referred to as one **mole** of the respective substance. One GAW of hydrogen (1.008 g of H) is one mole of hydrogen atoms. One mole of H$_2$ would be 2.016 g of H$_2$ molecules. One GFW of water (18.015 g of H$_2$O) would be one mole of water. Later on the mole will be defined in another way concerning the number of atoms or molecules in one mole of a substance. The relative weights of one mole of different substances are shown in Figure 4.2.

Frequently in chemistry it is necessary to know how many moles are represented by a given number of grams of substance, or to know the mass of substance that would represent a certain number of moles. The following examples will show you how to do these kinds of calculations.

Example 4.3

Sodium is a solid, metallic element that is soft enough to be cut with a knife. How many moles of sodium (GAW's of Na) are represented by 90.0 g of that element?

Using the approximate atomic weight of sodium we can write

$$1 \text{ mole Na} = 23.0 \text{ g Na}$$

This equation can be rearranged to form a conversion factor which, when multiplied by 90 g, will convert grams to moles and eliminate the gram units:

$$90.0 \text{ g Na} \times \underbrace{\frac{1 \text{ mole Na}}{23.0 \text{ g Na}}}_{\substack{\text{conversion} \\ \text{factor}}} = 3.9 \text{ mole Na}$$

Example 4.4

The most widely used analgesic (pain killer) in the world is aspirin. How many grams of aspirin, $C_8H_8O_3$, are present in 0.750 mole of the compound? The molecular weight of aspirin is 152 amu, so we can write

$$1 \text{ mole } C_8H_8O_3 = 152 \text{ g } C_8H_8O_3$$

or simply,

$$1 \text{ mole} = 152 \text{ g}$$

As in the previous example, rearrangement of this equation will form a conversion factor that, when multiplied by 0.750 mole, will convert moles to grams and eliminate the mole units:

$$0.750 \text{ mole} \times \frac{152 \text{ g}}{1 \text{ mole}} = 114 \text{ g of aspirin}$$

One gram atomic weight (one mole) of an element and one gram formula weight (one mole) of a compound have an important characteristic in common. Each contains an identical number of the species referred to in its name. For example, one mole of any element contains 6.02×10^{23} atoms of that element. One mole of a compound contains 6.02×10^{23} formula units or molecules of that compound. Notice in each case that one mole represents 6.02×10^{23} units (molecules, atoms, or formula units) of the substance. A more satisfactory definition of the mole would be a quantity of a substance that contains 6.02×10^{23} atoms, molecules, or formula units of that substance.

The number of particles in one mole is so important in chemistry that it is given a special name, **Avogadro's number**, in honor of the Italian physicist Amedeo Avogadro (1776–1856):

$$\text{Avogadro's Number} = 6.02 \times 10^{23} \frac{\text{atoms}}{\text{GAW}} \text{ or } \frac{\text{molecules}}{\text{GFW}} \text{ or } \frac{\text{formula units}}{\text{GFW}}$$

Table 4.2 shows the meaning of Avogadro's number for one mole of six different substances.

TABLE 4.2 The Mole and Avogadro's Number			
Substance	Atomic or formula weight	Mass of one mole	Contained in one mole
N	14 amu	14 g	6.02×10^{23} N atoms
N_2	28	28 g	6.02×10^{23} N_2 molecules
CH_4	16	16 g	6.02×10^{23} CH_4 molecules
NaCl	58.5	58.5 g	6.02×10^{23} NaCl formula units
Na^+	23	23 g	6.02×10^{23} Na^+ ions
Cl^-	35.5	35.5 g	6.02×10^{23} Cl^- ions

Example 4.5

How many molecules are present in 88.0 g of carbon dioxide, CO_2? The GFW of CO_2 is 44.0 g.

One mole (44.0 g) of CO_2 will contain Avogadro's number of molecules:

$$44.0 \text{ g } CO_2 = 6.02 \times 10^{23} \text{ molecules}$$

This equation can be written as a conversion factor to be used to convert grams of CO_2 to molecules of CO_2:

$$88.0 \text{ g} \times \frac{6.02 \times 10^{23} \text{ molecules}}{44.0 \text{ g}} = 1.2 \times 10^{24} \text{ molecules}$$

Check Test Number 2

1. Sodium hydroxide is a caustic substance used in drain cleaners and in the manufacture of soap. How many grams of NaOH represent 0.80 mole of that compound?
2. One kilogram of water, H_2O, represents what number of moles of water? How many molecules of water are present in 1 kg of water?
3. How many oxygen atoms are in one molecule of hydrogen peroxide, H_2O_2? In one mole of H_2O_2?

Answers:

1. $0.80 \text{ mole} \times \dfrac{40.0 \text{ g}}{\text{mole}} = 32.0 \text{ g NaOH}$

2. 1 kg = 1000 g $\qquad$ $1000 \, \cancel{g} \times \dfrac{1 \text{ mole}}{18 \, \cancel{g}} = 55.5 \text{ mole } H_2O$

$$55.5 \, \cancel{\text{mole}} \times \dfrac{6.02 \times 10^{23} \text{ molecules}}{1 \, \cancel{\text{mole}}}$$

$$= 3.34 \times 10^{25} \text{ molecules}$$

3. The formula shows two oxygen atoms per molecule. Since there are 6.02 $\times \, 10^{23}$ molecules in one mole, the number of oxygen atoms in one mole of H_2O_2 is twice Avogadro's number; 1.2×10^{24} atoms.

Percent Composition of Compounds

Every sample of a pure compound contains the same elements in a definite weight ratio. You may recall this fact as the Law of Definite Composition. In 1 g of water there is 0.1119 g of hydrogen and 0.8881 g of oxygen. The composition of a compound can be stated in terms of the weight percent of each element in its formula. The result is called the **percent composition** of the compound. The percent by weight of an element is determined by dividing the weight of that element in a sample by the weight of the sample, then multiplying by 100% to convert the decimal to percent. In terms of some element A,

$$\text{percent A in sample} = \frac{\text{grams of A}}{\text{grams of sample}} \times 100\%$$

Using the data given above for water:

$$\%H = \frac{0.1119 \text{ g}}{1.0000 \text{ g}} \times 100\% = 11.19\%$$

$$\%O = \frac{0.8881 \text{ g}}{1.0000 \text{ g}} \times 100\% = 88.81\%$$

What do these figures mean? In terms of hydrogen, they mean that 11.19% of the weight of one molecule of water is hydrogen, or on a larger scale, that 11.19% of any size sample of water is hydrogen. For example, in 1000 g of water, there would be 111.9 g of hydrogen, with the remainder of the sample (88.81%) being 888.1 g of oxygen.

The percent composition of a compound can also be determined from its formula. The sample size is taken as its GFW, and the amount of each element present in the sample is equal to its contribution in grams to the GFW. The following example demonstrates this method.

Example 4.6

Calculate the percent composition of calcium chloride from its formula, $CaCl_2$.

First the gram formula weight of $CaCl_2$ must be determined. For convenience let us use the approximate gram atomic weights of the elements:

$$1 \times GAW\ Ca = 1 \times 40.1\ g = \quad 40.1\ g$$

$$2 \times GAW\ Cl = 2 \times 35.5\ g = \underline{\quad 71.0\ g}$$

$$111.1\ g = GFW\ CaCl_2$$

Note that in 111.1 g of $CaCl_2$ there are 40.1 g of Ca and 71.0 g of Cl. The percent composition can be calculated from these values:

$$\%\,Ca = \frac{40.1\ g}{111.1\ g} \times 100\% = 36.1\%$$

$$\%\,Cl = \frac{71.0\ g}{111.1\ g} \times 100\% = 63.9\%$$

Check Test Number 3

Calculate the percent composition by weight of methane, CH_4.

Answer:

$$\%\,C = \frac{12.0\ g}{16.0\ g} \times 100\% = 75.0\%; \quad \%\,H = \frac{4.0\ g}{16.0\ g} \times 100\% = 25.0\%.$$

Balancing Chemical Equations

Earlier you learned how equations are used to describe chemical reactions since they show what substances are consumed and which ones are produced. But for an equation to be totally correct, it must be balanced. A **balanced equation** is one that has the same number of atoms of each element on both sides. A balanced equation obeys the Law of Conservation of Mass since an identical mass of each element is present at the beginning (in the reactants) and at the end (in the products). The equation describing the decomposition of calcium carbonate (limestone) is already balanced when the formulas for the reactant and products are written out:

reactant   products

$$CaCO_3 \xrightarrow{\Delta} CaO + CO_2$$

1Ca ... balances 1Ca
1C .... balances 1C
3O .... balances .... 3O (1 in CaO,
                                 2 in CO_2)

If an identical number of each kind of atom does not appear on both sides after the formulas are written out, then the equation is not balanced, and though it shows what chemical changes take place, it violates the Law of Conservation of Mass.

Equations are balanced by placing whole-number **coefficients** before the formulas and symbols of the reactants and products to selectively change the number of each species. A coefficient of 2 placed before a formula will double the number of atoms of each kind; $2CO_2$ represents two carbon atoms and (2×2) four oxygen atoms. It is incorrect to change the subscript numbers of a formula to increase the number of atoms. This changes the identity of the compound to something not really involved in the reaction and which may not even exist. For example, $2CO_2$ is correct, C_2O_4 is not. Consider the reaction of magnesium metal with hydrochloric acid (HCl), a reaction which forms magnesium chloride ($MgCl_2$) and hydrogen gas (H_2):

$$HCl + Mg \longrightarrow MgCl_2 + H_2$$

A quick examination shows that the equation is not balanced. There are two hydrogens and two chlorines on the product side, but only one of each on the reactant side. The equation can be balanced by placing a coefficient of 2 before the formula for hydrochloric acid:

coefficient $2\,HCl + Mg \longrightarrow MgCl_2 + H_2$

2H .... balances 2H
2Cl .... balances 2Cl
1Mg .... balances 1Mg

To emphasize a point made earlier, it would be wrong to balance the equation this way:

cannot change $H_2Cl_2 + Mg \rightarrow MgCl_2 + H_2$ (WRONG)
subscripts

The species H_2Cl_2 is not hydrochloric acid, and the equation does not describe the reaction it is supposed to describe.

Figure 4.3 illustrates the balancing of the equation for the formation of water (H_2O) from hydrogen (H_2) and oxygen (O_2).

Often, equation balancing is done by trial and error, changing one coefficient at a time until each element is brought into balance. Equations can be balanced correctly and quickly if you follow the method outlined below:

1. Make sure the correct symbols and formulas are used for each reactant and product.

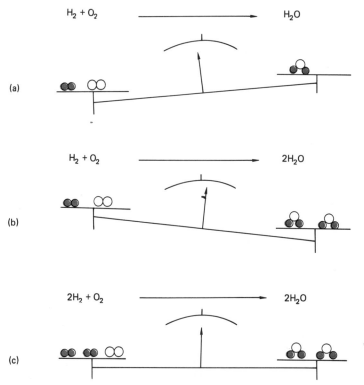

Fig. 4.3. A balanced equation has an identical mass of each element on both sides of the arrow. (a) One molecule of H_2 plus one molecule of O_2 weighs more than one molecule of H_2O. The equation is not balanced. (b) Doubling the number of H_2O molecules gives the same number of oxygen atoms and the same mass of oxygen on both sides. But, the equation is not balanced in terms of hydrogen. (c) Doubling the number of H_2 molecules will make the number of hydrogen atoms and the mass of hydrogen equal on both sides. The equation is now balanced.

2. Count the number of atoms of each type on both sides, and determine which are not balanced.

3. Balance one element at a time, placing whole number coefficients before the substances containing that element.

   a. It is sometimes easier if you balance oxygen and hydrogen last.
   b. Sometimes, as you bring one element into balance another will be shifted out of balance in the process. You will continually need to watch for this.
   c. Once balanced, the set of coefficients used in the equation should be the smallest whole numbers that give the correct balance.

4. Check to see if the equation is balanced correctly when you think you are finished. Remember, every equation that describes an actual chemical change can be balanced, though some may take longer than others.

Here are some examples showing how equations are balanced.

Example 4.7

Balance the following equations:

a. $$CH_4 + O_2 \rightarrow CO_2 + H_2O$$

Carbon is balanced but oxygen and hydrogen are not. Since hydrogen is contained in a single compound on each side, it might be easier to balance it first by placing a 2 before the formula of water:

$$CH_4 + O_2 \rightarrow CO_2 + 2H_2O$$

Oxygen can now be balanced by placing a 2 before O_2, the symbol for oxygen. A quick check will show that the equation is balanced:

$$CH_4 + 2O_2 \longrightarrow CO_2 + 2H_2O \quad \text{(balanced)}$$

$$
\begin{aligned}
1C &\ldots\ldots \text{balances} \ldots\ldots 1C \\
4H &\ldots\ldots \text{balances} \ldots\ldots 4H \\
4O &\ldots\ldots \text{balances} \ldots\ldots 4O
\end{aligned}
$$

b. $$Fe + O_2 \rightarrow Fe_2O_3$$

First, balance iron:

$$2Fe + O_2 \rightarrow Fe_2O_3$$

To balance oxygen, a coefficient of 2 is placed before Fe_2O_3 and, at the same time, a 3 before O_2:

$$2Fe + 3O_2 \rightarrow 2Fe_2O_3$$

Now oxygen is balanced, but iron is no longer balanced. This can be corrected by changing the coefficient of Fe from 2 to 4. The equation is now balanced. Check each element to verify this:

$$4Fe + 3O_2 \rightarrow 2Fe_2O_3 \qquad \text{(balanced)}$$

Check Test Number 4

Balance the following equations:
a. $HgO \rightarrow Hg + O_2$
b. $H_2 + N_2 \rightarrow NH_3$
c. $KClO_3 \rightarrow KCl + O_2$

Answers:
a. $2HgO \rightarrow 2Hg + O_2$
b. $3H_2 + N_2 \rightarrow 2NH_3$
c. $2KClO_3 \rightarrow 2KCl + 3O_2$

The Meaning of the Balanced Equation

A balanced equation can be interpreted in several ways. Four ways of reading the equation that describes the formation of ammonia from hydrogen and nitrogen are shown below:

$$3H_2 \quad + \quad N_2 \quad \rightarrow \quad 2NH_3$$

$3H_2$	N_2	$2NH_3$
3 molecules	1 molecule	2 molecules
3 moles	1 mole	2 moles
$3 \times (6.02 \times 10^{23})$ molecules	$1 \times (6.02 \times 10^{23})$ molecules	$2 \times (6.02 \times 10^{23})$ molecules
6 grams	28 grams	34 grams

The terms listed beneath each species represent different interpretations for the symbols and formulas used in the equation, and note how important the coefficients are in each one. In terms of moles and grams, the balanced equation tells you that *three* moles of hydrogen (6.0 g of H_2) will react completely with *one* mole of nitrogen (28.0 g of N_2) to produce *two* moles of ammonia (34.0 g of NH_3). The balanced equation is important not only because it can show what species are involved in a reaction, but because it can show how much of each reactant and product is involved. We can sum up the meaning of the balanced equation this way: *A balanced equation is a mole statement, and it shows the mole ratios of all reactants and products.* Let us show what this statement means by using the balanced equation that describes the reaction of hydrogen with oxygen to produce water:

$$2H_2 + O_2 \rightarrow 2H_2O$$

2 moles   1 mole   2 moles

The mole ratios shown by the balanced equation are as follows.

1. The ratio of H_2 and O_2:
   The balanced equation shows that two moles of H_2 react with one mole of O_2, and this can be written as a ratio:

$$\frac{2 \text{ moles } H_2}{1 \text{ mole } O_2} \quad \text{or} \quad \frac{1 \text{ mole } O_2}{2 \text{ moles } H_2}$$

2. The ratio of H_2 and H_2O:

$$\frac{2 \text{ moles } H_2}{2 \text{ moles } H_2O} \quad \text{or} \quad \frac{2 \text{ moles } H_2O}{2 \text{ moles } H_2}$$

3. The ratio of O_2 and H_2O:

$$\frac{1 \text{ mole } O_2}{2 \text{ moles } H_2O} \quad \text{or} \quad \frac{2 \text{ moles } H_2O}{1 \text{ mole } O_2}$$

Each mole ratio states the number of moles of one substance that are needed to produce or consume a certain number of moles of another. We cannot prepare two moles of water unless we use up one mole of oxygen (ratio 3) and two moles of H_2 (ratio 2). Mole ratios are very useful in calculations that involve balanced chemical equations. Remember though, that correct ratios can only be written if the equation is balanced.

Calculations Using Balanced Equations

Since balanced equations tell us the number of moles of each substance involved in the reaction, it is not a difficult task to calculate the amounts of each substance that are consumed or formed. Consider the reaction between sodium and chlorine which forms sodium chloride:

$$Na + Cl_2 \rightarrow NaCl$$

Suppose we wanted to know how many moles of NaCl will form if 2.50 moles of Cl_2 are reacted with sodium. Let us solve this problem in a step-by-step sequence that can be followed for all problems of this type.

Step 1.   Balance the equation. The only way you can get the correct mole ratios is from the balanced equation:

$$2Na + Cl_2 \rightarrow 2NaCl$$

Step 2.   Determine what is being sought, and what is given. Then using the balanced equation, write down the mole ratio involving these substances with the substance being sought in the numerator, and the substance given in the denominator. This mole ratio acts as a conversion factor.

Sought: moles of NaCl
Given: 2.50 moles of Cl_2
Mole ratio: $\dfrac{\text{moles of sought}}{\text{moles of given}}$ or $\dfrac{\text{2 moles NaCl}}{\text{1 mole } Cl_2}$

Step 3.   Use the mole ratio of the sought and given substances, and the given number of moles of chlorine to calculate the number of moles of NaCl that can be produced. The general equation for this calculation is

moles of sought substance =

$$\text{moles of given substance} \times \underbrace{\frac{\text{moles of sought}}{\text{moles of given}}}_{\substack{\text{obtained from the} \\ \text{balanced equation}}}$$

Placing the quantities from step 2 in this equation will provide the number of moles of NaCl:

$$\text{moles of NaCl} = 2.50 \, \cancel{\text{moles } Cl_2} \times \frac{2 \text{ moles NaCl}}{1 \, \cancel{\text{mole } Cl_2}} = 5.00 \text{ moles}$$

So we see that 5 moles of NaCl can be prepared if we react 2.50 moles of Cl_2 with sodium. An additional calculation step might be involved if instead of moles, the mass of NaCl is desired. All you would need to do is to convert the number of moles of NaCl to grams using the GFW of the compound:

$$1 \text{ GFW NaCl} = 1 \text{ mole NaCl} = 58.5 \text{ g}$$

$$\text{mass of NaCl} = \frac{\text{g NaCl}}{1 \text{ mole}} \times \text{moles NaCl}$$

$$= \frac{58.5 \text{ g}}{1 \text{ mole}} \times 5.00 \text{ moles}$$

$$\text{mass of NaCl} = 292 \text{ g}$$

The following examples will demonstrate the use of balanced equations in chemical calculations.

Example 4.8

Magnesium hydroxide, $Mg(OH)_2$, is used in several antacid preparations to remove excess hydrochloric acid from the stomach. How many moles of hydrochloric acid can be consumed by 0.10 mole (5.8 g) of $Mg(OH)_2$? The equation is

$$Mg(OH)_2 + HCl \rightarrow MgCl_2 + H_2O \text{ (unbalanced)}$$

1. Balance the equation:

$$Mg(OH)_2 + 2HCl \rightarrow MgCl_2 + 2H_2O$$

2. Sought: moles of HCl
    Given: 0.10 mole $Mg(OH)_2$
    The mole ratio of sought over given from the balanced equation is:

$$\frac{2 \text{ moles HCl}}{1 \text{ mole Mg(OH)}_2} \quad \begin{array}{l} \longleftarrow \text{ sought} \\ \longleftarrow \text{ given} \end{array}$$

3. Calculate the moles of sought substance:

$$\text{moles HCl} = \text{moles Mg(OH)}_2 \times \frac{2 \text{ moles HCl}}{1 \text{ mole Mg(OH)}_2}$$

$$= 0.10 \text{ mole} \times \frac{2 \text{ moles}}{1 \text{ mole}} = 0.20 \text{ mole HCl}$$

So, 5.8 g of $Mg(OH)_2$ can remove 0.20 mole of HCl from the stomach. In terms of mass, 0.20 mole of HCl is

$$0.20 \text{ mole} \times \frac{36.5 \text{ g}}{\text{mole}} = 7.3 \text{ grams of HCl}$$

Example 4.9

Potassium reacts violently when placed in water, producing potassium hydroxide (KOH) and hydrogen gas (H_2).

$$K + H_2O \rightarrow KOH + H_2 \text{ (unbalanced)}$$

How many grams of hydrogen, H_2, can be produced if 11 g (0.28 mole) of potassium reacts with water?

1. Balance the equation

$$2K + 2H_2O \rightarrow 2KOH + H_2$$

2. Sought: grams of H_2
   Given: 11 g of K, which is 0.28 mole of K
   Mole ratio of sought over known from the balanced equation:

$$\frac{1 \text{ mole } H_2}{2 \text{ moles K}}$$

3. Calculation of the number of moles of H_2:

$$\text{moles of } H_2 = 0.28 \text{ mole K} \times \frac{1 \text{ mole } H_2}{2 \text{ mole K}} = 0.14 \text{ mole}$$

Converting moles of H_2 to grams:

$$1 \text{ mole } H_2 = 2.0 \text{ g } H_2$$

$$\text{mass } H_2 = 0.14 \text{ mole} \times \frac{2.0 \text{ g}}{\text{mole}} = 0.28 \text{ g}$$

So, 0.28 g of H_2 can be produced when 11 g of potassium reacts with water.

Check Test Number 5

1. How many moles of barium oxide (BaO) will react with three moles of hydrochloric acid?

$$BaO + HCl \rightarrow BaCl_2 + H_2O \text{ (unbalanced)}$$

2. How many grams of oxygen are needed to produce 102 g (1 mole) of aluminum oxide (Al_2O_3)?

$$Al + O_2 \rightarrow Al_2O_3 \text{ (unbalanced)}$$

Answers:

1. $BaO + 2HCl \rightarrow BaCl_2 + H_2O$

$$3 \text{ moles HCl} \times \frac{1 \text{ mole BaO}}{2 \text{ moles HCl}} = 1.5 \text{ mole BaO}$$

2. $4Al + 3O_2 \rightarrow 2Al_2O_3$

$$\text{moles of } O_2 = 1 \text{ mole } Al_2O_3 \times \frac{3 \text{ mole } O_2}{2 \text{ moles } Al_2O_3} = 1.5 \text{ mole}$$

$$\text{mass of } O_2 = 1.5 \text{ mole} \times \frac{32 \text{ g}}{\text{mole}} = 48 \text{ g}$$

Terms

Several important terms appear in Chapter 4. You should understand the meaning of each:

atomic weight	Avogadro's number
atomic mass unit	mole
formula weight	percent composition of a compound
molecular weight	balanced equation
gram atomic weight	mole ratios in a balanced equation
gram formula weight	coefficient

Questions

Answers to the starred questions appear at the end of the text.

1. What is the importance of the carbon-12 atom in the atomic weight scale?

*2. Why are atomic weights considered relative weights?

3. How do the terms formula weight and gram formula weight differ?

*4. What is the relationship between Avogadro's number and the mole?

5. Why does a balanced equation obey the law of conservation of mass?

6. Complete the following:
   *a. One mole of chlorine atoms (Cl) = _____ atoms

   *b. One mole of chlorine molecules (Cl_2) = _____ molecules

   *c. One mole of chlorine atoms = _____ grams

   d. One mole of chlorine molecules = _____ grams

   e. One mole of chlorine molecules = _____ atoms

*7. Interpret the following equation in terms of molecules, moles, and grams of each species:

$$H_2 + I_2 \rightarrow 2HI$$

*8. Which term, formula weight or molecular weight, can be applied to a greater number of species? Why?

9. Balance the following equations:
   a. $Na + H_2O \rightarrow NaOH + H_2$
   b. $H_2 + P_4 \rightarrow PH_3$
   c. $Ag_2O \rightarrow Ag + O_2$
   d. $FeO + Al \rightarrow Fe + Al_2O_3$
   e. $P_4 + I_2 \rightarrow PI_3$
   f. $C_2H_6 + O_2 \rightarrow CO_2 + H_2O$
   g. $Cl_2 + KI \rightarrow I_2 + KCl$
   h. $NaHCO_3 + HCl \rightarrow NaCl + H_2O + CO_2$
   i. $Al + CuSO_4 \rightarrow Cu + Al_2(SO_4)_3$
   j. $MgO + HCl \rightarrow MgCl_2 + H_2O$
   k. $CuS + O_2 \rightarrow CuO + SO_2$
   l. $Mg + O_2 \rightarrow MgO$
   m. $FeCl_3 + Cu \rightarrow CuCl_2 + Fe$
   n. $Ca + H_2O \rightarrow Ca(OH)_2 + H_2$
   o. $TiCl_4 + H_2O \rightarrow TiO_2 + HCl$
   p. $HF + SiO_2 \rightarrow SiF_4 + H_2O$
   q. $CaC_2 + H_2O \rightarrow C_2H_2 + Ca(OH)_2$
   r. $Al + H_2SO_4 \rightarrow Al_2(SO_4)_3 + H_2$
   s. $BaCl_2 + Na_2SO_4 \rightarrow BaSO_4 + NaCl$
   t. $S + O_2 \rightarrow SO_3$

10. Calculate the number of moles of substance in each of the following (use approximate atomic weights).
    *a. 20.0 g of CH_4
    b. 100 g of H_2SO_4
    *c. 1×10^{23} molecules of H_2O
    d. 35.5 g of Cl_2
    *e. 3500 g of $C_{12}H_{22}O_{11}$
    f. 6.02×10^{25} atoms of iron

11. Calculate the number of grams in each of the following quantities.
    *a. 0.50 mole of HCl
    b. 2.10 moles of Au
    *c. 6.02×10^{20} molecules of UF_6
    d. 2.0 moles of $Al_2(SO_4)_3$
    e. 1.00×10^{25} atoms of H
    *f. 10 GFW's of BaO

*12. Calculate the percent composition of ethyl alcohol, C_2H_6O.

13. Which compound contains the highest percentage of gold?
   a. $AuCl_3$
   b. Au_2O_3
   c. $AuPO_4$

14. Determine the mole ratios of all species in each of the following equations:
   *a. $2B + 3F_2 \rightarrow 2BF_3$
   *b. $2Cu + O_2 \rightarrow 2CuO$
   c. $2Na + 2HCl \rightarrow 2NaCl + H_2$

15. Write balanced equations for each of the following.
   *a. Methane and chlorine react to form chloroform ($CHCl_3$) and HCl.
   *b. Propane (C_3H_8) burns in oxygen to form carbon dioxide and water.
   c. Silver oxide (Ag_2O) decomposes when heated to form silver and oxygen.
   d. Sulfuric acid (H_2SO_4) reacts with aluminum hydroxide [$Al(OH)_3$] to form aluminum sulfate [$Al_2(SO_4)_3$] and water.

*16. How many moles of hydrochloric acid can form if 2.00 moles of hydrogen react with chlorine?

$$H_2 + Cl_2 \rightarrow HCl$$

17. How many moles of oxygen are consumed if 0.50 mole of zinc is converted to zinc oxide? How many moles and how many grams of zinc oxide form?

$$Zn + O_2 \rightarrow ZnO$$

*18. How many grams of sodium hydroxide would react with 73.0 g of hydrochloric acid?

$$NaOH + HCl \rightarrow NaCl + H_2O$$

*19. How many carbon atoms are in 18.0 g of glucose, $C_6H_{12}O_6$? The formula weight of glucose is 180 amu.

*20. In living green plants, the process of photosynthesis converts carbon dioxide and water to glucose. How many grams of carbon dioxide are needed to produce 1 g of glucose?

$$CO_2 + H_2O \rightarrow C_6H_{12}O_6 + O_2$$

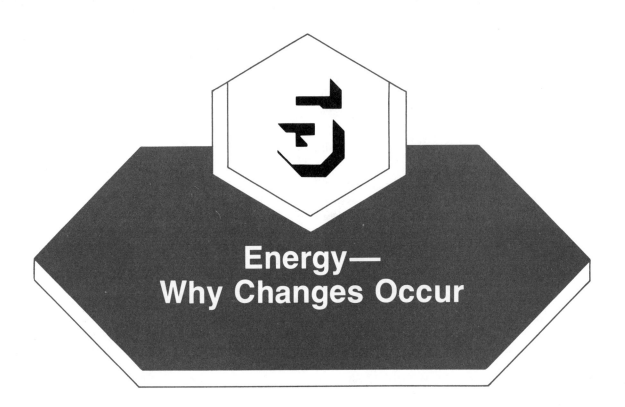

Energy—Why Changes Occur

By this time you have learned several facts about matter and the kinds of changes it undergoes. An important part of any physical or chemical change is that it also involves energy. In fact, if energy changes did not take place, it would be impossible for physical or chemical changes to occur or for life to exist. Chemistry must therefore be concerned with both matter *and* energy. In this chapter we will be primarily concerned with energy in the form of heat that accompanies chemical and physical changes. By doing so, it is hoped that you will sense the importance of energy in all the changes that take place around and in you every day.

Objectives

By the time you finish this chapter, you should be able to do the following.

1. Define energy and work and show how they are related.
2. Explain the Law of Conservation of Energy, and how it applies to energy changing form.
3. Describe the different forms of energy including kinetic and potential energy.

4. Describe the electromagnetic spectrum and characterize its different parts in terms of wavelength and ability to damage tissue.

5. Explain how a chemical compound can be regarded as a potential energy source.

6. Describe the relationship between the average kinetic energy of molecules and temperature.

7. Describe the units used to measure heat energy and the operation of a calorimeter.

8. Explain what is meant by exothermic and endothermic processes.

9. Explain, in terms of energy, what all processes must have in common to occur.

10. Describe how the fuel or energy value of a food is determined and give the average fuel values for fats, carbohydrates, and proteins.

Energy

What is energy? Unlike matter it does not have properties that we can easily sense like color or odor, but it can be defined in terms of what its presence allows to be done. **Energy** is possessed by a body if it has the ability to change the position or arrangement of a sample of matter, that is, if it has the ability to do **work.** When work is done, energy is transferred from the object doing the work to the object being moved, whether on the very small scale of molecules or the larger scale of our world. Energy is required for a nurse to lift a patient onto a bed, since work is being done as the patient is moved from one place to another. During the process, energy is transferred from the nurse to the patient. Stirring a solution also requires energy, since the liquid is set into motion as it receives energy from the person doing the stirring. In fact, if the solution is stirred hard enough and long enough, the energy put into it can cause it to become very hot, and even boil.

As another example, suppose you allow an ice cube to melt in your hand. As the ice begins to melt, your hand becomes cold. Both effects are due to a single occurrence, the flow of energy, in the form of heat, out of your hand and into the ice cube. The water molecules are moved from their fixed positions in the ice crystals and become free to move about in the liquid. Energy is absorbed and molecules are moved. The removal of heat energy from your hand, which you sense as a cooling sensation, causes changes in the arrangement of the molecules in the tissue, and if the cooling were allowed to continue, damage not unlike that from frostbite would occur—all because energy is being taken away.

Scientists have been unable to find a single physical, chemical, or biological process that allows energy to be acquired by one object except at the expense of that possessed by another. Furthermore, though energy can be transferred from one place to another, and can change form, it can be neither created nor destroyed in the process. This last statement is an important observation known as the **Law of Conservation of Energy.** In any process then, from the melting of wax to a nuclear explosion, it is possible to account for *all* the energy involved in it. Considering both the law of conservation of energy and the law of conservation of mass, it is apparent that in ordinary chemical and physical changes neither mass nor energy is created or destroyed, though either may change form. Also, it is a fundamental rule of nature that we cannot use heat energy with 100% efficiency. Only a fraction of the energy consumed ever goes into doing work. Our best heat engines operate at about a 40% efficiency, meaning that 60% of the energy is transferred to the environment while only 40% is used to do work (see Figure 5.1).

When we speak of energy we also need to speak of work, since as energy is used, the two are linked together. We can recognize that work is being done if we see something move, like the rustling leaves of a tree, a fast train, or a soaring baseball. But work is also done in our bodies, much of it on the molecular scale, and it consumes energy just as any other kind of work does. The energy for this *biological work* is supplied by the foods we consume. As food is digested, it is broken down into smaller molecules, transported across the intestinal membrane and eventually sent to other parts of the body. Often the movement of digested material to cells and tissues must be done against a

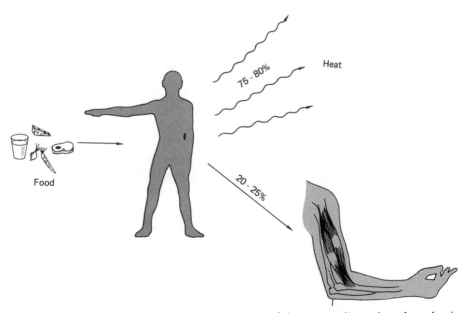

Fig. 5.1. The human body uses about 20–25% of the energy it receives from foods to do biological work. The rest is lost in the form of heat.

resistance to the flow, and this requires energy since *transport work* must be done to move against the resistance. This movement is known as active transport. In the cells, large molecules, such as proteins that are necessary for the growth of the cell, are continuously being synthesized from smaller molecules. This *biosynthesis work* requires energy. When a muscle contracts, work is being done as the muscle tissues draw together. As you move your arms, legs, or eyes, *mechanical work* is being done, which also requires energy.

The Different Forms of Energy

Though energy cannot be created or destroyed, it can change form. As a lightbulb burns, electrical energy is converted into heat energy and light energy. Since we use energy in all its forms, it would be useful to briefly describe some of them.

1. *Heat energy* is obtained from the burning of fuels and from the sun. Most energy is ultimately lost to the environment as heat. Heat is related to the motion of atoms and molecules in matter.
2. *Mechanical energy* is that of machinery in motion. A moving arm, leg, or car possesses mechanical energy. A moving leg can give energy to a football by sailing it through the air with a good kick.
3. *Electrical energy* is generated by friction (leather shoes on a wool rug), the motion of electrical conducting materials in a magnetic field, and in certain chemical reactions (batteries).
4. *Chemical energy* is possessed by a compound because of the kinds and arrangement of atoms in its molecules. Upon reaction, some of this energy can be released if the products of the reaction possess lesser amounts of energy. Some of the chemical energy contained in gasoline is released as it is burned and converted to carbon dioxide and water. Chemical energy is discussed in more detail later on in this chapter.
5. *Light energy* or *electromagnetic energy* can be thought of as energy traveling as waves through space at a speed of 3×10^{10} cm/sec, commonly called the speed of light. A wave is represented in Figure 5.2, and the wavelength λ (lambda) is the distance between two adjacent crests. Light that has a short wavelength is more energetic and damaging to tissue than that with a longer wavelength. Wavelengths are often stated in centimeters, angstroms (1 Å = 10^{-10} m), or nanometers (10^{-9} m). The electromagnetic spectrum is shown in Figure 5.3. It may be surprising to realize that the many different kinds of radiation you have heard about are all part of the same kind of energy. Whether called waves (microwaves) or rays (cosmic rays), the only difference between the different categories of light is the

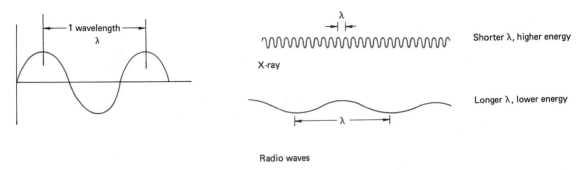

Fig. 5.2. One wavelength, λ, is the distance between adjacent crests of a light wave. The energy of electromagnetic radiation increases as the wavelength becomes shorter. X rays have very short wavelengths, are very energetic, and and can cause cell damage. Radio waves have longer wavelengths, are of low energy, and to the best of our knowledge, do not cause cell damage.

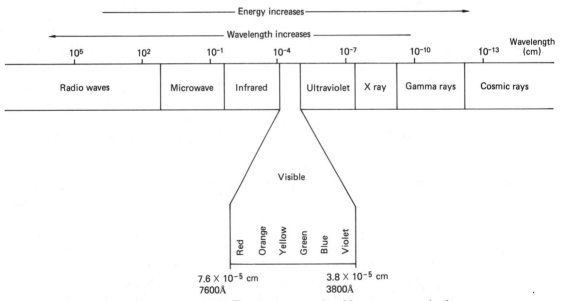

Fig. 5.3 The electromagnetic spectrum. The human eye is able to sense only the relatively narrow visible region of the spectrum.

wavelength of the radiation. Only a small portion of the spectrum is visible to the naked eye, the *visible* region, and it includes the familiar colors of the rainbow. *Infrared* (ir) radiation can also be sensed by our bodies as heat. The warmth we feel from the sun is due to its infrared radiation being absorbed by our skin. Looking at the other region of the spectrum, *radiowaves* have very long wavelengths and are of low energy. To the best of our knowledge, they do not cause damage to tissue. *Microwave* radiation of the proper wavelength is able to increase the motion of water molecules in food, thus heating them evenly throughout as they are cooked. At other wavelengths, microwave radiation is used in radar and communications. *Ultraviolet* light (uv) has shorter wavelengths than visible light, and being more energetic can cause molecules in the skin to break up or change form. This is what ultimately causes a suntan or a sunburn. Bacteria can be destroyed by ultraviolet light, a property that prompts its use in sterilization units. Ultraviolet radiation also provides an important benefit since it is involved in the production of vitamin D in the skin. *X rays* have short wavelengths and therefore are very energetic. Though x rays can cause severe tissue damage, very brief exposures for diagnostic purposes are considered worth the risk, since a great deal of information can be obtained about internal organs and bones without surgery. The effects of x rays on tissue are cumulative over time, so exposure must be limited and carefully controlled. Cancer cells are more sensitive to x rays than normal cells, so therapeutic exposure is used to treat certain forms of cancer. Cancer cells are destroyed preferentially. Astronauts must be protected from the very

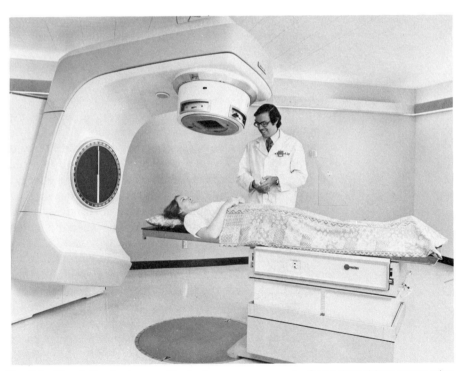

Fig. 5.4 X radiation is also used to destroy cancerous cells. Special machines are used to concentrate a constantly moving x-ray beam on the affected area while keeping the exposure of healthy tissue to a minimum. (Varian Associates, Palo Alto, California)

energetic *gamma* and *cosmic* rays from the sun, or severe cell damage could be fatal after only a brief exposure.

The earth's atmosphere filters out most of the short-wavelength, high-energy radiation from the sun, protecting life on the surface from its destructive effects (Figure 5.5).

Kinetic Energy and Potential Energy

Energy in its various forms can be classified as being either kinetic or potential energy. An object possesses **kinetic energy** if it is in motion. A javelin hurling through the air and a moving car both have kinetic energy by virtue of their motion, and both can do work if they strike another object. A javelin rips up the ground when it lands, doing work (Figure 5.6, p. 90). A moving car would be doing work if it wandered off the road and demolished a mailbox. The water molecules in a beaker of water also possess kinetic energy. Though we cannot see them, the molecules are in constant, rapid motion and thus possess kinetic energy.

The amount of kinetic energy possessed by an object can be determined if its mass and velocity are known. The equation that relates these terms is

$$\text{kinetic energy} = \tfrac{1}{2}(\text{mass}) \times (\text{velocity})^2$$
$$\text{K.E.} = \tfrac{1}{2}\,m\,v^2$$

The equation shows that the kinetic energy of an object is directly related to its mass (m) and to the square of its velocity (v^2, which is equal to $v \times v$). If the mass of a moving object is doubled, its kinetic energy is doubled, but if the velocity is doubled, its kinetic energy is increased fourfold. A car traveling at ten miles per hour has four times as much kinetic energy as it would have at five miles per hour, and it could do four times as much work if it ran into a parked car. Water molecules have very small masses, yet boiling water possesses a large amount of kinetic energy because the molecules move about at very high velocities.

It is not difficult to see how the kinetic energy of a moving car could be determined. Its mass and velocity can be measured rather easily. But how is it possible to measure the kinetic energy of the molecules in a glass of water? As it turns out, the **average kinetic energy** of a collection of molecules can be measured by taking the temperature of the sample. The heat energy contained in a beaker of water is responsible for the motion of the molecules. As more heat flows into the water, the average velocity of the molecules increases (and so does the average kinetic energy) and the temperature goes up. This is an important point to remember: the temperature of a sample of matter is directly related to the average kinetic energy of the molecules that make it up.

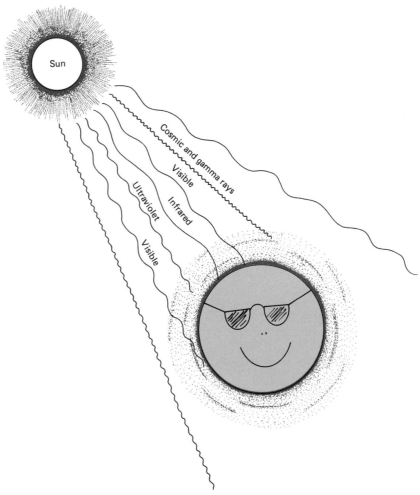

Fig. 5.5. The earth's atmosphere filters out most of the energetic, short-wavelength radiation from the sun, protecting life on the surface from its harmful effects. The atmosphere acts like sunglasses for the earth.

Before we go any further we should examine what is meant by "average kinetic energy." Suppose we had a collection of 100 water molecules. If we could measure the velocity of each one at some instant, we would find some that would be moving extremely fast, and some moving extremely slow, but most would have velocities somewhere between these extremes. The average of these individual velocities can be used to calculate the average kinetic energy of the sample (Figure 5.7). There is enough energy in boiling water at 100°C due to the motion of the molecules to cook an egg, destroy bacteria, or scald tissue. In each case, work is being done.

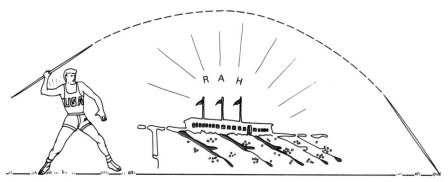

Fig. 5.6. A javelin sailing through the air has kinetic energy due to its motion. Its kinetic energy allows it to do work as it rips up the ground when it lands.

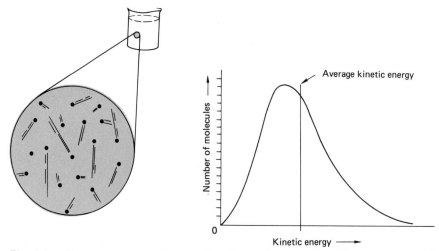

Fig. 5.7. The temperature of a sample of water is proportional to the average kinetic energy of the molecules in it. Some will be moving at very high velocities and have very large kinetic energies. Others will have very low velocities and very small kinetic energies. A plot showing the number of molecules at a given instant with a certain kinetic energy, shows most having kinetic energies between the extremes, with the average being to the right of the peak of the curve.

Potential energy is possessed by an object by virtue of its position, condition, or composition. It is stored energy that can be made to do work as it is converted to kinetic energy. A car stopped at the top of a hill has potential energy by virtue of its position. As it rolls down the hill to a point of lower potential energy due to the effect of gravity, its potential energy is converted to kinetic energy, the energy of motion (Figure 5.8). A tightly wound clock spring has potential energy because of its wound condition. It is converted to kinetic

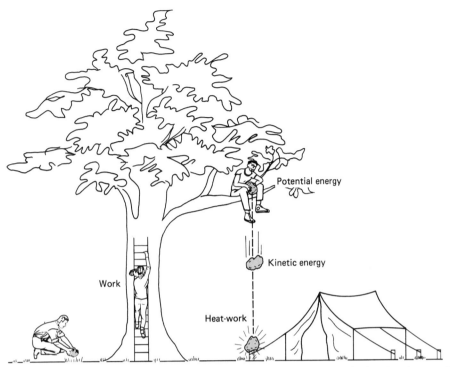

Fig. 5.8. Sometimes a little ingenuity is needed to drive a tent stake into hard ground. The potential energy of a rock is increased as it is carried up to a convenient branch. The energy gained by the rock comes from the work done to raise it to the branch. When the rock is dropped, potential energy is converted to kinetic energy, the energy of motion. When it strikes the stake, the kinetic energy is converted into heat and work as the stake moves deeper into the ground.

energy as it unwinds and moves the clock mechanism. Sucrose (table sugar) possesses potential energy by virtue of its composition. When it is burned in the body (metabolized), part of its stored energy is released and is used to do biological work and generate heat, thus maintaining life.

Check Test Number 1

1. Indicate which of the following represents kinetic energy and which represents potential energy.
   a. blood rushing through an artery
   b. a pound of coal
   c. a running electric motor
   d. boiling water
   e. water held behind a hydroelectric dam

2. Arrange the following kinds of light in sequence from long wavelength to short wavelength.
   a. red light
   b. infrared
   c. microwaves
   d. radiowaves
   e. blue light
   f. x rays

Answers:

1. (a), (c), and (d) are kinetic; (b) and (e) are potential.
2. Radiowaves have the longest wavelength; then microwave, infrared, red light, blue light, and x rays, the shortest of them all.

The Measurement of Heat Energy

You know that heat energy is required to warm a cup of cold water to make hot coffee. But, do you know how much heat is required? The heat produced by a single match would not do, but that from a box of matches probably could. In chemistry, medicine, and the dietary sciences, the measurement of heat is quite important since it is part of all chemical and physical changes both in the laboratory and in the body.

The unit of heat energy is the **calorie**, the amount of heat needed to raise the temperature of one gram of water one degree Celsius. It is sometimes called the small calorie, and is abbreviated cal. The calorie is a small unit of energy and is often replaced by the *kilocalorie*, a more convenient unit, which is 1000 times larger than the calorie. It is sometimes called the large calorie and is abbreviated kcal or Cal (capital C). One kilocalorie will raise the temperature of one kilogram of water one degree Celsius.

The amount of heat needed to raise the temperature of one gram of any substance one degree Celsius is called the **specific heat** (SH) of that substance. From the definition of the calorie, the specific heat of water would be 1.00 cal/g °C. The unit of specific heat can be read as "calories per gram degree," and it clearly describes the function of specific heat, being the number of calories involved when one gram of a substance changes temperature by one degree (see Figure 5.9). Water has a relatively high specific heat value, which means it can absorb or release quite a bit of heat without undergoing wide temperature changes. This makes water a good substance to use in hot packs since it can store and release a large amount of heat without changing its temperature very much. Since the body is mostly water, it takes relatively large amounts of heat to change our body temperature from the normal 37°C. The specific heats of several substances are listed in Table 5.1. Metals characteristically have small specific heats.

TABLE 5.1 Specific Heats of Several Substances	
Substance	Specific heat $\dfrac{cal}{g\,°C}$
Water	1.00
Alcohol	0.58
Mineral oil	0.50
Aluminum	0.21
Iron	0.11
Silver	0.06
Gold	0.03

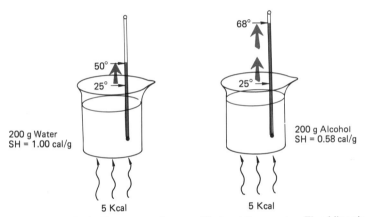

Fig. 5.9. Alcohol has a smaller specific heat than water. Five kilocalories of heat energy can raise the temperature of 200 g of alcohol 43°C, from 25° to 68°C, but would raise the temperature of 200 g of water by only 25°C, from 25°C to 50°C.

It is not difficult to calculate the amount of heat needed to raise the temperature of a quantity of matter if its specific heat is known. The amount of heat energy Q will depend on the mass of substance, its specific heat, and the size of the temperature change in degrees Celsius (ΔT). The equation that relates these terms for some substance A is

$$\text{quantity of heat} = (\text{mass of } A) \times (\text{SH of } A) \times (\Delta T \text{ of } A)$$
$$Q = m_A \times \text{SH}_A \times \Delta T_A$$

The units of the three terms cancel to leave calorie, the needed unit for Q. The following example will show how the equation can be used.

Example 5.1

How many calories of heat energy are needed to raise the temperature of 700 g of water from 25.0°C to 40.0°C? The specific heat of water is found in Table 5.1, and the mass of water and the temperature change are given in the question. We need to solve for Q.

$$\Delta T = 40.0° - 25.0° = 15.0°C$$

$$Q = M_{water} \times SH_{water} \times \Delta T_{water}$$

$$= 700 \, g \times 1.00 \frac{cal}{g°C} \times 15.0°C$$

$$= 10,500 \text{ cal or } 10.5 \text{ kcal}$$

If the question in Example 5.1 would have been asked in terms of the amount of heat *released* if 700 g of water cooled from 40.0° to 25.0°C, the number of calories would have been the same. If 10.5 kcal of heat energy was needed to increase the temperature 15.0°C, 10.5 kcal would be evolved as the temperature decreased by the same amount.

Heat of Fusion and Vaporization

When ice melts to form liquid water, heat energy is absorbed. During the melting process, the energy that is absorbed is used to free water molecules from the crystalline ice structure and so does *not* cause an increase in the kinetic energy of the molecules. This means that the temperature does *not* change, and it remains constant at 0°C (see Figure 5.10). The amount of heat needed to change one gram of a solid at its melting point to one gram of liquid at the same temperature is called its **heat of fusion** (melting). For water at 0°C, the heat of fusion is 80 cal/g. It takes 80 cal to change 1 g of ice at 0°C to one gram of water at 0°C:

1 g ice at 0°C + 80 cal → 1 g water at 0°C

(the heat of fusion)

For the reverse process, 80 cal must be removed from 1 g of water at 0°C to form 1 g of ice at the same temperature. Cold packs containing ice are often used to remove heat from inflammed tissue or to reduce a high fever, since each gram of ice can absorb a considerable amount of heat as it melts.

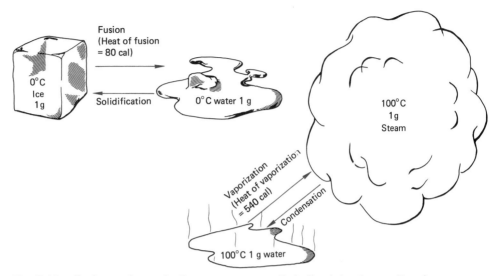

Fig. 5.10. Fusion and vaporization are processes that absorb heat as matter changes from one state to another. For water, the heat of fusion is 80 cal/g, and the heat of vaporization is 540 cal/g. The reverse processes, solidification and condensation, evolve the same amounts of heat. Note that temperature does not change in these processes.

Example 5.2

How much heat is absorbed by 150 g of ice in a cold pack as it melts at 0°C? The heat of fusion of ice is 80 cal/g:

$$Q = 150 \; g \times 80 \; \frac{cal}{g}$$

$$= 12{,}000 \; cal \; or \; 12 \; kcal$$

As a matter of comparison, if cold water was used to absorb the same amount of heat, it would require 2400 g (over 5 lb) of water if it were to undergo a 5°C temperature change. It certainly would be more convenient for a patient to manage 150 g of ice in a cold pack than over 2 kg of cold water!

The amount of heat absorbed when 1 g of a liquid at its boiling point is converted to 1 g of vapor at the same temperature is its **heat of vaporization.** For water, the heat of vaporization is 540 cal/g:

$$1 \; g \; water \; at \; 100°C + 540 \; cal \rightarrow 1 \; g \; steam \; at \; 100°C$$

(the heat of vaporization)

Since the heat of vaporization of water is so large, steam has a much greater energy content than liquid water at the same temperature. For this reason, surgical instruments are sterilized in an atmosphere of steam rather than in boiling water, since as each gram of steam condenses on the instrument's surface, 540 calories of energy are released to kill bacteria.

Check Test Number 2

1. How much heat is released if 10 g of steam condense to liquid water at 100°C?
2. How much heat is required to bring 500 g of water at 20°C to the boiling temperature, 100°C?

Answers:

1. 540 cal are released if 1 g of steam condenses, so if 10 g condense, the amount of heat will be

$$10 \, \cancel{g} \times 540 \, \frac{cal}{\cancel{g}} = 5400 \text{ cal}$$

2. $Q = (m)(SH)(\Delta T)$, $\Delta T = 100°C - 20°C = 80°C$, $SH = 1.00 \, \dfrac{cal}{g°C}$

$$Q = (500 \, \cancel{g})(1.00 \, \frac{cal}{\cancel{g}\cancel{°C}})(80°\cancel{C})$$

$$= 40,000 \text{ cal}$$

$$= 40 \text{ kcal}$$

Energy and Chemical Changes

Chemical reactions that produce heat energy as they take place are called **exothermic** reactions (see Figure 5.11):

reactants → products + heat   (exothermic)

Many reactions are exothermic, such as the burning of methane or coal, the burning of glucose (a sugar from carbohydrates) in the body, and the formation of HCl from its elements:*

$$CH_{4(g)} + 2O_{2(g)} \rightarrow CO_{2(g)} + 2H_2O_{(l)} + 213 \text{ kcal}$$
$$H_{2(g)} + Cl_{2(g)} \rightarrow 2HCl_{(g)} + 44.1 \text{ kcal}$$

---

*The subscripts (s), (l), (g) stand for solid, liquid, and gas indicating the physical state of each substance in the reaction. If a reactant or product was in a different state than that indicated, the amount of heat produced would be affected.

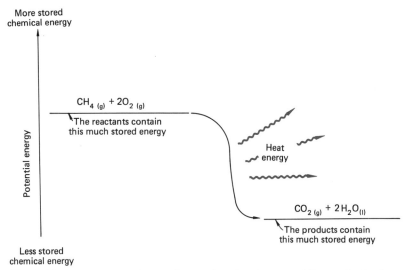

Fig. 5.11. An exothermic reaction evolves heat energy. The products of an exothermic reaction contain less stored chemical energy than do the reactants.

Chemical reactions that consume or absorb heat energy as they proceed are called **endothermic** reactions (see Figure 5.12):

$$\text{reactants} + \text{heat} \rightarrow \text{products} \quad \text{(endothermic)}$$

Endothermic reactions require a constant input of heat energy as they take

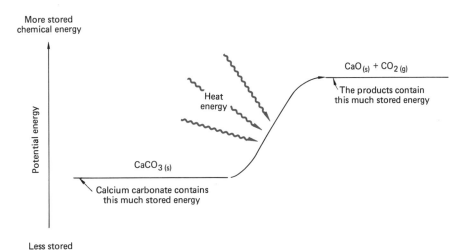

Fig. 5.12. An endothermic reaction absorbs heat energy. The products of an endothermic reaction contain more stored chemical energy than do the reactants.

place. If the energy supply is cut off, the reactions stop. The decomposition of calcium carbonate is an endothermic reaction:

$$CaCO_{3(s)} + 42.6 \text{ kcal} \rightarrow CaO_{(s)} + CO_{2(g)}$$

The amount of heat evolved or absorbed during a chemical change is measured using a **calorimeter.** The calorimeter shown in Figure 5.13 is used to measure the heat produced when a substance is burned in an atmosphere of pure oxygen. The reaction is carried out in a sealed chamber, which is submerged in a known mass of water. Since energy is conserved, the heat produced by the exothermic combustion reaction flows into the water, raising its temperature. Since the mass of water and its temperature change are known, the quantity of heat evolved by the reaction can be calculated:†

heat produced by the reaction = heat absorbed by the water

$$Q_{\text{reaction}} = Q_{\text{water}}$$

$$Q_{\text{reaction}} = m_{\text{water}} \times SH_{\text{water}} \times \Delta T_{\text{water}}$$

Example 5.3

One gram of glucose was burned in a calorimeter, raising the temperature of 500 g of water from 22.0°C to 29.6°C. How much heat is produced when one gram of glucose is converted to carbon dioxide and water under these conditions?

$$Q_{\text{reaction}} = Q_{\text{water}}$$

$$\Delta T = 29.6°C - 22.0°C = 7.6°C$$

$$Q_{\text{reaction}} = 500 \text{ g} \times 1.00 \frac{\text{cal}}{\text{g} \cdot °C} \times 7.6°C$$

$$= 3800 \text{ cal or } 3.8 \text{ kcal (exothermic)}$$

If the combustion of 1 g of glucose produces 3.8 kcal, 2 g would produce twice as much heat, 7.6 kcal. This points out an important feature of chemical changes, that the amount of heat involved in a change is directly proportional to the mass of reactant consumed. This also holds true for physical changes.

---

†In actual practice, small corrections are made to account for the heat absorbed by the calorimeter itself, and for that which is lost to the environment due to imperfect insulation. We will neglect these here.

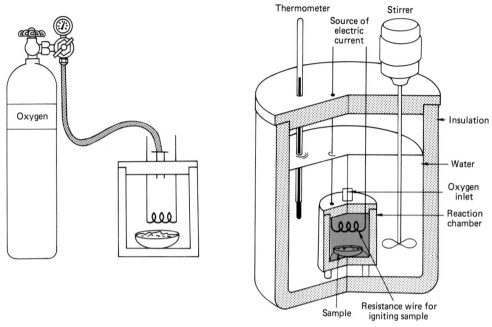

Fig. 5.13. A calorimeter is used to measure quantities of heat. A sample of combustible material is placed in the reaction chamber, which is filled with pure oxygen. Once immersed in a known amount of water, the sample is electrically ignited and burned. The amount of heat produced is determined by measuring the change in temperature of the water bath.

In Example 5.2 it was shown that if 80 calories of heat were required to melt 1 g of ice, then 12,000 calories would be required to melt 150 g of ice (150 g × 80 cal/g = 12,000 cal).

Check Test Number 3

1. One-half gram of carbon was burned in a calorimeter to form carbon dioxide. The temperature of 400.0 g of water around the reaction chamber changed from 25.0°C to 34.8°C. How much heat is produced in this reaction per 0.50 g of C and per mole of C?

Answer:

1. $\Delta T = 34.8°C - 25.0°C = 9.8°C$; $SH = 1.00\ \dfrac{cal}{g°C}$; $m = 400.0\ g$

   $Q = m \times SH \times \Delta T = 400.0\ \cancel{g} \times 1.00\ \dfrac{cal}{\cancel{g}\cancel{°C}} \times 9.8°\cancel{C} = 3920\ cal$

      $= 3.92$ kcal per 0.50 g of C

   1 mole C = 12.0 g C

The heat produced per mole would be

$$\frac{3.92 \text{ kcal}}{0.50 \text{ g}} \times \frac{12.0 \text{ g}}{1 \text{ mole}} = 94.1 \frac{\text{kcal}}{\text{mole}}$$

Chemical reactions take place in and around us all the time, and like many things we tend to take them for granted. But there is one important question that should be asked. Why do chemical reactions occur? The details of the answer lie in the complexities of thermodynamics, the study of energy and work, but it comes down to a single, fundamental principle: *Chemical reactions will take place if, in the process, there is a net loss of energy to the environment*. The same requirement holds for physical changes as well. After all, water left to its own devices only runs downhill, from higher to lower potential energy. As it flows, there is a net loss of energy as work is done moving rocks and soil and perhaps a waterwheel along the way.

At first glance, it is not difficult to see why exothermic chemical reactions take place, since they release energy which is ultimately lost to the environment. But what about endothermic reactions? Since they absorb energy as they take place, it would seem that nature is running uphill energywise, forming products that contain more energy than did the reactants. But remember, an endothermic reaction cannot take place unless energy is given to it by some energy source. So when we consider the case of an endothermic reaction, we must also take into account the reaction that is coupled with it that produces the needed energy. Only in this way can the real energy change or the *net* energy change for the process be determined. It is something like pushing a wagon up hill. The potential energy of the wagon increases as it moves up the hill, but only at the expense of the energy of the person doing the pushing. Since the wagon cannot move up hill by itself, we must take the energy changes of both the pusher and the wagon into account in order to determine the *net* energy change.

Consider the decomposition of calcium carbonate, an endothermic reaction. It is known that 42.6 kcal of heat energy is required to decompose one mole of $CaCO_3$, limestone, to calcium oxide and carbon dioxide. The decomposition will not take place without added energy. Suppose the energy is supplied by the combustion of methane. Because heat energy cannot change form or be transferred to another place with 100% efficiency, we would have to generate more than 42.6 kcal of heat from methane for every mole of limestone decomposed. This results in an important observation, that more energy *must be produced* by the exothermic reaction *than is absorbed* by the endothermic one, resulting in a *net* energy loss to the environment. Though a reaction can by itself be considered endothermic, it can take place only if it is coupled to another process to produce an overall net loss of energy. Remember, no physical or chemical change can take place without energy ultimately being transferred to the environment. Nature only runs downhill energywise.

The energy that exists in compounds comes from the sun, our ultimate source of energy. In *photosynthesis*, a series of reactions which comprise an endothermic process produces glucose and oxygen from carbon dioxide and water in green plants. The energy needed for the process comes from reactions on the sun, and is stored as potential energy in plants. Once the plants are eaten by man or animal, the stored energy is released in a process called *cellular respiration*, in which the glucose is converted back to carbon dioxide and water, releasing the stored energy (see Figure 5.14). In the human body, only about 20% to 25% of the released energy is used to do biological work. The rest escapes to the environment as heat. Even our bodies cannot use energy with a high efficiency. Without a constant input of energy, our bodies would be unable to do the necessary life-maintaining work, and would soon cease to function. We are alive today because of energy-producing reactions that occured 93 million miles away on the sun.

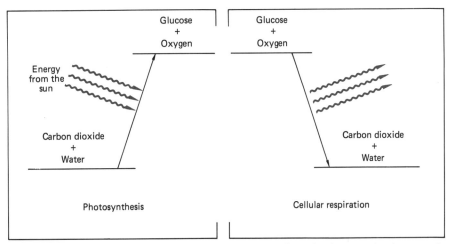

Fig. 5.14. Energy from the sun is captured by green plants in photosynthesis, as carbon dioxide and water are converted into glucose. When plants are eaten and digested the stored energy released as glucose is changed back to carbon dioxide and water in cellular respiration.

Energy and Nutrition

People who diet to lose weight pay careful attention to the number of "calories" in the foods they eat. In nutrition, the term calorie refers to the energy of the kilocalorie. To avoid confusion the abbreviation for the nutritional calorie is written with a capital C, Cal, to distinguish it from the small calorie. The nutritional calorie is the amount of heat in kilocalories produced when a given amount of food is burned in a calorimeter, and it indicates the value of that food as a fuel in the body. For example, if the amount of heat produced when one gram of cheddar cheese was burned was 4700 calories, its caloric value would be listed as 4.7 Cal per gram. In the body, then, each gram of cheddar cheese eaten would supply that much energy.

Most of the energy required by our bodies is supplied by carbohydrates and fats. When carbohydrates are metabolized, the end result is the transport of glucose by the blood to cells where it reacts with oxygen to produce carbon dioxide, water, and heat energy. Carbohydrates produce, on the average, about 4 kcal of energy per gram (4 Cal/g). Fats produce more energy per gram than carbohydrates, averaging about 9 kcal per gram (9 Cal/g). Energy is stored in the body as fat (potential energy), and it serves that purpose well since it has a higher energy content per gram and is not soluble in water, a requirement for the storage of a substance in the body. The metabolism of proteins produces an average of about 4 kcal of energy per gram (4 Cal/g) in the body. Only a very small amount of the protein that enters the body is used for energy though. Rather, it ends up as new cells, skin, hair, organ walls, and so forth. Only during starvation is protein used in any appreciable amount for energy.

The composition and caloric values of several foods are listed in Table 5.2.

TABLE   5.2
Composition and Caloric Value of Common Foods

Food	Approximate composition* (%)			Caloric value (Cal) per gram
	Protein	Fat	Carbohydrate	
Apples (raw)	0.4	0.5	13	0.6
Cheese (cheddar)	28	37	4	4.7
Eggs	13	10	0.7	1.6
Bread (white, enriched)	9	3	52	2.8
Hamburger	22	30	0	3.6
Milk	3.3	4	5	0.7
Peanuts	26	39	22	5.6

*Percentages will not total 100% because of the varying amounts of water in foods.

The amount of energy required per day by an individual varies according to age, weight, and the amount of work performed. An adult at rest and in a warm environment requires about 1500 Cal per day. Doing an average amount of work, the requirement increases to from 2500 to 3500 Cal per day. If more calories are taken in than needed, the excess is stored in the body as fat.

Not only should a person take in the proper number of calories each day, but they should be provided as fats, proteins, and carbohydrates. The *recommended dietary allowance* (RDA) for optimum nutrition requires 45% of the daily caloric intake to be carbohydrates, 40% fats and the remainder protein.

Example 5.4

___

If the optimum caloric intake for a person was 2000 Cal, and if it was to be supplied according to the RDA values above, how many grams of fat, carbohydrate, and protein would be required in the daily diet? (Assume for the sake of simplicity that protein will be entirely converted to energy.)

45% of the 2000 calories should be from carbohyrates, 40% from fats, and 15% from protein.
Cal from carbohydrates = 2000 Cal × 0.45 = 900 Cal
Cal from fats = 2000 Cal × 0.40 = 800 Cal
Cal from proteins = 2000 Cal × 0.15 = 300 Cal

Knowing that the average caloric content of carbohydrates and proteins is 4 Cal/g, and of fats 9 Cal/g, the amount of each type of food can be calculated:

$$\text{grams of carbohydrates} = 900 \; \cancel{Cal} \times \frac{1 \text{ g}}{4 \; \cancel{Cal}} = 225 \text{ g}$$

$$\text{grams of fat} = 800 \; \cancel{Cal} \times \frac{1 \text{ g}}{9 \; \cancel{Cal}} = 89 \text{ g}$$

$$\text{grams of protein} = 300 \; \cancel{Cal} \times \frac{1 \text{ g}}{4 \; \cancel{Cal}} = 75 \text{ g}$$

___

The stored energy in glucose is much greater than that needed by the cells at any instant for usual activities, and if it were released all at one time, a very large part of it would be wasted as heat. So in the body the burning of glucose is coupled with other reactions that allows a more efficient, controlled transfer of energy in the cell.

Terms

___

Several important terms appeared in Chapter 5. You should understand the meaning of each of the following:

energy

work

law of conservation of energy

biological work

electromagnetic spectrum

calorie

kilocalorie

nutritional calorie

specific heat

ΔT

ultraviolet light

visible light

infrared light

wavelength

kinetic energy

potential energy

heat of fusion

heat of vaporization

exothermic change

endothermic change

photosynthesis

cellular respiration

Questions

Answers to the starred questions appear at the end of the text.

*1. What must a body possess if it is to have the ability to do work?

2. Which of the following represents work being done?
   a. pushing a wheelchair
   b. pouring water from a pitcher into a glass
   c. leaning against a brick wall
   d. cooking an egg in boiling water
   e. separating red capsules from blue capsules

*3. If a pan of boiling water is set out on a table to cool, what happens to the heat energy it contains?

4. Explain why we are so concerned about the world's energy supply when the law of conservation of energy states that energy can never be destroyed?

5. List three kinds of biological work that are done in our bodies, and briefly describe each process.

*6. Describe in general terms how the temperature of a beaker of water is related to the motion of the water molecules.

7. Indicate one wavelength (λ) on the figure.

*8. How is the wavelength of electromagnetic radiation (light) related to its energy and its ability to do cellular damage?

9. Which of the following are known to cause damage to living tissue?
   a. cosmic rays   c. ultraviolet light
   b. radiowaves   d. gamma rays

*10. Which of the following represent kinetic energy and which represent potential energy?
   a. wind blowing across a lake
   b. gun powder
   c. a bottle of I.V. solution suspended above a patient
   d. the I.V. solution that is flowing into a patient's vein
   e. a diver poised on the end of a diving board
   f. a "set" mousetrap

*11. How many small calories are contained in one nutritional calorie?

12. Why does the temperature of a piece of ice remain constant (at 0°C) as it absorbs heat energy and melts to form liquid water (at 0°C)?

*13. Which of the following are exothermic changes, and which are endothermic?
  a. $2HgO + heat \rightarrow 2Hg + O_2$
  b. $CaO + H_2 \rightarrow Ca(OH)_2 + heat$
  c. $N_2 + 3H_2 \rightarrow 2NH_3 + heat$
  d. $Na_{(s)} + heat \rightarrow Na_{(l)}$

14. Describe the operation of a calorimeter that is designed to burn substances in pure oxygen.

15. Ultimately, after all things are considered, what can you say about the *net* energy change for any process?

*16. If the energy contained in the products of a reaction is less than that in the reactants, is the reaction exothermic or endothermic?

17. What is the recommended dietary allowance of fats, protein, and carbohydrates in terms of our percent intake of Calories?

*18. Is protein usually used as an energy source by the body?

*19. What is the difference, in terms of energy storage and energy release, between photosynthesis and cellular respiration?

*20. How many kilocalories of heat energy are required to raise the temperature of 1.00 kg of water from 25°C to 80°C? SH water = 1.00 cal/g°C.

21. The specific heat of copper is 0.09 cal/g°C. How much heat will flow out of a 100 g copper casting as it cools from 400°C to 25°C?

*22. How many kilocalories of heat can be absorbed by 500 g of ice at 0°C as it melts to form water at the same temperature? The heat of fusion for water is 80 cal/g.

23. How many grams of water at 100°C can be converted into steam with 150 kcal of heat? The heat of vaporization for water is 540 cal/g.

*24. When one gram of ammonium chloride (NH_4Cl) is dissolved in 100 g of water, the temperature falls by 0.65°C. What would be the final temperature of a solution prepared by dissolving 12 g of NH_4Cl in 100 g of water initially at 25.0°C?

*25. Which of the following would undergo the greatest temperature change if 100 g of each absorbed 0.50 kcal of heat energy?
  a. alcohol; SH = 0.58 cal/g °C
  b. benzene; SH = 0.41 cal/g °C

26. One gram of fudge was burned in a calorimeter containing 600 g of water. The temperature of the water bath changed from 23.2°C to 30.0°C. What is the fuel value per gram of fudge?

*27. How many grams of fat could produce enough heat energy to equal a daily caloric need of 2800 Cal? What is this equivalent to in pounds?

28. When one mole of methane (CH_4) is burned, 213 kcal of heat is released. How many moles and how many grams of methane would produce one million kcal of heat when burned?

*29. How many grams of limestone, $CaCO_3$, can be decomposed to CaO and CO_2 if it absorbs 4.26×10^4 kcal of heat? One mole of $CaCO_3$ can be decomposed with 42.6 kcal of energy.

The Structure of the Atom, and the Periodic Table

Once it was realized that rocks and water and air were actually composed of atoms, a new age of chemistry was at hand. Everything was composed of atoms. And there were different kinds of atoms, one kind for each element. An aluminum atom was not like a sulfur atom, and each had its own set of properties to prove it.

But with the acceptance of the existence of atoms came a host of new questions. One of the most fundamental of these asked what caused the atoms of one element to be so different from those of another element? Why are aluminum atoms and sulfur atoms so different? A good part of the answer is in this chapter, and to learn it we will need to look into the atom to see how it is put together. Only then will we be able to understand why there are differences, as well as similarities, among the elements.

Near the end of the 19th century scientists acquired the ability to look indirectly into the atom. What they saw caused another revolution in chemistry, which carried over into medicine as well. We will begin a discussion of the atom with the facts that began the revolution and end the chapter surveying some of the reasons for the similarities and differences between the elements.

Objectives

By the time you finish this chapter you should be able to do the following.

1. Describe the three principal particles that make up the atom, and describe where each is located within the atom.
2. Define atomic number and mass number, and, given both for an atom, tell how many electrons, protons, and neutrons it contains.
3. Define isotope, and be able to recognize isotopes of an element.
4. Describe the changes that take place in an atom that cause it to emit light energy.
5. Describe the Bohr atom and state the maximum number of electrons that can be in each of the first three energy levels.
6. Describe the relation between the "orbit" of an electron about the nucleus and the "energy level" of that electron.
7. Draw the atomic structure of the first 20 elements, show the arrangement of electrons in the energy levels, and be able to indicate which are the valence electrons.
8. Write the electron-dot symbol for the elements in groups I through VIII.
9. Describe the Periodic Table in terms of its groups and periods.
10. Describe what is meant by a periodic property, and indicate how the sizes of atoms and their ionization energies change within the second and third periods.

The Particles that Form the Atoms

You have already learned that the atom is an incredibly small particle of matter; yet it is composed of still smaller fragments collectively called **subatomic particles.** The three principal subatomic particles are the electron, the proton, and the neutron.

The **electron** (e^-) is the lightest subatomic particle and has a mass of 9.107×10^{-28} g. It bears the smallest unit of negative electrical charge ever measured, and for simplicity it is symbolized as $1-$ (one negative charge).

The **proton** (p^+) is about 1840 times heavier than the electron with a mass of 1.672×10^{-24} g. The proton bears an electrical charge that is equal in size to that on the electron but positive in sign. It is therefore symbolized $1+$ (one positive charge). The principal difference between a negative charge and a positive charge is the way they act toward each other and toward other charges of the same sign. Positive and negative charges are attracted to each other, but they are repelled by charges of the same sign. You may remember the old saying that goes "like charges repel, and unlike charges attract."

The **neutron** (n) bears no electrical charge; it is neutral. It has a mass slightly larger than the proton, 1.675×10^{-24} g.

The approximate mass of the subatomic particles can be stated in atomic mass units, the relative weight unit of the atomic weight scale. The neutron and proton have a mass of about 1 amu, and the approximate mass of the electron can be remembered as 0 amu.

Within the atom, the relatively massive protons and neutrons are together in a very small, dense body called the **nucleus.** The nucleus is in the center of the atom, and it bears a positive charge equal to the number of protons it contains. The electrons are distributed around the nucleus, forming what can be visualized as a spherical cloud of negative electrons that encapsulates the nucleus. The nucleus is only about one 100-thousandth the size of the electron cloud. To put this into perspective: if the nucleus were the size of a golf ball, the distance across the atom would be more than two miles. You can see then, that the very small electrons occupy a great deal of space, and the nucleus occupies only a small fraction of the volume of an atom, as shown in Figure 6.1.

TABLE 6.1 The Subatomic Particles					
Name	Location in the atom	Symbol	Charge	Mass (g)	Approximate mass in amu
Electron	outside the nucleus	e^-	1−	9.107×10^{-28}	0
Proton	nucleus	p^+	1+	1.672×10^{-24}	1
Neutron	nucleus	n	0	1.675×10^{-24}	1

Atomic Numbers and Mass Numbers

We are now ready to learn why an atom of aluminum is so different from an atom of sulfur in its chemical behavior. They are both composed of the same subatomic particles, but the number of protons, electrons, and neutrons in an aluminum atom is different from the number of each in a sulfur atom. Ultimately, it is the number of protons in the nucleus that determines the identity of an atom. The number of protons in the nucleus is given by the **atomic number** of the element, and for a neutral atom, the atomic number also equals the number of electrons around the nucleus:

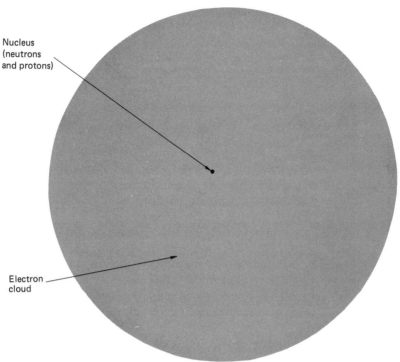

Nucleus
(neutrons
and protons)

Electron
cloud

Fig. 6.1. The nucleus is only a small fraction of the size of the atom, though it contains nearly all of its mass.

$$\text{atomic number} = \text{number of protons in the nucleus}$$
$$= \text{number of electrons around the nucleus}$$

Each element has its own atomic number which is different from that of all the other elements. The atomic number of hydrogen is 1 ($1p^+$, $1e^-$), for helium it is 2 ($2p^+$, $2e^-$), and as you move across the Periodic Table (Figure 6.5) the atomic number increases by one for each of the subsequent elements. Since aluminum has an atomic number of 13, all aluminum atoms have 13 protons in the nucleus and 13 electrons about it. Sulfur has an atomic number of 16, so a sulfur atom has 16 protons surrounded by 16 electrons.

The number of neutrons in a nucleus can be determined from the mass number of the atom. The **mass number** is equal to the sum of the number of neutrons plus the number of protons in the nucleus:

$$\text{mass number} = \text{number of protons} + \text{number of neutrons}$$

An atom of helium with two neutrons and two protons in the nucleus has a mass number of 4. A carbon atom with six neutrons and six protons has a mass number of 12. Symbols that indicate the mass number and atomic number of an atom can be written using superscripts and subscripts on the left side of the symbol:

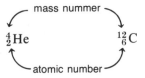

The mass number can also be indicated in the name of the element, such as helium-4 and carbon-12. The number of neutrons in the nucleus equals the difference between the mass number (protons + neutrons) and the atomic number (protons) of the atom:

number of neutrons = mass number − atomic number

The mass number of an atom also equals the *approximate* atomic weight of the atom. Both the neutron and proton have an approximate atomic weight of 1 amu, so the atomic weight of helium-4 ($^4_2$He) would be about 4 amu since it contains two neutrons and two protons. The electrons do not have a significant effect on the atomic weight since they have a much smaller mass.

The number of subatomic particles that make up three different atoms is shown in Table 6.2. Note that the number of protons and neutrons need not be equal.

TABLE 6.2 The Composition of Three Atoms					
			Number of:		
Atom	Atomic number	Mass number	e^-	p^+	n
$^{27}_{13}$Al	13	27	13	13	14
$^{238}_{92}$U	92	238	92	92	146
$^1_1$H	1	1	1	1	0

nuclear particles

Isotopes

As knowledge of the atom increased, it became clear that even though all atoms of a given element had the same chemical properties, they all did not have the same weight. Each atom of an element had the same number of protons and electrons, but they did not all have the same number of neutrons. As a result, the atoms had different mass numbers. Atoms with the same atomic numbers but different mass numbers are called **isotopes** of one another. Isotopes, then, are atoms of the same element that differ in weight because they contain a different number of neutrons.

A typical example is chlorine. There are two isotopes of chlorine found in nature, chlorine-35 ($^{35}_{17}Cl$) and chlorine-37 ($^{37}_{17}Cl$). Approxmately three out of every four chlorine atoms are chlorine-35, which contains two less neutrons than chlorine-37:

$$^{35}_{17}Cl \qquad 17e^-, 17p^+, 18n \qquad\qquad ^{37}_{17}Cl \qquad 17e^-, 17p^+, 20n$$

The relative number of atoms of each isotope in nature is called the **natural abundance** of the isotope. The atomic weight of chlorine-35 is approximately 35 amu, and that of the heavier chlorine-37 atom is 37 amu. The atomic weight you see on the Periodic Table for chlorine is about 35.5, an average of these values that takes into account the relative number of atoms of each isotope found in nature. Since three out of four chlorine atoms have an atomic weight of 35 amu, and the fourth 37 amu, the average atomic weight of the four atoms is the sum of the four values divided by 4:

	35 amu
3 atoms of $^{35}_{17}Cl$ weigh	35 amu
	35 amu
1 atom of $^{37}_{17}Cl$ weighs	37 amu
	———
all 4 atoms weigh	142 amu

$$\text{average atomic weight of chlorine atoms} = \frac{142\ \text{amu}}{4\ \text{atoms}} = 35.5\ \frac{\text{amu}}{\text{atom}}$$

The atomic weight of carbon is the average value of its isotopes, carbon-12 ($^{12}_{6}C$), carbon-13 ($^{13}_{6}C$), and carbon-14 ($^{14}_{6}C$). You may recall in Chapter 4 that the atomic weight scale is based on carbon-12, which is assigned an atomic weight of exactly 12.00 amu. The average atomic weight for carbon that appears on the Periodic Table is slightly larger than 12.00 since about 1% of all the carbon atoms in nature are heavier isotopes. The atomic weights of all the elements given on the periodic table are average values of their naturally oc-curring isotopes.

Isotopes exist for all the elements. There are three isotopes of hydrogen. The most abundant form of hydrogen is $^{1}_{1}H$, which is sometimes called protium to distinguish it from the less abundant isotopes $^{2}_{1}H$ (deuterium) and $^{3}_{1}H$ (tritium).

Certain isotopes of some elements are not very stable and in time they disintegrate, a property known as radioactivity. Carbon-14 and tritium are both unstable isotopes, as are those of uranium. In fact, of the five known isotopes of fluorine, only one is stable, fluorine-19 ($^{19}_{9}F$).

Check Test Number 1

1. Fill in the missing values:

Atom	At. No.	Mass No.	Number of:		
			e^-	p^+	n
(a) $^{23}_{11}Na$	___	___	___	___	___
(b) ___	15	___	___	___	16

2. Which of the following pairs are isotopes?
   (a) $^{36}_{18}X$, $^{40}_{18}X$ (b) $^{40}_{18}X$, $^{40}_{19}X$ (c) $^{36}_{18}X$, $^{40}_{19}X$

3. What is the approximate atomic weight of sodium-23?

Answers:
   1. (a) $^{23}_{11}Na$, 11, 23, 11, 11, 12;
      (b) $^{31}_{15}P$, 15, 31, 15, 15, 16
   2. (a) same atomic numbers, different mass numbers
   3. 23 amu (11 p^+ and 12 n each contributing 1 amu)

The Arrangement of the Electrons

At this point we need to focus our attention on the electrons that surround the nucleus. The chemical properties of an atom depend not only on the number of electrons it has, but also on the way they are arranged in space and in energy.

It may seem surprising, but much of what is known about the arrangment of electrons in an atom came from the study of light. If a sample of hydrogen gas is heated to a high temperature or has an electric discharge passed through it, it glows and emits a reddish light. If this light is passed through a prism, it separates into four narrow bands of light, each of a different color and wavelength (λ). This set of four colored bands is called the **visible emission spectrum** of hydrogen, which is shown in Figure 6.2. Other elements treated the same way also produced an emission spectrum. If sunlight is passed through the prism you would not see individual lines of color; rather there would be a continuous spread of the colors of the rainbow from red to violet, quite unlike the result obtained with hydrogen.

In Chapter 5 you learned that light is a form of energy, and that light of longer wavelengths is of lower energy than light of shorter wavelengths. The orange line in the emission spectrum of hydrogen is of lower energy than the green

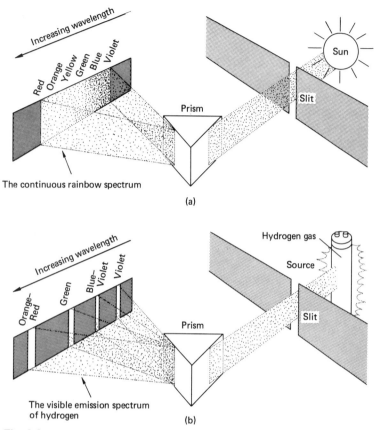

Fig. 6.2. (a) When sunlight is passed through a prism, it separates to form the continuous spectrum of the rainbow. (b) Visible light emitted from hydrogen atoms heated to a high temperature will form a spectrum containing four distinct bands of light. This is the visible emission spectrum of hydrogen.

line, and of the four, the violet line has the highest energy. In 1913, a Danish physicist, Neils Bohr, was able to interpret the spectrum of the hydrogen atom in terms of a definite arrangement of energy levels in the atom that could be occupied by electrons. The result of his work was the first successful model of the atom. Let us examine this model and later show how it can produce the lines of an emission spectrum.

The Bohr Atom   The only way Bohr could explain the existence of lines of definite energy in the spectrum of hydrogen was to conclude that there were fixed **energy levels** that the electron could occupy. Each energy level corresponded to a circular **orbit** or path that the electron could follow as it traveled around the nucleus. In a sense, Bohr's atom looked much like a small solar system, with the nucleus at the center and the electron moving in an orbit of definite radius and energy.

The orbit nearest the nucleus represents the lowest energy level, and each larger orbit represents a successively higher energy.

Each energy level (or orbit) is identified by a number symbolized n, which can equal 1, 2, 3, 4, ... The number is called a **quantum number**. The lowest energy level, or smallest orbit, has an n value of 1, for the next higher level n equals 2, and so forth. Sometimes letters are also used to identify each level, beginning with K for $n=1$ and continuing alphabetically as shown in Table 6.3. For several reasons, quantum numbers are used more often than letters to describe the energy levels.

TABLE   6.3
Labels for the Energy Levels of the Atoms

Energy level	1st	2nd	3rd	4th	5th
Quantum number—n	1	2	3	4	5
Letter designation	K	L	M	N	O

A diagram of the atom showing the orbits along with another showing the corresponding energy levels appears in Figure 6.3.

You may be wondering how the line spectrum of hydrogen can be connected

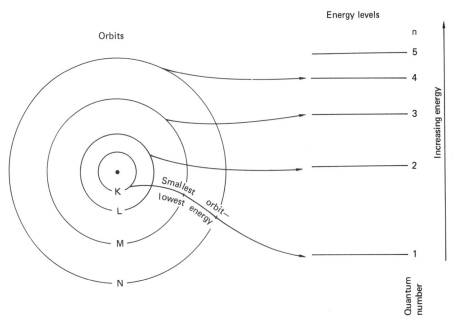

Fig. 6.3. Each orbit of the Bohr atom represents the definite level of potential energy. The smallest orbit (K) has the lowest energy, the next larger orbit (L) has a higher energy, and so forth.

to the model of the atom that Bohr proposed. It can be explained in terms of the fixed energy levels and the movement of an electron from one level to another. In an atom, if an electron is in a higher energy level, say the $n=3$ level, and then drops to a lower energy level, say the $n=2$ level, the electron will lose an amount of energy exactly equal to the energy difference between the $n=3$ and $n=2$ levels. In atoms, this lost energy is emitted from the atom as light of a definite wavelength, and it will produce a single line in the spectrum of the atom. (See Figure 6.4.) It is important to realize that the electron can only experience a change in its energy *exactly* equal to the difference between any two energy levels. Any other change for an electron in an atom is "forbidden" by the laws of nature. Electrons can also gain energy and be "excited" to a higher energy level. But again the energy gained is exactly equal to the energy separation between the lower and higher energy levels.

The changes that cause the four lines in the visible emission spectrum of hydrogen are described in Table 6.4. In each case the energy emitted is exactly equal to the difference between the higher energy level and the lower energy level.

A remarkable feature of the Bohr model of the hydrogen atom is that it can be used as a model for atoms that contain many electrons. The electrons in these atoms exist in definite energy levels too, and every different element will display a characteristic series of lines in its emission spectrum that acts like a fingerprint for the element. Just as in hydrogen, the spectrum of lines results from electrons in the excited atoms dropping to lower energy levels and emitting light in the process.

An important clinical test used in all hospital laboratories measures the amount of sodium ion (Na^+) and potassium ion (K^+) in blood serum. These ions are determined by measuring the *amount* of light they emit as they are carefully aspirated into the high-temperature environment of a flame. Sodium emits yellow light when excited in a flame. A sample of serum that contains more sodium ions than another will emit a brighter (more intense) yellow light.

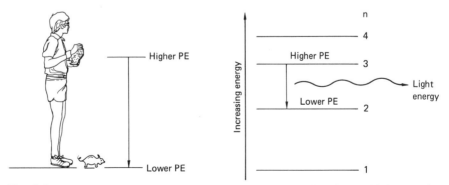

Fig. 6.4. Whenever an object, be it a rock or an electron, falls from a higher to a lower potential energy level, it loses an amount of energy exactly equal to energy difference between the two levels. With electrons in atoms, the lost energy is emitted as light.

	TABLE 6.4	
Electron Changes that Cause the Visible Hydrogen Spectrum		
Electron moves from	Wavelength of energy emitted (Å)	Line color
level 3 → level 2	6564	orange-red
level 4 → level 2	4863	green
level 5 → level 2	4342	blue-violet
level 6 → level 2	4103	violet

By carefully measuring the brightness of the light it is possible to know the amount of sodium ion in the serum sample. Potassium emits red-violet light and it can be measured in the same way. Since each ion emits light in a different region of the visible spectrum, sodium and potassium can be determined simultaneously.

The picture of the atom is now a bit more detailed, but a few additional facts remain concerning the number of electrons that can populate each energy level. The maximum number of electrons that can be placed in any energy level is $2n^2$, where n is the quantum number of the level. In the first level, $n=1$, there can be a maximum of $2 \times (1)^2$ or 2 electrons. In the second level, $n=2$, the maximum number is $2 \times (2)^2$ or 8 electrons, and in the third level, $n=3$, $2 \times (3)^2$ or 18 electrons. This is summarized in Table 6.5. An atom in its lowest state of energy will have its electrons in the lowest energy levels possible, without ever exceeding the maximum number of electrons allowed in each. The lowest energy arrangement of electrons in an atom is called the **ground-state electronic configuration** of the atom.

	TABLE 6.5	
Maximum Number of Electrons in the Energy Levels of an Atom	
Energy level n	Maximum number of electrons $2 \times n^2$
1	2
2	8
3	18
4	32

The Structures of Atoms

We can draw the structures of atoms using diagrams that show the electrons in each energy level. The nucleus can be symbolized showing its protons enclosed in a circle. Though the neutrons are not shown, realize they will be part of the nucleus. The structure of the hydrogen atom (At. No. = 1) with its single electron is represented two different ways below. Each shows the single electron in the lowest energy level. For convenience let us use the abbreviated structure on the right for other atoms.

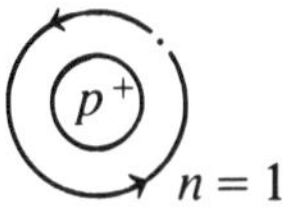

H
(At. No. = 1)

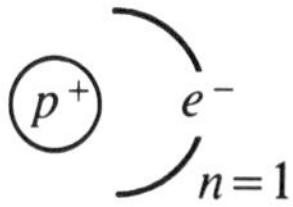

Helium would fill the first energy level with two electrons:

He
(At. No. = 2)

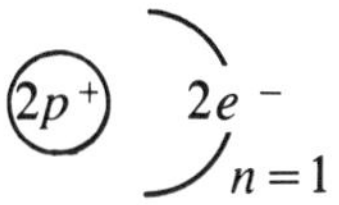

The three electrons in lithium require one electron to be placed in the second energy level after the first level is filled:

Li
(At. No. = 3)

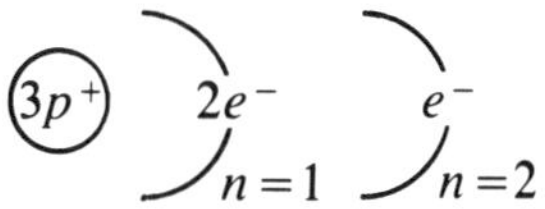

The next seven elements, beryllium through neon, will each have one more electron in the second level than the preceding element. With neon, the second shell is completely filled with eight electrons:

Be
(At. No. = 4)

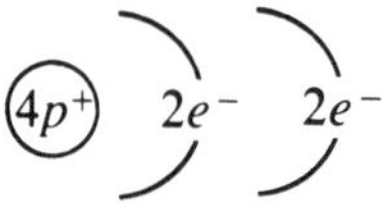

B
(At. No. = 5)

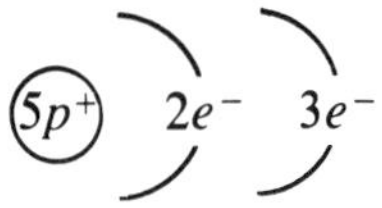

C
(At. No. = 6)

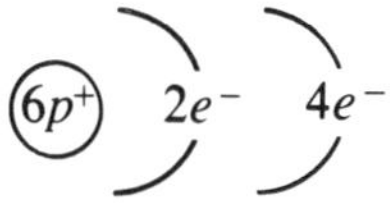

N
(At. No. = 7)

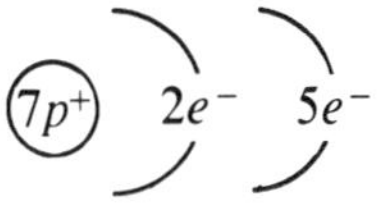

O
(At. No. = 8)

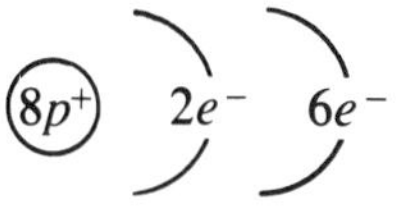

F
(At. No. = 9)

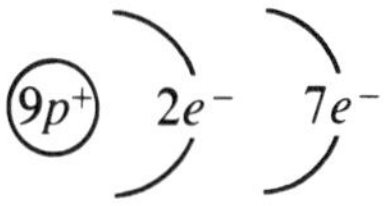

Ne
(At. No. = 10)

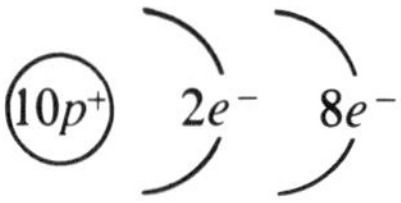

The sodium atom, with eleven electrons, brings the third energy level into use:

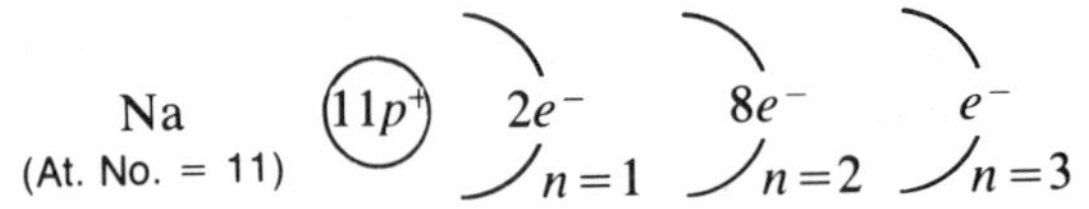

Na $11p^+$ $2e^-$ $8e^-$ e^-
(At. No. = 11) $n=1$ $n=2$ $n=3$

The next seven elements, magnesium (At. No. 12) through argon (At. No. 18), add additional electrons to the third level to give it a total of eight. Though the third level can actually hold 18 electrons, there is a *temporary pause* at eight electrons, as the next two electrons enter the fourth level for potassium (At. No. 19) and calcium (At. No. 20):

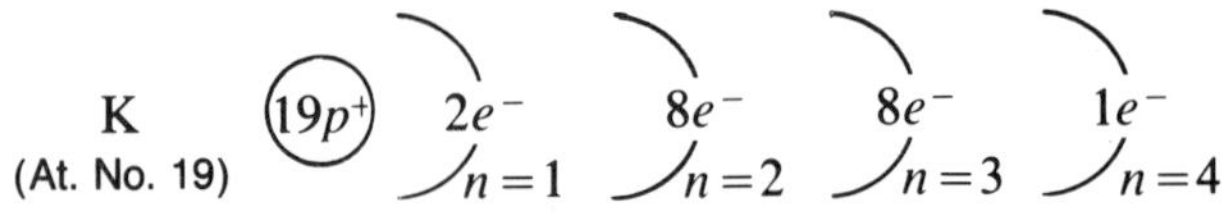

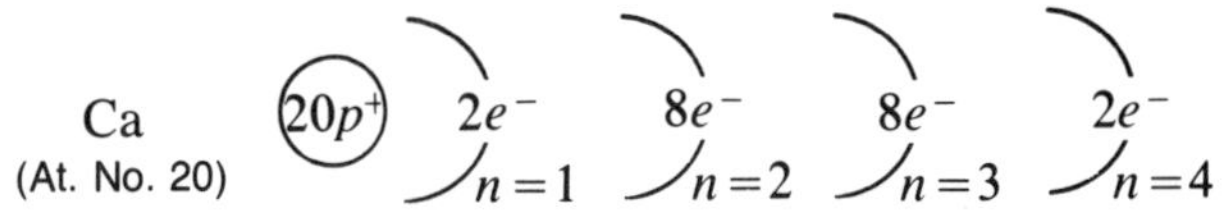

K $19p^+$ $2e^-$ $8e^-$ $8e^-$ $1e^-$
(At. No. 19) $n=1$ $n=2$ $n=3$ $n=4$

Ca $20p^+$ $2e^-$ $8e^-$ $8e^-$ $2e^-$
(At. No. 20) $n=1$ $n=2$ $n=3$ $n=4$

From this point on, electrons continue to add to the third level until the maximum of 18 electrons has been reached in zinc (At. No. 30):

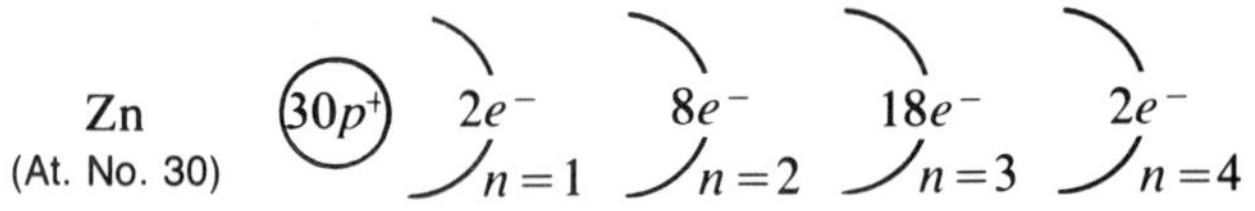

Zn $30p^+$ $2e^-$ $8e^-$ $18e^-$ $2e^-$
(At. No. 30) $n=1$ $n=2$ $n=3$ $n=4$

The Valence Electrons

Not every electron in an atom is directly involved in its chemistry. For most of the elements only those electrons in the outermost level are significant. The outermost electrons are often called the **valence electrons** of an atom to emphasize their importance. There is one valence electron in lithium, the one in the second level (the outermost level). The one valence electron in sodium is in the third level. It might seem strange that only those electrons furthest removed from the nucleus are the ones that most affect its chemistry, but when you realize these electrons are essentially on the outside of the atom and not as tightly held to the nucleus as the others, it seems quite reasonable.

Check Test Number 2

1. Draw the electronic configurations of aluminum (At. No. 13) and sulfur (At. No. 16). Point out the valence electrons in each.
2. Briefly explain how light can be emitted from an atom.

Answers:

1.

Al $(13p^+)$ $2e^-$ $8e^-$ $(3e^-)$ valence electrons

$n=1$ $n=2$ $n=3$

S $(16p^+)$ $2e^-$ $8e^-$ $(6e^-)$ valence electrons

$n=1$ $n=2$ $n=3$

2. When an electron excited to a higher energy level returns to a more stable lower energy level, the loss in energy of the electron is emitted as light energy. The energy of the light exactly equals the difference in energy between the two energy levels.

The Periodic Table of the Elements

The similarity in chemical behavior of certain elements was recognized long before people knew about electrons and protons. Sodium and potassium have nearly identical chemical properties, and chlorine and bromine are very similar also. During the mid 1800's, Lothar Meyer, a German chemist, and Dimitri Mendeleev, a Russian chemist, independently arranged the known elements into a table in order of their increasing atomic weights. After a few adjustments, elements with similar chemical properties appeared at regular intervals along the series. A property or set of properties that repeat at definite intervals is said to behave in a **periodic** way. The seasons, the tides, and our heart beat are periodic since they repeat at regular intervals of time. The significance of the periodic behavior of the elements was clearly demonstrated by Mendeleev when he predicted the existence of three elements that were as yet undiscovered and even went on to predict many of their properties with uncanny accuracy. To honor him and his achievement, element 101 is named Mendelevium.

The modern *Periodic Table of the elements* is much more extensive than Mendeleev's, and it appears inside the front cover of this text and in Figure 6.5. In the modern table, elements are arranged in order of increasing atomic number, not atomic weight. In 1914, a British physicist named Henry Mosley showed that the periodic properties of the elements were a function of their atomic numbers, a fact that has come to be known as the **periodic law**.

In the Periodic Table, each element occupies a single box which contains its symbol, atomic number, and atomic weight. Some tables include still more information in each box.

Key:
- 1 — Atomic number
- H — Symbol
- 1.00797 — Atomic weight

Period	I	II												III	IV	V	VI	VII	VIII
1	1 H 1.00797																		2 He 4.0028
2	3 Li 6.939	4 Be 9.0122												5 B 10.811	6 C 12.01115	7 N 14.0067	8 O 15.9994	9 F 18.9984	10 Ne 20.183
3	11 Na 22.9898	12 Mg 24.312												13 Al 26.9815	14 Si 28.086	15 P 30.9738	16 S 32.064	17 Cl 35.453	18 Ar 39.948
4	19 K 39.102	20 Ca 40.08	21 Sc 44.956	22 Ti 47.90	23 V 50.942	24 Cr 51.996	25 Mn 54.9380	26 Fe 55.847	27 Co 58.9332	28 Ni 58.71	29 Cu 63.54	30 Zn 65.37		31 Ga 69.72	32 Ge 72.59	33 As 74.9216	34 Se 78.96	35 Br 79.909	36 Kr 83.80
5	37 Rb 85.47	38 Sr 87.62	39 Y 88.905	40 Zr 91.22	41 Nb 92.906	42 Mo 95.94	43 Tc (99)	44 Ru 101.07	45 Rh 102.905	46 Pd 106.4	47 Ag 107.870	48 Cd 112.40		49 In 114.82	50 Sn 118.69	51 Sb 121.75	52 Te 127.60	53 I 126.9044	54 Xe 131.30
6	55 Cs 132.905	56 Ba 137.34	57 °La 138.91	72 Hf 178.49	73 Ta 180.948	74 W 183.85	75 Re 186.2	76 Os 190.2	77 Ir 192.2	78 Pt 195.09	79 Au 196.967	80 Hg 200.59		81 Tl 204.37	82 Pb 207.19	83 Bi 208.980	84 Po (210)	85 At (210)	86 Rn (222)
7	87 Fr (223)	88 Ra (226)	89 †Ac (227)	104 Rf (257)	105 Ha (260)														

°	58 Ce 140.12	59 Pr 140.907	60 Nd 144.24	61 Pm (147)	62 Sm 150.35	63 Eu 151.96	64 Gd 157.25	65 Tb 158.924	66 Dy 162.50	67 Ho 164.930	68 Er 167.26	69 Tm 168.934	70 Yb 173.04	71 Lu 174.97
+	90 Th 232.038	91 Pa (231)	92 U 238.03	93 Np (237)	94 Pu (242)	95 Am (243)	96 Cm (247)	97 Bk (247)	98 Cf (249)	99 Es (254)	100 Fm (253)	101 Md (256)	102 No (253)	103 Lr (257)

Fig. 6.5.   The modern Periodic Table.

The horizontal rows of elements are called **periods**. The first three periods are shorter than the others and are called the short periods. The other four are called the long periods. The first period contains only two elements, the second and third each have eight, the fourth and fifth each have 18. The sixth and seventh periods contain an even larger number of elements, but to keep the table from becoming too wide, elements number 57–71 (the lanthanides) and 89–103 (the actinides) are placed beneath the table. With the exception of the first period, each begins with elements that are metals and ends with elements that are nonmetals.

The columns of elements are called **groups** or **families**. Elements within each group will have many similar chemical properties. Several groups in the Periodic Table are identified by a Roman numeral. The Roman numeral not only stands for the group number, but it also indicates the number of electrons in the outermost energy level of each atom in the group, that is, the number of valence electrons. Each element in group I has *one* valence electron, and each element in group V has *five* valence electrons. The metals in group I are collectively called the **alkali metals**, those in group II the **alkaline earth** metals, and those in group VII are called the **halogens**. The elements in group VIII are called the

rare gases or the **noble gases**. They are particularly unreactive elements, and except for helium each has eight electrons in the outermost shell.

Figure 6.6 shows the division of the periodic Table into classes of elements. The similarity in the number of valence electrons for the alkali metals and the halogens is shown in Table 6.6.

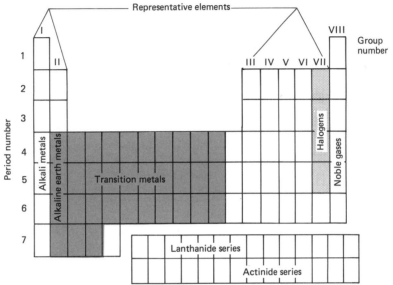

Fig. 6.6. Classes of elements in the Periodic Table.

TABLE 6.6
Electronic Configurations of Elements in Groups I and VII*

| | Group I | | | | | | Group VII | | | | |
| | Energy level | | | | | | Energy level | | | | |
Element	1	2	3	4	5	Element	1	2	3	4	5
Li	2	(1)				F	2	(7)			
Na	2	8	(1)			Cl	2	8	(7)		
K	2	8	8	(1)		Br	2	8	18	(7)	
Rb	2	8	18	8	(1)	I	2	8	18	18	(7)

*Circled numbers indicate valence electrons.

The elements in groups I through VII are called the **representative elements,** and some of them are very important in living systems. Fats and carbohydrates are composed of only three of these elements, carbon, hydrogen, and oxygen. Proteins are made up of these three plus nitrogen and sulfur.

The three rows containing 10 elements in the 4th, 5th, and 6th periods that bridge groups II and III are the **transition metals.** Not only are they similar to elements within their own groups, but they are very often similar to other transition metals within their period. Table 6.7 shows the electronic configurations of some transition elements. The changing number of electrons in the $n=3$ level is masked somewhat by the constant pair of electrons in the $n=4$ level. Because electrons from both the 3rd and 4th levels are involved, the chemistry of a given transition metal can be quite varied. Small amounts of several transition metals are necessary in the body for life to exist. Iron is involved in hemoglobin, the oxygen carrier in the bloodstream. Cobalt, manganese, zinc, and copper are found in different enzymes that control the progress of certain biochemical reactions. Chromium is believed to be involved with the use of insulin by the body. Still other transition metals like mercury and silver are quite toxic since they can interfere with life-sustaining processes.

TABLE   6.7
The Electronic Configuration of Four Transition Metals

| Element | Symbol | Atomic number | Energy level, $n =$ | | | |
			1	2	3	4
Manganese	Mn	25	2	8	13	2
Iron	Fe	26	2	8	14	2
Cobalt	Co	27	2	8	15	2
Nickel	Ni	28	2	8	16	2

Check Test Number 3

1. Why are the chemical properties of elements within a group usually quite similar?
2. What do the elements within a period have in common in terms of their electronic structures?

Answers:
1. Elements within a group have the same number of valence electrons, and the valence electrons are most important in their chemistry.
2. The outermost electrons of each element in a given period are in the same energy level. 1st period—1st energy level, 2nd period—2nd energy level, and so forth.

Electron-Dot Symbols for the Elements

Because the chemistry of an element is closely related to the number of electrons in its outermost energy level, there are times when it is useful to include them in the symbol. The electrons can be represented by dots (·), placed above, below, and to either side of the symbol. The electron-dot symbol for carbon should show that it has four valence electrons:

 each dot stands for one valence electron

It is easy to write an electron-dot symbol for the representative elements (those in groups I through VIII), since the number of valence electrons is the same as the group number. The electron-dot symbols for the elements of the first three periods are given in Table 6.8. As you move across the second and third periods note how the first four dots are placed singly on each side of a symbol before they are doubled up in group V. This is a good convention to follow when writing electron-dot symbols. We will find these symbols to be quite useful in Chapter 8 when we discuss how atoms react with each other to form compounds.

TABLE 6.8
Electron-Dot Symbols of the First 18 Elements

Period \ Group	I	II	III	IV	V	VI	VII	VIII
1	H·							·He·*
2	Li·	·Be·	·Ḃ·	·Ċ·	·N̈·	·Ö·	:F̈·	:N̈e:
3	Na·	·Mg·	·Äl·	·S̈i·	·P̈·	·S̈·	:C̈l·	:Är:

*Though helium (He) is in group VIII, its electron-dot symbol shows only two electrons, the maximum number allowed in the first energy level.

Check Test Number 4

Expand the list of elements in Table 6.8 by writing the electron-dot symbols for: K, Ca, As, Se, Br, I, and Xe.

Answer:

$$K· , ·Ca· , ·Äs· , ·S̈e· , :B̈r· , :Ï· \text{ and } :Ẍe:$$

Periodic Properties of the Elements

The arrangement of elements into groups and periods in the periodic table offers several advantages, since it frees us from having to view each element as a completely different entity, with a completely different set of properties.

Instead, you can learn general properties of each group of elements which in turn apply to each element in it. There are also trends that appear within the periods that help us learn more about the elements in them.

Those properties of elements that generally increase or decrease as you move across a period are called **periodic properties.** For example, the metallic character (shiny surface, conduction of electricity, strength) of the element in a period decreases from left to right, and the period ends with a decidedly nonmetallic rare gas. Except for the first period (H and He) this trend is seen to repeat again and again in each period, so one could think of the metallic property as being a periodic property. You might think that something as general as a "metallic property" is too vague, and in some ways it is. But there are properties that are more specific that show the same kind of trends, and which can be related to the metal to nonmetal trend. These are the sizes of the atoms and their ionization energies.

The Size of Atoms

An atom can be thought of as a very small sphere, like a marble. The size of this sphere can be described by its radius. Since atoms are so small, the radius is usually given in angstroms (1 Å = 10^{-10}m). The atomic radii (the radius of atoms) of several elements appear in Figure 6.7. Notice how the radii *decrease* from left to right across a period, and *increase* going down a group. The decrease in size across a period is caused by the addition of one proton to the nucleus of each succeeding atom. The step-by-step increase in the positive charge on the nucleus draws the valence electrons in closer each time and atoms get progressively smaller. Atoms increase in size passing down a group because the outermost electrons are in successively higher energy levels, which represent larger orbits. Of course, the positive charge on the nucleus also increases in each element as you go down a group, but the increased attraction for the outermost electron that this would cause (reducing the size of the atom) is offset by the inherently larger size of each successive outermost energy level.

Overall, the largest atoms are found in the lower left-hand corner of the Periodic Table and the smallest ones are in the upper right-hand corner.

The Ionization Energy of Atoms

The **ionization energy** of an atom is the amount of energy needed to completely remove an electron from an atom. It provides a measure of how tightly an atom holds on to its electrons. The removal of an electron from an atom is called **ionization.** The ionization energies are measured once the atoms are vaporized and in the gaseous state. The equation describing the process for sodium is

$$\text{ionization}$$
$$\text{Na}_{(g)} + \text{energy} \rightarrow \text{Na}^+_{(g)} + e^-$$

Removal of a negative electron from the neutral sodium atom will form a sodium **ion** with a 1+ charge.

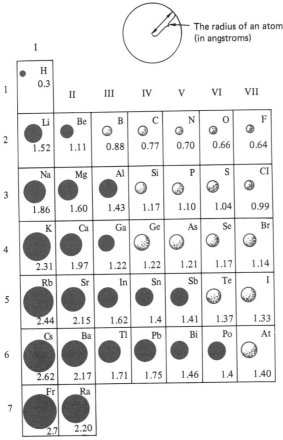

Fig. 6.7. The size of an atom can be described by its radius. Atomic radii tend to decrease across a period and increase down a group.

The more tightly an atom holds on to its valence electrons, the larger will be the ionization energy. Across a period, the atoms get smaller and smaller because the electrons are held increasingly more tightly by the atoms. So we would expect the ionization energies to increase as the atoms get smaller. This trend is seen in Figure 6.8, which graphically displays the ionization energies for the first 20 elements. Note that ionization energies are stated in kilocalories per mole, such as 118 kcal/mole for sodium. This means that 118 kilocalories of energy are needed to remove one electron from each of 6.02×10^{23} (1 mole) gaseous sodium atoms.

You can see how the ionization energies increase across a period in Figure 6.8. The group I metals have the lowest values (Li=124, Na=118) and the group VIII elements have the highest values (Ne=497, Ar=362). The minor "jags" seen in the curve as it traverses the second and third periods are caused by

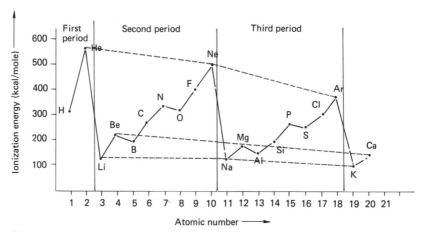

Fig. 6.8. The ionization energies for the elements generally increase across a period and decrease down a group.

slight increases in stability of some elements due to their electron arrangements. We need not be concerned with these minor effects here. Also note how the ionization energies trend to lower values as you pass down a group. This is most easily seen for the metals of group I and the rare gases of group VIII. This trend agrees with the increasing size of atoms going down a group.

The relatively low ionization energies for the elements on the left side of the Periodic Table are responsible for many of the properties characteristic of metals. The higher values for the elements on the right side of the Periodic Table are responsible for many of the properties characteristic of the nonmetals. Also, the trends in ionization energy begin to explain why the elements toward the bottom of groups III, IV, and V show increasingly metallic properties compared to those at the top. The elements toward the bottom have lower ionization energies.

Overall, then, we see that elements in the lower left-hand corner of the Periodic Table have the lowest ionization energies, and those in the upper right-hand corner the highest. You can remember this trend by relating it to the change in the size of atoms presented earlier.

Check Test Number 5

1. List the three elements given below in order of increasing size of their atoms.

Si   Mg   Cl

2. Do metals or nonmetals generally have the larger ionization energies?

Answers:
1. Mg is largest, Si next, and Cl the smallest. This is their order in the second period.
2. Nonmetals have larger ionization energies than metals.

The Modern Picture of the Atom (Optional)*

In the years following the Bohr theory of the atom, it became increasingly clear that the electron could not travel about the nucleus in an orbit of definite size. Though an electron would be in a definite energy level identified by a quantum number, the location of the electron had to be described in terms of **probability.** Those regions of space where the probability of finding an electron of a particular energy is high are called **orbitals.** An orbital is not the same as an orbit. An orbital can be thought of as a small volume of space in which the electron can be found most of the time. The shape and size of the orbitals are obtained from complex mathematical calculations, but the important point is that each orbital can hold a maximum of two electrons.

The number of orbitals in an energy level is equal to its quantum number squared, n^2. Thus, in the first energy level there is one ($1^2 = 1$) orbital, so the first level can hold a maximum of two electrons, the same number as in the Bohr model of the atom. There are four orbitals in the second energy level ($2^2 = 4$), and they can hold a maximum of eight electrons, again just as in the Bohr atom. Let us take a look at the shapes of the orbitals in the first two energy levels.

The one orbital in the first energy level is shaped like a sphere that surrounds the nucleus. This orbital is depicted in Figure 6.9. It is called the $1s$ orbital ($1 = 1$st energy level, $s = $ spherical shaped orbital). The region in which the electron would be found 90% of the time would be within this sphere. It might be very near the nucleus at any instant or far from it, but nearly all the time it would be in the spherical region.

Two kinds of orbitals appear in the second level, one is an s orbital like that in the first level only larger in size, and the others are called p orbitals. Though the $2s$ orbital is spherical, the three $2p$ orbitals are each shaped like "dumbbells" with two lobes, one on either side of the nucleus. In Figure 6.10, the nucleus is placed at the center (origin) of an x, y, z axis system, and each p orbital is located on one of the three axes. The p orbital on the z axis is labeled $2p_z$, the orbital along the y axis is $2p_y$, and the orbital on the x axis is $2p_x$.

*Omission of this section will not result in a loss of continuity.

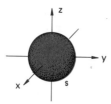

Fig. 6.9.   The spherical probability picture of an s orbital. The nucleus is at the origin of the axis system.

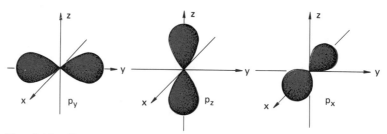

Fig. 6.10.   The p orbitals are described by dumbbell-shaped probability pictures. There are three p orbitals in a set, and each is aligned along one of the three axes, x, y, or z.

The third energy level contains nine orbitals ($3^2=9$); one s orbital ($3s$), three p orbitals ($3p$), and five orbitals called d orbitals ($3d$). We will not be concerned with the shapes of the d orbitals, but just like all orbitals, each one can hold a maximum of two electrons. The orbitals of the first three energy levels of an atom are summarized in Table 6.9.

TABLE   6.9
Orbitals and Electrons in the First Three Energy Levels of an Atom

Energy level	Level quantum number (n)	Number of orbitals (n^2)	Kinds of orbitals	Symbol of orbitals	Maximum number of electrons
1st	1	1	s (one)	1s	2 = 2
2nd	2	4	s (one) p (three)	2s 2p	2 6 } = 8
3rd	3	9	s (one) p (three) d (five)	3s 3p 3d	2 6 10 } = 18

The energy of the orbitals in the first three levels of an atom increases in the order shown below:

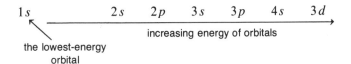

$1s$ — the lowest-energy orbital

$2s$ $2p$ $3s$ $3p$ $4s$ $3d$

increasing energy of orbitals

An easy way to remember the sequence is to write out the orbital notations in the pattern shown below, then starting at the $1s$ orbital, follow the arrow to succeeding higher energy ones:

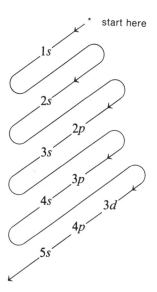

The electronic configuration of an atom can be written using orbital notations. Remember, just as we saw with the Bohr atom configurations, electrons fill the lowest-energy levels first, then the next lowest, and so on. A shorthand method of showing the number of electrons in an orbital, or set of identical orbitals (such as the p orbitals), is shown below for (a) two electrons in a $1s$ orbital and (b) three electrons in a set of $2p$ orbitals:

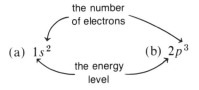

the number of electrons

(a) $1s^2$ (b) $2p^3$

the energy level

The way you could write the electronic configuration for lithium (At. No.=3) is shown below. The way the structure is described in the Bohr model is also shown for comparison.

Li $3p^+$ $2e^-$ $1e^-$ Li $1s^2\ 2s^1$
         $n=1$ $n=2$

Bohr notation   Modern notation

The electronic configurations for the first 11 elements are given in Table 6.10. Note how each orbital or set of identical orbitals is filled with the maximum number of electrons before the next higher energy orbital is used. As in the Bohr atom, the electrons in the highest energy level (the outermost electrons) are the **valence electrons.**

TABLE 6.10 The Electronic Configurations of the First 11 Elements. The Valence Electrons are Underlined		
Element	Atomic number	Electronic configuration
H	1	$\underline{1s^1}$
He	2	$\underline{1s^2}$
Li	3	$1s^2\ \underline{2s^1}$
Be	4	$1s^2\ \underline{2s^2}$
B	5	$1s^2\ \underline{2s^2\ 2p^1}$
C	6	$1s^2\ \underline{2s^2\ 2p^2}$
N	7	$1s^2\ \underline{2s^2\ 2p^3}$
O	8	$1s^2\ \underline{2s^2\ 2p^4}$
F	9	$1s^2\ \underline{2s^2\ 2p^5}$
Ne	10	$1s^2\ \underline{2s^2\ 2p^6}$
Na	11	$1s^2\ 2s^2\ 2p^6\ \underline{3s^1}$

The modern picture of the atom may seem more complicated, but it is in better agreement with our modern knowledge of nature. Still, the Bohr model of the atom continues to serve us well since it presents many of the important facts about atoms in an easily understood way.

Terms

Several important terms appeared in Chapter 6. You should understand the meaning of each one.

subatomic particle   Periodic Table
electron   periodic law

proton

neutron

nucleus

orbit

energy level

atomic number

mass number

isotope

emission spectrum

electronic configuration

valence electrons

electron-dot symbol

quantum number

group

period

alkali metal

halogen

transition metal

representative element

periodic property

atomic radius

ionization energy

†orbitals

†s orbital

†p orbital

Questions

Answers to the starred questions appear at the end of the text.

1. Where are the protons, neutrons, and electrons located in the atom?

*2. What is the charge on the electron, the proton, and the neutron? How do the masses of these subatomic particles compare?

*3. What does the atomic number of an element equal? What does the mass number of an atom equal?

*4. Complete the chart:

Element	Atomic number	Mass number	Number of		
			e^-	p^+	n
boron	——	11	——	——	——
——	3	——	——	——	4
——	——	——	——	12	12

*5. What are isotopes? How are they similar, how do they differ? Which of the following would represent isotopes?

(a) $^{14}_{6}C$, $^{12}_{6}C$;   (b) $^{14}_{6}C$, $^{14}_{7}N$;   (c) $^{1}_{1}H$, $^{2}_{1}H$

†Optional terms.

6. Is the atomic weight of chlorine on the Periodic Table in any way connected to the natural abundance of the isotopes of that element?

7. Briefly describe the model of the atom proposed by Neils Bohr in 1913.

8. Compare the size of the orbits of electrons in an atom to their energies.

*9. If an electron moved from a higher energy orbit to a lower energy orbit, what would happen to the energy lost by the electron? How much energy would be lost by the electron?

*10. Draw the first three energy levels of an atom and label each with the correct quantum number. Which is the lowest energy level? What is the maximum number of electrons that can be in each of these energy levels?

11. Draw pictures of the following atoms showing the number of protons in the nucleus and the arrangement of electrons in the energy levels.
   a. Boron (At. No. 5)   **c.** Argon (At. No. 18)
   *b. Aluminum (At. No. 13)   ***d.** Potassium (At. No. 19)

12. What is wrong with each of the following?

*a. Li $(3p^+)$ $3e^-$ $_{n=1}$ b. P $(15p^+)$ $2e^-$ $_{n=1}$ $8e^-$ $_{n=2}$ $6e^-$ $_{n=3}$
     (At. No. 3)   (At. No. 15)

c. O $(8p^+)$ $6e^-$ $_{n=2}$ $2e^-$ $_{n=1}$
    (At. No. 8)

13. What are the valence electrons in an atom? Where are they located?

*14. Write the electron-dot symbols for: hydrogen, chlorine, neon, sulfur, oxygen, nitrogen, and sodium.

15. In what order are the elements arranged in the Periodic Table? What is meant by the terms "group" and "period" in reference to the Periodic Table?

*16. Of what use are the group numbers (I through VIII) found on the Periodic Table? What is the collective name given the elements in groups I through VII?

*17. Give an example of: (a) an alkali metal, (b) a halogen, (c) a transition metal, (d) a rare gas, (e) an element in group III, (f) an element in the first period.

18. Why do representative elements in the same group have similar chemical properties?

19. In which energy level are the valence electrons located for the elements in the second period?

*20. What is meant by the term "periodic property"?

21. How do the sizes of the atoms of the elements change in going from left to right across a period? How do they change going down a group?

*22. Which atom would be largest, which smallest, of these five? F, N, Rb, Be, Na.

23. What is the ionization energy of an atom? How do the ionization energies change in going from left to right across a period?

*24. Generally, how do the ionization energies of metals compare with those of non-metals? Which element in each period has the highest ionization energy?

*25. †What is an orbital? How many electrons can be put in an orbital?

*26. †Compare the shape of an *s* orbital with that of a *p* orbital. In what way is the 2*s* orbital different from the 1*s* orbital?

*27. †How many *p* orbitals are in the second energy level? What is the principal difference between them?

*28. †Give the electronic configurations of the following elements using orbital notation. Underline the valence electrons.
  a. Li
  b. C
  c. F

The Nucleus—Radioactivity and Radiation

By now you have learned quite a lot about the atom. You know about the protons and neutrons in the nucleus and that the electrons around it have much to do with the chemistry of the atom. A lot of emphasis is placed on the electrons in a chemistry course, so much so that we sometimes take the nucleus for granted. So before we get too far along, we need to pause and consider the nucleus and some of the things that can go on there that can affect people's health.

In 1896 Henri Becquerel, a French physicist, accidentally discovered that uranium ore was capable of exposing photographic film if placed nearby. Becquerel was aware of the same effect being caused by x rays, which were discovered the year before, and he assumed some kind of x-ray-like radiation was being produced by the uranium ore that exposed the film. After further experiments, he was convinced that the radiation came from uranium itself, by a process that did not in any way resemble a chemical reaction. Several years later, the process which produced radiation was explained as a decay or breaking apart of the nucleus of the uranium atom. The breaking apart of nuclei with the production of radiation came to be known as radioactivity.

Soon after Becquerel's discovery, it was learned that the radiation from uranium could damage tissue and even cause death. Becquerel himself received a radiation "burn" on his chest from carrying around a vial of uranium ore in his shirt pocket. Eventually people learned how to control this kind of radiation so it could be used to our advantage. Medical specialists are very concerned with the effects of radiation on living systems since it can be very harmful. But if used with care it can be a formidable weapon against cancer and an invaluable aid in diagnosis.

We will start the discussion of the nucleus by describing the kinds of radiation produced by radioactivity, how it affects tissue, and then how it can be controlled and used to fight disease.

Objectives

By the time you finish this chapter you should be able to do the following:

1. Describe the three kinds of radiation that accompany radioactive decay.
2. Write equations describing nuclear reactions that produce alpha, beta, and gamma radiation.
3. Describe the difference between natural radioactivity and artificial radioactivity.
4. Describe what is meant by ionizing radiation, and explain how it can cause damage to a cell.
5. Explain the terms: biological half-life, radioactive half-life, and effective half-life.
6. Describe how radiation is detected using a Geiger counter, film badge, and scintillation detector.
7. Describe the units used to measure radiation: curie, roentgen, rad, and rem.
8. Describe three precautions you can take to avoid or minimize exposure to radiation.
9. Describe how radiation can be used in cancer therapy.
10. Describe how radioactive isotopes can be used to diagnose diseases of the internal organs.

Radiation and Nuclear Equations

Several elements found in nature exist only as naturally radioactive atoms. Generally, these elements will have an atomic number greater than 83 (bismuth) and will exist in several isotopic forms. These radioactive isotopes, commonly called **radioisotopes,** spontaneously **decay** producing **radiation** and often isotopes of other elements, which may also be radioactive.

There are three kinds of radiation that can accompany radioactive decay: alpha radiation, beta radiation, and gamma radiation. Sometimes only one form of radiation is emitted, but some radioisotopes produce two kinds simultaneously. Each can be symbolized by the Greek letter for which it is named; α (alpha), β (beta), and γ (gamma).

Alpha radiation consists of a stream of particles that each contain two protons and two neutrons. Alpha particles, as they are often called, are just like helium nuclei and have a mass of 4 amu and a 2+ electrical charge. The alpha particle itself is not radioactive, it is simply the product of a radioactive disintegration:

alpha particle $=$ helium nucleus

$$\alpha \quad = \quad \left(\begin{array}{c} 2p^+ \\ 2n \end{array}\right)$$

Alpha particles are not able to penetrate the skin much deeper than about 0.07 mm, the thickness of the protective layer of dead cells. Light clothing, paper, or a few centimeters of air will absorb alpha radiation and serve as protection from an external source. Alpha radiation can be dangerous, though, if an alpha source is brought into the body by mouth or breathed in through the lungs.

When a radioisotope decays emitting an alpha particle, a new element is formed, called a **daughter,** which has a mass number four units less and an atomic number two units less than the **parent** isotope. A nuclear reaction can be expressed in equation form just as a chemical reaction can. The symbols of the parent and daughter isotopes are written with the correct mass numbers and atomic numbers, and the alpha particle is symbolized as a helium nucleus, $^4_2$He. This symbol is identical to that used for the helium atom, but in nuclear reactions it represents only the composition of the helium nucleus, the alpha particle. Uranium-238 decays to produce thorium-234 and an alpha particle. The nuclear equation is

$$^{238}_{92}\text{U} \quad \rightarrow \quad ^{234}_{90}\text{Th} \quad + \quad ^4_2\text{He}$$

Parent	Daughter	Alpha
nucleus	nucleus	particle
uranium-238	thorium-234	

At this point a few things should be said about writing nuclear equations. Two rules must be followed:

1. The *sum* of the *mass numbers* on both sides of the equation must be equal.
2. The *sum* of the *atomic numbers* on both sides of the equation must be equal.

When both rules are obeyed, the nuclear equation is said to be balanced. In the previous equation, the mass number of the uranium isotope (238) equals the sum of the mass numbers of the thorium isotope (234) plus that of the alpha particle (4). The atomic number of uranium (92) equals the sum of the atomic numbers of thorium (90) plus that of the alpha particle (2).

$$\text{mass numbers: } 238 = (234 + 4)$$
$$^{238}_{92}U \qquad ^{234}_{90}Th + {}^{4}_{2}He$$
$$\text{atomic numbers: } 92 = (90 + 2)$$

Other nuclear reactions that produce alpha radiation are given below. Notice how the rules for balancing nuclear equations are obeyed. The first describes the decomposition of thorium-234 to radium-230, and the second describes the change of polonium-210 to lead-206:

$$^{234}_{90}Th \rightarrow {}^{230}_{88}Ra + {}^{4}_{2}He$$

$$^{210}_{84}Po \rightarrow {}^{206}_{82}Pb + {}^{4}_{2}He$$

Beta radiation consists of streams of high-speed electrons emitted from the nuclei of decaying radioisotopes. Being electrons, beta particles have an electrical charge of 1− and a small mass, usually approximated as 0 amu. Because of the small size of a beta particle, it can penetrate skin to a depth of about 4 mm, but not deep enough to reach the internal vital organs. Beta radiation can travel several thousand centimeters in air and can cause radiation burns on the skin, a condition not unlike a severe sunburn. Heavy clothing, thick cardboard, and inch-thick wood provide protection from external beta radiation, but just as with alpha radiation, a beta source carried inside the body can be very harmful.

beta particle = high-speed electron from the nucleus

$$\beta \qquad = \qquad \qquad e^-$$

When a beta particle is emitted from a nucleus, the net effect is the loss of one neutron and the appearance of one proton. Though an oversimplification, and not entirely correct, you can think of beta emission resulting from the breaking apart of a neutron into a proton and a high-speed electron (the beta particle). In nuclear equations the beta particle is symbolized $_{-1}^{0}e$ with a mass number of zero and an assigned atomic number of −1. Beta emission will produce a daughter isotope with the same mass number as the parent but with an atomic number one unit larger. Carbon-14 decays with beta emission forming nitrogen-14:

$$^{14}_{6}\text{C} \rightarrow {}^{14}_{7}\text{N} + {}^{0}_{-1}e$$

Phosphorus-32 produces sulfur-32 and the beta particle:

$$^{32}_{15}\text{P} \rightarrow {}^{32}_{16}\text{S} + {}^{0}_{-1}e$$

Note in each equation that the sums of the mass numbers and atomic numbers are equal.

Gamma radiation is a form of electromagnetic radiation very much like x rays. Gamma radiation (or gamma rays) has very short wavelengths, is very energetic, and can penetrate deeply into the body and cause severe cell damage. Thick sheets of lead or concrete are required to shield gamma radiation. The principal difference between gamma rays and x rays is their source. Gamma rays come from the nucleus of decaying atoms and x rays result from certain energy changes of the electrons around the nucleus. Since gamma and x rays are similar in their effects on tissue, both will be discussed in later sections.

gamma ray = electromagnetic radiation like x rays

γ = $\sim\!\sim\!\sim\!\sim$

Gamma rays are usually, but not always, emitted along with alpha or beta particles in nuclear reactions. They have no mass nor electrical charge and in nuclear equations are symbolized γ. A new element will be produced with gamma emission *if* it is accompanied by alpha or beta emission. Two sources of gamma radiation used in medicine are cobalt-60 and technetium-99m. Cobalt-60 also produces a beta particle:

$$^{60}_{27}\text{Co} \rightarrow {}^{60}_{28}\text{Ni} + \underset{\substack{\text{Beta} \\ \text{particle}}}{{}^{0}_{-1}e} + \underset{\substack{\text{Gamma} \\ \text{ray}}}{\gamma}$$

Technetium-99m is called a **metastable isotope** as indicated by the *m* following the mass number. A metastable isotope contains some "extra" energy that can be released as gamma radiation when it converts to a more stable form of the *same* isotope:

$$^{99m}_{43}\text{Tc} \rightarrow {}^{99}_{43}\text{Tc} + \gamma$$

The characteristics of the three kinds of radiation are summarized in Table 7.1.

| | | | | TABLE 7.1 | |
| | | | Alpha, Beta, and Gamma Radiation | | |
Name	Symbol	Electrical charge	Approximate mass (amu)	Description	Penetrating ability
Alpha	α, ^4_2He	2+	4	nucleus of He atom	low
Beta	β, $^0_{-1}e$	1−	0	high-speed electron	moderate
Gamma	γ	0	0	high-energy electro-magnetic radiation, like x rays	very high

Check Test Number 1

1. List the three kinds of radiation that accompany radioactive decay in order of increasing penetrating power.
2. Write nuclear equations describing the following:
   a. Strontium-90 ($^{90}_{38}\text{Sr}$) decaying to form yttrium-90 ($^{90}_{39}\text{Y}$) plus a beta particle.
   b. Radon-222 ($^{222}_{86}\text{Rn}$) decaying to form polonium-218 ($^{218}_{84}\text{Po}$) plus an alpha particle.
   c. Lead-214 ($^{214}_{82}\text{Pb}$) decaying to form bismuth-214 ($^{214}_{83}\text{Bi}$) plus a beta particle and a gamma ray.

Answers:
1. In order of increasing penetrating power: alpha, beta, gamma.
2. a. $^{90}_{38}\text{Sr} \rightarrow ^{90}_{39}\text{Y} + ^{0}_{-1}e$
   b. $^{222}_{86}\text{Rn} \rightarrow ^{218}_{84}\text{Po} + ^{4}_{2}\text{He}$
   c. $^{214}_{82}\text{Pb} \rightarrow ^{214}_{83}\text{Bi} + ^{0}_{-1}e + \gamma$

Natural and Artificial Radioactivity

About 20 radioactive isotopes are found in nature, such as those of uranium and radium. These natural radioisotopes are said to display **natural radioactivity.** However, most of the radioisotopes used in medicine are produced artificially in nuclear reactors and huge machines called accelerators. These "artificial" radioisotopes display what is then called **artificial radioactivity.** Whether a radioisotope is found in nature or is made artificially, it will have the same properties; the only difference is in the origin.

A common method used to prepare radioisotopes involves bombarding a stable target atom with neutrons (symbolized $_0^1n$) in a nuclear reactor. Once the neutron penetrates the target nucleus and is captured by it, a gamma ray is emitted leaving the new radioactive isotope behind. Radioactive cobalt-60 is prepared from nonradioactive cobalt-59 in this way:

$$_{27}^{59}\text{Co} + _0^1n \quad \rightarrow \quad _{27}^{60}\text{Co} + \gamma$$

Nonradioactive   Radioactive

Isotopes of nearly every element have been prepared this way. In many cases adding a particle to the nucleus will produce a different element, a process known as **transmutation.** Phosphorus-32 and iodine-131, two radioisotopes used in medicine, are prepared by transmutation processes. A proton, $_1^1\text{H}$, is also formed in the first reaction.

$$_{16}^{32}\text{S} + _0^1n \rightarrow _{15}^{32}\text{P} + _1^1\text{H} + \gamma$$

Radioactive

$$_{52}^{130}\text{Te} + _0^1n \rightarrow _{53}^{131}\text{I} + _{-1}^0e + \gamma$$

Radioactive

Transmutation can also be made to occur using alpha particles, protons, electrons, or nuclei of smaller atoms in place of neutrons. The elements beyond uranium in the Periodic Table (those with atomic numbers above 92) are all prepared in transmutation reactions. They are not found in nature.

The Biological Effects of Radiation

Radiation can be extremely harmful to living systems because of its ability to cause chemical changes in tissue. Alpha particles, beta particles, gamma rays, and x rays interact with molecules in cells creating ions and highly reactive fragments called free radicals. Because of these effects, nuclear radiations and x rays are called **ionizing radiations.** Though alpha and beta particles are less penetrating than gamma or x rays, their mass and electrical charge cause them to be more effective in forming these reactive species. The ions and free radicals formed when a water molecule is fragmented by radiation are especially important since cells contain a large amount of water. Three ways in which radiation can affect a water molecule are shown in Figure 7.1. The ions and radicals can react with proteins, enzymes, and genetic material (DNA) in the cell with profound and even fatal results. Sometimes delayed cell division can occur, resulting in the formation of cell giants having a volume several hundred times larger than normal cells. Chemical changes in the genetic material in the

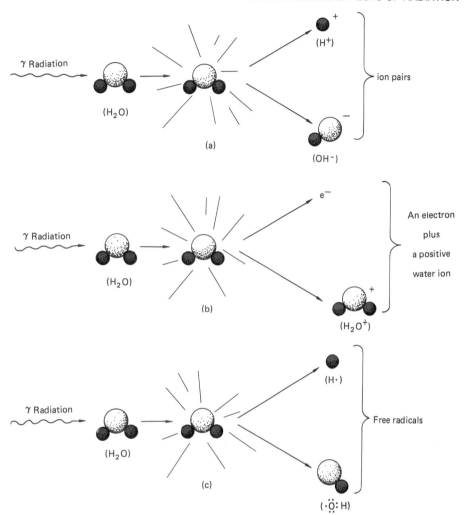

Fig. 7.1. Ionizing radiation can split water molecules into highly reactive fragments which can attack various components in cells and cause damage. Once the energetic radiation is absorbed by the molecule it can (a) split into a hydrogen ion H^+ and hydroxide ion OH^-; (b) lose an electron and become a positive water ion H_2O^+; or (c) split to form very reactive neutral fragments called free radicals.

cell nucleus can cause tumors, birth defects (if in embryos), cancer, and even mutations in future generations.

It does not require a large exposure to ionizing radiation to bring about these changes. Many people believe there is potential danger in even the lowest levels of exposure since the effects are cumulative, that is, the effect of each exposure is added to that of each preceding exposure. Over time, daily exposure to low levels of radiation can represent as serious a threat to health as a single large exposure. Experts generally agree that any exposure to radiation can cause some changes in a cell's genetic material, but long-term effects caused by exposure to small amounts of radiation are very difficult to measure. No one really knows if low-level exposure can cause nongenetic damage such as cancer. Some people believe any exposure is harmful while others believe there is a **threshold exposure level** below which nongenetic damage is unlikely. Radiation exposure below this threshold level would cause only slight damage, if any, which would be quickly repaired by the body. Because of the differing opinions, and lack of solid evidence either way, minimum allowed radiation exposure standards are arbitrarily set as low as reasonably possible. It is generally felt that any risk accompanying the use of x rays and radioisotopes for diagnostic purposes is greatly offset by the reward of an early diagnosis of disease.

Cells that are undergoing rapid division are more susceptible to radiation damage, such as those in lymphatic tissue, the intestinal tract, bone marrow, gonads, and especially those in the developing embryo. The eye is susceptible to cataract formation while nerve and muscle cells are more resistant to radiation damage. The rapidly multiplying cells of a cancer can be slowed in their growth or destroyed by radiation, causing it to be a powerful tool in cancer therapy.

Overexposure to nonlethal amounts of radiation either in a single dose or by slow accumulation over time may result in **radiation sickness.** The symptoms of radiation sickness are nausea, vomiting, prostration, reduced white cell count, diarrhea, hemorraghing, and anemia. Susceptibility to infectious disease is also increased owing to the reduced effectiveness of the body's immunity system. Exposure to lethal amounts of radiation will cause death in a matter of hours or weeks due to the failure of one or more vital functions.

At this point you may be having serious reservations about ever being around or working with nuclear radiation or x rays. It is important to be aware of the hazards of radiation, but it is equally important to realize that radiation can be handled safely and with great medical benefits. It is something like fire. Controlled and in its place, it is a tremendous benefit to all of us, but uncontrolled, it can be a devastating force.

Check Test Number 2

1. Write the nuclear equation describing the transmutation of gold-197 ($^{197}_{79}$Au) to gold-198 ($^{198}_{79}$Au) by the absorption of one neutron. Would gold-198 prepared in this way be considered a natural radioisotope?
2. What regions of the body are most susceptible to radiation damage?

Answers:
1. $^{197}_{79}$Au $+$ $^{1}_{0}n$ $\rightarrow$ $^{198}_{79}$Au $+$ γ. Gold-198 would be considered artificial.
2. Lymphatic cells, intestinal tract, bone marrow, gonads, and in general, any tissue undergoing rapid cell division.

Half-Life

Carbon-14 and phosphorus-32 are both radioactive and produce beta particles when they decay. But there is a major difference between them. Phosphorus-32 decays at a much faster *rate* than does carbon-14. The rate of decay of a radioisotope is described by its **half-life** ($t_{\frac{1}{2}}$), the time required for one half the radioactive atoms in a sample to decay. The half-life for phosphorus-32 is a relatively short 14 days, while for carbon-14 it is 5760 years.

Perhaps an example would describe the idea of half-life better. Suppose you started with 1000 atoms of phosphorus-32. Fourteen days later only 500 of them would remain, the other 500 would have decayed. In the next 14-day period, half of the remaining 500 atoms would decay and only 250 would remain. After a third half-life period, only 125 atoms would remain.

Radioisotopes used in the body for diagnostic purposes should have half-lives on the order of several days or less to minimize the patient's exposure to radiation. These short half-life radioisotopes are prepared artificially, since those found in nature characteristically have half-lives on the order of thousands or millions of years. Several radioisotopes with their half-lives are given in Table 7.2.

A second kind of half-life is important when radioisotopes are used in the body, and it is called the biological half-life of the isotope. The **biological half-life** measures the rate at which the body can eliminate the isotope through biological means. Both the biological and radioactive half-lives of an isotope need to be considered when determining how long a patient receiving radio-isotopes will remain radioactive. Both means of diminishing radioactivity are summed up in what is called the **effective half-life** of a radioisotope in the body. The effective half-life is always smaller than either the radioactive half-life or the biological half-life. An isotope with a long biological half-life should then have a short radioactive half-life to minimize exposure, and vice versa. The three half-lives of several medically important radioisotopes are given in Table 7.3.

TABLE 7.2 The Half-lives of Several Radioisotopes		
Isotope	Half-life $t_{1/2}$	Radiation
Iron-59	46.3 days	β, γ
Cobalt-60	5.2 years	β
Iodine-131	8 days	β
Radium-226	1590 years	α, γ
Uranium-235	800 million years	α, γ

TABLE 7.3 Radioactive, Biological, and Effective Half-Lives for Some Medically Important Isotopes				
Isotope	Use	Radioactive half-life	Total-body biological half-life	Effective half-life
Phosphorus-32	leukemia	14.3 days	257 days	13.5 days
Iodine-131	thyroid	8 days	138 days	7.6 days
Technetium-99m	several organs	0.25 day	1 day	0.2 day
Xenon-133	lungs	5 days	very short	very short

Detection of Radiation

The human body cannot sense the presence of moderate amounts of radiation, so a variety of tools have been developed for this purpose. Some are designed to measure the amount of radiation present at a given time while others measure the dose of radiation accumulated over several hours or weeks.

A familiar device used to detect the presence of ionizing radiation is the **Geiger–Müller counter**, or Geiger counter for short. It uses ions formed when radiation passes through a gas to complete an electrical circuit. The greater the intensity of radiation, the greater the number of ions that form and the greater the flow of electricity. A meter or audible "clicker" informs the user of the presence of radiation. The Geiger counter is described in Figure 7.2, p. 146.

Photographic film can be "exposed" by radiation. The **film badge** (Figure 7.3), worn by people who work around radiation, contains several layers of film sandwiched between thin sheets of cadmium metal in a plastic holder. Once developed, the amount of exposure is measured by specialists and translated into a dose quantity of radiation received by the wearer. A cumulative record is kept for the individual and if minimum safety levels of radiation are ever exceeded, the individual is moved to another work area away from radiation sources for a specified time.

A **dosimeter** performs the same function as the film badge, measuring doses of radiation received over a given period of time. It is first charged with electricity and then slowly discharges as radiation strikes it. Dosimeters have an advantage over film badges since the absorbed dose can be monitored daily by peering through a small window to see the position of an indicator on a scale. Dosimeters are usually not considered as reliable as film badges.

An important detector used to locate and measure radiation in the body is the **scintillation camera.** It contains a thin layer of crystals that emit flashes of light when struck by radiation. The light is then converted into electrical signals and recorded. Scintillation cameras are used in machines that scan the body to locate and measure radiation, and a typical instrument is shown in

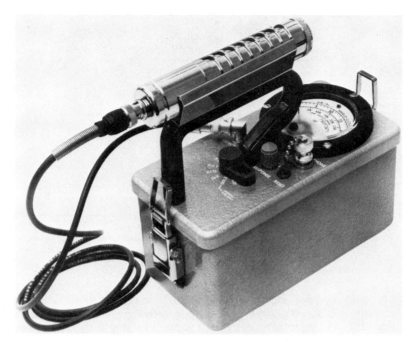

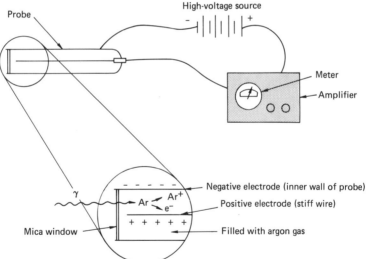

Fig. 7.2. The Geiger–Müller counter is used to detect and measure radiation. Radiation entering the window of the probe ionizes argon atoms forming Ar^+ and electrons. The positive argon ions are attracted to the negative electrode (the inside wall of the probe) and the electrons are attracted to the positive electrode (a wire in the center of the tube). When the electrons and ions reach the electrodes, a small electric current will flow, which is amplified and indicated on a meter. The greater the intensity of the radiation, the greater the electric current. (Picker Corporation, Cleveland, Ohio)

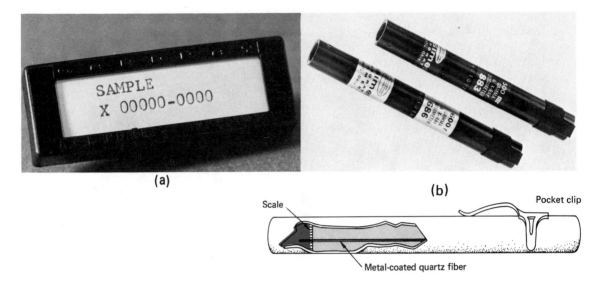

(a)

(b)

Scale

Pocket clip

Metal-coated quartz fiber

Fig. 7.3. The film badge (a) and pocket dosimeter (R.S. Landauer, Jr., and Company, Glenwood, Illinois) (b) are used to monitor exposure to radiation. The deflection of a quartz fiber in the dosimeter is proportional to the amount of ionizing radiation that has passed through it, just as is the degree of exposure of the film in the film badge. (Dosimeter Corporation of America, Cincinnati, Ohio)

Figure 7.4. These instruments produce x-ray-like pictures of the tissue that is emitting the radiation. (Photographs of brain scans appear in Figure 7.6.)

Units for Measuring Radiation

When radiation is used in therapy or diagnosis, it is imperative that the administered dose is large enough to bring about the desired result without subjecting the patient to undue exposure. Over the years, several units have been established for describing amounts of radioactivity or radiation. Some are more useful than others for measuring the effects of radiation on tissue.

The **curie** (Ci) is used to measure the amount of radioactivity in a radioactive source. One curie equals 37 billion disintegrations per second. It represents a large amount of activity, so units of millicurie ($1 \text{ mCi} = 10^{-3}$ Ci) and microcurie ($1 \text{ } \mu\text{Ci} = 10^{-6}$ Ci) are often used. The curie does not provide any information as to the kind of radiation emitted by the sample (α, β, or γ) or its effect on tissue.

The **roentgen** (r, pronounced rĕnt́gĕn) is used to describe the energy of x-ray and gamma radiation. One roentgen of radiation will produce about 2 billion pairs of ions when passing into one cubic centimeter of air at 0°C and standard

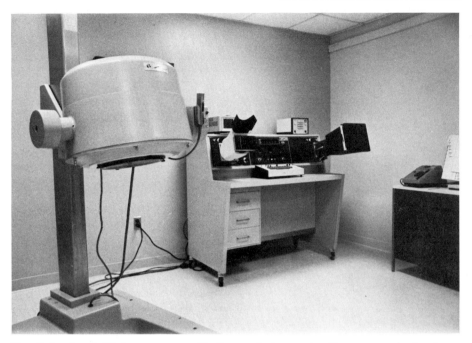

Fig. 7.4. A scintillation camera with the control console. The camera is the large cylinderical unit mounted on the stand to the left of the control console. The camera "lens" at the base of the cylinder is positioned over the organ to be scanned. (Courtesy of Isographics Inc., Stone Mountain, Georgia.)

atmospheric pressure. The roentgen is not a large amount of radiation since a single gamma ray can produce thousands of ions. However, an exposure to 600–700 roentgens would be fatal for most people. Unfortunately the roentgen only describes the effect of radiation on air and not bone or soft tissue. In reality, different materials absorb radiation to varying degrees.

The *r*adiation *a*bsorbed *d*ose, called the **rad** (D), takes the nature of the absorbing material into account, and it is used for all forms of ionizing radiation. One rad of radiation will generate 100 ergs of energy in one gram of absorbing tissue. You may not be familiar with the erg as a unit of energy, but 100 ergs is not very much energy, about two-millionths of a calorie. However, 100 ergs does represent a significant amount of absorbed radiation. One hundred rads absorbed by the body will cause radiation sickness, and 600 rads would be fatal for most people. Typical doses of radiation absorbed in routine x-ray examinations are: chest x ray: 0.04 to 1 rad, gastrointestinal examination: 1 rad, and an arm or leg x ray: 0.25 to 1 rad. An exposure of one roentgen of x-ray or gamma radiation is roughly equivalent to an absorbed dose of one rad for humans.

A further refinement in the measurement of the absorbed dose of radiation takes into account the effect of each kind of radiation on tissue. One rad of

TABLE 7.4 Units of Radiation Measurement	
Unit	Description
Curie (Ci)	An amount of radioactive material that undergoes 37 billion disintegrations per second.
Roentgen (r)	The amount of gamma or x radiation that will produce about 2×10^{10} ion pairs in 1 cc of dry air.
Radiation absorbed dose—rad (D)	An amount of radiation absorbed by tissue that produces 100 ergs of energy per gram of tissue. For x rays and gamma radiation, 1 roentgen of exposure is a dose of about 1 rad.
Roentgen equivalent for man (rem)	Considers the effect of different kinds of ionizing radiation (α, β, γ, and x rays) on tissue. The number of rems equals the dose in rads multiplied by factors that relate the ability of a particular kind of radiation to cause damage in tissue. The cumulative effects of radiation can be expressed in rems.
LD_{50}-30 day	The dose required by each member of a population to kill 50% of them within 30 days.

alpha radiation is ten times more dangerous to humans than one rad of beta, gamma, or x radiation. To compensate for this, the **relative biological effectiveness** (RBE) of alpha radiation is assigned a value of 10, and the others a value of 1. Multiplying the RBE of the radiation by the absorbed dose in rads gives a new unit of absorbed dose called the **dose equivalent** (DE). Sometimes other factors are also included to compensate for the way different organs absorb radiation. The unit of the DE is the **rem**, derived from *r*oentgen *e*quivalent for *m*an. One rem of any radiation will cause the same biological effect in humans as the absorption of one roentgen (or one rad) of x rays.

Cumulative doses of radiation are expressed in rems. An individual who received 1 rem of x radiation and 0.5 rem of alpha radiation has received a total dose equivalent of 1.5 rem.

The effect of radiation on living organisms can also be expressed in terms of the dose that would be fatal for 50% of the exposed population within 30 days. The required lethal dose can be symbolized LD_{50}-30 days. It is estimated to be 500 rems for humans. It is 700 rems for rats.

The units of radiation measurement are summarized in Table 7.4.

Radiation Safety

It should be clear to you that ionizing radiation can be harmful. Anyone working around radiation must take precautions to minimize exposure. All of us are subject to the **natural background** radiation that comes to us from the sun, radioactivity in the earth, and pollution. The average dose from this background is about 0.17 rem per year. The National Council on Radiation Protection and Measurements sets a limit of 0.5 rem per year from all sources for the average U. S. citizen, and individuals who work with radiation should not exceed a whole-body dose of 5 rem per year.

Any area in which radiation is used should be identified with the nuclear radiation symbol, a purple three-bladed figure on a yellow background seen in Figure 7.5. Often patients who have received diagnostic radioisotopes will be identified with the same symbol and will need special treatment. Since much of an administered radioisotope dose may be excreted in the days immediately following a test, all excrement must be retained for proper disposal. In some cases, visitation of the patient must be limited to brief periods.

If you are working around radiation, there are three ways to keep exposure to a minimum: (1) keep distance between you and the radiation source; (2) use adequate shielding; (3) minimize exposure times. A few words would be in order about each.

Radiation travels away from the source in all directions and in straight lines. The intensity of the radiation diminishes as the distance from the source increases. Radiation is like light from a candle in this sense. Near the candle the light is more intense and you can read by it, but further away the light becomes so dim (less intense) it is not able to illuminate the page and reading becomes difficult. If you *double* your distance from a radiation source, the radiation intensity is cut to *one-fourth* the original amount. *Tripling* the distance cuts it to *one-ninth* the original amount.

Fig. 7.5. The caution symbol for radiation is a magenta figure on a yellow background. Below the symbol the words "Radioactive Materials," "Radiation Area," or "High Radiation Area" indicate the nature of the radiation in the area.

Proper shielding will absorb radiation. Thin sheets of metal or inch-thick wood will stop alpha and beta radiation, but gamma and x rays require fairly thick sheets of lead or concrete walls. Radiation labs will have shielded areas for the protection of technicians and patients as well as portable floor-standing shields, lead-lined aprons, and body drapes. Heavy lead containers are necessary to store radioisotopes safely, and all associated syringes, cups, and dishes must be properly shielded also.

Keeping the time of exposure to radiation to a minimum will also reduce the risk of harm. If you are working with radioisotopes, know the procedures thoroughly beforehand so they can be carried out with dispatch.

A further precaution should be mentioned. Never smoke, drink, or eat around radioactive materials. Ingestion of radioisotopes by mouth can be quite dangerous and it is unnecessary.

Use of Radiation in Therapy

Radiation is successfully used in the treatment of cancer. Cancer cells, just like normal cells, can be destroyed or slowed in their growth when exposed to radiation. The treatment is further enhanced by the generally faster rate of repair of normal cells compared to cancer cells when damaged by radiation.

Because of their ability to penetrate the body, gamma radiation and x radiation are both used to destroy tumors. The principal source of gamma radiation is cobalt-60. A 75–200 Ci sample of cobalt-60 is housed in a lead chamber and positioned far enough from the patient to provide an average dose of about 0.002 roentgen per hour. In practice, the radiation source is mounted on a machine that rotates it around the patient so that the cancerous tissue is constantly exposed to the gamma rays while surrounding healthy tissue receives a reduced dose. The technique is called **rotational teletherapy**.

Cesium-137 has also been used as a source of gamma radiation. It has a longer half-life than cobalt-60 and needs replacement less often. Also the gamma radiation from cesium-137 is lower in energy than that from cobalt-60, an advantage for treatment of head and neck conditions.

Radioisotopes can be encapsulated in small metal tubes or needles for placement in affected tissue. Both radium-226 and cobalt-60 have been used this way and the technique is called **implantation therapy**. Small plaques of radioactive beta emitters can be placed on the surface of the skin or eye to treat superficial tumors. Penetration can reach from a fraction of a millimeter to four or five millimeters depending on the choice of isotope. This technique is called **surface therapy**.

Some elements naturally accumulate in certain parts of the body, and it provides the basis for **internal therapy**. Radioactive iodine-131 will accumulate in the thyroid gland, and it can be used to treat cancer of the thyroid and hyperthyroidism. Phosphorus-32 accumulates in bone tissue and it can be used to treat leukemia, a cancer affecting white blood cells. The source of leukemia is located in the marrow of the bones, and the progress of the disease can be reduced by the beta radiation emitted by the decaying phosphorus-32 located there.

Use of Radioisotopes in Diagnosis

The use of radioisotopes as a diagnostic tool has grown rapidly since the mid-1950's, and a new medical discipline called nuclear medicine has evolved.

A radioisotope, once introduced into the body, will continue to decay and emit radiation. The radiation (usually gamma radiation) can be detected by a scintillation camera, and the location and movement of the radioisotope can be monitored. In practice, the pattern of radiation from the body is translated into a photograph which shows an organ or region of the body much like what is seen on an x-ray photograph. However, the image obtained with radioisotopes sometimes allows detection of pathological conditions that are difficult or even impossible to find with x-ray techniques. Sometimes radioisotopes will tend to concentrate in the region of a tumor and produce a "hot spot" in the image. This is seen with brain tumors. Other times, as in the liver or thyroid, radioisotopes will concentrate in normal tissue and a tumor appears as a region of low radiation called a "cold spot." Brain scans are seen in Figure 7.6.

Radioisotopes are not administered as pure elements for diagnostic purposes. Rather, they are incorporated into water-soluble compounds that can be given orally or intravenously. Often the compound will have a high affinity for a particular organ or region of the body causing the radioisotope to be concentrated there. A compound that contains a radioactive atom is said to be **tagged**. Red blood cells can be tagged with radioactive chromium-51 and injected into the blood stream to monitor blood flow or red blood cell survival. Iodine-131 is administered as a solution of tagged sodium iodide (NaI) for thyroid studies. Technetium-99m is used as a solution of sodium pertechnetate (NaTcO$_4$) for organ studies, or technetium pyrophosphate (Tc$_2$P$_2$O$_7$) for bone studies. Because of its high visibility and short effective half-life, technetium-99m is widely used for diagnostic studies.

Several diagnostic uses of radioisotopes are described in Table 7.5.

TABLE   7.5
Diagnostic Uses of Radioisotopes In the Body

Area of body	Isotope and how administered	Use
Brain	technetium-99m (I.V.)	detects and locates brain tumors, stroke
	iodine-131 (inj. into cerebrospinal fluid)	detects fluid buildup on brain, blockage of movement of cerebrospinal fluid
Thyroid	iodine-131 (oral)	determines rate of iodine uptake by thyroid: rapid uptake indicates hyperthyroid, reduced uptake, hypothyroid conditions
	iodine-131 or technetium-99m (I.V.)	determine shape and size of thyroid; locate abnormal masses in the gland
Lung	xenon-133 (inhaled)	determines if lungs fill properly; locates regions of reduced ventilation, tumors
	substances tagged with iodine-131 or technetium-99m (I.V.)	substance gathers in lungs and helps locate blood clots
Heart	technetium-99m (I.V.)	can determine cardiac output, size and shape of heart, locate myocardial infarcts, holes between chambers
Liver–Spleen	technetium-99m (I.V. as sulfur colloid)	determines size and shape of liver and spleen, location of tumors
	copper-64 (I.V.)	measures metabolic function of the liver to aid diagnosis of Wilson's disease (accelerated absorption of copper by the body)
Pancreas	selenium-75 (I.V.)	determines size and shape of pancreas and extent of abnormal masses
Kidney	technetium-99m, iodine-131, or chromium-51 (I.V.)	determine renal function, location of cysts; a common follow-up procedure for kidney transplant patients
Bone	technetium-99m, barium-131 (I.V.)	allows early detection of the extent of bone tumors and active sites of rheumatoid arthritis
Blood	red blood cells tagged with chromium-51 (I.V.)	determine blood volume in the body, survival rate of red blood cells
	technetium-99m (I.V.)	determines blood flow

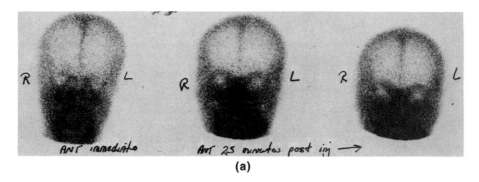

(a)

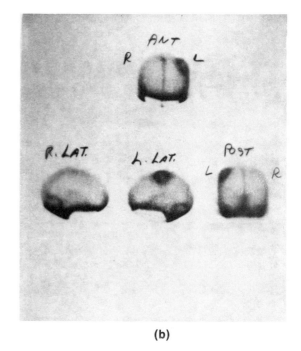

(b)

Fig. 7.6. Scans obtained in two different brain studies showing the distribution of Tc^{99m} in the blood system about the brain. In (a) the distribution of the radioisotope appears as white regions that outline the shape of the brain. The brain appears normal since "hot spots" do not appear on or within the brain volume either immediately or 25 minutes after injection of the radioisotope. In (b) a concentration of the radioisotope on the left hemisphere of the brain locates an abnormal growth. In this set of scans, the "hot spot" appears as a dark region because the scans were prepared for viewing in a different way than those in (a). Ant—anterior; Post—posterior. (Courtesy of Isographics Inc., Stone Mountain, Georgia.)

Terms

Several important terms were introduced in Chapter 7. You should understand the meaning of the following:

radioactivity	effective half-life
radioactive decay	Geiger counter
alpha radiation	film badge
beta radiation	curie
gamma radiation	roentgen
natural radioactivity	rem
artificial radioactivity	rad
transmutation	LD_{50}-30 day
ionizing radiation	background radiation
radiation sickness	teletherapy
half-life	internal therapy
biological half-life	implantation therapy

Questions

Answers to the starred questions appear at the end of the text.

*1. What part of an atom is changing in a radioactive disintegration?

 2. What are the three kinds of radiation that are products of radioactive decay?

*3. Complete the following chart to describe the three kinds of nuclear radiation:

Radiation	Symbol	Charge	Approximate mass (amu)	Penetrating ability
_____	_____	2+	_____	_____
_____	$\beta, -_1^0 e$	_____	_____	_____
_____	_____	_____	_____	very high

 4. Why are gamma rays and x rays considered to be similar? Considering their source, how are they different?

*5. What is ionizing radiation?

 6. What can happen to a water molecule when exposed to ionizing radiation?

*7. What tissues in the body are more susceptible to damage by ionizing radiation?

 8. Which element could be represented by the nuclear symbol, $_{82}^{210}X$?

9. Write the nuclear symbol for each isotope listed below:
   *a. carbon-14   c. radium-226
   *b. uranium-238   d. cobalt-60

10. Write the nuclear equation for each of the following nuclear reactions. The symbols for some elements are in parentheses.
    *a. Uranium-238 decaying to form thorium-234 (Th) plus an alpha particle.
    b. Cobalt-60 decaying to form nickel-60 plus a beta particle and a gamma ray.
    *c. Phosphorus-32 decaying to form sulfur-32 plus a beta particle.
    d. Xenon-133 decaying to form cesium-133 plus a beta particle and a gamma ray.
    *e. Radium-226 (Ra) decaying to produce radon-222 (Rn) plus an alpha particle and a gamma ray.

11. What is transmutation?

12. Write nuclear equations for each of the following transmutation reactions:
    *a. Uranium-238 plus a neutron (^1_0n) forming uranium-239 plus a gamma ray.
    b. Nitrogen-14 plus a neutron forming carbon-14 plus a proton (^1_1H).

*13. The radioactive half-life of xenon-133 is 5 days. If you started with 1×10^5 atoms of Xe-133, how many would remain after 30 days?

14. If in one hour the number of radioactive bismuth-214 atoms in a sample fell from 2×10^4 to 2.5×10^3, what is the half-life of bismuth-214?

*15. Considering half-life information only, which of the following hypothetical radio-isotopes would *not* be suitable for internal diagnostic use?

Isotope	Radioactive $t_{1/2}$	Biological $t_{1/2}$
A	12 hours	6.5 days
B	13.6 years	640 days
C	43.1 weeks	6.8 years
D	6.1 weeks	2.3 days

16. How do the radioactive half-lives of artificial radioisotopes compare with those of natural radioisotopes?

*17. Which "unit" used for measuring radiation would best fit each of the following:
    a. tabulating cumulative doses of radiation
    b. exposure to x rays or gamma rays only
    c. the amount of radioactivity in a sample
    d. the probable dose of radiation that would be lethal for a particular animal

18. How does radiation affect photographic film so that the film badge can be used to measure absorbed doses of radiation?

19. Briefly describe how a Geiger–Müller counter works.

20. Describe the caution sign that is used to mark radioactive materials and radiation areas.

*21. If you were 6 feet from a radiation source, would the radiation intensity be greater, the same, or less at a spot 12 feet from the source? If you think the intensity will change, predict by how much.

22. What could you use to shield yourself from the following radiations?
    a. gamma rays
    b. alpha rays
    c. beta rays
    d. x rays

*23. If you work around radiation, what three precautions should you take to minimize exposure?

24. Why can ionizing radiation be used to treat cancer?

*25. Why is iodine-131 used for thyroid studies?

26. Why is it necessary to retain the urine and stool of patients who have been administered diagnostic radioisotopes?

27. What does it mean to "tag" a molecule with a radioactive isotope?

Atoms in Combination—
Ionic and Covalent
Compounds

In Chapter 6 you learned that an atom of each of the 106 elements has its own unique identity, the result of the number of protons in its nucleus. Yet this relatively small number of elements can, through combination with one another, form millions of compounds.

The human body is a collection of thousands of compounds. Some are rather simple like water and sodium chloride while others are very complex like the proteins, nucleic acids, and fats. But no matter how simple or complex compounds may be, they all come into being because of certain fundamental interactions between two or more atoms. These interactions, which involve the valence electrons of each atom, result in the formation of chemical bonds which join the atoms together. If atoms combine with the formation of ions, then an ionic compound forms with the ions held together by ionic bonds. On the other hand, if atoms combine to form molecules, the result is a covalent compound and the atoms are held together by covalent bonds. Both kinds of compounds will be studied in this chapter.

The understanding of chemical bonding and the formation of compounds is a very important part of any chemistry course. You should do your best to understand each section before going on to the next. The discussion begins with the octet rule, which will help you to predict formulas of many compounds. Then ionic compounds and covalent compounds will be examined with special emphasis on the bonds that join the atoms together. You will learn how to predict the shapes of covalent molecules after you learn how to represent molecules with Lewis structures. Finally the way ionic and covalent molecules are named will be described.

Objectives

By the time you finish this chapter you should be able to do the following:

1. State the octet rule and tell how it is used to predict the formulas of compounds.
2. Describe and compare the ionic bond and the covalent bond.
3. Predict whether the bond formed between two elements would likely be ionic or covalent.
4. Compare some of the properties of ionic and covalent compounds.
5. Predict the formulas of ionic compounds using the octet rule and the charges on ions as guides.
6. Predict the formulas of covalent compounds using the octet rule as a guide.
7. Write equations using electron-dot symbols to describe the formation of ionic and covalent compounds.
8. Write the name of an ionic or covalent compound given its formula, or write the formula given its name.
9. Draw Lewis structures for covalent molecules or polyatomic ions given the formulas and skeleton structures.
10. Denne electronegativity and describe the trends in values across the Periodic Table.
11. Describe the following kinds of covalent bonds: single, double, triple, polar, nonpolar, and coordinate.
12. Predict the shapes of simple covalent molecules from their Lewis structures, and give the probable bond angles.

The Octet Rule

With the exception of helium, atoms of the rare gases have eight electrons (an octet of electrons) in their highest occupied energy level. Helium has two electrons which completely fill the first energy level, the $n=1$ level. The rare gases are chemically unreactive, that is, they are very stable atoms. Except for a very few compounds of xenon and krypton, stable compounds of these elements do not exist.

TABLE 8.1
The Stable Electronic Configurations of the Rare Gases

	Energy level (n)					Electrons in highest occupied energy level	Electron-dot symbol
	1	2	3	4	5		
He	②					2	·He·
Ne	2	⑧				8	:Ne:
Ar	2	8	⑧			8	:Ar:
Kr	2	8	18	⑧		8	:Kr:
Xe	2	8	18	18	⑧	8	:Xe:

As chemistry developed, it became clear that there was a relationship between the stability of an atom or ion and the number of electrons in its highest occupied energy level. If that energy level contained an octet of electrons, the atom or ion turned out to be unusually stable. This observation, which holds true for most of the representative elements, led to the development of the octet rule. The **octet rule** states that atoms of the representative elements, when in chemical reactions, will lose, gain, or share the necessary number of electrons so that the highest occupied energy level will end up with an octet of electrons. Stated another way, an atom in a reaction will lose, gain, or share the necessary number of electrons to give it the same electronic configuration as the nearest rare gas. This last statement brings the stable helium configuration into the picture even though it does not have an octet of electrons.

The octet rule will be a useful guide in the study of ionic and covalent compounds. There are exceptions to the rule though, and they will be discussed later in the chapter. Also, the rule does not apply to transition metals since factors other than numbers of electrons can influence their stability in compounds.

Check Test Number 1

Write the ground-state electronic configuration of each ion given below, and identify the rare gas with the identical configuration. Identify the electrons in each ion that represent the octet.

(a) F^- (b) S^{2-} (c) Na^+ (d) Ca^{2+}

Answers: $n = 1\ 2\ 3$

   a. F^- ($10e^-$)   2 ⑧   same as Ne
   b. S^{2-} ($18e^-$)   2 8 ⑧   same as Ar
   c. Na^+ ($10e^-$)   2 ⑧   same as Ne
   d. Ca^{2+} ($18e^-$)   2 8 ⑧   same as Ar

Ionic Compounds —The Sodium Chloride Case

Ionic compounds are composed of ions, both positive ions (cations) and negative ions (anions). They are the most common type of compound that forms when metals react with nonmetals. Perhaps the most familiar ionic compound is sodium chloride, NaCl, a white solid also known as table salt. It forms when sodium and chlorine are brought together and reacted. The equation describing the reaction of sodium atoms with chlorine atoms* to form sodium chloride is

$$Na + Cl \rightarrow NaCl$$

But let us look at the reaction more closely to see what happens to the electrons in each atom. The Bohr pictures of these elements show the arrangement of the 11 electrons of sodium and the 17 electrons of chlorine:

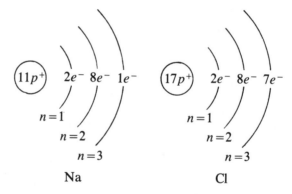

It may not be obvious at this point, but sodium and chlorine are near ideal mates for reaction. Sodium is a group I metal, and its single valence electron is not tightly held by the atom. Its relatively low ionization energy verifies this (Figure 6.8). Chlorine is a group VII nonmetal, and the relatively high ionization energy of chlorine indicates that the seven valence electrons are held quite tightly by the atom. Both elements are only one electron away from achieving stable electronic configurations like those of the rare gases. The *loss* of the single valence electron by sodium will produce a positive sodium ion, Na^+, with an electronic configuration identical to that of neon. The ion has an octet of electrons in the second energy level which, for the ion, is now the highest occupied energy level. Chlorine can achieve the stable argon configuration if

---

*Pure chlorine actually exists as molecules containing two chlorine atoms, Cl_2. In this discussion we will consider the reaction to be between sodium and chlorine atoms. The atoms are formed as the Cl_2 molecules break apart during the reaction.

it *gains* one electron and becomes a negative chloride ion, Cl^-. The added electron completes an octet of electrons in the third energy level.

Now you can see why these elements are well mated for reaction. Sodium, acting like a typical group I metal, has a loosely held electron to lose. Chlorine, acting like a typical group VII nonmetal, has a high attraction for an additional electron.

The reaction between these elements is the transfer of an electron from sodium to chlorine causing the formation of two stable ions in the process:

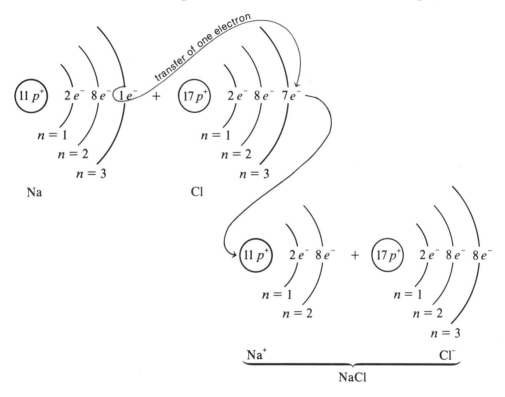

The reaction could just as easily have been described using electron-dot symbols (Table 6.8). The electron-dot symbol for chlorine represents an atom with seven electrons and one vacancy in the valence level. With nonmetals especially, you should see the dot symbols both in terms of electrons present *and* vacancies:

Now some things need to be said about the product of the reaction. In actual practice, a reaction between sodium and chlorine may involve trillions of atoms which become trillions of ions, half of them positive, half negative. These positive and negative ions are strongly attracted to each other by virtue of their opposite charges. The ions pack together in a highly organized crystalline structure that allows the attractive forces to be at a maximum while reducing the repulsive forces between ions of the same charge to a minimum. In the sodium chloride crystal, each Na^+ is surrounded by six Cl^-, and each Cl^- is in turn surrounded by six Na^+. The crystal is overall electrically neutral, since there is one negative charge for each positive charge. The highly ordered arrangement of ions in the crystal is called a **lattice,** and the crystal lattice of NaCl is shown in Figure 8.1. If you examine crystals of sodium chloride with the naked eye or with a magnifying glass (shake some table salt in your hand and examine it next time you eat lunch), they will look like little cubes, often fused into one another, with right-angle corners. The cubic shape of the crystals is the result of the way the ions are arranged in the crystal. Other ionic compounds

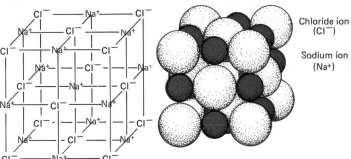

Fig. 8.1. The sodium chloride crystal lattice. (Courtesy of the American Museum of Natural History.) Each Na^+ is surrounded by six Cl^- and each Cl^- is surrounded by six Na^+.

may have crystals with a different shape since both the size and charge of the positive and negative ions will greatly affect how they will be arranged in the crystals.

The **formula** of sodium chloride, or any ionic compound for that matter, represents the smallest whole number ratio of the different ions in the crystal. Since there is one Na^+ for each Cl^-, the formula is NaCl. Note that the charges on the ions are *not* included in the formula. There are no discrete molecules in ionic crystals. It is not possible to identify small groups of ions in a crystal that act as independent units as molecules would. So when it is necessary to speak of some fundamental part of an ionic compound that is related to its composition, the formula unit is used. The **formula unit** is given by the formula of the ionic compound. One formula unit of NaCl is one Na^+ and one Cl^-.

The Ionic Bond

The **ionic bond** is the net force of attraction between ions of opposite charge in a crystal. Said another way, the ions in a crystal of an ionic compound are held together by ionic bonds. The ionic bonds form as the positive and negative ions pack together to form the crystal. As the bonds form there is a release of energy, mostly in the form of heat, that is responsible for the stability of the crystal. The amount of energy released, which can be quite large, is related to the strength of the ionic bonds. Because the ions are strongly held together, crystals of ionic compounds are very hard and brittle, and they usually have high melting temperatures.

Another name that is sometimes used instead of ionic bonding is electrovalent bonding, and ionic bonds can be called electrovalent bonds.

Check Test Number 2

1. Why is it incorrect to speak of "molecules of NaCl" in the sodium chloride crystal?
2. Why are the positive and negative ions arranged in a highly organized way in crystals of ionic compounds?
3. What is the ionic bond?

Answers:
1. No one Na^+ is associated with only one Cl^-; every ion is associated with several ions and is part of the crystal as a whole.
2. Only in a highly organized structure of ions can the attractive forces between the ions of opposite charge be at a maximum so the crystal is as stable as possible.
3. The ionic bond is the net force of attraction between ions of opposite charge.

Ions and the Periodic Table

Before we proceed to discuss the formation of other ionic compounds, several comments need to be made about ions. When a sodium atom loses an electron and becomes a sodium ion, it is important to realize that the only similarity between the two species is the composition of the nucleus. The ion has a set of properties different from those of the atom. And because the sodium ion has the same electronic configuration as the rare gas neon, it does not mean it will behave like a neon atom. The same kind of differences exist for the chlorine atom and its ion. They are different species, and an ion should never be referred to as an atom.

Ions do not have the same size as the atoms from which they are formed (see Figure 8.2). A positive ion (a cation) is invariably *smaller* than the parent atom because of the loss of one or more electrons. The electrons remaining in the ion are held more tightly by the nucleus since its positive charge exerts an influence over fewer electrons. A negative ion (an anion) is invariably *larger* than its parent atom. The increased number of electrons are not able to be held as tightly by the nucleus.

The elements at the extremes of the Periodic Table, that is, those in groups

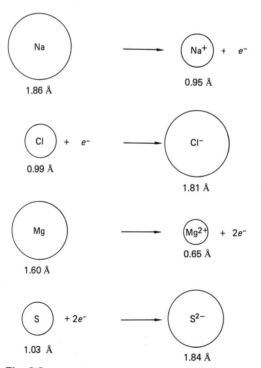

Fig. 8.2.   Positive ions are invariably smaller than their parent atoms and negative ions are invariably larger than their parent atoms.

I, II, VI, and VII, readily form ions with rare gas configurations. All the metals in group I form ions with a 1+ charge. Hydrogen is unique since it commonly forms a 1+ ion, but it can also form a 1− ion, the hydride ion, with the helium configuration. The metals in group II form ions with a 2+ charge. The nonmetals in group VI form ions with a 2− charge, and those in group VII form ions with a 1− charge. Aluminum, the most common element in group III forms an ion with a 3+ charge. Since the elements in group IV and V would need to lose or gain several electrons to reach rare gas configurations, stable ions of these elements are not common. The metals near the bottom of these groups do form positive ions, but for our purposes we will not be concerned with them here. The transition metals all form positive ions, and many form two or more ions of different charge. Copper forms two ions, Cu^+ and Cu^{2+}. Iron forms Fe^{2+} and Fe^{3+}. The charges on some of the other ions of the transition metals are found in Figure 8.3.

Two terms are commonly used to describe the loss or gain of electrons by atoms or ions. The loss of one or more electrons by a species is called **oxidation.** Sodium is **oxidized** as it loses one electron to become the sodium ion. The gain of one or more electrons by a species is called **reduction.** Chlorine is **reduced** as it gains one electron to form the chloride ion. Oxidation and reduction always occur simultaneously in chemical reactions since an atom can lose an electron only if another species is present to accept it. Oxidation and reduction are discussed in greater detail in Chapter 11.

All the ions discussed up to this point have been **simple ions,** that is, ions formed from individual atoms. But you may recall from Chapter 3 (Table 3.7) that several **polyatomic ions** are commonly encountered. Polyatomic ions are species carrying a charge that contain two or more atoms bonded together. Most polyatomic ions are anions and to refresh your memory, several common polyatomic ions are again listed in Table 8.2. You should learn the name, composition, and charge of each of these.

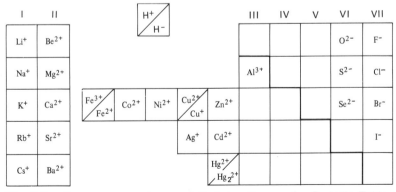

Fig. 8.3. Some common ions and their location on the Periodic Table. Notice that hydrogen can form both a 1+ ion and a 1− ion, the hydride ion.

TABLE 8.2 Some Common Polyatomic Ions	
NH_4^+	ammonium ion
OH^-	hydroxide ion
NO_3^-	nitrate ion
CN^-	cyanide ion
HCO_3^-	hydrogen carbonate ion (also called the bicarbonate ion)
CO_3^{2-}	carbonate ion
SO_4^{2-}	sulfate ion
PO_4^{3-}	phosphate ion
HPO_4^{2-}	hydrogen phosphate ion
$H_2PO_4^-$	dihydrogen phosphate ion
$C_2H_3O_2^-$	acetate ion

Check Test Number 3

1. Referring to the Periodic Table, predict the charge on the ion formed from each of the following elements.
   a. Ca   d. S
   b. K   e. Al
   c. Br   f. O
2. In the reaction between sodium and chlorine to form NaCl, which element is oxidized? Which is reduced?
3. From memory, give the name of each of the following polyatomic ions.
   a. NH_4^+ d. SO_4^{2-}
   b. OH^- e. CO_3^{2-}
   c. NO_3^- f. $C_2H_3O_2^-$

Answers:
1. (a) Ca^{2+}, (b) K^+, (c) Br^-, (d) S^{2-}, (e) Al^{3+}, (f) O^{2-}
2. Na is oxidized, Cl is reduced.
3. (a) ammonium ion, (b) hydroxide ion, (c) nitrate ion, (d) sulfate ion, (e) carbonate ion, (f) acetate ion

Predicting Formulas of Ionic Compounds

A great deal of what has been said about the formation of sodium chloride can be applied to the formation of other ionic compounds as well. Generally, metals and nonmetals react with the transfer of electrons forming ions of opposite charge. The ions will organize themselves in a definite ratio (so there is an equal number of + and − charges) to form a crystal. The formula of the compound is the simplest ratio of the ions in the crystal. The formulas of many ionic compounds can be predicted by first determining the ions that each reacting element will form and then combining them in the smallest ratio that will give an equal number of + and − charges. The formula is then written using this ratio of ions.

In the following examples different metals and nonmetals are selected to form an ionic compound, and the formula of the compound is to be predicted. A convenient procedure to follow is outlined below.

Step 1: Determine the ion formed by the metal.

Step 2: Determine the ion formed by the nonmetal. (Use the octet rule or Figure 8.3 in steps 1 and 2.)

Step 3: Determine the smallest whole number ratio of positive to negative ions so the charges sum to zero.

Step 4: Write the formula using the ratio of ions found in step 3.

Example 8.1

$$Ca + O \rightarrow \ ?$$

Step 1: Calcium is a group II metal and will form the stable ion, Ca^{2+}.

Step 2: Oxygen is a group VI nonmetal and will form the stable ion, O^{2-}.

Step 3: The simplest ratio of ions that will provide an equal number of + and − charges is

$$1 \ Ca^{2+} \ \text{to} \ 1 \ O^{2-}$$

Step 4: The formula with this ratio is *CaO, calcium oxide.*

The reaction between calcium and oxygen can be written using electron-dot symbols to show the movement of electrons from calcium to oxygen:

$$\cdot Ca\cdot \ + \ \ddot{\ddot{O}} \ \longrightarrow \ Ca^{2+}, \ :\ddot{O}:^{2-}$$
$$(CaO)$$

Example 8.2

$$Ca + Br \rightarrow \ ?$$

Step 1:   As in Example 8.1, calcium will form Ca^{2+}.

Step 2:   Bromine is in group VII, and will form Br^-.

Step 3:   The simplest ratio of ions that will provide an equal number of + and − charges is

$$1\ Ca^{2+}\ \text{to}\ 2\ Br^-$$

Step 4:   The formula of the product will be *CaBr$_2$*, calcium bromide.

Using electron-dot symbols the equation would be

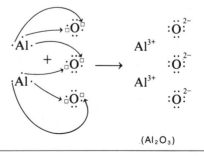

$$\text{(CaBr}_2\text{)}$$

Example 8.3

$$Al + O \rightarrow ?$$

Step 1:   Aluminum, a group III metal, can lose three electrons to form Al^{3+}.

Step 2:   Oxygen forms O^{2-}.

Step 3:   The simplest ratio would be

$$2\ Al^{3+}\quad \text{to}\quad 3\ O^{2-}$$

(a total of 6+)   (a total of 6−)

Step 4:   The formula is then, Al_2O_3, aluminum oxide.

Using electron-dot symbols, the equation would be

$$\text{(Al}_2O_3\text{)}$$

The formulas of ionic compounds that involve polyatomic ions are determined by choosing the simplest ratio of ions that provides an equal number of positive and negative charges, just as was done in the previous examples. The formula of the compound involving K^+ and SO_4^{2-} would be K_2SO_4, potassium sulfate. Two K^+ are needed to offset the 2− charge on the sulfate ion. The compound formed from the ammonium ion NH_4^+ and the sulfide ion S^{2-} would be $(NH_4)_2S$. The ammonium ion is enclosed in parentheses to avoid confusion with the subscript numbers. The formula would be read as "N-H-4 taken twice, S." Parentheses are commonly used in formulas that include two or more polyatomic ions of a particular type, such as $Ca(OH)_2$, $Al_2(SO_4)_3$, and $Cu(NO_3)_2$. Let us do one more example. What is the formula of the compound formed from Ca^{2+} and PO_4^{3-}? In order to get the same number of positive and negative charges in the formula, you will need three Ca^{2+} and two PO_4^{3-} (6+ and 6−). The compound would be $Ca_3(PO_4)_2$, calcium phosphate.

Now try your skill at predicting formulas in Check Test Number 4.

Check Test Number 4

1. Predict the formulas of the ionic compounds formed when the following pairs of elements react:
   a. Na, I   d. Rb, O
   b. Ca, F   e. Ba, S
   c. Al, Br   f. Ag, Cl
2. Predict the formulas of the compounds formed when the following pairs of ions are combined:
   a. Li^+, CO_3^{2-} c. Ca^{2+}, NO_3^-
   b. NH_4^+, SO_4^{2-} d. Fe^{3+}, SO_4^{2-}

Answers:

1. a. NaI   d. Rb_2O
   b. CaF_2 e. BaS
   c. $AlBr_3$ f. AgCl
2. a. Li_2CO_3 c. $Ca(NO_3)_2$
   b. $(NH_4)_2SO_4$ d. $Fe_2(SO_4)_3$

Naming Ionic Compounds

Up to this point the names of several ionic compounds have been given without much explanation, and you might think each compound has its own peculiar name. But fortunately, such is not the case. As the number of known compounds grew, chemists devised systematic methods to name them. This systematic approach to nomenclature (the naming of things) is based on sets of rules that, when followed, will allow the formula of a compound to be written if its **systematic name** is known and vice versa. It also means that a particular compound will be known by the same name to all chemists. But still today, there are some very common compounds that are often called by their **common names.** For example, NaCl is commonly called table salt or salt instead of sodium chloride, its systematic name. Because of their continued use, common names for certain compounds will be written in parentheses alongside the systematic names when they appear.

Let us begin with the systematic rules that are used for naming *binary* ionic compounds, that is, ionic compounds composed of only two elements. The rules for naming binary covalent molecules will appear later on in this chapter.

A. Ionic compounds containing only two elements

Rule 1.   The name of the metal appears first in the name of the compound, just as the symbol of the metal is first in its formula. It is usually not capitalized.

Rule 2.   The name of the nonmetal follows that of the metal, and the name of the nonmetal in binary compounds ends in *-ide*. The -ide ending is added to the stem of the name of the element as shown below.

Element	Stem	Ion	Name in compound
Fluorine	fluor-	F^-	fluoride
Chlorine	chlor-	Cl^-	chloride
Bromine	brom-	Br^-	bromide
Iodine	iod-	I^-	iodide
Oxygen	ox-	O^{2-}	oxide
Sulfur	sulf-	S^{2-}	sulfide

Rule 3.   If the metal commonly appears in compounds in only *one* charge state [such as the metals in group I (1+), group II (2+), and Al^{3+}, Zn^{2+}, Cd^{2+}, and Ag^+], the name of the binary ionic compound is formed by following the name of the metal with that of the nonmetal using the -ide ending as described in rule 2. The ammonium ion (NH_4^+) is also treated as an ion of fixed charge. The following examples all contain positive ions of fixed charge.

NaCl	sodium chloride	(table salt)
NaF	sodium fluoride	
LiBr	lithium bromide	
Na_2O	sodium oxide	
K_2S	potassium sulfide	
NH_4Cl	ammonium chloride	(sal ammoniac)
$AlCl_3$	aluminum chloride	
ZnO	zinc oxide	
$(NH_4)_2S$	ammonium sulfide	
CaO	calcium oxide	(lime)
Al_2O_3	aluminum oxide	(alumina)

Suppose you were given the name of one of the compounds listed above and asked to write its formula. To do this, you need to determine the charges on both the ions so they can be combined in the correct ratio just as was done earlier. If the formula of calcium fluoride is needed, recall that calcium (group II) forms Ca^{2+} and fluorine (group VII) forms F^-. You would need two F^- for one Ca^{2+} to form a neutral formula, CaF_2.

Rule 4. If the metal commonly forms ions of more than one charge, such as iron (Fe^{2+} and Fe^{3+}), then the size of the positive charge is written as a Roman numeral in parentheses after the name of the metal.

Though many metals fall into this category, we will only be concerned with three of them here: iron (Fe^{2+}, Fe^{3+}), copper (Cu^+, Cu^{2+}), and mercury* (Hg_2^{2+}, Hg^{2+}). There is an older scheme used to indicate the different ions of these three metals that you should know. It uses the stem of the Latin name for the metal and adds an *-ous* or *-ic* ending to indicate the lower or higher positive charge, respectively. Both schemes are shown below, though the systematic name is preferred.

Ion	Systematic name	Older name
Fe^{2+}	iron (II)	ferrous
Fe^{3+}	iron (III)	ferric
Cu^+	copper (I)	cuprous
Cu^{2+}	copper (II)	cupric
Hg_2^{2+}	mercury (I)	mercurous
Hg^{2+}	mercury (II)	mercuric

The following examples show why it is important to indicate the size of the positive charge on the metal in compounds containing these elements. If you simply called a compound "iron chloride," you would not be able to clearly identify it as $FeCl_2$ or $FeCl_3$.

	Systematic name	Older name
$FeCl_2$	iron (II) chloride†	ferrous chloride
$FeCl_3$	iron (III) chloride	ferric chloride
Cu_2O	copper (I) oxide	cuprous oxide
CuO	copper (II) oxide	cupric oxide
Hg_2Cl_2	mercury (I) chloride	mercurous chloride
$HgCl_2$	mercury (II) chloride	mercuric chloride

*Hg_2^{2+} is a diatomic mercury ion. You can think of it as being two Hg^+ ions joined together by a covalent bond: Hg^+-Hg^+.

†Read as "iron-two chloride," "iron-three chloride," "copper-one oxide," and so forth.

B. Ionic compounds containing polyatomic ions

Rule 5. Ionic compounds that contain polyatomic anions are named with the metal first followed by the name of the anion.

$Ca(OH)_2$	calcium hydroxide	(slaked lime)
NaOH	sodium hydroxide	(lye)
$NaHCO_3$	sodium hydrogen carbonate or sodium bicarbonate	(baking soda)
$FePO_4$	iron (III) phosphate	
$Mg(OH)_2$	magnesium hydroxide	(milk of magnesia)
KCN	potassium cyanide	
$CaCO_3$	calcium carbonate	(limestone)
$Mg(C_2H_3O_2)_2$	magnesium acetate	
$Fe_2(SO_4)_3$	iron (III) sulfate	
Na_2HPO_4	sodium hydrogen phosphate	
Hg_2SO_4	mercury (I) sulfate	

Ionic compounds containing the hydroxide ion (OH^-), the cyanide ion (CN^-), and the ammonium ion (NH_4^+) have the -ide endings of binary compounds even though they contain more than two elements. Though these are exceptions, they should not present a problem since compounds containing these ions are common.

Check Test Number 5

1. Correctly name the following compounds.
   a. MgO   d. K_2O
   b. AlF_3 e. LiI
   c. CuS   f. Al_2S_3
2. Correctly name the following compounds.
   a. $BaSO_4$ d. $HgSO_4$
   b. $Ca(HCO_3)_2$ e. $RbNO_3$
   c. $Fe(OH)_3$ f. $AlPO_4$
3. Write the correct formula for each of the following compounds.
   a. ammonium bromide
   b. barium nitrate
   c. copper (II) sulfate
   d. calcium hydroxide

Answers:
1. (a) magnesium oxide, (b) aluminum fluoride, (c) copper (II) sulfide, (d) potassium oxide, (e) lithium iodide, (f) aluminum sulfide
2. (a) barium sulfate, (b) calcium hydrogen carbonate or calcium bicarbonate, (c) iron (III) hydroxide, (d) mercury (II) sulfate, (e) rubidium nitrate, (f) aluminum phosphate
3. (a) NH_4Br, (b) $Ba(NO_3)_2$, (c) $CuSO_4$, (d) $Ca(OH)_2$

Covalent Compounds

You have learned that reactions between metals and nonmetals will produce ionic compounds composed of positive and negative ions neatly arranged in crystalline structures. The ions form as electrons are transferred from metal atoms to nonmetal atoms. But there are many materials that are not composed of ions, such as water, carbon dioxide, gasoline, and the plastics. Most of these compounds are formed when nonmetals react with each other, and they exist as neutral molecules containing two or more atoms. The atoms are joined together by pairs of electrons shared between them. This **shared electron-pair bond** is called a **covalent bond**. Compounds whose atoms are joined by covalent bonds are called **covalent compounds**.

The hydrogen molecule, H_2, is the simplest example of a covalent species. It is formed by the combination of two hydrogen atoms. Since neither atom has a greater tendency to gain or lose electrons than the other (they are identical in all respects), each can acquire a stable rare gas configuration by *sharing* their electron with the other. The shared electron pair joins the atoms together and gives *each* two electrons in their valence level (the $n=1$ level). In a sense, each has attained the stable configuration of the helium atom.

The joining of two hydrogen atoms to form the molecule can be described using the classic Bohr pictures of the atoms or using the more convenient electron-dot symbols. Small ×'s and dots are used to represent the valence electrons on different atoms, though in practice it is not possible to tell one electron from another once in a bond.

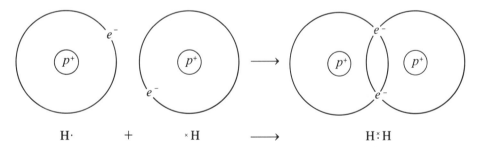

H· + ×H ⟶ H×H

Two hydrogen atoms bonded together in a molecule represent a more stable condition (that is, they are at a lower energy) than the separated atoms. The formation of the covalent bond adds stability to the atoms, and the energy released as the bond forms is called the **bond energy** (Figure 8.4). For the hydrogen molecule, the bond energy is 103 kcal per mole. This means that 103 kcal of energy is released for every 6.02×10^{23} (1 mole) molecules formed. This is also the amount of energy needed to "break" one mole of molecules apart into the separated atoms. Bond energies are used to state the stength of chemical bonds. The larger the bond energy, the stronger the bond.

You may wonder how the formation of a covalent bond can increase the

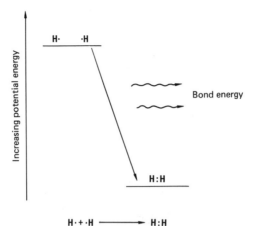

Fig. 8.4. The energy released upon the formation of the covalent bond is equal to its bond energy. Two hydrogen atoms in a molecule are more stable (are at a lower energy) than two separated hydrogen atoms. Bond formation is a downhill process, energywise.

stability of two hydrogen atoms. A detailed answer can become quite involved, but perhaps an approach that compares the attractions and repulsions between electrons and nuclei in the atoms and the molecule might be useful. In the hydrogen molecule, each electron is attracted to two nuclei and each nucleus is attracted to two electrons. This gives a total of four attractions in the molecule as shown in Figure 8.5. There are also two repulsions in the molecule, but they are relatively weak. One is between the two positive nuclei and the other is between the two negative electrons. In each isolated hydrogen atom, there is only one attraction, that between the one electron and the nucleus. So in two hydrogen atoms there are two attractions and no repulsions. Comparing the two atoms with the molecule, then, shows that there is a net increase in at-

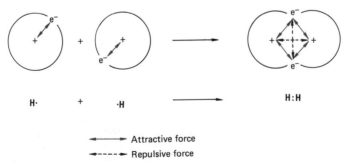

Fig. 8.5. The stability of the covalent bond is due to the net increase in the forces of attraction in the molecule compared to the separated atoms. In each atom there is a single attractive force, but in the molecule there are four stronger forces of attraction and two weaker repulsive forces.

tractions once the atoms form the molecule, and this makes the molecule more stable.

Now let us look at some other elements to see how they form molecules. When two fluorine atoms combine to form the fluorine molecule, each atom shares one valence electron, an "unpaired" electron, with the other to form a *shared pair* of electrons. This gives each atom an octet. The shared electrons are counted in the octet of both atoms:

$$:\ddot{F}\cdot \ + \ _{x}^{x}\!\overset{x}{\underset{x}{F}}\!x \ \longrightarrow \ :\overset{..}{\underset{..}{F}}\overset{x}{\underset{x}{x}}\overset{x}{\underset{x}{F}}x$$

the unpaired electrons (F_2)

Each atom in the fluorine molecule has four pairs of electrons in its valence level. Those not involved in the bond are called **unshared electron pairs** or **nonbonding pairs**.

$$:\overset{..}{\underset{..}{F}}:\overset{..}{\underset{..}{F}}:\ \longleftarrow \text{unshared pairs}$$
(nonbonding pairs)

shared pair ──┘
(bonding pair)

The shared electron pairs in a covalent molecule are commonly represented with a dash (–), drawn between the atoms.

$$H{-}H \qquad :\overset{..}{\underset{..}{F}}{-}\overset{..}{\underset{..}{F}}:$$

A single pair of electrons shared between two atoms is called a **single bond**. The other halogens (the group VII elements) also exist as diatomic molecules employing single bonds:

$$:\overset{..}{\underset{..}{Cl}}{-}\overset{..}{\underset{..}{Cl}}: \qquad :\overset{..}{\underset{..}{Br}}{-}\overset{..}{\underset{..}{Br}}: \qquad :\overset{..}{\underset{..}{I}}{-}\overset{..}{\underset{..}{I}}:$$

(Cl_2) (Br_2) (I_2)

The structures used to represent the hydrogen molecule and the halogen molecules are called **Lewis structures**. In 1916 an American chemist, G. N. Lewis, introduced the concept of the shared electron pair bond or covalent bond. He also introduced the octet rule. Lewis structures show how the valence electrons of each atom in a molecule are distributed. They are very useful to chemists since they show how atoms are bonded together in molecules. They also show unshared electron pairs which can play an important role in determining the shape of a molecule as you will see later.

Hydrogen can form a large number of covalent compounds with other nonmetals. In each case, hydrogen is joined to the other element by a single bond. Hydrogen forms a compound with each of the halogens:

$$\text{H—}\ddot{\text{F}}\text{:} \qquad \text{H—}\ddot{\text{C}}\text{l:} \qquad \text{H—}\ddot{\text{B}}\text{r:} \qquad \text{H—}\ddot{\text{I}}\text{:}$$

Hydrogen combines with oxygen to form water. Oxygen, a group VI element, has two unpaired electrons and thus needs two electrons to complete its octet. This requires combining with two hydrogen atoms. The Lewis structure of water shows that oxygen forms two bonds, one to each hydrogen:

$$\text{H}\colon\!\ddot{\text{O}}\colon\!\text{H} \longrightarrow \text{H}\!:\!\ddot{\text{O}}\!:\!\text{H} \qquad or \qquad \text{H—}\ddot{\text{O}}\text{—H}$$
$$(\text{H}_2\text{O})$$

The other elements in group VI form similar compounds such as H_2S and H_2Se. Hydrogen combines with nitrogen to form ammonia. Nitrogen, a group V element, with its three unpaired electrons requires three additional electrons to achieve an octet, so three hydrogen atoms are required. The Lewis structure of ammonia shows that nitrogen forms three bonds:

$$\text{H}\colon\!\text{N}\colon\!\text{H} \longrightarrow \text{H}\!:\!\ddot{\text{N}}\!:\!\text{H} \qquad or \qquad \text{H—}\ddot{\text{N}}\text{—H}$$
$$\text{H} \qquad\qquad\qquad \text{H} \qquad\qquad\qquad\qquad \text{H}$$
$$(\text{NH}_3)$$

Analogous compounds are formed by the other elements in group V such as PH_3 and AsH_3.

You should be able to predict the formula of the simplest compound formed between carbon and hydrogen. Carbon is a group IV element, $\cdot\dot{\text{C}}\cdot$, and hydrogen has one electron to share, $H\times$. Carbon can achieve an octet if it combines with *four* hydrogen atoms to form CH_4, methane:

$$\begin{array}{ccccc} \text{H} & & \text{H} & & \text{H} \\ \text{H}\colon\!\text{C}\colon\!\text{H} & \longrightarrow & \text{H}\!:\!\text{C}\!:\!\text{H} & or & \text{H—C—H} \\ \text{H} & & \text{H} & & \text{H} \\ & & (\text{CH}_4) & & \end{array}$$

You should also expect other members of group IV to form compounds with similar formulas, such as SiH_4.

Multiple Bonds

Many atoms can share more than one pair of electrons with another atom resulting in **multiple bonds** between two atoms. The oxygen molecule is made up of two atoms each donating two electrons to the bond. This results in *two* shared pairs of electrons and an octet about each atom:

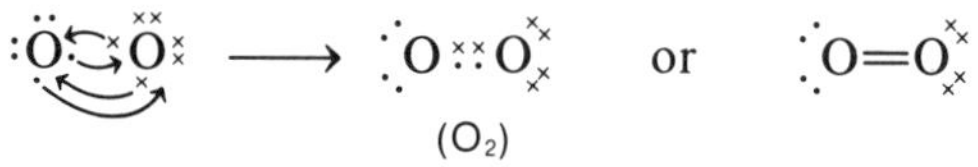

$$:\overset{\times\times}{\underset{..}{O}}\overset{..}{\underset{..}{O}}: \longrightarrow :\overset{..}{O}::\overset{\times}{\underset{\times}{O}}: \quad \text{or} \quad :\overset{..}{O}=\overset{\times}{\underset{\times}{O}}:$$
$$(O_2)$$

The oxygen molecule has a **double bond**, since the number of shared pairs is double that in a single bond.

The Lewis structure of nitrogen molecule, N_2, shows a **triple bond** between the two nitrogen atoms, the result of sharing three pairs of electrons. Since each nitrogen atom needs three electrons to complete its octet, each shares three electrons with the other, for a total of six shared electrons:

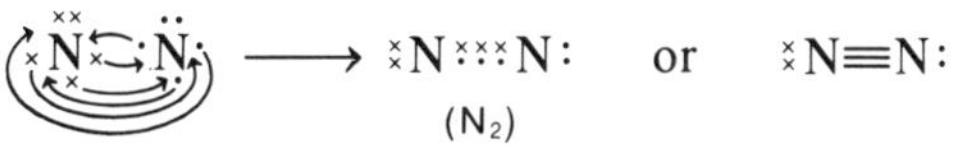

$$\overset{\times\times}{(\overset{\times}{.}N\overset{}{.}\overset{}{.}N\overset{.}{.})} \longrightarrow \overset{\times}{.}N\overset{\times\times\times}{.}N: \quad \text{or} \quad \overset{\times}{.}N\equiv N:$$
$$(N_2)$$

Multiple bonding between two atoms rarely exceeds the three shared pairs of the triple bond. The bond energies of single and multiple bonds are not the same. Comparing bonds between the same elements, triple bonds are the strongest, and double bonds are stronger than single bonds. The bond energies for the carbon–carbon bonds point this out: $-C\equiv C-$ (230 kcal/mole), $\diagup C=C\diagdown$ (125 kcal/mole) and $-\overset{|}{\underset{|}{C}}-\overset{|}{\underset{|}{C}}-$ (85 kcal/mole). Other molecules that display multiple bonds are shown below:

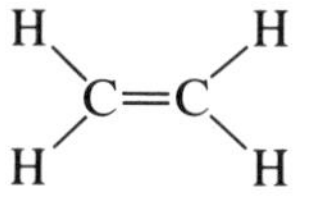

$$\overset{H}{\diagdown}C=C\overset{H}{\diagup}_{\diagdown H} \qquad H-C\equiv C-H \qquad H-\overset{}{\underset{|}{C}}=O$$
$$\overset{H}{\diagup}\qquad\qquad\qquad\qquad\qquad\qquad H$$

ethylene   acetylene   formaldehyde
(used to make   (a fuel)   (used to preserve
polyethylene)   tissue samples)

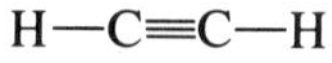

If you go back and study each of the molecules discussed so far, you will note that each element forms a definite number of covalent bonds. Hydrogen forms one bond, oxygen forms two bonds, nitrogen three bonds, and carbon, four. The bonding patterns that each of these common nonmetals can have are summarized in Table 8.3.

The information summarized in Table 8.3 can be used to draw Lewis structures for covalent compounds of these elements if their formulas are known. For example, the Lewis structure of chloroform, $CHCl_3$, can be drawn once we figure out how the five atoms can be joined so each uses the number of bonds shown in the table. The formula of chloroform shows there are

3   $:\overset{..}{\underset{..}{Cl}}-$

1   $H-$

1   $-\overset{|}{\underset{|}{C}}-$ (or one of the other bond patterns shown for C)

TABLE 8.3
Covalent Bonding Patterns for Some Common Elements

Element	Group	Number of bonds	Bond patterns	
H·	I	1	H—	one single bond
:$\ddot{\text{X}}$·	VII	1	:$\ddot{\text{X}}$—	one single bond

X=F, Cl,Br,I

Element	Group	Number of bonds	Bond patterns	
·$\ddot{\text{O}}$·	VI	2	—$\ddot{\text{O}}$—	two single bonds
			:$\ddot{\text{O}}$=	one double bond
·$\dot{\ddot{\text{N}}}$·	V	3	—$\overset{\mid}{\ddot{\text{N}}}$—	three single bonds
			—$\ddot{\text{N}}$=	one single, one double bond
			:N≡	one triple bond
·$\dot{\text{C}}$·	IV	4	—$\overset{\mid}{\underset{\mid}{\text{C}}}$—	four single bonds
			$>$C=	two single, one double bond
			=C=	two double bonds
			—C≡	one single, one triple bond

The only way the atoms can be arranged so each element forms the indicated number of bonds is

$$\text{H}-\overset{\displaystyle :\ddot{\text{Cl}}:}{\underset{\displaystyle :\ddot{\text{Cl}}:}{\text{C}}}-\ddot{\text{Cl}}:$$

This is the Lewis structure for chloroform. No other arrangement of atoms would satisfy the bonding requirements of the three elements. Here are some other examples.

Example 8.4

Predict the Lewis structure of hydrogen peroxide, H_2O_2. Hydrogen peroxide is a common antiseptic.
  The formula shows four atoms are present in the molecule:

$$2 \quad \text{H}—$$

$$2 \; —\ddot{\text{O}}— \quad (\text{or} = \ddot{\text{O}}\!:)$$

Since oxygen must use two bonds, the only possible arrangement of the four atoms is

$$\text{H}—\ddot{\text{O}}—\ddot{\text{O}}—\text{H}$$

Any other arrangement would either leave one atom out or use too many bonds to oxygen or hydrogen.

Example 8.5

Draw the Lewis structure for ethane, C_2H_6. The formula tells us we have

$$6 \quad \text{H}—$$

$$2 \; —\overset{\textstyle |}{\underset{\textstyle |}{\text{C}}}—$$

If these eight atoms are to be joined together in a single molecule, the two carbon atoms must be joined and the unused bonds then filled with hydrogen.

$$—\overset{\textstyle |}{\underset{\textstyle |}{\text{C}}}—\overset{\textstyle |}{\underset{\textstyle |}{\text{C}}}— \quad \longleftarrow$$

This leaves six bonds open to join the six hydrogen atoms.

$$\text{H}—\overset{\textstyle \text{H}}{\underset{\textstyle \text{H}}{\text{C}}}—\overset{\textstyle \text{H}}{\underset{\textstyle \text{H}}{\text{C}}}—\text{H}$$

The Lewis structure for ethane shows there is a carbon–carbon single bond in the molecule. This might not have been predicted from the formula.

Example 8.6

Draw the Lewis structure for carbon dioxide, CO_2. The formula shows that we have

$$1 \quad -\overset{|}{\underset{|}{C}}- \quad \text{(or other bond pattern)}$$

$$2 \quad -\overset{..}{\underset{..}{O}}- \quad \text{(or} =\overset{..}{O}\!\!:)$$

We know that carbon must form four bonds, and that each oxygen atom must form two bonds. The solution is to use carbon as $=C=$ and each oxygen as $=\overset{..}{O}$.

$$:O= \qquad =C= \qquad =\overset{..}{O}:$$
$$\underbrace{}$$
$$:\overset{..}{O}=C=\overset{..}{O}:$$

This structure satisfies the required number of bonds for both C and O (that is, it satisfies the octet rule), and we see that carbon dioxide has two double bonds.

Now try your skill at drawing Lewis structures for some other covalent molecules in Check Test Number 6.

Check Test Number 6

Draw Lewis structures for the following using as a guide the number of bonds each element can form.
(a) NF_3, (b) Cl_2CO, (c) HCN

Answers:

(a) $:\overset{..}{\underset{..}{F}}-\overset{..}{\underset{|}{N}}-\overset{..}{\underset{..}{F}}:$
$\quad\quad\;\; :\overset{..}{\underset{..}{F}}:$

(b) $:\overset{..}{\underset{..}{Cl}}:$
$\quad\quad\quad\;\; \overset{}{\underset{:\overset{..}{\underset{..}{Cl}}:}{}} C=\overset{..}{O}:$

(c) $H-C\equiv N:$

Another method to derive Lewis structures that is usable for a much larger number of species (including polyatomic ions) will be presented later in this chapter.

Polar and Nonpolar Covalent Bonds

Covalent bonds are shared pairs of electrons, and you may wonder if the electrons are shared equally by the atoms or if they are shared unequally. The answer depends on the nature of the atoms sharing the electrons.

Electrons shared by identical atoms, as in H_2, O_2, or N_2, are shared equally.

$$H \mid H \qquad H \mid :F$$

Fig. 8.6.   The covalent bond in the hydrogen molecule is nonpolar. The electrons in the bond are shared equally and are no closer to one atom than the other. The covalent bond in hydrogen fluoride is polar. The electrons in the bond are closer to F because of its greater attraction for electrons.

Neither atom has a greater tendency to attract the electrons than the other since they are identical. Covalent bonds in which electrons are shared equally are called **nonpolar covalent bonds.** The electron pair shared by hydrogen and fluorine in HF is not shared equally by both atoms. The atoms are not alike, and fluorine exerts a greater attraction for the shared electrons than hydrogen. This causes the shared pair to be held closer to fluorine. The electrons are still shared, but shared unequally. A covalent bond in which the electrons are shared unequally is a **polar covalent bond;** the bond in HCl is also a polar bond; the electrons are attracted more strongly by chlorine. Generally, covalent bonds between unlike atoms are polar bonds. (See Figure 8.6.) The affect of polar covalent bonds on the properties of covalent compounds will be discussed in Chapter 10.

The Electronegativity of Atoms

The tendency for an atom to attract the electrons shared in a bond to itself is described by its **electronegativity.** The more strongly the atom attracts these electrons, the greater is its electronegativity. The electronegativities of elements are stated as numbers. The most electronegative element, fluorine (it has the greatest attraction for electrons of all the elements), is assigned a value of 4.0. Cesium, one of the least electronegative elements (it has a very low attraction for electrons), has a value of 0.7. All the other elements have electronegativities between these extremes. The electronegativity values for several common elements are given in Figure 8.7. You do not need to memorize the values for these elements, but you should recognize that elecronegativities generally increase from left to right in a period and decrease going down a group. The nonmetals have relatively high electronegativities while the metals have low values.

Electronegativity values can be used to determine the relative polarity of a covalent bond. The greater the *difference* in the electronegativities of two bonded atoms, the more polar will the bond be. The hydrogen–fluorine bond in HF is quite polar. The difference between the electronegativities of hydrogen and fluorine is 1.9 (F=4.0, H=2.1). The covalent bond in HCl is less polar than

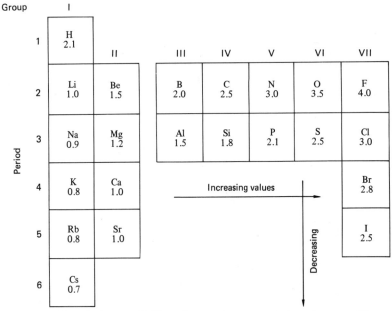

Fig. 8.7.   The electronegativities of several representative elements. Electronegativities generally increase across a period and decrease down a group.

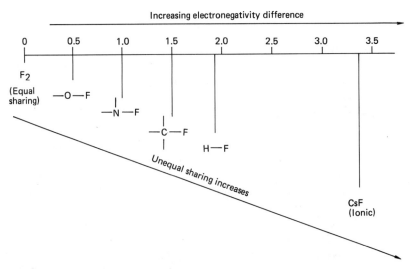

Fig. 8.8.   The polarity of a covalent bond will increase as the difference in electronegativity between two bonded atoms increases. A large difference in electronegativity results in an ionic bond.

that in HF since the electronegativity difference is only 0.9 (Cl=3.0, H=2.1). The covalent bond in the hydrogen molecule is nonpolar because both atoms have the same electronegativity and the difference is, of course, zero. Generally if the difference in the electronegativities of two elements is greater than about 2.0, it is likely they will form an ionic compound when they react. In a sense, the ionic bond represents the extreme case of polarity in a bond. (See Figure 8.8.)

Check Test Number 7

1. For each pair indicate which bond would be more polar. You may wish to refer to Figure 8.7.

    a. $-\overset{|}{\underset{|}{C}}-H$, $-\overset{|}{\underset{|}{C}}-F$ b. $-S-Cl$, $-\overset{|}{N}-Cl$ c. $H-I$, $H-Br$

2. Using the trends in electronegativities in the periodic table, predict which element of each pair is more electronegative.
    a. C, N
    b. O, S
    c. Br, Cl
    d. Ca, Be
    e. Br, Se
    f. I, Sr

Answers:
1. The pair of atoms with the greater electronegativity difference will have the more polar bond:

    (a) $-\overset{|}{\underset{|}{C}}-F$,   (b) $-S-Cl$,   (c) $H-Br$

2. (a) N, (b) O, (c) Cl, (d) Be, (e) Br, (f) I

Naming Covalent Compounds

The names of several covalent compounds have already appeared in the previous sections, and just as with ionic compounds, there are formal schemes used to name them. One of these schemes is particularly descriptive and it is used to name binary covalent compounds, those that contain only two elements. This method will be described here. Other schemes, which can be used to name more complicated compounds, will be introduced in later chapters as they are needed.

When using the descriptive scheme, the name of the first element in the formula is written in full followed by the name of the second element but with the -ide ending added to its stem. Additionally, the number of atoms of each element in the formula is indicated using Greek prefixes. These prefixes are listed in Table 8.4.

Let us show how the scheme works by naming CCl_4, a covalent compound with one carbon and four chlorine atoms in the molecule. The name of the first element, carbon, will appear first in the name followed by the name of the second element properly modified with the -ide ending, chloride. The prefix tetra- will be used before chloride to indicate there are four chlorine atoms in the formula. The name of CCl_4 would be carbon tetrachloride. Notice the name of the first element is separated from the rest of the name by a single space.

CCl_4

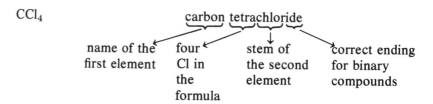

The prefix mono- was omitted from the name since the name of the element itself indicates one atom of it in the formula. The mono- prefix is sometimes used to emphasize the difference between compounds composed of the same elements such as carbon monoxide (CO) and carbon dioxide (CO_2).

TABLE 8.4 Prefixes Used to Indicate Numbers of Atoms	
Prefix	Number of atoms
mono-	1 (usually omitted)
di-	2
tri-	3
tetra-	4
penta-	5
hexa-	6
hepta-	7
octa-	8

The names of several binary covalent compounds are given below. Notice how the numerical prefixes describe the formula of the compound.

SO_2	sulfur dioxide
SO_3	sulfur trioxide
N_2O	dinitrogen oxide (laughing gas)
NO_2	nitrogen dioxide
N_2O_5	dinitrogen pentoxide
PCl_5	phosphorus pentachloride
SiO_2	silicon dioxide
S_2Cl_2	disulfur dichloride
CS_2	carbon disulfide
H_2O	dihydrogen oxide (water)

Because some compounds are so familiar to us, we often forgo their scientific names and use common names instead. For that reason, most chemists would call H_2O water and not dihydrogen oxide, and NH_3 is usually called ammonia and not nitrogen trihydride. It should also be added that the prefixes in Table 8.4 are *not* used in the names of the ionic compounds unless they appear in the name of a polyatomic ion such as in the dihydrogen phosphate ion, $H_2PO_4^-$.

Check Test Number 8

Correctly name the following binary covalent compounds.
a. CBr_4 b. PF_3 c. N_2O_4 d. SCl_2
e. SF_6 f. B_2O_3 g. NO   h. OF_2

Answers:

a. carbon tetrabromide   e. sulfur hexafluroide
b. phosphorus trifluoride   f. diboron trioxide
c. dinitrogen tetroxide   g. nitrogen oxide
d. sulfur dichloride   h. oxygen difluoride

The Coordinate Covalent Bond

In each example of a covalent bond studied up to this point, half of the shared electrons came from each of the atoms joined by the bond. Each fluorine atom provided one of the electrons in the shared pair in the flourine molecule. But there are cases in which *both* electrons in a bond come from only one of the two atoms. Covalent bonds in which both of the shared electrons are provided by only one atom are called **coordinate covalent bonds.** The only difference between an ordinary covalent bond and a coordinate covalent bond is the way they are formed. After either bond is formed, it behaves like any other covalent bond. An example of coordinate covalent bonding is the formation of the ammonium ion from ammonia and a hydrogen ion. The ammonia molecule provides the two electrons for the bond between the nitrogen atom and H^+:

$$H-\overset{\overset{\displaystyle H}{|}}{\underset{\underset{\displaystyle H}{|}}{N}}{:}H^+ \longrightarrow \left[H-\overset{\overset{\displaystyle H}{|}}{\underset{\underset{\displaystyle H}{|}}{N}}-H\right]^+$$

$$(NH_4{}^+)$$

Coordinate covalent bonds are found in many species, such as the polyatomic ions. But it is important to remember that they are just like any other covalent bond that involves the sharing of two electrons. Its principal distinction is the single atom source of the electron pair that forms the covalent bond.

Covalent Bonds in Polyatomic Ions

Polyatomic ions are composed of two or more atoms joined together by covalent bonds, and the bonded atoms carry an electrical charge. Though you have already been introduced to several ions of this type, let us look at the Lewis structure of three of them.

A typical polyatomic ion is the hydroxide ion, OH^-. Its Lewis structure shows an octet of electrons about oxygen with a pair of electrons shared with hydrogen to form a covalent bond:

$$\left[:\overset{..}{\underset{..}{O}}:H\right]^- \quad \text{or} \quad \left[:\overset{..}{\underset{..}{O}}-H\right]^-$$

The ion bears a 1− charge because it contains one more electron than the total number of protons in the two nuclei ($10e^-$, $9p^+$). This "extra" electron is used by oxygen to complete its octet. The very poisonous cyanide ion, CN^-, has a triple bond between the carbon and nitrogen atoms, as seen in its Lewis structure:

$$\left[:C\equiv N:\right]^-$$

The acetate ion, $C_2H_3O_2^-$, contains several atoms joined by covalent bonds. The "extra" electron is acquired to complete the octet on one of the oxygen atoms, and it gives the ion a 1− charge:

$$\left[\begin{array}{c} \text{H} \\ | \\ \text{H}\!-\!\text{C}\!-\!\text{C} \\ | \\ \text{H} \end{array} \underset{\ddots\ddot{\text{O}}:}{\overset{\ddot{\text{O}}:}{\diagdown}} \right]^{-}$$

Polyatomic ions are stable species as evidenced by their ability to go through many chemical reactions without being destroyed.

Deriving Lewis Structures

Earlier the Lewis structures of several molecules were derived by simply combining the unpaired electrons on two or more atoms to form covalent bonds and octets of electrons. This approach is quite straightforward and it works well for many molecules, but for others (and the polyatomic ions) determining Lewis structures can become more difficult. Part of this difficulty stems from the use of coordinate covalent and ordinary covalent bonds in the same species. This means that some atoms can donate pairs of electrons to form more bonds than that predicted solely from the numbers of unpaired electrons. For this reason another method that can be used to derive Lewis structures is presented here. It uses a four-step procedure to distribute valence electrons about the atoms so each atom has an octet of electrons.

Step 1: Draw the skeleton structure of the molecule or ion, joining the bonded atoms with single bonds.

Skeleton structures show which atoms in a molecule or ion are joined by covalent bonds. The skeleton structure for CO_2 would be O—C—O. It shows that both oxygen atoms are bonded to carbon. The information needed to draw skeleton structures will be given in most examples and problems.

Step 2: Total the valence electrons on all the atoms. If it is an anion, *add* one electron to the total for each negative charge; if a cation, *subtract* one electron from the total for each positive charge.

Step 3: From the electron total in step 2, subtract two electrons for each single bond drawn in the skeleton structure.

Step 4: Distribute the remaining electrons around the skeleton structure as unshared *pairs* to give each atom an octet. Hydrogen only gets two electrons.

In some cases there may be too few electrons available to give each atom an octet when all electrons are used as unshared pairs. In these cases, you will need to convert one or more single bonds to double or triple bonds by converting unshared pairs to shared pairs. The new shared pair then counts toward the octet of both atoms. For example, any attempt to draw the Lewis structure for carbon dioxide without using double bonds would lead to structures that do not have octets on all atoms. The eight pairs of valence electrons might be distributed like this:

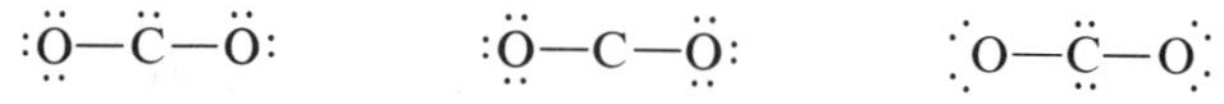

$$:\ddot{O}-\ddot{C}-\ddot{O}: \qquad :\ddot{O}-C-\ddot{O}: \qquad \ddot{O}-\ddot{C}-\ddot{O}$$

But shifting two electron pairs to form double bonds will allow each of the three atoms to have an octet:

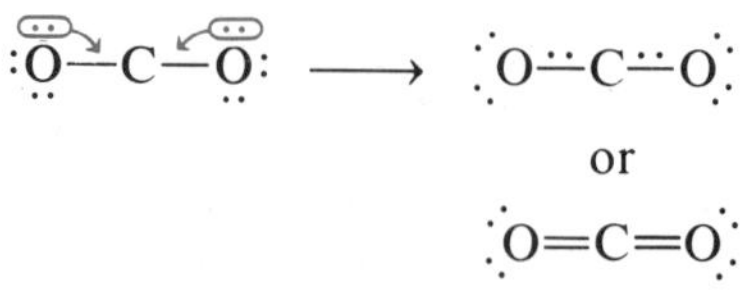

$$:\ddot{O}-C-\ddot{O}: \longrightarrow :O \ddot{=} C \ddot{=} O:$$

or

$$:\ddot{O}=C=\ddot{O}:$$

The following examples will show how to use this method for deriving Lewis structures.

Example 8.7

Derive the Lewis structure for the sulfate ion, SO_4^{2-}. Each oxygen atom is bonded to sulfur.

Step 1:   Draw the skeleton structure from the information given:

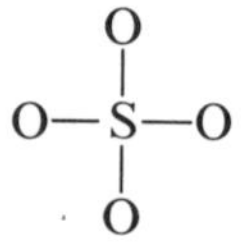

$$\begin{array}{c} O \\ | \\ O-S-O \\ | \\ O \end{array}$$

Step 2:   Total valence electrons:

$$
\begin{aligned}
4\,O &= 4 \times 6e^- &&= 24e^- \\
1\,S &= 1 \times 6e^- &&= 6e^- \\
2- \text{ charge on ion} &= 2 \text{ more electrons} &&= 2e^- \\
\hline
\text{Total} &&&= 32e^-
\end{aligned}
$$

Step 3:   Subtract electrons already used in skeleton structure:

$$
\begin{aligned}
& 32e^- \text{ (total)} \\
-\; & 8e^- \text{ (4 single bonds)} \\
\hline
& 24e^- \text{ (12 pairs of electrons)}
\end{aligned}
$$

Step 4:   Distribute remaining electrons to give each atom an octet:

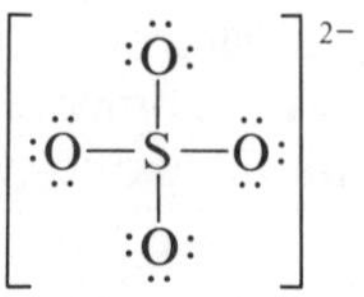

$$
\left[\begin{array}{c} :\ddot{O}: \\ | \\ :\ddot{O}-S-\ddot{O}: \\ | \\ :\ddot{O}: \end{array} \right]^{2-}
$$

The remaining 24 electrons are distributed as 12 unshared pairs about the oxygen atoms. A check shows an octet about each atom and a total of 32 electrons (16 pairs), so this would be a reasonable Lewis structure for SO_4^{2-}.

It may seen unusual to see each oxygen atom bonded to sulfur with a single bond, but a pair of electrons is donated to each oxygen atom and it shares this with sulfur to complete its octet. That single pair of electrons is symbolized as a single bond.

Example 8.8

Derive the Lewis structure for sulfur dioxide, SO_2. Each oxygen is bonded to sulfur.

Step 1:   Skeleton structure:

$$O\text{—}S\text{—}O$$

Step 2:   Valence electron total:

$$2\,O = 2 \times 6e^- = 12e^-$$
$$\underline{1\,S = 1 \times 6e^- = 6e^-}$$
$$\text{Total} = 18e^-$$

Step 3:   Subtract electrons used as bonds:

$$18e^- \text{ (total)}$$
$$\underline{-\ 4e^- \text{ (2 single bonds)}}$$
$$14e^- \text{ (7 pairs of electrons)}$$

Step 4.   Distribute electrons to form octets:
   Using the electrons as unshared pairs will not give each atom an octet. One attempt may look like this $:\ddot{O}\text{—}\ddot{S}\text{—}\ddot{O}:$, with sulfur short two electrons. Depending on which oxygen atom is chosen to provide a pair of electrons for a double bond, two similar Lewis structures are possible:

A check shows either would be a satisfactory Lewis structure for SO_2.

Resonance

Experiments have shown that neither Lewis structure derived for sulfur dioxide in Example 8.8 accurately represents the *true* structure of SO_2. Both Lewis structures show one single and one double bond in the molecule, yet experiments show that both sulfur–oxygen bonds are identical, having properties between those of single and double bonds. How can this be? The answer to this dilemma may seem like chemical "sleight of hand" at first glance, but SO_2 displays the property of **resonance**, which makes it impossible to use any single Lewis structure to accurately represent the molecule. The actual structure of SO_2 is assumed to be an average of the two structures derived in Example 8.8. It is called a **resonance hybrid** since it is a hybrid of both Lewis structures, and the sulfur–oxygen bonds are neither single nor double bonds but something between these extremes. A resonance hybrid cannot be accurately drawn, but it can be represented in an approximate way as an average of the two Lewis structures:

$$:\ddot{O}=\ddot{S}-\ddot{O}: \rightleftharpoons \left\{:\ddot{O}\text{---}\ddot{S}\text{---}\ddot{O}:\right\} \rightleftharpoons :\ddot{O}-\ddot{S}=\ddot{O}:$$

Resonance
hybrid

Whenever it is possible to draw two or more Lewis structures for a molecule or ion that differ *only* in the location of one or more multiple bonds, then the true structure will be a resonance hybrid. The species displays resonance. The true structure of the nitrate ion, NO_3^-, is a resonance hybrid formed from contributions from each of the following:

$$\left[\begin{array}{c} \ddot{O} \\ \parallel \\ N \\ \diagup \ \diagdown \\ :\ddot{O} \quad \ddot{O}: \end{array}\right]^- \quad \left[\begin{array}{c} :\ddot{O}: \\ | \\ N \\ \diagup \ \diagdown \\ :\ddot{O} \quad \ddot{O}: \end{array}\right]^- \quad \left[\begin{array}{c} :\ddot{O}: \\ | \\ N \\ \diagup \ \diagdown \\ :\ddot{O} \quad \ddot{O}: \end{array}\right]^-$$

Resonance adds stability to a molecule or ion.

Check Test Number 9

1. Derive Lewis structures for the following:

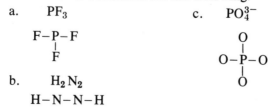

    a. PF_3 c. PO_4^{3-}

$$\begin{array}{c} F-P-F \\ | \\ F \end{array} \qquad \begin{array}{c} O \\ | \\ O-P-O \\ | \\ O \end{array}$$

    b. H_2N_2

$$H-N-N-H$$

2. Draw three Lewis structures that would be suitable for SO_3, sulfur trioxide. Would SO_3 be expected to display resonance?

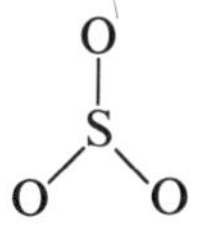

Answers:

1. a $:\ddot{F}-\ddot{P}-\ddot{F}:$,   b $H-\ddot{N}=\ddot{N}-H$,   c $\left[\begin{array}{c} :\ddot{O}: \\ | \\ :\ddot{O}-P-\ddot{O}: \\ | \\ :\ddot{O}: \end{array}\right]^{3-}$

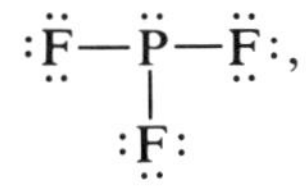

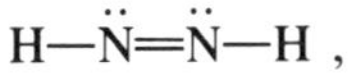

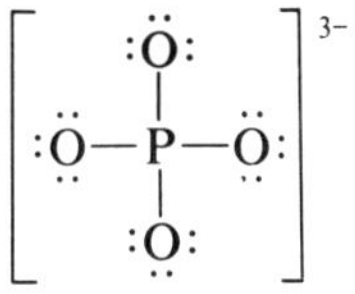

2.

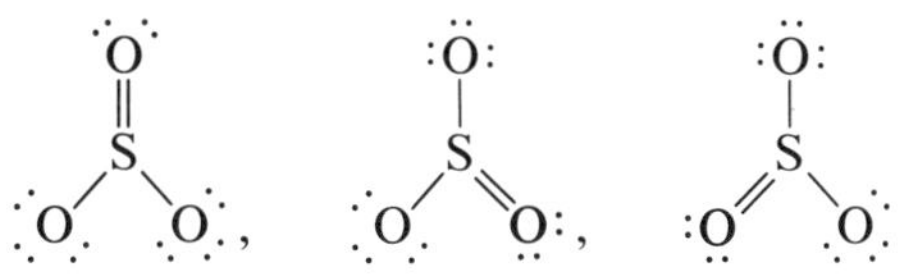

The three structures differ only by the location of one double bond, so SO_3 would be predicted to display resonance.

Exceptions to the Octet Rule

Although most of the molecules and polyatomic ions you will encounter in this course obey the octet rule, there are some fairly common ones that do not. One such species is nitrogen dioxide, NO_2, a molecule that has an odd number of valence electrons. A reasonable Lewis structure for NO_2 is shown below. Notice that nitrogen does not have an octet of electrons:

$$2\,O = 2 \times 6e^- = 12e^-$$

$$\underline{1\,N = 1 \times 5e^- = 5e^-}$$

$$\text{valence electron total} = 17e^-$$

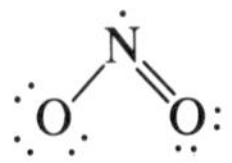

Boron trifluoride, BF_3, is a stable molecule with only six electrons about the boron atom. Boron is a group III element and it uses its three valence electrons to bond to the three fluorine atoms:

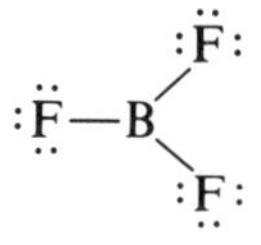

There are also species that have more than an octet of electrons about an atom. Two such compounds are phosphorus pentafluoride, PF_5, and sulfur hexafluoride, SF_6:

$$
\begin{array}{c}
\text{F} \\
| \\
\text{F—P—F} \\
\diagup \quad | \\
\text{F} \quad \text{F}
\end{array}
\qquad
\begin{array}{c}
\text{F} \\
\text{F} \diagdown | \diagup \text{F} \\
\text{S} \\
\text{F} \diagup | \diagdown \text{F} \\
\text{F}
\end{array}
$$

(10e– about P)   (12e– about S)

The Shapes of Molecules

The physical and chemical properties of a molecule are not only determined by the kinds of atoms that make it up and how they are joined together, but also by the shape of the molecule, that is, the way the atoms are arranged in space. For this reason you should become acquainted with the common shapes of molecules and how to predict them.

By now you can appreciate how useful Lewis structures are for showing the bonding arrangements in molecules. But their value goes beyond this since they can also be used to predict the shapes of molecules. This is because Lewis structures show both shared and unshared pairs of valence electrons in molecules, and the valence electron pairs to a large extent govern how the atoms will be arranged in space.

Each pair of electrons, whether shared or unshared, in the valence level of an atom acts as if it were a small cloud of negative charge. These clouds (or electron pairs) mutually repel each other because of their negative charge. The repulsion forces them to become oriented about the atom so they are located as far from each other as possible. As an example, consider the BeH_2 molecule. The Lewis structure of BeH_2 is H:Be:H (it does not obey the octet rule). There are two pairs of electrons about beryllium, the central atom of the molecule. The **central atom** of a molecule is the one to which the other atoms are attached. Only one possible orientation of the two electron pairs will reduce their mutual repulsion to a minimum, and that will locate them on opposite sides of the beryllium atom. Since each electron pair is shared with hydrogen, it means the hydrogen atoms will be on opposite sides also. This gives the BeH_2 molecule a **linear shape,** with the three atoms lying along a straight line. If the molecule is drawn using dashes for the bonds, the angle between the two Be–H bonds is seen to be 180°, the linear bond angle:

H—Be—H

180°

The angle between adjacent bonds in a molecule is called a **bond angle.** Bond angles are commonly used to describe the shapes of molecules. Molecular shapes are also described by **bond lengths,** the distances between the nuclei of two bonded atoms. The shape of diatomic molecules like HCl and H_2 can be described by their bond lengths since they do not have a bond angle. We will be more concerned with bond angles than bond lengths here.

If the central atom of a molecule is surrounded by three electron pairs, they will be oriented to the corners of an equilateral triangle with the central atom in its center. Boron trifluoride, BF_3, is such a molecule. The shape of BF_3 is described as being **trigonal planar.** All four atoms are in the same plane and all bond angles are 120°, the trigonal bond angle:

BF_3

Four pairs of electrons about an atom will be oriented toward the corners of a regular tetrahedron, a four-sided figure with triangular faces. Methane, CH_4, has four pairs of electrons about carbon and the molecule has a **tetrahedral shape** with all bond angles being $109\frac{1}{2}°$, the tetrahedral angle. The carbon atom is in the center of the tetrahedron:

CH_4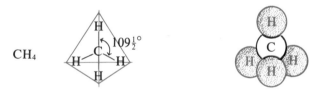

The preferred orientations for two, three, and four pairs of electrons about an atom are summarized in Table 8.5. There are also preferred orientations for five or more pairs of electrons, but we will not be concerned with them.

It was said earlier that the shape of a molecule is described by the location

TABLE 8.5			
Preferred Orientation of Electron Pairs About an Atom			
Number of pairs	Preferred orientation	Bond angle	Example molecules
2	linear	180°	BeH_2, BeF_2
3	trigonal planar	120°	BF_3, BCl_3
4	tetrahedral	$109\frac{1}{2}°$	CH_4, CCl_4

of its atoms (or more precisely, the nuclei of the atoms) in space. With BeH_2, BF_3, and CH_4 each electron pair on the central atom is in a bond, and the shape of the molecule is the same shape as that adopted by the electron pairs. However, if one or more of the electron pairs is an unshared pair, the molecular shape will not be the same as that of the electron pairs themselves. Both ammonia and water have four pairs of electrons about the central atoms and they will be oriented in a tetrahedral shape. But the atoms in NH_3 are in a **pyramid shape** and the atoms in H_2O form a molecule with a **bent shape** as shown in Figure 8.9. The bond angles in NH_3 are 107° and in H_2O, 105°. These are slightly less than the tetrahedral angle predicted for four pairs of electrons, but whenever one or two of the four electron pairs are unshared pairs, the bond angles are found to be slightly less than the ideal $109\frac{1}{2}°$.

Double and triple bonds have about the same affect on the shapes of molecules as single bonds. They can be regarded, as far as shapes are concerned, as a

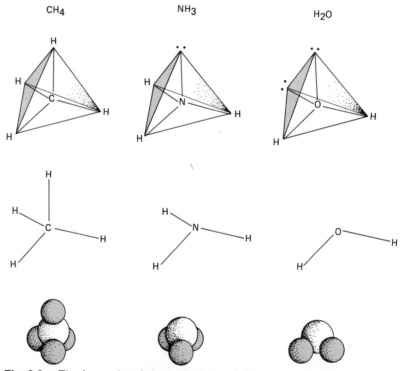

Fig. 8.9.   The four pairs of electrons about the central atom in CH_4, NH_3, and H_2O are oriented toward the corners of a regular tetrahedron. The shape of each molecule is determined by the location of the atoms. CH_4 is tetrahedral, NH_3 is a pyramid, and H_2O is bent. Though each molecular shape is different, all are the result of the orientation of four pairs of electrons about the central atom.

pair of electrons. The shape of the sulfur dioxide molecule, SO_2, can be considered as the result of repulsions between an unshared pair, a single bond pair, and a double bond about sulfur. The electrons act "effectively" as three pairs of electrons oriented to the corners of an equilateral triangle. Considering only the atoms, the molecule is bent with a predicted bond angle of 120° (or slightly less due to the unshared pair):

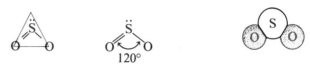

You may recall that sulfur dioxide displays resonance, but either Lewis structure for the molecule can be used to predict its structure.

Predicting Molecular Shapes

Molecular shapes can be predicted from the Lewis structures of most molecules or polyatomic ions. We will restrict the predictions to species that have an easily identified central atom, since it is the orientation of the electron pairs about the central atom that determines the shape. In later chapters the shapes of larger molecules will be discussed.

The first thing to do when attempting to predict the shape of a molecule or ion is to draw its Lewis structure and locate the central atom. Second, count the "effective" number of electron pairs: unshared pair=1, single bond=1, multiple bond=1 (effectively). Then determine the preferred orientation of the electron pairs as listed in Table 8.5; then name the shape outlined by the atoms: linear, bent, pyramid, trigonal planar, or tetrahedral.

The Lewis structure for sulfur dichloride, SCl_2, shows four pairs of electrons about sulfur, and they would be directed toward the corners of a tetrahedron. The molecule will have a bent shape and the bond angle should be less than $109\frac{1}{2}°$:

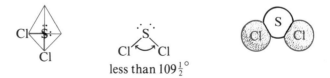

less than $109\frac{1}{2}°$

The Lewis structure of the nitrate ion, NO_3^-, shows two single and one double bond about nitrogen. Together they behave effectively as three electron pairs, so they will point to the corners of an equilateral triangle:

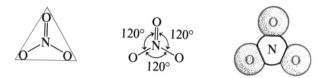

The predicted shape of the ion will be trigonal planar and all bond angles will be equal to 120°. Now try your hand at predicting shapes in Check Test Number 10.

Check Test Number 10

Predict the shapes and bond angles of the following.
(a) H_2S, (b) PO_4^{3-}, (c) PF_3

Answers:
  a. bent, with a bond angle less than $109\frac{1}{2}°$
  b. tetrahedral, all bond angles are $109\frac{1}{2}°$
  c. pyramid, bond angles less than $109\frac{1}{2}°$

Comparing Ionic and Covalent Compounds

The kind of bonding that exists in a compound, whether ionic or covalent, will have a great affect on the properties of the compound. There are many compounds that contain both covalent and ionic bonds in the formula unit, such as sodium acetate or potassium hydroxide. In terms of classification, these compounds would be considered ionic. Generally, the presence of an ionic bond (or ions) in the compound places it in the ionic class, even if there are many covalent bonds in the ions.

sodium acetate $(NaC_2H_3O_2)$ potassium hydroxide (KOH)

Perhaps the simplest way to present the differences that exist between ionic and covalent compounds is to summarize their characteristics in tabular form for comparison. This is done in Table 8.6. It should be added that many compounds have properties that overlap into both categories.

TABLE 8.6 Characteristics of Ionic and Covalent Compounds		
	Ionic compounds	Covalent compounds
Kind of particles present	positive and negative ions	molecules
Type of bonding	ionic	covalent
Physical state at room temperature (about 25°C)	all are solids	can be either solids, liquids, or gases
Nature of the solid compounds	crystalline; hard, brittle crystals with a relatively high melting temperature	can be crystalline or amorphous (without organized structure); crystals are often soft, easily crushed, and melt at relatively low temperatures
Behavior when dissolved in water	if soluble in water, will dissociate into the ions; the solutions can conduct electricity	if soluble in water, will remain in molecular form (barring reaction with water); the solution will not conduct electricity

Terms

Several important terms appeared in Chapter 8. You should understand the meaning of each.

octet rule
ionic compound
ionic bond
formula unit
systematic name
common name
oxidation
reduction
covalent compound
lattice
linear shape

covalent bond
shared pair
unshared pair
single bond
double bond
triple bond
bond energy
Lewis structure
nonpolar covalent bond
skeleton structure
pyramid shape

polar covalent bond
electronegativity
coordinate covalent bond
resonance
resonance hybrid
molecular shape
bond angle
bond length
tetrahedral shape
trigonal planar shape
bent shape

Questions

Answers to the starred questions appear at the end of the text.

1. For each of the elements listed, give the ion it would form to achieve an octet of electrons in the highest occupied energy level.

   *a. Li   d. S
   *b. O   e. I
   *c. Ba   f. Al

* 2. Which of the following are ionic compounds? Which are covalent compounds?

   a. $BaCl_2$ d. H_2S
   b. MgF_2 e. NaBr
   c. CS_2 f. $CaCO_3$

   3. Using electron-dot symbols write an equation describing the reaction between sodium and oxygen to form sodium oxide, Na_2O.

*4. What is the ionic bond?

*5. Classify each of the following as either oxidation or reduction.

   a. $Na \rightarrow Na^+ + 1e^-$ c. $O + 2e^- \rightarrow O^{2-}$
   b. $Ca \rightarrow Ca^{2+} + 2e^-$ d. $Cu^{2+} + e^- \rightarrow Cu^+$

6. Predict the formula of the compound formed when the following pairs of elements react.

   *a. Ca, Br   d. Al, F
   b. Ag, S   *e. Sr, O
   *c. K, O   f. Ba, Cl

7. Predict the formula of the compound formed when the following pairs of ions are combined.

   *a. Li^+, PO_4^{3-} d. Hg_2^{2+}, Cl^-
   b. Sr^{2+}, NO_3^- *e. Na^+, $C_2H_3O_2^-$
   *c. Al^{3+}, SO_4^{2-} f. NH_4^+, SO_4^{2-}

8. Write the correct formula for each of the following compounds.

   *a. calcium iodide   *g. aluminum sulfide
   b. ammonium chloride   h. copper (II) acetate
   *c. silver cyanide   *i. potassium nitrate
   d. mercury (II) oxide   j. sodium bicarbonate
   *e. magnesium sulfate   *k. calcium hydrogen phosphate
   f. iron (III) hydroxide   l. barium carbonate

9. Give the correct name for each of the following ionic compounds.

   *a. $CaCl_2$ *e. $KHCO_3$ *i. Ag_2CO_3
   b. $NaNO_3$ f. SrS   j. $Ba(C_2H_3O_2)_2$
   *c. Fe_2O_3 *g. CuI   *k. KCN
   d. ZnO   h. $Al(OH)_3$ l. $(NH_4)_3PO_4$

10. Write the correct formula for the following, and give the systematic name for each one.

    *a. ferric sulfide   d. ferrous oxide
    b. cuprous chloride   *e. cupric nitrate
    *c. mercuric oxide   f. mercurous chloride

11. What is the difference between the bond in NaCl and Cl_2?

12. Why do the elements O, Cl, and N exist as diatomic molecules?

13. What are double bonds? Triple bonds?

***14.** Draw the Lewis structures for the compounds of hydrogen with Br, P, B, and C.

***15.** Draw the Lewis structures for the most likely compounds formed between the following pairs of elements.
    a. N, F   **c.** C, S
    b. Cl, Br   **d.** P, Cl

16. Which of the following bonds would be polar covalent bonds?
    a. H—H   **c.** :C≡N:⁻   **e.** :O̤=O̤:
    b. NaBr   **d.** H—C̤l:

***17.** Arrange each group of elements from left to right in order of increasing electronegativity.
    a. Cl, P, S
    b. Mg, K, C, F
    c. P, Al, Ca

18. What property of an element is described by its electronegativity?

***19.** In what general region of the Periodic Table do you find the most electronegative elements?

***20.** Which is the most polar covalent bond: F—F, H—Br, $-\overset{|}{\underset{|}{C}}-Cl$, or $-\overset{|}{N}-H$? Why?

21. What is the relationship between the strength of a chemical bond and the number of electrons shared in the bond?

***22.** Using the fact that carbon forms four bonds and hydrogen one bond in compounds, derive the Lewis structures for each compound listed.
    a. C_3H_8
    b. C_2H_4
    c. C_2H_2

23. What is the difference between an ordinary covalent bond and a coordinate covalent bond?

***24.** Draw Lewis structures for each of the following.

    a. CO_3^{2-}, $O-\overset{|}{\underset{O}{C}}-O$ **c.** O_2^{2-} (peroxide ion), O—O

    b. PH_3, $H-\overset{|}{\underset{H}{P}}-H$ **d.** CO (carbon monoxide), C—O

25. What is wrong with each of these Lewis structures?

a. SO_3, :Ö—S̈—Ö:
              |
             :O:

c. NO_2^-, [:Ö—N̈—Ö:]⁻

b. CH_4O

H
|
H—C=O:
|
H

d. SiF_4

26. Identify each bond angle as being either the tetrahedral angle, the trigonal angle, the linear angle, or neither of the three.
    a. 105°
    b. 120°
    c. 90°
    d. 180°
    e. $109\frac{1}{2}°$
    f. 60°

*27. Predict the shape of each of the following species.

a.
H—C=O
     |
     H

c.
     H
     |
:C̈l—C—C̈l:
     |
     H

e.
:F̈—N̈—F̈:
     |
    :F̈:

b.
:C̈l—P̈—C̈l:
     |
   :Br:

d.
: Ö—S=O:
     |
   :O:

28. Correctly name each of the following binary covalent compounds.
    *a. SF_4
    b. SF_6
    *c. P_2O_5
    d. CI_4
    *e. NCl_3
    f. BrCl
    *g. PI_3
    h. $AsCl_3$
    *i. $AsCl_5$
    j. N_2O_4

29. Write the formulas of these compounds.
    *a. boron tribromide
    b. phosphorus pentafluoride
    *c. hydrogen chloride
    d. carbon tetrachloride
    e. selenium dichloride

30. Identify each of the following as a characteristic property of either an ionic compound or a covalent compound.
    *a. hard, brittle crystals
    b. can be solids, liquids, and gases
    *c. exists as molecules
    d. can evaporate
    *e. soft crystals
    f. forms ions in water
    *g. solutions do not conduct electricity
    h. ionic bonds
    *i. solids have relatively low melting temperature
    j. all are solids

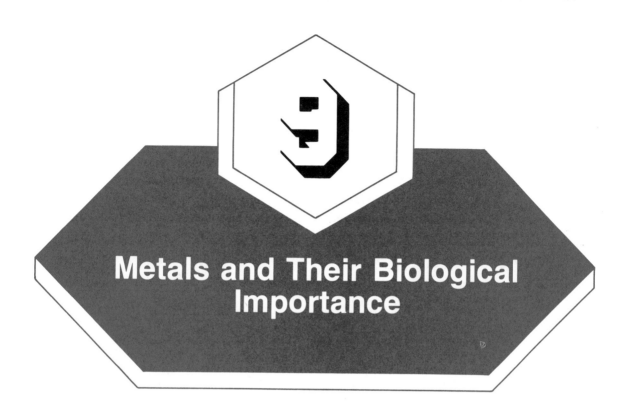

Metals and Their Biological Importance

Metals account for about $2\frac{1}{2}\%$ of the total weight of the body, and their presence is extremely important in nearly all facets of body chemistry. For example, the body of an average adult contains about 4 g of iron, but this small amount is vital for the absorption, transport, and storage of oxygen. Many enzymes, the biological catalysts in the body, require the presence of certain metals if they are to function at all. Carbonic anhydrase is such an enzyme. It has a very large molecular weight, on the order of about 30,000, yet this huge molecule requires the presence of one zinc ion (Zn^{2+}) buried within it so it can function. Without the zinc ion, it would be an inactive chunk of protein. The muscles and nervous system would not function without the optimum ratios of sodium, potassium, and other metals.

This chapter is primarily designed to point out the important biological role of metals in the body. But because metals are important to us in other ways too, there will be brief discussions of metallic bonding and alloys, mixtures of metals designed to have special properties. The chapter will then conclude with a survey of ten important metals in the body.

Objectives

By the time you finish this chapter you should be able to do the following:

1. Describe the electron-sea model of metallic bonding.
2. Describe what an alloy is and identify some important alloys.
3. Describe the general composition of dental amalgam and dental gold.
4. Describe the three general forms in which metals are incorporated into the chemistry of the body.
5. List ten important metals in the body and give at least one important function of each.

The Metallic Bond

In the previous chapter, two main types of bonding were described: ionic bonding, which results from the complete transfer of electrons from one atom to another; and covalent bonding, the result of electron sharing between atoms. Metals present a third kind of bonding that in some ways can be considered a mixture of both ionic and covalent effects.

A piece of iron, sodium, or gold is made up of millions of atoms arranged in a crystalline structure. But unlike ionic or covalent solids, metals can conduct electricity, can be pounded into different shapes (malleability), and drawn into wires (ductility). Any theory of bonding in metals that describes how metal atoms are attracted to one another should be consistent with these unique properties of metals.

One theory of the **metallic bond** that is reasonably successful and straightforward is the **electron-sea model** (see Figure 9.1). Other theories also exist, but they are beyond the scope of this text. The outermost energy levels of metal atoms are only partially full, usually containing one, two, or three electrons. In a metallic crystal, the outermost energy levels of atoms can overlap or mix with like energy levels on adjacent atoms on all sides. The result is the formation of a large, three-dimensional energy level that spans millions of atoms. Since the outermost electrons (valence electrons) are usually not tightly held by metal

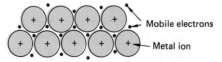

Fig. 9.1.   The electron-sea model for metallic bonding proposes the attraction of positive metal ions for the sea of highly mobile electrons.

atoms (the relatively low ionization energies of metals verify this), one or more of the valence electrons can readily leave an atom and become part of the large "common" energy level and drift throughout the crystal. The metal atoms, having lost the valence electrons, exist as ions imbedded in the highly mobile sea of electrons. It is the net attraction between the positive metal ions and the negative sea of electrons that results in the **metallic bond** that holds the crystal together. Earlier the metallic bond was described as a mixture of both ionic and covalent interactions since, at once, there is a sharing of valence electrons with other atoms and an electrostatic attraction between the electron sea and the metal ions in it.

The highly mobile sea of electrons accounts for the high electrical conductivity (electricity is a flow of electrons) of metals. The fact that metals are malleable and ductile can be understood in terms of the relative ease with which the electron sea insulates and substantially reduces any repulsive forces between the metal ions as they move by one another.

Alloys

An **alloy** is a mixture of elements, mostly metals, that has the characteristic properties of metals. They are most often prepared by melting different metals together to form a homogeneous liquid solution. The solution of metals is then cooled and solidified. The alloying process allows the properties of metals to be modified to meet specific needs. For example, pure gold (24 karat gold) is too soft to be used in jewelry or dental restorations. A pure gold crown on a molar would quickly become deformed and require replacement. But an alloy of gold, silver, copper, and perhaps small amounts of other metals, retains the color and corrosion resistance of gold while being hard enough to withstand the pressures encountered in chewing food. Metal pins, joints, and plates implanted in the body must not only be strong, but must resist corrosion and be compatible with bone and soft tissue. Two alloys that have been developed to meet these requirements along with several others are described in Table 9.1.

Alloys of mercury, a liquid metal at room temperature, are called **amalgams**. Amalgams can be prepared by simply dissolving suitable metals in a small amount of mercury at room temperature. The so called "silver" dental fillings are amalgams of silver, tin, copper, and zinc dissoved in a small amount of mercury. The amalgam can be packed into a prepared cavity and shaped while it is still soft. After 3 to 5 minutes the amalgam is moderately hard and reaches full hardness in about 24 hours.

You might be alarmed at the thought of filling teeth with an alloy that contains mercury, since mercury is widely mentioned as being one of the most toxic metals. In dental amalgam, mercury is present as elemental mercury (mercury atoms). Studies have shown that small amounts of ingested mercury metal,

TABLE 9.1
Composition and Uses of Several Alloys

Name of alloy	Composition by weight	Uses
Brass	Cu (60–85%) Zn (15–40%)	hardware fittings (corrosion resistant, easily worked)
Stainless steel	Fe (about 80%) Cr (about 18%) Ni (about 1%) C (about 0.5%)	laboratory utensils, sinks, and pans
Stainless steel for surgical implants	Fe (62–71%) Cr (17–20%) Ni (10–14%) Mo (2–4%)	surgical implants for joints, support pins, and plates
Vinertia alloy	Co (67%) Cr (27%) Mo Ni }(6%) Mn	surgical implants for joints
Dental amalgam	Ag (70%) Sn (maximum of 29%) Cu (2%) Zn (0–2%) Hg (2%)	tooth filling material (easily shaped, hardens in a matter of hours)
Dental gold	Au (70%) Ag (15%) Cu (about 10%) Pd Pt }(about 5%) Zn	tooth restorations, crowns, and bridges
Sterling silver	Ag (93%) Cu (7%)	jewelry, tableware

such as one might receive following the filling of a tooth, are excreted in the urine in a few days as mercury atoms. However, it is also possible that mercury metal can be oxidized in the body to form Hg_2^{2+}, a toxic ion of mercury. This ion may become incorporated into soluble mercury compounds that may enter the bloodstream. Once in the blood, these mercury compounds may attack brain cells, causing permanent damage. On the other hand, if a small amount of mercury is released into the mouth as Hg_2^{2+}, it rapidly combines with the chloride ion in saliva to form Hg_2Cl_2, a compound that is only very slightly soluble. In this form, mercury does not seem to represent as serious a threat to health, and it is eventually excreted. Mercury has been used in tooth fillings for years, and it has not been regarded as a hazard. There may be some serious questions raised about this belief in the future.

Though mercury in dental fillings does not seem to represent an immediate health hazard, it should be added that dental personnel should take precautions to avoid inhaling mercury vapors. Mercury spills will evaporate in time and it has been noted that once brought *into the lungs*, mercury atoms can become converted into toxic mercury compounds and enter the bloodstream.

Metals in the Body

Metals are present in the body as positive ions, and not as atoms. This is an important point to remember. When someone speaks of "calcium in the body," they are not speaking of metallic calcium but calcium in the form of its ion, Ca^{2+}.

There are three principal ways in which metal ions are incorporated into the chemical makeup of the body.

1. Metal ions existing as ions in solution. Blood serum and the fluids in and around cells are water solutions that contain many different substances dissolved in them. Some metals, notably sodium and potassium, are present in these fluids as free ions, Na^+ and K^+. It should be added that anions (such as Cl^- and HCO_3^-) are also present since, in solutions, each positive charge must be balanced by a negative charge.

2. Metal ions in structures composed of aggregates of positive and negative ions (usually associated with protein). The major portions of bone and teeth are composed of Ca^{2+}, PO_4^{3-}, and OH^- ions in a solid crystalline structure of simple formula: $Ca_5(PO_4)_3OH$. In both bones and teeth, this white solid is dispersed throughout a protein matrix that imparts a degree of flexibility to the structure.

A protein-bound ionic aggregate of Fe^{3+}, OH^-, and $H_2PO_4^-$ ions is used to store iron in the liver, spleen, and other areas of the body. The spherical, ionic aggregate is called **ferritin** and each aggregate can contain as many as 2500 Fe^{3+} ions and be as much as 25% iron by weight. The aggregate opens up to release iron as it is required by the body. A second iron storage unit called **hemosiderin** is much like ferritin, except it can contain up to 45% iron.

3. Metal ions in structures that bind them either loosely or firmly to covalent molecules. Some of the most intensively studied biochemical species are those containing metal ions bonded to relatively large covalent molecules. Perhaps the best-known example of this kind of metal ion involvement is Fe^{2+} in **heme**, the oxygen-carrying unit of hemoglobin (see Figure 9.2). The heme unit is a flat structure of carbon, nitrogen, oxygen, and hydrogen with an iron (II) ion firmly held in the center by bonds to four nitrogen atoms. The heme unit is in turn bonded to a large protein unit (globin) to form hemoglobin

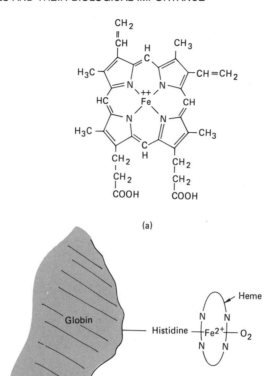

(a)

(b)

Fig. 9.2.   (a) Heme. (b) Heme is attached to a large protein called globin to form hemoglobin. The histidine unit of the globin molecule bonds to the Fe^{2+} of heme. Oxygen is carried on the opposite side of the heme unit by joining with Fe^{2+}.

(Chapter 22). The function of hemoglobin is to transport oxygen from the lungs through the bloodstream to other parts of the body. It does this by attaching an oxygen molecule to the Fe^{2+} ion in the heme unit. The oxygen molecule is then released elsewhere in the body. The heme unit is also found in myoglobin, the oxygen storage unit in cells. Chlorophyll, the green pigment in plants, looks similar to the heme group except a magnesium ion, Mg^{2+}, replaces the iron ion. And in a similar fashion, a cobalt ion, Co^{2+}, is incorporated in the vitamin B_{12} molecule. Metal ions are also found in many enzymes. In some, it is located at the "active site" of the catalyst, firmly bound to protein. In others, the metal ion is less firmly held and assists in the operation of the enzyme in a less direct way.

Of course, a given metal ion is not necessarily found exclusively in only one of the three categories. For example, calcium is found as free Ca^{2+} in serum, as a protein-bound ion in enzymes and serum, and in solid bone and teeth; and sodium is found as free ions in body fluids, combined with other ions in bone, and weakly associated with certain enzymes.

The Biologically Important Metals

Ten metals are especially important in the chemistry of the body. Four are representative elements: sodium (Na), potassium (K), calcium (Ca), and magnesium (Mg). The remaining six are transition elements: iron (Fe), cobalt (Co), copper (Cu), manganese (Mn), zinc (Zn), and molybdenum (Mo). The transition metals are needed in trace quantities by the body (less than 0.01% of body weight), but even in minute amounts, each is vital.

Metals possess certain chemical properties that make them particularly suited for biochemical processes: (1) They can form compounds with proteins and other compounds. (2) Some metals exhibit catalytic behavior that is utilized in several enzymes. (3) Several metals, especially the transition metals, can form ions of different charge, such as iron (Fe^{2+}, Fe^{3+}) and copper (Cu^{+}, Cu^{2+}), and this allows additional flexibility in their biochemical roles. For example, electrons can be transferred from one molecule to another through compounds containing iron ions, such as the cytochromes. The Fe^{3+} ion in cytochrome can gain an electron from one molecule and become Fe^{2+}. Then, the Fe^{2+} ion can transfer the electron to a different molecule and become Fe^{3+} once again.

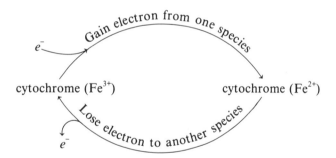

The descriptions of 10 biologically important metals that follow do not detail the actual chemical processes in which each is involved. That could quickly become exceedingly complicated. Rather, some of the roles played by each metal in the body will be pointed out so you can better appreciate their importance.

Sodium and Potassium	Normal range in plasma	Percent of body weight in a normal adult
	Na^{+} 310−334 mg/100 ml	0.10%
	K^{+} 13.7−19.6 mg/100 ml	0.22%

Sodium and potassium ions are widely distributed throughout the body. Sodium ion is the principle positive ion in blood plasma and extracellular fluid, that is, the fluid that surrounds cells. It is also found within cells, but for each sodium ion in a cell, there are about 14 on the outside. The diference in concentration in large part governs the flow of water into and out of cells. Sodium is also involved in the mechanism by which potassium ion, glucose (blood sugar), and amino acids (the building blocks of proteins) are transported into cells.

Potassium ions are the most abundant positive ions inside the cells and are found at considerably lower concentrations in extracellular fluid. Potassium ions are required by cells for efficient protein synthesis and the metabolism of glucose which provides energy for the cell.

Both potassium and sodium ions, working together, are involved in the transmission of nerve impulses. A specific concentration balance is necessary for maximum nerve sensitivity and muscle control. Sodium ions are required for muscle contraction, and it is known that potassium ion permits the heart muscle to relax between beats.

The loss of potassium ion, which may occur in severe diarrhea, can cause the potassium level in cells to drop so low that the rhythm of the heartbeat will become irregular and general muscle weakness will occur. Fortunately, the lost ions can be readily replaced through medication or diet, and the condition reversed.

Sodium is obtained in the diet principally from table salt, baking soda ($NaHCO_3$), and green leafy vegetables. Sources of potassium are legumes, green leafy vegetables, whole grains, meats, and nuts.

Calcium and Magnesium

Normal range in serum	Percent of body weight in a normal adult
Ca^{2+} 8.5–10.5 mg/100 ml	2.0%
Mg^{2+} 1.8–3.0 mg/100 ml	0.04%

Both calcium and magnesium are present in the body as ions, Ca^{2+} and Mg^{2+}. Calcium ion is needed for the formation of bone and teeth, and growing children require up to 1.5 g of calcium per day for this purpose. It is also required for milk formation during lactation, and for maintenance of a regular heartbeat. Calcium ion is important for adequate blood clotting, and may be administered to patients for this purpose. On the other hand, certain drugs can be used to reduce the level of calcium ion in blood to minimize clotting, but if the serum calcium level falls too low, muscular twitching and, eventually, convulsions can occur. Chronic low calcium levels can bring about the removal of calcium from bone which can eventually lead to degradation of the bone, and bone bending (osteomalacia) can result. There is also evidence that muscle contraction (in response to nerve stimulation) involves a flow of calcium ions, and withdrawal of these ions accompanies muscle relaxation.

Magnesium ions are required during all stages of protein synthesis in cells. They are present at a relatively high concentration in cells, and are associated with and stabilize DNA and RNA structures. The metabolism of glucose for energy, and the transfer of this energy across cell membranes, also requires the presence of magnesium ions. Energy transfer in brain, nerve, and muscle cells is particularly sensitive to the presence of magnesium ion. Magnesium also is important for proper nerve transmission and muscle contraction. However, if the concentration of magnesium is allowed to get too high, as can occur if it is administered directly into the bloodstream, it can cause depression and anaesthesia and has been used in this way to reduce convulsions. In plants, magnesium is an important part of chlorophyll.

Magnesium is obtained from nuts, beans, peanut butter, whole grains, green leafy vegetables, and dairy products. Calcium is readily obtained from cheese, milk, and green vegetables.

Iron

Normal range in serum	Percent of body weight in a normal adult
Fe^{2+} 0.05–0.15 mg/100 ml	0.005%

Total iron binding capacity: 0.25–0.41 mg/100 ml

Iron is present in the body as Fe^{2+} and Fe^{3+} ions. About 70% of the iron is found in hemoglobin, the oxygen carrier in the blood. About 15% is stored (ferritin, hemosiderin) in the liver, spleen, kidneys, and bone marrow. The remainder is either in myoglobin in cells or is found in plasma. Blood plasma (that which remains once the blood cells are removed from whole blood) contains less than 0.1% of the body's iron, but the plasma iron concentration is very important to good health. Plasma iron levels are highest early in the morning and lowest in midafternoon, and they are usually about 10 to 15% lower at all times for women. If severe bleeding, giving birth, or menstruation increases the need for iron beyond that supplied in the diet, anemia can occur. Often, anemia can be corrected by administration of iron salts, such as iron (II) sulfate ($FeSO_4$), since Fe^{2+} is readily absorbed in the intestinal tract. But a second requirement must also be met. There must be sufficient iron-binding capacity in plasma to hold the iron once it is absorbed. In plasma, iron is bound to a protein called transferrin, and if it is unable to hold the needed amount of iron, iron salts alone will not cure anemia. Normally, between 20 and 55% of the total iron-binding capacity (TIBC) of transferrin is utilized at any one time, and the determination of the TIBC is part of the standard procedure used to diagnose anemia.

Iron ions in the cytochromes are the principal electron carriers in the body, since iron can cycle between the 2+ and 3+ ions as an electron is gained or lost.

Good sources of iron in the diet are organ meats (especially liver), shellfish, beans, green leafy vegetables, and egg yolks.

Manganese

Normal range in serum	Percent of body weight in a normal adult
Mn^{2+} 8.0 × 10^{-5} – 2.6 × 10^{-4} mg/100 ml	0.000 1%

In the body, manganese exists primarily as Mn^{2+} and Mn^{3+} ions. Though manganese is required only in trace amounts, it is vital for the operation of several enzymes and is essential for normal bone structure. Mn^{2+} can serve in place of Mg^{2+} in the transfer of energy across cell membranes. Low blood manganese levels have been reported in children suffering convulsions.

Manganese is obtained in the diet from nuts, beans, fruits, cereals, grain, and green leafy vegetables.

Cobalt

Normal range in serum as vitamin B_{12}	Percent of body weight in a normal adult as Co
$1.5 \times 10^{-2} - 7.5 \times 10^{-2}$ mg/100 ml	0.000 004%

The most important biological compound of cobalt is vitamin B_{12} (cobalamin) in which cobalt exists as the Co^{2+} ion. In fact, research suggests that cobalt is effective in the body only when present in the B_{12} vitamin or one of its derivatives. Vitamin B_{12} is required in the formation of hemoglobin, and deficiencies cause pernicious anemia, a rare condition that affects protein synthesis. Vitamin B_{12} is stored in sufficient quantities in the liver to meet bodily needs for three to five years.

The human body cannot synthesize vitamin B_{12}, and so it must be obtained from the foods we eat, notably lean meat, dairy products, and eggs. But vitamin B_{12} cannot be absorbed through the intestinal wall unless it encounters hydrochloric acid (gastric acid) and the ''intrinsic factor'' enroute to the intestines. The intrinsic factor is a protein that joins to the vitamin so it can pass through the intestinal wall.

Cobamide, a B_{12} derivative, is important in the synthesis of amino acids in the body, which are then used to make proteins. Vitamin B_{12} and its derivatives are considered to be some of the most remarkable substances found in the body chemistry.

Copper

Normal range in serum	Percent of body weight in a normal adult
$8.5 \times 10^{-2} - 1.1 \times 10^{-1}$ mg/100 ml	0.000 4%

Copper in the form of the Cu^+ and Cu^{2+} ions is found combined with large molecules in the liver, brain, heart, and kidneys. Though rare, severe copper deficiency leads to degradation of the sheath around the spinal cord, weakening of certain blood vessel walls, decoloration of hair, reduced hemoglobin production, and inhibition of the energy-producing reactions in cells. A large excess of copper ion can produce severe mental illness and even death, though in some cases drugs can be administered that are able to remove copper ion from tissues to reverse these effects. Wilson's disease is a rare congenital condition characterized by a progressive accumulation of copper in the kidney, liver, and brain. The disease causes liver and kidney failure and severe disorders of the nervous system. If not treated (by leaching excess copper from tissue) the disease is fatal.

In mollusks (snails, crabs, octopus), the so called blue-blooded animals, oxygen is carried by a copper-containing protein, hemocyanin.

Good sources of copper in the diet are raisins, shellfish, organ meats, nuts, legumes, and vegetable oils.

Zinc

Normal range in serum	Percent of body weight in a normal adult
Zn^{2+}, 0.050–0.15 mg/100 ml	0.0025%

Zinc is present in the body as Zn^{2+}. Zinc ions are needed for the process that allows carbon dioxide to be absorbed by the blood and eliminated in the lungs. Zinc stabilizes insulin, the hormone required for the transfer of glucose into the cells. Zinc is also involved in one of the key chemical reactions of vision, and seems to accelerate the healing of wounds. There is some evidence that an excess of zinc in relation to copper in the body is related to increased disease of the arteries. The optimum ratio of zinc to copper is unknown.

A reserve of zinc is maintained in skin and bone, though movement of zinc ion from one tissue to another seems limited. For this reason a daily intake of zinc is needed, and good sources of this metal are nuts, red meat and shellfish.

Molybdenum

Normal range in serum	Percent of body weight in a normal adult
not known	Mo 0.000 02%

Molybdenum has been identified as a component of several enzymes, and so it is believed to have an essential role in the body. Molybdenum is stored in the liver, and one molybdenum-containing enzyme is believed to be involved in the release of iron from ferritin. Molybdenum is also involved in the transfer of electrons between biochemical species, since its ions can display varying charge.

Molybdenum is obtained in the diet from nuts, red meat, and shellfish.

Terms

Several new terms were introduced in Chapter 9. You should understand the meaning of the following.

metallic bond
electron-sea model

alloy
amalgam

Questions

Answers to the starred questions appear at the end of the text.

1. Describe the electron-sea model for metallic bonding.

*2. What feature of the electron-sea model accounts for the high electrical conductivity of metals?

*3. a. What is an alloy?
   b. Why are alloys used in certain applications instead of pure metals?

4. What is an amalgam?

5. What precautions should be taken when working with mercury metal to avoid damage to one's health?

6. List the three categories that describe how metal ions are incorporated into the body chemistry.

*7. What metal is found in (a) heme, (b) chlorophyll, and (c) vitamin B_{12}?

8. What is the most abundant metal in the body?

*9. What is the principal metal ion found in blood plasma?

10. What is the most abundant metal ion found in cells?

*11. Which metal ion is important for proper blood clotting?

12. Which two metal ions work together in the transfer of nerve impulses?

*13. Which metal ion has been used to reduce convulsions?

14. In what species is most of the body's iron found at any one time?

*15. How is the total iron-binding capacity in plasma related to the treatment for anemia?

16. What metal is associated with Wilson's disease?

*17. In what kinds of compounds is cobalt used by the body?

18. How many grams of zinc would be in a normal 160-lb adult? (1 lb = 454 g)

Gases, Liquids, Solids, and the Forces between Molecules

The three common states of matter, gas, liquid, and solid, were first introduced in Chapter 3, and the energy changes that accompany the conversion of one state to another (fusion, vaporization) were described in Chapter 5. Several other aspects of the three states will be examined in this chapter so you will have a better understanding of matter and how it behaves.

The major difference between solids, liquids, and gases when viewed at the molecular level is the closeness and organization of the molecules. Water molecules in ice are very close together and highly organized in the crystalline solid. In liquid water, the molecules are still close together, but because they can move around one another, the degree of organization is much lower. In steam, the molecules are widely separated with no organization at all.

You might wonder why some substances are solids at room temperature (about 21°C), while others are liquids or solids. The answer lies in the size of the forces of attraction that exist between molecules. If the forces are very large, a substance will most likely be a solid, and if very weak, it will probably be a gas. If the attractive forces are intermediate, it may be a liquid at room temperature.

The chapter begins with the study of gases. After examining a theory that describes gases at the molecular level, several physical laws that relate the volume of a quantity of gas to its pressure and temperature are presented. Then, the forces of attraction that can exist between molecules are described. The chapter concludes with a discussion of liquids and solids.

Objectives

By the time you finish this chapter you should be able to do the following:

1. State the four postulates of the kinetic molecular theory and show how each is important in describing the properties of a gas.
2. Describe how a gas exerts pressure.
3. Describe and use: Boyle's law, Charles' law, Dalton's law, and the ideal gas law.
4. State the standard conditions of temperature and pressure.
5. State the volume of one mole of gas at STP, and relate it to Avogadro's law and the density of a gas.
6. Describe how the blood gases exchange between the air, blood, and tissue.
7. Describe London forces, dipole–dipole forces, and hydrogen bonds, and predict which would be expected in a sample of a compound given its structural formula.
8. Describe evaporation, condensation, the equilibrium vapor pressure, and the normal boiling point of a liquid.
9. Describe the four kinds of crystalline solids that are commonly found in nature.

The Kinetic Molecular Theory of Gases

Many years ago scientists developed a model to describe a so-called "ideal" gas. The model, with modification, can also be applied to liquids and solids, but for now we will be more concerned with what it says about gases. This model is called the **kinetic molecular theory** (KMT), and it can be stated as a series of postulates, that is, as assumptions that are taken without proof. The KMT provides an excellent picture of what a gas is like in terms of the molecules that make it up, and over the years it has needed only slight modification to be consistent with new discoveries made concerning the behavior of matter.

The work "kinetic" implies motion, and the KMT describes a gas as a collection of molecules in motion. You will recall that kinetic energy is the energy of a molecule due to its motion (Chapter 5). We will use the term molecule to represent the individual particles that make up a gas, though in helium or argon the particles are atoms not molecules. Let us examine the postulates of the KMT to see what they say about gases.

1. A gas consists of molecules that are far apart from each other. Compared to the size of the molecules, the space between them is very large.

A gas is mostly empty space. Only about 0.1% of the volume of helium in a balloon at standard pressure is occupied by helium atoms; 99.9% of the volume is empty space. Gases can be compressed since only the space between molecules is getting smaller.

2. The molecules of a gas are in continuous, rapid motion. They move in straight lines and change direction only when they collide with each other or the wall of the container.

At room temperature the average speed of an oxygen molecule is about 1000 miles an hour. Some are moving faster and some slower than this. The pressure exerted by a gas is the result of the collision of molecules with the walls of its container.

3. When two molecules collide, there is no net loss of kinetic energy, though kinetic energy may be transferred from one molecule to the other.

The importance of the third postulate may not be apparent right away, but it means that molecules in motion will stay in motion. They will not all eventually slow down and fall to the bottom of the container. Collisions that keep the *total* amount of kinetic energy possessed by the molecules unchanged are called "elastic" collisions.

4. The average kinetic energy of the molecules is affected only by the temperature of the gas.

If the temperature of a gas increases, the average kinetic energy of the molecules increases and they travel at a higher average speed ($KE = \frac{1}{2}mv^2$). If the temperature decreases, the molecules move slower since they have less kinetic energy. It is interesting that only temperature affects the kinetic energy of molecules. The temperature of a gas is really a measure of its average kinetic energy. Furthermore, no matter what gas is studied, O_2, CO_2, or He, if they are all at the same temperature, they all have the same average kinetic energy, Figure 10.1.

It is interesting to note that the KMT provides all the information needed to derive the gas laws that will be presented later on. The method used to do this is mathematical, and will not be presented here. But it does show that the four postulates of the KMT must be fundamentally correct.

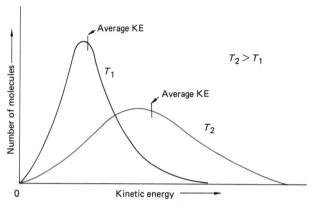

Fig. 10.1.   The average kinetic energy of a collection of molecules increases as temperature increases. Notice at either temperature some molecules have very low kinetic energy while others have very high kinetic energy.

Pressure

Gases exert pressure. The word "pressure" implies a push of some kind and gases do push against the walls of the container that holds them. The pressure is the result of millions of collisions of rapidly moving molecules with the container walls. Each collision exerts a tiny force against the wall, and the total force of these collisions over a given area (a square centimeter or square inch) is the **pressure** of the gas:

$$\text{pressure} = \frac{\text{force of molecular collisions}}{\text{area receiving the force}}$$

The pressure of a gas can be measured with a **barometer**, a device invented by the Italian scientist Torricelli in 1643. A barometer similar to that used by Torricelli is shown in Figure 10.2. This kind of barometer is commonly used to measure the pressure of the atmosphere. The long glass tube, which is sealed shut at the upper end, rests in an open dish of mercury. A column of mercury is supported in the tube by the pressure of the atmosphere pushing down on the mercury in the dish. The downward pressure exerted by the mercury column equals the pressure of the atmosphere. There is no air *in* the tube above the mercury column (and therefore no air pressure pushing down on the column), so the length of the mercury column turns out to be a measure of the pressure exerted by the atmosphere. At sea level, the average distance from the surface of the mercury in the dish to the top of the column is 760 mm. This is taken as a pressure of 760 mm Hg ("760 millimeters of mercury"). The unit of **mm Hg** is commonly used for pressure. In honor of Torricelli, a pressure equivalent

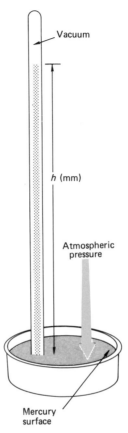

Vacuum

h (mm)

Atmospheric
pressure

Mercury
surface

Fig. 10.2. A mercury barometer. The column of mercury in the tube is supported by the pressure of the atmosphere. Atmospheric pressure equals the height of the column (h) measured from the mercury surface in the dish to the top of the column.

to 1 mm Hg is called a **torr**, so 760 mm Hg could be stated as 760 torr. Another unit of pressure is the atmosphere (atm), which is equal to 760 torr. Two atmospheres would then equal 2 atm × 760 torr/atm or 1520 torr. Perhaps you are familiar with another common unit used for pressure, **pounds per square inch**. One atmosphere equals 14.7 1b/sq. in.

$$760 \text{ mm Hg} = 760 \text{ torr} = 1 \text{ atmosphere} = 14.7 \frac{\text{lb}}{\text{sq. in.}}$$

One can easily convert a pressure given in one set set of units to another using the above relationships. If you were given a pressure of 740 torr and needed to convert it to the equivalent pressure in atmospheres, use the fact that 760 torr = 1 atm:

$$740 \, \text{torr} \left(\frac{1 \, \text{atm}}{760 \, \text{torr}} \right) = 0.974 \, \text{atm}$$

We will use the units of torr and atmosphere in this chapter.

There are other devices that can be used to measure pressures of gases. Respirators, autoclaves, and pressure reduction valves on cylinders of gas use mechanical devices that display pressure on a gauge. These are easier to use than mercury barometers, though the dials may be calibrated in mm Hg, torr, atmospheres, or pounds per square inch. Electronic devices are sometimes used to measure very low pressures, on the order of fractions of a torr.

Gases and the Gas Laws

When we speak of a "gas" we are referring to a substance that is completely in the gas phase at ordinary temperatures and pressures, such as oxygen, nitrogen, or carbon dioxide. Each gas has a unique set of chemical properties that are determined by the elements that make it up and the way the atoms are joined together. But many (though not all) of the physical properties of *all* gases are very nearly the same. For example, the volume occupied by one mole of carbon dioxide (44 g of CO_2) is the same as that occupied by one mole of oxygen (32 g of O_2) if both volumes are measured at the same temperature and pressure. Also, the effect of changes in temperature or pressure on the volume occupied by a sample of gas is essentially the same for all gases.

Because of the similarity in the physical behavior of gases, several laws have been discovered that can be applied to any gas. They are called the gas laws, and they relate how the volume occupied by a gas is affected by changes in pressure, temperature, and the amount of gas in the sample, which is usually expressed in moles.

Boyle's Law— Volume and Pressure

One of the first physical laws of nature was determined by Robert Boyle in 1662. Boyle studied how the volume of a confined sample of gas varied with the pressure exerted on it. As the pressure increased, the volume decreased and vice versa. The results of his work are summarized as a law that bears his name.

> **Boyle's law:** The volume of a fixed sample of any gas varies *inversely* with the pressure of the gas as long as the temperature of the gas does not change.

The "inverse" relationship of volume and pressure is an important part of Boyle's law, and perhaps it can be best explained with an example. Suppose you had 8 l of air in a container at a pressure of one atmosphere. Then you compressed the air by reducing the volume of the container to 4 l, just half the original volume. The pressure exerted by the air at a volume of 4 l would now be 2 atm, just twice as great as it was at first. By reducing the volume to half, the pressure doubles. This is the inverse relationship. If the volume had been reduced to 2 l, just one-fourth the original volume, the pressure would have quadrupled to 4 atm. Of course the temperature of the air would have to remain constant during this time, as required by Boyle's law. The inverse relationship of pressure and volume is also shown in Figure 10.3, and a relatively simple mathematical equation for Boyle's law can be derived from the volumes and pressures in the figure. In each case (A, B, and C), the product of $P \times V$ equals the same number, that is, a constant with units of liter-atmosphere:

	P	*V*	*P* × *V*
A	1 atm	8 l	8 l atm
B	2 atm	4 l	8 l atm
C	4 atm	2 l	8 l atm

Since, for a given quantity of gas, $P \times V$ = constant over a wide range of pressures and volumes, it is possible to write an equation relating the pressure and volume of a gas at one condition (P_1, V_1) to the corresponding terms at some second condition (P_2, V_2). The equation would hold only if the temperature of the gas remains constant:

$$P_1 V_1 = P_2 V_2 \qquad \text{(at constant temperature)}$$

If you know any three terms in the equation, you can solve for the fourth. For example if you know the volume of a gas at one pressure, you can calculate the volume it would have at a different pressure, as long as the temperature is constant. Let us do some examples using Boyle's law.

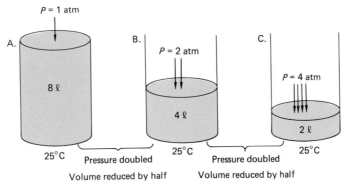

Fig. 10.3. Boyle's law in operation. If the pressure on a gas doubles, while the temperature is held constant, the volume of the gas is reduced by half. The pressure exerted by the gas equals that applied to the gas.

Example 10.1

Oxygen gas can be supplied to hospitals in steel cylinders under high pressure. A 20-l cylinder contains oxygen at a pressure of 130 atm. If the gas leaked out of the cylinder, what volume would this oxygen occupy at a pressure of 1 atm? The gas is measured at the same temperature in each case.

initial condition: $P_1 = 130$ atm; $V_1 = 20$ l

final condition: $P_2 = 1$ atm; $V_2 = ?$

$$P_1 V_1 = P_2 V_2$$

$$(130 \text{ atm}) (20 \text{ l}) = (1 \text{ atm}) V_2$$

solving for V_2:

$$V_2 = \frac{(130 \text{ atm})}{(1 \text{ atm})} (20 \text{ l}) = 2600 \text{ l}$$

The answer is reasonable. The oxygen should occupy a much larger volume at 1 atm. Notice how the units of pressure cancel out. Any pressure units (or volume units) can be used as long as both known terms use the same units so they cancel.

Example 10.2

Certain autoclaves use a sterilizing gas to destroy microorganisms. How many times can an autoclave that has a volume of 15 l be filled with gas at a pressure of 3 atm from a 10-l tank of gas that is initially at a pressure of 150 atm? All values are measured at the same temperature. First we will need to determine the volume the gas would occupy at 3 atm.

initial condition: $P_1 = 150$ atm, $V_1 = 10$ l

final condition: $P_2 = \quad 3$ atm, $V_2 = ?$

$$V_2 = \frac{P_1}{P_2}(V_1) = \frac{(150 \text{ atm})}{(3 \text{ atm})} (10 \text{ l})$$

$$V_2 = 500 \text{ l}$$

Then, if each time the autoclave is filled requires 15 l of gas (at 3 atm), the number of times the autoclave can be filled is

$$500 \text{ l} \left(\frac{1 \text{ time}}{15 \text{ l}} \right) = 33.3 \text{ times} = 33 \text{ times}$$

The act of breathing, inhaling and exhaling air, is the result of pressure–volume changes in the body. As you inhale (inspire), the chest cavity expands in volume causing a momentary reduction of the pressure of air in the lungs. This would be predicted by Boyle's law. Air will flow into the lungs until the pressure equals that of the atmosphere. As you exhale (expire), the chest cavity contracts in volume, momentarily increasing the pressure of the air in the lungs above that of the atmosphere. Air then flows out of the lungs, from a region of higher to one of lower pressure. Breathing demonstrates an important fact about gases. Gases will spontaneously flow from a region of higher pressure to one of lower pressure. The flow will continue until the pressures become equal. This principle will be examined again in more detail later on in this chapter.

Charles' Law— Volume and Absolute Temperature

If the pressure on a gas remains constant (so *it* will not cause a volume change) an increase in temperature of *any* gas will cause its volume to increase, and a decrease in temperature will cause its volume to decrease.

In 1787, Jacques Charles, a French scientist, reported his findings on the relationship between the volume of a gas and its temperature. He found that a temperature change of 1°C would change the volume of a gas by an amount equal to 1/273 of its volume at 0°C (both volumes measured at the same pressure). If, for example, a quantity of oxygen had a volume of 1000 ml at 0°C, its volume would *increase* by about 3.7 ml if its temperature rose to 1°C.

$$\text{volume increase} = \frac{1}{273}(1000 \text{ ml}) = 3.7 \text{ ml}$$
$$(0° \rightarrow 1°C)$$

The oxygen would occupy 1003.7 ml. If, on the other hand, the gas was cooled to −10°C, the 10-degree decrease in temperature would *decrease* its volume by 37 ml, and it would occupy only 963 ml:

$$\text{volume decrease} = \frac{10}{273}(1000 \text{ ml}) = 37 \text{ ml}$$
$$(0° \rightarrow -10°C)$$

If this idea is carried to very large temperature changes, one would soon conclude that decreasing the temperature of the gas to −273°C would reduce its volume to zero, and it would disappear:

$$\text{volume decrease} = \frac{273}{273}(1000 \text{ ml}) = 1000 \text{ ml}$$
$$(0° \rightarrow -273°C)$$

But this does not happen with real gases. They become liquids before they reach −273°C, and the observations made by Charles on gases would no longer apply.

There is an important point that must be made about the temperature of −273°C (−273.15°C, to be exact). Several different experiments have led people to regard this temperature as the lowest temperature possible. It is called the **absolute zero** of temperature. The **absolute temperature scale** or **Kelvin scale,** as it is often called, begins at absolute zero (0°K, "zero degrees Kelvin") and proceeds upward. One Kelvin degree represents the same temperature change as one Celsius degree, so the two temperature scales can be related by a simple equation, which was first introduced in Chapter 2:

$$\text{Kelvin temperature} = \text{Celsius temperature} + 273°$$

$$°K = °C + 273°$$

So 0°C is 273°K, and 0°K is −273°C.

The reason for delving into the Kelvin temperature scale at this point will soon be apparent. It turns out that the relation between the volume of a gas and its temperature discovered by Charles in 1787 is most conveniently stated using temperatures in degrees Kelvin. **Charles' law** can be stated as follows:

> The volume of a fixed quantity of gas varies directly with its absolute temperature (°K), as long as the pressure on the gas remains constant.

Charles' Law is a "direct" relationship between volume and temperature. If the temperature in degrees Kelvin doubles, the volume of the gas doubles. If the Kelvin temperature triples, the volume triples. Of course, as stated in the law, each volume must be measured under identical conditions of pressure. The volume–temperature behavior of a gas at constant pressure is shown in Figure 10.4.

The volume–temperature information displayed in Figure 10.4 can be used to develop a mathematical equation for Charles' law. In each case the quotient of V/T (T in °K) equals a constant:

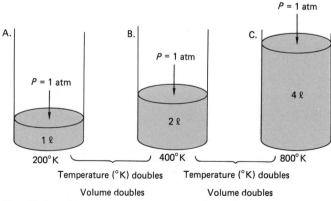

Fig. 10.4. Charles' law in operation. If the absolute temperature (°K) doubles, while pressure is held constant, the volume doubles.

	V	T	V/T
A	1 l	200°K	0.005 l/°K
B	2 l	400°K	0.005 l/°K
C	4 l	800°K	0.005 l/°K

We can then equate the V/T term measured under one condition, V_1/T_1, to that measured under a different condition, V_2/T_2, for a gas at constant pressure:

$$\frac{V_1}{T_1} = \frac{V_2}{T_2} \quad \text{(pressure held constant)}$$

This is the convenient form of Charles' law. If the volume of a gas is known at one temperature, the volume it would occupy at a different temperature can be calculated. It is important to remember that all temperatures must be stated in degrees Kelvin.

Example 10.3

A balloon filled with helium has a volume of 3.5 l at a temperature of 23°C. What would the volume be at 60°C? Assume the pressure is constant. Both temperatures must first be converted to degrees Kelvin:

$$\text{initial temperature} = T_1 = (23° + 273°) = 296°K$$

$$\text{initial volume} = V_1 = 3.5 \, l$$

$$\text{final temperature} = T_2 = (60° + 278°) = 338°K$$

$$\text{final volume} = V_2 = ?$$

$$\frac{V_1}{T_1} = \frac{V_2}{T_2}$$

Solving for V_2:

$$V_2 = \frac{T_2}{T_1}(V_1) = \frac{338°\cancel{K}}{296°\cancel{K}}(3.5 \, l) = 4.0 \, l$$

Example 10.4

At what temperature would a 100-l volume of methane gas, measured at 22°C, occupy 250 l? Both volumes are to be at the same pressure.

$$\text{initial temperature} = T_1 = 22° + 273° = 295°K$$

$$\text{initial volume} = V_1 = 100 \, l$$

$$\text{final temperature} = T_2 = \, ?$$

$$\text{final volume} = V_2 = 250 \, l$$

Using Charles' law to find T_2:

$$T_2 = \frac{V_2}{V_1}(T_1) = \frac{250 \, l}{100 \, l}(295°K) = 740°K$$

In degrees Celsius = 740° − 273° = 467°C
At 467°C, the methane would occupy 250 l.

Check Test Number 1

1. A helium-filled weather balloon had a volume of 5000 l at 750 torr. Assuming the temperature remains constant, what would be the volume of
   the balloon at an elevation where the pressure is 250 torr?
2. A sample of carbon dioxide has a volume of 2.50 l at 25°C. What volume
   would it occupy at 100°C? Assume pressure remains constant.

Answers

1. Boyle's law: $P_1 V_1 = P_2 V_2$; $V_2 = 5000 \, l \left(\dfrac{750 \text{ torr}}{250 \text{ torr}} \right) = 15000 \, l$

2. Charles' law: $\dfrac{V_1}{T_1} = \dfrac{V_2}{T_2}$; $V_2 = 2.50 \, l \left(\dfrac{373°K}{298°K} \right) = 3.13 \, l$

   $T_1 = 25° + 273° = 298°K$
   $T_2 = 100° + 273° = 373°K$

Standard Temperature and Pressure   The volume of any gas is affected by its temperature and pressure as described by Boyle's law and Charles' law. In order to compare the amounts of any two gases in terms of their volumes, they must both be measured at the *same* temperature and pressure. As a matter of convenience, people have chosen a specific temperature and pressure as standards for comparing gas volumes. The **standard temperature** is 0°C (273°K) and the **standard pressure** is 760 torr (1 atm).

standard temperature and pressure (STP):

0°C and 760 torr

or

273°K and 1 atm

Though gas volumes are often measured at pressures and temperatures other than STP, they can be converted to standard conditions using Boyle's law, Charles' law, or the combined gas law.

The Combined Gas Law   Boyle's law and Charles' law can be combined in a single equation so the effect of simultaneous changes in pressure and temperature on the volume of a gas can be handled easily. The **combined gas law** is

$$\frac{P_1 V_1}{T_1} = \frac{P_2 V_2}{T_2}$$

You should be able to recognize the Boyle's law relationship of P and V and the Charles' law relationship of T and V in the equation. If a volume change takes place at constant temperature, then $T_1 = T_2$ and temperatures cancel, leaving Boyle's law. If pressure remains constant, then $P_1 = P_2$, and they cancel, leaving Charles' law. The following example problem uses the combined gas law to determine the volume of a gas at STP.

Example 10.5
_____

One mole of carbon dioxide occupies a volume of 40.0 l at 200°C and 737 torr. What volume would one mole of carbon dioxide occupy at STP?

initial conditions: $V_1 = 40.0\,l$

$P_1 = 737$ torr

$T_1 = 200° + 273° = 473°K$

final conditions: $V_2 = ?$

$P_2 = 760$ torr

$T_2 = 0° + 273° = 273°K$

The combined gas law can be rearranged to solve for V_2, the volume at STP.

$$V_2 = (V_1)\left(\frac{T_2}{T_1}\right)\left(\frac{P_1}{P_2}\right)$$

$$V_2 = (40.0 \text{ l})\left(\frac{273°\text{K}}{473°\text{K}}\right)\left(\frac{737 \cancel{\text{torr}}}{760 \cancel{\text{torr}}}\right)$$

$$V_2 = 22.4 \text{ l}$$

Molar Volume and Avogadro's Law
The volume occupied by one mole of any gaseous substance is its **molar volume,** and at STP the molar volume of *any* gas is 22.4 l.

$$\text{molar volume at STP} = 22.4 \text{ l}$$

One mole of nitrogen ($N_2 = 28$ g), one mole of hydrogen ($H_2 = 2$ g), and one mole of sulfur trioxide ($SO_3 = 80$ g) all occupy the same volume, 22.4 l, at STP as shown in Figure 10.5.

Avogadro recognized the relationship between the volumes of gases and the mass (or number of moles) of those gases many years ago. The relationship is called **Avogadro's law:**

> Equal volumes of gases measured under the same conditions of temperature and pressure contain the same number of moles (or molecules) of gas.

Avogadro's law says that the volume occupied by a gas is "directly" related to the number of moles of gas. If one mole of oxygen occupies 22.4 l at STP,

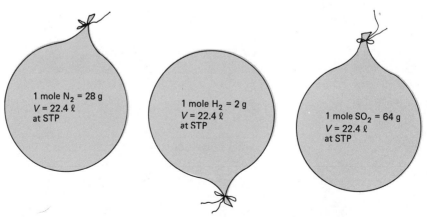

1 mole $N_2 = 28$ g
$V = 22.4$ ℓ
at STP

1 mole $H_2 = 2$ g
$V = 22.4$ ℓ
at STP

1 mole $SO_2 = 64$ g
$V = 22.4$ ℓ
at STP

Fig. 10.5.   One mole of any gas occupies the same volume at STP: 22.4 l.

then two moles would occupy (2 moles × 22.4 $\frac{1}{\text{mole}}$) or 44.8 l at the same temperature and pressure. Notice that volumes are compared under the same conditions of temperature and pressure.

The *density* of a gas at STP can be determined by dividing the weight of one mole of the gas by its volume at STP, 22.4 l. Gas densities are commonly reported in units of grams per liter. The density of hydrogen gas (H_2) at STP is

$$\text{Density of } H_2 \text{ at STP} = \frac{2.02 \text{ g } H_2}{1 \text{ mole}} \times \frac{1 \text{ mole}}{22.4 \text{ l}} = 0.0902 \text{ g/l}$$

The density of hydrogen is less than that of air (1.25 g/l), so a hydrogen-filled balloon will be buoyed up in air and rise. For the same reason a cork floats on water because its density is less than that of water.

The density of any gas is directly related to its molecular weight. One mole of helium has a mass of 4.00 g, nearly twice that of a mole of hydrogen. The density of helium at STP is nearly twice that of hydrogen,

$$\text{Density of He at STP} = \frac{4.00 \text{ g He}}{1 \text{ mole}} \times \frac{1 \text{ mole}}{22.4 \text{ l}} = 0.178 \text{ g/l}$$

Check Test Number 2

1. Calculate the density of sulfur dioxide gas (SO_2) at STP.
2. Three liters of oxygen gas and three liters of hydrogen gas both exist at the same temperature and pressure.
   (a) Which contains the greater number of moles of gas?
   (b) Which contains the greater number of grams of gas?

Answers:

1. Density $= \dfrac{64 \text{ g } SO_2}{1 \text{ mole}} \times \dfrac{1 \text{ mole}}{22.4 \text{ l}} = 2.86$ g/l at STP
2. (a) Avogadro's law says that identical volumes at the same T and P have the same number of moles of gas.
   (b) The mass of O_2 is greater than the mass of H_2. Each volume contains the same number of molecules, but oxygen molecules are heavier (32 g/mole) than hydrogen molecules (2 g/mole).

The Ideal Gas Law   You have already seen how Boyle's law, Charles' law, and Avogadro's law relate the volume occupied by a gas to pressure, temperature, and the number of moles of gas. The three laws can be combined in a single equation that relates the four terms conveniently. The equation is commonly called the **ideal gas law:**

$$PV = nRT$$

Because of its versatility, the ideal gas law is one of the most useful equations describing gases.

The symbol R in the equation is called the **ideal gas law constant.** If pressure is in **atmospheres,** volume in **liters,** the quantity of gas given in **moles,** and temperature used in degrees **Kelvin,** the value is

$$R = 0.0821 \frac{1\,atm}{mole\,°K}$$

The value of the constant is the same for all gases, but you must be certain the units used for P, V, T, and n agree with those of R. Pressure must be in atmospheres, volume in liters, temperature in °K, and n in moles.

The following examples apply the ideal gas law.

Example 10.6

Derive the value of the ideal gas law constant R from the fact that one mole of any gas occupies 22.4 l at 1 atm and 273°K (STP).

$$R = \frac{PV}{nT} = \frac{(1\,atm)\,(22.4\,l)}{(1\,mole)\,(273°K)}$$

$$R = 0.0821 \frac{1\,atm}{mole\,°K}$$

Example 10.7

In certain surgical procedures the abdominal cavity is inflated with CO_2 to provide working area for the surgeon. How many moles of CO_2 are used when 3.0 l of the gas occupy the cavity at 37°C and 1.0 atm?

Temperature must be expressed in Kelvin degrees: $37° + 273° = 310°K$

Rearranging the ideal gas law to solve for n yields

$$n = \frac{PV}{RT} = \frac{(1.0\,atm)\,(3.0\,l)}{(0.0821\,l\,atm/mole\,°K)\,(310°K)}$$

$$n = 0.089\,mole\,CO_2$$

Example 10.8

As a matter of convenience large hospitals often buy liquified oxygen instead of high-pressure oxygen gas in cylinders. As the liquified oxygen vaporizes, the oxygen gas is piped throughout the hospital. How many liters of oxygen gas at 22°C and 1.0 atm could be obtained from 160 kg of liquid oxygen?

First determine the number of moles of O_2 in 160 kg of O_2, then use the ideal gas law to determine the volume at 1 atm and 22°C (295°K).

$$\text{moles of } O_2 = 160 \text{ kg} \left(\frac{1000 \text{ g}}{1 \text{ kg}}\right)\left(\frac{1 \text{ mole } O_2}{32 \text{ g}}\right) = 5000 \text{ moles}$$

$$V = \frac{nRT}{P} = \frac{(5000 \text{ moles}) (0.0821 \text{ l atm/mole °K}) (295°K)}{(1 \text{ atm})}$$

$$V = 121,097 \text{ l or about } 121,000 \text{ l}$$

Check Test Number 3

1. What volume would 2.00 moles of oxygen occupy at 2.20 atm and 0°C?
2. What would be the pressure of hydrogen gas stored in a steel cylinder that contained 10.0 moles of H_2 in a volume of 10.0 l at 25°C?

Answers:

$$1. \ V = \frac{nRT}{P} = \frac{(2.00 \text{ mole}) (0.0821 \text{ l atm/mole °K}) (273°K)}{2.20 \text{ atm}} = 20.4 \text{ l}$$

$$2. \ P = \frac{nRT}{V} = \frac{(10.0 \text{ mole}) (0.0821 \text{ l atm/mole °K}) (298°K)}{10.0 \text{ l}} = 24.4 \text{ atm}$$

Dalton's Law—Pressures of Gaseous Mixtures   Suppose you had a 1-l container of oxygen gas at a pressure of one atmosphere, and a second container of identical size filled with nitrogen gas at a pressure of one atmosphere. Then suppose further that you transferred all the oxygen from its container to the nitrogen container. What do you think the total pressure of the mixture would be? Before we answer that question, let us examine the mixture. The nitrogen gas still occupies a volume of 1 l and it still exerts a pressure of 1 atm. The oxygen, even though it is now mixed with nitrogen, also occupies a volume of 1 l and so it will exert a pressure of 1 atm. To answer the questions then, the total pressure (P_T) of the mixture would be the sum of the nitrogen and oxygen pressures, or 2 atm. The pressure exerted by each individual gas in a mixture of gases is called its **partial pressure** (p), and the total pressure is simply the sum of the partial pressures, see Figure 10.6.

This observation concerning the pressure of a mixture of gases is stated in **Dalton's law**:

The total pressure exerted by a mixture of gases is the *sum* of the pressure each gas would exert if it were present alone under the same conditions.

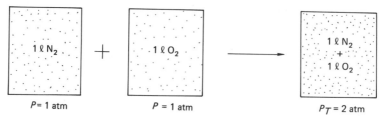

Fig. 10.6. Dalton's law in operation. If 1 l of oxygen and 1 l of nitrogen, each at a pressure of 1 atm, are combined in a 1-l container, each gas still exerts a pressure of 1 atm, but the total pressure of the mixture is 2 atm. The total pressure is the sum of the partial pressures of each gas in the mixture.

Mathematically, Dalton's law is

$$P_T = p_A + p_B + p_C + \cdots$$

where p_A, p_B, p_C, etc., are partial pressures of gases A, B, C, etc.

The pressure exerted by the atmosphere is the sum of the partial pressures of the gases that make it up. The compositions of clean, dry (water-free) air is given in Table 10.1, along with the associated partial pressures. When the atmospheric pressure is measured with a barometer, it is the total pressure of the mixture:

$$P_{atm} = p_{N_2} + p_{O_2} + p_{Ar} + p_{all\ else}$$

Before leaving this section, a point needs to be made concerning the relationship between the partial pressures of gases in mixtures and the movement of the individual gases from one mixture to another. Suppose you had two gaseous mixtures of oxygen and carbon dioxide at the same total pressure separated by a porous membrane. The membrane separates the mixtures but

TABLE 10.1 Composition of the Atmosphere near Sea Level		
Gas	Percent by volume	Partial pressure (torr)
Nitrogen	78.09	593.5
Oxygen	20.95	159.1
Argon	0.93	7.1
Carbon dioxide	0.032	
Neon	0.0018	0.3
Helium	0.00052	
Methane	0.00015	

allows gases to pass through and in this sense acts like a cell wall or blood capillary in the body. Assume that one of the mixtures contains oxygen at a partial pressure of 100 torr and carbon dioxide at a partial pressure of 50 torr. The other mixture contains oxygen at a partial pressure of 50 torr and carbon dioxide at a partial pressure of 100 torr, as shown in Figure 10.7.

The partial pressure of a gas is a measure of how hard it "pushes" to pass through the membrane. In medicine the partial pressure of a gas is called its **tension**. Since the partial pressure of oxygen is greater in one mixture than in the other, oxygen can be thought of as pushing harder on one side of the membrane than on the other. As a result, there will be a *net* flow of oxygen from the higher partial pressure region to the lower partial pressure region. The flow will continue until the partial pressure of oxygen is the same on either side of the membrane. Likewise, there will be a *net* flow of carbon dioxide from the higher to lower region of its partial pressures until they too become equal. After a period of time, both mixtures will have an identical composition, with identical partial pressures for both oxygen and carbon dioxide, and there will no longer be a *net* movement of either gas across the membrane.

It is interesting that the components of a mixture of gases act independently of the other gases in the mixture. Its net movement is determined only by the differences in its own partial pressure. Differences in partial pressures of oxygen and carbon dioxide are responsible for the migration of oxygen from the lungs to cells in the body and the movement in the reverse direction of carbon dioxide.

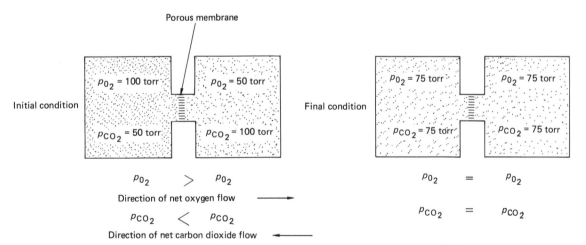

Fig. 10.7. Two different gaseous mixtures separated by a porous membrane will, in time, achieve identical compositions. The direction of the net flow of a gas is determined by the difference in its partial pressures on either side of the membrane. Oxygen flows from the left chamber (p_{O_2} = 100 torr) into the chamber on the right (p_{O_2} = 50 torr). Carbon dioxide flows in the opposite direction as dictated by its partial pressures. The net flow will continue until the partial pressures become equal for each gas in both chambers.

Check Test Number 4

Assume the total pressure exerted by four gases in an inhalation anesthetic is 750 torr. The partial pressure of O_2 in the mixture is 200 torr; for N_2, 480 torr; and for water vapor, 25 torr. What would be the partial pressure of the anesthetic gas?

Answer:

$$P_T = p_{O_2} + p_{N_2} + p_{H_2O} + p_{anesthetic}$$

$$750 \text{ torr} = 200 \text{ torr} + 480 \text{ torr} + 25 \text{ torr} + p_{anesthetic}$$

$$p_{anesthetic} = 45 \text{ torr}$$

The Blood Gases—CO_2 and O_2

Every living cell in the body carries out cellular respiration to generate energy. The process is complex but it can be summarized in the familiar equation in which glucose reacts with oxygen to produce carbon dioxide, water, and energy:

$$C_6H_{12}O_6 + 6O_2 \rightarrow 6CO_2 + 6H_2O + energy$$

During respiration, oxygen is consumed and carbon dioxide is produced. Both gases are transported through the bloodstream and for this reason they are called the blood gases in medicine.

The exchange of these gases between the body and the atmosphere takes place in the lungs across thin, porous membranes that enclose the alveoli. Each alveolus is in intimate contact with a fine network of capillaries, and it is here that gas exchange takes place. The direction a gas will flow across the alveoli-capillary membrane is determined by the difference in the partial pressures (tensions) of the gas on either side of the membrane system. The net flow of a gas will be from a region of higher to one of lower partial pressure of that gas. The partial pressures of oxygen and carbon dioxide in the atmosphere, alveolar air, blood, and tissue are listed in Table 10.2.

The figures in Table 10.2 reveal that venous blood returning to the lungs has a relatively low oxygen pressure (40 torr) and an elevated carbon dioxide pressure (46 torr) compared to alveolar air. So, as venous blood passes through the capillary network around the alveoli, oxygen flows into the blood and carbon dioxide flows out (see Figure 10.8). This gas exchange converts venous blood to oxygen-rich arterial blood that is pumped throughout the body by the heart.

Once arterial blood enters the fine capillary system in tissue, gas exchange again occurs as dictated by the relative oxygen and carbon dioxide partial

	p_{O_2} (torr)	p_{CO_2} (torr)
TABLE 10.2 Partial Pressures of Oxygen and Carbon Dioxide		
Atmosphere	157	0.3
Alveolar air	100	40
Venous blood	40	46
Arterial blood	100	40
Tissue	40 (or less)	46

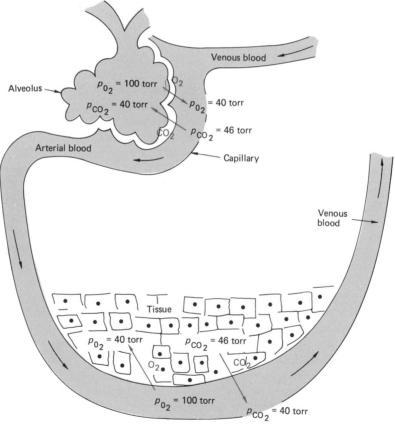

Fig. 10.8. The movement of oxygen and carbon dioxide into and out of the alveoli, blood, and tissue is largely controlled by differences in the partial pressures of these gases. Oxygen enters the blood and carbon dioxide leaves at the alveoli in response to partial pressure differences. The direction of flow is reversed in tissue, again as governed by partial pressure differences.

pressures. Oxygen diffuses from arterial blood (p_{O_2} = 100 torr) into tissue where the oxygen pressure is 40 torr or less (depending on how strenuously the tissue is being used, as in a muscle for example). At the same time, carbon dioxide passes from the tissue (p_{CO_2} = 46 torr) into the arterial blood (p_{CO_2} = 40 torr). Arterial blood is converted into venous blood at this point and is recirculated back to the lungs to start the process all over again.

The partial pressure of oxygen in alveolar air is significantly increased in oxygen therapy. An atmosphere that contains from 40 to 50 percent oxygen is administered to patients using an oxygen tent, facial mask, or nasal catheter. The elevated partial pressure of oxygen (approximately 350 torr) increases the oxygenation of blood passing through the lungs. Atmospheres containing higher concentrations of oxygen can cause irritation to the respiratory tract, but they are used for short periods of time when necessary.

Forces between Molecules

There certainly must be some kind of attractive force that exists *between* molecules, otherwise why would they cling together to form liquids or solids? The forces between molecules are called **intermolecular forces**, and the size of these forces determines, to a large extent, whether a substance will be a solid, liquid, or gas at room temperature and pressure.

Perhaps the best place to start this discussion is to reexamine the nature of the forces of attraction that exist in ionic compounds. This might seem a bit strange at this point since molecules do not exist in ionic compounds, but there is a parallel in the nature of the attractive forces between ions and those between molecules.

Earlier, in Chapter 8, you were told that ionic compounds (such as NaCl or KNO_3) were all solids at room temperature because of the **electrostatic attractions** that exist between the positive and negative ions that are packed tightly together in the crystal. In a similar manner, though not to the same degree, neutral molecules are attracted to each other by electrostatic forces too. Just as with ions, the attractive forces become significant only when the molecules are quite close together as in liquids or solids. Even though a molecule is overall neutral, that is, it does not bear a positive or negative charge, some molecules possess a permanent separation of charge as a result of the unequal sharing of electrons in polar covalent bonds. A good example of this kind of molecule is HCl, hydrogen chloride. Since chlorine is more electronegative than hydrogen (Cl = 3.0, H = 2.1), the shared pair of electrons in the bond are, on the average, held closer to chlorine. This causes the chlorine end of the molecule to bear a small negative charge symbolized δ−. The hydrogen end therefore bears a small positive charge of equal size, δ+. The Greek delta, δ, symbolizes a "partial charge," that is, one that is smaller in size than that on an electron or proton.* The HCl molecule, then, acts like an electric **dipole**. A dipole is a body with a negative end and a positive end. We can symbolize the dipole as an elongated body showing the partial charges on either end. This emphasizes the existence of the dipole in a molecule *and* that the molecule is *polar*:

$$\delta + \left(\text{H}\rule{1cm}{0.4pt}\text{Cl} \right) \delta -$$

Arrows ($+\!\!\longrightarrow$) are also used to indicate the polarity of a bond or a molecule. The arrowhead points to the $\delta-$ end, and the crossed end, to the $\delta+$ end of the dipole:

$$\begin{array}{c} +\!\!\longrightarrow \\ \text{H}\!\!-\!\!\text{Cl} \end{array}$$

If polar molecules are close together, the negative end of one will be attracted to the positive end of another. The force of attraction between dipoles is called the **dipole–dipole force**. Some of the ways in which HCl dipoles can interact are shown in Figure 10.9.

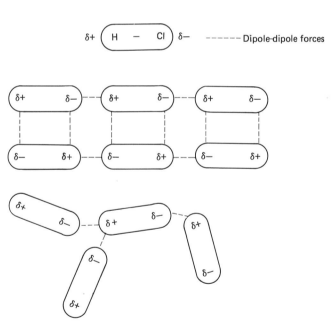

Fig. 10.9. Dipole–dipole forces between HCl molecules as they might appear in solid HCl and liquid HCl. In each, the δ- end of one molecule is attracted to the $\delta+$ end of another.

*The partial charges are the result of unequal electron sharing. Do not think that electrons or protons are being cut up into smaller pieces that then have smaller charges.

Some other molecules that exist as dipoles are H_2O, ICl, and NH_3. Each has polar covalent bonds, and the shapes of the molecules allow the polarity of the bonds to add together to give each molecule an overall polarity ($\longmapsto$):

You might want to review the electronegativities of the common elements (Figure 8.7) to understand why the polarity of each bond is drawn as it is.

Dipole–dipole forces are not at all as strong as the covalent bonds within molecules. But they are large enough to affect the physical properties of compounds.

The molecules of many compounds do not possess permanent dipoles and so do not engage in dipole–dipole interactions. Some of these kinds of molecules are H_2, I_2, O_2, CCl_4, BF_3, and CO_2. The diatomic molecules (H_2, I_2, and O_2) do not have polar bonds since electrons are shared by identical atoms. The bonds in CCl_4, BF_3, and CO_2 are polar, but the shapes of these molecules cause the polarity of the bonds to cancel to zero. The symmetry of these molecules creates a "standoff" in a kind of two-, three-, or four-way tug-of-war. Since no polar bond is the winner, the molecule is *nonpolar* overall:

There are intermolecular forces that exist between nonpolar molecules (those without a *permanent* dipole). This second kind of force is called the **London force** or **dispersion force**. London forces arise as the result of the formation of "instantaneous dipoles" in molecules. These instantaneous dipoles form by the rapid shifting of the electrons in the molecule from side to side. The heavy nuclei do not move, only the electrons, with the result that one end of the molecule is richer in electrons for an instant. This electron-rich end of the molecule then becomes $\delta-$, and the other end, $\delta+$. Even though the dipole exists for only an instant, it induces adjacent molecules to "flex" their electrons to form instantaneous dipoles so that the $\delta-$ end of one is against the $\delta+$ end of the other. In this condition there is an electrostatic attraction between the molecules that lasts for an instant, as shown in Figure 10.10. In the next instant, the "instantaneous dipoles" will reform in another direction reestablishing the attraction between molecules. In any solid or liquid, where the molecules are close together, there are millions upon millions of these short-lived dipoles forming and reforming every instant. Though the individual London forces are

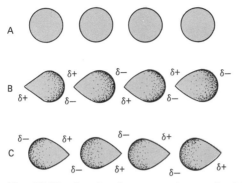

Fig. 10.10. London forces are the result of the formation of "instantaneous" dipoles in molecules. In A, the electrons are equally distributed in the molecules and no attractive forces are present. In B, the electrons have shifted to one side, causing one end of the molecule to be richer in electrons (δ-) and the opposite end poorer in electrons (δ+). Attractive forces then exist between the molecules. In C, the electrons have shifted to the opposite side in the next instant reestablishing the attractive forces. London forces are important when molecules are close together as in liquids and solids.

weak, their presence everywhere in a liquid or solid causes them to be quite significant.

London forces exist between *every* kind of species that has one or more electrons: atoms, molecules, or ions. It is the only kind of intermolecular force that exists between nonpolar molecules. Organic compounds that contain only carbon and hydrogen (hydrocarbons) are essentially nonpolar, so the only attractive force between these molecules is the London force. Polar molecules will have dipole–dipole *and* London forces attracting them together.

The third kind of intermolecular force is called **hydrogen bonding**. It is the strongest of the three forces, but it only occurs in molecules that contain a hydrogen atom bonded to either oxygen, nitrogen, or fluorine:

$$\text{---O---H} \qquad\qquad \text{---N---H} \qquad\qquad \text{F---H}$$

Nitrogen, oxygen, and fluorine are very electronegative elements, and the electron pair they share with hydrogen is drawn so far away from hydrogen that the bond is extremely polar. The small hydrogen atom is left with a substantial partial positive charge, and the other atom (O, N, or F) with a substantial partial negative charge:

$$\overset{\delta^-}{O}\text{---}\overset{\delta^+}{H} \qquad\qquad \text{---}\overset{\delta^-}{N}\text{---}\overset{\delta^+}{H} \qquad\qquad \overset{\delta^-}{F}\text{---}\overset{\delta^+}{H}$$

Hydrogen bonds form between the δ+ hydrogen in one molecule and the δ- oxygen, nitrogen, or fluorine in another, Figure 10.11. The dotted lines in Figure

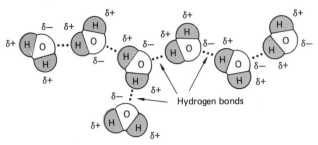

Fig. 10.11. Hydrogen bonds between water molecules. The hydrogen atom in one molecule is strongly attracted to the oxygen atom in an adjacent molecule.

10.11 represent the hydrogen bonds between water molecules. Hydrogen bonding is much like a dipole–dipole interaction, but it is unique in that it is considerably stronger than usual dipole–dipole forces.

Hydrogen bonding in water is responsible for its abnormally high boiling temperature. Comparison of the boiling points of the hydrogen compounds of the elements in group VI, Figure 10.12, shows a decreasing trend in boiling temperatures as you pass from H_2Te to H_2Se to H_2S. You would think the decreasing trend would continue with H_2O, but this is not so. Hydrogen bonding in water holds the molecules together so strongly that a much higher temperature (therefore much higher kinetic energy) is needed to separate them to form steam. Ammonia and hydrogen fluoride also have abnormally high boiling temperatures compared to the hydrogen compounds of their respective groups, again because of hydrogen bonding.

More will be said about the importance of hydrogen bonding in certain biologically important molecules in later chapters.

The three intermolecular forces, dipole–dipole, London, and hydrogen bonding, are sometimes as a group referred to as **van der Waals forces** in honor of Johannes van der Waals, who postulated the idea of intermolecular forces in 1873.

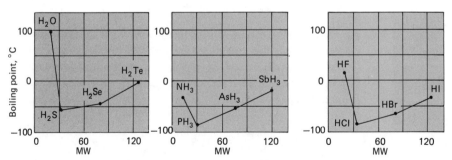

Fig. 10.12. Because of hydrogen bonding the boiling points of H_2O, NH_3, and HF are abnormally high compared to those of the hydrogen compounds of the other elements in each group.

Check Test Number 5

Indicate the type(s) of intermolecular forces that you would expect to find in the following:

a. Br_2 e. $S=C=S$ (linear)
b. CO   f. $CH_3-CH_2-CH_3$
c. NH_3 g. HF
d. CBr_4 (tetrahedral)   h. H_3C-OH (methyl alcohol)

Answers:
a. London
b. London, dipole–dipole
c. London, dipole–dipole, H bonding
d. London
e. London
f. London
g. London, dipole–dipole, H bonding
h. London, dipole–dipole, H bonding

Liquids

Water, ethyl alcohol, and motor oil are three common materials that are liquids at room temperature. The molecules in these liquids are very close together and touching each other. For this reason, liquids are not able to be compressed like gases since a great deal of energy would be needed to force the molecules much closer together. However, the molecules in liquids are in constant motion and continually move around each other. The forces of attraction between the molecules are large enough to keep them together, but not so large as to restrict their movement in the liquid. No two liquids have the same size forces of attraction between molecules and this causes them to have different physical properties like boiling and melting points, viscosity, and surface tension.

The **viscosity** of a liquid is a measure of its resistance to flow. Generally, liquids that are made up of relatively small molecules with weaker intermolecular forces have low viscosities because the molecules can flow past each other easily and quickly. Liquids that are made up of relatively large molecules, which may have larger intermolecular forces, have higher viscosities. The molecules are less able to move past each other easily and so they flow slower. A motor oil contains large, long-chain molecules that intertwine and are highly attracted to one another. They cannot move past each other quickly so many oils flow slowly at room temperature, that is, they have higher viscosities.

Usually, the viscosity of a liquid decreases as its temperature increases. The increased kinetic energy of the molecules at higher temperatures allows some

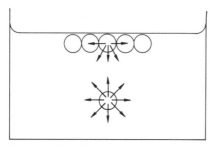

Fig. 10.13.   The surface tension of a liquid. Molecules at the surface of a liquid experience forces drawing them together along the surface and into the liquid. The surface acts like a thin, invisible skin stretched across the liquid. A molecule deep within the liquid experiences forces in all directions, unlike those at the surface.

of the attractive forces between molecules to be overcome. Molecules can then move by each other more easily.

The **surface tension** of a liquid is another property related to the intermolecular forces in liquids. The surface of a liquid acts like a very thin elastic "skin" that is stretched across it. The surface tension is a measure of the strength of this "skin." Molecules at the surface of a liquid experience attractive forces that draw them together across the surface and inward toward the bulk of the liquid. A molecule on the surface, then, has a "one-sided" attraction for the molecules around it, which causes them to form the thin skin on the surface. A molecule buried within the liquid, on the other hand, experiences attractive forces in all directions, as shown in Figure 10.13. It is the surface tension that causes droplets of water to have a near spherical shape when spilled on a plastic table top. Also, the surface tension of water is strong enough that small water bugs can scoot across its surface moving on top of the water, not floating in it.

The magnitude of the surface tension of a liquid is related to the magnitude of the intermolecular forces in the liquid. Various substances, called **surfactants,** can be added to liquids to reduce their surface tension. Soap will reduce the surface tension of water and enhance the ability of a soap solution to clean greasy dirt. Surfactants can also be used in inhalation therapy to reduce the surface tension of solutions used in atomizers. With reduced surface tension, smaller droplets of liquid are formed in the aerosol that are able to carry medication to the smaller regions of the alveoli on the lungs. Surfactants also aid loosening mucus in the lungs so it can be coughed up and expelled.

Evaporation, Condensation, and Vapor Pressure   If a glass of water is placed on a table and left alone, eventually the water will disappear. In our common experience we say the water has evaporated. **Evaporation** is the process by which molecules escape from the surface of a liquid to the gas phase:

$$\text{liquid} \xrightarrow{\text{evaporation}} \text{gas}$$

Once in the gas phase, they can move far from the liquid and never return. Let us take a closer look at evaporation from the point of view of the molecules in the liquid.

The molecules in a glass of water at room temperature are in constant motion, that is, they possess kinetic energy. Some are moving very fast (they have high KE) while some are moving slower (low KE). The attractive forces between water molecules hold them together as a liquid. Some molecules, though, will have enough kinetic energy to break away from their neighbors and, if near the surface, escape from the liquid, that is, evaporate. If the temperature of the liquid is raised, a larger fraction of the molecules will be able to escape, and the liquid will evaporate faster.

Since the high-energy molecules are the ones that leave, the average kinetic energy of those left behind in the liquid will decrease. This means the temperature of the liquid will drop and it will become cooler. Evaporation is therefore a cooling process, a fact you can demonstrate by rubbing alcohol on your hand and letting it evaporate. Evaporation is an imporant process in maintaining normal body temperature on a hot day.

Now suppose another glass of water is placed on a table, but this time the glass is covered with a tight-fitting lid. Evaporation will again occur but now the gaseous molecules cannot escape. As the number of molecules in the gas phase increases, the likelihood that they will collide with the surface of the liquid and be recaptured increases also. The movement of molecules from the gas phase to the liquid is called **condensation**:

$$\text{gas} \xrightarrow{\text{condensation}} \text{liquid}$$

(Liquids and solids are sometimes referred to as the "condensed states" of matter.) Eventually a point will be reached where the number of molecules *leaving* from the liquid each second equals the number *returning* to the liquid each second. The rates of the opposing changes are then equal, as shown in Figure 10.14.

The condition in which opposing changes occur at the same rate is called **dynamic equilibrium.** It can be symbolized in equations with opposing arrows

$$\text{liquid} \rightleftharpoons \text{gas}$$

(the condition of equilibrium)

At equilibrium, the total number of molecules in the gas phase is constant. For every molecule that leaves the liquid one returns, and even though evaporation and condensation continue, the number of molecules in the gas phase does not change.

The **vapor pressure** of a liquid is equal to the pressure exerted by its vapor when measured at equilibrium. Molecules of the liquid once in the gas phase will exert pressure by virtue of their collisions with the walls of the container.

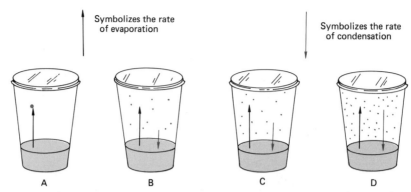

Fig. 10.14.   A liquid in a closed container will reach a state of dynamic equilibrium between the liquid and vapor phases. The rate of evaporation does not change, but as the number of molecules in the vapor phase increases (A through C), the rate of condensation increases. Eventually the rates of evaporation and condensation become equal (D), and the liquid–vapor equilibrium is established.

The magnitude of the vapor pressure of a liquid depends *only* on its temperature. As the temperature increases, the number of molecules in the gas phase at equilibrium increases, so its vapor pressure will increase also. The effect of temperature on the vapor pressure of four different liquids is shown in Figure 10.15. At any given temperature the liquid with the lowest vapor pressure is the one with the greatest intermolecular forces (H_2O), and vice versa. Diethyl ether has the highest vapor pressure at 20°C because it has the lowest intermolecular forces. It is the most **volatile** of the four liquids.

The Normal Boiling Point

As the temperature of a liquid increases, eventually its vapor pressure will equal standard atmospheric pressure, 760 torr. The temperature at which this occurs is called the **normal boiling point** of the liquid. In Figure 10.15, the normal boiling point of diethyl ether is 34.6°C and of water, 100°C.

A heated liquid will boil whenever its vapor pressure equals the pressure over its surface (which could be much less than 760 torr). At very high elevations, such as the top of Mt. Everest, the atmospheric pressure is much less than standard pressure and water boils around 80°C. Though this would be the boiling point at that pressure, remember that the normal boiling point is that at 760 torr. Boiling points of liquids in tables and reference books are, unless otherwise indicated, normal boiling points.

Just as the boiling point of a liquid is lowered if the pressure above its surface is lowered, the boiling point will exceed the normal value at pressures above standard pressure. This is the principle behind high-pressure autoclaves used to sterilize surgical equipment. The pressure in the sealed autoclave is raised to the point where water boils around 130°C. Bacteria cannot long survive in an environment of 130°C steam. The boiling point of water in pressure cookers used in the home can be as high as 110°C. At this temperature, foods cook about twice as fast as at 100°C.

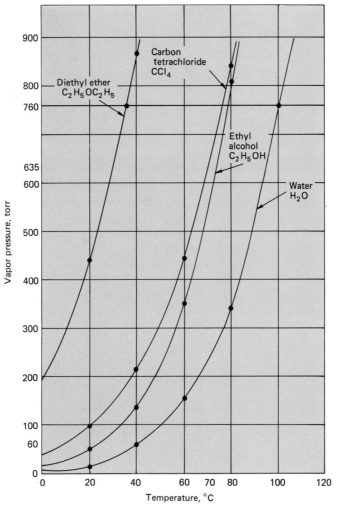

Fig. 10.15. The equilibrium vapor pressure of a liquid increases as the temperature increases. The normal boiling point of a liquid is the temperature at which its vapor pressure equals 760 torr.

Solids

The particles that make up solids (atoms, molecules, or ions) are attracted to each other by relatively strong forces. These attractions hold the particles close together (usually closer than in liquids) in definite, highly organized arrangements that cause solids to have fixed volumes and shapes. Yet, the particles possess kinetic energy but it is insufficient to overcome the forces of attraction that hold each in its place. Each particle just vibrates back and forth about its place in the solid. It cannot readily move past its neighbors. If the temperature of a solid is raised, the vibrations become more violent and eventually a temperature will be reached where the particles have enough kinetic energy to overcome some of the attractive forces and it will melt. The temperature at which a solid is converted to a liquid is its **melting point.**

Most pure solids are **crystalline,** that is, they exist as crystals with a definite shape. You may recall the term used to describe the highly ordered arrangement of particles in a crystaline solid, the **lattice** of the crystals. Some solids do not have a highly organized arrangement of particles. These noncrystalline solids are called **amorphous solids.** Some common amorphous solids are glass, rubber, tar, and polyethylene. The particles that compose some amorphous solids can move past each other fairly easily, as demonstrated by the stretching of a rubber band. Amorphous solids usually soften over a fairly wide temperature range, then melt. Crystalline solids usually melt quickly when the melting point is reached.

Crystalline solids can be classified into four groups, depending on the kind of particles that make them up and the kinds of forces that hold the particles together.

Ionic solids have aready been discussed in Chapter 8. The positive and negative ions that make them up are strongly held together by electrostatic forces. Because the attractive forces are strong, ionic compounds have high melting points (in the hundreds of degrees C), are hard, brittle crystals, and they may or may not be soluble in water. Ionic compounds are not soluble in nonpolar, organic liquids like benzene, gasoline, or carbon tetrachloride.

Molecular solids are composed of individual covalent molecules held together in the crystal by the intermolecular forces described earlier. The force of attraction between the molecules is much less than that between ions in ionic solids. As a result, molecular solids have lower melting points and the crystals are relatively soft and easily crushed. They are usually not soluble to any great extent in water, but will dissolve in organic liquids. Examples of molecular solids (and the forces that hold the molecules together) are I_2 (London forces); ice (H bonds, dipole–dipole, and London forces); and ICl (dipole–dipole and London forces).

Some molecular solids have the property of passing from the solid phase directly into the gaseous phase *without* becoming a liquid during the process. This is called **sublimation:**

$$\text{solid} \xrightarrow{\text{sublimation}} \text{gas}$$

Several common compounds sublime at room temperature and pressure: Mothballs (para-dichlorobenzene or naphthalene) sublime at room temperature, and when placed in clothing the fumes permeate the fabric.

Covalently bonded solids are composed of atoms that are covalently bonded to each other to form a hard, rigid three-dimensional network of atoms that is

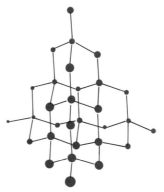

Fig. 10.16. The structure of diamond, a covalently bonded solids. Each carbon atom is bonded to four others arranged tetrahedrally about it.

really one large molecule, a macromolecule. Diamond (C) and quartz (SiO_2) are covalently bonded solids. In diamond, each carbon atom is covalently bonded to four other atoms arranged in a tetrahedron about it. In quartz, each silicon atom (Si) is bonded to four oxygen atoms, and each oxygen to two silicon atoms. The structure of diamond appears in Figure 10.16, and you can see how this arrangement of atoms would form a rigid solid that would be very hard. Covalently bonded solids characteristically have hard crystals that melt only at very high temperatures (usually over 1500°C). They are not soluble in any of the common solvents.

The last of the four types of crystalline solids is the **metallic solid.** They are the metals, and they are composed of organized arrangements of atoms held together by metallic bonds (Chapter 9). Some metallic solids are very soft and can be cut with a knife (Na, K, Ca), while others are very hard (W and Fe). Most metals can be deformed by hammering or drawn into wires without shattering the crystal. They are not soluble in the common solvents (unless they react with the solvent, as sodium does when placed in water).

Terms

Several important terms appeared in Chapter 10. You should understand the meaning of each of the following.

kinetic molecular theory   partial pressure
pressure   intermolecular force
barometer   London force
mm Hg   dipole–dipole force
torr   hydrogen bonding

atmosphere	surface tension
standard pressure	surfactant
standard temperature	equilibrium vapor pressure
Boyle's law	normal boiling point
Charles' law	evaporation
combined gas law	condensation
Avogadro's law	sublimation
ideal gas law	amorphous solid
R	molecular solid
Dalton's law	covalently bonded solid
viscosity	

Questions

Answers to the starred questions appear at the end of the text.

1. Compare gases, liquids, and solids in terms of the closeness of molecules and their organization.

*2. Describe how a gas exerts pressure on the walls of its container.

*3. How is the kinetic energy of the molecules in a gas affected by temperature?

4. What are the four postulates of the kinetic molecular theory?

5. Describe how a mercury barometer works, and give four units that can be used to express pressure.

*6. Twelve liters of oxygen at a pressure of 350 torr would occupy what volume if its pressure increased to 760 torr while temperature remained constant?

7. What pressure is required to compress 10 l of air at 1 atm to 2.3 l while holding temperature constant?

*8. A sample of helium has a volume of 5.0 l at 27°C and 1 atm of presure. What volume would it occupy at −25°C and 1 atm of pressure?

9. What volume would 44 g of CO_2 occupy at 0°C and 760 torr?

*10. What is the density of CO_2 at STP?

11. A sample of nitrogen gas has a volume of 14.0 l at 27°C and 740 torr. What volume would it occupy at STP?

12. Use the ideal gas law to determine the following:
   *a. The number of moles of O_2 in 100 l of O_2 at 100°C and 2.00 atm.
   b. The pressure exerted by 64 g of SO_2 at 50°C in a 10-l container.
   c. The volume of 0.25 mole of CO_2 at 500°C and 0.10 atm.

13 What is the total pressure of a mixture of CO_2 at 0.30 atm, O_2 at 0.15 atm, and N_2 at 0.90 atm?

*14. Two flasks, A and B, are connected by a tube. Flask A contains helium at a pressure of 0.50 atm and hydrogen at a pressure of 0.25 atm. Flask B contains helium at a pressure of 0.25 atm and hydrogen at a pressure of 0.50 atm. Into which flask will helium flow? Into which flask will hydrogen flow?

15. Using the data in Table 10.2, describe the movement of oxygen from the alveoli to tissue, and the reverse flow of carbon dioxide in terms of differences in partial pressures of the two gases.

16. What kind(s) of intermolecular attractive forces would you predict would be present in the following species:

   *a. Br_2
   b. CBr_4
   *c. H_2O

   d. CH_3-NH_2
   *e. CO_2
   f. HBr

   *g. $CH_3CH_2CH_2CH_2CH_3$
   h. CH_3CH_2-OH
   *i. H_2S (bent molecule)

17. What is the effect of increasing temperature on the viscosity and surface tension of a liquid?

18. What is the purpose of a surfactant?

*19. How does a temperature increase affect the equilibrium vapor pressure of a liquid? What is the normal boiling point of a liquid?

20. Describe how the rates of evaporation and condensation are related in a condition of dynamic equilibrium. Why is the equilibrium condition termed "dynamic"?

*21. What is sublimation?

22. What are the four classes of crystalline solids?

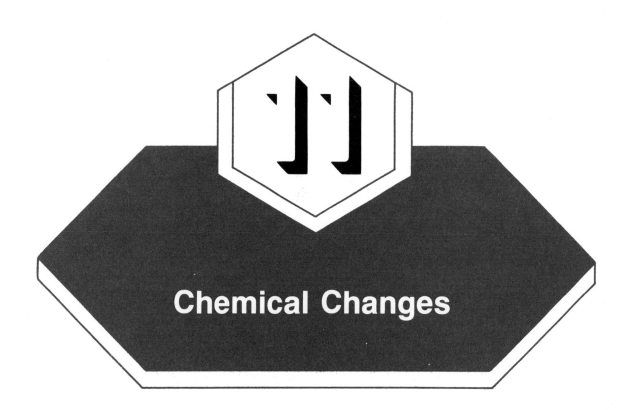

Chemical Changes

Life is sustained by chemical changes. The food we eat is produced through a series of chemical changes in plants and animals. Digestion of that food requires chemical changes just as does metabolism, a series of reactions that use the products of the digestive process. Every movement of a muscle, transmission of a nerve impulse, or division of a cell is the result of specific chemical reactions. In earlier chapters you have been shown how equations are used to symbolize chemical changes, and how reactions are accompanied by the release or absorption of energy. Other aspects of chemical changes will be discussed in this chapter.

Because of the large number of known chemical reactions, the chapter will begin with a discussion of a few general reaction types. By classifying reactions according to their similarities, it will be easier to discuss reactions as they appear in later chapters. Special emphasis will be placed on oxidation–reduction reactions. These reactions involve the transfer of electrons from one species to another, and they are very important in many body processes. Some chemical reactions take place very quickly, while others are very slow. The factors that influence the speed of chemical reactions will also be presented. The chapter will conclude with a description of reversible chemical reactions, and an introduction to the condition of equilibrium in chemical reactions.

Objectives

By the time you finish this chapter you should be able to do the following:

1. List four types of chemical reactions and give examples of each.
2. Define oxidation, reduction, oxidizing agent, and reducing agent.
3. Assign oxidation numbers to each atom in a molecule or ion.
4. Identify the reduced and oxidized species in an oxidation–reduction reaction.
5. Define what is meant by the "rate of a chemical reaction."
6. Describe the three requirements that must be met if two species are to react in a chemical change.
7. Describe how a catalyst influences the rate of a chemical reaction.
8. Describe how the temperature and the concentration of reactants influence the rate of a reaction.
9. Describe the state of chemical equilibrium.

Types of Chemical Reactions

Though thousands of different chemical reactions are known, there are relatively few general reaction types. Let us look at four of them along with several examples of each type.

1. Combination Reactions (Synthesis)

As the name implies, **combination reactions** combine two or more elements or compounds to produce a new compound. A general equation for the combination reaction is

$$\underbrace{A + B}_{\text{Elements or compounds}} \longrightarrow \underset{\text{A new compound}}{AB}$$

Examples:

$$C + O_2 \longrightarrow CO_2$$
$$CaO + CO_2 \longrightarrow CaCO_3$$
$$H_2 + Cl_2 \longrightarrow 2HCl$$

2. Decomposition Reactions

Decomposition reactions are those in which a compound is decomposed to form (a) two or more simpler compounds, (b) the elements that composed the compound, or (c) both simpler compounds and elements.

A **simpler compound** is one with fewer atoms in its formula than the decomposed compound. Often, decomposition requires heating, which is indicated by the delta sign (Δ) written over the arrow. A general equation for a decomposition reaction that requires heating is

$$AB \xrightarrow{\Delta} \underbrace{A + B}$$

Compound   Simpler compounds
or elements

Examples:

$$2HgO \xrightarrow{\Delta} 2Hg + O_2$$
$$CaCO_3 \xrightarrow{\Delta} CaO + CO_2$$
$$2KClO_3 \xrightarrow{\Delta} 2KCl + 3O_2$$

3. Single-Replacement Reactions

Single-replacement reactions take place between elements and compounds. One element replaces another in a compound to produce a new compound plus the replaced element. These reactions usually take place with both compounds in solution; however, the elements may or may not be in solution. As the general equation shows, one element replaces another, producing a new compound:

$$A + BC \longrightarrow AC + B$$

Element   Compound   New Compound   Replaced element
(replaced by A)

Examples:

$$Zn + CuSO_4 \longrightarrow ZnSO_4 + Cu$$
$$Cu + 2AgNO_3 \longrightarrow Cu(NO_3)_2 + 2Ag$$
$$Ca + H_2SO_4 \longrightarrow CaSO_4 + H_2$$
$$Br_2 + 2KI \longrightarrow 2KBr + I_2$$

4. Double-Replacement Reactions

In **double-replacement reactions** two compounds react to produce two different compounds. The reactions usually take place in solution, and most always one of the products is either an evolved gas, an insoluble solid (a precipitate), or a soluble covalent species. The general equation shows that there is an exchange of partners in going from reactants to products:

$$\underbrace{AB + CD} \longrightarrow \underbrace{AD + CB}$$

Two compounds   Two new compounds

Examples:

$$HCl \; + \; NaOH \; \longrightarrow \; NaCl \; + \; H_2O$$
$$BaCl_2 \; + \; Na_2SO_4 \; \longrightarrow \; 2NaCl \; + \; BaSO_4\downarrow \quad (\downarrow \text{ indicates a precipitate})$$
$$HNO_3 \; + \; NaHCO_3 \; \longrightarrow \; NaNO_3 \; + \; H_2CO_3$$
$$\llcorner\!\!\longrightarrow \; H_2O \; + \; CO_2$$

(Once formed, H_2CO_3 may decompose to produce H_2O and CO_2, which is evolved as a gas.)

Check Test Number 1

Classify each reaction as either combination, decomposition, single replacement, or double replacement:

a. $(NH_4)_2Cr_2O_7 \; \xrightarrow{\;\Delta\;} \; 4H_2O + N_2 + Cr_2O_3$
b. $2Cu + O_2 \; \longrightarrow \; 2CuO$
c. $Pb(NO_3)_2 + 2KI \longrightarrow PbI_2 + 2KNO_3$
d. $2Fe + 6HCl \; \longrightarrow \; 2FeCl_3 + 3H_2$

Answers:
(a) decomposition, (b) combination, (c) double replacement, and (d) single replacement.

Oxidation–Reduction Reactions

In many cases, a reaction that would fit into one of the four categories listed above could also be described as an **oxidation–reduction** reaction. Oxidation–reduction reactions, or **redox** reactions as they are conveniently called, involve the transfer of electrons from one species to another. The substance that loses electrons is said to be **oxidized**, and that which gains electrons is **reduced**. Oxidation and reduction always occur simultaneously. An atom, ion, or molecule cannot lose electrons unless another atom, ion, or molecule is there to accept them. The species that loses electrons is called the **reducing agent** in a redox reaction. The reducing agent causes another substance to be reduced by giving electrons to it. The reduced substance is then called the **oxidizing agent**, since by accepting electrons it causes something to be oxidized.

The language may seem a bit confusing at first, but notice the cause-and-effect relationship of the terms: An oxidizing agent is reduced. A reducing agent is oxidized. Then remember that *o*xidation is the loss of electrons and reduction is just the opposite.

In order to help you recognize if a chemical change is an oxidation–reduction reaction, it is necessary to introduce the concept of the **oxidation number** of an atom. Every atom in a molecule or ion can be arbitrarily assigned an oxidation number which indicates whether the atom had lost or gained electrons as it became part of a molecule or ion. In many cases, electrons are not completely lost or gained by an atom, as is the case in a covalent molecule. In these cases, the oxidation number is a measure of the electrons that are partially lost or gained in polar covalent bonds, bonds that share electrons unequally between atoms. If an atom has electrons removed from it, either partially or totally, a positive (+) oxidation number is assigned to it. If electrons are gained, either partially or totally, the atom is assigned a negative (−) oxidation number. Because these oxidation numbers can change for certain atoms in a redox reaction, they help us keep track of electron transfers.

Since oxidation numbers relate the shift of electrons toward or away from an atom in a compound, something that is not always easily determined, they will be assigned using a set of rules.

Rule 1. The oxidation number of a free element is zero. The oxidation number of Cl is 0, whether it is a free atom or in the chlorine molecule Cl_2.

Rule 2. In compounds, the oxidation number of a group I metal is 1+, and the oxidation number of a group II metal is 2+. Sodium is 1+ in NaCl, Na_2O, and Na_2CO_3. Calcium, a group II metal, has an oxidation number of 2+ in CaO, $CaCl_2$, and $CaSO_4$. The metals in groups I and II exist as ions in their compounds. As a general rule, *the oxidation number of any monatomic ion equals the charge on that ion.* So, the oxidation number of chlorine as a chloride ion, Cl^-, is 1−. A monatomic ion is one derived from a single atom. They are also called simple ions.

Rule 3. The usual oxidation number for oxygen in its compounds is 2−.*

Rule 4. The usual oxidation number for hydrogen in its compounds is 1+.*

Rule 5. The oxidation numbers for other atoms in a molecule or polyatomic ion are determined after the above rules have been applied, so that (a) for neutral compounds, the sum of all oxidation numbers equals *zero*, and (b) for polyatomic ions, the sum of all oxidation numbers equals the *charge on the ion.*

Let us use these rules to assign oxidation numbers to atoms in molecules and ions. Rules 2, 3, and 4 will be used to assign values to monatomic ions, oxygen, and hydrogen. Then, rule 5 will be used to determine the oxidation number of the remaining elements.

Example 11.1

What is the oxidation number of nitrogen in nitrogen dioxide, NO_2?

---

*Both oxygen and hydrogen can assume oxidation numbers other than 2− and 1+, respectively. In peroxides, such as H_2O_2 or Na_2O_2, oxygen is 1−. If hydrogen is combined with metals, as in CaH_2 or $LiAlH_4$, its oxidation number is 1−. Since these are not commonly seen oxidation numbers for these elements, we will not be concerned with them in this discussion.

From rule 3 we know that each oxygen is 2−, and since there are two oxygen atoms in the molecule, the total for oxygen is 2x (2−) or 4−.

We also know that the sum of the oxidation numbers of all the atoms must equal zero (rule 5a). This can be stated in an equation which can be used to determine the oxidation number of nitrogen:

$$NO_2$$

N		O_2	
(Ox. No. of N)	+	2 (Ox. No. of O)	= 0
(N)	+	2(2−)	= 0
(N)	+	(4−)	= 0
(N) = 4+			

The oxidation number of N is 4+.

Example 11.2

What is the oxidation number of carbon in sodium carbonate, Na_2CO_3?

The oxidation number of sodium, a group I metal, is 1+, and oxygen is 2−. Knowing that the sum of the oxidation numbers for every atom must equal zero in the compound, we can develop an equation to determine the oxidation number of carbon.

$$Na_2CO_3$$

Na_2		C		O_3	
2(Na)	+	(C)	+	3(O)	= 0
2(1+)	+	(C)	+	3(2−)	= 0
(2+)	+	(C)	+	(6−)	= 0
		(C)	=	4+	

The oxidation number of carbon is 4+.

As you can see, the equations used to determine the oxidation number of an atom are solved using algebra. If you need to brush up on your algebra skills, the math review in the Appendix might be helpful.

Example 11.3

What is the oxidation number for phosphorus in the phosphate ion, PO_4^{3-}?

Knowing that oxygen is 2−, and that the sum of all oxidation numbers must equal the charge on the ion (rule 5b), we can write

$$PO_4{}^{3-}$$

$$\overbrace{\qquad\qquad}$$

$$\begin{array}{ll} P & O_4{}^{3-}\end{array}$$

$$\begin{aligned}
(P) + 4(O) &= 3- \qquad \text{(The charge on the}\\
(P) + 4(2-) &= 3- \qquad \text{ion is 3$-$)}\\
(P) + (8-) &= 3-\\
P &= 5+
\end{aligned}$$

Example 11.4

What is the oxidation number of chromium (Cr) in the dichromate ion, $Cr_2O_7{}^{2-}$?

We will first determine the total oxidation number for $2(Cr)$. Half the total will be assigned to each chromium atom.

$$Cr_2O_7{}^{2-}$$

$$\overbrace{\qquad\qquad}$$

$$\begin{array}{ll} Cr_2 & O_7{}^{2-}\end{array}$$

$$\begin{aligned}
2(Cr) + 7(O) &= 2- \\
2(Cr) + 7(2-) &= 2- \qquad \text{(The charge on the ion)}\\
2(Cr) + (14-) &= 2- \qquad \text{(Total for 2Cr)}\\
2(Cr) &= 12+ \\
Cr &= 6+
\end{aligned}$$

Before proceeding, it is important that the meaning of the oxidation number is clear to you. In the first example, nitrogen in NO_2 was assigned an oxidation number of $4+$. This does *not* mean that nitrogen is present in the NO_2 molecule as a nitrogen ion with a charge of $4+$. Nor does the $2-$ oxidation number of oxygen mean it is present as the oxide ion (O^{2-}). The oxidation numbers in molecules and polyatomic ions only relate the shift of electrons toward or away from an atom because of unequal electron sharing in polar covalent bonds. For this reason, chlorine in the chlorine molecule is assigned an oxidation number of zero. The electrons in the chlorine–chlorine bond are shared equally, and neither atom ends up with a net loss or gain of electrons.

Now try your hand at assigning oxidation numbers to elements in Check Test Number 2.

Check Test Number 2

Assign the correct oxidation number to the underlined element in each of the following: (a) $\underline{S}O_2$, (b) $Na_3\underline{P}O_4$, (c) $Ba\underline{S}O_4$, (d) $\underline{N}O_3^-$, (e) $\underline{N}H_4^+$, (f) $H_6\underline{C}_2O$.

Answers:

a. $(S) + 2(2-) = 0;\ S = 4+$

b. $3(1+) + (P) + 4(2-) = 0;\ P = 5+$

c. $(2+) + (S) + 4(2-) = 0; S = 6+$
d. $(N) + 3(2-) = 1-; N = 5+$
e. $(N) + 4(1+) = 1+; N = 3-$
f. $6(1+) + 2(C) + (2-) = 0; 2(C) = 4-, C = 2-$

Once you know how to assign oxidation numbers to elements, you can use them to determine what is being oxidized and what is being reduced in redox reactions. If the oxidation number of an element increases (becomes more positive) in a reaction, the element is being oxidized. If the oxidation number decreases, the element is reduced in the reaction.

$$\text{Oxidation}$$
$$\text{(Increasing oxidation number)}$$
$$\xrightarrow{\hspace{3cm}}$$

Oxidation number: $\quad \cdots 3- \quad 2- \quad 1- \quad 0 \quad 1+ \quad 2+ \quad 3+ \cdots$

$$\xleftarrow{\hspace{3cm}}$$
$$\text{(Decreasing oxidation number)}$$
$$\text{Reduction}$$

The reaction of calcium metal with water is both a single-replacement reaction and a redox reaction:

$$Ca + 2H_2O \longrightarrow Ca(OH)_2 + H_2$$

The oxidation number of calcium changes from zero for the free element to 2+ in $Ca(OH)_2$. Calcium is oxidized; its oxidation number increases from zero to 2+. The element that is reduced is hydrogen. In water, hydrogen has an oxidation number of 1+, and it decreases to zero in H_2:

$$\text{Oxidation}$$
$$(0) \xrightarrow{\hspace{3cm}} (2+)$$
$$Ca + 2H_2O \longrightarrow Ca(OH)_2 + H_2$$
$$(1+) \xrightarrow{\hspace{3cm}} (0)$$
$$\text{Reducton}$$

The synthesis of sodium chloride from its elements is also a redox reaction. Sodium is oxidized and chlorine is reduced:

$$\text{Oxidation}$$
$$(0) \xrightarrow{\hspace{3cm}} (1+)$$
$$2Na + Cl_2 \longrightarrow 2NaCl$$
$$(0) \xrightarrow{\hspace{3cm}} (1-)$$
$$\text{Reduction}$$

Mercury(II) oxide can be decomposed with heating to form mercury and oxygen. Mercury is reduced and oxygen is oxidized (a change in oxidation number from 2− to zero for oxygen is an algebraic increase, therefore oxidation):

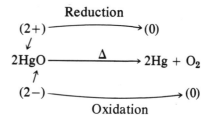

Oxidation numbers can also be used to determine if a reaction proceeds *without* oxidation–reduction. For example, the decomposition of calcium carbonate is not a redox reaction, since no element experiences a change in its oxidation number. Calcium stays 2+, carbon 4+, and oxygen 2−:

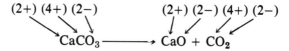

Check Test Number 3

Identify the element that is oxidized and the element that is reduced in the following redox reactions:
a. $Cu + 2AgNO_3 \rightarrow 2Ag + Cu(NO_3)_2$
b. $H_2 + Cl_2 \rightarrow 2HCl$
c. $FeO + H_2 \rightarrow Fe + H_2O$

Answers:

a. Cu, oxidized:
$$\overset{(0)}{Cu} \rightarrow \overset{(2+)}{Cu(NO_3)_2}$$

  Ag, reduced:
$$\overset{(1+)}{AgNO_3} \rightarrow \overset{(0)}{Ag}$$

b. H_2, oxidized:
$$\overset{(0)}{H_2} \rightarrow \overset{(1+)}{HCl}$$

  Cl_2, reduced:
$$\overset{(0)}{Cl_2} \rightarrow \overset{(1-)}{HCl}$$

c. H_2, oxidized:
$$\overset{(0)}{H_2} \rightarrow \overset{(1+)}{H_2O}$$

  Fe, reduced:
$$\overset{(2+)}{FeO} \rightarrow \overset{(0)}{Fe}$$

Oxidation and reduction can also be described in terms that do not directly require following changes in oxidation numbers. These alternate methods are frequently useful when dealing with large biological molecules that contain many atoms. Assigning oxidation numbers to each atom in these large molecules becomes a tedious job, so alternate methods are desirable.

1. A substance is **oxidized** if it either gains oxygen atoms or loses hydrogen atoms in a reaction.

2. A substance is **reduced** if it either loses oxygen atoms or gains hydrogen atoms in a reaction.

Notice that these descriptions let us speak of a *compound* being either oxidized or reduced as opposed to a single atom or atoms within the molecule. In cellular respiration, glucose reacts with oxygen to form carbon dioxide, water, and energy. Glucose is oxidized since it gains oxygen to form the products:

$$C_6H_{12}O_6 + 6O_2 \longrightarrow 6CO_2 + 6H_2O + energy$$

The oxidation of methyl alcohol, CH_3OH, to formaldehyde, H_2CO, in the liver involves the loss of two hydrogen atoms from the alcohol. The reduction reaction, which occurs simultaneously, involves an enzyme. Only the oxidation is shown in the equation:

$$CH_3OH \xrightarrow{\text{enzyme}} H_2CO + 2H$$

The conversion of succinic acid to fumaric acid in the body is also an oxidation reaction, since two hydrogen atoms are lost by succinic acid. The enzyme is reduced. Only the oxidation reaction is shown.

Succinic acid   Fumaric acid

As you can see, there are several ways to describe oxidation and reduction. Whichever method we choose to use in a given case is more a matter of convenience, since each is equally correct. Ultimately, though, the description of oxidation and reduction in terms of changes in oxidation number is fundamental, and it provides the foundation for the other methods. The descriptions of oxidation and reduction are summarized below:

Oxidation	Reduction
increase in oxidation number	decrease in oxidation number
loss of electrons	gain of electrons
gain of oxygen atoms	loss of oxygen atoms
loss of hydrogen atoms	gain of hydrogen atoms

The Rates of Chemical Reactions

Some chemical reactions take place very quickly, like the burning of gasoline in an automobile engine or the exposure of photographic film in a high-speed camera. Others, like the rusting of iron in moist weather, take place slowly. The speed at which a chemical reaction occurs is commonly called the **rate** of the reaction. In an engine, gasoline burns at a high rate. Iron rusts at a low rate in moist weather.

The rate at which chemical reactions take place in the body is of great importance to good health. Not only must the body pick up oxygen in the lungs to maintain cellular respiration, but it must pick it up fast enough to meet the oxygen needs of the cells. Also, the complex chemical reactions responsible for nerve transmission and muscle action must be fast enough so you can react quickly if you accidently touch a hot object.

Chemists describe the rate of a reaction in terms of how fast a reactant is consumed or how fast a product is formed. If at some point in a reaction, 0.0010 mole of a compound is consumed in one minute, the rate at that point could be expressed in terms of that compound as 0.0010 mole/minute. For example, the rate at which hydrogen and iodine react to form hydrogen iodide can be determined by measuring the amount of iodine (I_2) that is consumed each second or minute at some point during the reaction:

$$H_2 + I_2 \longrightarrow 2HI$$

How Chemical Reactions Take Place

If two species are to react, be they molecules, atoms, or ions, certain conditions must be met. First, the two species must come into contact with each other in a collision. Second, the collision must take place with at least some minimum amount of energy if a reaction is to occur. Third, in many cases the species must be oriented in a particular way as they collide so the proper atoms contact each other. Let us look at each of these conditions in some detail so the importance of each is understood.

It seems reasonable to expect that two molecules must collide and contact each other if they are to react. Ammonia, NH_3, and hydrogen chloride, HCl, are both gases that react to form ammonium chloride, NH_4Cl, a white solid. But, if the two gases are not allowed to mix together so the molecules can interact, the reaction will be impossible:

$$NH_{3(g)} + HCl_{(g)} \longrightarrow NH_4Cl_{(s)}$$
$$(NH_4^+, Cl^-)$$

The fact that molecules need to collide with a certain amount of energy is perhaps not as obvious as the first condition. If two species are to react, with the making and breaking of bonds, the valence shell electrons on both must be forced together so electrons can rearrange themselves as needed. It takes energy to force the negatively charged clouds of electrons into each other, because the like charges repel one another.

The minimum amount of energy that must be involved in the collision so that a reaction can take place is called the **activation energy** for the reaction. The activation energy represents an energy ''hill'' or ''barrier'' that must be crossed if reactants are to form products. The energy needed to cross this barrier comes primarily from the kinetic energy (the energy of motion) of the colliding molecules. In Chapter 5, potential energy diagrams were used (Figures 5.11, 5.12) to show that the amount of heat absorbed or produced in a chemical reaction equalled the difference in the potential energies of the reactants and products. These diagrams can be modiified to show the energy barrier for a reaction, as in Figure 11.1. The difference between the potential energy (stored chemical energy) of the reactants and the highest point on the curve is equal to the activation energy for the reaction. Collisions between molecules that do not have enough energy to pass over the barrier, that is, cannot meet the activation energy, will not result in reaction. The collision would not be ''effective.'' In a sense, the activation energy can be viewed as a mountain that separates two cities. To get from one city to the other, you must have enough energy to first cross over the mountain. The difference between the potential energy of the

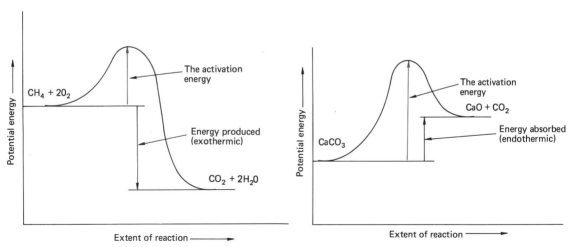

Fig. 11.1. The energy pathway for the conversion of reactants to products. The activation energy for a reaction is the energy difference between the reactants and the maximum point on the curve. The difference in energy separating reactants and products is the energy produced or absorbed by the reaction.

reactants and products equals the amount of energy released or absorbed in the reaction, as was shown in Chapter 5. Whether a reaction is exothermic or endothermic, there is an activation energy barrier that must be crossed if the reaction is to take place. Later on you will learn how the rate of a reaction is affected by the size of the activation energy.

Finally, except for very simple species, molecules must collide in such a way that the atoms that are to join together contact each other. In the reaction between HCl and NH_3, the hydrogen end of the HCl molecule most collide with the nitrogen atom of the NH_3 molecule if NH_4Cl is to form. Any other orientation in the collision would not be an "effective" collision, even if the collision was very energetic and met the activation energy requirement:

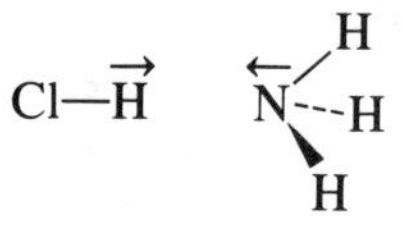

The effective collision 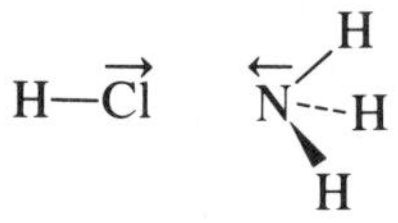   An ineffective collision

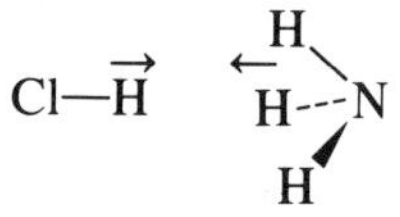

An ineffective collision 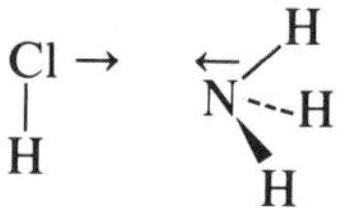   An ineffective collision

Simple species like atoms or monatomic ions can collide with any orientation since they are spherical particles. There is no orientation requirement for the reaction of two hydrogen atoms to form the hydrogen molecule:

The Activated Complex

At this point you may wonder what happens once two molecules collide with the right amount of energy and correct orientation. The molecules combine to form a high-energy, relatively unstable species called an activated complex. The **activated complex** is a species that is neither reactant nor product, but is intermediate between the two, and in it, bonds can both break and form. Energywise, the activated complex is at the highest point in the potential energy curve for a reaction, as shown in Figure 11.2. The activated complex is not something that can be isolated in a reaction—it is too unstable. Once the complex forms, it can fall apart to form the reaction products, or it may fall apart in a different way to reform the reactants.

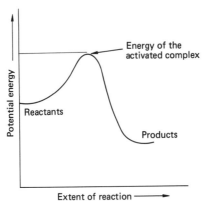

Fig. 11.2. The activated complex is a high-energy intermediate in a reaction. During the brief lifetime of the complex, chemical bonds are made and broken as reactants are converted to products. The potential energy of the complex equals the energy of the highest point along the pathway from reactants to products.

Factors That Influence the Reaction Rate

There are four principal factors that determine how fast a chemical reaction will proceed:

1. the nature of the reactants
2. the temperature
3. the concentration of the reactants
4. the presence of a catalyst

The Nature of the Reactants
At first glance, the "nature of reactants" might seem a bit vague, but there are several characteristics that a reactant may possess that can have a marked effect on the rate with which it can react. Some of these characteristics are (a) whether the reactant is an ion or a neutral atom or molecule, (b) the shape and size of the species, (c) the polarity of the reactant, (d) the polarity and strength of bonds involved in the reaction, and (e) how the reactant interacts with the solvent if it is in solution.

Reactions between ions of opposite charge are nearly always faster than those between neutral species when carried out under similar conditions. The opposite charges on reacting ions cause them to be attracted to each other, which increases the rate at which they can react, This kind of attraction does not occur with neutral species, though polar molecules can experience weak attractions for each other if they are close together. The shape and size of a reactant are both going to affect the likelihood of reactant molecules colliding together with the proper orientation. Generally, large, complex molecules are less likely to collide with proper orientation than small molecules or atoms. Reactions that involve the breaking of covalent bonds are often somewhat slow, but generally, weaker bonds will be broken easier and faster than stronger bonds. The role of the solvent in a reaction can affect the rate in many ways. More will be said about the way a solvent interacts with the dissolved species in Chapter 12, but certain reactions can only be carried out conveniently in solution, especially those between ions.

In reactions, the size of the activation energy barrier largely depends on the nature of the reacting species. Generally, reactions between oppositely charged ions in solution have low activation energies. Reactions between neutral molecules have higher activation energies.

Temperature   A reaction can be made to take place at a faster rate if the temperature is increased. At higher temperatures, the reacting molecules will be traveling at higher speeds because of their greater average kinetic energy. As a result, they will collide more frequently and the collisions will be more energetic (they hit harder). As the number of collisions each second increases, the opportunity for reaction increases. The increased energy of the collisions increases the chances that the required activation energy will be supplied so products can form.

As a general rule for reactions that involve the breaking of covalent bonds, a 10°C rise in temperature will double the reaction rate. There are many exceptions to this rule, but it does provide a rough estimate of the importance of a temperature change on reaction times.

Concentration of Reactants   Increasing the concentration of one or more of the reactants also increases the rate of the reaction. When the concentration of a reactant is increased, a larger number of molecules is made to occupy the same volume. Because the number of molecules is greater, there will be more collisions each second, which causes the rate to increase. On the other hand, if the concentrations of the reactants are lowered, the reaction will proceed at a slower rate. This is because the number of collisions will be fewer each second. There is no simple "rule of thumb" that can be used to predict how much change will be seen in a reaction rate due to concentration changes. This usually must be determined by careful experimental studies. But rates will increase if concentrations of one or more reactants increase, and decrease if concentrations decrease.

Catalysts   For many reactions there are substances that can be added to the mixture of reactants to increase the reaction rate. These substances are commonly called **catalysts**. A catalyst is a substance that increases the rate of a chemical reaction but in doing so is not consumed during the reaction. Once the reaction is over, the catalyst can be recovered unchanged. A catalyst increases the rate of reaction by providing an alternate, easier "pathway" for the conversion of reactants to products. There are several ways a catalyst can change the reaction

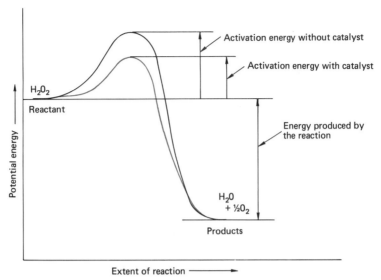

Fig. 11.3. A catalyst lowers the activation energy of a reaction, thereby increasing the rate of reaction. For the decomposition of one mole H_2O_2, the activation energy without the catalyst is 18.0 kcal; with a catalyst it is 13.5 kcal.

pathway, but in each case there is an overall lowering of the activation energy. The effect of a catalyst on the activation energy of the decomposition of hydrogen peroxide, H_2O_2, is shown in Figure 11.3. The catalyst in this reaction is the iodide ion, I^-, and the activation energy is only three-fourths of what it would be if the catalyst were not used. The catalyst is usually written above the arrow in the equation for the reaction:

$$2H_2O_2 \xrightarrow{\quad I^- \quad} 2H_2O + O_2$$

By lowering the overall activation energy, a larger fraction of the collisions between molecules will be able to meet the lowered energy requirement for effective collisions. More products can then form each second.

Frequently, catalysts are employed so that reactions can be carried out at temperatures lower than what would be needed if a catalyst were not used. For example, oxygen gas can be produced by decomposing potassium chlorate, $KClO_3$, by heating it to a high temperature. Usually a temperature of about 425°C is needed to produce oxygen at a convenient rate:

$$2KClO_3 \xrightarrow[\quad 425°C \quad]{} 2KCl + 3O_2$$

If a small amount of manganese (IV) oxide, MnO_2, a catalyst for the reaction, is first mixed with $KClO_3$, oxygen can be produced at the same rate at only 275°C:

$$2KClO_3 \xrightarrow[275°C]{MnO_2} 2KCl + 3O_2$$

The thousands of life-sustaining reactions in the body are catalyzed by special proteins called enzymes (Chapter 24). Without enzymes, the reactions could not take place fast enough at normal body temperature to maintain life. There are thousands of different enzymes in the body, and each one has a specific role in body chemistry.

One last word about catalysts. A catalyst cannot cause a reaction to occur that would not take place otherwise. It can only affect the rate of reactions that are possible when no catalyst is present.

Check Test Number 4

1. State whether each of the following would increase or decrease the rate of a chemical reaction: (a) increasing the concentration of one reactant but not the other, (b) lowering the temperature, (c) decreasing the concentration of one of the reactants, (d) adding a suitable catalyst.
2. What two changes could you make to bring about an increase in the number of collisions each second between reactant molecules?

Answers
1. (a) increase, (b) decrease, (c) decrease, (d) increase.
2. Increase the concentration on one or more reactants and increase the temperature.

Reversible Reactions

Up to this point, chemical reactions have been regarded as processes that take place only in one direction, from left to right as the equation is written. The substances on the left side of the equation have always been called the **reactants,** and those on the right side, **products.** But some reactions can occur in either direction depending on the circumstances. For example, take the reaction between hydrogen and iodine which forms hydrogen iodide. The temperature of the reaction is high enough so that iodine is in the gaseous state, so (g) subscripts are used in the equation to show that all species are gases, and the reaction takes place in a sealed container.

$$H_{2(g)} + I_{2(g)} \longrightarrow 2HI_{(g)}$$
$$\underrightarrow{\text{forward direction}}$$

The combination of H_2 and I_2 to produce HI can be called the **forward reaction,** since the equation is read in the forward direction, from left to right. After some

HI is produced, an interesting event begins to take place. Hydrogen iodide begins to react with itself, producing H_2 and I_2, just the reverse of the forward reaction. If we write the equation with the same order of substances as was used above, we can symbolize the **reverse reaction** by reversing the direction of the arrow:

$$H_{2(g)} + I_{2(g)} \longleftarrow 2HI_{(g)}$$

$$\underline{\text{reverse direction}}$$

We say this reaction occurs in the "reverse" direction only because we would need to read the equation from right to left. It is common practice to write equations for **reversible reactions** using two opposing arrows ($\rightleftharpoons$) in a single equation:

$$H_{2(g)} + I_{2(g)} \rightleftharpoons 2HI_{(g)}$$

More will be said about this reaction in the following section dealing with chemical equilibrium.

In reversible reactions, the terms reactant and product can be applied to each substance in the equation. Hydrogen iodide is a product in the forward reaction, but a reactant in the reverse reaction.

A very important reversible reaction is that between hemoglobin in the blood and oxygen. As the blood circulates through the lungs, hemoglobin picks up oxygen and becomes oxyhemoglobin. Oxyhemoglobin is then transported to other parts of the body where the reverse reaction occurs as it loses oxygen and becomes hemoglobin once again. This reversible reaction can be written more conveniently with words since the formulas are very complex:

$$\text{hemoglobin} + \text{oxygen} \rightleftharpoons \text{oxyhemoglobin}$$

There are many reactions known that are considered to be reversible, yet there are many other reactions that for all practical purposes are irreversible. You may wonder what determines whether or not a reaction will be reversible. The answer requires that we return to activation energies. If a reaction is to be significantly reversible, the activation energies for the forward and reverse reactions should not be greatly different. This is shown in Figure 11.4 for the reversible hydrogen iodide reaction. The difference in the two activation energies is only about 2.5 kcal, a small amount of energy compared with the activation energies of the forward and reverse reactions, about 40 kcal. Compare this with the case of a typical irreversible reaction, the combustion of methane, CH_4:

$$CH_4 + 2O_2 \longrightarrow CO_2 + 2H_2O$$

The difference between the activation energies of the forward and reverse

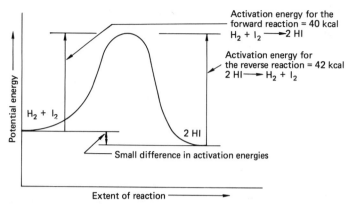

Fig. 11.4. If a reaction is to be significantly reversible, the activation energies of the forward and reverse reactions should be similar.

reactions is about 192 kcal, a large amount of energy. Once methane reacts with oxygen, the likelihood that the reaction products will combine to form CH_4 and $2O_2$ is remote. Carbon dioxide and water remain as products.

Chemical Equilibrium

You have already been introduced to the concept of equilibrium in Chapter 10 in terms of the equilibrium vapor pressure of liquids. The condition of equilibrium was reached when the rate at which the liquid evaporated equalled the rate at which gaseous molecules condensed to the liquid (see Figure 10.14). An important point of that discussion was that equilibrium was attained only when the rates of the opposing changes (evaporation and condensation) became equal. This same idea of opposing changes occurring at equal rates can also be applied to reversible chemical reactions, since they too can exist in a state of equilibrium.

The equilibrium state, as it applies to chemical reactions, is called **chemical equilibrium**. Let us return to the reversible hydrogen iodide reaction and use it to describe how a chemical reaction can reach a point of equilibrium.

If an equal number of moles of gaseous H_2 and I_2 are placed in a sealed container held at 200°C, they will begin to react. At first, since only H_2 and I_2 are present, only the forward reaction takes place, producing HI:

$$H_{2(g)} + I_{2(g)} \longrightarrow 2HI_{(g)}$$

The rate at which HI forms will depend on the concentrations of both reactants. Since H_2 and I_2 are consumed in the forward reaction, their concentrations

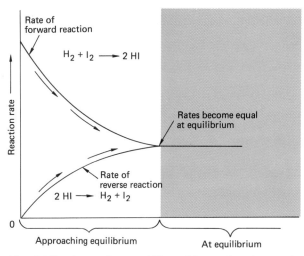

Fig. 11.5. As a mixture of H_2 and I_2 react and come to equilibrium with HI, the rate of the forward reaction decreases while the rate of the reverse reaction increases. Eventually, the two rates become equal and the reaction is then at equilibrium.

continually decrease and, as a result, the rate of the forward reaction steadily decreases. During this time, the HI molecules that are produced begin to react, producing H_2 and I_2 in the reverse reaction:

$$H_{2(g)} + I_{2(g)} \longleftarrow 2HI_{(g)}$$

At first the concentration of HI is low, and the rate of the reverse reaction is slow, but as the concentration of HI increases, the rate of the reverse reaction increases steadily. With the rate of the forward reaction decreasing and the rate of the reverse reaction increasing, eventually a point is reached where the rates become equal. This is shown in Figure 11.5. Whenever the rates of the forward and reverse reactions are equal in a reversible reaction, it exists in **chemical equilibrium:**

$$H_{2(g)} + I_{2(g)} \rightleftharpoons 2HI_{(g)}$$

At equilibrium, HI is consumed in the reverse reaction as fast as it is produced in the forward reaction. The concentration of HI has reached a maximum and it will remain constant. It cannot get any higher since every time a molecule of HI is produced, another is consumed. The concentrations of H_2 and I_2 remain constant at equilibrium for the same reasons. They are consumed as fast as they are formed. The changes in the concentrations of H_2, I_2, and HI as equilibrium is reached are shown in Figure 11.6. Chemical equilibrium is a dynamic state. Both reactions are taking place and molecules are made and consumed each instant.

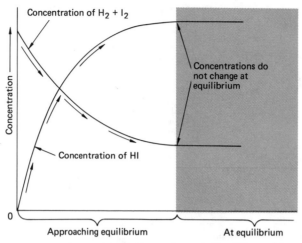

Fig. 11.6. As a mixture of equal amounts of H_2 and I_2 reacts and comes to equilibrium with HI, the concentration of each species continues to rise or fall until equilibrium is reached. From then on, they remain constant.

Terms

Several important terms were introduced in Chapter 11. You should under-stand the meaning of each of the following:

combination reaction	reaction rate
decomposition reaction	activation energy
single-replacement reaction	effective collision
double-replacement reaction	activated complex
oxidation	catalyst
reduction	reversible reaction
redox reaction	forward reaction
oxidizing agent	reverse reaction
reducing agent	chemical equilibrium
oxidation number	

Questions

Answers to the starred questions appear at the end of the text.

1. Classify each of the following reactions as either combination, decomposition, single replacement, or double replacement:

   *a. $2H_2 + C \longrightarrow CH_4$

   b. $Ca(OH)_2 + H_2SO_4 \longrightarrow CaSO_4 + 2H_2O$

   *c. $2Au_2O_3 \xrightarrow{\Delta} 4Au + 3O_2$

   *d. $CaCl_2 + Na_2C_2O_4 \longrightarrow CaC_2O_4 + 2NaCl$

   e. $Cl_2 + 2KBr \longrightarrow 2KCl + Br_2$

   *f. $2Fe + 6HNO_3 \longrightarrow 2Fe(NO_3)_3 + 3H_2$

   g. $2Al + 3Cl_2 \longrightarrow 2AlCl_3$

   *h. $CuO + 2HCl \longrightarrow CuCl_2 + H_2O$

   i. $Zn + CuCl_2 \longrightarrow ZnCl_2 + Cu$

   *j. $BaO + H_2O \longrightarrow Ba(OH)_2$

   k. $2Fe(OH)_3 \xrightarrow{\Delta} Fe_2O_3 + 3H_2O$

   *l. $CuO + SO_2 \longrightarrow CuSO_3$

2. What is the oxidation number of the underlined element in each of the following?
   *(a) $C\underline{O}_2$, *(b) $\underline{Br}_2$, (c) $Na_2\underline{C}O_3$, *(d) $H\underline{Cl}O_3$, (e) $H_2\underline{C}O$, *(f) $\underline{S}O_4^{2-}$, (g) $H\underline{C}O_3^-$, *(h) $H_2\underline{S}O_3$.

*3. Which of the following reactions are not redox reactions?

   a. $Na_2O + H_2O \longrightarrow 2NaOH$

   b. $Cu + 4HNO_3 \longrightarrow Cu(NO_3)_2 + 2H_2O + 2NO_2$

   c. $KOH + HCl \longrightarrow KCl + H_2O$

   d. $2Zn + O_2 \longrightarrow 2ZnO$

   e. $Ni(OH)_2 \xrightarrow{\Delta} NiO + H_2O$

4. In each of the following identify which element is oxidized and which is reduced?

   *a. $4Na + O_2 \longrightarrow 2Na_2O$

   b. $Mg + H_2SO_4 \longrightarrow MgSO_4 + H_2$

   *c. $4FeO + O_2 \longrightarrow 2Fe_2O_3$

   *d. $2NaClO_3 \xrightarrow{\Delta} 2NaCl + 3O_2$

   e. $Fe_2O_3 + 3H_2 \xrightarrow{\Delta} 2Fe + 3H_2O$

5. Four compounds are listed below along with a product that each may form in a redox reaction. Indicate whether the compound is oxidized or reduced.

Compound	Product
*a. CH_3OH	H_2CO
*b. H_2O_2	H_2O
c. $H_2C{=}CH_2$	$H_3C{-}CH_3$
d. CH_3CHO	CH_3COOH

6. List the four factors that can influence the rate of a chemical reaction.

*7. What is meant by the term "activation energy" for a chemical reaction?

8. What effect does a catalyst have on the rate of a chemical reaction?

*9. How does a catalyst affect the activation energy of a chemical reaction?

10. What species are the biological catalysts in the body?

*11. If two molecules are to react, they must collide with each other. Indicate how each of the following will affect the number of collisions per second between reactant molecules: (a) increasing the temperature, (b) adding a catalyst, (c) increasing the concentration of one or more reactants.

12. Why would you expect reaction A to be faster than reaction B? Both take place in water at 25°C.

$$A: \quad Ba^{2+} + SO_4^{2-} \longrightarrow BaSO_4\downarrow$$

$$B: \quad 2H_2O_2 \longrightarrow 2H_2O + O_2$$

*13. One million molecules of a given reactant are consumed each second in a reaction at 20°C. Assuming the reaction rate doubles for each 10°C increase in temperature, how many molecules will be consumed each second at 30°C? At 40°C?

14. Which of the following equations describe reversible reactions?
   a. $CO_2 + H_2O \rightleftharpoons H_2CO_3$
   b. $4Al + 3O_2 \longrightarrow 2Al_2O_3$
   c. $HCl + NaOH \longrightarrow NaCl + H_2O$
   d. $2SO_2 + O_2 \rightleftharpoons 2SO_3$

*15. What can be said about the rates of the forward and reverse reactions for a chemical reaction at equilibrium?

16. Why do the concentrations of all substances in a reversible reaction remain constant once at equilibrium?

Water, Solutions, And Colloids

Water is the most abundant compound on the surface of the earth and, not surprisingly, in our bodies as well. Water is so much a part of our daily lives we often take many of its unique properties for granted. Water has the lowest formula weight (FW H_2O = 18) of any compound that is a liquid at room temperature. It is the only common liquid that expands when it freezes. This makes the density of ice less than that of water, so ice will float on its surface. This single fact keeps ponds, lakes, and rivers from freezing completely solid in the winter killing aquatic life in the process. The ice layer across the surface insulates the liquid water beneath it from the freezing temperatures.

Water also has an abnormally high specific heat (1.0 cal/g °C), heat of fusion (80 cal/g), and heat of vaporization (540 cal/g) compared to other liquids (Chapter 5). Water is capable of dissolving many substances, forming solutions that may be vital to the life-sustaining processes in the body or that may simply allow a substance to be handled more conveniently. Water is a remarkable compound, and we cannot live without it.

The chapter will begin with a discussion of the importance of water in the body. Then the nature of water as a solvent will be described along with several methods of expressing the concentration of a dissolved substance in a solution. Certain physical properties of solutions will be described with special attention given to osmosis and dialysis. Finally, we will examine colloids, mixtures that in some ways resemble solutions but are quite different in others.

Objectives

By the time you finish this chapter you should be able to do the following:

1. Describe the three principal regions of the body in which water is found.
2. Describe how water is used in the body.
3. Describe the characteristics of solutions and compare them with those of colloids and suspensions.
4. Express the concentration of a dissolved substance in a solution in terms of weight/volume percent and molarity.
5. Describe how to prepare a solution of a given percent or molar concentration.
6. Define the following terms used to describe solutions: dilute, concentrated, saturated, unsaturated, solute, and solvent.
7. Describe the effect of a dissolved substance on the vapor pressure, boiling point, freezing point, and osmotic pressure of a solution compared to the pure solvent.
8. Describe osmosis and dialysis and compare the effect of an isotonic, hypertonic, and hypotonic solution on red blood cells.
9. Describe two ways of purifying water for laboratory use.

The Water in Our Bodies

Between 55% and 65% of the weight of the average adult is water, and for infants, water accounts for about 80% of body weight. Water is essential to life and if you suffered a loss of about 10% of your body water through illness, injury, or starvation, you would be severely dehydrated and your well-being would be seriously affected. A 20% loss of body water is usually fatal. Infants are even more susceptible to the adverse affects of water loss. A 5% to 10% loss of body water by an infant will bring on the symptoms of severe dehydration. Frequently, the loss of water is accompanied by the loss of sodium ion (Na^+). For this reason people who do heavy work in hot weather frequently take salt tablets to replace the sodium ion lost in perspiration.

Water performs several important functions in the body that can be broadly classed as temperature regulation, transport, lubrication, and hydrolysis. The regulation of the temperature of the body is largely controlled by the evaporation of water from the skin and through the lungs. A fever increases the loss of water through the skin by an estimated 50–75 ml for each degree of fever (°F) in a 24-hr period. The fact that water does not have to undergo wide temperature variations as it absorbs or releases heat (water has a large specific heat, 1.0 cal/gram °C) allows it to stabilize the temperature of the body. The movement of oxygen, carbon dioxide, nutrients, foods, and wastes within the body takes place in the blood and the fluids in and around cells. The lubrication of joints and internal organs is necessary if they are to move freely and without pain. The major joints in the body are bathed in a viscous, slippery synovial fluid which is largely water. Water is the principal reaction medium for chemical reactions within cells and in the digestive processes. Water is also consumed in some of these reactions, as well as being produced in others. The general name for any reaction that uses water as a reactant is **hydrolysis**. The reactions that break down large protein molecules and fat molecules in digestion are hydrolysis reactions. They will be discussed in later chapters.

Most of the water in the body is located within cells, and it is called **intracellular water**. About 40% of your body weight is intracellular water (about 25 l of water). The spaces between cells also contain large amounts of water, and it is called **interstitial** or **extracellular water**. About 15% of your weight is extracellular water, and the movement of nutrients and wastes between the cells and the circulatory system takes place through the extracellular water. Most of the remaining water is present in blood plasma, the intravascular fluid in arteries, veins, and capillaries. The blood in the average adult represents about 5% of the weight of the body, see Figure 12.1. Of course, the water in

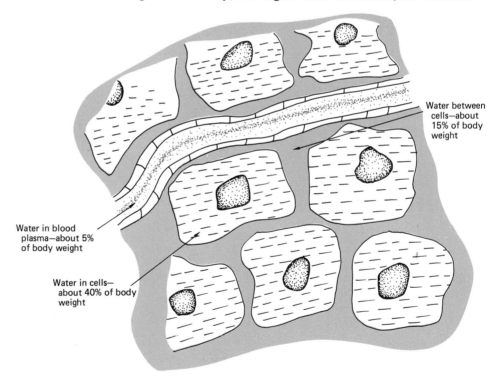

Water between cells—about 15% of body weight

Water in blood plasma—about 5% of body weight

Water in cells— about 40% of body weight

Fig. 12.1. Approximately 60% of your body weight is water. The water within cells, intracellular water, represents about 40% of your body weight. That between cells, extracellular water, totals about 15% of body weight. Most of the remainder, about 5% of body weight, is in blood plasma.

and around cells is not just pure water, there are many substances dissolved in it. Many of these substances are ionic, and the average composition of intracellular, extracellular, and plasma fluids in healthy adults is given in Table 12.1. In each type of fluid, there is an equal number of positive and negative charges. The column in Table 12.1 headed mEq/l (milliequivalents per liter) lists concentration values that take the charges of ions into account. Note that the sum of all the cation concentrations in mEq/l equals the sum for all the anions within each of the three types of fluids. Though it will not be obvious to you at this point, this is the same as saying there is a negative charge for every positive charge in each solution. Notice that the principal ion in intracellular fluid is K^+, while in extracellular fluid and plasma it is Na^+.

For a healthy person, the amount of water taken into the body each day should equal the amount lost. Our principal sources of water are the liquids we drink and the food we eat. A smaller amount of water is produced within the body in metabolism. Water is lost through the skin and lungs as well as through the kidneys (urine) and the intestinal tract (feces). The average adult will consume about 2500 ml of water each day, but the amount can vary widely from individual to individual. The abnormal retention of water by the body is called **edema**, and it causes a swelling of tissue and frequently a puffiness in the face. Edema can be brought on by excessive sodium (Na^+) retention in the extracellular fluid due to kidney disease or by heart or liver disease.

TABLE 12.1
Composition of Intracellular, Extracellular, and Plasma Fluids

	Intracellular			Extracellular			Plasma		
	ion	mg/l	mEq/l	ion	mg/l	mEq/l	ion	mg/l	mEq/l
Cations	K^+	6260	160	Na^+	3340	145	Na^+	3270	142
	Mg^{2+}	425	35	K^+	155	4	K^+	195	5
	Na^+	230	10	Ca^{2+}	60	3	Ca^{2+}	100	5
				Mg^{2+}	24	2	Mg^{2+}	24	2
Anions	HPO_4^{2-}	6720	140	Cl^-	4080	115	Cl^-	3730	105
	anionic protein	—	55	HCO_3^-	1830	30	HCO_3^-	1465	25
				organic acids	—	5	anionic protein	—	16
	HCO_3^-	490	8	HPO_4^{2-}	95	2	organic acids	—	6
	Cl^-	70	2	SO_4^{2-}	50	1	HPO_4^{2-}	95	2
				anionic protein	—	1	SO_4^{2-}	50	1

TABLE 12.2 Typical Water Intake and Loss for an Adult				
Intake			Loss	
liquids	1000 ml		lungs	400 ml
solid foods	1200 ml		skin	400 ml
metabolism	300 ml		kidneys	1500 ml
			intestinal	200 ml
	2500 ml			2500 ml

Typical daily water intake and loss figures for a healthy adult are given in Table 12.2.

Solutions

Many of the substances you encounter every day are solutions. You may not think of drinking water as a solution, but it is. Drinking water contains small amounts of dissolved air and minerals. Air is also a solution made up of nitrogen, oxygen, water vapor, and other gases. Gold jewelry is a solution of gold and copper. In medicine, solutions of drugs are used for injections, and liquids that are administered intravenously may be solutions of dextrose (glucose) and ionic substances dissolved in water.

As you can see, solutions can be gases, liquids, or solids, but for our purposes we will be more interested in those that are liquids, since they are by far the most common kinds of solutions used in medicine. Several kinds of solutions are listed in Table 12.3.

A **solution** can be defined as a homogeneous mixture of two or more substances. Being homogeneous, the composition of a solution is the same every-

TABLE 12.3 Types of Solutions with Examples		
Type	Physical state	Examples
gas in a liquid	liquid	carbonated beverages (CO_2 in H_2O)
liquid in a liquid	liquid	rubbing alcohol, antifreeze
solid in a liquid	liquid	sugar water, salt water
gas in a gas	gas	air, some anesthetic gases
gas in a solid	solid	hydrogen absorbed in titanium
liquid in a solid	solid–liquid	dental amalgam, certain flexible plastics
solid in a solid	solid	jewelry gold, brass, alloys

where throughout its volume. As a mixture, the relative amounts of each substance are not necessarily fixed, but rather can be varied by adding more of one or the other.

If a spoonful of sucrose, common table sugar, is dissolved in water, a solution is formed. Water is called the **solvent** since sucrose is dissolved in it. Sucrose is the **solute** since it is the substance being dissolved. The terms solvent and solute are commonly used to describe solutions. Generally, the solvent is the substance present in the greater amount, and the solute in the lesser amount. Solutions that use water as the solvent are called **aqueous solutions.**

There are several characteristics that are common to all liquid solutions. Let us look at them as they relate to aqueous solutions:

1. Solutions are clear and transparent. You can see through them and particles of solute are not visible at all.

2. Solutions may be colorless or colored depending on the solute. A solution of sucrose or salt in water is colorless, but a solution of copper (II) sulfate in water is a beautiful blue color due to the presence of the copper ion, Cu^{2+}.

3. The solute is present as individual positive and negative ions or as neutral molecules. Soluble ionic compounds like NaCl enter the solution as sodium and chloride ions:

$$NaCl_{(s)} \rightarrow Na^+_{(aq)} + Cl^-_{(aq)}$$

The subscript (aq) written after each ion indicates they are dissolved in water—in an aqueous solution. Substances that produce ions when they dissolve are called **electrolytes**, since their solutions can conduct an electric current. Any soluble ionic compound is an electrolyte. Potassium nitrate, KNO_3, is an electrolyte since it exists as ions in solution, $K^+_{(aq)}$ and $NO^-_{3(aq)}$. Sugars and alcohols are molecular substances, and when dissolved in water they exist as neutral molecules. Solutes that are present in solutions as neutral molecules are called **nonelectrolytes**, since their solutions will not conduct an electric current. Certain highly polar covalent molecules, like HCl, interact so strongly with water that they dissociate into ions. A solution of HCl in water is a mixture of ions in water, $H^+_{(aq)}$ and $Cl^-_{(aq)}$. More will be said about the events that take place as ionic and covalent compounds dissolve in water in the next section.

4. The solute will not settle out of a solution upon standing for any length of time. The constant motion of the solvent molecules and the solute particles, because of their kinetic energy, keeps the solution homogenous.

5. The solute cannot be separated from the solvent by simple filtration through filter paper. Both solute and solvent are present as small, individual species that readily pass through the pores in filter paper. Both water and dissolved substances are able to pass through most all cell membranes too.

The amount of solute that can dissolve in a given volume of solvent depends on the nature of the solute and the solvent (how well they can accommodate each other), the temperature of the solution, and, if the solute is a gas, the pressure exerted by the gas over the solution.

The degree of solubility of a substance in water is usually expressed as the grams of solute that can dissolve in 100 g of water. Sucrose is very soluble in water. You can dissolve about 200 g of sucrose in 100 g of water at room temperature. Even though the amount of sucrose exceeds the amount of water in the solution, water is still considered the solvent since sucrose dissolves in it. Barium sulfate, the substance used in the "barium milkshake" that is drunk during a gastrointestinal x-ray examination, has a very low solubility in water. Only about 0.00030 g of $BaSO_4$ will dissolve in 100 g of water at room temperature. For all practical purposes, substances like barium sulfate are regarded as being insoluble in water, though in reality they are very slightly so.

In most cases, but not all, the amount of a given solid substance that can dissolve in 100 g of water increases as the temperature of the solution increases. For example, at 0°C, you can dissolve 13.3 g of KNO_3 in 100 g of water, but at 100°C, 247 g of KNO_3 will dissolve. The opposite is true if the solute is a gas. The solubility of oxygen decreases as the temperature of the solution increases. At 0°C, 100 g of water can dissolve 0.007 g of O_2. At 50°C, it will dissolve 0.003 g of O_2, and at 100°C, the solubility is essentially zero.

Pressure will also affect the solubility of a gas in a liquid. If the pressure of a gas in contact with a liquid is increased, the amount of gas that will dissolve in the liquid is increased also. It is as if the pressure exerted by the gas forces it to enter the liquid. Carbonated beverages are made by dissolving carbon dioxide in aqueous solutions using gas pressures in excess of 1 atm.

Any solution that contains all of the dissolved solute that it can hold at a given temperature is called a **saturated solution**. When we say that the solubility of NaCl in water at 25°C is 36.5 g salt per 100 g of water, we are giving the composition of a saturated solution. If more NaCl is added, it will simply fall to the bottom of the solution and will not change the amount of NaCl dissolved in the solution.

A saturated solution in contact with undissolved solute is another kind of dynamic equilibrium system. The undissolved solute is actually going into solution, but at the same time and at the same rate, dissolved solute is coming out of solution forming crystals, Figure 12.2. This can be represented with an equation for a saturated NaCl solution in contact with solid NaCl; the opposing arrows indicate equilibrium:

$$NaCl_{(s)} \rightleftharpoons Na^+_{(aq)} + Cl^-_{(aq)} \text{ (saturated solution)}$$

Every time a sodium ion enters the solution, one leaves the solution. The same is true for the chloride ions. Any solution that is allowed to remain in contact with undissolved solute will become a saturated solution in time.

Fig. 12.2   There is a dynamic equilibrium between dissolved solute in a saturated solution and undissolved solute. In A, NaCl is added to a beaker of water and the rate at which it dissolves is indicated by the black arrow. In B, NaCl is still dissolving at the same rate, but the NaCl concentration in the solution is high enough that it begins to return to the crystal at an appreciable rate indicated by the colored arrow. In C, the concentration of NaCl in solution is so high, the rate of its return to the crystal equals the rate at which it dissolves. The solution is then saturated.

Solutions that contain smaller amounts of solute than that needed to be saturated are known as **unsaturated solutions**. Ten grams of NaCl in 100 g of water at 25°C would be an unsaturated solution.

Water as a Solvent

Water has been called the "universal solvent," not because it is the most common liquid in our lives, but because it can dissolve varying amounts of many compounds and at least small amounts of most. Of course there is no such thing as a universal solvent, that is, one that could dissolve everything, but water has exceptional solvent qualities and it can dissolve many polar molecules, those with polar groups, and ionic compounds.

Water is a polar liquid, because the water molecules are dipoles. You will recall the water molecule has a bent shape, and the polar covalent hydrogen–oxygen bonds cause the entire molecule to act as a dipole:

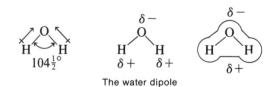

The water dipole

The negative end of the water dipoles will be attracted to the positive ends of polar molecules or groups dissolved in it, and vice versa. Water also engages in hydrogen bonding, not only with itself but also with other substances dissolved in it that contain — O — H or — $\overset{|}{N}$ — H groups. The polarity of water coupled with its hydrogen bonding ability enhance the solvent power of water because they allow water to accomodate polar molecules and ions in solution. If a liquid is unable to interact with a dissolved species to accomodate and hold it in solution, it cannot be a good solvent for that substance. A good example of how water can accomodate a dissolved substance is seen in the sodium chloride solution.

As sodium chloride dissolves in water, the ions enter the liquid and immediately become surrounded by water molecules. Each sodium ion is encased in a layer of water molecules with the negative ends of the water dipoles closest to the positive ion. Each chloride ion is also surrounded by water molecules but with the positive ends of the dipoles closest to the negative ion, as shown in Figure 12.3. The layer of water dipoles about each ion creates an environment similar to that experienced by the ion in the crystal. Positive ions are in contact with the negative ends of the water dipoles, and negative ions are in contact with the positive ends of the dipoles. Water accommodates the ions in solution, and oppositely charged ions are kept apart and insulated from one another by the water molecules about them.

The interaction of a molecule or ion with the solvent is called **solvation.** If the solvent is water, a more specific term, **hydration,** is often used. Ions are

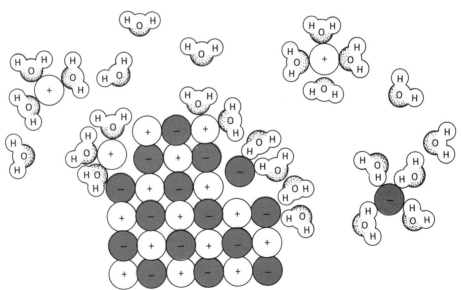

Fig. 12.3. As an ionic compound dissolves in water, each positive and negative ion becomes surrounded by a layer of water molecules as it leaves the crystal. The water dipoles are oriented so the negative ends are closer to the positive ions, and the positive ends of the dipoles are oriented closer to the negative ions. The ions are hydrated in solution.

hydrated in water, and if this were not so ionic compounds would not be soluble in it. Some ionic compounds are only very slightly soluble in water because the ions are so strongly attracted to each other in the crystal that they cannot be readily removed to form a solution. Very often, compounds that have large positive (2+, 3+) and large negative (2−, 3−) charges on the ions will not be very soluble in water.

Though sodium chloride is soluble in water, it is not soluble in carbon tetrachloride, CCl_4, or gasoline because these nonpolar solvents cannot solvate the ions and keep them apart in solution. If the ions would enter the liquid to form a solution, they would quickly come together and reform the NaCl crystal. Oil and other nonpolar compounds are not soluble in water because nonpolar molecules cannot interact well with the polar water molecules. They are "squeezed out" of water because the water molecules can interact better with themselves.

Many alcohols and sugars are soluble in water and are present in solution as molecules. Ethyl alcohol, CH_3CH_2OH, is a polar molecule as well as being one that can hydrogen bond with water. It can be readily accommodated by water. Sugars contain many polar — O—H groups that can interact with water through hydrogen bonding.

Ethyl alcohol with hydrogen bonds to water

Glucose with hydrogen bonds to water

Polar molecules that are not capable of hydrogen bonding with water can still be solvated through the attraction of water dipoles for the $\delta+$ and $\delta-$ ends of the solute molecule dipole, Figure 12.4. If the solute is an extremely polar molecule (HCl, HBr, or HI), the dipole–dipole interactions can be so strong that the solute molecule will break apart and exist in solution as ions, Figure 12.5.

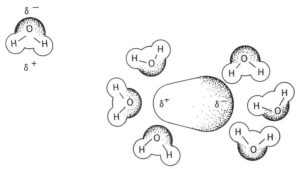

Fig. 12.4. The interactions between water molecules and polar covalent molecules is largely responsible for their solubility in water. Notice how the polar water dipoles are arranged and attracted to the $\delta+$ and $\delta-$ ends of the polar molecule.

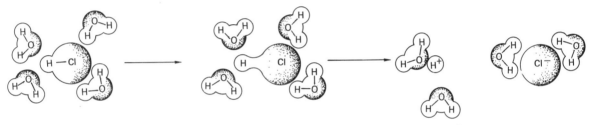

Fig. 12.5. The interaction of the water and HCl dipoles is so large, the HCl molecule dissociates into ions in solution.

Expressing the Concentration of Solutions

There are several ways to express the concentration of a solute in solution. The most useful methods clearly indicate the amount of solute (in grams or moles) in a given amount of solution or solvent. Several of these methods will be described since they are widely used in the health care field. Yet, solutions are also described in a relative way as being either concentrated or dilute. A **concentrated** solution contains a larger amount of solute in a given volume of solution. A **dilute** solution contains a smaller amount of solute in a given volume of solution. These terms are useful since they allow the concentrations of two solutions to be compared in a general way. For some acids and bases, the terms have specific meanings and they indicate definite concentrations, but these are exceptions to the usual meaning of the terms.

Percent by Weight or Volume   There are three methods for stating concentration in terms of the percent of solute in a solution. The most widely used of the three is the **weight/volume percent, % (w/v).** The numerical value of a weight/volume percent equals the number of grams of solute present in 100 ml of solution.

$$\% \,(\text{w/v}) \text{ of solute} = \frac{\text{grams of solute}}{\text{volume of solution in ml}} \times 100\%$$

Because of its common use in medicine, we will examine the weight/volume percent in more detail than the other percent schemes.

A 5%(w/v) glucose solution contains 5 g of glucose in 100 ml of solution. A 0.05% (w/v) solution of an antihistamine contains 0.05 g of antihistamine in 100 ml of solution. Frequently, with commercially prepared solutions, the solute concentration is simply stated as "percent" without the qualification that it is a weight/volume percent. Generally, it is safe to assume that the percent is on a weight/volume basis, but if you need to know this with certainty you should check with the manufacturer.

The following examples show how weight/volume percent concentrations are used in calculations. Note the units, g/ml, are not included in the % (w/v) term.

Example 12.1

What is the %(w/v) of potassium iodide, KI, in 800 ml of a solution that contains 30.0 g of KI?

$$\% \,(\text{w/v}) \text{ KI} = \frac{\text{g KI}}{\text{ml of soln}} \times 100\%$$

$$= \frac{30.0 \text{ g}}{800 \text{ ml}} \times 100\%$$

$$= 3.75\%(\text{w/v})$$

There are 3.75 g of KI in each 100 ml of solution.

Example 12.2

How many grams of sodium chloride are needed to prepare 1000 ml (1 liter) of 0.90%(w/v) solution of NaCl in water? (This is an isotonic NaCl solution.)

$$0.90\%(\text{w/v}) \text{ is equal to } \frac{0.90 \text{ g NaCl}}{100 \text{ ml solution}}$$

This ratio can be used as a conversion factor to determine the number of grams of NaCl present in 1000 ml of solution:

$$\text{grams of NaCl needed} = 1000 \text{ ml} \left(\frac{0.90 \text{ g}}{100 \text{ ml}} \right) = 9.0 \text{ g}$$

The solution is prepared by dissolving 9.0 g of NaCl in a volume of water slightly less than 1000 ml, then adding additional water until the volume is brought up to 1000 ml. Thorough mixing is necessary to ensure a homogeneous solution.

Example 12.3

How many milliliters of a 2.0%(w/v) glucose solution would contain 40 g of glucose?

$$2.0\%(w/v) \text{ is equal to } \frac{2.0 \text{ g glucose}}{100 \text{ ml solution}}$$

Again, the ratio can be used as a conversion factor to convert grams of glucose to the volume of solution needed:

$$\text{volume of solution} = 40 \text{ g} \left(\frac{100 \text{ ml}}{2.0 \text{ g}} \right) = 2000 \text{ ml}$$

2000 ml of a 2.0%(w/v) solution would contain a total of 40 g of glucose.

A common concentration term used in clinical laboratory work when reporting glucose, calcium ion, or other components of blood serum, urine, or cerebrospinal fluid is the **milligram percent** (mg %). It is a weight/volume percent equal to the number of milligrams of the component in 100 ml of the analyzed solution. Eighty milligrams of glucose in 100 ml of serum would be stated as 80 mg% glucose.

The concentration of a solution expressed as a **volume/volume percent,** %(v/v), is equal to the volume of solute in 100 volumes of solution. About the only place you would be likely to come across this method of expressing concentrations is in alcohol–water solutions. If a rubbing alcohol solution (isopropyl alcohol in water) is stated as a 70% solution, it contains 70 ml of alcohol in 100 ml of solution. Though it is frequently omitted, the concentration should be stated as 70%(v/v) or 70% by volume to be perfectly correct.

The third concentration scheme using percent is that based entirely on weight.

The **weight/weight percent,** %(w/w), is equal to the number of **grams** of solute in 100 g of solution. Weight percents are generally more useful for mixtures of solids since they are conveniently measured by weight. Liquid solutions, on the other hand, are usually measured by volume, which is responsible for the greater use of the weight/volume percent.

For solutions that are extremely dilute, it is frequently cumbersome to express concentrations on a percent basis. As an alternative, units of **parts per million** (ppm) or **parts per billion** (ppb) are commonly used. The "parts" can be in units of weight or volume, whichever is more convenient in a given application. In terms of weight, one part per million equals one gram of solute in one million grams of solution. As a matter of comparison, the percent figures are really stating "parts per hundred" of solute in a solution. Fluorinated drinking water contains 1 ppm of fluoride ion (F^-). Using parts per million on a weight/volume basis, this is equivalent to one gram of F^- in one million milliliters of water or one milligram of F^- in one liter of water. This is a very dilute solution, and if expressed in percent terms would be equal to 0.0001%(w/v). In a glass of fluorinated drinking water, about 250 ml, there would be only 0.00025 g of fluoride ion.

The concentrations of pollutants in the environment are frequently stated in parts per million or parts per billion. Though these imply very low concentrations, many in the medical and scientific community are concerned because long-term exposure to even low levels of certain pollutants may be potentially dangerous.

Molarity   A concentration scheme that is more useful to chemists is **molarity**, since it expresses the amount of solute in terms of moles. Since elements and compounds react in definite mole ratios (given by the coefficients in balanced equations), molar concentrations are especially appropriate when solutes are used in chemical reactions. The symbol for molarity or molar concentration is M, and the molarity of a solution can be expressed in an equation in terms of the total number of moles of solute and the volume of the solution in liters:

$$M = \frac{\text{moles of solute in the solution}}{\text{volume of the solution in liters}} = \frac{\text{moles}}{\text{liter}}$$

Note the units of molarity are moles per liter, though they are replaced by the symbol M when stating the molar concentration of a solution. At other times, when doing calculations, it may be convenient to restate M as moles per liter so you can check if the proper units cancel.

Perhaps the easiest way to become familiar with the molar concentration scheme is to show its use with a few examples. The following examples are typical of calculations based on molarity.

Example 12.4

What is the molarity of a solution that contains 120 g of sodium hydroxide, NaOH, in 2.0 l of solution?

You are given the volume of the solution, but you need the number of moles of NaOH.

$$1 \text{ mole NaOH} = 40 \text{ g of NaOH}$$

$$\text{moles of NaOH} = 120 \, \cancel{g} \left(\frac{1 \text{ mole}}{40 \, \cancel{g}} \right) = 3.0 \text{ moles}$$

$$M = \frac{\text{moles NaOH}}{V \text{ soln in liters}} = \frac{3.0 \text{ moles}}{2.0 \text{ l}} = 1.5M$$

The molarity of the solution reveals that each liter of the solution contains 1.5 moles (60 g) of NaOH.

Example 12.5

How many milliliters of a 2.0M sulfuric acid solution, H_2SO_4, should be taken to obtain 0.10 mole of the acid?

First, you can restate the concentration of the solution in terms of moles per liter and use this as a conversion factor:

$$2.0M = \frac{2.0 \text{ moles } H_2SO_4}{1 \text{ l of solution}}$$

$$\text{Volume of solution containing } 0.10 \text{ mole of } H_2SO_4 = 0.10 \, \cancel{\text{mole}} \left(\frac{1 \text{ l}}{2.0 \, \cancel{\text{mole}}} \right)$$

$$= 0.050 \text{ l}$$

$$= 50 \text{ ml of solution}$$

50 ml of 2.0M solution of H_2SO_4 contains 0.10 mole of H_2SO_4.

Example 12.6

What volume of a 1.25M solution of glucose, $C_6H_{12}O_6$, can be prepared with 135 g of glucose?

First, determine the number of moles of glucose you have:

$$1 \text{ mole } C_6H_{12}O_6 = 180 \text{ g of } C_6H_{12}O_6$$

$$\text{moles glucose} = 135\,\cancel{g}\left(\frac{1 \text{ mole}}{180\,\cancel{g}}\right) = 0.750 \text{ mole}$$

A 1.25M solution contains $\dfrac{1.25 \text{ moles glucose}}{1\,\text{l}}$

Using the number of moles of glucose you have, and the number of moles that are present in 1 l of a 1.25M solution, calculate the volume of solution you could prepare with only 0.750 mole of glucose.

$$\text{volume of solution} = 0.750\,\cancel{\text{mole}}\left(\frac{1\,\text{l}}{1.25\,\cancel{\text{mole}}}\right)$$

$$= 0.600\,\text{l}$$

You could prepare 0.600 l of a 1.25M glucose solution using 135 g of glucose. The correct way to prepare this solution is to first dissolve the glucose in a volume of water less than 0.600 l, then add additional water, with stirring, until the solution reaches a volume of 0.600 l. A graduated cylinder (Chapter 2) can be used to prepare the solution, see Figure 12.6.

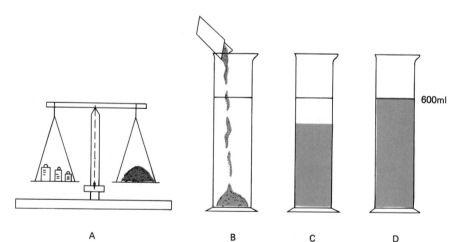

| A | B | C | D |

Fig. 12.6. The procedure for preparing 600 ml of a 1.25M solution of glucose in water. A: Accurately weigh out 135 g of glucose. B: Add the glucose to a clean graduated cylinder with a capacity of at least 600 ml. C: Dissolve the glucose in 400 to 500 ml of pure water to prepare a solution that is more concentrated than you desire; then, D: Add more water to dilute the solution until the volume reaches the 600 ml mark on the graduated cylinder. Stir the solution well as it is being prepared. Only by dissolving the glucose in a lesser volume of water, then diluting it to the volume desired can you be certain the final volume is that of the solution and not just the solvent.

Example 12.7

How many grams of sodium chloride are in 0.800 l of a 0.400M NaCl solution?

First, solve for the number of moles of NaCl:

$$\text{moles of NaCl} = 0.800\,\cancel{l}\left(\frac{0.400\text{ mole}}{1\,\cancel{l}}\right)$$

$$= 0.320\text{ mole NaCl}$$

$$1\text{ mole NaCl} = 58.5\text{ g of NaCl}$$

$$\text{g NaCl} = 0.320\,\cancel{\text{mole}}\left(\frac{58.5\text{ g}}{1\,\cancel{\text{mole}}}\right) = 18.7\text{ g}$$

In the previous examples the molarities of solutions were rewritten in terms of the moles of solute per liter of solution, and then used as conversion factors. Though it was perhaps not initially evident, the example problems were solved using rearranged forms of the defining equation for molarity. To determine the number of moles of solute in a given volume of solution of known molarity, you can use the equation in this form:

$$\text{moles of solute} = (M)\,(\text{volume of solution in liters})$$

To determine the volume of a solution of known molarity that contains a certain number of moles of solute, use

$$\text{volume of solution in liters} = \frac{\text{moles of solute}}{M}$$

Either approach is satisfactory, but the conversion factor approach may give you a better understanding of calculations involving molar concentrations.

Check Test Number 1

1. One-half liter (0.500 l) of an aqueous solution of sodium hydroxide contains 60 g of NaOH. State the concentration of NaOH in terms of
   a. %(w/v)
   b. molarity
2. How many milliliters of a 1.5%(w/v) NaCl solution contains 0.30 g of NaCl?

Answers:

1. a. $\%(w/v)\ NaOH = \dfrac{60\ g\ NaOH}{500\ ml} \times 100\% = 12\%(w/v)$

   b. 60 g of NaOH = 1.5 moles of NaOH

$$M = \frac{1.5\ mole}{0.50\ l} = 3.0M$$

2. $1.5\%(w/v) = \dfrac{1.5\ g\ NaCl}{100\ ml}$

$$\text{volume of solution} = 0.30\ g\left(\frac{100\ ml}{1.5\ g}\right) = 20\ ml$$

Dilution of Solutions

In order to save storage space in hospitals and laboratories, solutions are often prepared and stored at concentrations much higher than those used in various procedures. Portions of these concentrated solutions, or "stock solutions" as they are called, are then diluted to prepare the required solutions. As solvent is added to a given volume of a concentrated solution, the volume of the solution increases and the concentration of the solute decreases—but the actual amount of solute (in moles or grams) remains constant. This fact allows equations to be written in terms of the volumes and concentrations of the concentrated solution and the dilute solution that is to be prepared. If molar concentrations are used, the equation is

$$V_{conc} \times M_{conc} = V_{dil} \times M_{dil}$$

$$V_c M_c = V_d M_d$$

If weight/volume percents are used:

$$V_{conc} \times \%(w/v)_{conc} = V_{dil} \times \%(w/v)_{dil}$$

$$V_c\ \%_c = V_d\ \%_d$$

The volumes can be expressed in any unit desired, as long as they are used consistently in a calculation and refer only to the volumes of the solutions.

With these equations you can calculate the volume of a concentrated solution that must be diluted to prepare a certain volume of a dilute solution, as shown in the following examples.

Example 12.8

What volume of $12M$ hydrochloric acid must be diluted to prepare 600 ml of $0.10M$ HCl?

Rearranging the dilution equation using molar concentrations, solve for V_c, the volume of concentrated HCl:

$$V_c = \frac{V_d\,M_d}{M_c} = \frac{(600\ ml)\,(0.10M)}{(12M)} = 5.0\ ml$$

5.0 ml of $12M$ HCl, when diluted to 600 ml by addition of water, will produce the $0.10M$ solution desired.

Example 12.9

How many milliliters of a $10.0\%(w/v)$ solution of glucose is required to prepare 1.0 l of a $4.0\%(w/v)$ solution?

$$V_c = \frac{V_d\,\%_d}{\%_c} = \frac{(1000\ ml)\,(4.0\%)}{(10.0\%)} = 400\ ml$$

400 ml of a $10.0\%(w/v)$ solution when diluted to 1000 ml will produce the $4.0\%(w/v)$ solution desired.

Check Test Number 2

1. What volume of a $6.0M$ CaCl$_2$ solution is needed to prepare 3.00 l of a $0.50M$ solution?
2. What volume of a $0.90\%(w/v)$ NaCl solution can be prepared from 25.0 ml of a $12.0\%(w/v)$ NaCl solution?

Answers:

1. $V_c = \dfrac{V_d\,M_d}{M_c} = \dfrac{(3.00\ l)\,(0.50M)}{(6.0M)} = 0.25\ l$

2. $V_d = \dfrac{V_c\,\%_c}{\%_d} = \dfrac{(25.0\ ml)(12.0\%)}{(0.90\%)} = 333\ ml$

The Physical Properties of Solutions

If you placed a glass containing a solution of glucose in water next to a glass of pure water, you would be hard pressed to see any difference between them. Both would be colorless, clear, and transparent. Though a visual examination would not reveal any differences, other experiments would show that the solution and the solvent do have different physical properties. Four of these properties are especially important, and as a group they are called the colligative properties of solutions. **Colligative properties** are those that depend only on the concentration of solute particles in the solution and not on the identity of the solute. The four colligative properties concern the effect of solute concentration on the vapor pressure, boiling point, freezing point, and osmotic pressure of a solution. Each of these properties will be discussed as they apply to dilute, aqueous solutions of nonvolatile solutes, such as sugars and ionic compounds in water.

The Vapor Pressure of Solutions

Solutions have *lower* vapor pressures than do the pure solvents at a given temperature. The more concentrated a solution, the lower will its vapor pressure be compared to the pure solvent. Particles of the solute (molecules or ions) reduce the ability of the solvent molecules to escape from the liquid and enter the gas phase. One way to explain this effect is to consider the composition of the layer of molecules at the surface of a solution. In pure water, every molecule at the surface is a solvent molecule, and each one has the ability to escape if it has enough kinetic energy. In the solution, some of the nonvolatile solute particles take the place of solvent molecules at the surface, reducing the number of solvent molecules that are in position to escape from the liquid. With fewer molecules able to escape, the vapor pressure of the solution will necessarily be less than that of the solvent. A comparison of the vapor pressures of pure water and a solution of glucose in water is shown in Figure 12.7.

The Boiling Points of Solutions

The boiling point of a solution is *higher* than that of the pure solvent alone. The more concentrated a solution the higher will be its boiling point compared to that of the pure solvent. This is a natural consequence of the reduced vapor pressures of solutions. Because the vapor pressure of a solution is lower compared to that of the solvent itself, the solution must be brought to a higher temperature to get its vapor pressure to equal atmospheric pressure—the point where boiling occurs. The elevation of the boiling point of a solution compared to the solvent is shown in Figure 12.7. The size of the increase in the boiling point of a solution compared to the boiling point of the solvent is proportional to the number of moles of solute particles in one kilogram (1000 g) of solvent,* A solution containing one mole of glucose (180 g of $C_6H_{12}O_6$) in 1000 g of water has a normal boiling point of 100.5°C, about one-half degree higher than pure water. Two moles of glucose in 1000 g of water would have twice the increase, and the normal boiling point would be raised to 101°C. Each mole of glucose molecules per kilogram of water elevates the boiling point by about 0.5°C. One mole of NaCl in 1000 g of water will raise the boiling point of the solution the

*The **molal** concentration of a solution is equal to the number of moles of solute per kilogram of solvent. Concentrations expressed in molality are convenient when describing colligative properties, but they will not be used here because of their limited application.

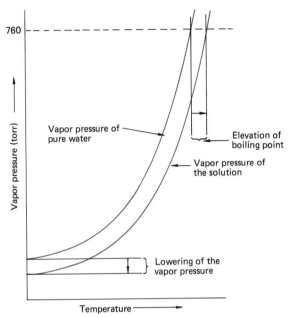

Fig. 12.7. The effect of dissolved solute on the vapor pressure and boiling point of water. The vapor pressure of the solution is less than that of pure water at any temperature. The boiling point of the solution is greater than that of pure water.

same amount as two moles of glucose. This is because NaCl dissociates into ions, one mole of Na^+ and one mole of Cl^-, which equals two moles of solute particles, so the solution boils at 101°C. This is consistent with the definition of a colligative property, that is, one that depends on the concentration of solute particles, be they ions or molecules. One mole of glucose produces one mole of glucose molecules, but one mole of NaCl produces two moles of ions when dissolved in water.

The effect of a solute on the freezing point of a solution compared to that of the pure solvent is even more dramatic, since the temperature changes per mole of solute are larger.

The Freezing Points of Solutions

The freezing point of a solution (the temperature at which the solution freezes) is *lower* than that of the pure solvent alone. The more concentrated a solution the lower will be its freezing point. Pure water freezes at 0°C, but a solution containing one mole of glucose in 1000 g of water freezes at −1.9°C. The freezing point is lowered 1.9°C for each mole of solute particles in 1000 g of water. One mole of NaCl in 1000 g of water produces a solution that freezes at −3.8°C (2 × −1.9°C). Remember, one mole of NaCl produces two moles of ions, so the freezing point is lowered by twice the amount caused by one mole of glucose molecules.

There are several practical uses made of the effect of a solute on the freezing point of water. The engine cooling system in an automobile can be filled with water, but in "freezing" weather it could freeze and damage the engine. To solve this problem, a solution of ethylene glycol ($HOCH_2 - CH_2OH$) in water is used, which remains liquid at temperatures far below the freezing point of water alone. Salt is spread on snow-covered sidewalks and streets during the winter to form solutions that remain liquid at temperatures below the freezing point of water.

Of course, there are other solvents besides water, and their vapor pressures and freezing points are lowered and their boiling points raised as solutes are dissolved in them. One difference, though, is the *amount* of change that occurs in the freezing and boiling points compared to water solutions of the same concentration. For example, the boiling point of benzene, an organic solvent, is raised by 2.5°C and its freezing point is lowered by 5.1°C if one mole of solute is dissolved in 1000 g of benzene. As you can see, these changes are greater than those for water.

Osmosis and Osmotic Pressure

Before you can appreciate the nature of the fourth colligative property, osmotic pressure, you need to learn about the phenomenon of osmosis. Cell membranes, the walls of blood vessels, and the lining of the digestive tract are **semipermeable membranes**, that is, they only allow small molecules and ions to pass through them while barring the passage of larger species. The proper movement of water, nutrients, and other small molecules across these membranes is necessary if life is to exist.

A semipermeable membrane can be thought of as a sheet riddled with tiny holes, much like a sieve. Small molecules can pass through these holes but larger ones cannot. An **osmotic membrane** is one that will only allow water molecules to pass through, and this represents the simplest form of transport across a semipermeable membrane. **Osmosis** is defined as the net movement of water through a semipermeable membrane from a more dilute solution (or pure water) into a more concentrated solution. Notice that osmosis is the net movement of water from one liquid into another through a membrane that separates the two liquids.

Osmosis can be demonstrated by separating an aqueous glucose solution from a quantity of pure water with an osmotic membrane, as shown in Figure 12.8. Water can pass through the membrane in either direction, but it will pass *into* the glucose solution at a faster rate than it can leave. This is because the "concentration" of water molecules is greater in pure water than in the glucose solution. Let us think for a minute about the concentration of water molecules and how this can affect the rate of their movement through the membrane. If a water molecule is to pass through a hole in the membrane, it must collide with that hole as it moves about. On the pure water side, every molecule that strikes a hole is, of course, a water molecule. And the number of molecules that can pass through the membrane each second (the rate of water flow) is proportional

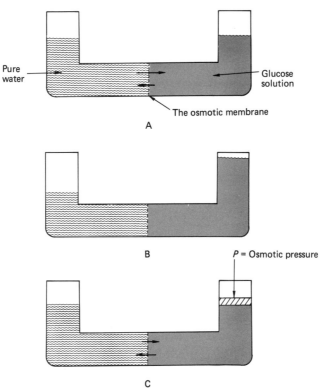

Fig. 12.8. Osmosis is the net flow of water from one liquid to another through an osmotic membrane. In A, a glucose solution is separated from pure water by an osmotic membrane. Water can pass through the membrane in either direction, but it passes into the glucose solution at a faster rate than it flows in the opposite direction as indicated by the arrows. In B, as osmosis takes place, the volume of the glucose solution increases and that of the water decreases. The osmotic flow of water into the solution can be stopped by applying pressure to the glucose solution. In C, the pressure required to bring the osmotic flow to a halt is the osmotic pressure of the glucose solution.

to the number of "water–hole" collisions each second. Now, compare this to the solution side of the membrane. Here only a fraction of the collisions with the membrane holes involve water molecules; the rest involve glucose molecules. As a result, there are fewer opportunities for water molecules to leave this solution each second. The overall result is that water molecules will flow out of the pure water into the solution faster than they can flow in the opposite direction.

The net flow of water across a semipermeable membrane can be stopped by applying pressure on the solution into which the water is flowing. The pressure that is needed to just stop the flow of water is called the **osmotic pressure** of the solution. The amount of pressure required is determined only by the concentration of solute particles (molecules or ions) in the solution. Since only the concentration of solute is important, and not the nature of the solute, osmotic pressure is classed as a colligative property. The osmotic pressure of a 2% glucose solution is twice as great as that for a 1% glucose solution.

As stated in the definition, osmosis will also occur if a dilute solution and a concentrated solution are separated by a semipermeable membrane. The net flow of water will be from the dilute solution (higher water concentration) into the concentrated solution (lower water concentration). The flow will continue until the concentrations of the two solutions become equal. At this point each solution will have the same osmotic pressure.

Osmosis plays an important role in the chemistry of the body since the movement of materials into and out of cells takes place through the semipermeable cell walls.* A cell contains many substances dissolved in an aqueous medium with larger, undissolved species dispersed or suspended in it. The aqueous medium within the cell will therefore exert an osmotic pressure. If it is equal to the osmotic pressure of the extracellular fluid that surrounds the cell, a net flow of water into or out of the cell will not occur. On the other hand, if the osmotic pressures are *not* the same, there will be a flow of water from the solution of lower osmotic pressure *into* the solution of higher osmotic pressure. The language might be confusing here, but when we speak of water flowing from a place of lower osmotic pressure into one of higher osmotic pressure, it is the same as saying water is flowing from a more dilute solution (higher water concentration) into a more concentrated one (lower water concentration). The relationship between the osmotic pressure of the solution in red blood cells and that of their environment has been known for many years, and it provides an excellent example of the importance of osmotic pressure in body fluids.

Red blood cells contain hemoglobin dispersed in an aqueous solution that is enclosed in a semipermeable membrane (the wall of the cell). The osmotic pressure exerted by the contents of the cell is the same as that of a 0.9%(w/v) solution of NaCl. Red cells bathed in a 0.9% NaCl solution (physiological saline) undergo no visible change. There is no net flow of water into or out of the cell because the osmotic pressures are the same. Solutions with the same osmotic pressure are said to be **isotonic**. A 5%(w/v) glucose (also called dextrose) solution is isotonic with red blood cells too. Intravenous solutions are prepared so they are isotonic with red blood cells.

If red blood cells are placed in a solution of lower osmotic pressure, such as a more dilute NaCl solution, the net flow of water into the cells will cause them to rapidly balloon and rupture. The rupturing of red blood cells is called **hemolysis**, an event that will occur if they are placed in a **hypotonic** solution, that is, one with lower osmotic pressure than the solution in the red blood cell.

---

*The walls of cells are actually dializing membranes that allow water and other small species to pass through them. If we focus on just the flow of water, we can think of them as osmotic membranes in this discussion. Dialysis is discussed later on in this chapter.

If a hypotonic solution is administered intravenously, the resulting hemolysis could have serious consequences for the patient.

If red blood cells are placed in a solution of higher osmotic pressure, a **hypertonic** solution, water will flow out of the cells and they will collapse. The collapsing of a red blood cell is called **crenation**. Administration of a hypertonic solution intravenously, unless done in a special way, can have serious effects for a patient. Seriously ill patients who must receive all nourishment intravenously are frequently given hypertonic solutions so the volume of fluid taken into the body is kept down to a level the body can manage, usually about 3 l a day. If the hypertonic solution is slowly injected into a large artery that carries a large volume of blood, it can be quickly diluted by the blood to a level that is isotonic with red blood cells. This is done by releasing the solution into the superior vena cava, a large vessel leading to the heart, through a catheter inserted in the subclavian vein.

The effects of isotonic, hypotonic, and hypertonic solutions on red blood cells is shown in Figure 12.9.

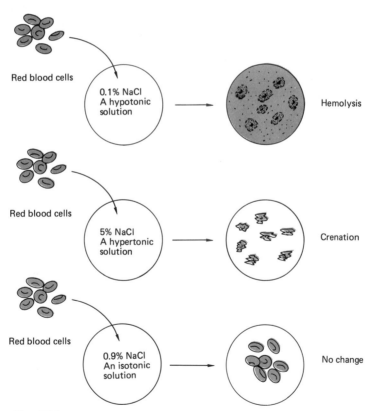

Fig. 12.9. The effects of hypotonic, hypertonic, and isotonic solutions on red blood cells.

Colloids

Up to this time we have dealt only with true solutions, homogeneous mixtures in which the solute is in the form of individual ions or molecules. The solute particles in solutions are very small, on the order of 0.05–0.25 nanometers (1 nm = 10^{-9} m) in diameter. They cannot be seen with either an optical or electron microscope, and they cannot be filtered out, nor do they settle out upon standing.

Colloids are not true solutions, though both share some of the same characteristics. **Colloids** are made up of larger particles ranging in size from approximately 1–100 nm in diameter that are "dispersed" in a solvent. The solvent is called the dispersing medium. A colloid is characterized only by the size of the particles and not by their composition. Colloidal particles are aggregates of several hundred molecules that are so large they cannot pass through semipermeable membranes. A **colloidal dispersion** (do not call them solutions) may or may not be transparent, but like a solution the particles cannot be removed by filtration nor do they settle out of the dispersion upon standing. Colloidal dispersions do exert a modest osmotic pressure. A beam of light passing through a transparent colloidal dispersion will be visible, a phenomenon known as the **Tyndall effect**. The beam is invisible in a true solution, Figure 12.10. There are eight different kinds of colloids, and they are listed in Table 12.4. Notice that the dispersing medium can be either a solid, a liquid, or a gas.

The large biological molecules within cells are of colloidal size, and they are dispersed in aqueous solutions as sols (see Table 12.4). The plasma proteins in blood albumin and the globulins are all of colloidal size and dispersed as sols. Within cells, the proteins are present as sols that are somewhat rigid and semisolid. Except for our bones and teeth, our bodies are largely colloidal in nature.

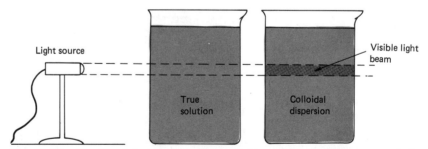

Fig. 12.10. The Tyndall effect. A beam of light passing through a true solution is not visible from the side because the dissolved particles are very small and cannot reflect the light. When the light beam passes through a colloidal dispersion, a visible beam is seen because colloidal particles are large enough to reflect the light. The Tyndall effect can be used to distinguish a solution from a colloid.

TABLE   12.4
Types of Colloids

Dispersing medium	Dispersed substance (colloid particles)	Type of colloid	Examples
gas	liquid	liquid aerosol	fog, aerosol spray
	solid	solid aerosol	smoke
liquid	gas	foam	whipped cream
	liquid	liquid emulsion	mayonnaise, milk
	solid	sol, gel	protoplasm, jellies
solid	gas	solid foam	foam rubber, marshmallows
	liquid	solid emulsion	cheese, butter
	solid	solid sol	pearls, some alloys

If particles that are much larger than those that form colloids, say larger than 100 nm, are dispersed in water, they will be visible to the naked eye and will settle out in time. Mixtures of this type are called **suspensions**. Like colloids, suspensions also show the Tyndall effect, but they are heterogeneous, show no osmotic pressure, and the suspended particles can be separated by simple filtration. You can form a suspension by vigorously mixing fine sand with water, but eventually the sand will settle to the bottom of the container. Red blood cells, white cells, and platelets are present in blood as a suspension. The pumping of the blood by the heart keeps them in suspension. Many liquid antacids are suspensions of either $Al(OH)_3$ or $Mg(OH)_2$ or both in water. They must be shaken well before use. The barium milkshake mentioned earlier is also a suspension. Solutions, colloids, and suspensions are compared in Table 12.5.

TABLE   12.5
Comparing Solutions, Colloids, and Suspensions

	Solutions	Colloids	Suspensions
Particle diameter	0.05–0.25 nm	1–100 nm	100 nm
Visibility of particles	invisible	visible with electron microscope	visible to naked eye
Settling of particles on standing	no	no	yes
Remove particles by filtration	no	no	yes
Pass through dialysis membrane	yes	no	no
Osmotic effect	high	low	none
Tyndall effect	no	yes	yes

Dialysis

Dialysis uses a semipermeable membrane to separate dissolved ions and molecules in a solution from larger colloidal particles that are dispersed in the solution. Dialysis resembles osmosis in many respects, except in dialysis both solute and solvent pass through the **dialyzing membrane**. The net flow of a given molecule or ion is from the solution in which its concentration is greater into the solution that is more dilute in that species. The flow will continue until the concentration of that dissolved species is the same on either side of the membrane.

The artificial kidney machine uses dialysis to remove toxic substances from the blood of people suffering from kidney failure. The purification of blood by dialysis is called **hemodialysis**. The patient's blood is pumped through a long, coiled cellophane tube that is immersed in a specially prepared solution of $NaCl$, $NaHCO_3$, KCl, and glucose, Figure 12.11. The concentration of each substance in the solution, called the dialysate, is adjusted so that normal concentrations of Na^+, K^+, Cl^-, HCO_3^-, and glucose are maintained in the blood as toxic materials are removed. Remember, an ion will not pass out of the blood into the dialysate if the concentration of the ion is the same in both. The condition of the dialysate must be monitored frequently to ensure that the toxic substances do not accumulate to levels that would bring them back into the blood. At this point the dialysate must be replaced with a fresh solution. An individual with severe kidney impairment may require two or three hemodialysis treatments per week, and each treatment may requre from four to six hours.

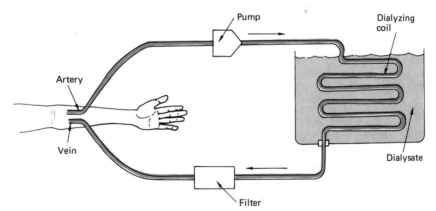

Fig. 12.11. Hemodialysis. The artifical kidney removes wastes from blood by dialysis. The patient's blood is pumped through a long, coiled cellophane dialysis tube that is immersed in a carefully prepared solution called the dialysate. Metabolic wastes pass through the dialysis membrane from the blood into the dialysate. Loss of essential ions from the blood is prevented by keeping the concentration of each one the same in the dialysate as in the blood.

The Purification of Water

Water as it comes from the tap is rarely pure enough for use in medical or chemical laboratories. In most areas of the country the principal contaminants are dissolved salts of calcium, magnesium, and iron which are in solution as Ca^{2+}, Mg^{2+}, Fe^{3+}, HCO_3^-, and SO_4^{2-} ions. Other ions might also be present, but these are the ones that are generally found in what is called "hard" water. The metal ions in hard water react with soap to form an insoluble, gray-white scum that deposits on fabric, porcelain, and skin (Chapter 20). Two methods are commonly used in laboratories to obtain water that is relatively free of these ions. One is distillation, a process based on the fact that ionic impurities are not volatile and will remain behind as water is boiled away. The other is ion exchange, a technique that replaces the unwanted positive and negative ions with hydrogen and hydroxide ions that then combine to form water.

Distillation involves heating a quantity of impure water to boiling and directing the water vapor into a cooled tube (the condenser) where it returns to the liquid and is then collected in a clean vessel. A typical distillation apparatus is shown in Figure 12.12. In every-day language it is called a "still." The dissolved ionic impurities are not volatile (they do not evaporate), and they remain behind as the water boils away. The water vapor is therefore free of these impurities,

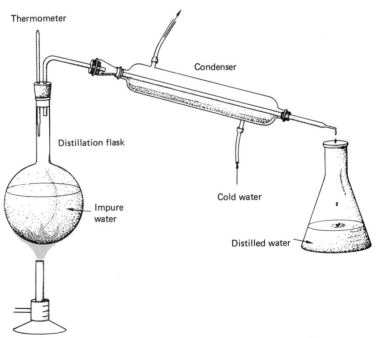

Thermometer

Condenser

Distillation flask

Impure water

Cold water

Distilled water

Fig. 12.12. A typical laboratory distillation apparatus used to prepare distilled water.

and when condensed it will form water that is pure enough for accurate clinical work. Water purified by distillation is called **distilled water**. Distillation can also be used to separate mixtures of two or more volatile liquids based on the differences in their boiling points. The petroleum industry uses distillation to separate crude oil into useful products (Chapter 15).

Very pure water can be obtained by the **ion exchange** method. The process works like this. Impure water is first passed through a column of a synthetic cation exchange resin that removes all the metal cations and replaces them with hydrogen ions, $H^+_{(aq)}$. For each positive charge removed, one is replaced in the form of a hydrogen ion. After the metal ions are removed, the water is then passed through a column of a synthetic anion exchange resin which replaces the dissolved anions with hydroxide ions, $OH^-_{(aq)}$. For each negative charge removed from solution, one hydroxide ion is released to replace it. During the ion exchange process, an equal number of $H^+_{(aq)}$ and $OH^-_{(aq)}$ ions are released and they quickly react to form water:

$$H^+_{(aq)} + OH^-_{(aq)} \rightarrow H_2O$$

The beauty of the ion exchange method is that ionic impurities end up being replaced with water molecules. Water purified in this way is called **deionized water**, or D.I. water for short. It has a very low concentration of ionic impurities, but unlike distilled water, it may not be free of bacteria because the water is never boiled. The apparatus used in ion exchange is relatively simple, and it is shown in Figure 12.13.

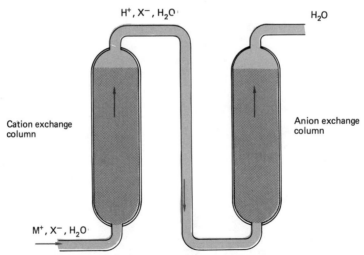

Fig. 12.13. Removal of ionic impurities from water by ion exchange. As water passes through the cation exchange column, metal ions are replaced by hydrogen ions, H^+_{aq}. In the anion exchange column, hydroxide ions, OH^-_{aq}, replace the other anions. The H^+_{aq} and OH^-_{aq} ions react to form water. Water purified in this way is essentially free of ionic impurities.

Water that is to be used for drinking must be treated chemically to destroy disease-causing bacteria. Most often this is done by adding from 0.2 to 1.0 g of chlorine to each 1000 l of water that passes through a water treatment plant. At this level, enough chlorine remains in the water to destroy any bacteria that may enter while it is distributed throughout a city. In recent years ozone, O_3, a very reactive form of oxygen, has been used to disinfect water. It is more expensive than chlorine but it is more effective at destroying microorganisms. Also, it does not impart an objectionable taste to water.

Terms

Several important terms appeared in Chapter 12. You should understand the meaning of each of the following.

intracellular water	molarity
extracellular water	parts per million (ppm)
solution	colligative property
solvent	osmosis
solute	osmotic pressure
aqueous solution	semipermeable membrane
electrolyte	dialysis
nonelectrolyte	hemodialysis
saturated solution	isotonic solution
unsaturated solution	hypotonic solution
solvation	hypertonic solution
hydration	hemolysis
concentrated	crenation
dilute	colloid
weight/volume %	suspension
weight/weight %	Tyndall effect
volume/volume %	distillation
milligram %	ion exchange

Questions

Answers to the starred questions appear at the end of the text.

*1. How does a homogeneous solution differ from a compound?
2. Why is water a good solvent for many ionic compounds?

*3.   What disadvantages are there in expressing the concentration of a solution as being unsaturated, concentrated, or dilute?

4.   Compare the characteristics of solutions and colloids with respect to (a) size of solute particles, (b) passage of solute through a dialyzing membrane, (c) separation of the solute by filtering through filter paper, (d) a beam of light passing through each.

*5.   What solute species (molecules, ions) are present in the following solutions?
     a. ethyl alcohol in water
     b. KNO_3 in water
     c. HCl in water

*6.   How do osmosis and dialysis differ in terms of the permeability of the membrane that separates the two solutions?

7.   Give the concentration of each of the following solutions in terms of weight/volume percent, %(w/v):
     *a. 3.0 g of $NaNO_3$ in 200 ml of solution
     *b. 25 g of KBr in 500 ml of solution
     c. 100 g of NaCl in 1.0 l of solution
     d. 1.5 g of $CaCl_2$ in 0.050 l of solution

*8.   How many milliliters of a 5.0M glucose solution must be used to prepare 250 ml of a 1.5M glucose solution?

9.   How many milliliters of a 3.5%(w/v) solution of NaCl are needed to obtain
     *a. 10 g of NaCl
     b. 5.5 g of NaCl
     c. 0.50 g of NaCl

*10.   Comparing a 5% aqueous glucose solution with pure water, which has the higher
     a. boiling point
     b. vapor pressure
     c. freezing point
     d. osmotic pressure

11.   Describe how you would prepare 500 ml of a 0.20M solution of NaCl in water.

12.   Indicate with a check mark which solution(s)

	3% glucose	5% glucose	2% NaCl
is (are) isotonic with red blood cells			
is (are) hypotonic with red blood cells			
will cause hemolysis of red blood cells			
will cause crenation of red blood cells			

*13.   A 1% NaCl solution and a 2% NaCl solution are separated by an osmotic membrane:
     a. Which solution has the higher osmotic pressure?
     b. Into which solution will water move?

*14.   Which solution in each pair of aqueous solutions will have the *greater* osmotic pressure?
     a. 1M NaCl, 1M glucose
     b. 0.5M NaCl, 1M glucose
     c. 1M glucose, 1M ethyl alcohol

15. Calculate the molarity of the following solutions:
   *a. 10.0 g of NaOH in 2.00 l of solution.
   *b. 80 g of glucose, $C_6H_{12}O_6$, in 900 ml of solution.
   c. 300 g of NaCl in 1000 ml of solution.
   d. 1.0 g of $CaCl_2$ in 1.3 l of solution.
   e. 6.3 g of HNO_3 in 700 ml of solution.

16. How many grams of each solute is needed to prepare
   *a. 1.00 l of a 0.500M solution of KCl?
   *b. 500 ml of a 2.30M solution of NaCl?
   c. 100 ml of a 0.010M solution of glucose, $C_6H_{12}O_6$?
   d. 10.0 l of a 0.85M solution of NaOH?

17. How many grams of solute are in
   *a. 300 ml of a 1.0%(w/v) KCl solution?
   *b. 50 ml of a 4.5%(w/v) NaCl solution?
   c. 1200 ml of a 12.0%(w/v) sucrose solution?
   d. 0.50 l of a 0.90%(w/v) NaCl solution?

18. Calculate the weight/weight percent, %(w/w), concentration of the following solution:
   *a. 10 g of NaCl in 200 g of water.
   b. 0.350 g of KI in 30.0 g of water.
   c. 500 g of glucose in 1000 g of water.

19. What volume of a 1.5M solution could be prepared from 35.0 g of KOH?

*20. A 2.3 l volume of a 0.50M sucrose solution was diluted by addition of 800 ml of water. What is the concentration of the new solution?

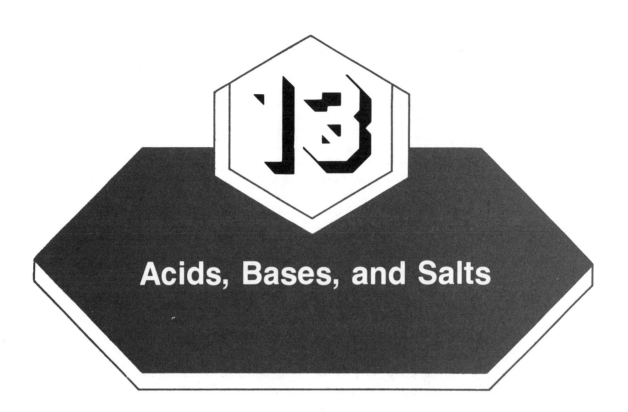

Acids, Bases, and Salts

The terms acid and base have been used for hundreds of years. Originally, "acid" was used to refer to substances with a sour taste. A "base" was a substance capable of destroying an acid, and in the process, forming a new substance called a salt. Chemistry, just as medicine, has improved over the years, and acids and bases are better understood today. They are versatile compounds, and find use in virtually every major industrial process. Moreover, they are of vital importance in the chemistry of the body. Medical technologists and respiratory therapists speak in terms of the acid–base balance of the body. Physicians with ulcer patients are concerned about the gastric acid produced in the stomach. Diabetics are concerned not only with the presence of sugar in the urine, but also whether the urine is acidic or basic. Certainly, if you are interested in health and health care, you must learn about acids and bases, two of the most important classes of compounds.

The chapter begins with descriptions of both acids and bases, their properties, and common examples of each. Neutralization, the reaction of acids with bases, will be described next, to emphasize the similarity in neutralization reactions. You have probably heard the term pH used many times. Since it is commonly used to describe solutions of acids and bases, it too will be described here. Buffers are solutions designed to regulate pH, so we will talk about them and their importance in the body. Another solution concentration term will be introduced: normality. It is frequently used for solutions of acids and bases, and is especially convenient in titration, the analysis of acidic or basic solutions.

Objectives

By the time you finish this chapter, you should be able to do the following:

1. Describe an acid and base in terms of the Arrhenius theory.
2. Describe the hydronium ion, and show how it can be symbolized.
3. List at least three properties of both acids and bases.
4. Describe the difference between a strong acid and a weak acid.
5. Given an acid and base, write the formula and ionic and net-ionic equations for the neutralization reaction.
6. Given the formula of a salt, write the formulas of the acid and base required to make it.
7. Write the equation describing the ionization of water, and define an acidic, basic, and neutral solution in terms of the relative hydroxide and hydronium ion concentrations.
8. Use the ion-product equation for water to relate hydronium ion concentrations to hydroxide ion concentrations and vice versa.
9. Define pH, and describe the pH scale.
10. Given the hydronium ion concentration of a solution, calculate the pH, and vice versa.
11. Define pOH and relate it to the pH of an aqueous solution.
12. Describe how a buffer operates to regulate the pH of a solution.
13. Describe the three major blood buffer systems.
14. Describe acidosis and alkalosis and the way these conditions may come about.
15. Given the formula of an acid or base, calculate its gram equivalent weight.
16. Define normality, and determine the normality of an acid or base solution.
17. Describe the procedure of titration, and show how it can be used to determine acid or base concentrations.

The Arrhenius Theory of Acids and Bases

Why do so many different compounds behave in such similar ways that they can all be classed as acids? And why do other compounds show different kinds of properties in common, which allow them to be classed as bases? What do acids have in common with each other, and what do bases have in common?

Over the past 200 years, several attempts have been made to answer these questions, and theories of acids and bases came and went. In 1887, Svante

Arrhenius, a Swedish chemist, proposed a theory that described acids and bases in terms that are still used today:

An **acid** is any substance that produces hydrogen ions, H^+, when dissolved in water.

Hydrogen chloride gas would be considered an acid, since when dissolved in water it dissociates to form hydrogen and chloride ions. The subscript (aq) indicates the ions are in an aqueous solution.

$$HCl_{(g)} \xrightarrow[\text{water}]{\text{in}} H^+_{(aq)} + Cl^-_{(aq)}$$

A solution of hydrogen chloride in water is called hydrochloric acid. Another name for the hydrogen ion is proton, since it is just the nucleus of a hydrogen atom, a proton.

A **base** is any substance that produces hydroxide ions, OH^-, when dissolved in water.

Sodium hydroxide, NaOH, is a base. When dissolved in water, the ionic solid enters solution as sodium and hydroxide ions.

$$NaOH_{(s)} \xrightarrow[\text{water}]{\text{in}} Na^+_{(aq)} + OH^-_{(aq)}$$

Several years after Arrhenius described an acid in terms of the hydrogen ion in water, it became evident that hydrogen ions did not exist in water as free, independent particles. Rather, they were closely associated with water molecules, forming species that can be symbolized H_3O^+, an ion formed from a hydrogen ion and a water molecule ($H^+ + H_2O = H_3O^+$). The H_3O^+ species is called the **hydronium ion.** More complex ions involving two, three, or four molecules of water with a hydrogen ion undoubtedly also exist, but the symbol H_3O^+ is used as a simple way to indicate the involvement of water with the hydrogen ion. The hydronium ion can also be symbolized as $H^+_{(aq)}$. Though it is perhaps not as descriptive as H_3O^+, it does represent the hydrogen ion as it exists in water. The hydronium ion can be symbolized in the following ways:

$$\left[H\!:\!\overset{\cdot\cdot}{\underset{\cdot\cdot}{O}}\!:\!H \atop H \right]^+ \qquad \left[H\!-\!\overset{}{\underset{|}{O}}\!-\!H \atop H \right]^+ \qquad H_3O^+ \qquad H^+_{(aq)}$$

The fact that hydronium ions exist in water instead of individual hydrogen ions does not make the Arrhenius theory obsolete. We could easily update the definition of an acid as a substance that produces hydronium ions when dissolved in water.

The Properties of Acids

All acids have certain properties in common that are due to the hydronium ions they form when dissolved in water. Solutions of acids have a sour taste, though you should not try to verify this fact in the laboratory. The common acids used in laboratories are sufficiently concentrated to cause chemical burns on skin. Instead, the sour taste of acids can be experienced in lemon and grapefruit juice, which contain citric acid. Acids are also able to change the color of certain organic dyes. One such dye is litmus, a substance obtained from plants. An acid will change the color of blue litmus to red. Small strips of paper treated with litmus (litmus paper) can be used to quickly check to see if a solution is acidic. If a drop of the solution placed on blue litmus causes it to turn red, the solution is acidic.

Acids are very reactive compounds, a property due to the reactivity of the hydronium ion. Acids react with bases to form salts and water. The reaction is called **neutralization** since, as they react, the properties of both the acid and base disappear, and they are neutralized. Hydrochloric acid and sodium hydroxide react to form sodium chloride (a salt) and water:

$$HCl_{(aq)} + NaOH_{(aq)} \rightarrow NaCl_{(aq)} + H_2O$$
acid + **base** $\rightarrow$ **salt** + **water**

Because neutralization reactions are important in acid–base chemistry, they will be discussed later on in more detail.

Acids also react with metal oxides (Na_2O, CaO, etc.) to form salts and water:

$$Na_2O_{(s)} + 2HCl_{(aq)} \rightarrow 2NaCl_{(aq)} + H_2O$$
metal oxide + **acid** $\rightarrow$ **a salt** + **water**

When carbonate or bicarbonate salts are combined with acids, the CO_3^{2-} or HCO_3^- ions join with hydrogen ions to form carbonic acid, H_2CO_3. Carbonic acid is not very stable, and most of it will decompose forming water and carbon dioxide gas:

$$H_2CO_{3(aq)} \rightleftharpoons H_2O + CO_{2(g)}$$

If a solution of sodium carbonate, Na_2CO_3, or sodium bicarbonate, $NaHCO_3$, is treated with hydrochloric acid, bubbles of carbon dioxide quickly form, rise to the top, and escape. The solution "fizzes" as the bubbles pop through the surface.

$$Na_2CO_{3(aq)} + 2HCl_{(aq)} \rightarrow 2NaCl_{(aq)} + H_2O + CO_{2(g)}$$

$$NaHCO_{3(aq)} + HCl_{(aq)} \rightarrow NaCl_{(aq)} + H_2O + CO_{2(g)}$$

The "fizz" that occurs when certain over-the-counter antacids are dropped into water is caused by the reaction of sodium bicarbonate with acidic substances that are also in the tablet. They do not react with each other until they are brought into solution as the tablet begins to dissolve in water.

Acids react with many metals to yield hydrogen gas and a salt of the metal. Zinc will react with hydrochloric acid to produce zinc chloride and hydrogen gas:

$$Zn_{(s)} + 2HCl_{(aq)} \rightarrow ZnCl_{2(aq)} + H_{2(g)}$$

metal + acid   →   **salt**   **+ hydrogen gas**

Some metals like gold, mercury, silver, copper, and platinum do not react with acids to produce hydrogen gas. Others like lead, tin, and nickel react slowly, while zinc, aluminum, and magnesium react quickly. Acids should not be stored in or handled with objects made of these metals. The metals in Group I (Li, Na, K, Rb, and Cs) are so reactive that they produce violent reactions with acids. In fact, these metals can even react with cold water to produce hydrogen:

$$2Na_{(s)} + 2H_2O \rightarrow 2NaOH_{(aq)} + H_{2(g)}$$

These reactions too are violent, and they must be carried out with great care.

Five Common Acids

There are many acids used in chemistry and medicine, but a few of them are more widely used than the others: hydrochloric acid, sulfuric acid, phosphoric acid, nitric acid, and acetic acid.

Hydrochloric acid, $HCl_{(aq)}$, is prepared by dissolving hydrogen chloride gas in water. The polar HCl molecules interact so strongly with the polar water molecules that they dissociate into ions, as shown in Figure 12.5 in the previous chapter:

$$HCl_{(g)} + H_2O \rightarrow H_3O^+ + Cl^-_{(aq)}$$

Hydrochloric acid is a **monoprotic acid,** a term used to describe acids that have only *one* hydrogen in the molecule that is able to ionize in water. This acid can be purchased commercially as 12 molar (12*M*) hydrochloric acid, which is referred to as "concentrated" HCl. The more concentrated solutions of HCl may slowly release fumes of hydrogen chloride gas upon standing in an open container in a warm room. Inhalation of these fumes produces a burning, choking sensation, and prolonged exposure can cause inflammation and ulceration of the respiratory tract.

A dilute solution of HCl, about 0.1*M*, is secreted in the stomach and is an important part of gastric juice. The mucous lining of the stomach protects tissue from the corrosive action of the acid.

Concentrated solutions of hydrochloric acid can cause severe chemical burns on skin and permanent damage or blindness if gotten in the eyes. Dilute solutions of the acid are less hazardous to skin, though they can produce dermatitis. Any acid that is gotten on the skin or in the eyes should be flushed away quickly with a large amount of running water. If a concentrated solution of HCl is ingested, it may cause damage to the mucous membranes, the esophagus, and the stomach. Unless treated promptly, circulatory collapse and even death could occur.

Sulfuric acid, H_2SO_4, is prepared by dissolving sulfur trioxide gas in water to form a liquid that is nearly pure H_2SO_4 (the rest is water) at a concentration of 18*M* H_2SO_4:

$$SO_{3(g)} + H_2O \rightarrow H_2SO_{4(l)}$$

Concentrated sulfuric acid (18*M*) is an oily, colorless liquid that has a high affinity for water. If spilled on the skin it will quickly remove water from tissue causing a painful chemical burn. Ingestion of sulfuric acid can cause the same damage to any tissue it contacts. Even a small amount of concentrated sulfuric acid can be fatal if ingested.

Sulfuric acid is a **diprotic acid,** since each H_2SO_4 molecule has *two* hydrogens that can ionize in solution. Diprotic acids dissociate, or ionize, in two steps. One hydrogen ion comes off in the first step, and the other in the second step. To simplify the equations, $H^+_{(aq)}$ will be used to symbolize the hydronium ion:

$$H_2SO_{4(aq)} \rightarrow H^+_{(aq)} + HSO^-_{4(aq)} \qquad \text{(1st Step)}$$

$$HSO^-_{4(aq)} \rightleftarrows H^+_{(aq)} + SO^{2-}_{4(aq)} \qquad \text{(2nd Step)}$$

Notice that the second step is written as an equilibrium ($\rightleftarrows$). The HSO^-_4 ions produced in the first step are only partially dissociated—that is, only a fraction of them come apart at any one time to form sulfate and hydronium ions.

More sulfuric acid is manufactured in the U.S. each year than any other chemical—billions of pounds per year. Most is used by industry: For example, the liquid in automobile batteries is sulfuric acid.

Phosphoric acid, H_3PO_4, is prepared by dissolving solid P_4O_{10} in water producing a thick, syrupy, colorless liquid that is $15M$ in H_3PO_4:

$$P_4O_{10\,(s)} + 6H_2O \rightarrow 4H_3PO_{4\,(aq)}$$

Phosphoric acid is a **triprotic acid**, dissociating in three steps when dissolved in water. Notice that each step is written as an equilibrium. Phosphoric acid only partially dissociates in solution:

$$H_3PO_{4\,(aq)} \rightleftharpoons H^+_{(aq)} + H_2PO^-_{4(aq)} \qquad \text{(1st Step)}$$
$$H_2PO^-_{4(aq)} \rightleftharpoons H^+_{(aq)} + HPO^{2-}_{4(aq)} \qquad \text{(2nd Step)}$$
$$HPO^{2-}_{4(aq)} \rightleftharpoons H^+_{(aq)} + PO^{3-}_{4(aq)} \qquad \text{(3rd Step)}$$

Phosphoric acid is used to make fertilizers, detergents, dental cements, and, because of its pleasant acid taste, it is used in many carbonated soft drinks. Concentrated solutions of phosphoric acid can irritate skin and the mucous membranes. These solutions are not as hazardous to tissue as sulfuric acid.

Phosphoric acid derivatives are involved in many vital processes in the chemistry of the body, as you will see when you study metabolism.

Nitric acid, HNO_3, is produced by dissolving nitrogen dioxide gas in water. Nitrogen oxide, NO, is also produced in the reaction, but it can be easily removed, leaving a concentrated acid solution that is $16M$ in HNO_3:

$$3NO_{2(g)} + H_2O \rightarrow 2HNO_{3\,(aq)} + NO_{(g)}$$

Concentrated solutions of nitric acid give off choking fumes that can cause chronic bronchitis and damage to lung tissue if inhaled over a prolonged period. Nitric acid is a *monoprotic acid*:

$$HNO_{3\,(aq)} \rightarrow H^+_{(aq)} + NO^-_{3(aq)}$$

Nitric acid will cause yellow stains on the skin by reacting with proteins in the skin. Nitric acid can cause severe burns to skin, eyes, and internal organs.

Acetic acid, $HC_2H_3O_2$ or CH_3COOH, is an organic acid with a pungent, vinegarlike odor. It may be prepared from ethyl alcohol or other suitable organic compound. It can be purchased as the pure acid (no water present), which is $17M$ in $HC_2H_3O_2$. Another name for the pure acid is glacial acetic acid, since it freezes at 17°C to form a glacierlike mass of solid acid.

Acetic acid is a *monoprotic* acid. Though there are four hydrogen atoms in the acid molecule, only one, the hydrogen bonded to oxygen, ionizes in water. The other three are strongly held by carbon:

Only this hydrogen can ionize in solution

Because only one of the four hydrogen atoms is able to ionize, the formula of acetic acid can be written $HC_2H_3O_2$, setting one hydrogen off from the others. Writing the formula as CH_3COOH also sets one hydrogen off, but additionally, the —COOH part of the formula shows that the compound is an organic acid. You will learn why when you study organic chemistry. The dissociation of acetic acid in water can be written using either type of formula:

$$HC_2H_3O_{2(aq)} \rightleftharpoons H^+_{(aq)} + C_2H_3O^-_{2(aq)}$$
$$CH_3COOH_{(aq)} \rightleftharpoons H^+_{(aq)} + CH_3COO^-_{(aq)}$$

The dissociation of acetic acid is also written as an equilibrium, since only a small fraction of the molecules are dissociated into ions at any one time.

Dilute solutions of acetic acid can be ingested without problem; in fact, vinegar is a 4%–5% solution of acetic acid in water. The sour taste of vinegar accents food in ways many people enjoy. Ingestion of concentrated solutions of acetic acid can be hazardous though, damaging tissue in the mouth and gastrointestinal tract. Circulatory collapse, uremia, and death can result in extreme cases.

Strong and Weak Acids

An acid can be classified as being either a strong acid or a weak acid depending on the degree to which it is dissociated in solution. A **strong acid** is one that is completely or nearly completely dissociated into ions in solution. Both hydrochloric acid and nitric acid are strong acids since they exist in solution almost totally as ions. Sulfuric acid is also a strong acid. The first step of the dissociation of H_2SO_4 is essentially complete, forming a solution of $H^+_{(aq)}$ and $HSO^-_{4(aq)}$ ions. The second step of the dissociation is not nearly as extensive as the first. Less than 10% of the HSO^-_4 ions are dissociated at any one time.

A **weak acid** is one that dissociates only to a small degree in solution. It exists mostly as acid molecules. The most common weak acid is acetic acid, $HC_2H_3O_2$. In a $0.10M$ solution of acetic acid, only 13 of every 1000 acid molecules are dissociated into hydronium and acetate ions. The remaining 987 acid molecules hold together and are in equilibrium with the ions. Remember, an equilibrium is a condition in which two opposing changes occur at the same rate. In the case of weak acids, molecules of the acid are constantly dissociating to form ions. But at the same time, and at the same rate, the ions are recombining to form acid molecules. Every time an acid molecule dissociates, one is formed. Weak acid equilibria are important in the chemistry of the blood as you will see later on.

Phosphoric acid is a moderately weak acid. When dissolved in water, about 24% (24 out of 100) of the H_3PO_4 molecules dissociate to form $H_{(aq)}^+$ and $H_2PO_4^-$ ions. Less than 0.1% of the $H_2PO_4^-$ ions dissociate. As you can see, nearly all of the hydronium ion in solution is produced by the first step of the dissociation. The second and third steps make a negligible contribution.

One more thing must be said about strong and weak acids. The terms "strong" and "weak" refer to the extent to which the acids dissociate in solution and *not* the concentration of the acid. A $0.01 M$ solution of hydrochloric acid is a dilute solution of a strong acid, and a $10 M$ solution of acetic acid is a concentrated solution of a weak acid.

The strengths of several acids are given in Table 13.1.

The Properties of Bases

Bases have many properties in common since they all form hydroxide ions when dissolved in water. A dilute solution of a base feels slippery on the skin, like soapy water. The base actually dissolves a small amount of tissue to form a slippery solution. Basic solutions have a bitter taste. They can also change the color of some organic dyes. A solution of a base will change the color of litmus from red to blue, just the opposite color change caused by acids. Acids and bases do act contrary to each other in many instances, and the effect on the color of litmus is just one example of this. The way acids and bases change the color of litmus can be remembered this way: a *B*ase turns litmus to *B*lue, and an acid does just the opposite.

TABLE   13.1
Common Strong and Weak Acids

Acid	Formula	Classification
Hydrochloric acid	HCl	strong
Nitric acid	HNO_3	strong
Sulfuric acid	H_2SO_4	strong
Phosphoric acid	H_3PO_4	moderate
Acetic acid	$HC_2H_3O_2$	weak
Carbonic acid	H_2CO_3	weak
Boric acid	H_3BO_3	weak

Fats and oils can be broken down to form soap and glycerine by heating them in aqueous solutions of sodium or potassium hydroxide. This important reaction is called **saponification**, and it is described in Chapter 20. Solutions of these bases are also able to react with aluminum and zinc metal, dissolving the metal and forming hydrogen gas. Obviously these bases should not be placed in containers made of these metals. The most important chemical property of bases is that they react with acids to form salts and water. The fact that acids and bases neutralize each other again points out the contrary or opposite nature of each kind of substance.

Some Common Bases—Strong and Weak

Two of the most common bases are sodium hydroxide (lye), NaOH, and potassium hydroxide, KOH. They are solid, ionic compounds that are very soluble in water. As they dissolve, the metal ions and hydroxide ions in the crystal separate and enter the solution:

$$NaOH_{(s)} \xrightarrow[\text{water}]{\text{in}} Na^+_{(aq)} + OH^-_{(aq)}$$

$$KOH_{(s)} \xrightarrow[\text{water}]{\text{in}} K^+_{(aq)} + OH^-_{(aq)}$$

Solutions containing very high concentrations of hydroxide ion, around $20M$, can be prepared from these compounds. Because of the high hydroxide ion concentration, these solutions are corrosive to all tissue, and if ingested they can cause violent pain in the throat and stomach, vomiting, and gastric bleeding. If not immediately fatal, severe scarring of the esophagus can occur that may constrict the passage of fluids and solid foods. Many common household drain cleaners contain sodium hydroxide, and care must be taken in their use and storage. A small child could be severely injured by swallowing only a teaspoonful of any of them.

Calcium hydroxide, $Ca(OH)_2$, and magnesium hydroxide, $Mg(OH)_2$, are also common bases, but neither is very soluble in water. They cannot be used to prepare solutions that have high hydroxide ion concentrations. A saturated solution of $Ca(OH)_2$ is about $0.04M$ in hydroxide ion, and one of $Mg (OH)_2$ is only $0.0003M$ in hydroxide ion. Basic solutions this dilute are not hazardous to tissue. For this reason, suspensions of magnesium hydroxide in water, called "milk of magnesia," are used internally as both an antacid and a laxative.

Calcium hydroxide suspensions and solutions are used as astringents to reduce discharges from tissue. A calcium hydroxide solution is also called "limewater."

Another term that has been used to describe these compounds is **alkali**, and solutions of these bases are **alkaline** solutions. Both of these terms have been used for many years and you should be familiar with them, though the terms "base" and "basic solution" will be used most often in this text.

Sodium hydroxide, potassium hydroxide, and calcium and magnesium hydroxide are all considered **strong bases**, since they are completely dissociated in solution into ions. Even though $Ca(OH)_2$ and $Mg(OH)_2$ form only dilute solutions because of their low solubility in water, that which is in solution is completely dissociated.

Another basic solution that is often encountered is one of ammonia in water, called aqueous ammonia. Ammonia, NH_3, is a gas at room temperature and it is very soluble in water. In solution, it reacts with water to a slight extent to form ammonium ions and hydroxide ions:

$$NH_{3(aq)} + H_2O \rightleftharpoons NH_{4(aq)}^+ + OH_{(aq)}^-$$

Most of the ammonia is present in solution as ammonia molecules. Only about 13 of every 1000 molecules will have reacted with water at any one time to form the ions. The equation is written as an equilibrium, since the forward reaction and the reverse reaction are taking place constantly at the same rate. Aqueous ammonia is a **weak base**, since only a small fraction of the dissolved ammonia molecules are involved in the formation of ions at any time.

The odor of ammonia is well known to most people. Concentrated aqueous ammonia, which is obtained commercially as a $15M$ solution, produces a strong, suffocating ammonia odor when exposed to the air. Inhalation of a small amount of ammonia over a brief period of time does not represent a hazard, but prolonged inhalation of the concentrated fumes can cause edema of the respiratory tract as plasma fluid enters and fills the alveoli and intervening spaces. Ammonia vapors are used as a stimulant (smelling salts) to revive persons who have fainted.

Aqueous ammonia is sometimes called ammonium hydroxide, a name that implies the existence of a compound with the formula NH_4OH. Such a compound has never been isolated though, so it is really more correct to use the name aqueous ammonia.

Neutralization Reactions

The most important chemical property of both acids and bases is their ability to react with each other. The reaction is called **neutralization,** and the products of a neutralization reaction are a salt (an ionic compound) and water (a covalent compound). The characteristic properties of the acid and the base are lost in the reaction, and for that reason we say they have neutralized each other. Some observations that would verify the loss of acidic and basic properties are shown in Figure 13.1.

Earlier, the reaction of hydrochloric acid with sodium hydroxide was used as an example of neutralization. Just as we had done in the past, the equation was written using the complete formulas of each reactant and product. Equations using complete formulas for all substances are called **formula equations**:

$$HCl_{(aq)} + NaOH_{(aq)} \rightarrow NaCl_{(aq)} + H_2O$$

A formula equation

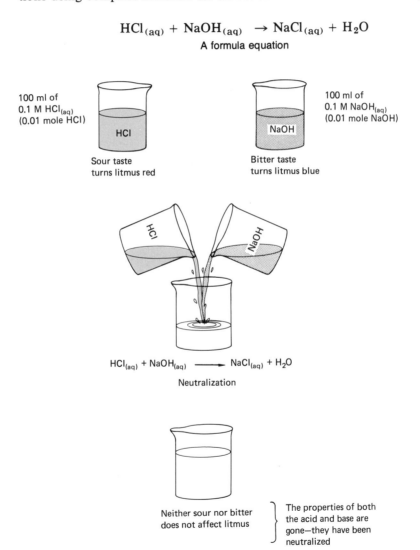

100 ml of
0.1 M HCl$_{(aq)}$
(0.01 mole HCl)

HCl

Sour taste
turns litmus red

100 ml of
0.1 M NaOH$_{(aq)}$
(0.01 mole NaOH)

NaOH

Bitter taste
turns litmus blue

$$HCl_{(aq)} + NaOH_{(aq)} \longrightarrow NaCl_{(aq)} + H_2O$$

Neutralization

Neither sour nor bitter
does not affect litmus

The properties of both
the acid and base are
gone—they have been
neutralized

Fig. 13.1. Acid–base neutralization. The properties of the acid and the base disappear when they react.

Formula equations are very useful, but in many cases they do not describe the true condition of each substance as it exists in solution. The acid, the base, and the salt are present in solution as ions—they are completely dissociated. This fact can be incorporated into the neutralization equation by rewriting the formula equation showing these compounds as pairs of ions. An equation that represents each dissociated compound as ions in solution is called an **ionic equation.** For simplicity, the hydronium ion is again symbolized as $H^+_{(aq)}$:

$$\overbrace{H^+_{(aq)} + Cl^-_{(aq)}}^{HCl_{(aq)}} + \overbrace{Na^+_{(aq)} + OH^-_{(aq)}}^{NaOH_{(aq)}} \rightarrow \overbrace{Na^+_{(aq)} + Cl^-_{(aq)}}^{NaCl_{(aq)}} + H_2O$$

An ionic equation

Notice that $Na^+_{(aq)}$ and $Cl^-_{(aq)}$ appear on both sides of the ionic equation. This means they exist as ions in solution both before *and* after the reaction. They are not directly involved in the neutralization. The ionic equation can be simplified by canceling them out of the ionic equation:

$$H^+_{(aq)} + \cancel{Cl^-_{(aq)}} + \cancel{Na^+_{(aq)}} + OH^-_{(aq)} \rightarrow \cancel{Na^+_{(aq)}} + \cancel{Cl^-_{(aq)}} + H_2O$$

By doing so, we are left with the "heart" of the neutralization reaction, and this equation is called a **net-ionic equation:**

$$H^+_{(aq)} + OH^-_{(aq)} \rightarrow H_2O$$

A net-ionic equation

Notice what the net-ionic equation is telling us about neutralization. The reaction between an acid and a base is really $H^+_{(aq)}$ combining with $OH^-_{(aq)}$ to form water. The identical reaction takes place no matter what acid or base is used. The sodium and chloride ions are called **spectator ions.** They act as spectators in a sense, watching the action from a distance. If the reaction solution is evaporated to dryness, solid, crystalline NaCl can be obtained, showing that it is a product of the reaction.

Formula, ionic, and net-ionic equations are given for two other neutralization reactions below.

Formula: $HNO_{3(aq)} + KOH_{(aq)} \rightarrow KNO_{3(aq)} + H_2O$
Ionic: $H^+_{(aq)} + NO^-_{3(aq)} + K^+_{(aq)} OH^-_{(aq)} \rightarrow K^+_{(aq)} + NO^-_{3(aq)} + H_2O$
Net-ionic: $H^+_{(aq)} + OH^-_{(aq)} \rightarrow H_2O$

Formula: $2HCl_{(aq)} + Ca(OH)_{2(aq)} \rightarrow CaCl_{2(aq)} + 2H_2O$
Ionic: $2H^+_{(aq)} + 2Cl^-_{(aq)} + Ca^{2+}_{(aq)} + 2OH^-_{(aq)} \rightarrow Ca^{2+}_{(aq)} + 2Cl^-_{(aq)} + 2H_2O$
Net-ionic: $2H^+_{(aq)} + 2OH^-_{(aq)} \rightarrow 2H_2O$

One medical application of neutralization is the use of antacids to remove excess hydrochloric acid from the stomach. Emotional stress or unusual eating habits can cause the secretion of a larger volume of acid in the stomach than is needed for digestion. This can be quite uncomfortable for some people. Antacids are made up of compounds that react with hydrochloric acid. *Maalox* contains magnesium hydroxide, $Mg(OH)_2$, and aluminum hydroxide, $Al(OH)_3$, in an aqueous suspension or in tablets. Both compounds can react with stomach acid, but because of their low solubility in water, neither will cause the stomach contents to become basic. Another common antacid, *Amphojel,* uses only $Al(OH)_3$. Calcium carbonate, $CaCO_3$, and magnesium carbonate, $MgCO_3$, appear in other antacid preparations. You may recall that carbonates react with acid to produce carbon dioxide and water.

Check Test Number 1

Write the formula, ionic, and net-ionic equations for the reaction of sodium hydroxide and nitric acid. The acid, base, and salt are completely dissociated in water.

Answer:

Formula: $HNO_{3(aq)} + NaOH_{(aq)} \rightarrow NaNO_{3(aq)} + H_2O$

Ionic: $H^+_{(aq)} + NO^-_{3(aq)} + Na^+_{(aq)} + OH^-_{(aq)} \rightarrow Na^+_{(aq)} + NO^-_{3(aq)} + H_2O$

Net-ionic: $H^+_{(aq)} + OH^-_{(aq)} \rightarrow H_2O$

Salts

Neutralization reactions are said to produce a salt and water. Perhaps you have wondered about the term "salt," since it does not seem to represent any particular compound (except in common usage to mean NaCl). **Salt** is a collective term used to describe the ionic compound *formed* in an acid–base neutralization reaction. A salt consists of a positive ion (or ions) that comes from the base and a negative ion (or ions) that comes from the acid. You will recall that these are the spectator ions in a neutralization reaction. Sodium chloride is considered a salt since it will be formed if NaOH and HCl react:

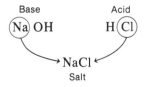

Likewise, potassium nitrate, KNO_3, and calcium chloride, $CaCl_2$, are classed as salts:

Diprotic and triprotic acids can form salts that may not have all the ionizable hydrogens replaced by metal ions. Depending on the amount of sodium hydroxide that reacted with sulfuric acid, two salts are possible:

$$H_2SO_{4(aq)} + NaOH_{(aq)} \rightarrow NaHSO_{4(aq)} + H_2O$$
$$H_2SO_{4(aq)} + 2NaOH_{(aq)} \rightarrow Na_2SO_{4(aq)} + 2H_2O$$

Phosphoric acid can form three different sodium salts, NaH_2PO_4, Na_2HPO_4, and Na_3PO_4. It is not difficult to figure out what acid and what base must be reacted to produce a particular salt. Just remember the positive ion comes from the base and the negative ion comes from the acid. Calcium sulfate, $CaSO_4$, would be formed if $Ca(OH)_2$ and H_2SO_4 react.

Of course, salts can be made in other ways too, not just in acid–base neutralization reactions. Three other ways salts can be made are

$$acid + metal \rightarrow salt + H_2$$
$$acid + metal\ carbonate \rightarrow salt + H_2O + CO_2$$
$$acid + metal\ bicarbonate \rightarrow salt + H_2O + CO_2$$

The way salts are named was presented in Chapter 8 in the discussion of ionic compounds. You may wish to review how these compounds are named.

Check Test Number 2:

Give the formulas of the acid and base that would produce each of the following salts in a neutralization reaction.

a. Na_2SO_4 d. $Ca(NO_3)_2$
b. K_3PO_4 e. $Mg_3(PO_4)_2$
c. $NaC_2H_3O_2$ f. $(NH_4)_2SO_4$

Answers:

a. $NaOH + H_2SO_4$ d. $Ca(OH)_2 + HNO_3$
b. $KOH + H_3PO_4$ e. $Mg(OH)_2 + H_3PO_4$
c. $NaOH + HC_2H_3O_2$ f. $NH_{3(aq)} + H_2SO_4$

The Ionization of Water

Perhaps you have already recognized how important water has been in the way acids and bases are defined. Acids and bases were described as compounds that formed either one or the other part of a water molecule, namely, $H^+_{(aq)}$ or $OH^-_{(aq)}$, when dissolved in water. We know these ions can be thought of as parts of a water molecule because they combine to form water in neutralization reactions. Water will play still another role in our discussion of acids and bases, particularly in the way we describe a solution as being either acidic or basic.

It may surprise you at first, but a sample of pure water is not entirely made up of water molecules. A very few of them, about one in 500 million, are dissociated into hydronium ions, and hydroxide ions. The ions exist in equilibrium with the undissociated water molecules:

$$H_2O \rightleftharpoons H^+_{(aq)} + OH^-_{(aq)}$$

You might be puzzled to see an equation that shows water molecules coming apart to form $H^+_{(aq)}$ and $OH^-_{(aq)}$ ions, since you just learned that the opposite reaction takes place in a neutralization reaction. Actually, there is not an inconsistency here. Think of it this way. An acid and base will react to produce water, but the water that is produced (along with that in the solution) is a compound that is slightly dissociated, as shown by the equation above.

In pure water at room temperature, the concentration of hydronium ion is $0.0000001 M$, a small but significant value. Numbers this small are more conveniently written in exponential form, so the hydronium ion concentration would be $1 \times 10^{-7} M$. Since one hydroxide ion is formed for each hydronium ion in pure water, the hydroxide ion concentration is also $1 \times 10^{-7} M$. Neither ion has a greater concentration than the other, so pure water is neither acidic nor basic—it is neutral.

A **neutral solution** can then be defined as one in which the concentrations of $H^+_{(aq)}$ and $OH^-_{(aq)}$ are equal. Enclosing these symbols in square brackets, [ ], forms an expression that means "molar concentration" of the species. A neutral solution, then, is one in which

$$[H^+_{(aq)}] = [OH^-_{(aq)}]$$
Neutral solution

This is read as: The molar concentration of $H^+_{(aq)}$ is equal to the molar concentration of $OH^-_{(aq)}$.

In any aqueous solution, the concentrations of both ions are related in a see-saw fashion. If the concentration of $H^+_{(aq)}$ goes up, the concentration of $OH^-_{(aq)}$ goes down, and vice versa. This inverse relationship between the two concentrations can be expressed in a single equation.

$$[H^+_{(aq)}]\,[OH^-_{(aq)}] = K_w \text{ (a constant for water)}$$

The value of the constant can be determined by substituting the concentrations of the two ions in the neutral solution into the equation and multiplying:

$$[H^+_{(aq)}]\,[OH^-_{(aq)}] = (1 \times 10^{-7})\,(1 \times 10^{-7}) = 1 \times 10^{-14}$$

$$K_w = 1 \times 10^{-14}$$

The result is called the **ion-product equation** for water, and it can be used for any solution in which water is the solvent. K_w is called the **ion-product constant:**

$$[H^+_{(aq)}]\,[OH^-_{(aq)}] = 1 \times 10^{-14} = K_W$$

If a small amount of acid is added to a sample of pure water, the solution becomes acidic, and the $H^+_{(aq)}$ concentration increases to a value *larger* than $1 \times 10^{-7}M$. As the $H^+_{(aq)}$ concentration goes up, the $OH^-_{(aq)}$ concentration decreases until the product of the two concentrations again equals 1×10^{-14}, the ion-product constant. An **acidic solution** can be defined as one in which the hydronium ion concentration is greater than the hydroxide ion concentration:

$$[H^+_{(aq)}] > [OH^-_{(aq)}]$$
Acidic solution

A small amount of base added to pure water will increase the hydroxide ion concentration while decreasing the hydronium ion concentration. The product of the two ion concentrations will still be 1×10^{-14}. In a **basic solution,** the hydroxide ion concentration is greater than the hydronium ion concentration.

$$[OH^-_{(aq)}] > [H^+_{(aq)}]$$
Basic solution

The ion-product equation can be used to determine the concentration of the hydronium ion if the concentration of the hydroxide is known and vice versa. This is done in the following examples. You may wish to consult the Appendix to review how exponential numbers are used in calculations.

Example 13.1

The hydronium ion concentration in a sample of black coffee is $1 \times 10^{-5}M$. What is the $[OH^-_{(aq)}]$?
First, let us rearrange the ion product equation to solve for $[OH^-_{(aq)}]$:

$$[OH^-_{(aq)}] = \frac{1 \times 10^{-14}}{[H^+_{(aq)}]}$$

You are given that $[H^+_{(aq)}] = 1 \times 10^{-5}$.

$$[OH^-_{(aq)}] = \frac{1 \times 10^{-14}}{1 \times 10^{-5}}$$

$$[OH^-_{(aq)}] = 1 \times 10^{-9}$$

Example 13.2

The hydroxide ion concentration in household ammonia is $0.001M$. What is the $[H^+_{(aq)}]$?

$$[H^+_{(aq)}] = \frac{1 \times 10^{-14}}{[OH^-_{(aq)}]}$$

For $[OH^-_{(aq)}]$, $0.001M$ is $1 \times 10^{-3}M$.

$$[H^+_{(aq)}] = \frac{1 \times 10^{-14}}{1 \times 10^{-3}}$$

$$[H^+_{(aq)}] = 1 \times 10^{-11}$$

Check Test Number 3

Would the following concentrations represent an acidic, basic, or neutral solution?

a. $[H^+_{(aq)}] = 1 \times 10^{-9}$

b. $[OH^-_{(aq)}] = 1 \times 10^{-3}$

c. $[OH^-_{(aq)}] = 1 \times 10^{-7}$

d. $[H^+_{(aq)}] = 1 \times 10^{-2}$

Answers:

		$[OH^-_{(aq)}]$		$[H^+_{(aq)}]$
a.	basic	1×10^{-5}	$>$	1×10^{-9}
b.	basic	1×10^{-3}	$>$	1×10^{-11}
c.	neutral	1×10^{-7}	$=$	1×10^{-7}
d.	acidic	1×10^{-12}	$<$	1×10^{-2}

pH—Another Way to Express $H^+_{(aq)}$ Concentration

When dealing with solutions that have very low hydronium ion concentrations, such as blood, urine, intravenous solutions, and those used in studies of plant and animal systems, it becomes awkward to express units of concentration in terms of molarity. If you had a solution that was $0.00001M$ in $H^+_{(aq)}$, it would be cumbersome to say to someone that the solution is "one one hundred thousandth molar" or "one times ten to the minus five molar" ($1 \times 10^{-5}M$) in hydronium ion. The natural instinct, then, is to develop a simpler way of expressing hydronium ion concentration. This was done in 1909, by a Danish biochemist, S. P. L. Sørensen, who proposed the use of the pH unit. It is related to the hydronium ion concentration by the following equation:

$$pH = -\log [H^+_{(aq)}]$$

There is one term in this equation that you may not be familiar with, log, which is an abbreviation for logarithm.

The logarithm of a number is equal to the exponent or power to which 10 must be raised so that it equals that number. This may be easier to understand if we use an example. Suppose you want the log of 100. The number 100 can be expressed as 1×10^2 or just 10^2 in exponential notation:

$$100 = 1 \times 10^2 = 10^2$$

Notice, 10 is raised to the power of 2, that is, the exponent that must be used on 10 so it equals 100 is 2. The log of 100 is then 2, the value of the exponent:

$$100 = 10^2 \leftarrow \text{The exponent}$$

$$\log 100 = \log 10^2 = 2$$

The log of any number that can be expressed in the form 1×10^n is equal to the value of n (including the $+$ or $-$ sign of n). The log of 0.01 would equal -2:

$$\log 0.01 = \log 10^{-2} = -2$$

The logarithms of several numbers are given below.

Number	Exponential form	Log	Number	Exponential form	Log
10	10^1	1	0.1	10^{-1}	-1
1,000	10^3	3	0.001	10^{-3}	-3
1,000,000	10^6	6	0.000001	10^{-6}	-6
1,000,000,000	10^9	9	0.000000001	10^{-9}	-9

Now, let us get back to pH. To determine the pH of a solution of known hydronium ion concentration, do the following:

1. Write the molar concentration of the hydronium ion in exponential form, 10^n.
2. The log of $[H^+_{(aq)}]$ is equal to n.
3. Then multiply n by -1, to obtain the pH of the solution.

Let us do some example problems dealing with pH.

Example 13.3

What is the pH of the solution described earlier in this section that was $0.00001M$ in hydronium ion?

$$[H^+_{(aq)}] = 1 \times 10^{-5} = 10^{-5}$$

$$pH = -\log [H^+_{(aq)}]$$

$$pH = -\log [10^{-5}]$$

$$= -(-5)$$

$$pH = 5$$

It is certainly easier to simply say the pH of the solution is 5 than saying "one times ten to the minus five molar in hydronium ion."

Example 13.4

What is the pH of pure water, which has a hydronium ion concentration of $1 \times 10^{-7}M$?

$$[H^+_{(aq)}] = 10^{-7}$$

$$pH = -\log [10^{-7}]$$

$$= -(-7)$$

$$pH = 7$$

Neutral water has a pH of 7.

Example 13.5

The hydroxide ion concentration of household ammonia, $NH_{3(aq)}$, is $1 \times 10^{-3}M$. What is the pH of this solution?

Since pH is related to $[H^+_{(aq)}]$, we first must convert the hydroxide ion concentration to hydronium ion concentration. This is done with the ion-product expression:

$$[H^+_{(aq)}] = \frac{1 \times 10^{-14}}{[OH^-_{(aq)}]}$$

$$= \frac{1 \times 10^{-14}}{1 \times 10^{-3}}$$

$$[H^+_{(aq)}] = 1 \times 10^{-11} = 10^{-11}$$

$$pH = -\log [10^{-11}]$$

$$= -(-11)$$

$$pH = 11$$

In the previous examples, pH values of 5, 7, and 11 were obtained for acidic, neutral, and basic solutions, respectively. A pH of 7 represents a neutral solution. Any pH value *less than 7* represents an *acidic solution*. As the pH values get smaller the solutions become increasingly acidic, that is, they have increasingly larger hydronium ion concentrations. Any pH value *greater than 7* represents a *basic solution*. As the pH values get larger, the solutions become increasingly basic or alkaline. Notice how the pH values get smaller as the hydronium ion concentrations gets larger. Concentrated acid solutions have low pH's while concentrated base solutions have large pH's. The **pH scale** for aqueous solutions is shown in Figure 13.2, along with the hydronium and hydroxide ion concentrations that correspond to each number from 1 to 14. A difference of one pH unit anywhere along the pH scale represents a 10-fold

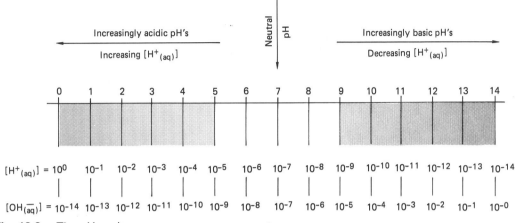

Fig. 13.2. The pH scale.

change in hydronium ion concentration. A difference of two pH units is a 100-fold (10×10-fold) change in hydronium ion concentration. The pH of several common solutions appears in Table 13.2.

Several pH values in Table 13.2 are not whole numbers. For example, the pH of lemon juice is given as 2.3, a number between 2 and 3. What must the hydronium ion concentration be if the pH is 2.3? It is not difficult to determine the approximate hydronium ion concentration knowing that the pH is between 2 and 3. A pH of 2 corresponds to a concentration of $10^{-2}M$, and a value of 3 corresponds to a concentration of $10^{-3}M$ for the hydronium ion. So in lemon juice the hydronium ion concentration is between 10^{-2} and $10^{-3}M$. The pH of milk is given as 6.4, a number between 6 and 7. The hydronium ion concentration of milk is then between 10^{-6} and $10^{-7}M$.

On the other hand, if you desire to state the approximate pH of a solution that is $5 \times 10^{-3}M$ in hydronium ion, a concentration between 1×10^{-2} and $1 \times 10^{-3}M$, the pH of the solution would be between 2 and 3.

TABLE 13.2			
The pH of Several Common Solutions			
Solution	pH	Solution	pH
0.1M HCl	1.0	Saliva	6.2–7.4
Gastric juice	1.4	Pure water	7.0
Lemon juice	2.3	Blood	7.35–7.45
Carbonated beverage	3.0–4.0	Bile	7.8–8.6
Black coffee	5.0	Milk of magnesia	10.5
Milk	6.4	Household ammonia	11
Urine	5.5–7.0	0.1M NaOH	13

We are not going to delve into logarithms to the extent required to state pH values in decimal numbers. Often it is sufficient to know just the range of hydronium concentration corresponding to a decimal pH, or to give the pH range corresponding to a hydronium concentration that cannot be expressed simply as $1 \times 10^n M$.

Check Test Number 4

1. Give the pH of each of the following solutions:
   a. Orange juice with $[H^+_{(aq)}]$ of $1 \times 10^{-4} M$.
   b. A household cleaner with $[H^+_{(aq)}]$ of $1 \times 10^{-10} M$.
   c. A soap solution with $[OH^-_{(aq)}]$ of $1 \times 10^{-6} M$.
2. What is the range of hydronium ion concentration of a solution with a pH of 8.3?

Answers:
   1. (a) pH = 4 (b) pH = 10; (c) $[H^+_{(aq)}] = 10^{-8}$, pH = 8.
   2. Between 10^{-8} and $10^{-9} M$ corresponding to the pH range from 8 to 9.

Measuring pH

In actual practice, the pH of a solution is measured directly much more often than it is calculated from the hydronium ion concentration. A common part of a urinalysis is to measure the pH of the specimen. Two methods are commonly employed to measure pH. The first uses **indicators**, substances (dyes) that change from one color to another in response to the hydronium ion concentration (or pH) of a solution. Litmus is an indicator, and you may recall it is red in acidic solutions, where the hydronium ion concentration is high, and blue in basic solutions where it is low. The color of litmus changes from red to blue over a pH range of from about 5 to 8. There are hundreds of indicators known, and each one produces a particular color change (red to yellow, blue to orange, etc.) over a particular narrow pH change. The range of change may be at low, intermediate, or high pH values.

The most convenient use of indicators as a means of measuring pH is **pH paper**, a name that is commonly used to describe strips of paper treated with various combinations of indicators. A drop of solution of unknown pH touched to a strip of pH paper will cause a color change. One kind of pH paper turns blue if the pH of the solution is around 11–12 (strongly basic). It turns green at a pH around 8–9 (mildly basic), yellow-orange at a pH of 5–6 (mildly acidic), and red at 1–2 (strongly acidic). Indicator paper, as it can also be called, is easy to use, and it is not very expensive. The paper just described measures approximate pH's over a wide range, but other kinds are available that measure very narrow ranges of pH. There is one problem though. If the solution to be measured is highly colored, the color of the solution itself may interfere with

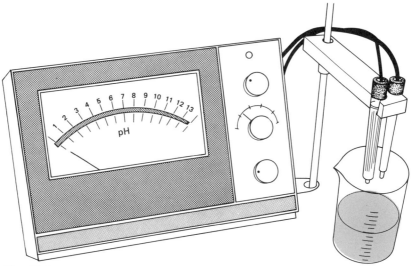

Fig. 13.3. A pH meter.

the color change of the indicators and accurate measurements may be impossible.

The pH of a urine sample is routinely measured using indicators. A plastic stick to which is bonded a small square of test paper is dipped into the specimen and removed. The color of the test paper is compared, after a few seconds, to a series of colors on a chart. Each color in the series corresponds to a certain pH. Once a color match is made, you have the pH of the urine sample (see Figure 21.1).

The second way that is used to measure pH is with a **pH meter**, an electronic device that is capable of very accurate pH measurements. A pH meter has two electrodes that are immersed in the solution to be tested. The electrodes contain chemicals that respond to the hydronium ion concentration of the solution, and through certain chemical reactions within the electrodes, an electrical signal is generated. The size of the signal is related to the hydronium ion concentration, and it is converted by the electronics to a pH value. The pH is then read directly from the meter, either on a digital readout or on a pH scale with a movable needle pointing to the pH value. An advantage of a pH meter is that it can be used in highly colored solutions without difficulty. One type of pH meter is shown in Figure 13.3.

pOH

Just as it is possible to use pH to express hydronium ion concentrations, a term called pOH can be used to express hydroxide ion concentrations:

$$pOH = - \log [OH^-_{(aq)}]$$

The pOH is not as widely used as pH, but you should be aware of its existence. The pH of an aqueous solution and the pOH of the same solution always add to 14:

$$pH + pOH = 14$$

If you know the pH of a solution, you can use this equation to calculate its pOH. If the pH is 5, then the pOH is (14 − 5) or 9. Also, if you are given the hydroxide ion concentration of a solution but need to calculate its pH, it may be easier to first find the pOH, then convert it to pH. This is done in the following example.

Example 13.6

The hydroxide ion concentration of a solution is $1 \times 10^{-3} M$. Calculate the pH of the solution.

$$[OH^-_{(aq)}] = 10^{-3} M$$

$$pOH = - \log [OH^-_{(aq)}]$$

$$pOH = -\log [10^{-3}]$$

$$= - (-3)$$

$$pOH = 3$$

Now, convert pOH to pH:

$$pH = 14 - pOH$$

$$pH = 14 - 3 = 11$$

Buffers—Keeping the pH Constant

If the vital chemical processes in biological systems are to take place properly, the pH of the solutions in which they occur must be held very close to some optimum value, usually around 7. You are aware of the life-sustaining role of blood carrying oxygen from the lungs to all parts of the body. The pH of blood plasma must be maintained close to 7.4 (very slightly basic) if it is to be an efficient oxygen carrier. The normal pH range for plasma is between 7.35 and 7.45. If, because of starvation or disease, the pH goes below 7.0 (severe acidosis) or above 7.8 (severe alkalosis), blood loses its ability to carry oxygen and death follows. But even before the pH reaches these fatal limits, symptoms of acidosis or alkalosis appear which can be quite serious. Both conditions will be discussed later on in this chapter.

As you can see, the control of pH must be an important function of body chemistry. Acidic and basic substances continually enter the bloodstream, yet they are absorbed without causing wide swings in pH. The control of pH is carried out by substances in the blood that can react with both acids and bases while maintaining a near constant pH. These substances allow blood to act as a buffer, a kind of chemical shock absorber. Before describing the blood buffer system, let us first look into the way a buffer operates.

A **buffer** is a solution that is able to maintain a near constant pH when small amounts of acid or base are added to it. Buffer solutions, as they are also called, contain either a weak acid *and* one of its salts (such as $HC_2H_3O_2$ + $NaC_2H_3O_2$ or H_2CO_3 + $NaHCO_3$), or a weak base *and* a salt of the weak base ($NH_{3(aq)}$ + NH_4Cl). Both components of a buffer work together to neutralize added acid or base. The best way to see how it is done is to study a typical example, a buffer made of acetic acid and sodium acetate. If each is present at nearly the same concentration, this buffer will hold the pH of the solution close to 4.7.

Acetic acid dissociates slightly in solution forming hydronium ions and acetate ions. Additional acetate ion is also present from sodium acetate, and the three species exist in solution in equilibrium. The pH of the solution is determined by the concentration of $H^+_{(aq)}$ that can exist in equilibrium with the larger concentrations of acetic acid molecules and acetate ions:

$$HC_2H_3O_{2(aq)} \rightleftharpoons H^+_{(aq)} + C_2H_3O^-_{2(aq)}$$

The fact that these three species exist in equilibrium with each other is very important in the operation of a buffer. To learn why, we need to reveal another fact about chemical equilibrium beyond that presented in Chapter 11.

In 1884, a French chemist, Henri Louis LeChâtelier (pronounced Le Shat-ē-ā), summarized his observations on equilibrium this way: If a stress is applied to a system in equilibrium, the system will change to minimize the effect of that stress. This is known as **LeChâtelier's principle**. In our case, the "system" is the acetic acid/sodium acetate buffer, and the "stress" will be either an increase or decrease in the hydronium ion concentration. Let us see how each kind of stress is handled by the buffer.

If the hydronium ion concentration in a buffer decreases, as it would if a strong base were added to it, more acetic acid would quickly dissociate to replace most of the hydronium ion that was consumed. Let's look at this as a series of events to see why this happens.

1. Added base reacts with $H^+_{(aq)}$:

$$HC_2H_3O_{2(aq)} \rightleftharpoons H^+_{(aq)} + C_2H_3O_2^-_{(aq)} \nearrow H_2O$$

Added base: $OH^-_{(aq)}$

2. This decreases the $H^+_{(aq)}$ concentration, putting the equilibrium under stress:

$$HC_2H_3O_{2(aq)} \rightleftharpoons H^+_{(aq)} + C_2H_3O_{2(aq)}^-$$

3. To remove the stress, more $HC_2H_3O_{2(aq)}$ must dissociate to replace most of the lost $H^+_{(aq)}$—the forward reaction of the equilibrium momentarily takes over,

$$HC_2H_3O_{2(aq)} \rightarrow H^+_{(aq)} + C_2H_3O_{2(aq)}^-$$

until:

4. Equilibrium is again established with the $H^+_{(aq)}$ concentration nearly the same as it was before the base was added. The stress on the equilibrium has been minimized:

$$HC_2H_3O_{2(aq)} \rightleftharpoons H^+_{(aq)} + C_2H_3O_{2(aq)}^-$$

On the other hand, if the hydronium ion concentration increases, as would happen if a strong acid were added to the buffer solution, the excess hydronium ions will combine with acetate ions to form acetic acid molecules until equilibrium is reestablished. Since acetic acid molecules are only slightly dissociated in solution, the added acid is effectively removed. Let us look at this case step by step.

1. Added acid increases the $H^+_{(aq)}$ concentration, putting the equilibrium under stress.

$$HC_2H_3O_{2(aq)} \rightleftharpoons H^+_{(aq)} + C_2H_3O_{2(aq)}^-$$

2. The reverse reaction becomes momentarily important and $H^+_{(aq)}$ is removed from solutions as acetate ion combines with it to form acetic acid molecules. Only the reverse reaction can remove this stress:

$$HC_2H_3O_{2(aq)} \leftarrow H^+_{(aq)} + C_2H_3O_{(aq)}^-$$

The reverse reaction continues until:

3. Equilibrium is reestablished with the $H^+_{(aq)}$ concentration nearly the same as it was before the acid was added. The stress on the equilibrium has been minimized:

$$HC_2H_3O_{2(aq)} \rightleftharpoons H^+_{(aq)} + C_2H_3O_{2(aq)}^-$$

The Blood Buffers

The blood has three principal buffer systems that maintain the pH of plasma very close to 7.4. They are the carbonate buffer, the phosphate buffer, and the plasma proteins that provide a bufferlike action. Let us briefly describe each one to learn how the acids and bases that enter the bloodstream are prevented from causing large changes in pH.

The Carbonate Buffer
The carbonate buffer is the most important one in blood. It is composed of carbonic acid, H_2CO_3, and the bicarbonate ion, HCO_3^-, the anion of the acid (as would be supplied by a salt of the acid). The H_2CO_3/HCO_3^- ratio is one mole of H_2CO_3 to 20 moles of HCO_3^-. This ratio provides a hydronium concentration that corresponds to a pH of 7.4:

$$H_2CO_{3(aq)} \rightleftharpoons H^+_{(aq)} + HCO^-_{3(aq)}$$

The carbonate buffer operates in the same way as the acetic acid buffer to control pH, so let us just look at the step that removes or provides hydronium ions in the buffer action. As acids enter the blood, the added hydronium ions react with bicarbonate ions to form carbonic acid until equilibrium is restored. If the carbonic acid concentration gets too large, it dissociates into CO_2 and water:

$$H_2CO_{3(aq)} \leftarrow H^+_{(aq)} + HCO^-_{3(aq)}$$

As bases enter the blood, the added hydroxide ion reacts with the hydronium ion to produce water. Carbonic acid then dissociates to replenish the hydronium ion until equilibrium is again restored:

$$H_2CO_{3(aq)} \rightarrow H^+_{(aq)} + HCO^-_{3(aq)}$$

The Phosphate Buffer
The phosphate buffer is composed of the dihydrogen phosphate ion, $H_2PO_4^-$, a weak acid, and the hydrogen phosphate ion, HPO_4^{2-}, the anion of the weak acid. They exist in equilibrium in plasma:

$$H_2PO^-_{4(aq)} \rightleftharpoons H^+_{(aq)} + HPO^{2-}_{4(aq)}$$

As acids enter the blood, the added hydronium ion combines with HPO_4^{2-} to form $H_2PO_4^-$. The reverse reaction continues until the equilibrium is reestablished. Added base is consumed as the hydroxide and hydronium ions react to form water. More $H_2PO_4^-$ ion then dissoicates to replace the hydronium ion and restore the equilibrium. In both cases the hydronium ion concentration ends up essentially unchanged, and the pH stays nearly constant.

The Plasma Protein Buffer Plasma proteins are very large molecules that possess groups that can bind hydronium ions, removing them from solution, and other groups that can neutralize hydroxide ions. Plasma proteins are not buffers of the type we have been discussing, but they do provide a buffering action because of their ability to react with acids and bases. We will put off further discussion of the buffer action of proteins until Chapter 22.

Acidosis and Alkalosis

The normal pH of blood is between 7.35 and 7.45, and in this state the body exists in a condition of acid–base balance. If the blood pH falls below 7.35, a condition known as **acidosis** exists. This is a medical term since the entire life-sustaining range of pH is actually on the basic side of neutral. In acidosis, the blood is just less basic then normal. The principal effect of acidosis is depression of the central nervous system, which may lead to disorientation, coma, and if untreated, death. If the blood pH rises above 7.45, the blood is more alkaline than normal, and the condition is called **alkalosis**. The effects of alkalosis are weak and irregular breathing, severe muscle contractions, convulsions, and even death in extreme cases.

Acidosis and alkalosis are brought about by changes in the blood buffer systems. Normal blood pH is maintained when the bicarbonate ion concentration is 20 times the carbonic acid concentration. Any change in this 20 to 1 ratio will bring about a change in the blood pH. In acidosis, the HCO_3^-/H_2CO_3 ratio is less than 20 to 1, either because bicarbonate ion decreases or carbonic acid increases in concentration. In alkalosis, the HCO_3^-/H_2CO_3 ratio is greater than 20 to 1, caused by either an increase in bicarbonate ion or a decrease in carbonic acid concentration.

Acidosis and alkalosis can be brought on by changes in respiration, which can raise or lower the carbon dioxide concentration in the blood. The level of CO_2 in the blood is ultimately determined by the concentration of CO_2 in the alveoli. If CO_2 is allowed to build up in the lungs because of shallow breathing due to disease or some other reason, the partial pressure of CO_2 increases, forcing the blood to retain CO_2. The increased level of CO_2 in the blood will increase the carbonic acid concentration above normal levels:

$$CO_{2(g)} \quad + \quad H_2O \quad \longrightarrow \quad H_2CO_{3(aq)}$$

| An increase in CO_2 concentration | causes | An increase in H_2CO_3 concentration |

The additional carbonic acid dissociates, increasing the hydronium ion concentration, lowering the pH, and bringing on acidosis.

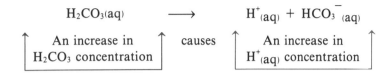

$$H_2CO_3(aq) \longrightarrow H^+_{(aq)} + HCO_3^-{}_{(aq)}$$

An increase in H_2CO_3 concentration ↑ causes ↑ An increase in $H^+_{(aq)}$ concentration

Because the acidosis is connected to a respiratory effect, it is called **respiratory acidosis.**

The body responds by increasing the rate and depth of breathing to reduce the concentration of CO_2 in the alveoli. The blood, then, will not retain the abnormally high level of CO_2. As the CO_2 level falls, the carbonic acid concentration will decrease, and as it does, the hydronium ion concentration will decrease also.

Respiratory alkalosis can be brought on by the opposite extreme in breathing. Rapid, deep breathing, such as may accompany a high fever or hysterical behavior, can lower the partial pressure of CO_2 in the alveoli. This allows an increased passage of CO_2 out of the blood and into the lungs. The reduced CO_2 level in blood causes the concentration of carbonic acid to fall. With less carbonic acid, the hydronium ion concentration becomes less and the blood pH rises. Alkalosis occurs:

$$CO_2(g) + H_2O \longleftarrow H_2CO_3(aq) \longleftarrow H^+_{(aq)} + HCO_3^-{}_{(aq)}$$

A decrease in CO_2 concentration ↓ causes ↓ A decrease in H_2CO_3 concentration ↓ which causes ↓ A decrease in $H^+_{(aq)}$ concentration ↓

The body responds by lowering the rate of breathing to bring the CO_2 level in the lungs back to normal. The blood then can retain the normal level of CO_2, which brings the carbonic acid concentration back up to normal, and with it, the hydronium ion concentration.

Since the body can control the depth and rate of breathing quickly, respiration is the principal means of keeping the blood at the correct pH.

Respiration is not the only cause of acidosis or alkalosis. All other causes are considered to be metabolic in origin. **Metabolic acidosis** is caused by an accumulation of acidic substances in the blood, which in turn causes a reduced bicarbonate ion concentration. It can also be caused by the direct loss of bicarbonate ion, as can occur in prolonged diarrhea. Several conditions can bring about metabolic acidosis: starvation—produces high levels of acidic metabolites; ingestion of acids; diabetes mellitus—acidic metabolites; kidney failure—acidic substances not excreted; and dehydration.

Metabolic alkalosis is the result of an abnormally high concentration of bicarbonate ion in the blood. It can be caused by an excessive loss of gastric acid in prolonged vomiting or because of an overdose of antacids. Kidney failure can also cause this condition.

The control of metabolic acidosis and alkalosis, to return the pH of blood to normal, is handled by the kidneys. In acidosis, the kidneys increase the level of bicarbonate ion by excreting acidic urine. In alkalosis, the kidneys excrete alkaline urine and bicarbonate ion, decreasing its concentration in the blood.

The Gram Equivalent Weight

Chemists often express quantities of pure substances in terms of moles of a compound. One mole of HCl is 36.5 g of HCl, and one mole of H_2SO_4 is 98 g of the acid. In Chapter 4, the mole was especially useful when dealing with balanced chemical equations. The acids and bases in the following equations react in a definite mole ratio:

$$HCl_{(aq)} + NaOH_{(aq)} \longrightarrow NaCl_{(aq)} + H_2O$$

1 mole   1 mole

$$H_2SO_{4(aq)} + 2NaOH_{(aq)} \longrightarrow Na_2SO_{4(aq)} + 2H_2O$$

1 mole   2 moles

One mole of NaOH neutralizes one mole of HCl, but two moles of NaOH are needed to neutralize one mole of H_2SO_4. This is because there are two replaceable hydrogens in H_2SO_4, but only one in HCl. In terms of the neutralizing power of the two acids, you can see that one mole of HCl is equivalent to one-half mole of H_2SO_4, since each quantity will neutralize the same amount (one mole) of NaOH:

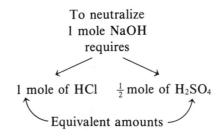

To neutralize
1 mole NaOH
requires

1 mole of HCl $\frac{1}{2}$ mole of H_2SO_4

Equivalent amounts

In the same way, one can determine equivalent amounts of bases:

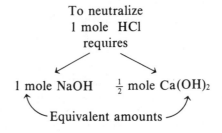

To neutralize
1 mole HCl
requires

1 mole NaOH $\frac{1}{2}$ mole $Ca(OH)_2$

Equivalent amounts

Since we can think in terms of equivalent amounts of acids and bases, chemists frequently describe them in terms of their gram equivalent weights. One **gram equivalent weight** of an **acid** is the number of grams of that acid required to neutralize one mole of $OH^-_{(aq)}$. The gram equivalent weight (GEW) of any acid can be determined this way:

$$GEW \ (acid) = \frac{\text{gram formula weight of the acid}}{\text{number of replaceable hydrogens in the formula}}$$

One **gram equivalent weight** of a **base** is the number of grams of that base required to neutralize one mole of $H^+_{(aq)}$. It can be determined this way for a base:

$$GEW \ (base) = \frac{\text{gram formula weight of the base}}{\text{number of hydroxide ions in the formula}}$$

The gram equivalent weights of several common acids and bases are determined below.

Acid	Gram formula weight		Number of replaceable hydrogen		GEW
HCl	36.5 g	÷	1	=	36.5 g
H_2SO_4	98 g	÷	2	=	49 g
$HC_2H_3O_2$	60 g	÷	1	=	60 g

Base	Gram formula weight		Number of hydroxide ions in formula		GEW
NaOH	40 g	÷	1	=	40 g
$Ca(OH)_2$	74 g	÷	2	=	37 g

Solutions containing either 36.5 g of HCl, 49 g of H_2SO_4, or 60 g of $HC_2H_3O_2$ will have the same capacity for neutralizing a base. They are equivalent amounts of acid. In fact, one GEW of an acid is commonly called an **equivalent** of acid. Also, 40 g of NaOH will neutralize the same amount of a given acid as 37 g of $Ca(OH)_2$ will. They are equivalent amounts of base in terms of neutralization. But perhaps the most useful feature of the gram equivalent weight is this: One GEW of any acid will neutralize just one GEW of any base. Or said another way, if the same number of equivalents of an acid and a base are combined, they will exactly neutralize each other. This important fact will be used when we discuss acid–base titrations.

Small amounts of acids and bases may be stated in terms of milliequivalents (mEq). One **milliequivalent** is one one-thousandth of an equivalent, 1 mEq = 0.001 equivalent.

Normality—Another Concentration Term

Up to now, the concentration of a solution of an acid or base has been given in terms of molarity (M), the number of moles of solute in one liter of solution. Another concentration scheme is also used for acids and bases. It expresses concentration in terms of the number of gram equivalent weights, or equivalents, in one liter of solution. Concentrations expressed in this way are called **normal** concentrations, and the **normality** (N) of a solution is defined this way:

$$\text{Normality} = \frac{\text{number of GEW of solute in solution}}{\text{volume of the solution in liters}}$$

or

$$= \frac{\text{equivalents of solute}}{\text{volume of the solution in liters}}$$

The units of normality are GEW/l or equivalents/l, but they are usually replaced by the symbol N. The following examples deal with normality.

Example 13.7

What is the normality of a sulfuric acid solution that contains 25 g of H_2SO_4 in 1.0 l of solution?

The volume of the solution is given, but you need to determine the number of gram equivalent weights or equivalents of H_2SO_4 in 25 g of the acid. It has already been shown that 1 GEW of H_2SO_4 = 49 g of H_2SO_4. This fact can be used to convert grams of acid to equivalents.

$$\text{number of GEW} = 25\,g\left(\frac{1\,\text{GEW}}{49\,g}\right) = 0.51\,\text{GEW}$$

$$N = \frac{\text{number of GEW}}{\text{volume of solution in liters}}$$

$$N = \frac{0.51\,\text{GEW}}{1.0\,l} = 0.51N$$

Example 13.8

How many equivalents of NaOH are in 500 ml of a 0.20N NaOH solution? The question can be answered by rearranging the normality equation and solving for the number of GEW. (0.20N can be written as 0.20 GEW/l and 500 ml is 0.50 l)

$$\text{number of GEW} = (N) \times (\text{volume of solution in liters})$$

$$\text{number of GEW} = 0.20 \frac{\text{GEW}}{\cancel{l}} \times 0.50 \cancel{l}$$

$$= 0.10 \text{ GEW}$$

In 500 ml of a 0.20 N NaOH solution there is 0.10 equivalent (4g) of NaOH.

The normality of a solution can be calculated quickly if the molarity of the solution is known. For acids, the normality is simply equal to the molarity times the number of replaceable hydrogens in the acid. For bases, the normality is the molarity times the number of hydroxide ions in the formula of the base.

Acids: $N = M \times$ (number of replaceable hydrogens)

Bases: $N = M \times$ (number of hydroxide ions in formula)

A 2M solution of $HCl_{(aq)}$ would also be a 2N solution, but a 2M solution of $H_2SO_{4(aq)}$ would be a 4N solution. A 2M solution of $NaOH_{(aq)}$ would be a 2N solution. A saturated solution of $Ca(OH)_{2(aq)}$ is 0.02M, but because there are two hydroxide ions in the formula, it would be a 0.04N solution. The normality of an acid or base solution will be either equal to or larger by a factor of 2, 3, etc., than the molarity. Since normal concentrations are based on equivalents of acids and bases, they are particularly useful when determining concentrations of these substances in solution. Before we get to that, though, try your hand at calculating normalities in the next Check Test.

Check Test Number 5

1. What is the normality of a sulfuric acid solution if 245 g of H_2SO_4 are in 2.0 l of solution?
2. What is the normality of a 5.0M sulfuric acid solution?

Answers:

1. $245 \cancel{g} \left(\dfrac{1 \text{ GEW}}{49 \cancel{g}} \right) = 5.0 \text{ GEW } H_2SO_4$

$N = \dfrac{5.0 \text{ GEW}}{2.0 \text{ l}} = 2.5N$

2. $10N$

Acid–Base Titration

Titration is a procedure used to determine the concentration of an acid or base in a solution. Suppose you needed to know the concentration of acid in a solution. Here is how it is done using titration. First, a known volume of the acid is carefully measured into a flask. A pipet can be used to do this, as shown in Figure 13.4. A few drops of an indicator solution are added to the acid. The indicator is carefully chosen so it changes color at the point where all the acid is just neutralized. Then a solution of base of known concentration is carefully added to the acid from a buret. The **buret** is a long glass tube calibrated in units of volume (milliliters) with a stopcock at the lower end. It is used to measure volumes of liquid delivered from it. The base is added until a point is reached where one drop causes the indicator to change color. At this point, all the acid has just been neutralized by the base. This is called the **equivalence point** of the titration, since the number of equivalents of acid neutralized exactly *equals* the number of equivalents of base added:

> At the equivalence point:
> number of equivalents of acid   =   number of equivalents of base

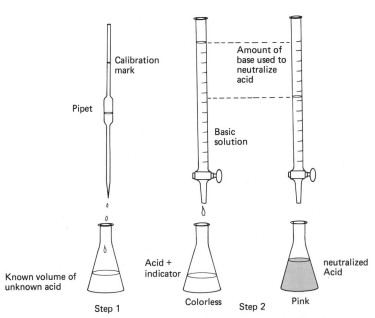

Fig. 13.4.   Titration of an acid with a base. Step 1: Obtain a known volume of the acid to be analyzed. A pipet is filled to the calibration mark with the acid solution and drained into a flask. A few drops of indicator are added. Step 2: The titration. A solution of base of known concentration is slowly added from a buret until the indicator just turns color. The volume of base required to reach the equivalence point is read from the buret.

The number of equivalents of base added can be calculated from the normality of the base, N_b, and the volume of base, V_b, using the equation that defines normality.

$$\text{number of equivalents of base} = V_b \times N_b$$

Likewise, the number of equivalents of acid neutralized equals the volume of the acid sample pipeted into the flask, V_a, times its normality, N_a. The normality of the acid is what is being sought in this titration.

$$\text{number of equivalents of acid} = V_a \times N_a$$

Since the equivalents of acid and base are equal at the equivalence point,

$$V_a \times N_a = V_b \times N_b$$

Knowing any three terms in the titration equation allows you to calculate the fourth. This equation holds true if an acid is titrated with a base, or if a base is titrated with an acid—just the reverse of the procedure described above. If the concentration of the base and the volume of base (read from the buret) are known, along with the volume of the acid added to the flask, the normal concentration of the acid can be determined using the titration equation:

$$N_a = \frac{V_b \times N_b}{V_a}$$

The volumes (V_a and V_b) may be expressed in liters or milliliters, but both volumes must use the same units. Since the volumes of solutions are usually not large in titrations, milliliters are commonly used. Let us do some example problems involving titration.

Example 13.9

A 20.0-ml solution of sulfuric acid required 45.0 ml of $0.20N$ NaOH to reach the equivalence point. What is the normality of acid solution?
You need to find N_a. You are given the following:

$$V_b = 45.0 \text{ ml}$$

$$N_b = 0.20N$$

$$V_a = 20.0 \text{ ml}$$

Substituting these into the titration equation allows the normality to be determined:

$$N_a = \frac{V_b N_b}{V_a} = \frac{(45.0 \text{ ml})(0.20N)}{(20.0 \text{ ml})}$$

$$N_a = 0.45N$$

Example 13.10

How many milliliters of $0.15N$ NaOH are required to just neutralize all the acid in 150 ml of $0.50N$ HCl?

You need V_b, and you are given: $V_a = 150$ ml

$$N_a = 0.50N$$
$$N_b = 0.15N$$

Rearrange the titration equation to solve for V_b, then substitute the known values into the equation and calculate:

$$V_b = \frac{V_a N_a}{N_b} = \frac{(150 \text{ ml})(0.50N)}{(0.15N)}$$

Check Test Number 6

1. What is the normality of an unknown base solution if 50.0 ml of the solution requires 35.0 ml of $0.30N$ hydrochloric acid to reach the equivalence point?
2. What volume of $2.0N$ sulfuric acid is required to neutralize all the base in 500 ml of $1.4N$ sodium hydroxide?

Answers:

1. $N_b = \dfrac{(35.0 \text{ ml})(0.30N)}{(50.0 \text{ ml})} = 0.21N$

2. $V_a = \dfrac{(500 \text{ ml})(1.4N)}{(2.0N)} = 350$ ml

Terms

Several new terms were introduced in Chapter 13. You should understand the meaning of each of the following.

acid	pOH
base	pH scale

hydronium ion

neutralization

strong acid

weak acid

ionic equation

net-ionic equation

salt

ion-product equation

pH

buffer

LeChâtlier's principle

acidosis

alkalosis

gram equivalent weight

normality

indicator

titration

equivalence point

Questions

Starred questions have answers given at the end of the text.

*1. To what ions did Arrhenius attribute the properties of acids? The properties of bases?

*2. Give the formulas and names for two strong acids and two strong bases.

*3. What is the hydronium ion? How is it symbolized in equations?

*4. List three properties of acids and three properties of bases.

*5. What is the principal difference between a strong acid and a weak acid?

6. Why is aqueous ammonia considered a weak base?

7. Classify each of the following acids as either monoprotic, diprotic, or triprotic.
   a. HCl c. HNO_3 e. H_2SO_4
   b. H_3PO_4 d. H_2CO_3 f. $HC_2H_3O_2$

*8. What is the first remedial action you should take if an acid or base is accidentally spilled on the skin?

*9. Use equations to show the two-step dissociation of sulfuric acid, H_2SO_4, in water.

*10. Concentrated sulfuric acid is described as being a dehydrating agent. What does this mean?

11. Complete and balance the following neutralization equations.
   *a. $2HCl + Ca(OH)_2 \rightarrow$
   *b. $HNO_3 + KOH \rightarrow$
   *c. $H_2SO_4 + 2NaOH \rightarrow$
   d. $H_3PO_4 + 2KOH \rightarrow$
   e. $H_2CO_3 + 2NaOH \rightarrow$
   f. $2HC_2H_3O_2 + Ca(OH)_2 \rightarrow$

12. Complete and balance the following equations that describe the chemical properties of acids.
   *a. $HCl + K_2O \rightarrow$
   *b. $HNO_3 + NaHCO_3 \rightarrow$
   *c. $HCl + Mg \rightarrow$
   d. $H_2SO_4 + BaO \rightarrow$
   e. $H_3PO_4 + 3NaOH \rightarrow$

13. Give the formula of the acid and base that must react to form each of the following salts.

Salt	Acid	Base
*KCl		
$BaSO_4$		
*$Ca(NO_3)_2$		
*$Ni(C_2H_3O_2)_2$		
Na_2HPO_4		

14. Write the formula, ionic, and net-ionic equations for the following reactions which occur in aqueous solution. All species are water soluble.
    *a. nitric acid + barium hydroxide → barium nitrate + water
    b. sulfuric acid + potassium hydroxide → potassium sulfate + water

15. Either the hydroxide or hydronium ion concentration is given for 5 different aqueous solutions below. Give the concentration of the ion not given.

	$[H^+_{(aq)}]$	$[OH^-_{(aq)}]$
*a.	10^{-3}	———
*b.	———	10^{-9}
*c.	10^{-10}	———
d.	———	10^{-2}
e.	10^{-7}	———

16. Calculate the pH of the following solutions given the hydronium or hydroxide ion concentrations.

$[H^+_{(aq)}]$	$[OH^-_{(aq)}]$
*a. 1×10^{-3}	*e. 1×10^{-4}
*b. 1×10^{-9}	*f. 1×10^{-11}
c. 1×10^{-2}	g. 1×10^{-6}
d. 1×10^{-11}	h. 1×10^{-13}

17. What is the hydronium ion concentration in each of the following solutions of known pH or pOH?
    *a. pH = 5   *e. pOH = 6
    *b. pH = 10   *f. pOH = 2
    c. pH = 14   g. pOH = 9
    d. pH = 2   h. pOH = 10

18. Give the range of hydronium ion concentration in solutions of the following pH.
    *a. 3.5
    *b. 6.8
    c. 11.6
    d. 8.3

19. Give the range of pH corresponding to the following hydronium ion concentrations.
    *a. $3 \times 10^{-4}M$ c. $4 \times 10^{-3}M$
    *b. $6 \times 10^{-8}M$ d. $7 \times 10^{-10}M$

*20. What is a buffer? What kinds of compounds are used to prepare buffers?

*21. What are the three major buffer systems in the blood? Which one of the three is considered most important.

*22. What is the normal pH range of blood? In what direction does the blood pH shift in acidosis? In alkalosis?

23. How can a change in respiration bring on acidosis? How can it bring about alkalosis?

*24. List two physiological causes of metabolic acidosis and metabolic alkalosis.

25. Calculate the gram equivalent weight of the following acids and bases.
   a. $Ba(OH)_2$ c. HNO_3 e. $HC_2H_3O_2$
   b. H_3PO_4 d. LiOH   f. H_2CO_3

26. Define normality.

*27. How many equivalents of HCl are in 600 ml of a $0.75N$ solution?

28. How many equivalents of NaOH are in 8.3 l of a $2.0N$ solution?

29. Determine the normality of the following solutions.
   a. $2M$ HCl   c. $0.02M$ $Ca(OH)_2$
   b. $1.5M$ H_2SO_4 d. $3.0M$ NaOH

*30. Calculate the normality of the following solutions.
   a. 73 g of HCl in 3.0 l of solution
   b. 100 g of HNO_3 in 1.50 l of solution
   c. 98 g of H_2SO_4 in 1.0 l of solution
   d. 80 g of NaOH in 1.2 l of solution
   e. 1.60 g of $Ca(OH)_2$ in 1500 ml of solution
   f. 1.5 moles of H_2SO_4 in 2.0 l of solution

31. Calculate the normality of these solutions.
   a. 120 g of HCl in 10 l of solution
   b. 80 g of HNO_3 in 3.0 l of solution
   c. 60.0 g of NaOH in 1200 ml of solution
   d. 6.50 mole H_2SO_4 in 4.00 l of solution
   e. 500 g of $HC_2H_3O_2$ in 2.0 l of solution
   f. 0.50 g of $Mg(OH)_2$ in 33 l of solution

*32. How many milliliters of $3.0N$ sodium hydroxide can be neutralized with 500 ml of $1.5N$ H_2SO_4?

*33. What is the meaning of the "equivalence point" in a titration?

*34. What is the function of the indicator in an acid–base titration?

*35. A 350-ml sample of acidic urine required 12.0 ml of $0.0010N$ NaOH to reach the equivalence point in a titration. What is the normality of the acidic components in this urine sample?

36. A 25.0-ml sample of acid required 41.5 ml of $0.10N$ sodium hydroxide to reach the equivalence point in a titration. What is the normality of the acid?

*37. How many milliliters of $0.75N$ hydrochloric acid are needed to just neutralize all the base in 1000 ml of $0.6N$ potassium hydroxide?

38. A 250-ml sample of an acid solution required 46.5 ml of $0.200N$ NaOH to reach the equivalence point in a titration. What is the normality of the acid?

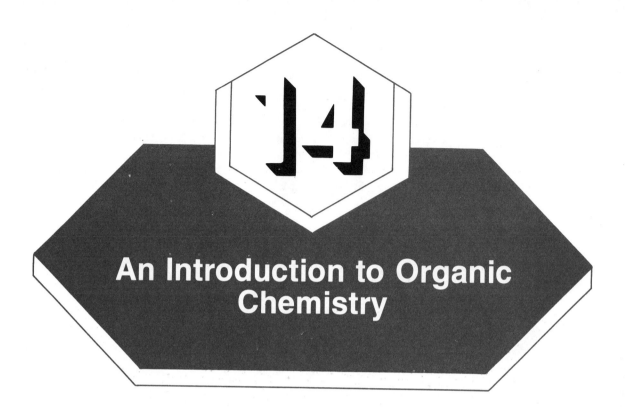

An Introduction to Organic Chemistry

About two hundred years ago the term "organic" was first used to describe substances found in or derived from living organisms. The organic compounds all contained a substantial amount of one particular element—carbon—and the study of these carbon-containing compounds was called **organic chemistry.** The study of compounds obtained from minerals, the atmosphere, and other non-living sources was called inorganic chemistry.

For many years people believed organic compounds could only be synthesized in living organisms, because only they could provide the certain necessary "vital force." As science advanced, the popularity of this belief waned, and it was eventually overturned in the 1840's when a German chemist, Fredrick Wöhler, synthesized urea (an organic compound found in urine) from ammonium cyanate (an inorganic salt) in the laboratory in the absence of a living organism. The inorganic compound simply rearranged to form an organic one:

$$NH_4OCN \xrightarrow{\Delta} H_2N-\overset{\displaystyle O}{\overset{\displaystyle \|}{C}}-NH_2$$

ammonium cyanate (inorganic) urea (organic)

Wöhler was not trying to disprove the idea of the "vital force," but several years after he published his work, its significance in terms of the vital force idea was recognized.

Today, organic compounds are regarded as those that contain carbon. But there are a few exceptions. The carbon compounds found in minerals are classed as inorganic, such as the carbonate and hydrogen carbonate salts (Na_2CO_3, $KHCO_3$, etc.). The oxides of carbon (CO and CO_2) and the carbides, which contain carbon combined with metals (CaC_2, Fe_3C, etc.), are also regarded as inorganic. All the other known carbon compounds, and there are over three million of them, are in the domain of organic chemistry, the largest single division of modern chemistry.

Though carbon is far from being the most abundant element in the earth's crust, its compounds are extremely important. Most of the substances that make up our bodies are organic compounds, compounds that are essential to life itself. The study of the chemistry of these important compounds is called biochemistry, an area that is closely related to organic chemistry. Food, plastics, fuels, pharmaceuticals, wood, fibers, and adhesives are either complicated mixtures of or pure organic compounds. The purpose of this chapter is to introduce you to some of the unique properties of carbon, and briefly survey the broad area of chemistry known as organic chemistry.

Objectives

By the time you finish this chapter you should be able to do the following:

1. Define the term "organic chemistry."
2. Describe the unique features of carbon that allow it to form such a large number of compounds.
3. Describe the spatial arrangement of the bonds about carbon in its characteristic bonding patterns found in organic compounds.
4. Write the structural formula for an organic compound given its condensed formula and vice versa.
5. Describe and name the common organic functional groups.

Carbon—The Exceptional Element

What is it about carbon that gives it such versatility that a major division of chemistry is devoted to the study of its many compounds? Though the answer is not simple, it would undoubtedly include the importance of the intermediate, middle-of-the-road properties of carbon. Carbon is about in the center of the second period, with its valence shell exactly half-filled with four electrons. It does not have a strong tendency to either gain or lose electrons to form ions, rather it shares its electrons with other atoms through covalent bonds. In organic compounds, there will always be four shared pairs of electrons (four bonds) about each carbon atom. Perhaps the best way to emphasize the unique properties of carbon is to summarize its versatility in forming compounds:

1. Carbon readily bonds to other carbon atoms forming straight chains, branched chains, or rings of carbon atoms. The carbon atom skeleton forms the structural framework of an organic molecule.

a. straight chains (a dash symbolizes a covalent bond):

$$H-\underset{\underset{H}{|}}{\overset{\overset{H}{|}}{C}}-\underset{\underset{H}{|}}{\overset{\overset{H}{|}}{C}}-\underset{\underset{H}{|}}{\overset{\overset{H}{|}}{C}}-\underset{\underset{H}{|}}{\overset{\overset{H}{|}}{C}}-\underset{\underset{H}{|}}{\overset{\overset{H}{|}}{C}}-\underset{\underset{H}{|}}{\overset{\overset{H}{|}}{C}}-\underset{\underset{H}{|}}{\overset{\overset{H}{|}}{C}}-\underset{\underset{H}{|}}{\overset{\overset{H}{|}}{C}}-H$$

b. branched chains:

Branches

c. rings:

2. Carbon can join to other atoms through single bonds (one shared pair of electrons), double bonds (two shared pairs of electrons), and triple bonds (three shared pairs of electrons). Carbon can have four different bonding patterns in its compounds (you may recall these from Chapter 8):

$$-\overset{|}{\underset{|}{C}}- \qquad \text{four single bonds}$$

$$\overset{\diagdown}{\underset{\diagup}{C}}= \qquad \text{two single bonds and one double bond}$$

$$=C= \qquad \text{two double bonds}$$

$$-C\equiv \qquad \text{one single bond and one triple bond}$$

The following molecules show carbon in these different bonding patterns:

$$H-\overset{\overset{\textstyle H}{|}}{\underset{\underset{\textstyle H}{|}}{C}}-H \qquad\qquad O=C=O$$

Methane   Carbon dioxide

Acrolein   Acetylene (ethyne)

3. Because of its intermediate properties, carbon can form strong bonds with several other nonmetal elements, principally hydrogen, oxygen, nitrogen, sulfur, phosphorus, and the elements of group VII, the halogens. Hydrogen is by far the most prevalent noncarbon element in organic compounds. Carbon can bond to each of these elements through single bonds, and in some cases through double or triple bonds. The way in which carbon can join to these elements is summarized in Table 14.1.

4. Since carbon bonds so readily with itself and to other nonmetals, as the number of atoms increases in a formula, the number of possible ways they can be joined together forming different structures increases also. Compounds with the same molecular formula but different structures are called **structural isomers** of one another. There are five known organic compounds with the formula C_6H_{14}, and each has the carbon atoms connected together in a different arrangement giving rise to five structural isomers. Since each structure is different, each isomer is a different compound with its own set of properties. More will be said about isomers in future chapters.

	TABLE 14.1	
	The Bonding of Carbon to Other Elements	
Element	Number of bonds to carbon	Bonding symbolized with dash structures
H	1	$H-\overset{\mid}{\underset{\mid}{C}}-$
O	1 or 2	$-O-\overset{\mid}{\underset{\mid}{C}}-,\quad O{=}C\overset{\diagup}{\diagdown}$
S	1 or 2	$-S-\overset{\mid}{\underset{\mid}{C}}-,\quad S{=}C\overset{\diagup}{\diagdown}$
N	1, 2, or 3	$-\overset{\mid}{\underset{\mid}{N}}-\overset{\mid}{\underset{\mid}{C}}-,-N{=}C\overset{\diagup}{\diagdown},\quad N{\equiv}C-$
P	1, 2, or 3	$-\overset{\mid}{\underset{\mid}{P}}-\overset{\mid}{\underset{\mid}{C}}-,-P{=}C\overset{\diagup *}{\diagdown},\quad P{\equiv}C-{}^{*}$
X (F, Cl, Br, I)	1	$X-\overset{\mid}{\underset{\mid}{C}}-$

*Not commonly found.

The Shapes of Organic Compounds

The three-dimensional shape of an organic molecule is primarily determined by the way the covalent bonds about each atom in the molecule are directed in space. Though the primary factors responsible for the shapes of molecules were discussed in Chapter 8, it would be useful to review how the bonds about carbon are arranged, and how these arrangements affect the structure of organic molecules.

1. If the carbon atom is bonded to four other atoms using *four single bonds,* each bond will be directed toward the corners of a regular tetrahedron. The angle between any two bonds in the tetrahedral arrangement is $109\frac{1}{2}°$. Methane, CH_4, has a **tetrahedral** shape as shown in Figure 14.1:

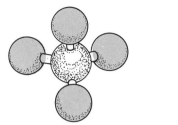

Fig. 14.1. Two models representing the methane molecule. The one on the left is a ball-and-stick model, and on the right, a space-filling model which gives a better indication of what the molecule actually looks like. Both show the tetrahedral arrangement of the four hydrogen atoms around the carbon atom.

The shape of a molecule that contains a chain of carbon atoms that have four single bonds can be viewed as a series of tetrahedra sharing corners. Though we often speak of a "straight chain" of carbon atoms, it actually has a zig-zag structure (see Figure 14.2).

2. If the carbon atom is bonded to three other atoms using *two single bonds and one double bond*, the three bonds will lie in the same plane and point to the corners of an equilateral triangle. The angle between the bonds is 120°.

Two molecules that have carbon atoms with this **trigonal planar** structure are formaldehyde and ethene:

$$
\underset{\text{H}}{\overset{\text{H}}{>}} \text{C}=\text{O}
\qquad
\underset{\text{H}}{\overset{\text{H}}{>}} \text{C}=\text{C} \underset{\text{H}}{\overset{\text{H}}{<}}
$$

Formaldehyde   Ethene

$CH_3CH_2CH_2CH_2CH_3$

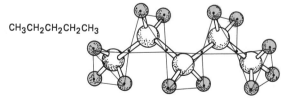

Fig. 14.2 The three dimensional shape of a straight-chain hydrocarbon is actually a zig-zag structure because of the tetrahedral arrangement of bonds about each carbon atom.

Formaldehyde and ethene are flat or planar molecules. The structure of ethene can be viewed as two equilateral triangles sharing a corner. The double bond holds both triangles in the same plane, as shown in Figure 14.3.

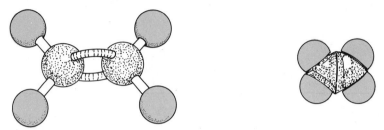

Fig. 14.3.   Ball-and-stick and space-filling models of ethene show its planar structure.

3. If the carbon atom is joined to two other atoms using *one single bond and one triple bond,* the bonds will be directed along a straight line and pointing in opposite directions. The same is true if the carbon atom is joined to two other atoms by *two double bonds.* The structure around these kinds of carbon atoms is said to be **linear,** and two examples are ethyne (acetylene), Figure 14.4, and carbon dioxide. The bond angle is 180°.

$$H—C≡C—H \qquad\qquad O=C=O$$

Ethyne   Carbon dioxide
(acetylene)

Fig. 14.4.   Ball-and-stick and space-filling models of ethyne (acetylene) show its linear structure.

Check Test Number 1

Describe how the bonds are arranged around the carbon atoms in each of the following compounds and give each bond angle.

(a)   (b)   (c)   H—C≡N

Answers:

    a. Planar, with all bond angles being 120°.

    b. Tetrahedral, and each bond angle is 109½°.

    c. Linear and the bond angle is 180°.

Writing Formulas for Organic Compounds

You have already learned how formulas are used to describe the composition of compounds. **Molecular formulas** give the actual number of atoms of each element in a molecule, though they do not show how the atoms are joined together.

Lewis formulas or **Lewis structures** are more revealing since they show not only what elements are present but also how the atoms are joined together in the molecule. They also show how the valence electrons are distributed around each atom. To refresh your memory, the Lewis formulas of two common organic compounds are shown below. Different symbols are used here for the valence electrons on adjacent atoms. Remember, carbon has four valence electrons, oxygen has six, and hydrogen one.

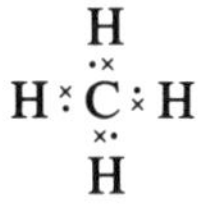

Methane 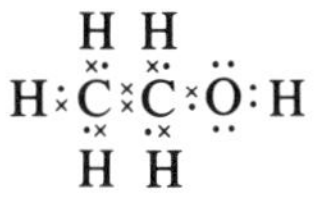   Ethyl alcohol

Organic compounds are very often represented using **structural formulas** or **dash structures** as they are sometimes called. A structural formula is very much like a Lewis formula, except that each pair of bonding electrons is symbolized using a dash (–) instead of dots or ×'s. Electron pairs not involved in bonding are usually omitted. The structural formulas of four organic compounds are shown below:

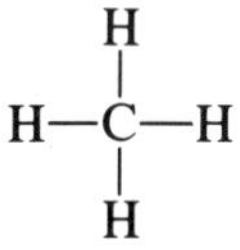

Methane 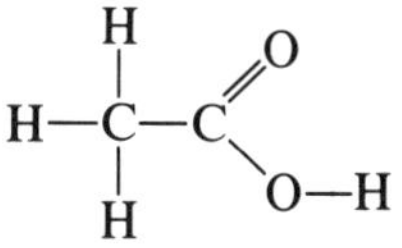   Acetic acid

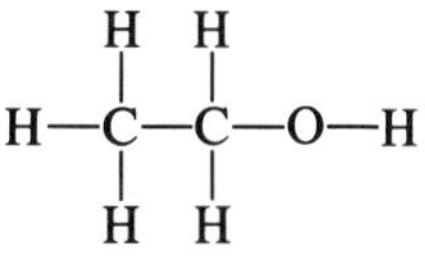

Ethyl alcohol 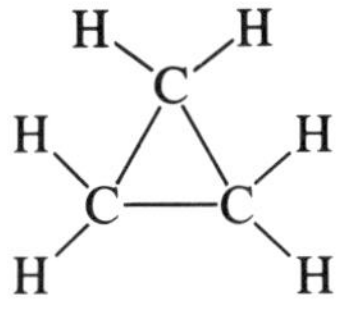   Cyclopropane

The **condensed formula** is an abbreviated way of writing a structural formula, and it is often used in textbooks to save space. The condensed formula for methane is CH_4. Even though it does not show the bonds or the arrangement of the atoms, the structure can be figured out since you know that carbon forms four bonds and hydrogen one.

Structural formulas can be condensed to varying degrees, ranging from those with only part of the formula condensed, to those that eliminate the dashes between atoms altogether. Certain chemically important groups of atoms called functional groups are always represented in condensed formulas by distinct arrangements of atoms: a carboxylic acid by –COOH, and an alcohol by –OH, for example. Functional groups are discussed later in this chapter, and the distinctive condensed formulas for each are given in Table 14.2. Care must be taken when writing condensed formulas so that only one possible structure can be derived from them. The following examples will demonstrate how structural formulas can be condensed to varying degrees:

Acetic acid 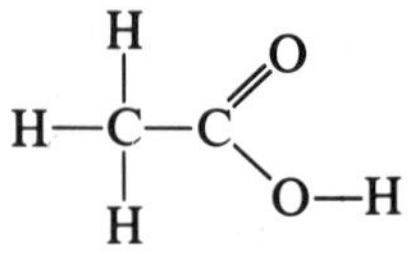
becomes

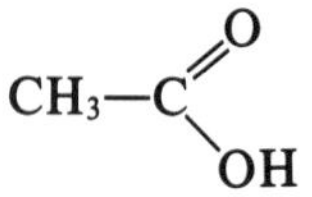

or CH_3COOH

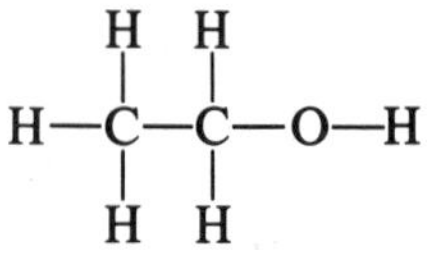

becomes

Ethyl alcohol $CH_3—CH_2—O—H$ or CH_3CH_2OH or C_2H_5OH

Cyclopropane 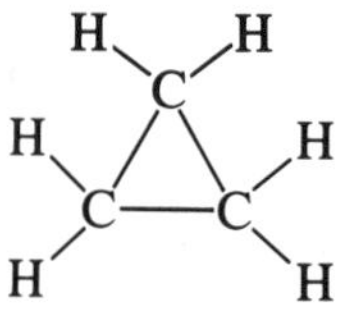
becomes

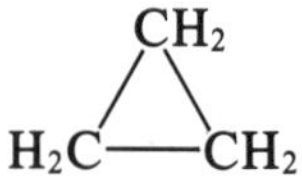

or

When rings of carbon atoms are symbolized using only lines, it is understood that a carbon atom is located at each corner with enough hydrogen atoms attached to it to completely fill its four bonds:

Isoprene (the building block for vitamin A),

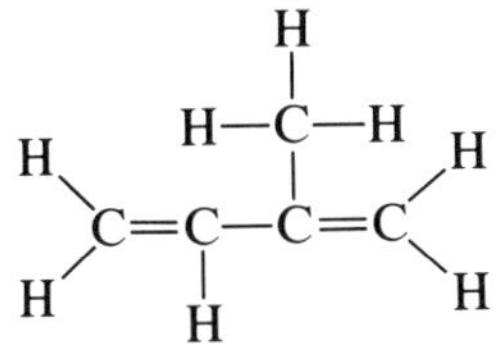

becomes

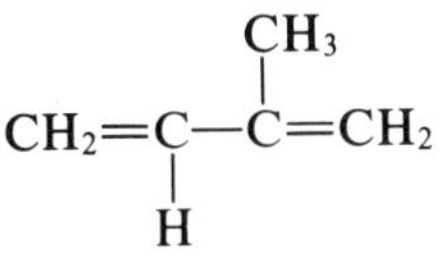

$$CH_2=C-\overset{CH_3}{\underset{H}{C}}=CH_2 \quad \text{or} \quad CH_2=CH\overset{CH_3}{C}=CH_2 \quad \text{or} \quad CH_2=CHC(CH_3)=CH_2$$

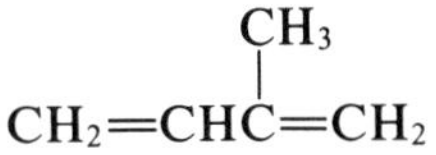

Check Test Number 2

The condensed formulas for three organic compounds are given below. Write each as a structural formula.

a. $CH_3CH(CH_3)CH_3$

b.  $\square$

c. $CH_2=CHCH=CH_2$

Answers:

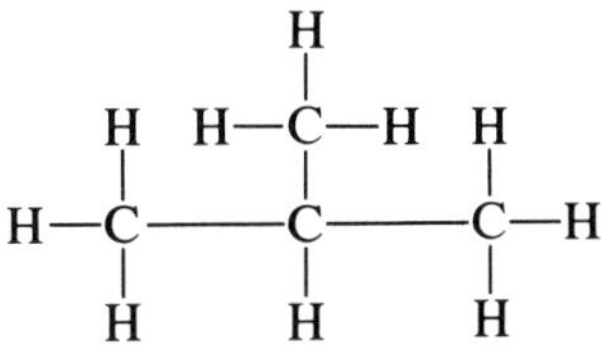

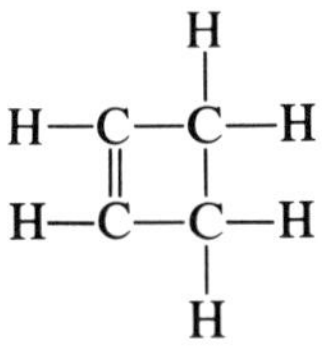

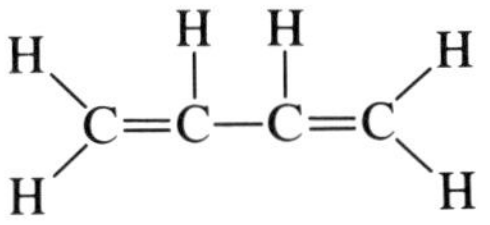

Modeling Organic Compounds

Sometimes it is not easy to visualize what a molecule actually looks like by just examining its structural formula. Because of this, chemists have devised different kinds of models to simulate the actual shapes of molecules. Two of these model types are used in the figures in this chapter. One is called a **ball-and-stick model,** and it uses small balls to represent atoms and short sticks or metal springs to represent the bonds between atoms. In a sense, they are chemical "tinker toys." A ball-and-stick model allows you to see how the atoms are arranged in space, and it also lets you see the angles between bonds easily. Though ball-and-stick models have several advantages, they do not provide the best picture of what a molecule really looks like.

A second kind of model, called the **space-filling model**, provides a more realistic picture of a molecule. The space-filling models seen in this chapter use sections of spheres to represent atoms. The sections are snapped together to form the model of a molecule. Because space-filling models position the atoms close together, as they are in actual molecules, they give a more accurate idea of what the molecule really is like. Different size spheres and sections of spheres are used to represent different size atoms. The bonds between atoms are not as easily seen as they are in the ball and stick models, but of course it is understood that the atoms are joined together by covalent bonds. Also the angles between bonds may not be as easily seen, but since space-filling models clearly show how the atoms are arranged in space the bond angles can be estimated.

As you can see, both kinds of models have their advantages and disadvantages. Generally, though, the space-filling models are more realistic than the ball-and-stick models.

The Three Major Types of Organic Compounds

Organic compounds can be separated into three categories based on the way the carbon skeleton of the compound is constructed. The categories are named aliphatic, aromatic, and heterocyclic.

The **aliphatic** group contains compounds with straight-chain, branched-chain, and certain cyclic arrangements (rings) of carbon atoms. If the compounds contain only carbon and hydrogen, they are called **hydrocarbons.** The aliphatics can have elements other than hydrogen bonded to carbon. **Aromatic** compounds all contain rings of carbon atoms, but unlike their cyclic aliphatic counterparts, they have an alternating arrangement of single and double bonds between the atoms in the ring. By far the most common aromatic compound is benzene. Benzene will be studied in Chapter 16. **Heterocyclic** compounds contain at least one ring that contains two or more different kinds of atoms. Besides carbon, the rings most commonly contain oxygen, nitrogen, or sulfur. Heterocyclic compounds will be described further as they appear in future chapters.

The Functional Groups of Organic Compounds

A **functional group** is an atom or group of atoms in a molecule that is responsible for many of its physical and chemical properties. Frequently it is the chemically reactive center of the molecule. For example, the hydroxyl group (–OH) is a functional group of a series of organic compounds known as alcohols.

$$CH_3OH \quad \text{methyl alcohol}$$
$$CH_3CH_2OH \quad \text{ethyl alcohol}$$
$$CH_3CH_2CH_2OH \quad \textit{n}\text{-propyl alcohol}$$

The chemical behavior of each of these alcohols is very similar because each contains the same functional group. If methyl alcohol undergoes a certain reaction, you can expect ethyl alcohol to undergo the same reaction because it has the same reactive center. Though the hydrocarbon part of the molecule of different alcohols will differ in size and shape, their properties will, in general, be determined by the nature of the functional group. The hydrocarbon portions, though not totally unimportant, have a lesser effect. Later on, when you study alcohols, you will really be studying the properties of the hydroxyl functional group, which can then be applied to alcohols in general.

Classifying compounds by their functional groups greatly simplifies the study of organic chemistry. Table 14.2 lists the common classes of functional groups. The symbol "R–," used in the general formulas, stands for any hydrocarbon group bonded to the functional group. R–OH could stand for methyl alcohol, ethyl alcohol, or any other alcohol. In some cases hydrogen can be substituted for the R group, and this is indicated by (H). You should begin to learn the class names and structural formulas of the functional groups right away.

TABLE   14.2
The Functional Groups in Organic Chemistry

Functional group	Functional group name	Class name of compounds	General formula of compounds	Condensed formula of compounds	Examples
$\diagup C=C \diagdown$	double bond	alkene	$R-\overset{\overset{H}{\mid}}{C}=\overset{\overset{H}{\mid}}{C}-R$ (H) (H)	$R-CH{=}CH-R$	$H_2C{=}CH_2$ ethene
$-C{\equiv}C-$	triple bond	alkyne	$R-C{\equiv}C-R$ (H) (H)	$R-C{\equiv}C-R$	$H-C{\equiv}C-H$ ethyne
$-OH$	hydroxyl	alcohol	$R-OH$	ROH	CH_3OH methyl alcohol
$-\overset{\mid}{\underset{\mid}{C}}-O-\overset{\mid}{\underset{\mid}{C}}-$ ether	ether	$R-O-R$	ROR	$H_3C-O-CH_3$ dimethyl ether
$\overset{O}{\underset{}{\parallel}}$ $\diagup C \diagdown$	carbonyl	aldehyde	$R-\overset{\overset{O}{\parallel}}{C}-H$ (H)	$RCHO$	$H_3C-\overset{\overset{\mid}{C}}{\underset{\underset{H}{\mid}}{}}{=}O$ acetaldehyde
		ketone	$R-\overset{\overset{O}{\parallel}}{C}-R$	$RCOR$	$H_3C-\overset{\overset{O}{\parallel}}{C}-CH_3$ acetone
$-\overset{\overset{O}{\parallel}}{C}\diagdown_{OH}$	carboxyl	carboxylic acid	$R-\overset{\overset{O}{\parallel}}{C}-OH$ (H) $\diagdown OH$	$RCOOH$	$H_3C-C\overset{\diagup O}{\diagdown_{OH}}$ acetic acid
$-\overset{\overset{O}{\parallel}}{C}\diagdown_{O-\overset{\mid}{C}-}$	ester	ester	$R-\overset{\overset{O}{\parallel}}{C}-O-R$ (H) $\diagdown O-R$	$RCOOR$	$H_3C-C\overset{\diagup O}{\diagdown_{OCH_2CH_3}}$ ethyl acetate
$-N-$	amino	amine	$R-\overset{\overset{}{\underset{\underset{H}{\mid}}{N}}}{}-H$ (primary)	RNH_2	H_3C-NH_2 methyl amine
			$R-\overset{\overset{}{\underset{\underset{H}{\mid}}{N}}}{}-R$ (secondary)	R_2NH	$H_3C-N-CH_3$ $\overset{}{\underset{H}{\mid}}$ dimethyl amine

TABLE 14.2 (cont.)
The Functional Groups in Organic Chemistry

Functional group	Functional group name	Class name of compounds	General formula of compounds	Condensed formula of compounds	Examples

			R–N–R (tertiary)
			     |
			     R
				R_3N	$H_3C–N–CH_3$
					     |
					     CH_3
					 trimethyl amine

$-\overset{\overset{O}{\|\|}}{C}-NH_2$	amide	amide	$R-\overset{\overset{O}{\|\|}}{C}-N-H_2$ (H)	$RCONH_2$	$H_3C-C\overset{\diagup O}{\diagdown NH_2}$ acetamide
$-NO_2$	nitro	nitro	$R-NO_2$	RNO_2	$H_3C–NO_2$ nitromethane
$-SH$	sulfhydryl	mercaptan	$R–SH$	RSH	$H_3C–SH$ methyl mercaptan
$-S-S-$	disulfide	disulfide	$R–S–S–R$	$RSSR$	$H_3C–S–S–CH_3$ dimethyl disulfide
$-X$ ($-F, -Cl,$ $-Br, -I$)	halo- (fluoro-, chloro-, bromo-, iodo-)	halide	$R–X$	RX	H_3CCl chloromethane

Check Test Number 3

1. Name the functional group(s) present in each of the following compounds.

    a. $CH_3CH_2OCH_3$

    b. $H_3C-C\overset{\diagup O}{\diagdown H}$

    c. $CH_3{=}C-CH_3$
            |
           CH_2

    d. $HS-CH_2-\overset{\overset{H}{\|}}{\underset{\underset{NH_2}{\|}}{C}}-C\overset{\diagup O}{\diagdown OH}$

    e. $H_3C-C\overset{\diagup O}{\diagdown NH_2}$

2. In which class should each of the following compounds be placed?

a. CH_3CH_2COOH

d. $CH_3-\overset{\overset{\displaystyle O}{\|}}{C}-CH_3$

e. CH_3CH_2OH

b. $CH_3-\overset{\overset{\displaystyle O}{\|}}{C}-O-CH_3$

c. $CH_3C\equiv CCH_3$

Answers:

1. (a) ether, (b) carbonyl, (c) double bond, (d) sulfhydryl, amino, carboxyl, (e) amide
2. (a) carboxylic acid, (b) ester, (c) alkyne, (d) ketone, (e) alcohol

Terms

Several important terms appeared in Chapter 14. You should understand the meaning of each.

hydrocarbon
straight chain
branched chain
cyclic hydrocarbon
condensed formula
functional group
alkene
alkyne
alcohol

ether
aldehyde
ketone
carboxylic acid
ester
amine
amide
mercaptan
halide

Questions

Answers to the starred questions appear at the end of the text.

1. Define "organic chemistry," and compare it with inorganic chemistry.
*2. Which of the following structures are incorrect?

a. 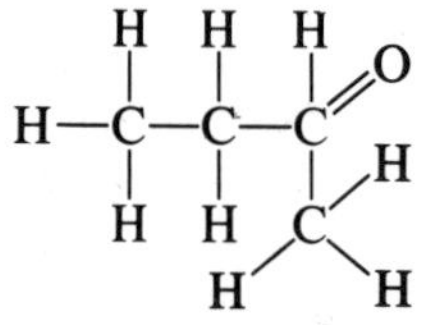
$$H-\overset{\overset{\displaystyle H}{|}}{\underset{\underset{\displaystyle H}{|}}{C}}-\overset{\overset{\displaystyle H}{|}}{\underset{\underset{\displaystyle H}{|}}{C}}-\overset{\overset{\displaystyle H}{|}}{\underset{\underset{\displaystyle C}{|}}{C}}\overset{\displaystyle O}{\diagup}_{H}$$

b.

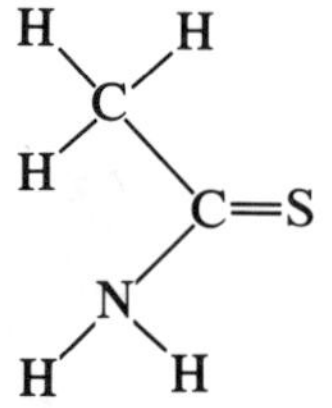

c.

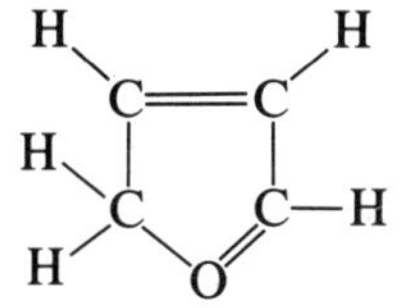

d. 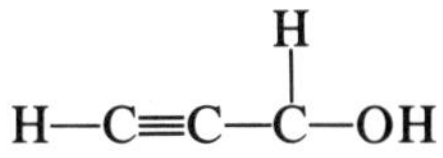 $H-C\equiv C-\overset{\overset{\displaystyle H}{|}}{C}-OH$

3. Draw Lewis structures for the follwing.
   a. $CH_3CH=CH_2$ b. CH_3COOCH_3 c. CH_3Cl

4. Draw the structural formulas for the following.
   a. $CH_3CH=CH_2$ f. CH_3OCH_3
   b. CH_3COOCH_3 g.
   c. CH_3Cl
   d. CH_3SH
   e. CH_2O

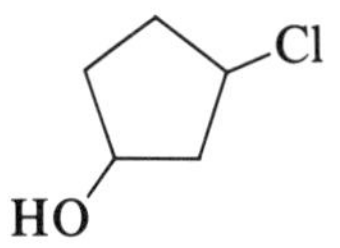

5. Give the name of each functional group in the following compounds:

*a. $CH_3CH=CHCH_3$

*g.

b.

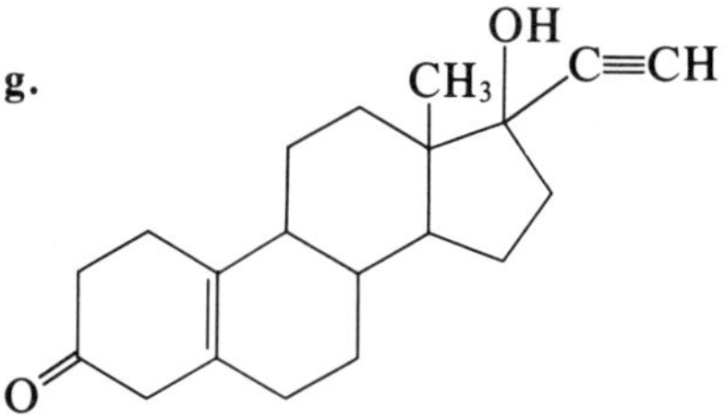

Norethynodrel (in birth control pills)

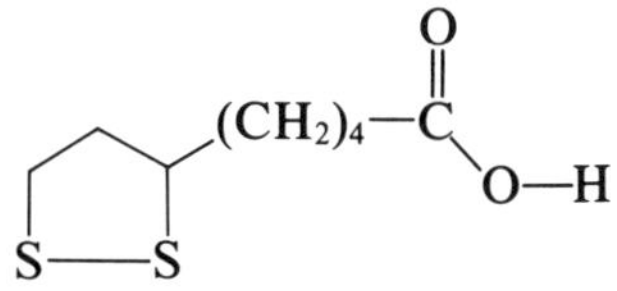

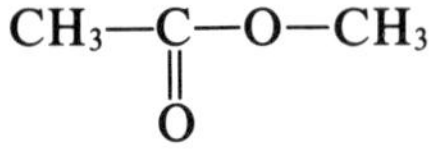

*c. $CH_3-\overset{\overset{\displaystyle O}{||}}{C}-O-CH_3$

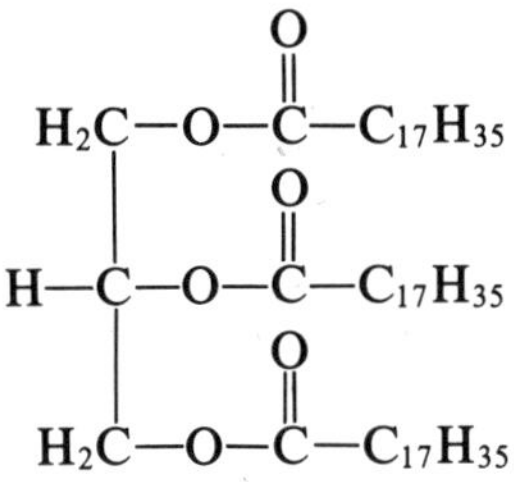

h.

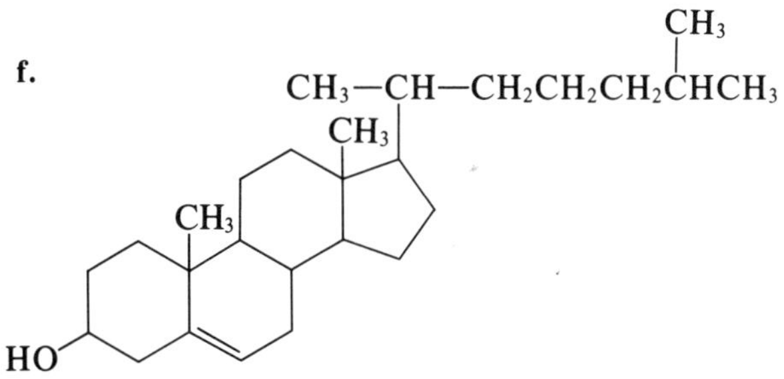

d. $CH_3CHOHCH_2COOH$

*e.

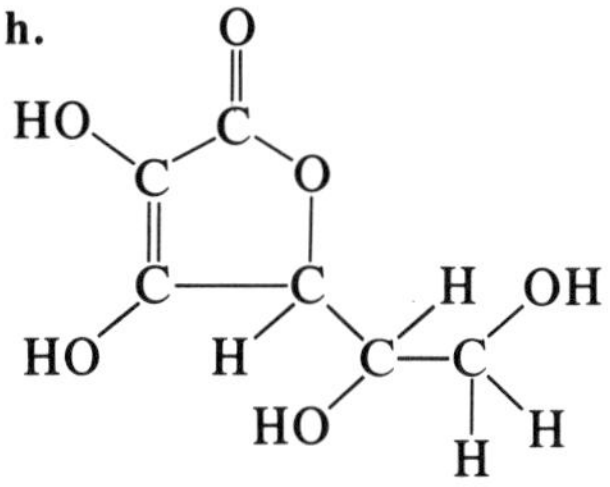

Vitamic C (ascorbic acid)

f.

$CH_3-\underset{\underset{\displaystyle CH_3}{|}}{CH}-CH_2CH_2CH_2\overset{\overset{\displaystyle CH_3}{|}}{C}HCH_3$

Cholesterol

6. In which class would each of the following compounds be placed? If there is more than one functional group, then identify the class for each functional group.

*a. $CH_3CH_2\overset{\overset{\displaystyle O}{\|}}{C}-CH_3$

b. $CH_3CH_2CH_2OH$

*c. 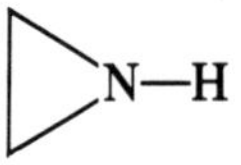 $\triangleright\!\!\!<^{N-H}$

d. $CH_3-C\overset{\displaystyle O}{\underset{\displaystyle NH_2}{\diagup}}$

*e. $CH_3\underset{\underset{\displaystyle OH}{|}}{CH}COOH$

f. $H_2C=CH-C\overset{\displaystyle O}{\underset{\displaystyle OCH_3}{\diagup}}$

*g. $HOCH_2CH_2CH_2\underset{\underset{\displaystyle NH_2}{|}}{CH}COOH$

h. H_2CO

*i. 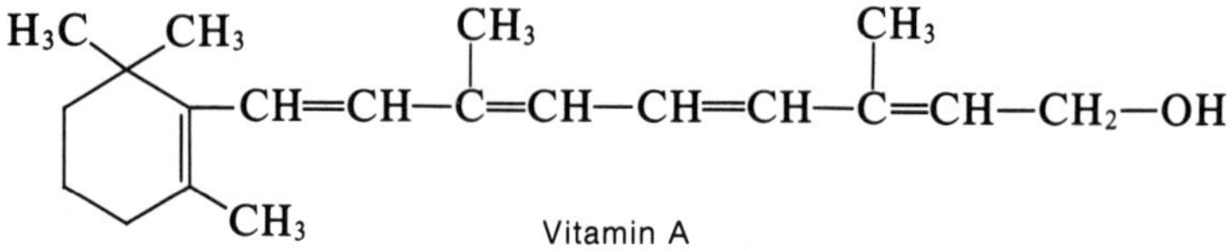   Vitamin A

H_3C CH_3 CH_3 CH_3
$-CH=CH-C=CH-CH=CH-C=CH-CH_2-OH$
CH_3

j. $CH_3CHBrCH_2CH=CH_2$

The Saturated Hydrocarbons

The saturated hydrocarbons are the simplest and least reactive class of organic compounds. They are composed only of carbon and hydrogen. The single feature that distinguishes the saturated hydrocarbons from other hydrocarbons is that every carbon atom is bonded to four other atoms by single, covalent bonds. There are no double or triple bonds between carbon atoms. The lack of multiple bonds together with the absence of other functional groups (Table 14.2) is largely responsible for their lack of chemical reactivity.

The saturated hydrocarbons are also called paraffins or alkanes. Paraffin is a name derived from Latin (*parum affinis*, little activity) which suggests their unreactive nature. Alkane, a term which is more commonly used, is derived from the scheme used to name these compounds. We will call the saturated hydrocarbons alkanes, as is most often done.

We encounter alkanes every day. Most of our petroleum fuels such as natural gas, gasoline, and fuel oil are alkanes. So are the lubricating oils and greases. In the health care field they are used as lubricants in the intestinal tract (mineral oil), as a base for creams and ointments, and as a protective coating on skin (petroleum jelly) to aid healing and prevent exposure to irritants.

The discussion will begin with a description of the first ten members of the alkane family and their structural isomers. Then, you will see how chemists name these compounds and their derivatives. Though the alkanes are not very reactive, some of the reactions they do undergo will be presented along with a description of their physical properties. The chapter will conclude with the cycloalkanes, saturated hydrocarbons with rings of carbon atoms.

Objectives

By the time you finish this chapter you should be able to do the following:

1. Explain why the alkanes are called saturated hydrocarbons.
2. Name and give the structural formulas for the first ten straight-chain alkanes.
3. Explain why the alkanes constitute a homologous series of compounds.
4. Determine if two compounds are structural isomers given their structural formulas.
5. Give the names and formulas of the most common alkyl groups.
6. Give the IUPAC name for an alkane or its derivative given its structural formula and vice versa.
7. Describe the physical and chemical properties of the alkanes.
8. Write a balanced equation for the combustion of any alkane.
9. Describe the details of the substitution reaction of Cl_2 with an alkane.
10. Name and draw the structural formulas for the first four cycloalkanes.
11. Describe the difference between the addition reaction and the substitution reaction of Cl_2 with cyclopropane.

The Alkanes

The **alkanes** are hydrocarbons in which each carbon atom is bonded to four other atoms through single bonds. Because each of the four bonds on every carbon atom is "filled," the alkanes are said to be saturated compounds. Frequently, the alkanes are called the **saturated hydrocarbons**. Alkanes can be either straight-chain or branched-chain molecules. The saturated hydrocarbons in which the carbon atoms are in rings are called **cycloalkanes**. The cycloalkanes will be discussed later on in this chapter.

Though the alkanes are not as important as some of the other classes of organic compounds, they deserve careful study since many other compounds can be considered derivatives of the alkanes. The first ten members of the straight-chain alkane family are listed in Table 15.1. You should learn the names, molecular formulas, and condensed structural formulas of each one. These names, with modification, will be used to name many other organic compounds.

TABLE 15.1
Straight-Chain Alkanes

Name	Molecular formula	Condensed structural formula	Boiling point (°C)	Density (g/ml)
Methane	CH_4	CH_4	−161	—
Ethane	C_2H_6	CH_3CH_3	−89	—
Propane	C_3H_8	$CH_3CH_2CH_3$	−42	—
Butane	C_4H_{10}	$CH_3CH_2CH_2CH_3$	0	—
Pentane	C_5H_{12}	$CH_3CH_2CH_2CH_2CH_3$	36	0.63
Hexane	C_6H_{14}	$CH_3CH_2CH_2CH_2CH_2CH_3$	69	0.66
Heptane	C_7H_{16}	$CH_3CH_2CH_2CH_2CH_2CH_2CH_3$	98	0.68
Octane	C_8H_{18}	$CH_3CH_2CH_2CH_2CH_2CH_2CH_2CH_3$	125	0.70
Nonane	C_9H_{20}	$CH_3CH_2CH_2CH_2CH_2CH_2CH_2CH_2CH_3$	151	0.72
Decane	$C_{10}H_{22}$	$CH_3CH_2CH_2CH_2CH_2CH_2CH_2CH_2CH_2CH_3$	174	0.73

The general formula for the open-chain (straight- and branched-chain) alkanes is C_nH_{2n+2}, where n is the number of carbon atoms in the molecule. The formula of the alkane with two carbon atoms, ethane, would be $C_2H_{2(2)+2}$, or C_2H_6. After the first four members of the series, the name of each compound uses a Greek prefix to indicate the number of carbon atoms in the molecule (see Table 8.6). This makes it easier to connect a name with a formula and vice versa. The -ane ending on each name means it is a saturated compound.

Notice in Table 15.1 that each member of the alkane series differs from the one just above or below it by a single $-CH_2-$ unit. Any series of compounds that displays a constant difference between its members is called a **homologous series.** Perhaps the most important characteristic of a homologous series is that each member possesses similar chemical properties. Each alkane behaves much like any other alkane in chemical reactions. Methane, propane, and hexane all react with oxygen to form the same products: carbon dioxide and water. The members of the other classes of organic compounds form homologous series too. Because of this, one only needs to study a few members of each class to learn the chemical properties of all its members.

Check Test Number 1

1. What would be the formula for the straight-chain alkane that has 30 carbon atoms?
2. What is a homologous series of compounds?

Answers:

1. $n=30$, $C_{30}H_{2(30)+2} = C_{30}H_{62}$
2. A homologous series of compounds is one in which the members differ from one another by some constant unit (one $-CH_2-$ group in the case of the alkanes).

Structural Isomers of the Alkanes

Structural isomers are compounds that have the same formula but different structures. They are different compounds, each with its own set of properties. The only thing they have in common is their formulas. There are two isomers with the formula C_4H_{10}, and they constitute the butane family. The common name for the straight-chain isomer is normal butane or just *n*-butane, and the common name for the branched-chain isomer is isobutane. Their different properties are reflected in their different boiling points.

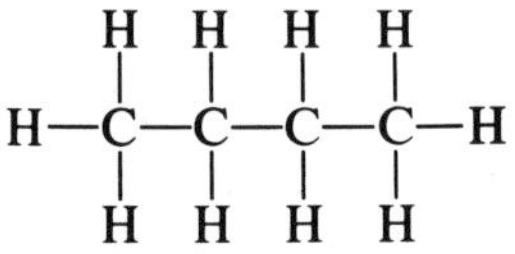

n-Butane (C_4H_{10})
Boiling point: 0°C

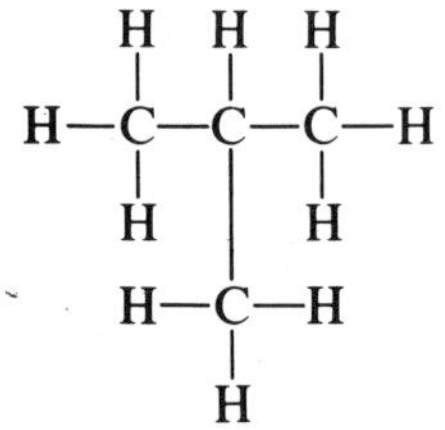

Isobutane (C_4H_{10})
Boiling point: –12°C

Models of these two isomeric compounds appear in Figure 15.1.

Going down the alkane series in Table 15.1, as the number of carbon atoms in each alkane increases, the number of possible isomers of each alkane increases also. There are three isomers of pentane (C_5H_{12}), five of hexane (C_6H_{14}), and nine of heptane (C_7H_{16}). In fact, the number of isomers for $C_{20}H_{42}$ is estimated to be over 360,000. For these very large alkanes, only a few of the isomers have been found in nature or synthesized in the laboratory, though in principle any isomer could be synthesized if needed.

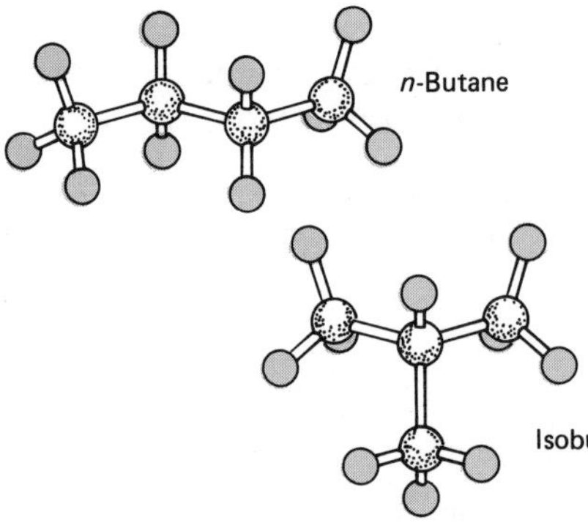

n-Butane

Isobutane

Fig. 15.1.   Ball-and-stick models of the two isomers of the butane family: *n*-butane and isobutane.

The common names of the three isomers of the pentane family are *n*-pentane, isopentane, and neopentane. Models of these isomers appear in Figure 15.2.

n-Pentane
Boiling point: 36°C

Isopentane
Boiling point: 28°C

Neopentane
Boiling point: 10°C

The preferred way to draw structural formulas of the alkanes is to have the longest chain of carbon atoms written horizontally as was done with *n*-pentane and isopentane. But it would be just as correct to draw the formula with the longest chain bent around into a U-shape or an L-shape or any other shape as long as the atoms are all correctly connected together. For example, all of the structural formulas that follow are correct for isopentane.

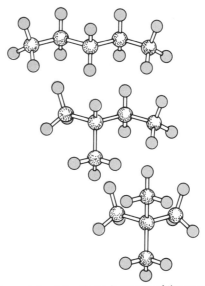

Fig. 15.2 Ball-and-stick models of the three isomers of the pentane family: *n*-pentane, isopentane, and neopentane.

$$H_3C-CH_2-\underset{\underset{\displaystyle CH_3}{|}}{CH}-CH_3 \quad H_3C-\underset{\underset{\displaystyle CH_3}{|}}{CH}-CH_2-CH_3 \quad \underset{\underset{\displaystyle CH_3}{|}}{\overset{\overset{\displaystyle CH_3}{|}}{CH_2}}-\underset{\underset{\displaystyle CH_3}{|}}{\overset{\overset{\displaystyle CH_3}{|}}{CH}} \quad \underset{\underset{\displaystyle H_3C-CH_2}{}}{\overset{\overset{\displaystyle CH_3}{|}}{CH}}-CH_3$$

     (a)   (b)   (c)   (d)

At first glance, they may not look like the same compound, and you might mistake them for isomers. If (b) were turned over end-for-end, it would be identical to (a). Atoms did not have to be disconnected and moved about to change (b) into (a) since they are the same compound. If they were isomers, you would have to break bonds and rearrange atoms to make the structural formulas the same. The carbon atoms are numbered to help you see that the molecule is just being turned over.

$$\overset{1}{CH_3}-\underset{\underset{\displaystyle \overset{5}{CH_3}}{|}}{\overset{2}{CH}}-\overset{3}{CH_2}-\overset{4}{CH_3} \quad \longrightarrow \quad \overset{4}{CH_3}-\overset{3}{CH_2}-\underset{\underset{\displaystyle \overset{5}{CH_3}}{|}}{\overset{2}{CH}}-\overset{1}{CH_3}$$

              (b)   Same as (a)

If the longest carbon chains in (c) and (d) were straightened out and written horizontally, they too would be quickly recognized as isopentane.

$$\underset{\underset{\displaystyle \overset{5}{CH_3}}{|}}{\underset{|3}{\overset{4}{CH_3}}} \quad \overset{1}{\underset{|2}{CH}} \longrightarrow \overset{4}{CH_3}-\overset{3}{CH_2}-\underset{\underset{\displaystyle \overset{5}{CH_3}}{|}}{\overset{2}{CH}}-\overset{1}{CH_3} \longleftarrow \underset{\underset{\displaystyle \overset{3}{CH_2}}{|}}{\underset{\overset{4}{CH_3}}{\overset{\overset{1}{CH_3}}{\underset{|2}{CH}}}}-\overset{5}{CH_3}$$

     (c)   Same as (a)   (d)

It must be emphasized that the four structures given for isopentane are *not* isomers. They look different only because we chose to draw the structural formula in different ways. Actually, the isopentane molecule can twist around forming shapes that do resemble the structural formulas given above. But it is still the same compound.

Check Test 2

1. Which of the following pairs of compounds are isomers?

   a. CH_3CH_2Cl, $ClCH_2CH_3$ b. $CH_3-\underset{\underset{\displaystyle CH_3-CH-CH_3}{|}}{CH}-CH_3$, $\underset{\underset{\displaystyle CH_3-CH_2}{}}{\overset{\overset{\displaystyle CH_3}{|}}{CH_2}}-\underset{\underset{\displaystyle CH_3}{}}{CH}$

c. CH_3—$\underset{\underset{CH_3-CH_2}{|}}{CH}$—$CH_3$, CH_3—$\underset{\underset{CH_3}{|}}{CH}$—$CH_2$—$CH_3$

d. H_2C—$\overset{CH_2}{\triangle}$—$CH_2$, $CH_3CH_2CH_3$

2. What would be the common names for the straight-chain isomers of hexane, pentane, and octane?

Answers:

1. a. identical compounds   c. identical compounds
   b. isomers   d. two different compounds since the formulas are different

2. *n*-hexane, *n*-pentane, *n*-octane

Naming Alkanes and Their Derivatives

There are thousands of alkanes, and each one of them has a name. The name is used to conveniently represent the compound in written form when structural formulas would be too cumbersome. Also, the name should supply just enough information about the compound so that its structural formula can be drawn if needed. The names of the ten alkanes in Table 15.1 are not difficult to remember, but as the compounds become larger and more complex, the need for a systematic scheme to name them is necessary.

The nomenclature system that is now used by all chemists for organic compounds is based on recommendations made by the International Union of Pure and Applied Chemistry (IUPAC). The IUPAC (pronounced "U-PAK") recommendations are presented as a series of rules that, when followed, will give names of compounds that will be understood by others who are familiar with the rules. Though the IUPAC system is widely accepted, common names are still often used for many compounds that are widely used everyday. Common names are frequently used in medicine, so you should make note of them as they are pointed out.

An important feature of the IUPAC system is the way various saturated hydrocarbon groups are regarded as substituents on the longest carbon chain of the molecule. A **substituent** is an atom or group of atoms that substitutes for a hydrogen atom on one of the carbon atoms in the longest chain. The saturated hydrocarbon substituents are called **alkyl groups**, and the name of the group is derived from the alkane with the same number of carbon atoms. The $-CH_3$ group is the methyl group. Its name is derived from methane by replacing the -ane ending with -yl. Notice the "open bond" on the methyl group. This is the point at which the carbon atom is attached to the longest carbon chain.

The methyl group (It occupies a site on the carbon chain that would otherwise belong to hydrogen. In a sense, it has been substituted for a hydrogen atom.)

Several common alkyl groups are listed in Table 15.2. If it is possible to relate two alkyl groups to a given alkane, each alkyl group is given a prefix (n-, iso-) to identify it. In general, the symbol R– is used to symbolize any alkyl group, and R–H would symbolize an alkane.

R—H $\xrightarrow[\text{hydrogen atom}]{\text{Remove one}}$ R—

An alkane   An alkyl group

Methane $\xrightarrow[\text{hydrogen atom}]{\text{Remove one}}$ Methyl group

The nomenclature rules that apply to alkanes and their saturated derivatives are given below. They will be expanded as needed in later chapters to cover other classes of organic compounds.

The Rules for Naming Alkanes and Their Derivatives

1. The names of all alkanes and their saturated derivatives end in -*ane*.

2. The longest continuous chain of carbon atoms (the main chain) is taken as the "parent compound" of the molecule. The molecule is named as a derivative of the parent compound. If the longest chain contains six carbon atoms, the molecule is named as a derivative of hexane.

TABLE 15.2 Four Common Alkyl Groups		
Alkyl group	Related alkane	Alkyl group name
—CH$_3$	methane	methyl
—CH$_2$CH$_3$	ethane	ethyl
—CH$_2$CH$_2$CH$_3$	propane	n-propyl
—C—H with CH$_3$ and CH$_3$	propane	isopropyl

3. The names of any alkyl groups attached to the main chain are placed *before* the name of the parent compound. If there are different alkyl groups, they can be listed in alphabetical order.

4. Nonhydrocarbon substituents are listed before the name of the parent compound with the alkyl groups and in alphabetical order.
   Common groups of this type are:

–F	fluoro	–I	iodo
–Cl	chloro	$-NO_2$	nitro
–Br	bromo		

5. The Greek prefixes, di-, tri-, tetra-, penta-, and so forth, are used to indicate the number of each kind of group attached to the main chain. If two methyl groups are present, they would appear in the name as dimethyl.

6. The position of each substituent on the main chain is indicated with a number. Each carbon in the main chain is numbered starting at the end that allows the lowest numbers to appear in the name.

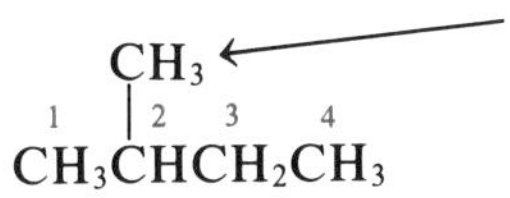

$$CH_3 \longleftarrow \quad \text{The methyl group is on carbon number 2,}$$
the lowest number that would identify that carbon atom.

$$\underset{1 \quad 2 \quad 3 \quad 4}{CH_3CHCH_2CH_3}$$

7. Commas are used to separate numbers.

$$\text{2,2,3-trimethylhexane}$$

8. Hyphens are used to separate numbers from words.

$$\text{3-chloropentane}$$

9. The name of the compound is written as one word, and the name of the "parent compound" always appears last.

The following examples apply these rules to develop the names for several compounds. Carefully note how each name is developed, and how commas and hyphens and prefixes are used.

Example 15.1

Give the IUPAC name for:

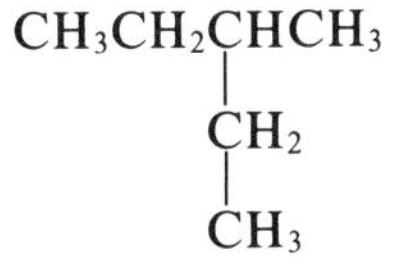

$$CH_3CH_2CHCH_3$$
$$|$$
$$CH_2$$
$$|$$
$$CH_3$$

The longest carbon chain contains five carbon atoms, so the "parent compound" is **pentane**.

$$\boxed{CH_3CH_2CH}\!\!-\!\!CH_3$$
$$\boxed{CH_2} \longleftarrow \text{Longest chain}$$
$$\boxed{CH_3}$$

You will need to make certain you choose the longest chain. The longest chain may be written in a straight line, but it might be in an L-shape or U-shape.

One methyl group is bonded to the pentane chain. In this example, the five carbon chain can be numbered from either end; the methyl group is on the number 3 carbon.

$$\overset{1}{CH_3}\!\!-\!\!\overset{2}{CH_2}\!\!-\!\!\overset{3}{CH}\!\!-\!\!\overset{}{\underparen{CH_3}} \longleftarrow \text{3-Methyl}$$
$$\overset{4}{CH_2}$$
$$\overset{5}{CH_3}$$

The IUPAC name is then:

3-methylpentane

Example 15.2

Give the IUPAC name for:

$$CH_3\!\!-\!\!CH_2\!\!-\!\!CH\!\!-\!\!CH_2\!\!-\!\!\overset{\displaystyle CH_3}{\underset{\displaystyle CH_3}{C}}\!\!-\!\!CH_3$$
$$\underset{\displaystyle Cl}{|}$$

The longest chain contains six carbon atoms, so the compound will be named as a derivative of **hexane**.

$$\overset{\displaystyle CH_3}{\boxed{CH_3\!\!-\!\!CH_2\!\!-\!\!\underset{Cl}{CH}\!\!-\!\!CH_2\!\!-\!\!\underset{CH_3}{C}\!\!-\!\!CH_3}} \longleftarrow \text{Longest chain}$$

There are two *methyl* groups (dimethyl) and a *chlorine* atom (chloro) attached to the hexane chain. The main chain must be numbered from right to left, so the smallest numbers appear in the name.

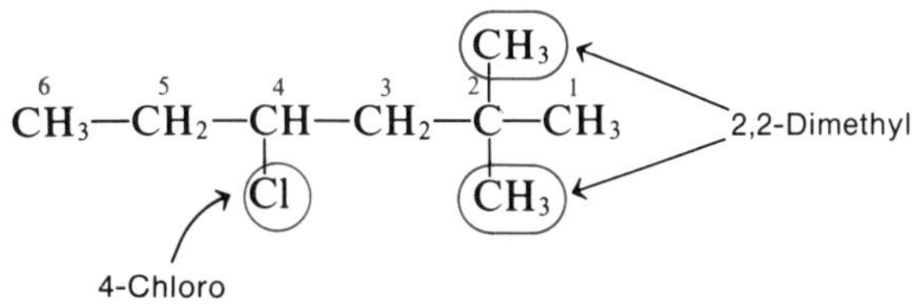

$$6 \quad 5 \quad 4 \quad 3 \quad 2 \quad 1$$
$$\text{CH}_3\text{—CH}_2\text{—CH—CH}_2\text{—C—CH}_3$$

with (CH₃) groups circled at top, 2,2-Dimethyl labeled; (Cl) circled, 4-Chloro labeled; (CH₃) circled at bottom.

The location of both methyl groups must appear in the name even if they are on the same carbon atom. The IUPAC name for the compound is then:

4-chloro-2,2-dimethylhexane

Example 15.3

Give the IUPAC name for:

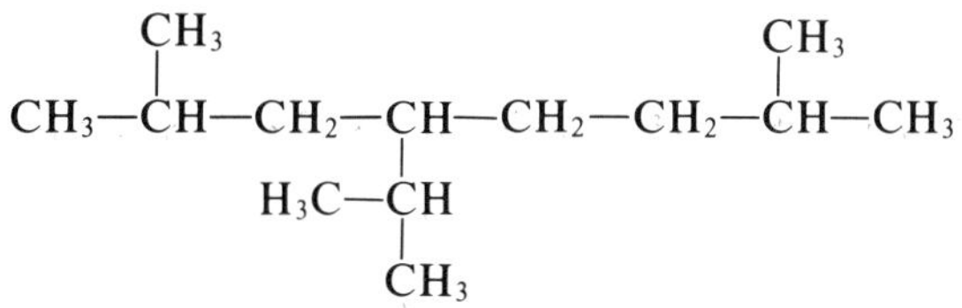

$$\text{CH}_3 \qquad\qquad\qquad \text{CH}_3$$
$$\text{CH}_3\text{—CH—CH}_2\text{—CH—CH}_2\text{—CH}_2\text{—CH—CH}_3$$
$$\text{H}_3\text{C—CH}$$
$$\text{CH}_3$$

The longest chain contains eight carbons, so it will be named as a derivative of **octane.** There are two *methyl* groups and one *isopropyl* group.
Numbering the carbons from left to right will give the smallest numbers: 2,7- dimethyl and 4-isopropyl. (Numbering in the reverse direction would give a set of larger numbers: 2,7-dimethyl and 5-isopropyl.) The IUPAC name would be:

2,7-dimethyl-4-isopropyloctane

The following examples are named correctly according to the rules given earlier. Incorrect names are also given to show some of the common errors that people make.

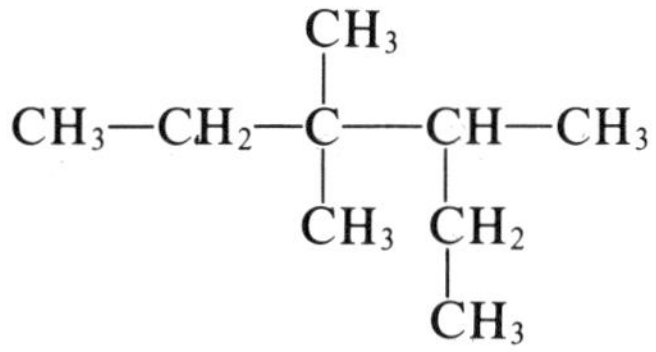

$$\text{CH}_3$$
$$\text{CH}_3\text{—CH}_2\text{—C—CH—CH}_3$$
$$\text{CH}_3 \quad \text{CH}_2$$
$$\text{CH}_3$$

Correct name: 3,3,4-trimethylhexane
Incorrect name: 2-ethyl-3,3-dimethylpentane (longest chain not chosen)

$$CH_3-\underset{\underset{\displaystyle CH_3}{|}}{\overset{\overset{\displaystyle CH_3}{|}}{C}}-CH_2-CH_3$$

Correct name: 2,2-dimethylbutane
Incorrect name: 2-dimethylbutane (two alkyl groups need two numbers)

$$CH_3-CH_2-\underset{\underset{\displaystyle CH_3}{\underset{\displaystyle |}{CH_2}}}{\overset{|}{CH}}-CH_2-CH_2-CH_3$$

Correct name: 3-ethylhexane
Incorrect name: 4-ethylhexane (longest chain numbered from wrong end)
3-*n*-propylpentane (did not use the longest chain)

$$\begin{array}{c} CH_3CH_2 \\ \\ CH_3CH_2 \end{array}\!\!\diagup\!\!\!\!\!\diagdown\, CH-\underset{\underset{\displaystyle CH_3}{|}}{CH}-CH_2-Br$$

Correct name: 1-bromo-3-ethyl-2-methylpentane
Incorrect name: 1-bromo-3,3-diethyl-2-methylpropane (incorrect choice of longest chain)

Check Test Number 3

1. What is the name of the alkyl group formed by removing one hydrogen atom from the end carbon in pentane. Draw the alkyl group.
2. Give the IUPAC names for:

a.
$$CH_3CH_2CH_2-\underset{\underset{\displaystyle CH_3}{|}}{\overset{\overset{\displaystyle Cl}{|}}{C}}-CH_3$$

b.
$$Br-\underset{\underset{\displaystyle CH_3}{|}}{\overset{\overset{\displaystyle CH_3}{|}}{CH}}-CH_2-\underset{\underset{\displaystyle CHCH_3}{\underset{\displaystyle |}{\overset{\displaystyle |}{CH_3}}}}{\overset{\overset{\displaystyle CH_3}{|}}{C}}-CH_2-\underset{\underset{\displaystyle CH_2}{\underset{\displaystyle |}{\overset{\displaystyle |}{CH_3}}}}{CH}-CH_2-CH_3$$

Answers:
1. *n*-pentyl, $-CH_2CH_2CH_2CH_2CH_3$
2. a. 2-chloro-2-methylpentane
   b. 2-bromo-6-ethyl-4-isopropyl-4-methyloctane

Structural Formulas from IUPAC Names

An IUPAC name contains all the necessary information needed to draw the structural formula of the compound. Suppose you wished to draw the structural formula of 1,1-dibromo-4-ethylhexane. Here is how it is done:

1. The parent compound is hexane, so a six-carbon atom chain is drawn and the carbon atoms are numbered from either end.

$$\overset{1}{C}-\overset{2}{C}-\overset{3}{C}-\overset{4}{C}-\overset{5}{C}-\overset{6}{C}$$

2. The substituent groups are joined to the appropriate carbon atoms:
   2 Br on carbon number 1
   1 CH_3CH_2- on carbon number 4

3. Complete the four bonds to each carbon by adding hydrogen.

Written in condensed form the formula is:

$$Br_2CH-CH_2-CH_2-\underset{\underset{CH_3}{\overset{|}{CH_2}}}{\overset{|}{CH}}-CH_2-CH_3$$

The structure of 2,2,4-trimethylpentane would be:

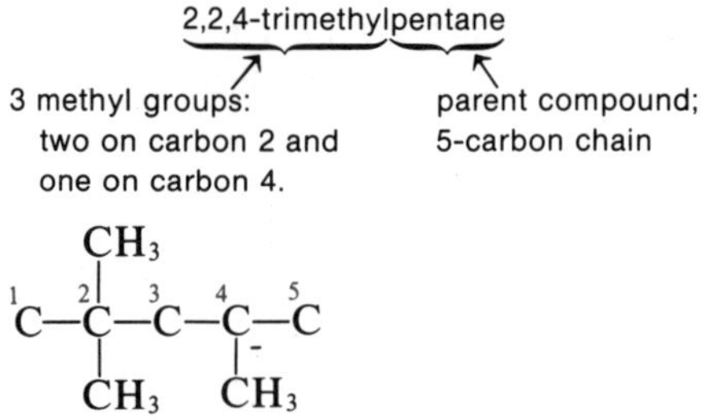

2,2,4-trimethylpentane

3 methyl groups: parent compound;
two on carbon 2 and   5-carbon chain
one on carbon 4.

$$
\begin{array}{c}
\overset{\displaystyle CH_3}{} \\
\overset{1}{C}-\overset{2}{C}-\overset{3}{C}-\overset{4}{C}-\overset{5}{C} \\
\underset{\displaystyle CH_3}{}\ \ \underset{\displaystyle CH_3}{}
\end{array}
$$

Adding the hydrogen, the condensed structural formula is:

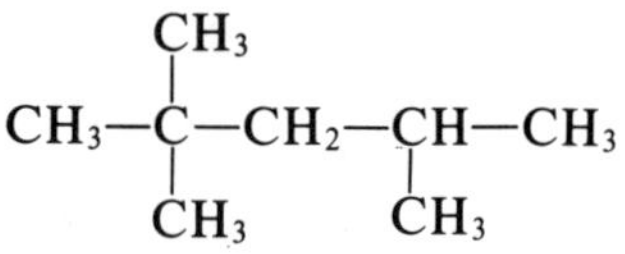

$$
\begin{array}{c}
CH_3 \\
| \\
CH_3-C-CH_2-CH-CH_3 \\
| \qquad\qquad | \\
CH_3 \qquad\ CH_3
\end{array}
$$

Check Test Number 4

Draw the condensed structural formulas for:
a. 1-bromo-3,4-dimethylpentane
b. 2,2,3,3-tetramethylhexane

Answers:

$$
\text{a.}\quad
\begin{array}{c}
CH_3 \\
| \\
BrCH_2-CH_2-CH-CH-CH_3 \\
| \\
CH_3
\end{array}
$$

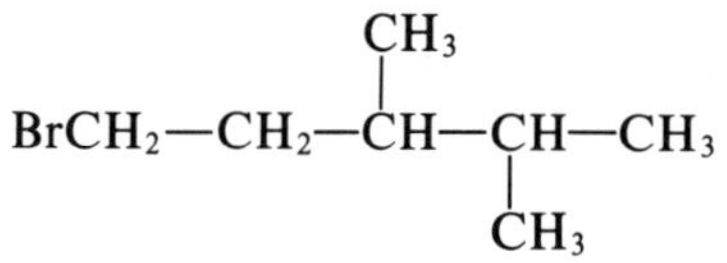

$$
\text{b.}\quad
\begin{array}{c}
CH_3\ \ CH_3 \\
|\qquad | \\
CH_3-C\!-\!\!-\!\!C-CH_2-CH_2-CH_3 \\
|\qquad | \\
CH_3\ \ CH_3
\end{array}
$$

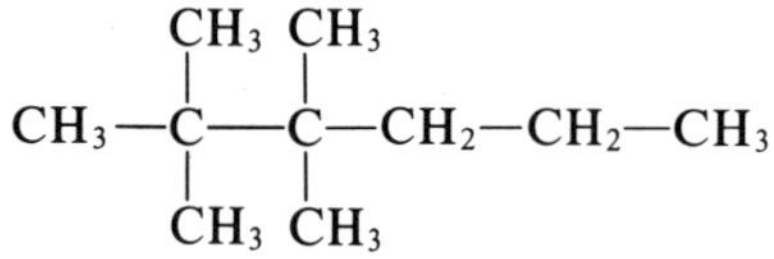

Sources of Organic Compounds

During the last 75 years our society has become increasingly dependent on a wide variety of organic compounds. Many of these compounds are in the alkane family. Alkanes are widely used as fuels, lubricants, and solvents. They are converted into other compounds that are used in the manufacture of synthetic fibers, structural plastics, drugs, and they have even been used in the manufacture of "synthetic food."

Some organic compounds are obtained from plants and animals, but most come either directly or indirectly from natural gas, petroleum, and coal. These underground sources are the remains of plant and animal tissue that, after millions of years in an environment of high pressures and temperatures, have been converted into simple organic compounds. Natural gas, as it comes from the ground, is a mixture of methane with lesser amounts of ethane, propane, and butane. Once methane is separated from the mixture, it is piped into homes and factories and used as a fuel. The remaining gaseous alkanes are used to make other organic chemicals or they too end up as fuels. Propane and butane are readily converted to liquids under pressure and stored in steel tanks as "bottle gas" or liquified petroleum gas (LPG).

Petroleum, or crude oil as it is often called, is of limited use as it comes directly from the ground. It is a thick, complicated mixture of organic compounds. Though its composition can vary from one location to the next, petroleum is principally a mixture of alkanes, cycloalkanes, and aromatic hydrocarbons along with smaller amounts of oxygen-, nitrogen-, and sulfur-containing compounds. In order to convert petroleum into useable materials, the mixture is separated into fractions by distillation. Distillation allows the components to be separated on the basis of their different boiling temperatures. As the petroleum is heated, the lower boiling components are removed first and stored for future use. Then, as the temperature increases, higher boiling fractions are obtained. The higher boiling fractions contain molecules with a larger number of carbon atoms. Because petroleum is separated into fractions that each contain several compounds that boil within a given temperature range, the process is called **fractional distillation**. The huge mixture of compounds in petroleum is "fractioned" into less complicated mixtures that boil over a given temperature range. Each fraction can be further purified if necessary, and pure compounds can be obtained.

Typical fractions obtained from the fractional distillation of petroleum are listed in Table 15.3.

Coal is an important source of many aromatic compounds. As coal-conversion

TABLE   15.3
Typical Petroleum Fractions from Crude Oil

Fraction	Alkanes present	Boiling range	Uses
Natural gas	C_1-C_4	below 30°C	fuels
Petroleum ethers	C_5-C_7	30–100°C	solvents
Gasoline	C_5-C_{10}	30–200°C	fuel
Kerosene, fuel oil	$C_{12}-C_{20}$	175–330°C	fuel, jet fuel
Lubricants	$C_{20}-C_{30}$	350°C and up	oils, grease
Paraffins	$C_{22}-C_{40}$	350°C and up	petroleum jelly, greases
Residue	$C_{25}-C_{50}$		paraffin wax, asphalt

technology grows and our petroleum reserves dwindle, coal may be our major source of organic compounds.

Physical Properties of the Alkanes

Alkanes are not soluble in water though they are soluble in organic solvents. The alkanes are nonpolar compounds, which accounts for their solubility in other nonpolar organic solvents. It is for this same reason that they are not soluble in highly polar liquids like water. The alkanes are less dense than water and, being insoluble, will float on its surface.

Pure alkanes are colorless, nearly odorless, and tasteless. This might seem contrary to your experience since gasoline is orange when it comes out of the pump and natural gas has a peculiar odor. Actually, dyes are added to gasoline if it contains tetraethyl lead, a toxic antiknock additive. Gasoline also contains aromatic compounds which are responsible for its characteristic odor. Sulfur compounds are added to natural gas to alert people in the event of a gas leak.

The boiling points and densities of the straight-chain alkanes increase in a regular way as the number of carbon atoms in the molecules increase (Table 15.1). This points out another useful feature of a homologous series, since it can also show trends in certain physical properties that parallel the size of the molecules.

Inhalation of pure methane, ethane, propane, or butane can produce a narcotic effect that blunts the senses, induces sleep, and in large quantities can produce complete insensibility and death. A mist of any of the lighter alkanes (C_5–C_{15}), when brought into the lungs, can cause severe damage to lung tissue by dissolving the fatty material in cell membranes. This reduces the elasticity of lung tissue, which makes it more difficult to expel fluids that accumulate there. A condition similar to pneumonia can develop which, in extreme cases, can be fatal. The lighter liquid alkanes can also dissolve fats and oils from the skin and a long-term exposure can cause a "chemical burn," complete with blisters, soreness, and loss of skin.

The heavier alkanes (C_{20}–C_{30}) are used in skin and hair lotions to replace natural oils lost in bathing. Mineral oil (purified C_{20}–C_{30}) is used as a laxative because of its lubricating property. Petroleum jelly is commonly used to protect the skin from exposure to irritants or to aid healing of minor skin problems. It is also used as a base for medicated salves.

Many organic and biochemical compounds that have a high proportion of alkyl structures in the molecules have some physical properties similar to those of the alkanes. Glyceryl tristearate is such a compound. It is a solid and a component of animal fat and, like the alkanes, it floats on water and is not soluble in it. It is soluble in many organic solvents. When pure, it is colorless, odorless, and tasteless. A large portion of the glyceryl tristearate molecule is made up of straight-chain alkyl groups. In large part they are responsible for its alkanelike physical properties.

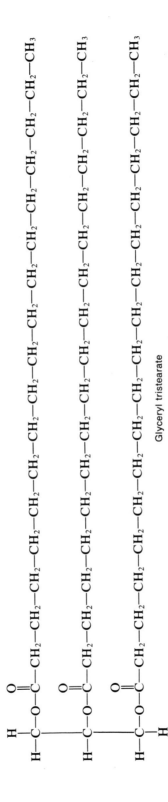

Glyceryl tristearate

The Chemical Properties of Alkanes

The alkanes are not very reactive chemically; in fact, they are the least reactive class of organic compounds. Because of their chemical inertness, they are often used as solvents for many reactive organic compounds.

Alkanes do undergo some very important reactions though. The most important of these is combustion. The **combustion** of an alkane is its rapid reaction with oxygen to produce carbon dioxide, water, and energy. We commonly say that a fuel "burns" though, rather than calling it combustion. These reactions are important because they produce the energy (both heat and light energy) that we use for heating, cooking, transportation, illumination, and the generation of electricity. The equations for the combustion of methane and octane show that identical products are formed in each case.

$$CH_4 + 2O_2 \rightarrow CO_2 + 2H_2O + \text{energy}$$

$$2C_8H_{18} + 25O_2 \rightarrow 16CO_2 + 18H_2O + \text{energy}$$

If the amount of oxygen is in short supply during combustion, carbon monoxide (CO) may form instead of carbon dioxide. This is called **incomplete combustion** and it commonly occurs to some extent whenever a fuel is burned.

$$2CH_4 + 3O_2 \rightarrow 2CO + 4H_2O + \text{energy}$$

Carbon monoxide is a toxic gas. If it enters the bloodstream, as it can when inhaled, it can form a moderately stable compound with hemoglobin and block its ability to carry oxygen. As the amount of "blocked" hemoglobin increases, the oxygen delivered to the cells may be reduced to the point where a person would become lethargic, sleepy, unable to think clearly, and perhaps even fall into a coma. In extreme cases death will follow. Even relatively small amounts of carbon monoxide can have a detrimental effect on the body. Cigarette smokers inhale carbon monoxide into their lungs with every puff. Heavy smoking can impair judgment and muscle control (smoker's nerves). The heart must work harder to get enough oxygen to the cells. For this reason, among others, people with heart disease should not smoke. Fortunately, the effect of small amounts of carbon monoxide can be reversed through application of pure oxygen or, in some cases, breathing clean air.

Alkanes also react with fluorine, chlorine, and bromine under the proper conditions. They do not directly react with iodine. The reactions with fluorine at room temperature are often violent and are of only limited use, but those with chlorine and bromine are more easily controlled and are used to produce chlorinated and brominated hydrocarbons. In these reactions, a halogen atom is substituted for (replaces) a hydrogen atom on the carbon chain. For this reason, they are called **substitution reactions**. Since both chlorine and bromine behave in the same way, we only need to study the chlorine reactions to learn some general facts about these reactions.

Methane and chlorine are both gases at room temperature. A mixture of the two gases will not react, though, unless it is heated to very high temperatures (400–500°C) or exposed to high-energy ultraviolet light. The high-energy condition is needed to break apart the chlorine molecule into chlorine atoms. The very reactive chlorine atoms can then attack the methane molecule to form chloromethane. You will not see the details of most of the reactions in this text, but it would be instructive for two reasons to study the reaction between methane and chlorine. First so you can see the "mechanism" that is involved as reactants are converted to products, and second so you can compare it to the reaction between sodium and chlorine that forms NaCl, which was described in Chapter 8.

The reaction between chlorine and methane to produce chloromethane can be described by a three-step process.

Step 1:   A chlorine molecule in the presence of U.V. light breaks apart into two chlorine atoms.

$$:\!\ddot{\text{C}}\text{l}:\!\ddot{\text{C}}\text{l}: \xrightarrow[\text{light}]{\text{U.V.}} :\!\ddot{\text{C}}\text{l}\cdot + \cdot\ddot{\text{C}}\text{l}:$$

Each chlorine atom is very reactive since it has a great desire to gain an electron to give it an octet of electrons (Chapter 8). It can do this by removing a hydrogen atom from methane to form HCl in the second step. The hydrogen and chlorine atoms share two electrons in a covalent bond.

Step 2:   A chlorine atom removes a hydrogen atom from methane forming HCl and a methyl radical.

$$:\!\ddot{\text{C}}\text{l}\cdot + \text{H}:\!\overset{\text{H}}{\underset{\text{H}}{\ddot{\text{C}}}}\!:\text{H} \longrightarrow \text{H}:\!\overset{\text{H}}{\underset{\text{H}}{\ddot{\text{C}}}}\!\cdot + \text{H}:\!\ddot{\text{C}}\text{l}:$$

Methyl radical

The methyl radical* is a very reactive species. Like the chlorine atom, it needs one more electron to complete an octet of electrons on carbon. It attacks a chlorine molecule in the third step of the reaction.

Step 3:   A methyl radical combines with one of the atoms of a chlorine molecule to form chloromethane, leaving a chlorine atom.

---

*In chemistry a radical is any species that has an odd number of electrons. The chlorine atom is also a radical; it has 17 electrons, an odd number. Radicals are very reactive.

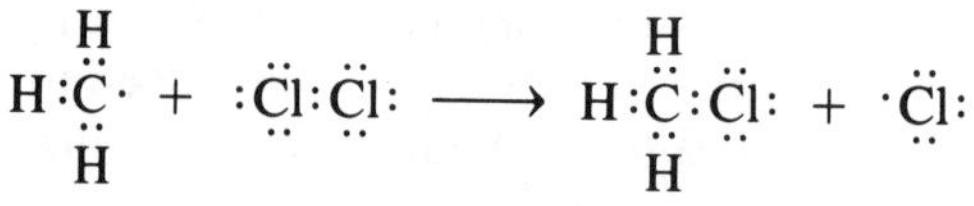

$$
\begin{array}{c}
\text{H} \\
\text{H} \textbf{:} \overset{\textbf{..}}{\text{C}} \textbf{·} \\
\text{H}
\end{array}
+ \;
\textbf{:} \overset{\textbf{..}}{\underset{\textbf{..}}{\text{Cl}}} \textbf{:} \overset{\textbf{..}}{\underset{\textbf{..}}{\text{Cl}}} \textbf{:}
\; \longrightarrow \;
\begin{array}{c}
\text{H} \\
\text{H} \textbf{:} \overset{\textbf{..}}{\text{C}} \textbf{:} \overset{\textbf{..}}{\underset{\textbf{..}}{\text{Cl}}} \textbf{:} \\
\text{H}
\end{array}
+ \;
\textbf{·} \overset{\textbf{..}}{\underset{\textbf{..}}{\text{Cl}}} \textbf{:}
$$

Chloromethane

The chlorine atom formed in the third step can react with a methane molecule as shown in the second step. It is interesting that for each chlorine atom used up in step 2, one is formed in step 3. Once the reaction gets started, it makes the chlorine atoms it needs to keep going.

The reaction between chlorine and methane certainly looks more complicated than the one between chlorine and sodium. There are no ions formed here, as there are when sodium chloride (NaCl) forms. The entire reaction involves neutral species. Also, the reaction with methane requires either light or heat to get it started, whereas the reaction with sodium starts at room temperature. The equation describing the overall reaction of chlorine and methane does not reveal the complicated steps that take place.

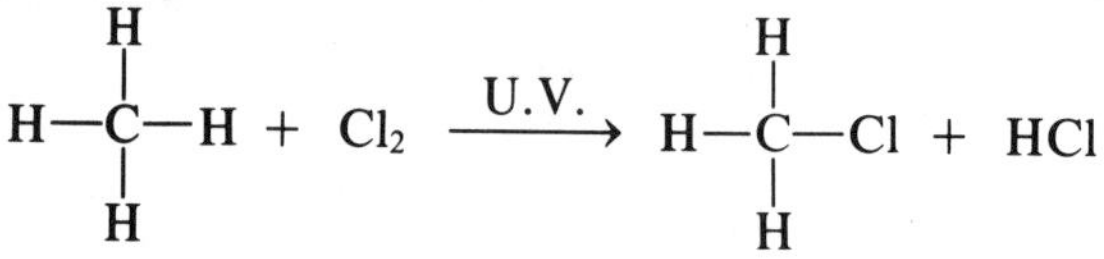

$$
\begin{array}{c}
\text{H} \\
| \\
\text{H} - \text{C} - \text{H} \\
| \\
\text{H}
\end{array}
+ \; Cl_2 \; \xrightarrow{\text{U.V.}} \;
\begin{array}{c}
\text{H} \\
| \\
\text{H} - \text{C} - \text{Cl} \\
| \\
\text{H}
\end{array}
+ \; HCl
$$

In actual practice, a mixture of chlorinated compounds is formed in this reaction. Chloromethane can react with chlorine too, and compounds containing two, three, and four chlorine atoms can form. The sequence of reactions is shown below; notice how each chlorinated methane is made from the one before it. Distillation can be used to separate the different products in the final mixture.

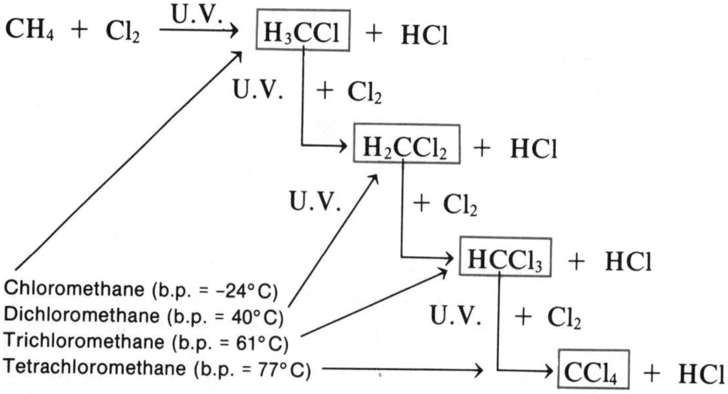

$$ CH_4 + Cl_2 \xrightarrow{\text{U.V.}} \boxed{H_3CCl} + HCl $$

U.V. $\quad$ + Cl_2

$\boxed{H_2CCl_2}$ + HCl

U.V. $\quad$ + Cl_2

$\boxed{HCCl_3}$ + HCl

U.V. $\quad$ + Cl_2

Chloromethane (b.p. = -24°C)
Dichloromethane (b.p. = 40°C)
Trichloromethane (b.p. = 61°C)
Tetrachloromethane (b.p. = 77°C)

$\boxed{CCl_4}$ + HCl

The chlorination of ethane and propane proceed by substitution just as was seen with methane, and a mixture of products is obtained in each case.

Several halogenated (halogen-containing) alkanes are listed in Table 15.4.

TABLE 15.4
Some Halogenated Alkanes

Compound	Common name	IUPAC name	Uses
H_3CCl	methyl chloride	chloromethane	topical anesthetic, refrigerant
H_2CCl_2	methylene chloride	dichloromethane	solvent, cleaning fluid
$HCCl_3$	chloroform	trichloromethane	solvent, cleaning fluid, has been used as inhalation anesthetic
CCl_4	carbon tetrachloride ("carbon tet")	tetrachloromethane	solvent, dry-cleaning fluid, fire extinguisher
CF_2Cl_2	Freon-12	dichlorodifluoromethane	aerosol propellant, refrigerant
CH_3CH_2Cl	ethyl chloride	chloroethane	topical anesthetic
$H-\overset{\displaystyle Br}{\underset{\displaystyle Cl}{C}}-\overset{\displaystyle F}{\underset{\displaystyle F}{C}}-F$	halothane	2-bromo-2-chloro-1,1,1-trifluoroethane	inhalation anesthetic

Though the properties of the halogenated alkanes allow them to be used in a variety of ways, they must be handled with care. The chlorinated alkanes can cause irrepairable damage to the liver, kidneys, and nervous system even if milliliter amounts are brought into the body. Obviously, these compounds should never be taken into the mouth. They should always be used in well-ventilated areas to minimize inhalation of their fumes, and solvent-resistant gloves should be worn to avoid absorption through the skin. Tetrachloromethane (CCl_4) can defat the skin, a condition that can lead to dermatitis.

Chloroform, carefully administered, was used as a general inhalation anesthetic for many years, but the difference between an anesthetic dose and a fatal dose is very small. The anesthetic property of chloroform is due to its depressant affect on the central nervous system. Fortunately, other halogenated alkanes that are known to be safer are available for anesthetic use.

Halogenated organic compounds are rarely found in nature, so they must be synthesized. Several have been found to be effective pesticides like DDT, lindane, and chlorodane. Unfortunately, many of these compounds are not biodegradable, and they build up in the soil and streams and eventually can enter our food supply. DDT is soluble in fatty tissue, and once it enters our body through the food we eat, it is slow to leave. Samples of human fatty tissue have been found to contain as much as 30 parts per million (0.013 g/1b of fat) of DDT. The long-term affect of DDT on humans is still in question, but to be on the safe side its use in the United States has been almost completely banned since 1972.

Check Test Number 5

1. Give the balanced equation for the complete combustion of propane, C_3H_8.
2. Why are the liquid alkanes often used as solvents for other organic compounds?
3. Why is ultraviolet light needed to start the reaction of chlorine with methane?

Answers:

1. $C_3H_8 + 5O_2 \rightarrow 3CO_2 + 4H_2O$
2. The alkanes are nonpolar and sufficiently unreactive that they can dissolve nonpolar organic compounds without reacting with them.
3. The U.V. light is needed to break apart the chlorine molecule to form the reactive chlorine atoms.

The Cycloalkanes

The **cycloalkanes** are saturated hydrocarbons that have the carbon atoms connected in rings. The general formula of the cycloalkanes is C_nH_{2n}. The cycloalkanes have two less hydrogen atoms than the open-chain alkanes with the same number of carbon atoms.

The simplest cycloalkane is cyclopropane. The cyclo- prefix is added to the name of the open-chain alkane with the same number of carbon atoms to indicate the ring structure.

Cyclopropane

C_3H_6

As was shown in Chapter 14, the condensed structure for cyclopropane is simply an equilateral triangle. It is understood that a $-CH_2-$ group is located at each corner of the triangle.

Symbolizes cyclopropane

Cyclopropane is a colorless, sweet-smelling gas at room temperature. When inhaled, it can produce unconsciousness in a matter of seconds. This has led to its use as a fast-acting anesthetic in emergency rooms and surgery. Unfortunately, cyclopropane forms an explosive mixture with air, and if used around electrical equipment the danger of fire or explosion is always present. As a result, cyclopropane has been replaced to a large extent by inflammable halogenated compounds, such as halothane, in recent years.

Other common cycloalkanes are cyclobutane, cyclopentane, and cyclohexane. Ball-and-stick models of the first three cycloalkanes appear in Figure 15.3.

Cyclobutane (C_4H_8)

$$H_2C\!-\!\!-\!\!-CH_2$$
$$|\qquad\qquad|$$
$$H_2C\!-\!\!-\!\!-CH_2$$

or ▢

Cyclopentane (C_5H_{10})

$$\begin{array}{c} H_2 \\ C \\ H_2C \qquad CH_2 \\ H_2C\!-\!\!-\!CH_2 \end{array}$$

or ⬠

Cyclohexane (C_6H_{12})

$$\begin{array}{c} H_2 \\ C \\ H_2C \qquad CH_2 \\ H_2C \qquad CH_2 \\ C \\ H_2 \end{array}$$

or ⬡

The four bonds about carbon in each cycloalkane should be arranged tetrahedrally in space if the molecule is to be as stable as an open-chain alkane. But in the flat cyclopropane and cyclobutane molecules, the bonds are forced into less ideal arrangements. Instead of bond angles of $109\frac{1}{2}°$ within the ring, they

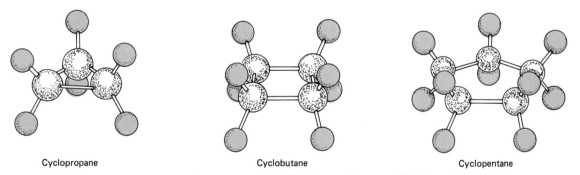

Cyclopropane   Cyclobutane   Cyclopentane

Fig. 15.3 Ball-and-stick models of cyclopropane (C_3H_6), cyclobutane (C_4H_8), and cyclopentane (C_5H_{10}).

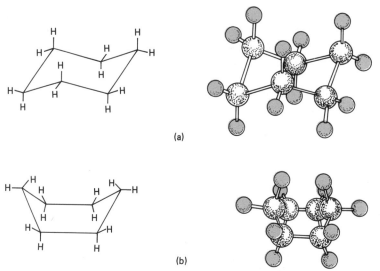

Fig. 15.4. (a) The chair form of cyclohexane, and (b) the boat form of cyclohexane.

are 60° and 90°. This introduces a strain in these rings that affects their chemical properties, as you will see later. The cyclohexane molecule eliminates this kind of strain by bending into shapes that resemble a chair or a boat. This allows the carbon atoms in the ring to maintain the tetrahedral arrangement of bonds, which makes the molecule more stable. The chair and boat forms of cyclohexane, as they are called, appear in Figure 15.4. The chair form is actually a little more stable than the boat form.

The location of groups or atoms substituted on the ring of a cycloalkane is indicated with numbers in the name of the compound. Just as with the open-chain alkanes, the carbon atoms are consecutively numbered so the smallest set of numbers appears in the name. If only one substituent is on the ring, the number is not included in the name.

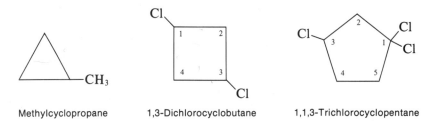

Methylcyclopropane   1,3-Dichlorocyclobutane   1,1,3-Trichlorocyclopentane

Cycloalkanes are present in petroleum, but only in small amounts. They all undergo substitution reactions with chlorine and bromine just as the open-chain alkanes do. Cyclopropane reacts with chlorine in the presence of U.V. light to produce chlorocyclopropane.

Chlorocyclopropane

The strain in the rings of cyclopropane and cyclobutane causes them to undergo ring-opening reactions to form straight-chain compounds. In order to break the ring, two atoms must be *added* to the molecule. Hydrogen will add to cyclopropane to form propane. Because atoms are added to the molecule, the reaction is called an **addition reaction**. Cyclobutane will add hydrogen to form *n*-butane in a similar manner. In the absence of U.V. light, cyclopropane can add chlorine to form 1,3-dichloropropane. If U.V. light is present, the substitution reaction will occur instead.

Terms

Several important terms appeared in Chapter 15. You should know the meaning of each of the following.

saturated hydrocarbon	structural isomers
alkane	IUPAC
paraffin	combustion
cycloalkane	incomplete combustion
substituent	substitution reaction
alkyl group	addition reaction
homologous series	fractional distillation

Questions

Answers to the starred questions appear at the end of the text.

*1. Describe the bonds formed by each carbon atom in a saturated hydrocarbon.

*2. Give two other names that are used for the saturated hydrocarbons.

3. Give the molecular formulas and the names of the first ten members of the alkane series.

*4. What would be the formula of the alkane that contains 14 carbon atoms?

5. Give the structural formulas for the structural isomers of pentane, and give the common name of the straight-chain isomer.

*6. Which of the following compounds are identical?

a.
$$H-\overset{\overset{\displaystyle H}{|}}{\underset{\underset{\displaystyle H}{|}}{C}}-\overset{\overset{\displaystyle H}{|}}{\underset{\underset{\displaystyle H}{|}}{C}}-\overset{\overset{\displaystyle H}{|}}{\underset{\underset{\displaystyle H}{|}}{C}}-\overset{\overset{\displaystyle H}{|}}{\underset{\underset{\displaystyle H}{|}}{C}}-Br$$

b.
$$H-\overset{\overset{\displaystyle Br}{|}}{\underset{\underset{\displaystyle H}{|}}{C}}-\overset{\overset{\displaystyle H}{|}}{\underset{\underset{\displaystyle H}{|}}{C}}-\overset{\overset{\displaystyle H}{|}}{\underset{\underset{\displaystyle H}{|}}{C}}-\overset{\overset{\displaystyle H}{|}}{\underset{\underset{\displaystyle H}{|}}{C}}-H$$

c.
$$H-\overset{\overset{\displaystyle H}{|}}{\underset{\underset{\displaystyle H}{|}}{C}}-\overset{\overset{\displaystyle H}{|}}{\underset{\underset{\displaystyle Br}{|}}{C}}-\overset{\overset{\displaystyle H}{|}}{\underset{\underset{\displaystyle H}{|}}{C}}-\overset{\overset{\displaystyle H}{|}}{\underset{\underset{\displaystyle H}{|}}{C}}-H$$

d.
$$Br-\overset{\overset{\displaystyle H}{|}}{\underset{\underset{\displaystyle H}{|}}{C}}-\overset{\overset{\displaystyle H}{|}}{\underset{\underset{\displaystyle |}{|}}{C}}-H$$
$$H-\overset{\overset{\displaystyle H}{|}}{\underset{\underset{\displaystyle H}{|}}{C}}-\overset{\overset{\displaystyle H}{|}}{\underset{\underset{\displaystyle H}{|}}{C}}-H$$

e.
$$H-\overset{\overset{\displaystyle H}{|}}{\underset{\underset{\displaystyle |}{|}}{C}}-\overset{\overset{\displaystyle H}{|}}{\underset{\underset{\displaystyle H}{|}}{C}}-H$$
$$Br-\overset{\overset{\displaystyle H}{|}}{\underset{\underset{\displaystyle H}{|}}{C}}-\overset{\overset{\displaystyle H}{|}}{\underset{\underset{\displaystyle H}{|}}{C}}-H$$

7. Name the following alkyl groups.
a. $-CH_3$
b. $-CH_2CH_2CH_2CH_2CH_3$
c. $-CH_2CH_2CH_3$
d. $-\overset{\overset{\displaystyle CH_3}{\diagup}}{\underset{\underset{\displaystyle CH_3}{\diagdown}}{C}}-H$
e. $-CH_2CH_3$

8. Would the cycloalkanes form a homologous series?

9. Give the IUPAC names for the following compounds.

*a. $CH_3CH_2\overset{\underset{\displaystyle |}{\underset{\displaystyle CH_3}{|}}}{CH}CH_2CH_2CH_3$

b.
$$\begin{array}{ccc} H_2C & \!\!\!\!\!-\!\!\!\!\! & CH_2 \\ | & & | \\ H_2C & \!\!\!\!\!-\!\!\!\!\! & CH_2 \end{array}$$

* c. $(CH_3)_3C—CH_2—C(CH_3)_3$

h.

$$CH_3\underset{\underset{CH_2}{|}}{\overset{\overset{CH_3}{|}}{C}}—CH_2—\underset{\underset{Br}{}}{\overset{\overset{CH_3}{|}}{C}}—CH_2CH_3$$
$$\underset{\underset{CH_3}{|}}{\underset{CH_2}{|}}$$

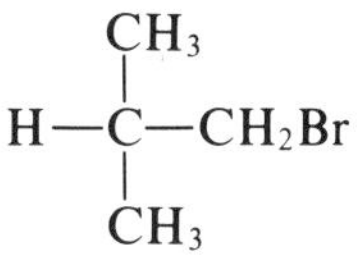

d.

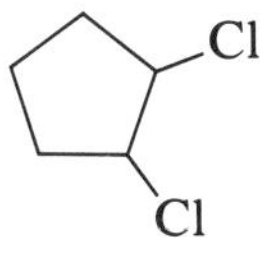

* e.

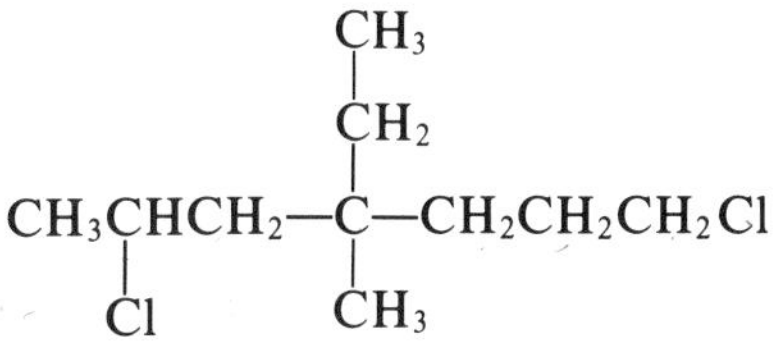

$$CH_3CHCH_2—\underset{\underset{CH_3}{|}}{\overset{\overset{\overset{CH_3}{|}}{CH_2}}{C}}—CH_2CH_2CH_2Cl$$
$$\underset{Cl}{|}$$

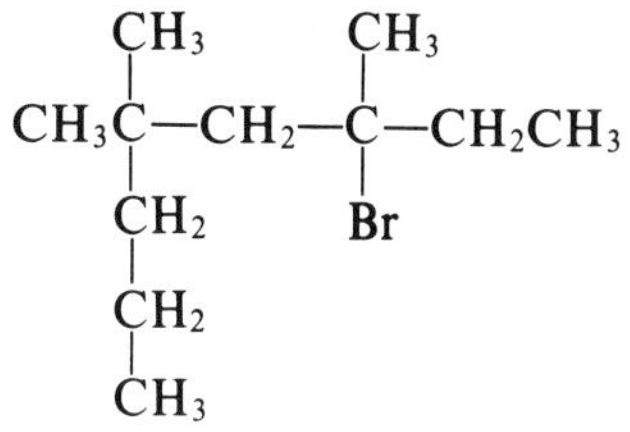

* i.

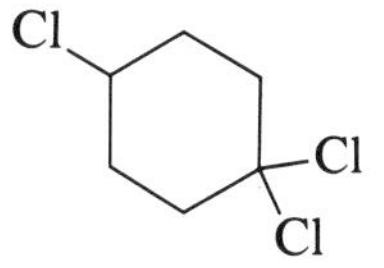

j. $CH_3CH_2CH_2CH_2CH_2CH_3$

* k. $HCCl_3$

f.

$$CH_3CH_2\overset{\overset{I}{|}}{C}H\overset{\overset{I}{|}}{C}H\overset{\overset{CH_3}{|}}{C}HCH_3$$

* g.

$$H—\underset{\underset{CH_3}{|}}{\overset{\overset{CH_3}{|}}{C}}—CH_2Br$$

10. Draw the structural formulas of the following compounds.
    *a. isobutane
    b. methylcyclohexane
    *c. 1,1,2-tribromocyclopentane
    d. *n*-hexane
    *e. 5-ethyl-3,3-dimethylheptane
    f. 1,3,5-trifluoropentane
    *g. 3-isopropylhexane
    h. dichloromethane

*11. Draw the structural formulas of the five isomers of hexane and give the IUPAC name for each one.

12. Which of the following describe typical physical properties of the alkanes.
    a. not soluble in water
    b. density greater than that of water
    c. tasteless
    d. boiling points increase as the molecular weights increase
    e. polar compounds
    f. obtained from petroleum
    g. densities decrease as the number of carbon atoms increase in the molecules
    h. colorless
    i. dissolve oils from the skin
    j. harmless if inhaled

*13. Give the equations for the complete combustion of C_3H_8, C_5H_{12}, and C_4H_8.

14. **a.** Why is carbon monoxide dangerous if inhaled?

   b. Under what conditions can carbon monoxide form during combustion of a fuel?

15. What is a substitution reaction?

*16. Write the step-wise equations for the reaction of chlorine, Cl_2, with methane in the presence of U.V. light to form (a) chloromethane, (b) dichloromethane, (c) trichloromethane, and (d) tetrachloromethane.

*17. The reaction of bromine, Br_2, with ethane produces the following compounds. Which are isomers?

   a. CH_3CH_2Br **d.** Br_3CCH_3 **g.** $Br_2CHCHBr_2$

   b. CH_3CHBr_2 **e.** $BrCH_2CHBr_2$ **h.** Br_3CCH_2Br

   c. $BrCH_2CH_2Br$ **f.** Br_2CHCBr_3 **i.** Br_3CCBr_3

*18. Which cycloalkane is used as an anesthetic? What hazard accompanies its use?

*19. Cyclopropane can react with chlorine gas both in the presence of U.V. light and in the dark. Give the equations for both reactions. Which is substitution and which is addition?

Alkenes, Alkynes, and the Aromatic Hydrocarbons

Many aliphatic organic compounds contain double or triple bonds between adjacent carbon atoms. They are called unsaturated compounds since they have the capacity to join additional atoms to the molecule in reactions that involve the multiple bond. Those organic compounds with a carbon–carbon double bond are called alkenes, and those with a carbon–carbon triple bond are called alkynes. Because these compounds are unsaturated, the alkenes and alkynes are chemically more reactive than alkanes. The unsaturated compounds are used in the preparation of many important drugs, solvents, plastics, and industrial chemicals.

The aromatic hydrocarbons comprise one of the three major types of organic compounds, the other two being the aliphatic and the heterocyclic compounds. The most important aromatic compound is benzene, C_6H_6, which has the six carbon atoms in a ring with a hydrogen atom joined to each carbon. The unique feature of aromatic compounds is the unusual bonding in the ring that joins the carbon atoms together and causes the ring to be very stable. Because of the unusual stability of the benzene ring, it can go through many different kinds of chemical reactions without being broken apart. Aromatic compounds do undergo chemical changes though, but they almost always involve the substitution of an atom or group for a hydrogen atom attached to the ring.

The chapter will begin with a discussion of the alkenes, followed by the alkynes, and concluding with the aromatic hydrocarbons. You will see how compounds of each kind are named as well as the kinds of chemical reactions they typically undergo.

Objectives

By the time you finish this chapter you should be able to do the following:

1. Describe the distinguishing bonding feature found in an alkene, an alkyne, and in an aromatic compound.
2. Name an alkene, an alkyne, or an aromatic compound given its structural formula and vice versa.
3. Describe *cis–trans* isomerism of alkenes.
4. Describe what is meant by an "addition" to an unsaturated compound.
5. Predict the product formed when hydrogen, a halogen, or a hydrogen halide adds to an alkene or an alkyne.
6. Describe several polymers that can be prepared from alkenes.
7. Describe the characteristic reactions of the aromatic hydrocarbons, and explain why they do not undergo typical addition reactions.
8. Describe the compounds that are known as polycyclic aromatic hydrocarbons.

The Alkenes

The **alkenes** contain at least one carbon–carbon double bond ($\text{C}=\text{C}$) in the molecule. Alkenes, unlike the alkanes, are **unsaturated compounds.** The general formula for an open-chain alkene that has one double bond is C_nH_{2n} (where n is a whole number equal to or greater than 2). An unsaturated compound that contains two carbon–carbon double bonds is called a **diene,** and if three double bonds are present, it is called a **triene.** Compounds in the alkene family are also called **olefins.**

The simplest alkene ($n=2$) is ethene, C_2H_4. The common name for ethene is ethylene. The -ene ending on the name is characteristic for alkenes. Ethene is a flat molecule, that is, all six atoms lie in the same plane. The bond angles are all 120°:

$$H \diagdown \quad \diagup H$$
$$C = C$$
$$H \diagup \quad \diagdown H$$

120°

120°

120°

Ethene (ethylene)

A model of ethene is shown in Figure 16.1. The double bond between the carbon atoms "locks" the molecule into a planar structure.

Several typical alkenes are listed in Table 16.1 with their IUPAC (International Union of Pure and Applied Chemistry) names. The presence of a double bond between two adjacent carbon atoms places each of these compounds in the alkene family.

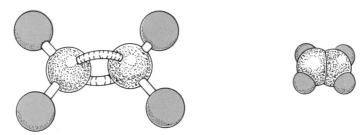

Fig. 16.1. Models of the ethene (ethylene) molecule. The double bond between the two carbon atoms locks the molecule in a planar shape.

	TABLE 16.1 Some Typical Alkenes		
Condensed formula	Molecular formula	IUPAC name	Boiling point (°C)
$CH_2 = CH_2$	C_2H_4	ethene	−104
$CH_3CH = CH_2$	C_3H_6	propene	−47
$CH_3CH_2CH = CH_2$	C_4H_8	1-butene	−6
$CH_3 - \underset{\underset{CH_3}{\vert}}{C} = CH_2$	C_4H_8	2-methylpropene	−7
$CH_3CH_2CH_2CH = CH_2$	C_5H_{10}	1-pentene	30
$CH_3 - \underset{\underset{CH_3}{\vert}}{CH} - CH = CH_2$	C_5H_{10}	3-methyl-1- butene	54
$CH_3CH_2 - \underset{\underset{CH_3}{\vert}}{C} = CH_2$	C_5H_{10}	2-methyl-1- butene	62
$CH_3CH_2CH_2CH_2CH = CH_2$	C_6H_{12}	1-hexene	63

Naming the Alkenes

Though several alkenes that have wide use are known by common names, the most suitable names for these compounds are derived from the recommendations of the International Union of Pure and Applied Chemistry (IUPAC). A set of rules that can be followed to derive the IUPAC names of alkenes are given below. These rules will also be used, with some modification, to name alkynes later on.

1. Examine the structural formula of the alkene and locate the longest carbon chain that contains the carbon–carbon double bond. This will be the "main chain" of the molecule:

$$
\begin{array}{l}
CH_3 \\
| \\
CH_2 \\
| \\
CH_3-CH-C{=}CH-CH_3 \\
| \\
CH_2 \\
| \\
CH_3
\end{array}
$$

The longest chain that contains the double bond

2. The main chain is given the name of the alkane (Table 16.1) with the same number of carbon atoms, except that the -*ane* ending of the alkane is changed to -*ene* for the alkene. In the example above, the main chain contains six carbon atoms, so the name of the parent compound is "hexene."

3. The carbon atoms in the main chain are consecutively numbered from the end *nearest* the double bond. This identifies the location of the double bond with the smallest possible number which is written in *front* of the name of the parent alkene and separated by a hyphen:

$$
\begin{array}{l}
CH_3 \\
| \\
CH_2 \\
\overset{3}{|} \\
CH_3-\overset{4}{CH}-\overset{3}{C}{=}\overset{2}{CH}-\overset{1}{CH_3} \\
\overset{5}{|} \\
CH_2 \\
\overset{6}{|} \\
CH_3 \quad \text{2-Hexene}
\end{array}
$$

4. Groups substituted on the main chain are named individually, positioned on the main chain with numbers, and written before the name of the parent

compound just as was done for the alkanes in Chapter 15. Looking at the hexene compound above, there is an ethyl group on carbon 3, and a methyl on carbon 4. The IUPAC name is

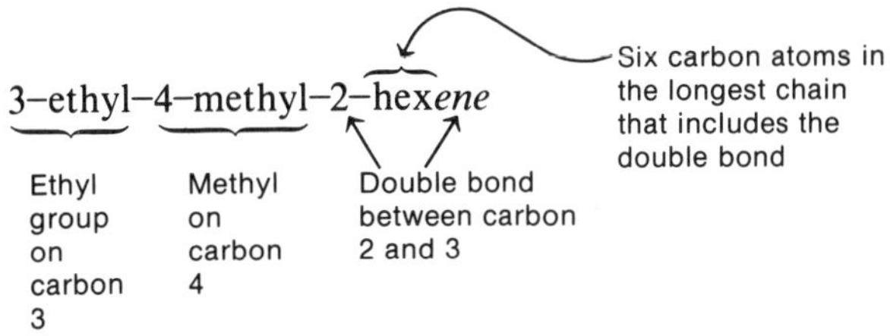

3–ethyl–4–methyl–2–hex*ene*

Ethyl group on carbon 3

Methyl on carbon 4

Double bond between carbon 2 and 3

Six carbon atoms in the longest chain that includes the double bond

5. If the molecule contains more than one double bond, then the location of each one must be identified with a number. If there are two double bonds, the name of the parent compound ends in *-diene,* and if there are three double bonds, the ending is *-triene.* Notice in the example below, only the *-ne* of the *-ane* ending of the parent alkane is dropped.

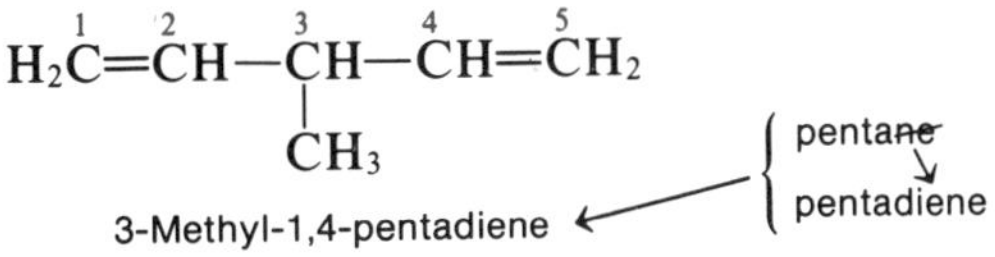

$$\overset{1}{H_2C}{=}\overset{2}{CH}{-}\overset{3}{CH}{-}\overset{4}{CH}{=}\overset{5}{CH_2}$$

$$CH_3$$

3-Methyl-1,4-pentadiene

pentane
pentadiene

The following examples show how the IUPAC names of other alkenes are derived.

Example 16.1

Give the IUPAC name for

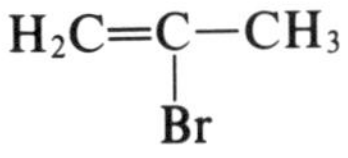

$$H_2C{=}\underset{Br}{C}{-}CH_3$$

The three-carbon chain that contains the double bond is named propene. A propene does *not* need to have the position of the double bond indicated with a number, since it can only be between carbon 1 and carbon 2. (Remember, the rules require us to locate the double bond with the smallest number possible. It would not be right to say it is between carbons 2 and 3.) The bromo group is on carbon 2, so the IUPAC name is

2-bromopropene

Example 16.2

Give the IUPAC name for

$$CH_3-CH-CH_2-CH=CH_2$$
$$|$$
$$Cl$$

The longest carbon chain that contains the double bond contains five carbons, so the name of the parent compound is pentene.

The chain is numbered from right to left so the double bond is at carbon 1, and the chloro group is at carbon 4. The IUPAC name is

4-chloro-1-pentene

Example 16.3

Give the IUPAC name for

$$CH_2=C-CH=C-CH=CH_2$$
$$|\qquad\quad|$$
$$CH_2\qquad CH_3$$
$$|$$
$$CH_3$$

The longest chain that contains all three double bonds has six carbon atoms; the name of the parent compound is hexatriene.

So that the location of the ethyl and methyl groups will use the lowest numbers, the main chain is numbered from left to right. (If *only* the double bonds were considered, the chain could have been numbered from either end, but to locate the alkyl groups with the smallest numbers, it is numbered from the left.) The IUPAC name is

2-ethyl-4-methyl-1,3,5-hexatriene

The carbon atoms in the ring of a **cycloalkene** are numbered so that the location of the double bond is always between carbon 1 and carbon 2. The locations of substituients on the ring are also indicated with numbers. The ring can be numbered in either a clockwise ($\curvearrowright$) or counterclockwise ($\curvearrowleft$) direction so that the smallest numbers are used for the substituents. If there is only one double bond in the ring, its location does not have to be indicated with a number—it is understood that it is between carbon 1 and carbon 2.

The correct IUPAC names for several alkenes and cycloalkenes are given below. Study them carefully as well as those in Table 16.1. Then name the compounds listed in Check Test Number 1.

$H_2C\!=\!CH\!-\!CH_2Cl$ 3-chloropropene

$\underset{\displaystyle H_2C\!=\!\underset{|}{C}\!-\!CH_3}{\overset{\displaystyle Cl}{}}$ 2-chloropropene

$\underset{H}{\overset{Cl}{\diagdown}}C\!=\!CH\!-\!CH_3$ 1-chloropropene

$H_2C\!=\!\overset{\displaystyle CH_3}{\underset{|}{C}}\!-\!-\!\overset{\displaystyle H}{\underset{|}{C}}\!=\!CH_2$ 2-methyl-1,3-butadiene

       3-methylcyclobutene

3,3,5-trichlorocyclohexene

1,4-cyclohexadiene

1,2-dimethylcyclopentene

cyclohexene

Check Test Number 1

Give the IUPAC name for each of the following compounds.

a. $CH_3\!-\!\underset{\underset{\textstyle CH_3}{|}}{CH}\!-\!CH_2\!-\!\underset{\underset{\textstyle CH_3}{|}}{C}\!=\!\underset{\underset{\textstyle CH_3}{|}}{C}\!-\!CH_3$

b. $CH_3CH\!=\!CHCH_2Br$

c. $CH_3\!-\!\underset{\underset{\textstyle Br}{|}}{C}\!=\!CH\!-\!CH\!=\!CHCl$

d.

Answers:
   a. 2,3,5-trimethyl-2-hexene
   b. 1-bromo-2-butene
   c. 4-bromo-1-chloro-1,3-pentadiene
   d. 4-bromo-1-methylcyclohexene (double bond is between carbons 1 and 2)

Cis–Trans Isomers of Alkenes

You have already learned about structural isomerism in Chapter 15, and how there can be two or more compounds with the same molecular formula but different structures. The structural isomers of butane, C_4H_{10}, are *n*-butane and isobutane (Figure 15.1).

Many alkenes display yet another kind of isomerism called **geometric isomerism,** or more specifically, *cis–trans* **isomerism.** It exists because the double bond between two carbon atoms in an alkene "locks" that part of the molecule around the double bond into a definite geometric arrangement. Unlike single bonds between carbon atoms, double bonds restrict free rotation and they hold the groups attached to the doubly bonded carbons in a definite arrangement. Because of this restricted rotation about a double bond, there are two geometric forms for 2-butene. Both isomers are shown below. The one on the left with the methyl groups on the same side of the molecule is called the *cis* isomer. The one on the right with the methyl groups on opposite sides is the *trans* isomer.

$$H_3C \diagdown \quad \diagup CH_3$$
$$C=C$$
$$H \diagup \quad \diagdown H$$

The *cis* isomer
cis-2-butene
b.p. = 4°C

$$H_3C \diagdown \quad \diagup H$$
$$C=C$$
$$H \diagup \quad \diagdown CH_3$$

The *trans* isomer
trans-2-butene
b.p. = 1°C

Cis-2-butene and *trans*-2-butene are actually different compounds by virtue of the different arrangement of atoms in space. The *cis* isomer cannot be changed to the *trans* isomer without breaking the double bond. Each isomer has different properties, as shown in their different boiling points. Models of the *cis* and *trans* isomers of 2-butene are shown in Figure 16.2.

Cis and *trans* isomers also exist for 2-bromo-2-butene. Here again we can use the positions of the two methyl groups to determine which isomer is *cis* and which is *trans:*

$$H \diagdown \quad \diagup Br$$
$$C=C$$
$$H_3C \diagup \quad \diagdown CH_3$$

Cis-2-bromo-2-butene

$$H_3C \diagdown \quad \diagup Br$$
$$C=C$$
$$H \diagup \quad \diagdown CH_3$$

Trans-2-bromo-2-butene

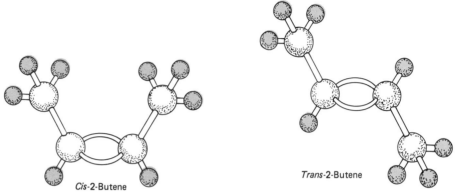

Fig. 16.2. *Cis*-2-butene and *trans*-2-butene are geometrical isomers of one another.

Not all alkenes display *cis–trans* isomerism. Only those alkenes that have two nonidentical atoms or groups on *each* of the double-bonded carbon atoms can exist as *cis–trans* isomers. Both 2-butene and 2-bromo-2-butene meet this requirement. Each has two different or nonidentical groups on *each* of the carbon atoms connected by the double bond. Ethene cannot have *cis–trans* isomers since both atoms bonded to each carbon are identical (they are both hydrogen). *Cis* and *trans* isomers do not exist for propene either, since identical atoms (hydrogen) exist on one of the carbon atoms in the double bond.

Because of identical atoms on this carbon, propene cannot exist as *cis-trans* isomers.

Check Test Number 2

1. Draw the *cis* and *trans* isomers of
   (a) 2-pentene and (b) 1,2-dichloroethene
2. Which of the following compounds would exhibit *cis–trans* isomerism?

    a. $CH_3-\overset{\overset{\displaystyle Cl}{|}}{C}=\overset{\overset{\displaystyle Cl}{|}}{C}-CH_3$

    b. $F_2C=CF_2$

    c. $CH_3-CH_2-CH=CH_2$

    d. $Cl-\overset{\overset{\displaystyle H}{|}}{C}=\overset{\overset{\displaystyle H}{|}}{C}-Br$

Answers:

1. (a)

cis

trans

(b)

2. (a) and (d) can exist as *cis–trans* isomers. (b) and (c) do not have two nonidentical groups on both double-bonded carbons.

Sources of Alkenes

Alkenes are not found in any substantial quantities in petroleum, yet petroleum provides the raw materials from which alkenes are made. The saturated alkanes in petroleum can be converted into alkenes by a process called **cracking.** When certain fractions of petroleum are vaporized and heated to very high temperatures (around 500°C) in the presence of certain catalysts, the larger saturated molecules are "cracked," that is, broken into smaller molecules. Depending on the conditions, many of these smaller molecules turn out to be alkenes, which are then separated and purified. A simple alkane like propane can produce both propene and ethene during the cracking process:

Butane can produce a mixture of four alkenes when it is cracked: 1-butene, 2-butene, propene, and ethene. Under different conditions high-temperature cracking can also be used to make automotive gasoline from heavier petroleum fractions by breaking the larger alkanes into the smaller ones used in gasoline.

Ethene is one of the most important members of the alkene family since it is used in the manufacture of many substances used in medicine, the home, and industry. Tons of ethene are made every day by cracking petroleum. At one time ethene gas was used as a general inhalation anesthetic since it produced unconsciousness quickly without the degree of postanesthetic nausea and discomfort that accompanied ether. Because of its flammability and the later discovery of better anesthetics, it is not used for this purpose today. Several years ago it was discovered that ethene is produced by fruits and vegetables as they ripen and that it accelerated the ripening process. Today ethene is used to control the ripening of tomatoes and a variety of citrus fruits. Since ethene is naturally produced by these fruits, the use of the same compound to artificially hasten ripening seems quite safe.

The Physical Properties of Alkenes

In many ways alkenes have physical properties much like the alkanes with the same number of carbon atoms. The small alkenes (ethene, propene, and butene) are gases at room temperature. Those containing from 5 to 18 carbon atoms (C_5–C_{18}) are liquids at this temperature, and those with more than 18 carbon atoms are solids. Alkenes are insoluble in water because of their generally low polarity. They are also less dense than water and, like the alkanes, will float on its surface.

The Chemical Properties of Alkenes

Alkenes undergo three major kinds of chemical reactions: (a) combustion—the rapid and visible reaction with oxygen to form carbon dioxide and water; (b) addition—reactions that add two atoms or groups of atoms across the double bond; and (c) polymerization—reactions that join many alkene molecules together to form long chains that can contain thousands of carbon atoms. Many common plastics are formed by the polymerization of alkenes. Let us examine each type of reaction.

Combustion   Alkenes, just like other hydrocarbons, combust (burn) to produce carbon dioxide, water, and energy. However, alkenes are rarely used as fuels since they are more valuable when used to make other essential chemicals. The equation showing the complete combustion of 1-octene, an alkene found in lemon oil, is given below:

$$CH_3(CH_2)_5CH = CH_2 + 12O_2 \rightarrow 8CO_2 + 8H_2O + energy$$

1-Octene

Addition Reactions of Alkenes   The most common kind of reaction for the alkenes is the addition reaction. Unlike the alkanes, which react by substituting one atom for another (a substitution reaction), alkenes react by adding two atoms or groups of atoms to the carbon–carbon double bond. This is called an addition reaction because atoms are "added" to the molecules and an unsaturated alkene is converted to a saturated product. An example of an addition reaction is that between chlorine and ethene:

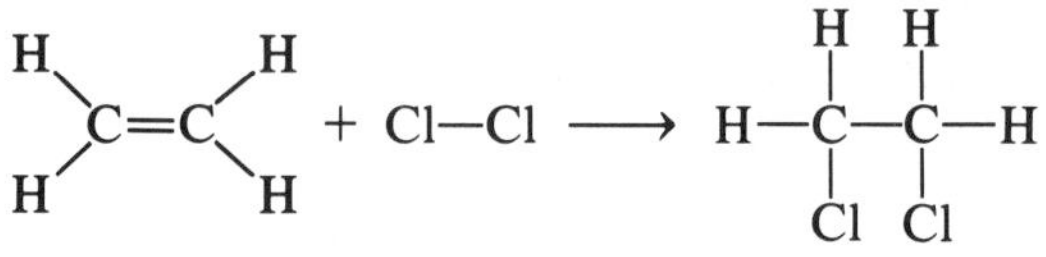

$$\begin{array}{c} \text{H} \\ \text{C}=\text{C} \\ \text{H} \end{array}\begin{array}{c} \text{H} \\ \\ \text{H} \end{array} + \text{Cl}-\text{Cl} \longrightarrow \text{H}-\overset{\overset{\displaystyle \text{H}}{|}}{\underset{\underset{\displaystyle \text{Cl}}{|}}{\text{C}}}-\overset{\overset{\displaystyle \text{H}}{|}}{\underset{\underset{\displaystyle \text{Cl}}{|}}{\text{C}}}-\text{H}$$

Ethene   1,2-Dichloroethane

One of the two bonds that form the double bond breaks allowing each carbon atom to join to one chlorine atom through a single covalent (shared electron pair) bond. The addition of two chlorine atoms to ethene is shown below in a way that depicts how the electrons of one bond of the double bond are made available to be shared with chlorine. This does not represent the actual "mechanism" of the reaction, though. It is only a simplified picture to describe the addition process.

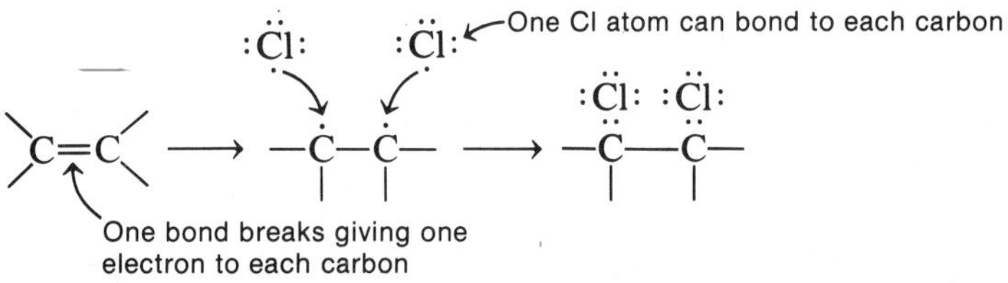

One Cl atom can bond to each carbon

One bond breaks giving one electron to each carbon

Bromine and iodine can also add to alkenes. The reaction with fluorine is often violent. The general name for these reactions that add halogen atoms to a molecule is **halogenation.**

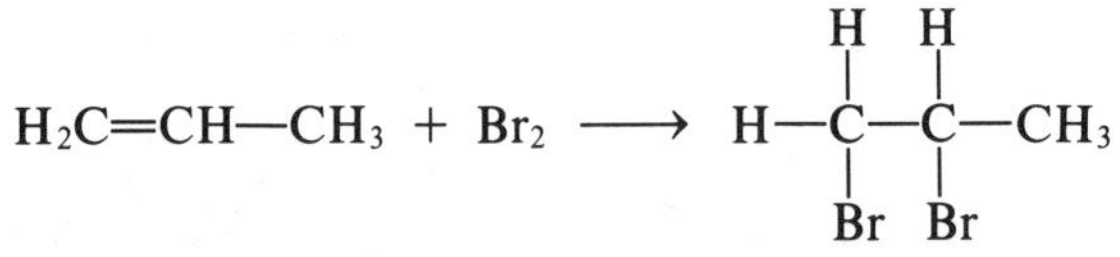

$$\text{H}_2\text{C}=\text{CH}-\text{CH}_3 + \text{Br}_2 \longrightarrow \text{H}-\overset{\overset{\displaystyle \text{H}}{|}}{\underset{\underset{\displaystyle \text{Br}}{|}}{\text{C}}}-\overset{\overset{\displaystyle \text{H}}{|}}{\underset{\underset{\displaystyle \text{Br}}{|}}{\text{C}}}-\text{CH}_3$$

1,2-Dibromopropane

$$\begin{array}{c} \text{H}_2\text{C}-\text{CH} \\ | \quad\quad || \\ \text{H}_2\text{C}-\text{CH} \end{array} + \text{I}_2 \longrightarrow \begin{array}{c} \text{H}_2\text{C}-\overset{\overset{\displaystyle \text{H}}{|}}{\text{C}}-\text{I} \\ | \quad\quad | \\ \text{H}_2\text{C}-\underset{\underset{\displaystyle \text{H}}{|}}{\text{C}}-\text{I} \end{array}$$

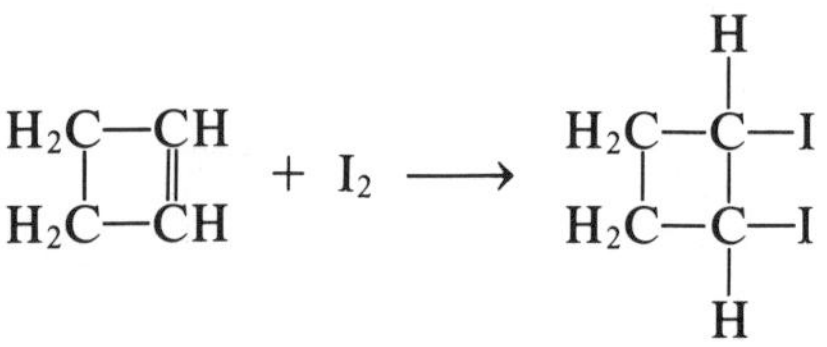

1,2-Diiodocyclobutane

Pure bromine is a red-brown liquid that is soluble in carbon tetrachloride (CCl_4). Dibromoalkanes that are produced when an alkene reacts with bromine are colorless. A common test used to detect the presence of a carbon–carbon double bond in a molecule is to add a drop of a solution of Br_2 in CCl_4 to a solution of the compound to be tested. If the compound is unsaturated, it will consume the bromine and the red-brown color will disappear. The bromine color test usually takes less than a minute.

Alkenes also add hydrogen to the double bond to form a saturated compound. A catalyst such as nickel (Ni), platinum (Pt), or palladium (Pd) is required, and the hydrogen gas is usually under high pressure. The addition of hydrogen to an unsaturated compound is called **hydrogenation.** The hydrogenation of propene produces propane:

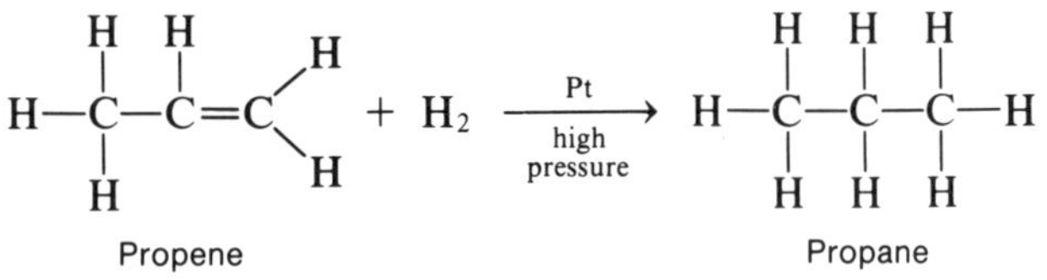

$$H-\overset{\overset{\displaystyle H}{|}}{\underset{\underset{\displaystyle H}{|}}{C}}-\overset{\overset{\displaystyle H}{|}}{\underset{\underset{\displaystyle H}{|}}{C}}=C\overset{\diagup H}{\diagdown H} + H_2 \xrightarrow[\text{high pressure}]{\text{Pt}} H-\overset{\overset{\displaystyle H}{|}}{\underset{\underset{\displaystyle H}{|}}{C}}-\overset{\overset{\displaystyle H}{|}}{\underset{\underset{\displaystyle H}{|}}{C}}-\overset{\overset{\displaystyle H}{|}}{\underset{\underset{\displaystyle H}{|}}{C}}-H$$

Propene   Propane

Oleomargarine and various cooking shortenings are prepared by partially hydrogenating vegetable oils. Vegetable oils contain many carbon–carbon double bonds and are often described as being polyunsaturated. As hydrogen is added to these double bonds, they become saturated, forming a product that can be solid or semisolid at room temperature. The hydrogenation of vegetable oils is described in more detail in Chapter 20.

Hydrogen chloride (HCl), hydrogen bromide (HBr), hydrogen iodide (HI), and water (HOH) can also add to carbon–carbon double bonds. With each of these reagents, hydrogen (–H) adds to one of the carbon atoms of the double bond, and the remainder of the molecule (–Cl, –Br, –I, or –OH) adds to the other. The addition of water requires the use of a strong acid like sulfuric acid (H_2SO_4) to catalyze the reaction.

The addition of these reagents to a double bond takes place in a definite way that is described by **Markovnikov's rule*:**

When a molecule of the general formula HX adds to a carbon–carbon double bond, the hydrogen joins to the carbon atom that has the greater number of hydrogens already bonded to it.

Note how the Markovnikov rule is followed in the following addition reactions:

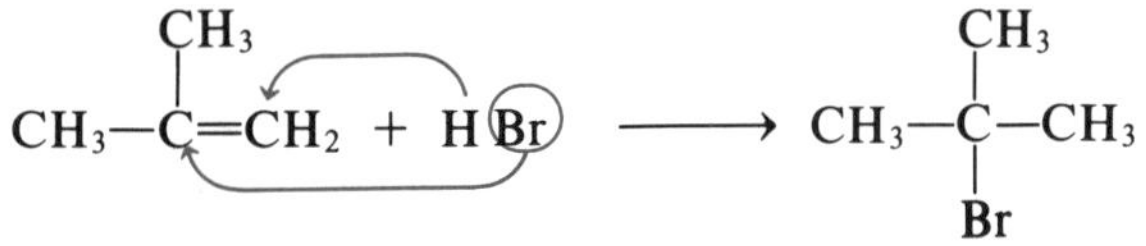

$$CH_3-\overset{\overset{\displaystyle CH_3}{|}}{C}=CH_2 + H\,Br \longrightarrow CH_3-\overset{\overset{\displaystyle CH_3}{|}}{\underset{\underset{\displaystyle Br}{|}}{C}}-CH_3$$

---

* Markovnikov was a Russian chemist who reported the facts that led to the rule in 1869.

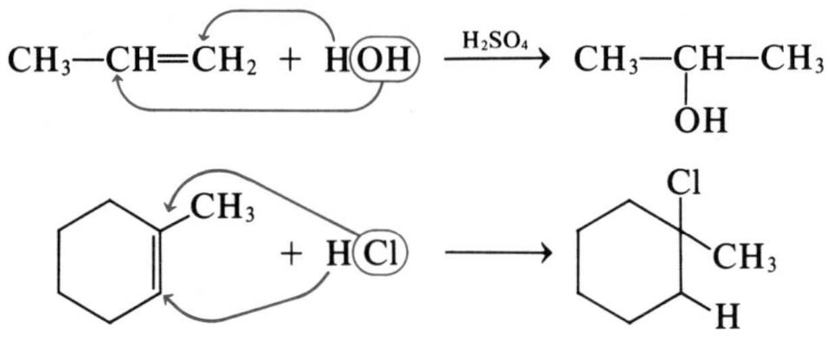

$$CH_3-CH=CH_2 + HOH \xrightarrow{H_2SO_4} CH_3-CH-CH_3$$
$$\underset{OH}{|}$$

Check Test Number 3

Write the structural formula of the product of each of the following reactions.

a. $CH_3-CH=CH_2 + Br_2 \longrightarrow$

b.  $+ H_2 \xrightarrow{Ni}$

c. $CH_3-\underset{\underset{CH_3}{|}}{C}=CH-CH_3 + HOH \xrightarrow{H_2SO_4}$

d. $CH_3-CH_2-CH=CH_2 + HCl \longrightarrow$

Answers:

a.

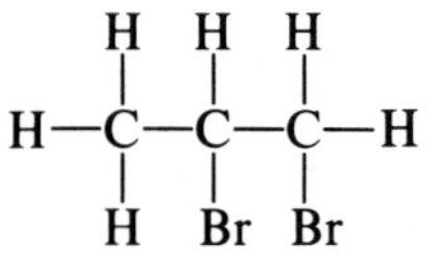

c.

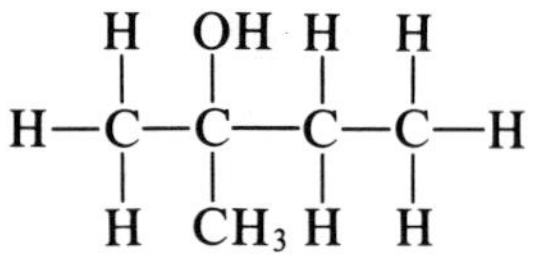

b.

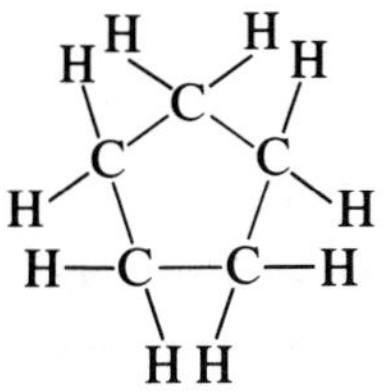

d.

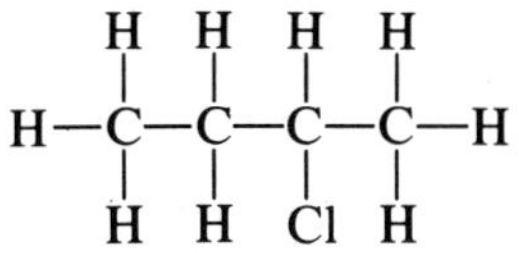

Polymerization of Alkenes   Many of the plastics that you encounter every day are made from alkenes. These plastics, which are more correctly called **polymers** (made of many parts), are very large molecules made from thousands of alkene molecules in a process called polymerization. The **polymerization** of an alkene joins many alkene units together to form the polymer. The individual alkene molecules used to make the polymer are called **monomers** (one part).

The most common plastic is polyethylene. It is formed by the polymerization of ethylene (ethene) in the presence of certain catalysts. Ethylene is the monomer and polyethylene is the polymer made from the monomer. The polymerization of three ethene molecules to form a short section of the polymer chain is shown below. Of course in practice, thousands of ethene molecules (monomers) are used to make each polymer chain.

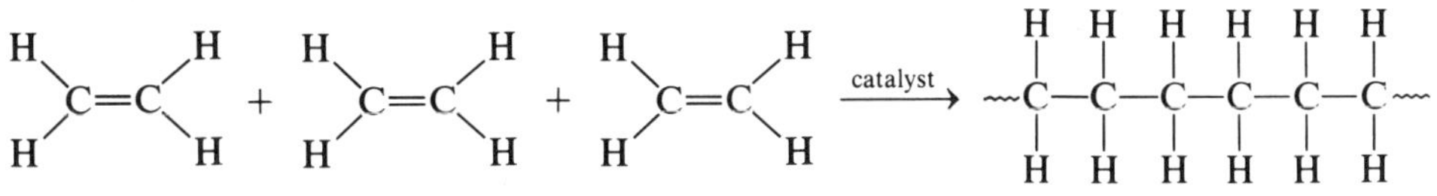

Ethene (ethylene)   Polyethylene

Notice that the polymerization of ethene is the addition of ethene molecules to one another across the carbon–carbon double bond in each molecule to form a chain of carbon atoms. The same thing occurs when other alkenes are polymerized to form other polymers. Several common polymers and the alkenes from which they are made are given in Table 16.2 (page 404).

Vinyl chloride, which is used to make polyvinylchloride, has been linked to a rare form of liver cancer. As a result, exposure to vinyl chloride has been greatly restricted by the government. Polyvinylchloride, on the other hand, is considered to be quite safe in this respect. This is not surprising though, because the polymer and the monomer are entirely different compounds.

The Alkynes

Alkynes are unsaturated compounds that contain a triple bond between two adjacent carbon atoms ($-C \equiv C-$). The general formula for open-chain hydrocarbons that contain one triple bond is C_nH_{2n-2}, where n is a whole number equal to or greater than 2. The simplest alkyne is then C_2H_2, ethyne. The common name for ethyne is acetylene. Notice how the common name for ethyne has an ending like that of an alkene. This is a good example of one kind of problem presented by some common names since the ending of acetyl*ene* would imply it is an alkene, which it is not. The ethyne molecule is linear, and the bond angles are 180°:

180°

$$H-C \equiv C-H$$

Ethyne
(acetylene)

A ball-and-stick model of ethyne is shown in Figure 16.3.

The IUPAC names for alkynes are derived in the same way names were derived for the alkenes, except that the name of an alkyne ends in -*yne*. The name of the parent compound comes from the longest chain of carbon atoms

TABLE 16.2
Several Common Polymers Formed From Alkenes

Monomer (Common name used)	Polymer	Uses
$H_2C{=}CH_2$ Ethylene	Polyethylene	plastic film, containers, toys
$H_2C{=}C{\overset{H}{\underset{Cl}{}}}$ Vinyl chloride	Polyvinylchloride (PVC)	vinyl film, water pipes, packaging
$(H_2C{=}C{\overset{H}{}}$ is the vinyl group)		
$H_2C{=}C{\overset{H}{\underset{C{\equiv}N}{}}}$ Acrylonitrile	Polyacrylonitrile	Orlon® and Acrilan® fibers used in clothing
$H_2C{=}C{\overset{H}{}}$ Styrene	Polystyrene	styrofoam, plastic sheeting, insulation
$H_2C{=}C{\overset{Cl}{\underset{Cl}{}}}$ Vinylidene chloride	Polyvinylidenechloride	Saran® wrap
$F_2C{=}CF_2$ Tetrafluoroethylene	Polytetrafluoroethylene	Teflon®

Fig. 16.3. Models of the linear ethyne (acetylene) molecule.

that contains the triple bond. The chain is numbered so that the position of the triple bond is indicated with the smallest number possible. The IUPAC names for several typical alkynes appear in Table 16.3.

The physical properties of the alkynes are similar to those of the alkanes and the alkenes. They are all insoluble in water, though they are soluble in many organic liquids. Ethyne, propyne, and 1-butyne are gases at room temperature. The higher members of the series are liquids or solids at room temperature.

Alkynes are not found in nature to any extent, and they are not of great biological importance in the body. However, carbon–carbon triple bonds do appear in certain synthetic oral contraceptives and in some antibiotics.

The Chemical Properties of Alkynes

Alkynes undergo combustion as do the other hydrocarbons. An important use of ethyne is as a fuel in oxyacetylene torches since its combustion can produce temperatures in excess of 2500°C. At these temperatures steel bars and sheets can be cut apart and welded together.

In many ways the chemical properties of alkynes are much like those of the alkenes. Alkynes and alkenes are both unsaturated compounds and both undergo addition reactions at the multiple bond. Alkynes can add either two or four atoms or groups to the triple bond. If only two atoms add, the product will be an alkene. If four atoms add, a saturated compound will be produced.

TABLE 16.3 Some Typical Alkynes			
Condensed formula	Molecular formula	IUPAC name	Boiling point (°C)
H–C≡C–H	C_2H_2	ethyne (acetylene)	−84
CH_3–C≡C–H	C_3H_4	propyne	−23
CH_3CH_2–C≡C–H	C_4H_6	1-butyne	9
CH_3–C≡C–CH_3	C_4H_6	2-butyne	27
$CH_3CH_2CH_2$–C≡C–H	C_5H_8	1-pentyne	40
CH_3CH_2–C≡C–CH_3	C_5H_8	2-pentyne	56

$$R—C\equiv C—R \begin{cases} +2A \longrightarrow R—\underset{\underset{A}{|}}{C}=\underset{\underset{A}{|}}{C}—R \quad \text{(alkene)} \\[2em] +4A \longrightarrow R—\underset{\underset{A}{|}}{\overset{\overset{A}{|}}{C}}—\underset{\underset{A}{|}}{\overset{\overset{A}{|}}{C}}—R \quad \text{(alkane)} \end{cases}$$

Hydrogen (H_2), the halogens (Cl_2, Br_2), or the hydrogen halides (HCl, HBr) can all add to the triple bond of an alkyne. If the reactions are carefully controlled, it is possible to add just two atoms to the triple bond, forming an alkene. The controlled hydrogenation (the addition of hydrogen) of 2-butyne in the presence of a special catalyst can produce 2-butene:

$$CH_3—C\equiv C—CH_3 \ + \ H_2 \ \xrightarrow[\substack{high \\ pressure}]{\substack{special \\ catalyst}} \ CH_3—\overset{\overset{H}{|}}{C}=\overset{\overset{H}{|}}{C}—CH_3$$

2-Butyne   2-Butene

Commonly the addition of hydrogen to an alkyne is not controlled and the saturated alkane is produced:

$$CH_3—C\equiv C—CH_3 \ + \ 2H_2 \ \xrightarrow[\substack{high \\ pressure}]{Ni} \ CH_3—\underset{\underset{H}{|}}{\overset{\overset{H}{|}}{C}}—\underset{\underset{H}{|}}{\overset{\overset{H}{|}}{C}}—CH_3$$

2-Butyne   Butane

Halogenation of an alkyne usually involves the addition of chlorine or bromine to the triple bond. The reactions with fluorine (F_2) are generally violent and those with iodine (I_2) produce compounds that are frequently unstable. With careful control, the addition of bromine or chlorine can be stopped after just two halogen atoms have added to the triple bond:

$$CH_3—C\equiv C—CH_3 \ + \ Br_2 \ \longrightarrow \ CH_3—\overset{\overset{Br}{|}}{C}=\overset{\overset{Br}{|}}{C}—CH_3$$

2-Butyne   2,3-Dibromo-2-butene

Further addition of halogen will produce the saturated compound:

$$CH_3—\overset{\overset{Br}{|}}{C}=\overset{\overset{Br}{|}}{C}—CH_3 \ + \ Br_2 \ \longrightarrow \ CH_3—\underset{\underset{Br}{|}}{\overset{\overset{Br}{|}}{C}}—\underset{\underset{Br}{|}}{\overset{\overset{Br}{|}}{C}}—CH_3$$

2,2,3,3-Tetrabromobutane

The addition of the hydrogen halides (HCl, HBr) proceeds in the same kind of two-stage reaction sequence. An additional consideration is that they will add according to Markovnikov's rule, that is, the hydrogen will go to the carbon atom that is richest in hydrogen. The addition of HCl to propyne proceeds as predicted by Markovnikov's rule:

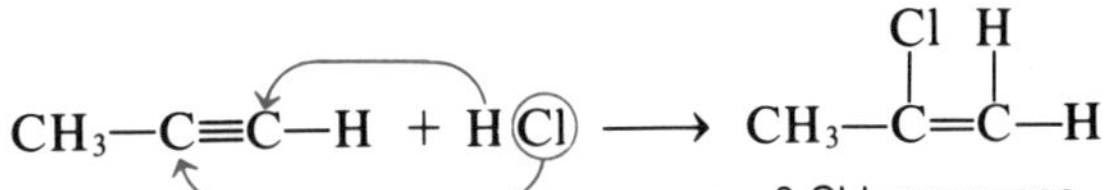

$$CH_3-C{\equiv}C-H \ + \ H Cl \longrightarrow CH_3-\overset{Cl}{\underset{}{C}}{=}\overset{H}{\underset{}{C}}-H$$

2-Chloropropene

Addition of a second molecule of HCl produces the saturated compound:

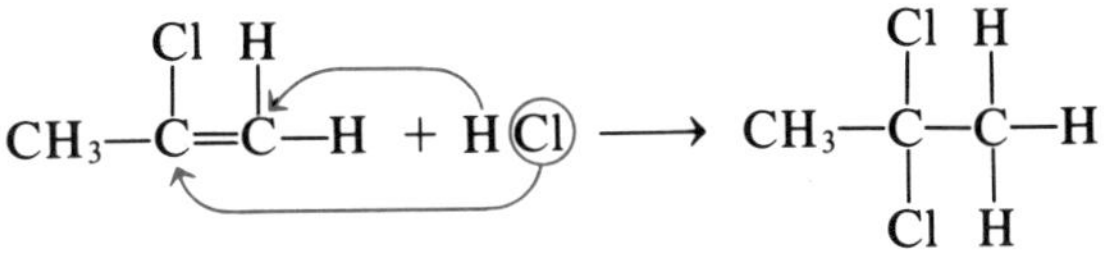

$$CH_3-\overset{Cl}{\underset{}{C}}{=}\overset{H}{\underset{}{C}}-H \ + \ H Cl \longrightarrow CH_3-\overset{Cl}{\underset{Cl}{C}}-\overset{H}{\underset{H}{C}}-H$$

2,2-Dichloropropane

Check Test Number 4

Give the formula and the name for the expected product in each of the following reactions.

a. $CH_3-C{\equiv}C-H \ + \ 2Cl_2 \rightarrow$

b. 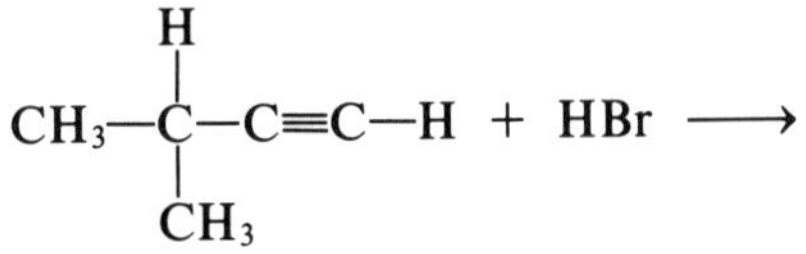 $CH_3-\overset{H}{\underset{CH_3}{C}}-C{\equiv}C-H \ + \ HBr \longrightarrow$

c. $CH_3CH_2CH_2-C{\equiv}C-H \ + \ 2H_2 \xrightarrow{Pt}$

Answers:

a. 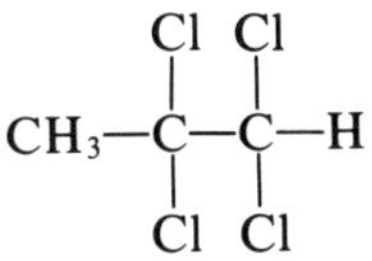 $CH_3-\overset{Cl}{\underset{Cl}{C}}-\overset{Cl}{\underset{Cl}{C}}-H$ 1,1,2,2-tetrachloropropane

b. 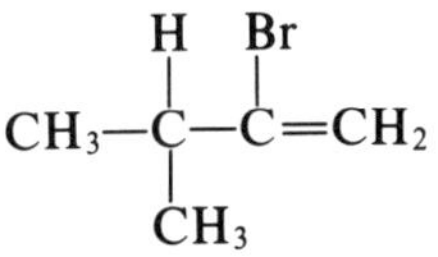 $CH_3-\overset{H}{\underset{CH_3}{C}}-\overset{Br}{\underset{}{C}}{=}CH_2$ 2-bromo-3-methyl-1-butene

c. $CH_3CH_2CH_2CH_2CH_3$ *n*-pentane

The Aromatic Hydrocarbons

The **aromatic hydrocarbons** include benzene and its derivatives and compounds that are benzenelike in their chemical properties. Perhaps the best way to learn about the aromatic hydrocarbons is to study benzene itself.

Benzene, C_6H_6, was first isolated by Michael Faraday in 1825, but many years were to pass before chemists were to feel confident that they understood why benzene had such unique properties. The low hydrogen/carbon ratio led chemists to believe that benzene must be a highly unsaturated compound with several double or triple bonds, yet benzene did not undergo the typical addition reactions with Br_2 or H_2 that were known for alkenes and alkynes. After further experiments it was clear that benzene and its derivatives constituted another major type of organic compound.

In 1865, August Kekulé, a German chemist, proposed that benzene must be a cyclic molecule with alternating single and double bonds between the carbon atoms in the ring. Furthermore, Kekulé proposed that the alternating single-double bond arrangement was in rapid oscillation around the ring causing the molecule to be resistant to addition of Br_2, HCl, and H_2. Kekulé represented the true structure of benzene as something intermediate between the two resonance forms shown below that are labeled A and B.

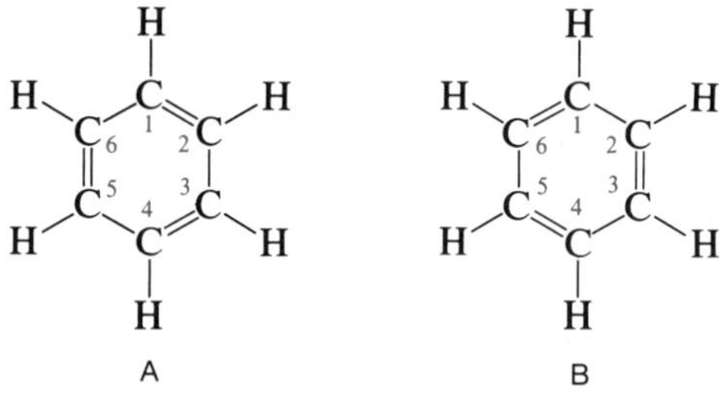

Kekulé structures
(resonance forms of benzene)

Modern chemists generally do not look at the benzene molecule in terms of a rapidly oscillating system of single and double bonds. Rather, the six electrons that would be responsible for one bond in each of the three double bonds are considered to be equally shared by all six carbon atoms at the same time as they freely move around the ring. As a result, the modern formula for the benzene ring is written as a hexagon of carbon atoms with a circle drawn in the center to symbolize the unique bonding in the ring. Of course, the Kekulé structures are also used, though it is understood that the alternating single-double bond arrangement only approximates the unusual nature of the bonding in the ring.

The benzene molecule is flat, with all twelve atoms lying in the same plane, as shown in Figure 16.4. It is insoluble in water but soluble in most nonpolar organic liquids. Though benzene and its derivatives were at one time called aromatic because of their generally pleasant odors, the term has taken on a new meaning which is used to describe compounds with ring structures which have the unique benzenelike bonding in the rings.

Benzene and other aromatic hydrocarbons can be obtained from the distillation of coal tar, a black, viscous, liquid obtained when coal is strongly heated in the absence of oxygen. Benzene was a widely used solvent in industry for years but its use has been curtailed by the discovery of its carcinogenic (cancer-causing) nature in test animals. It should be emphasized that even though benzene itself has been linked to cancer, there are many compounds that contain the benzene ring that are essential to life and do not have the potential to induce cancer. These essential compounds that contain the benzene ring are obtained from plant sources in our diet. Animals are unable to synthesize benzene rings in their bodies, but plants can do so.

The Chemical Properties of Aromatic Hydrocarbons

The characteristic reactions of the aromatic hydrocarbons are substitution reactions in which an atom or group is substituted for a hydrogen on the ring. Unlike the substitution reactions of the alkanes, those with aromatic compounds

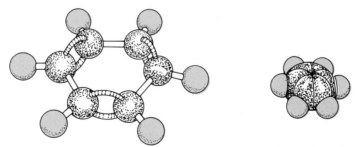

Fig. 16.4. Models of the planar benzene molecule. Though models can be constructed with alternating double and single bonds in the ring, they do not represent the true nature of the bonding between the six carbon atoms. Each carbon–carbon bond is identical to the others in the ring.

usually require a catalyst and do not need uv light or very high temperatures. The following are typical substitution reactions with benzene.

1. Halogenation—Cl or Br replacing H

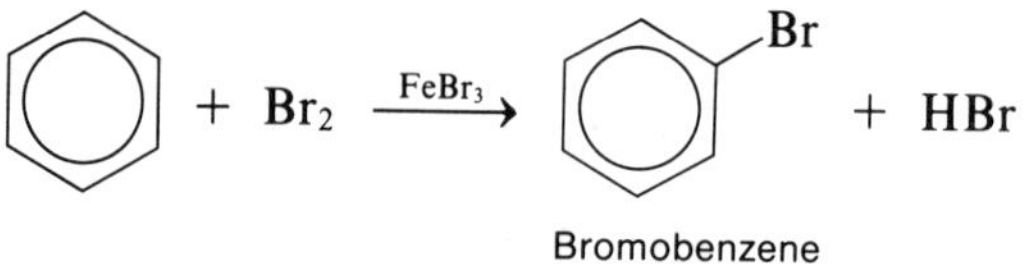

$$\text{benzene} + Br_2 \xrightarrow{\ FeBr_3\ } \text{C}_6\text{H}_5\text{Br} + HBr$$

Bromobenzene

2. Nitration—a nitro group ($-NO_2$) replacing H

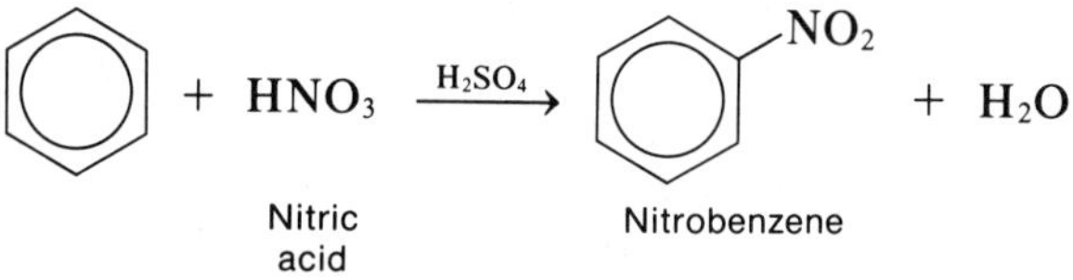

$$\text{benzene} + HNO_3 \xrightarrow{\ H_2SO_4\ } \text{C}_6\text{H}_5\text{NO}_2 + H_2O$$

Nitric
acid   Nitrobenzene

3. Sulfonation—a sulfonic acid group ($-SO_3H$) replacing H

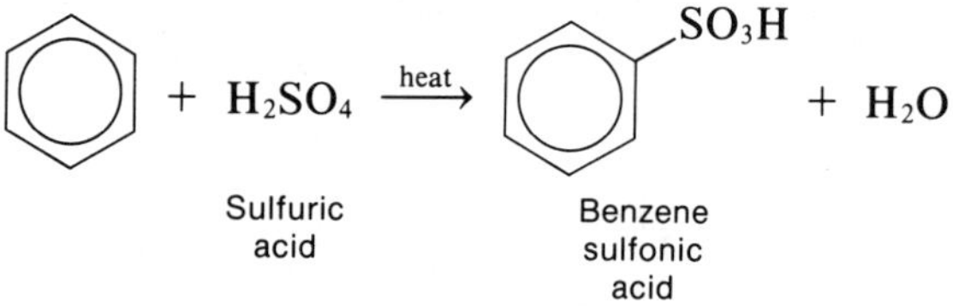

$$\text{benzene} + H_2SO_4 \xrightarrow{\ heat\ } \text{C}_6\text{H}_5\text{SO}_3\text{H} + H_2O$$

Sulfuric   Benzene
acid   sulfonic
                  acid

4. Alkylation—an alkyl group ($-R$) replacing H (the Friedel–Crafts reaction)

General reaction:

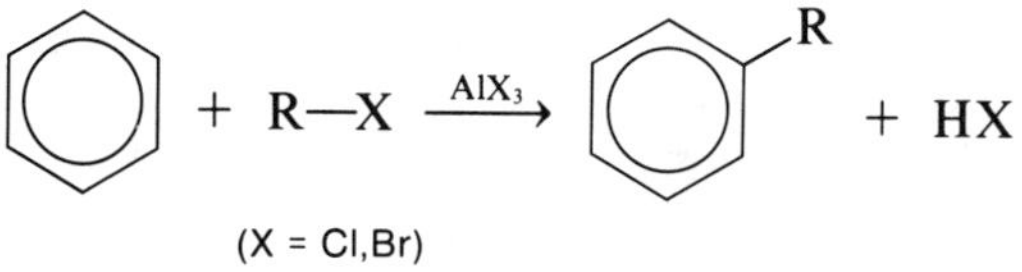

$$\text{benzene} + R-X \xrightarrow{\ AlX_3\ } \text{C}_6\text{H}_5\text{R} + HX$$

(X = Cl,Br)

Specific reaction:

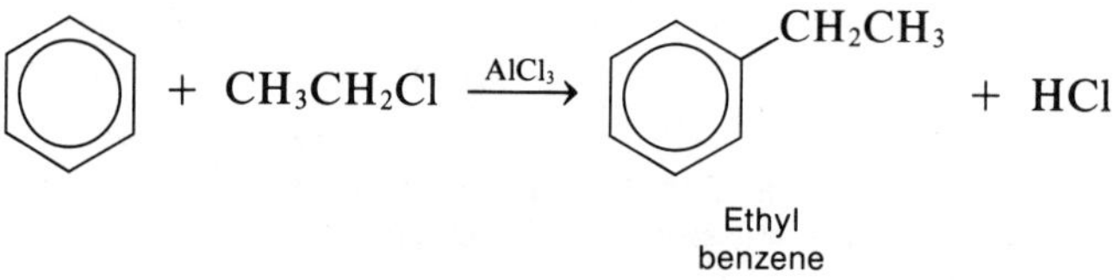

$$\text{benzene} + CH_3CH_2Cl \xrightarrow{\ AlCl_3\ } \text{C}_6\text{H}_5\text{CH}_2\text{CH}_3 + HCl$$

Ethyl
benzene

Naming the Derivatives of Benzene

Derivatives of benzene can have one or more of the hydrogens on the ring replaced by other atoms or groups. If only one hydrogen is replaced, as was done in each of the reactions described in the previous section, the IUPAC name of the product is simply the name of the substituting group added as a prefix to "benzene." The name is written as one word. Frequently, common names are used for certain benzene derivatives. Common names appear in parentheses below the IUPAC names in the following examples.

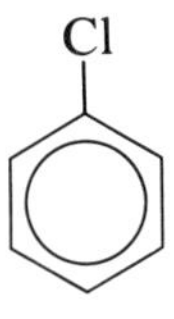

Chlorobenzene

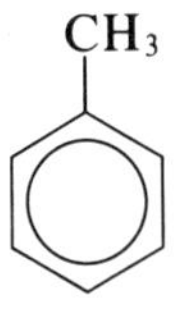

Methylbenzene
(toluene)

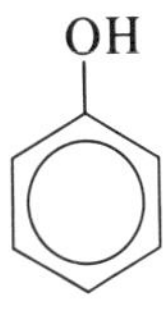

Hydroxybenzene
(phenol)

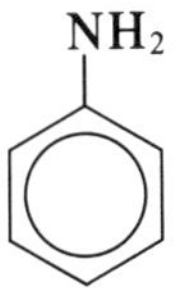

Aminobenzene
(aniline)

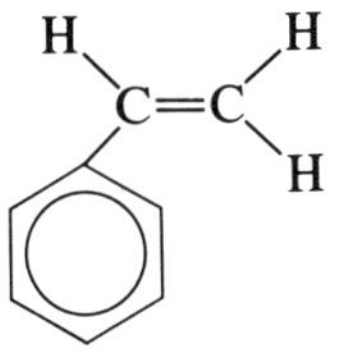

Vinyl benzene
(styrene)

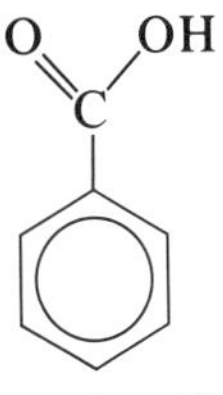

(Benzoic acid)

Though each of the compounds named above is drawn with the substituting group at the "top" of the ring, the group could have been placed at any of the other places on the ring. Any of the following structures would represent chlorobenzene.

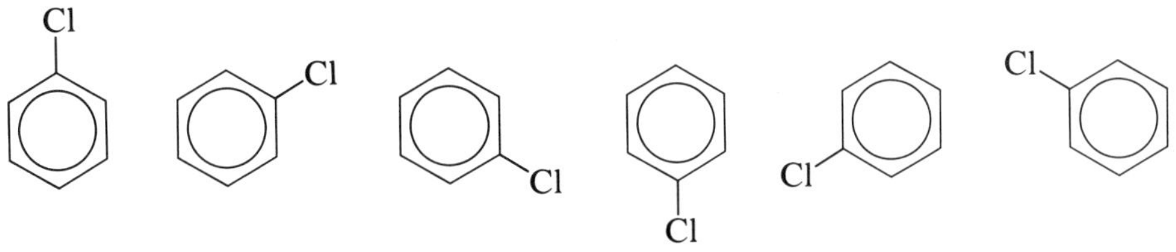

If two hydrogen atoms on the benzene ring are replaced by other groups, there are three possible ways to position them on the ring. This gives rise to three possible isomers. The three isomers of dibromobenzene are shown below. The common name for the isomer with both bromine atoms located on adjacent carbons is *ortho*-dibromobenzene, which is abbreviated as *o*-dibromobenzene.

If the two bromine atoms are separated by one carbon atom, it is called *meta*-dibromobenzene, or *m*-dibromobenzene. If the bromine atoms are on opposite sides of the ring, it is called *para*-dibromobenzene, or *p*-dibromobenzene.

Ortho-dibromobenzene   *Meta*-dibromobenzene   *Para*-dibromobenzene
o-dibromobenzene   *m*-dibromobenzene   *p*-dibromobenzene

An alternate way of naming the disubstituted benzenes is provided by the IUPAC system. Each carbon atom in the ring is numbered from 1 to 6 so that the smallest numbers appear in the name. This scheme is especially useful if two or more unlike groups are attached to the ring. The following examples use this method:

1,3-Dibromobenzene   1-Bromo-3-nitrobenzene   1-Amino-4-methylbenzene

A third method that is used to name the disubstituted compounds regards them as a derivative of some other aromatic compound and not benzene. Commonly they are named as derivatives of toluene, phenol, benzoic acid, aniline, or benzene sulfonic acid.

m-Bromotoluene   *p*-Chlorophenol   *o*-Bromoaniline

If three or more groups are substituted on the benzene ring, the compounds can be named by numbering the atoms in the ring as in the IUPAC system or as derivatives of other common aromatic compounds. Some of the following examples use both methods. Remember, the rings are numbered so the smallest set of numbers are used in the name.

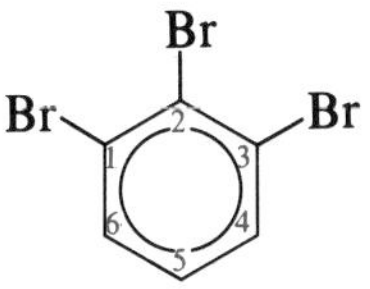

Br
Br——Br

1,2,3-Tribromobenzene

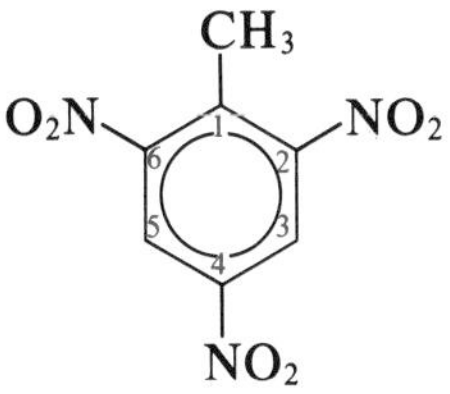

CH_3
O_2N——NO_2
NO_2

2,4,6-Trinitrotoluene
(—CH_3 is understood to be on carbon number 1)

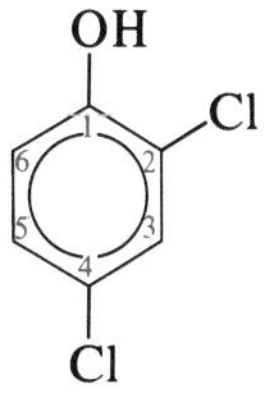

OH
Cl
Cl

2,4-Dichlorophenol
1-hydroxy-2,4-dichlorobenzene

Check Test Number 5

Draw the structural formula of each of the following compounds:
a. ethylbenzene
b. 1,3-dimethylbenzene
c. *p*-chloroaniline
d. 1,2,4-trinitrobenzene
e. 2,4-dibromotoluene
f. 2-ethyl-3-methylnitrobenzene

Answers:

a. 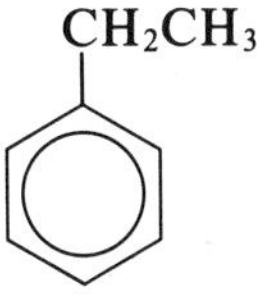 CH_2CH_3

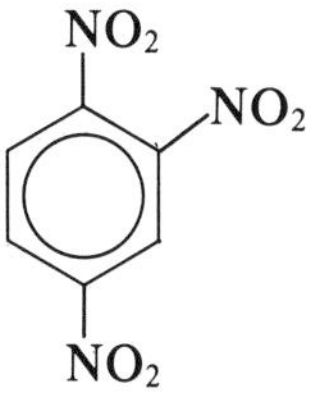

b. 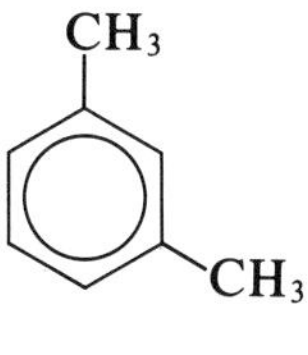 CH_3 CH_3

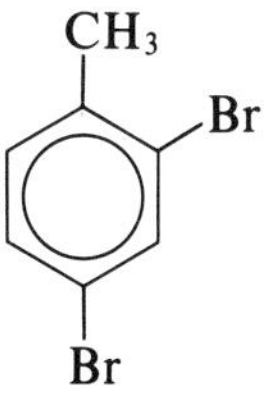

c. 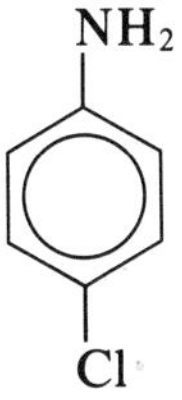 NH_2 Cl

d. NO_2 NO_2 NO_2

e. CH_3 Br Br

f. 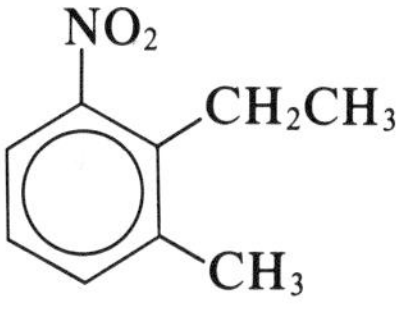 NO_2 CH_2CH_3 CH_3

Polycyclic Aromatic Compounds

The polycyclic aromatic compounds contain two or more benzene rings that are joined together by sharing ring carbon atoms. The simplest polycyclic aromatic compound is naphthalene, $C_{10}H_8$. Naphthalene is a white solid at room temperature and it has the odor of mothballs. (Both naphthalene and *p*-dichlorobenzene have been used to protect woolen clothing from the damage of moths.)

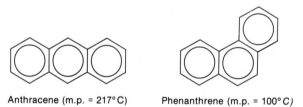

Naphthalene (m.p. = 80°C)

Two other polycyclic aromatic compounds are anthracene, $C_{14}H_{10}$, and phenanthrene, $C_{14}H_{10}$. Their condensed formulas show how three rings are fused, or joined, to form these large, flat aromatic molecules:

Anthracene (m.p. = 217°C)   Phenanthrene (m.p. = 100°C)

Several polycyclic aromatic compounds that contain more than three rings fused together have been found to cause cancer in test animals. One of the most familiar of these is 3,4-benzpyrene, $C_{20}H_{12}$. This carcinogenic compound is formed in the high-temperature environment of burning organic compounds. It is found in automobile exhausts, cigarette smoke, the smoke formed during the grilling of meat, and the burning of coal.

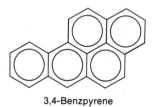

3,4-Benzpyrene

Terms

Several important terms appeared in Chapter 16. You should know the meaning of each of the following.

alkene

alkyne

aromatic hydrocarbon

polycyclic aromatic

olefin

unsaturated compound

geometric isomerism

cis–trans isomerism

diene

triene

Markovnikov's rule

monomer

polymer

polymerization

ortho, meta, para

halogenation

hydrogenation

Questions

Answers to the starred questions appear at the end of the text.

1. Why are alkenes and alkynes described as unsaturated compounds?

*2. What would be the formula for an alkene hydrocarbon with seven carbon atoms and one double bond? What would be the formula for an alkyne hydrocarbon that has five carbon atoms and one triple bond?

3. Give the correct IUPAC name for the following alkenes.

*a. $CH_3CH{=}CHCH_2$

b. $CH_3CHCH_2CH{=}CH_2$
　　　$|$
　　　CH_3

*c. $CH_2{=}CHCH_2CH{=}CH_2$

d. $CH_3{-}CH{-}C{=}CHBr$
　　　　$|$　　$|$
　　　CH_3　CH_3

*e.

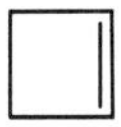

*f.

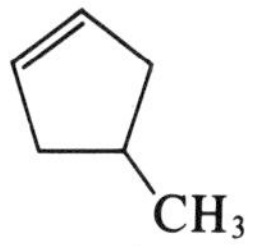

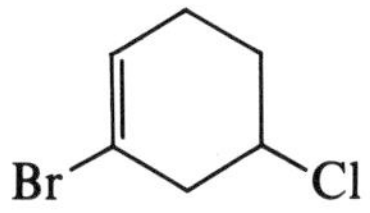

*h. $ClCH_2{-}CH{=}CCH_2CH_2CH_3$
　　　　　　　　　　$|$
　　　　　　　　　　Cl

g.

4. Why are there *cis* and *trans* isomers for 2-butene but not for 1-butene?

5. Draw the structural formulas for the *cis* and *trans* isomers of the following alkenes.

   a. $BrCH=CHBr$

   b. $BrCH=CHCH_3$

   c. $CH_3CH_2-CH-C=CHCH_3$
      $\quad\quad\quad\quad\;\;|\quad\;|$
      $\quad\quad\quad\;\;\;CH_3\;CH_3$

6. Give the structural formula for the product of each of the following addition reactions of alkenes.

   *a. $CH_2=CH_2 + H_2 \xrightarrow[\text{high pressure}]{\text{Pt}}$

   *b. $CH_3CH=CH_2 + HBr \longrightarrow$

   c. $CH_3CH_2\ CH=C(CH_3)_2 + Cl_2 \longrightarrow$

   d. 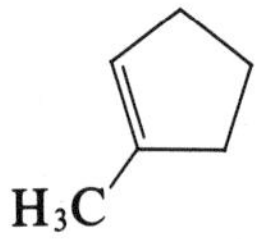 $+ HI \longrightarrow$

   e. $CH_3CH_2C=CHCH_3 + H_2O \xrightarrow{H_2SO_4}$
      $\quad\quad\quad\;\;|$
      $\quad\quad\quad CH_3$

7. Give the formula of the monomer that would be used to prepare the following polymers.

   *a. polyethylene$-CH_2-CH_2-CH_2-CH_2-$

   *b. polyvinylchloride$-CH_2-CHCl-CH_2-CHCl-$

   c. polystyrene $+CH_2-CH-CH_2-CH+$

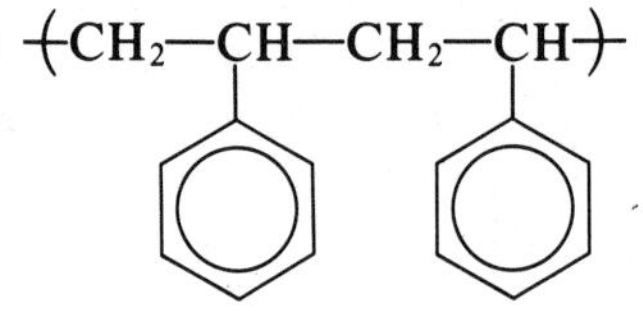

8. Describe the bromine test that is used to detect the presence of unsaturation in an organic compound.

9. Complete the following equations that describe reactions of alkynes.

   a. $H-C\equiv C-H + O_2 \xrightarrow{\text{combustion}}$

   b. $CH_3-C\equiv C-H + 1HBr \longrightarrow$

   c. $CH_3CH_2C\equiv CCH_2CH_3 + 2\ Br_2 \longrightarrow$

   d. 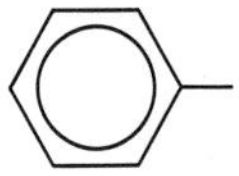 $-CH_2C\equiv CH + 2\ HCl \longrightarrow$

10. If the structural formula for benzene can be drawn with three carbon–carbon double bonds, why does it not undergo typical addition reactions like the alkenes?

*11. Indicate which dibromobenzene is *para*, *meta*, and *ortho*.

a. 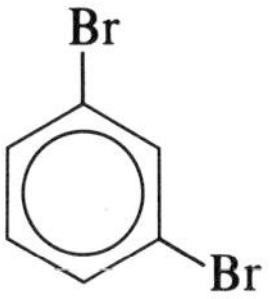   **b.** 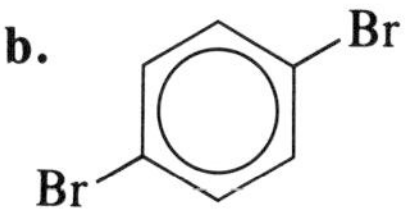   **c.**

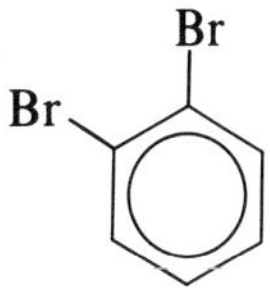

12. Complete the following equations.

*a. 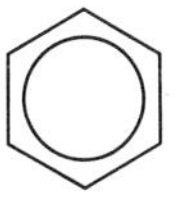 + CH_3Br $\xrightarrow{AlBr_3}$

b. 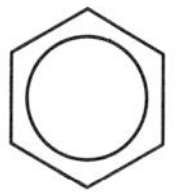 + HNO_3 $\xrightarrow{H_2SO_4}$

c. 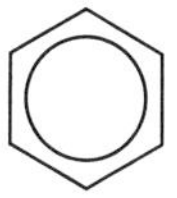 + Br_2 $\xrightarrow{FeBr_3}$

13. Name the following aromatic compounds.

*a.

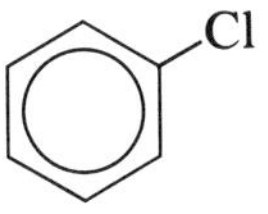

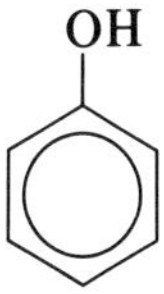

b.

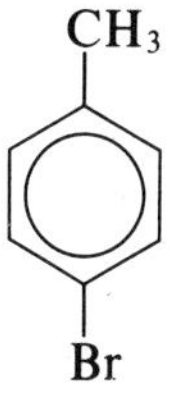

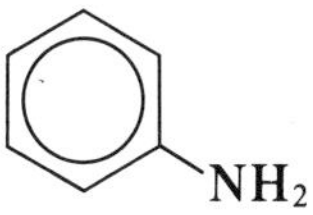

*c.

d.

*e.

f.

*g. *dynomite*

h.

*i.

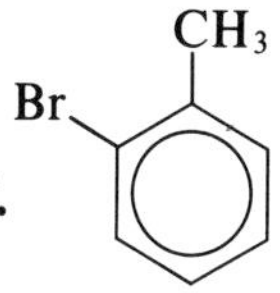

j. 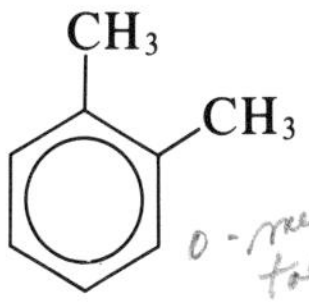 *o - methyl toluene*

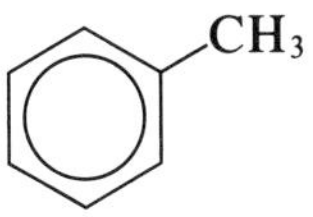

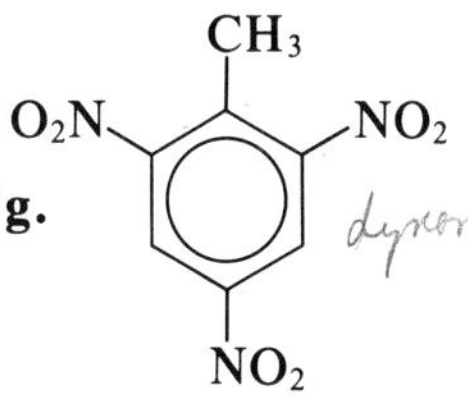

o - xylene

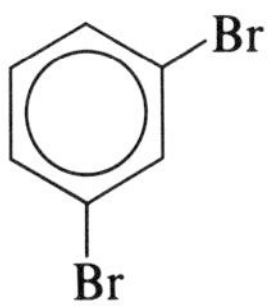

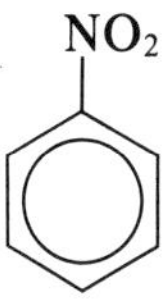

14. Give the structural formula for each of the following.
   *a. aniline   d. benzoic acid
    b. toluene   *e. phenol
   *c. p-dimethylbenzene   f. benzene sulfonic acid
15. Draw the structural formulas for napthalene and anthracene.

17

Alcohols, Phenols, and Ethers

Up to this point in the discussion of organic chemistry, we have been primarily concerned with hydrocarbons, compounds composed of only carbon and hydrogen. If we are to continue on with organic chemistry, it is time to include another element, oxygen. Specifically, we are going to be concerned with organic compounds that contain oxygen atoms bonded to two other atoms (carbon or hydrogen) by single bonds. There are two ways an oxygen atom can be incorporated into a hydrocarbon and still be joined to two other atoms. One way is to place it between a carbon atom and a hydrogen atom like this, $-\overset{\displaystyle |}{\underset{\displaystyle |}{C}}-O-H$. In doing so, a new class of compounds comes into being, the alcohols and phenols. The second way to include an oxygen atom is to place it between two carbon atoms, $-\overset{\displaystyle |}{\underset{\displaystyle |}{C}}-O-\overset{\displaystyle |}{\underset{\displaystyle |}{C}}-$. The result is another class of organic compounds, the ethers.

Alcohols, phenols, and ethers are important compounds in the area of health care, as you will see, and it important that you understand what they are and how they are used.

The chapter will begin with alcohols. The way they are classified, named, and prepared will be discussed along with their physical and chemical properties. The section on alcohols will end with a closer look at five alcohols that are important in daily life. Next, the phenols will be described, with special emphasis on their germicidal properties. Ethers will be discussed last. Their names, properties, and uses as anesthetics will be discussed.

Objectives

By the time you finish this chapter you should be able to do the following:

1. Identify a compound as an alcohol, phenol, or an ether given its condensed formula.
2. Identify an alcohol as being either primary, secondary, or tertiary given its structural formula.
3. Give the common name for at least five common alcohols and ethers.
4. Derive the IUPAC name for an alcohol given its structural formula.
5. Explain why ethers have lower boiling points and lower solubilities in water than do the comparable alcohols.
6. Describe the reactions used to prepare alcohols in the laboratory, specifically the hydration of alkenes and the reduction of carbonyls.
7. Give the product of a dehydration reaction of an alcohol under high- and low-temperature conditions.
8. Give the expected product of the controlled oxidation of a primary, secondary, and tertiary alcohol.
9. Describe the principal difference between a phenol and an aliphatic alcohol.
10. Give some uses of ethers and alcohols in the health care area.

Alcohols

The general formula for an **alcohol** is R–OH, where R– is an aliphatic group. The identifying feature of an alcohol is the **hydroxyl** functional group, –OH, bonded to carbon. In a sense, alcohols can be viewed as derivatives of water. If one of the hydrogen atoms in a water molecule is replaced by an aliphatic group (CH$_3$—, CH$_3$CH$_2$—, etc.), the result is an alcohol. The bent structure of the water molecule is retained in the alcohol, though formulas for alcohols are usually written with the –OH group in a straight line and not showing the bent molecular shape.

Water

An alcohol

Alcohols are divided into three subclasses depending on the number of alkyl groups attached to the carbon atom bearing the hydroxyl group. The subclasses are named **primary, secondary,** and **tertiary,** and they are identified by the symbols 1°, 2°, and 3°, respectively. If one alkyl group is joined to the hydroxyl-bearing carbon, it is a primary alcohol. If two alkyl groups are joined, it is a secondary alcohol, and if three alkyl groups are joined to the hydroxyl-bearing carbon, it is a tertiary alcohol. Examples of the three alcohol subclasses appear in Table 17.1.

TABLE   17.1
The Three Subclasses of Alcohols

Subclass	Symbol	General formula	Examples
Primary	1°	$R-\overset{\overset{H}{|}}{\underset{\underset{H}{|}}{C}}-OH$	$\boxed{CH_3}CH_2OH$ $\boxed{CH_3CH_2}CH_2OH$ $CH_3\,OH^*$
Secondary	2°	$R-\overset{\overset{H}{|}}{\underset{\underset{\boxed{R}}{|}}{C}}-OH$	$\boxed{CH_3}-\overset{\overset{OH}{|}}{\underset{\underset{H}{|}}{C}}-\boxed{CH_3}$ $\boxed{CH_3CH_2}-\overset{\overset{OH}{|}}{\underset{\underset{H}{|}}{C}}-\boxed{CH_3}$
Tertiary	3°	$\overset{\boxed{R}}{\underset{\boxed{R}}{R-C-OH}}$	$\overset{\boxed{CH_3}}{\underset{\boxed{CH_3}}{CH_3-C-OH}}$

*CH_3OH is methyl alcohol, the first member of the alcohol family, and it is considered a primary alcohol.

Check Test Number 1

Classify each of the following alcohols as either primary, secondary, or tertiary:

a. $CH_3CH_2\underset{\underset{\displaystyle CH_3}{|}}{C}HCH_2OH$

c. $CH_3-\underset{\underset{\displaystyle CH_3}{|}}{\overset{\overset{\displaystyle OH}{|}}{C}}-CH_2CH_3$

b. $CH_3-CH-\underset{\underset{\displaystyle CH_3\ H}{|}}{\overset{\overset{\displaystyle OH}{|}}{C}}-CH_3$

d. $\bigcirc\!\!\!\!\bigcirc-CH_2-\underset{\underset{\displaystyle CH_3}{|}}{\overset{\overset{\displaystyle CH_3}{|}}{C}}-OH$

Answers: (a) primary, (b) secondary, (c) tertiary, (d) tertiary

Naming the Alcohols

Common names are most often used for the first few members of the alcohol family. They are formed by first naming the alkyl group in the alcohol, and then adding the word "alcohol." The common names for three of the most widely used alcohols are

CH_3OH
Methyl alcohol

CH_3CH_2OH
Ethyl alcohol

$CH_3\overset{\overset{\displaystyle OH}{|}}{C}HCH_3$
Isopropyl alcohol

Some alcohols have more than one common name. Ethyl alcohol is also called grain alcohol, and methyl alcohol is called wood alcohol. Alcohols that are more complex than the first members of the family are more easily named by the IUPAC system.

The IUPAC rules for naming alcohols are similar in many ways to those used earlier for alkanes and alkenes. Let us go through the step-by-step process that will develop the IUPAC name for the following alcohol that contains several carbon atoms:

$$CH_3-\underset{\underset{\displaystyle CH_3}{\overset{\displaystyle |}{\underset{\displaystyle |}{CH_2}}}}{\overset{\overset{\displaystyle H}{|}}{C}}-\underset{\underset{\displaystyle CH_3}{|}}{\overset{\overset{\displaystyle OH}{|}}{C}}-CH_3$$

Step 1. Examine the structural formula and locate the longest chain of carbon atoms that has the –OH group bonded to it. This is the **main chain** of the alcohol.

$$CH_3-\underset{\underset{\displaystyle CH_3}{\overset{\displaystyle |}{\underset{\displaystyle |}{CH_2}}}}{\overset{\displaystyle H}{\overset{\displaystyle |}{C}}}-\underset{\underset{}{CH_3}}{\overset{\displaystyle OH}{\overset{\displaystyle |}{C}}}-CH_3$$

The longest chain containing the –OH group has five carbon atoms

Step 2. The main chain is given the name of the alkane (Table 15.1) with the same number of carbon atoms, except the *-e* ending is replaced by *-ol* to indicate that the compound is an alcohol. In our example, the main chain has five carbon atoms, so the name of the parent alcohol is *pentanol*.

Step 3. The carbon atoms in the main chain are consecutively numbered from the end *nearest* the hydroxyl group. This locates the hydroxyl group with the lowest possible number. This number is written in *front* of the name of the parent alcohol separated by a hyphen.

$$CH_3-\overset{\displaystyle H}{\underset{\underset{\underset{CH_3}{5|}}{\overset{4|}{CH_2}}}{\overset{3|}{C}}}-\overset{\displaystyle \boxed{OH}}{\underset{\underset{CH_3}{|}}{\overset{2|\ 1}{C}}}-CH_3$$

2-Pentanol

Step 4. Alkyl groups bonded to the main chain are then located and numbered, and their names are written in front of the name of the parent alcohol to give the complete IUPAC name. The same rules used for alkanes are used for the alkyl groups on alcohols.

$$CH_3-\overset{\displaystyle H}{\underset{\underset{\underset{CH_3}{5|}}{\overset{4|}{CH_2}}}{\overset{3|}{C}}}-\overset{\displaystyle OH}{\underset{\underset{CH_3}{|}}{\overset{2|\ 1}{C}}}-CH_3$$

2,3-Dimethyl-2-pentanol

If an alcohol contains two or more hydroxyl groups, special endings are used in the name. Alcohols of this type are known as polyhydric alcohols or polyalcohols.

Step 5. If the alcohol contains two –OH groups, it is a **diol**; three –OH groups, a **triol**, and so forth. The same IUPAC rules used to name all alcohols are followed, except that the *-e* ending of the name of the main chain is *not* dropped, and the ending, *-diol*, *-triol*, etc., is substituted for the *-ol* ending. Numbers before the name of the main chain are used to locate each –OH group.

$$\begin{array}{ccc}
\underset{\text{1,2-Ethanediol}}{\text{H}-\overset{\overset{\displaystyle \text{OH}}{|}}{\underset{\underset{\displaystyle \text{H}}{|}}{\text{C}}}-\overset{\overset{\displaystyle \text{OH}}{|}}{\underset{\underset{\displaystyle \text{H}}{|}}{\text{C}}}-\text{H}
& \qquad
\underset{\text{1,3-Propanediol}}{\text{H}-\overset{\overset{\displaystyle \text{OH}}{|}}{\underset{\underset{\displaystyle \text{H}}{|}}{\text{C}}}-\overset{\overset{\displaystyle \text{H}}{|}}{\underset{\underset{\displaystyle \text{H}}{|}}{\text{C}}}-\overset{\overset{\displaystyle \text{OH}}{|}}{\underset{\underset{\displaystyle \text{H}}{|}}{\text{C}}}-\text{H}
& \qquad
\underset{\text{1,2,3-Propanetriol}}{\text{H}-\overset{\overset{\displaystyle \text{OH}}{|}}{\underset{\underset{\displaystyle \text{H}}{|}}{\text{C}}}-\overset{\overset{\displaystyle \text{OH}}{|}}{\underset{\underset{\displaystyle \text{H}}{|}}{\text{C}}}-\overset{\overset{\displaystyle \text{OH}}{|}}{\underset{\underset{\displaystyle \text{H}}{|}}{\text{C}}}-\text{H}
\end{array}$$

Step 6. When naming nonaromatic cyclic alcohols by the IUPAC system, the carbon atom in the ring bonded to the –OH group is defined as carbon 1. This number is not used in the name unless two or more –OH groups are on the ring. Then, starting at carbon 1, the ring is numbered in either direction so that other groups are given the lowest numbers possible. The prefix **cyclo-** is used with the name of the alkane. The *-e* ending of the alkane is replaced by *-ol*. The structure and IUPAC name for menthol, a cyclic alcohol, is given below:

2-Isopropyl-5-methylcyclohexanol
(Menthol)

The structures and IUPAC names for several alcohols are given below. Study them carefully, then try to name the alcohols in Check Test Number 2.

CH_3OH	methanol
CH_3CH_2OH	ethanol
$CH_3CH_2CH_2OH$	1-propanol
$CH_3-\overset{\overset{\displaystyle \text{OH}}{|}}{\text{CH}}-CH_3$	2-propanol
$CH_3-\overset{\overset{\displaystyle \text{CH}_3}{|}}{\underset{\underset{\displaystyle \text{CH}_3}{|}}{\text{C}}}-\overset{\overset{\displaystyle \text{OH}}{|}}{\underset{\underset{\displaystyle \text{H}}{|}}{\text{C}}}-CH_2-CH_3$	2,2-dimethyl-3-pentanol

CH$_3$—C(Cl)(H)—CH$_2$CH$_2$OH   3-chloro-1-butanol

CH$_3$—C(H)(OH)—CH$_2$—CH$_2$OH   1,3-butanediol

cyclohexanol (with OH)

3,5-dichlorocyclohexanol (OH, Cl, Cl)

Check Test Number 2

Give the correct IUPAC name for the following alcohols:

a. CH$_3$CH$_2$CH$_2$CH$_2$OH

b. CH$_3$CHCH$_2$CHCH$_3$ (with CH$_3$, OH)

c. CH$_3$—C(OH)(CH$_3$)—CH$_3$

d. HOCH$_2$CH$_2$CHOH (with Cl)

e. 2,3-dimethylcyclopentanol structure (CH$_3$, CH$_3$, OH)

Answers: (a) 1-butanol, (b) 4-methyl-2-pentanol, (c) 2-methyl-2-propanol, (d) 1-chloro-1,3-propanediol, (e) 2,3-dimethylcyclopentanol.

Physical Properties of Alcohols

The common alcohols are all colorless liquids at room temperature. Those with a single –OH group usually have densities around 0.8 g/ml, while the diols and triols commonly have densities a little greater than 1 g/ml. Alcohols with a single –OH group are free-flowing liquids, but as the number of alcohol groups increases in the molecule, the liquids become more viscous (thicker) and do not flow as quickly.

The boiling points of alcohols are considerably higher than those of alkanes or ethers of similar molecular weight. For example, 1-butanol, diethyl ether, and pentane have nearly identical molecular weights, but the boiling point of the alcohol is over 80°C higher than either boiling point of the other two.

Compound	Formula	Molecular weight	Boiling point
1-Butanol	$CH_3CH_2CH_2CH_2OH$	74	118°C
Diethyl ether	$CH_3CH_2-O-CH_2CH_3$	74	36°C
Pentane	$CH_3CH_2CH_2CH_2CH_3$	72	35°C

The higher boiling points for alcohols arise from their ability to hydrogen bond to each other, something that ethers and alkanes are unable to do. Hydrogen bonding is a relatively strong intermolecular force of attraction between molecules, and more energy is required to separate the molecules in hydrogen-bonded liquids as they boil. Because of the higher energy requirement, they boil at higher temperatures. The reason compounds of similar molecular weight are compared is because each will have about the same size London force of attraction (Chapter 10):

The polar —OH group of an alcohol allows it to hydrogen bond.

Hydrogen bonds

A comparison of the physical properties of several alcohols with alkanes and ethers of similar molecular weight appears in Table 17.2. In each comparison, the alcohol has the highest boiling point. Also, notice that the first three alcohols are liquids, while the corresponding alkanes and ethers are gases. This too is due to the greater intermolecular forces that exist between alcohol molecules.

The ability of an alcohol to hydrogen bond is also responsible for the high solubility of the lower-molecular-weight alcohols in water. The methyl, ethyl, and propyl alcohols are soluble in water in any proportion. They are said to be **miscible** with water. However, as the size of the alkyl group of an alcohol increases, its solubility in water becomes less and less. The larger alkyl groups are not readily accommodated by water since they are nonpolar and do not interact well with water. Even though the alcohol group will hydrogen bond with water, an interaction that enhances solubility, it is not sufficient to hold the larger alkyl groups in solution. Generally, monohydroxyl alcohols (those with one –OH group) containing five or more carbon atoms are less soluble in water than those with fewer carbon atoms, as shown in Table 17.2.

Alcohols that contain two or more –OH groups are more soluble in water than the corresponding monohydroxyl alcohols. These alcohols are able to engage in more hydrogen bonding, giving them a greater solubility. Also, poly-

TABLE   17.2
Comparison of Certain Physical Properties of Alcohols, Ethers, and Alkanes
of Similar Molecular Weight

Compound	Condensed formula	Molecular weight	Physical state at 25°C	Boiling point (°C)	Solubility in water (g/100 ml)
Methanol	CH_3OH	32	liquid	65	all proportions
Ethane	CH_3CH_3	30	gas	−89	insoluble
Ethanol	CH_3CH_2OH	46	liquid	78	all proportions
Dimethyl ether	CH_3-O-CH_3	46	gas	−24	7.7
Propane	$CH_3CH_2CH_3$	44	gas	−42	insoluble
1-Propanol	$CH_3CH_2CH_2OH$	60	liquid	97	all proportions
Ethyl methyl ether	$CH_3-O-CH_2CH_3$	60	gas	8	moderately soluble
n-Butane	$CH_3CH_2CH_2CH_3$	58	gas	0	insoluble
1-Butanol	$CH_3CH_2CH_2CH_2OH$	74	liquid	118	7.9
Diethyl ether	$CH_3CH_2-O-CH_2CH_3$	74	liquid	35	6.1
n-Pentane	$CH_3CH_2CH_2CH_2CH_3$	72	liquid	36	insoluble
1-Pentanol	$CH_3CH_2CH_2CH_2CH_2OH$	88	liquid	138	2.3
Ethyl n-propyl ether	$CH_3CH_2-O-CH_2CH_2CH_3$	88	liquid	64	slightly soluble
n-Hexane	$CH_3CH_2CH_2CH_2CH_2CH_3$	86	liquid	69	insoluble
1-Hexanol	$CH_3CH_2CH_2CH_2CH_2CH_2OH$	102	liquid	157	0.6
Di-n-propyl ether	$CH_3CH_2CH_2-O-CH_2CH_2CH_3$	102	liquid	91	0.2
n-Heptane	$CH_3CH_2CH_2CH_2CH_2CH_2CH_3$	100	liquid	98	insoluble

alcohols have higher boiling points than monohydroxyl alcohols of similar molecular weight. The physical properties of four polyalcohols are given in Table 17.3 (p. 428). They are all very soluble in water.

The Preparation of Alcohols

The Hydration of Alkenes   Most secondary and tertiary alcohols are commercially prepared by the addition of water to alkenes. The addition of water to a molecule is called **hydration**. Ethyl alcohol is the only primary alcohol that may be prepared this

TABLE   17.3
Physical Properties of Four Polyalcohols

Polyalcohol	Common name	Structure	Molecular weight	Physical state	Boiling point (°C)	Solubility in g/100 ml
1,2-Ethanediol	ethylene glycol	$HOCH_2CH_2OH$	62	liquid	197	all proportions
1,2-Propanediol	1,2-propylene glycol	$CH_3\overset{\text{OH}}{\underset{\text{l}}{C}HCH_2OH}$	72	liquid	189	all proportions
1,2,3-Propanetriol	glycerol	$HOCH_2\overset{\text{OH}}{\underset{\text{l}}{C}HCH_2OH}$	92	liquid	290	all proportions
1,2,3,4,5,6-hexanehexol	sorbitol	$HOCH_2-CH-CH-CH-CH-CH_2OH$ (OH OH OH OH)	182	solid	295	83

way. The addition of water to the carbon–carbon double bond follows Markovnikov's rule: Hydrogen joins to the carbon already richest in hydrogen. The hydration reactions are catalyzed with sulfuric acid, H_2SO_4, and the general equation for the addition reaction is given below:

$$\underset{\text{Alkene}}{\overset{H}{\underset{R}{>}}C=C\overset{H}{\underset{R'}{<}}} + \underset{\text{Water}}{HOH} \xrightarrow{H_2SO_4} \underset{\text{Alcohol}}{R-\overset{H}{\underset{H}{C}}-\overset{H}{\underset{OH}{C}}-R'}$$

In the following specific examples, notice how Markovnikov's rule is followed as the secondary and tertiary alcohols are formed:

$$H_2C{=}CH_2 + HOH \xrightarrow{H_2SO_4} CH_3CH_2OH$$

Ethyl alcohol (1°)

$$CH_3-\overset{H}{\underset{}{C}}{=}CH_2 + HOH \xrightarrow{H_2SO_4} CH_3-\overset{H}{\underset{HO}{C}}-CH_3$$

2-Propanol (2°)

$$CH_3-\underset{\underset{CH_3}{|}}{C}{=}CH_2 + HOH \xrightarrow{H_2SO_4} CH_3-\overset{OH}{\underset{\underset{CH_3}{|}}{C}}-CH_3$$

2-Methyl-2-propanol (3°)

The addition of water to carbon–carbon double bonds is an important reaction in cells. During the metabolism of carbohydrates, fumaric acid is converted to malic acid by this process. In the body, the reaction is catalyzed by an enzyme:

$$HOOC-\overset{H}{\underset{H}{C}}=C-COOH + HOH \xrightarrow{enzyme} HOOC-\overset{OH}{\underset{H}{C}}-CH_2-COOH$$

Fumaric acid   Malic acid

Reduction of Carbonyls   The carbonyl group ($C = O$) in aldehydes, ketones, and carboxylic acids can be reduced by addition of hydrogen, in the presence of a catalyst, to form alcohols. Carboxylic acids and aldehydes are reduced to primary alcohols. Ketones produce secondary alcohols. Tertiary alcohols cannot be prepared this way. The reduction of a carboxylic acid first produces an aldehyde, which then is reduced to the alcohol:

$$R-\overset{O}{\underset{OH}{C}} + H_2 \xrightarrow{catalyst} R-\overset{O}{\underset{H}{C}} + H_2O$$

Carboxylic acid   Aldehyde

$$R-\overset{O}{\underset{H}{C}} + H_2 \xrightarrow{catalyst} R-\overset{H}{\underset{H}{C}}-OH$$

Aldehyde   1° Alcohol

$$\underset{R \quad R'}{\overset{O}{C}} + H_2 \xrightarrow{catalyst} R-\overset{OH}{\underset{H}{C}}-R'$$

Ketone   2° Alcohol

Here are some specific examples of the reduction of a carbonyl:

First step   Second step

$$CH_3\overset{O}{\underset{OH}{C}} + H_2 \xrightarrow{catalyst} CH_3-\overset{}{\underset{H}{C}}=O \xrightarrow[catalyst]{+ H_2} CH_3CH_2OH$$

Acetic acid   Acetaldehyde   Ethyl alcohol

$$H_3C \overset{\displaystyle \overset{O}{\|}}{\underset{}{C}} CH_3 + H_2 \xrightarrow{\text{catalyst}} CH_3 - \overset{\displaystyle OH}{\underset{\displaystyle H}{C}} - CH_3$$

Acetone   2-Propanol

The reduction of carbonyl compounds is an important reaction in biochemical systems. In the above examples hydrogen is supplied as hydrogen molecules, H_2, but in the body H atoms are supplied by the enzyme systems that catalyze the reactions.

Commercial Preparation of Methyl and Ethyl Alcohol

Methyl alcohol can be prepared by the reduction of formaldehyde, $H_2C=O$, but for many years it was prepared by the destructive distillation of wood. In this process, wood is heated to very high temperatures in the absence of air. As it decomposes, methyl alcohol is formed and removed by distillation. You can see why one of the common names for methyl alcohol is wood alcohol. Today, methyl alcohol is primarily prepared by the catalytic hydrogenation of carbon monoxide at high temperature and pressure:

$$CO + 2H_2 \xrightarrow[\Delta, \text{ pressure}]{\text{catalysts}} CH_3OH$$

Methyl alcohol

Two different processes are used to prepare ethyl alcohol commercially. If the alcohol is destined for industrial use, it is prepared by the hydration of ethene. On the other hand, if the alcohol is for human consumption in the U.S., as in alcoholic beverages, it must be prepared by the fermentation of grains. Grains contain sugars and starches that, in the presence of water and certain enzymes, are converted to ethyl alcohol and carbon dioxide:

$$\text{starch} \xrightarrow[+ \ H_2O]{\text{enzyme}} C_6H_{12}O_6 \xrightarrow{\text{enzyme}} 2CH_3CH_2OH + 2CO_2$$

Glucose   Ethyl alcohol

The fermentation of grains is one of the earliest known chemical processes. The enzymes that catalyze the reactions are obtained from yeast.

Fermentation can produce solutions that usually contain no more than about 20% (v/v) ethyl alcohol. At alcohol concentrations only a little higher than this, the enzymes become inactive and the process stops. Beers and wines can be made directly from the filtered fermentation solution. These beverages contain relatively low concentrations of alcohol, about 5% (v/v) for beers and 10–14% (v/v) for most wines. Beverages that contain higher concentrations of alcohol, such as whisky and rum, are prepared by distilling the filtered fermentation solution to obtain solutions as high as 95% (v/v) ethyl alcohol. Beverages usually go no higher than 50% (v/v) alcohol though. The "proof" of an alcoholic beverage is just twice the volume/volume per cent of alcohol that it contains. A whisky that is 50% (v/v) ethyl alcohol would be labeled 100 proof.

Check Test Number 3

1. Write the equation for the following reactions: (a) the hydration of 2-methyl-1-butene to produce an alcohol; (b) the preparation of ethyl alcohol from the reduction of acetaldehyde, $CH_3C = O$; (c) the reduction

    of the proper ketone to produce 2-propanol.
2. What is the % (v/v) of ethyl alcohol in an 86 proof whisky?

Answers:

1. a. $CH_3CH_2-C=CH_2 + HOH \xrightarrow{H_2SO_4} CH_3CH_2-\underset{\underset{CH_3}{|}}{\overset{\overset{OH}{|}}{C}}-CH_3$

    2-Methyl-1-butene

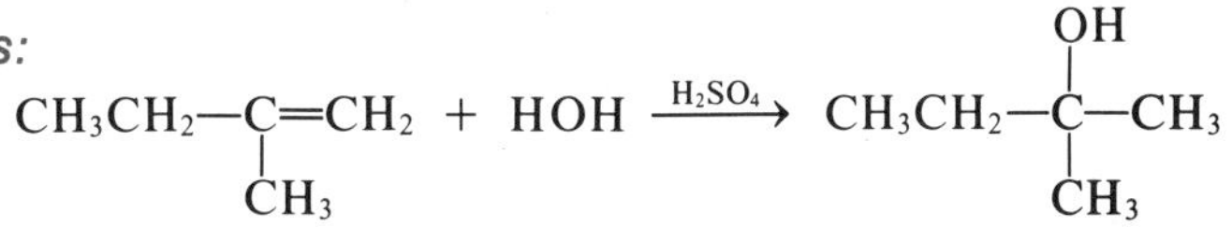

    b. $CH_3C=O + H_2 \xrightarrow{catalyst} CH_3CH_2OH$

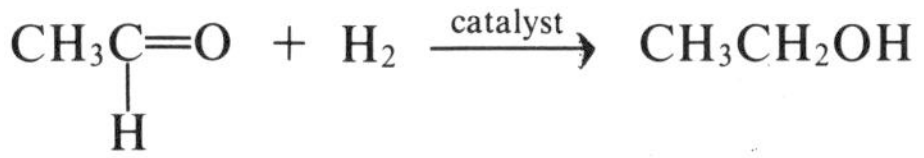

    c. $CH_3-\overset{\overset{O}{\|}}{C}-CH_3 + H_2 \xrightarrow{catalyst} CH_3-\underset{\underset{H}{|}}{\overset{\overset{OH}{|}}{C}}-CH_3$

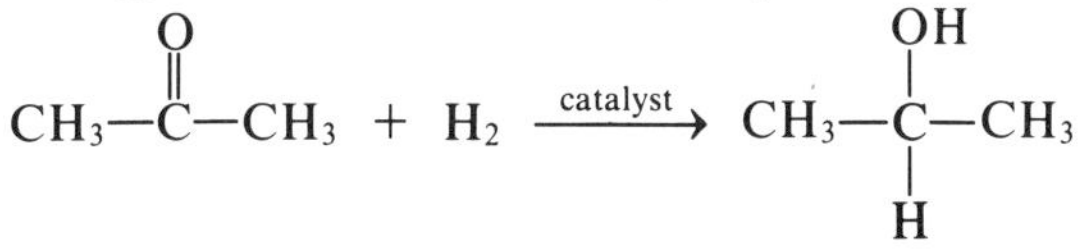

    Acetone   2-Propanol

2. 86 proof is 43% (v/v) ethyl alcohol.

The Chemical Properties of Alcohols

Dehydration of Alcohols

   The **dehydration** of an alcohol is the elimination of one molecule of water from a single alcohol molecule to produce an alkene ($>C=C<$) or from two alcohol molecules to produce an ether ($-\overset{|}{C}-O-\overset{|}{C}-$). The reactions are catalyzed by strong mineral acids such as sulfuric acid or phosphoric acid. Whether the product of the dehydration will be an alkene or an ether depends on the temperature of the reaction. At lower temperatures, around 130–140°C, the product will be an ether. At higher temperatures, 170–180°C, the product will be mostly alkene. Let us look at the reactions that produce alkenes first.

   Dehydration of primary alcohols at 180°C produces 1-alkenes:

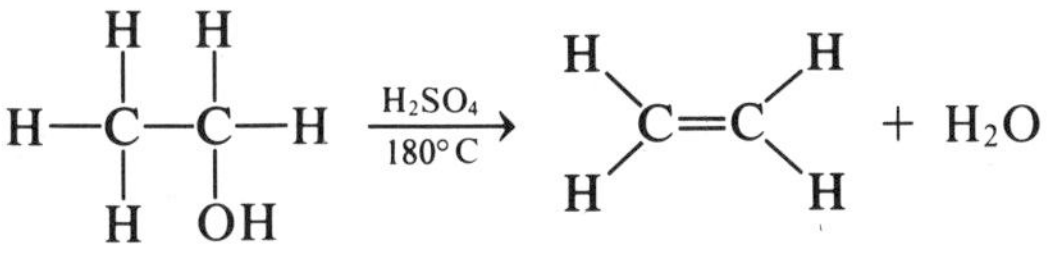

$$H-\underset{\underset{H}{|}}{\overset{\overset{H}{|}}{C}}-\underset{\underset{OH}{|}}{\overset{\overset{H}{|}}{C}}-H \xrightarrow[180°C]{H_2SO_4} \underset{\underset{H}{}}{\overset{\overset{H}{}}{C}}=\underset{\underset{H}{}}{\overset{\overset{H}{}}{C}} + H_2O$$

Ethyl alcohol (1°)   Ethene

$$CH_3-CH_2-\overset{\overset{\displaystyle H}{|}}{\underset{\underset{\displaystyle H}{|}}{C}}-\overset{\overset{\displaystyle H}{|}}{\underset{\underset{\displaystyle OH}{|}}{C}}-H \xrightarrow[180°C]{H_2SO_4} CH_3CH_2CH{=}CH_2 + H_2O$$

1-Butanol (1°)　　　　　　1-Butene

Dehydration of secondary and tertiary alcohols at higher temperatures usually produces a mixture of two alkenes, with one being produced in a greater amount than the other. For example, the dehydration of 2-butanol produces 2-butene as the major product and 1-butene as the minor product:

$$H-\overset{\overset{\displaystyle H}{|}}{\underset{\underset{\displaystyle H}{|}}{C}}-\overset{\overset{\displaystyle H}{|}}{\underset{\underset{\displaystyle OH}{|}}{C}}-\overset{\overset{\displaystyle H}{|}}{\underset{\underset{\displaystyle H}{|}}{C}}-\overset{\overset{\displaystyle H}{|}}{\underset{\underset{\displaystyle H}{|}}{C}}-H \xrightarrow[180°C]{H_2SO_4}$$

2-Butanol

$$\longrightarrow CH_3-\overset{\overset{\displaystyle H}{|}}{C}{=}\overset{\overset{\displaystyle H}{|}}{C}-CH_3 + H_2O$$

2-Butene
(major product)

$$\longrightarrow H_2C{=}\overset{\overset{\displaystyle H}{|}}{C}-CH_2CH_3 + H_2O$$

1-Butene
(minor product)

You may wonder why it is possible to form two different alkenes in a dehydration reaction, and why one is produced in a greater amount than the other. Let us answer this question by seeing how the water is eliminated. The hydrogen atom that combines with the –OH group can come from either carbon atom adjacent to the alcohol group. In 2-butanol, if it comes from carbon 1 and 1-butene is formed:

$$H-\overset{\overset{\displaystyle H}{|}}{\underset{\underset{\displaystyle \boxed{H}}{|}}{\underset{1}{C}}}-\overset{\overset{\displaystyle H}{|}}{\underset{\underset{\displaystyle \boxed{OH}}{|}}{\underset{2}{C}}}-\overset{3}{C}H_2\overset{4}{C}H_3 \xrightarrow[180°C]{H_2SO_4} H_2C{=}\overset{\overset{\displaystyle H}{|}}{C}-CH_2CH_3 + H_2O$$

1-Butene

If the hydrogen is removed from carbon 3, 2-butene is formed:

$$H_3\overset{1}{C}-\overset{\overset{\displaystyle H}{|}}{\underset{\underset{\displaystyle \boxed{OH}}{|}}{\underset{2}{C}}}-\overset{\overset{\displaystyle H}{|}}{\underset{\underset{\displaystyle \boxed{H}}{|}}{\underset{3}{C}}}-\overset{4}{C}H_3 \xrightarrow[180°C]{H_2SO_4} CH_3-\overset{\overset{\displaystyle H}{|}}{C}{=}\overset{\overset{\displaystyle H}{|}}{C}-CH_3 + H_2O$$

2-Butene

You can see how two different alkenes can form, but how can one predict which will be the major product? Whenever it is possible to form two different alkenes in a dehydration reaction, the major product will be the one with the *larger* number of alkyl groups on the double-bond carbon atoms. There are two alkyl groups about the double bond in 2-butene, but only one in 1-butene. 2-Butene is therefore the major product:

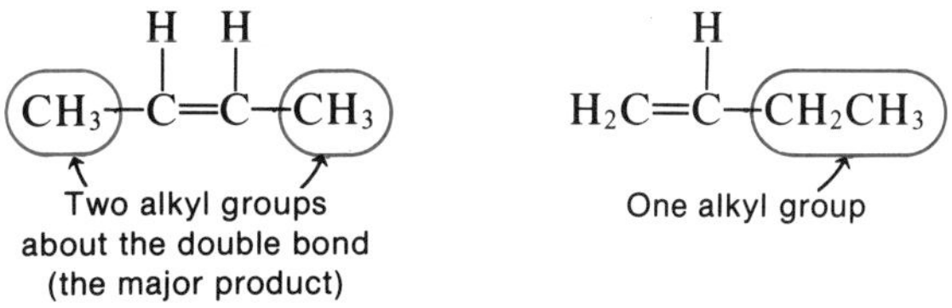

$$CH_3-C=C-CH_3 \qquad H_2C=C-CH_2CH_3$$

Two alkyl groups
about the double bond
(the major product)

One alkyl group

The dehydration of 2-methyl-2-butanol, a tertiary alcohol, produces two alkenes, with 2-methyl-2-butene as the major product. It has the larger number of alkyl groups about the double bond:

$$CH_3-\underset{OH}{\underset{|}{\overset{CH_3}{\underset{|}{C}}}}-CH_2CH_3 \xrightarrow[180°C]{H_2SO_4}$$

$$\longrightarrow CH_3-\overset{CH_3}{\overset{|}{C}}=\overset{H}{\overset{|}{C}}-CH_3 + H_2O$$

2-Methyl-2-butene
(major product)

$$\longrightarrow H_2C=\overset{CH_3}{\overset{|}{C}}-CH_2CH_3 + H_2O$$

2-Methyl-1-butene
(minor product)

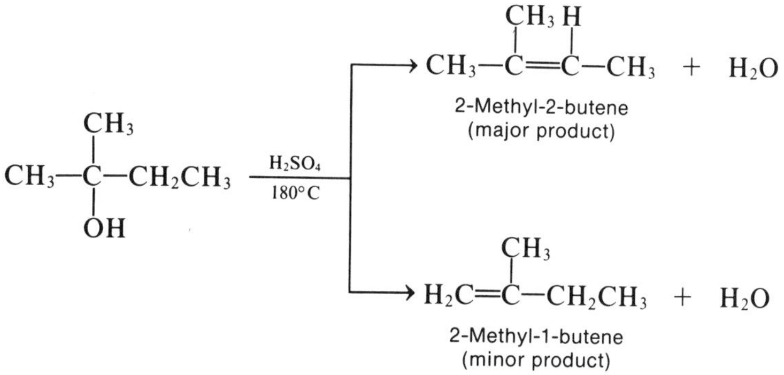

When dehydration reactions are carried out in the laboratory, it is common for secondary and tertiary alcohols to produce two different alkenes. But in the body, dehydration is strictly controlled by enzymes, and only one alkene is formed.

The preparation of ethers by the dehydration of alcohols involves the elimination of one molecule of water from two alcohol molecules. The reactions are carried out at lower temperatures than those used for alkene formation, but a strong acid is still required to catalyze the reactions. The general equation for the preparation of ethers is given below:

$$R-O\boxed{H \ + \ HO}R' \xrightarrow[140°C]{H_2SO_4} R-O-R' + H_2O$$

Alcohol   Alcohol   Ether

The two alcohols in the equation may be identical, or they may be different, as shown in the following examples:

$$CH_3O\boxed{H + HO}CH_3 \xrightarrow[140°C]{H_2SO_4} CH_3—O—CH_3 + H_2O$$

Methyl alcohol   Dimethyl ether

$$CH_3O\boxed{H + HO}CH_2CH_3 \xrightarrow[140°C]{H_2SO_4} CH_3—O—CH_2CH_3 + H_2O$$

Methyl   Ethyl   Methyl ethyl ether
alcohol   alcohol

The formation of ethers by simple dehydration is less prevalent in biological systems than the formation of alkenes. However, the synthesis of starch and cellulose in plants and glycogen in humans can be viewed as an ether synthesis. Starch, cellulose, and glycogen are polymers of glucose. As the polymers form, water is eliminated from two –OH groups on adjacent glucose molecules, joining them together through an ether link. The dehydration is controlled by an enzyme:

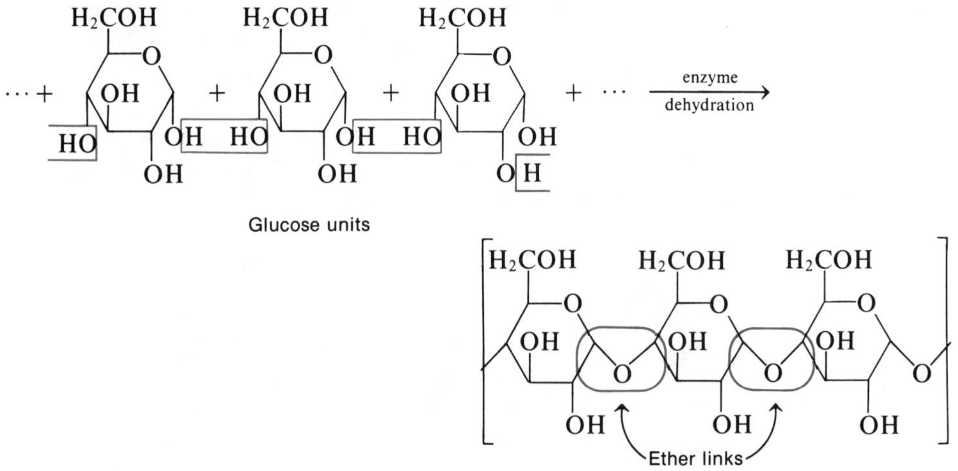

Glucose units

Starch

Check Test Number 4

1. Write the formula of the *major* product obtained when the following alcohols are dehydrated to form an alkene:

   a. $CH_3CH_2CH_2CH_2OH$

   b.

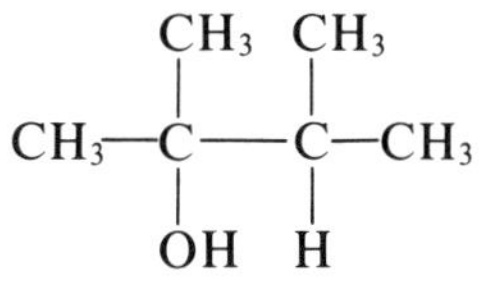

$$CH_3-\overset{\overset{\displaystyle CH_3}{|}}{\underset{\underset{\displaystyle OH}{|}}{C}}-\overset{\overset{\displaystyle CH_3}{|}}{\underset{\underset{\displaystyle H}{|}}{C}}-CH_3$$

   c.

$$CH_3-\overset{\overset{\displaystyle H}{|}}{\underset{\underset{\displaystyle CH_3}{|}}{C}}-\overset{\overset{\displaystyle OH}{|}}{\underset{\underset{\displaystyle H}{|}}{C}}-CH_3$$

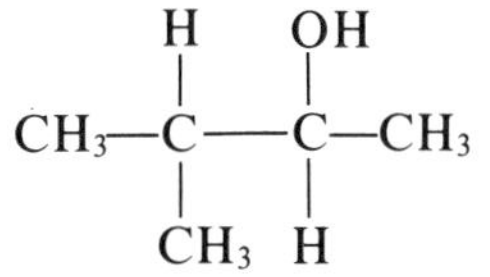

2. Write the formula for the ether formed when the following alcohols are dehydrated:

   a. $CH_3CH_2OH + HOCH_2CH_3 \xrightarrow{H_2SO_4}$

   b. $CH_3OH + HOCH_2CH_2CH_3 \xrightarrow{H_2SO_4}$

Answers:

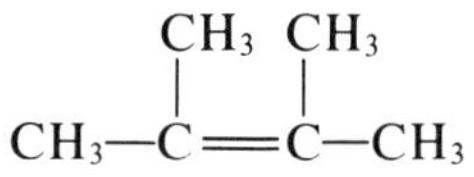

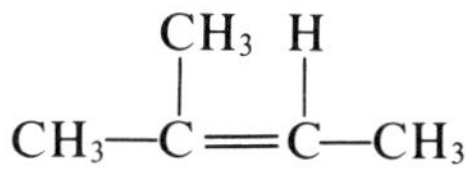

1. a. 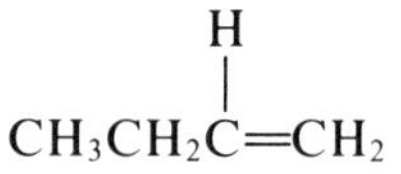 $CH_3CH_2\overset{\overset{\displaystyle H}{|}}{C}=CH_2$ 2. a. $CH_3CH_2-O-CH_2CH_3$

   b. $CH_3-\overset{\overset{\displaystyle CH_3}{|}}{C}=\overset{\overset{\displaystyle CH_3}{|}}{C}-CH_3$ b. $CH_3-O-CH_2CH_2CH_3$

   c. $CH_3-\overset{\overset{\displaystyle CH_3}{|}}{C}=\overset{\overset{\displaystyle H}{|}}{C}-CH_3$

The Oxidation of Alcohols

In Chapter 11 an oxidation reaction was described as one that either adds oxygen atoms to a molecule or removes hydrogen atoms from a molecule. Though oxidation can also be defined in other ways, this particular definition will be the most convenient one for us to use here.

Two common reagents that are used to oxidize alcohols (and other compounds) in the laboratory are potassium dichromate with sulfuric acid ($K_2Cr_2O_7-H_2SO_4$) and potassium permanganate with sulfuric acid

(KMnO$_4$–H$_2$SO$_4$). Because we will be more concerned with the organic product formed as the alcohol is oxidized, we will not include these oxidizing agents in equations. Instead, an oxidation reaction will be symbolized by simply writing [O] above the arrow in the equation: $\xrightarrow{[O]}$. The oxygen atom can be considered as coming from the oxidizing agent, and it may add to a molecule in an oxidation reaction, or it may combine with two hydrogen atoms removed from the molecule to form water. In biological systems, oxidations of alcohols are controlled by special enzymes, the alcohol dehydrogenases. If an oxidation equation describes a reaction in a cell, you only need to write "enzyme" over the arrow.

The oxidation of a primary alcohol produces an aldehyde (R$-\overset{\displaystyle O}{\overset{\displaystyle \|}{C}}-$H) which, unless quickly removed from the reaction medium, will be further oxidized to a carboxylic acid. Two H atoms are removed from the alcohol as it is oxidized to the aldehyde:

$$\begin{array}{c} H \\ | \\ H-C-OH \\ | \\ H \end{array} \xrightarrow{[O]} \begin{array}{c} \\ H-C=O \\ | \\ H \end{array} + H_2O$$

Methyl alcohol (1°)   Formaldehyde

Formaldehyde may then be oxidized to formic acid by gaining one O atom:

$$\begin{array}{c} H-C=O \\ | \\ H \end{array} \xrightarrow{[O]} \begin{array}{c} H-C=O \\ | \\ OH \end{array}$$

Formic acid

Enzymes can control the oxidation so it stops at the aldehyde. Ethyl alcohol is oxidized to acetaldehyde in the liver during metabolism:

$$\begin{array}{c} H \quad H \\ | \quad | \\ H-C-C-OH \\ | \quad | \\ H \quad H \end{array} \xrightarrow{enzyme} \begin{array}{c} H \\ | \\ H-C-C=O \\ | \quad | \\ H \quad H \end{array} + H_2O$$

Ethyl alcohol (1°)   Acetaldehyde

Secondary alcohols are oxidized to ketones (R$-\overset{\displaystyle O}{\overset{\displaystyle \|}{C}}-$R). The ketones do not undergo further oxidation. The alcohol loses two H atoms in the oxidation process:

$$\underset{\underset{\text{2-Propanol (2°)}}{}}{\overset{\text{OH}}{\underset{\overset{|}{H}}{\underset{|}{CH_3-C-CH_3}}}} \xrightarrow{\text{[O]}} \underset{\text{Acetone}}{\overset{\overset{\text{O}}{\|}}{CH_3-C-CH_3}} + H_2O$$

Tertiary alcohols do not undergo oxidation under normal laboratory conditions. To summarize the oxidation of alcohols we must remember that primary, secondary, and tertiary alcohols do not react in the same way:

$$1° \text{ alcohol} \xrightarrow{\text{[O]}} \text{aldehyde} \xrightarrow{\text{[O]}} \text{carboxylic acid}$$

$$2° \text{ alcohol} \xrightarrow{\text{[O]}} \text{ketone}$$

$$3° \text{ alcohol} \xrightarrow{\text{[O]}} \text{no reaction}$$

Ester Formation   Alcohols react with carboxylic acids to form **esters**. The reactions are catalyzed by strong acids. Though esters will be discussed in greater detail in Chapter 18, their preparation will be introduced now because alcohols are involved. The general equation for ester formation is

$$\underset{\text{Carboxylic acid}}{R-C\overset{\displaystyle O}{\underset{\boxed{OH}}{}}} + \underset{\text{Alcohol (1°, 2°, or 3°)}}{\boxed{H}\,O-R'} \xrightarrow{H^+} \underset{\text{Ester}}{R-C\overset{\displaystyle O}{\underset{O-R'}{}}} + H_2O$$

Acetic acid and ethyl alcohol react to form the ester ethyl acetate:

$$\underset{\text{Acetic acid}}{H_3C-C\overset{\displaystyle O}{\underset{\boxed{OH}}{}}} + \underset{\text{Ethyl alcohol}}{\boxed{H}\,O-CH_2CH_3} \xrightarrow{H^+} \underset{\text{Ethyl acetate (ester)}}{H_3C-C\overset{\displaystyle O}{\underset{O-CH_2CH_3}{}}} + H_2O$$

Esters may also be prepared from alcohols and inorganic acids, such as nitric acid (HNO_3), sulfuric acid (H_2SO_4), and phosphoric acid (H_3PO_4). Esters formed using an inorganic acid are usually called **inorganic esters** to distinguish them from organic esters made from carboxylic acids. An important inorganic ester is glyceryl trinitrate, commonly called nitroglycerine. Nitroglycerine is prepared from glycerol and nitric acid in the presence of sulfuric acid:

$$\begin{array}{ccc}
H_2C\!-\!O\!-\!\boxed{H \quad\quad HO}\!-\!NO_2 & & H_2C\!-\!ONO_2 \\
H\!-\!C\!-\!O\!-\!\boxed{H} + \boxed{HO}\!-\!NO_2 \xrightarrow{\;H_2SO_4\;} & H\!-\!C\!-\!ONO_2 + 3H_2O \\
H_2C\!-\!O\!-\!\boxed{H \quad\quad HO}\!-\!NO_2 & & H_2C\!-\!ONO_2
\end{array}$$

Glycerol	3 Nitric acid molecules	Glyceryl trinitrate (nitroglycerine)

Nitroglycerine is a pale yellow, oily liquid. It is a very powerful explosive, and will detonate with a mild shock. During the mid-1800's its use was accompanied by numerous fatal accidents because of its extreme sensitivity to shock. In 1866 the Swedish chemist Alfred Nobel discovered that nitroglycerine, when adsorbed on finely divided clay, formed an explosive that was very stable and would not easily detonate. The product was called dynamite, and Nobel's discovery made him a wealthy man. After his death, the bulk of his fortune was set aside, according to the instructions in his will, to recognize major contributions in peace, chemistry, medicine, literature, and other areas. The Nobel prizes are awarded annually from the proceeds of the Nobel Trust.

Nitroglycerine is also used as a drug. A 1.0% solution of nitroglycerine in ethyl alcohol is used as a heart stimulant. It dilates the smaller blood vessels and relaxes smooth muscles, which then causes a reduction in blood pressure and the associated pain of angina pectoris. Nitroglycerine can also be administered orally as tablets.

Phosphate esters are very important inorganic esters in biological systems. They are involved in the transfer and storage of energy in cells, and as necessary parts of many biological compounds, such as the nucleic acids DNA and RNA. When glucose is used by the cell for energy, it is first converted into a phosphate ester, glucose-6-phosphate:

Glucose-6-phosphate

Check Test Number 5

1. Write the equations for the oxidation of ethyl alcohol (both oxidation steps) and 2-butanol.
2. Write the structural formula of the ester formed in the reaction of acetic acid and methyl alcohol.

Answers:

1. $$H_3CCH_2OH \xrightarrow{[O]} H_3C{-}\underset{\underset{H}{|}}{C}{=}O \xrightarrow{[O]} H_3C{-}\underset{\underset{OH}{|}}{C}{=}O$$

$$H_3C{-}\underset{\underset{H}{|}}{\overset{\overset{OH}{|}}{C}}{-}CH_2CH_3 \xrightarrow{[O]} H_3C{-}\overset{\overset{O}{\|}}{C}{-}CH_2CH_3$$

2. $$H_3C{-}C{\Large\langle}{\overset{\displaystyle O}{\underset{\displaystyle O{-}CH_3}{}}}$$

Some Important Alcohols

Alcohols are some of the most common organic compounds encountered in daily living, both in the health care area and out. For that reason, five of the more important alcohols are described here in greater detail. Since each is more widely known by a common name, they will be used to identify each alcohol as it is discussed.

Methyl Alcohol: CH₃OH
IUPAC Name: Methanol
Other Common Name: Wood Alcohol

Methyl alcohol is used as a solvent in paints, shellacs, and in many industrial processes. About half of the methyl alcohol produced in the U.S. is oxidized to formaldehyde, which in turn is used in the manufature of plastics and other chemicals. Solutions of methyl alcohol in water are used as an antifreeze in automobiles. Methyl alcohol is toxic if taken internally. Once in the bloodstream it is carried to the liver and oxidized to formaldehyde and formic acid. As these highly toxic compounds accumulate, they are carried by the blood to all parts of the body. An early sign of methyl alcohol poisoning is paralysis of the optic nerve, resulting in blindness. Ingestion of as little as 10 ml of methyl alcohol can cause permanent blindness, and doses in excess of 30 ml are fatal. If detected early enough, methyl alcohol poisoning can be treated by administration of ethyl alcohol. Ethyl alcohol is preferentially oxidized in the liver, thus allowing methyl alcohol to be slowly eliminated through the kidneys.

Ethyl Alcohol: CH_3CH_2OH
IUPAC Name: Ethanol
Other Common Names: Grain Alcohol, Alcohol

Perhaps the most familiar member of the alcohol family is ethyl alcohol. In fact, the term ''alcohol'' as used in procedures and solutions usually means ethyl alcohol. Ethanol is widely used in industry as a solvent and as a starting material in the manufacture of many other chemicals. The excellent solvent properties of ethanol allow it to be used for solutions of drugs, certain pesticides, perfumes, and many polar and nonpolar organic compounds. Solutions using ethyl alcohol as the solvent are often called ''tinctures,'' such as ''tincture of iodine'' (I_2 in ethyl alcohol). Increasing amounts of ethyl alcohol are being used in gasohol, a 5–10% mixture of alcohol in gasoline for use in automobiles. A 70% (v/v) ethanol in water solution is used as a topical antiseptic to destroy microorganisms on the skin. Solutions containing higher concentrations of ethanol are not able to destroy microorganisms as well as the 70% solution.

Ingestion of moderate amounts of ethyl alcohol over a period of time can cause alcohol intoxication. Ethanol depresses the central nervous system, which leads to mental confusion, reduced muscle coordination, and eventual unconsciousness if taken in excess. Once ingested, ethanol quickly passes into the bloodstream and is carried to the liver. There it is rapidly oxidized to acetaldehyde, and then to acetic acid. The acetic acid is quickly converted to a form that is rapidly used by cells to produce energy. Since ethanol can be quickly metabolized to produce energy, it is sometimes administered to individuals in shock due to exposure to cold temperatures. Each gram of alcohol metabolized produces about 7 kcal of energy.

Ethyl alcohol destined for industrial use is altered in some way to prevent its use as a beverage. Once altered, it is referred to as **denatured alcohol.** Usually a small amount of methyl alcohol, benzene, or other toxic material is added to ''denature'' the alcohol, thus making it unfit for human consumption but without altering its excellent solvent property.

$$\text{OH}$$
$$|$$
Isopropyl Alcohol: $H_3C-CH-CH_3$
IUPAC Name: 2-propanol
Other Common Name: Rubbing Alcohol

Isopropyl alcohol is widely used as an industrial solvent. It is also used to prepare acetone and other organic compounds. Its principal medicinal use is as a rubbing alcohol for sponge baths and as an astringent for contracting blood vessels. Isopropyl alcohol is not absorbed through the skin, but if ingested, it is toxic.

Ethylene Glycol: $HOCH_2CH_2OH$
IUPAC Name: 1,2-Ethanediol
Other Common Name: Glycol

Ethylene glycol is a colorless, sweet-tasting liquid that is soluble in all proportions in water. Because of its high solubility in water and high boiling point, ethylene glycol is primarily used in antifreeze solutions in automobiles. Solutions of ethylene glycol in water can be prepared that remain liquid as low as -50°F. This diol is very toxic; in fact, it is as toxic as methyl alcohol if taken internally. It is oxidized in the liver to oxalic acid, a substance that depresses the central nervous system and destroys kidney cells:

$$HO{-}CH_2CH_2{-}OH \xrightarrow{enzyme} \underset{\text{Oxalic acid}}{HO{-}\overset{\overset{O}{\|}}{C}{-}\overset{\overset{O}{\|}}{C}{-}OH}$$

It is interesting to note that oxalic acid poisoning may also occur from eating very large quantities of rhubarb. The tart taste of rhubarb is due to the unusually high levels of oxalic acid that it contains.

$$\textit{Glycerol:} \quad H_2\overset{\overset{OH}{|}}{C}{-}\overset{\overset{OH}{|}}{C}H{-}\overset{\overset{OH}{|}}{C}H_2$$
IUPAC Name: 1,2,3-Propanetriol
Other Common Name: Glycerin, Glycerine

Glycerol is a nontoxic, viscous, sweet-tasting liquid that is soluble in all proportions in water. It is widely used as a moisturizer in hand creams, lotions, cosmetics, pharmaceuticals, and even in tobacco. When applied to skin or other surfaces, it forms a thin, flexible coat that retards the evaporation of moisture. As a result, skin retains moisture and remains supple. Glycerol is a by-product of soap manufacturing, which serves as a major source of this triol.

Thiols—Sulfur Analogs of Alcohols

Replacing the oxygen atom of an alcohol with a sulfur atom produces a class of compounds known as **thiols** or **mercaptans.** The –SH functional group is called the **sulfhydryl group.** The first three mercaptans are listed below along with their common names:

$$CH_3SH \quad \text{methyl mercaptan}$$
$$CH_3CH_2SH \quad \text{ethyl mercaptan}$$
$$CH_3CH_2CH_2SH \quad n\text{-propyl mercaptan}$$

Unlike alcohols which have pleasant odors, thiols have strong, disagreeable odors. In fact, one of the compounds responsible for the odor of skunks is n-butyl mercaptan, $CH_3CH_2CH_2CH_2SH$.

Many proteins contain the sulfhydryl group. When you study proteins you will see the important role played by these groups in determining the three-dimensional structures adopted by proteins.

Phenols

The **phenols** are aromatic alcohols which have the –OH group joined to a benzene ring. Though "phenol" is used to describe the family of aromatic alcohols, it is also the common name for the simplest aromatic alcohol:

OH

Phenol

The phenols are either solids or liquids at room temperature. Phenol itself is a colorless, crystalline solid with a characteristic odor reminiscent of hospital antiseptics. Since phenols had been known and studied for many years prior to the development of the IUPAC system of nomenclature, they are still often identified by their common names. For this reason, and because the IUPAC names can be quite complicated, common names will be used here.

About half of the phenol produced in the U.S. is used in the synthesis of phenolic resins, which are used in plastics, coatings, and waterproof adhesives. Phenol is also used to prepare dyes, pesticides, explosives, hormones, aspirin, and a host of other products.

Unlike the aliphatic alcohols discussed in the first part of this chapter, phenols are mildly acidic compounds. A dilute solution of phenol in water is known as **carbolic acid.** The acidic property of phenols is partially responsible for their use in antiseptics.

Phenols as Germicides

Some of the most widely used germicides are members of the phenol family. Over 100 years ago, Joseph Lister, an English surgeon, used a solution of phenol in water (carbolic acid) to prevent infection in wounds following surgery. Phenol was the first compound to be used extensively as an antiseptic, but because it causes burns when in contact with skin and because of its toxicity, other substances have been developed to take its place. Many of these are derivatives of phenol.

Hexyl resorcinol is a better antiseptic than phenol, and with fewer adverse side effects. At one time it was used as a urinary antiseptic, but in recent years antibiotics have become the drugs of choice. It is used to destroy parasitic worms. It is the active ingredient in several mouthwashes and throat lozenges.

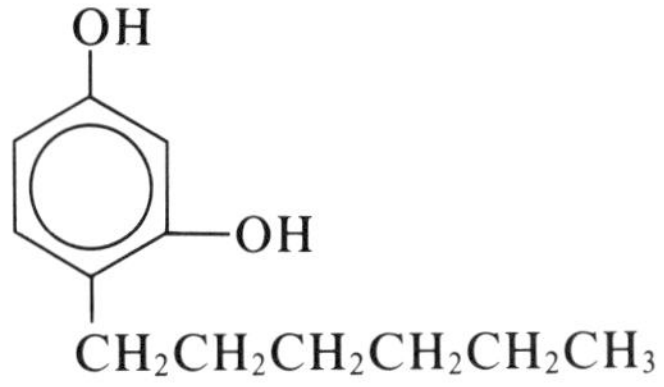

Hexyl resorcinol

Thymol has good antiseptic properties as well as a pleasant taste, so it is used in many mouthwashes and toothpastes. **Hexachlorophene** is another derivative of phenol that used to be in deodorants, toothpastes, germicidal soaps, and several cosmetic products. In 1972 hexachlorophene was linked to the apppearance of brain damage in infants bathed and powdered with products containing this antiseptic, and its use since that time has been greatly curtailed.

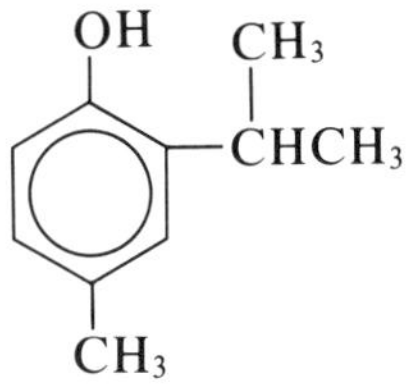

Thymol

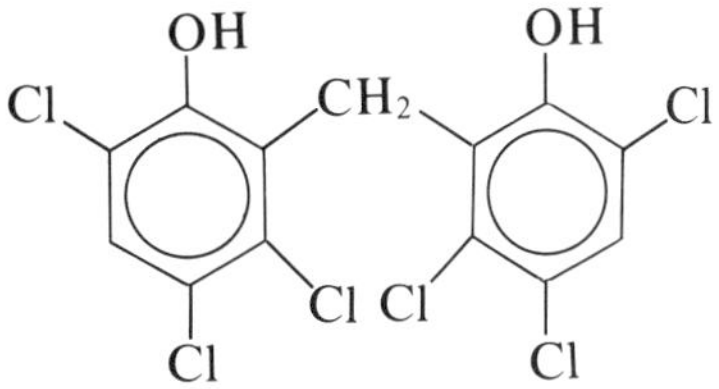

Hexachlorophene

Picric acid is used as an antiseptic in many burn ointments. It destroys microorganisms on contact and causes the coagulation of proteins in and around the burned area. The coagulated protein seals the wound and minimizes the loss of fluids and the risk of infection. Pure, crystalline picric acid is also used as an explosive.

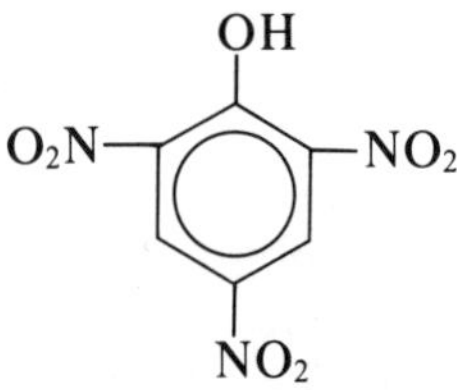

OH

O_2N— —NO_2

NO_2

Picric acid
(2,4,6-trinitrophenol)

The monomethyl derivatives of phenol are called **cresols,** and they are used extensively as disinfectants and wood preservatives. Dilute solutions of cresols are used in hospitals to destroy germs on floors, walls, and furniture. The antiseptic odor associated with hospitals is, in part, caused by the use of phenol and cresols as disinfectants.

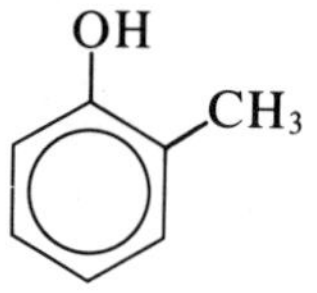

OH
CH_3
ortho-Cresol

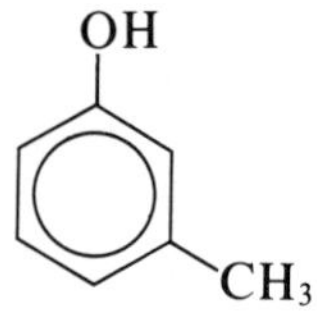

OH
CH_3
meta-Cresol

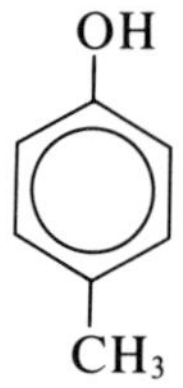

OH
CH_3
para-Cresol

Some phenols have properties that are less desirable than those of the previous examples. The rash caused by poison ivy and poison oak is due to a mixture of phenols known as **urushiol,** which is found in the sap of these plants. The phenols that compose urushiol differ in the number of double bonds in the alkyl group.

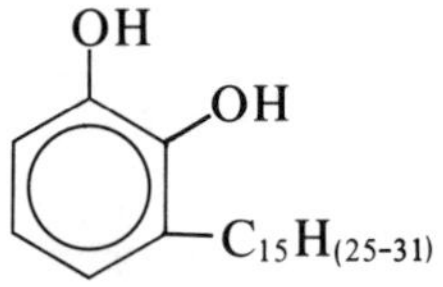

OH
OH
—$C_{15}H_{(25-31)}$

Urushiol

Other Important Phenols

There are a number of compounds that can be viewed as derivatives of phenol, though they contain other functional groups as well. One of them is **vanillin,** a substance that provides the pleasant vanilla flavor in foods. The principal mind-altering component in marijuana smoke is **tetrahydrocannabinol** (THC).

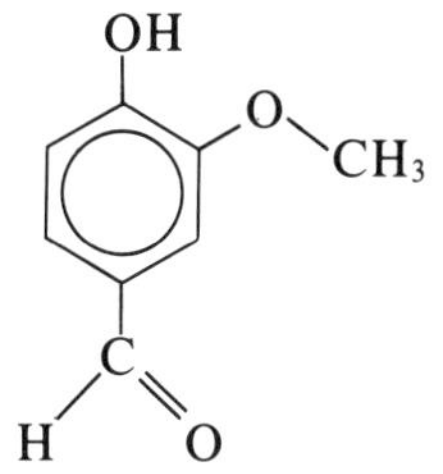

Vanillin

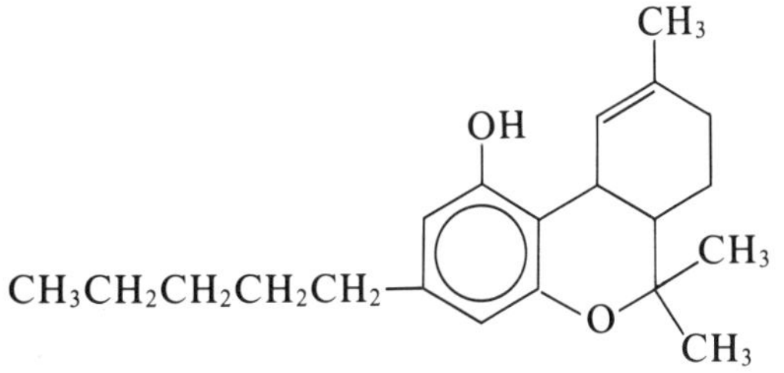

Tetrahydrocannabinol (THC)

Eugenol is the major component of clove oil, and it has both antiseptic and anesthetic properties. It is used in over-the-counter medicines to control the pain of toothaches. A mixture of eugenol and zinc oxide (ZnO) forms a plastic material known as ZOE, which is used to make dental impressions and temporary fillings.

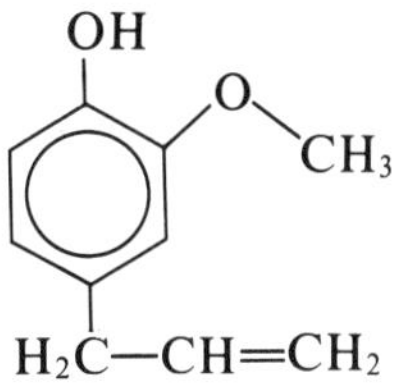

Eugenol

Diethylstilbestrol (DES) is a phenol derivative that has female sex hormone characteristics. At one time it was given to women during pregnancy to prevent miscarriages. Unfortunately, it was not until several years later that the full effect of the drug was known. An unusually large percentage of the daughters of these women developed cancer of the uterus, and as a result, the drug has been banned for use in humans. The drug is still allowed to be fed to cattle to promote growth and accelerate weight gain.

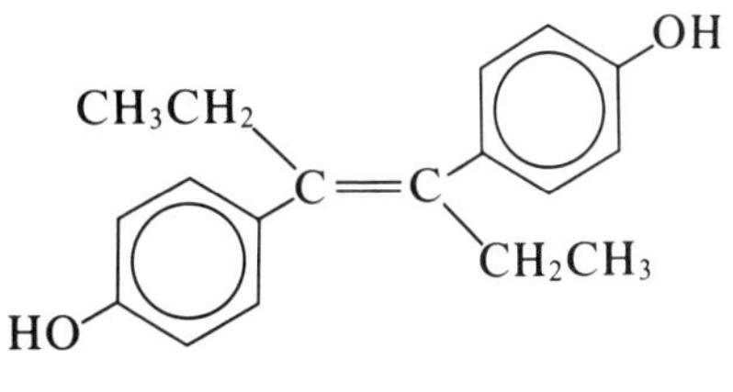

Diethylstilbestrol (DES)

BHA and **BHT** are two abbreviations that appear in small print on the labels of many processed foods. They are antioxidants, and prevent the reactions of oxygen with fats and oils that may produce materials with unpleasant tastes or smells. BHT stands for *b*utylated *h*ydroxy *t*oluene, and BHA represents the two isomers of *b*utylated *h*ydroxy *a*nisole.

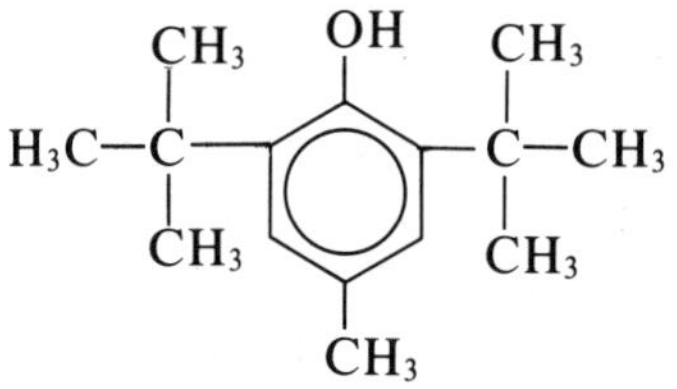

Butylated hydroxytoluene
(BHT)

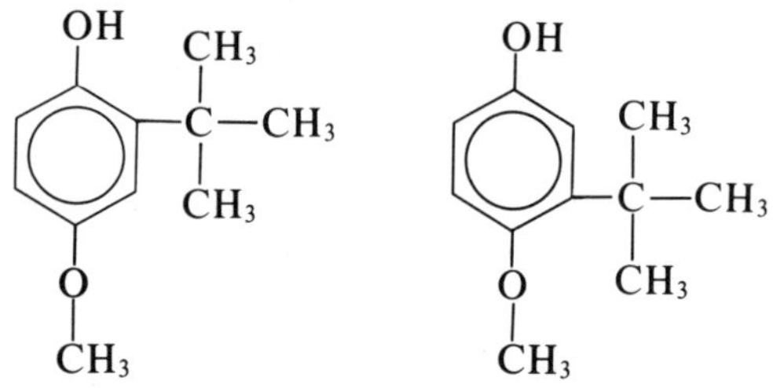

The butylated hydroxyanisoles
(BHA)

Ethers

The general formula for an ether is R−O−R, and the ether functional group is $-\overset{|}{\underset{|}{C}}-O-\overset{|}{\underset{|}{C}}-$. Like the alcohols, ethers can also be viewed as derivatives of water. In ethers, both hydrogen atoms are replaced by organic groups. Ether molecules have a bent shape, just as does the water molecule. Even though formulas for ethers are often written in a straight line, keep in mind that they are actually bent molecules.

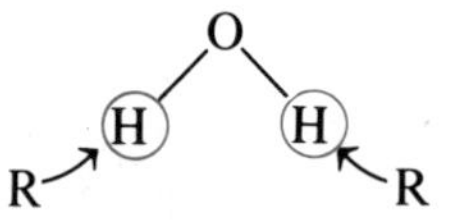

Water

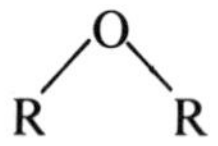

An ether

The two organic groups joined to the oxygen atoms in an ether may be identical, or they may be two different groups. Each of the following compounds is an ether. The oxygen atom in the ether functional group is indicated with an arrow.

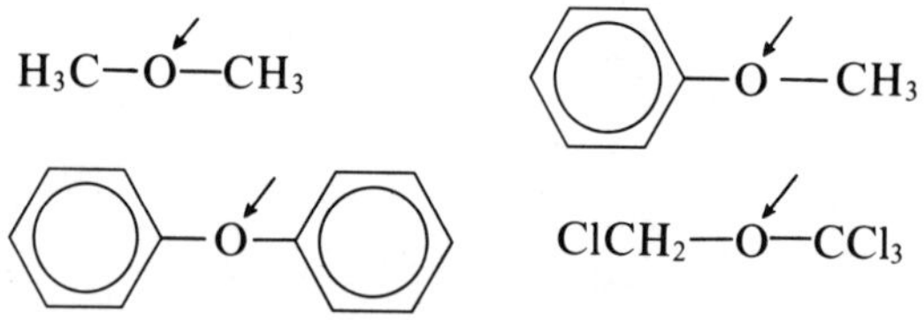

The synthesis of ethers by the controlled dehydration of alcohols was described in a previous section. To refresh your memory, the synthesis of diethyl ether is accomplished by the dehydration of ethyl alcohol using sulfuric acid as a catalyst:

$$\text{CH}_3\text{CH}_2\text{O}\boxed{\text{H} + \text{HO}}\text{CH}_2\text{CH}_3 \xrightarrow[140°]{\text{H}_2\text{SO}_4} \text{CH}_3\text{CH}_2-\text{O}-\text{CH}_2\text{CH}_3 + \text{H}_2\text{O}$$

Ethyl alcohol   Diethyl ether

Naming Ethers

An IUPAC system for naming ethers does exist, but since the more familiar members of the ether family are usually identified by their common names, it is reasonable to limit the discussion to that which is commonly done. The common names for ethers are formed by first naming each organic group, usually in alphabetical order, followed by the word "ether." If the two organic groups are identical, the prefix di- may be used as in dimethyl ether. Sometimes the prefix is omitted, and it is to be understood that both organic groups are then identical. The common names for several ethers are given below:

$\text{CH}_3-\text{O}-\text{CH}_3$	dimethyl ether or methyl ether
$\text{CH}_3-\text{O}-\text{CH}_2\text{CH}_3$	ethyl methyl ether
$\text{CH}_3-\text{O}-\text{CH}_2\text{CH}_2\text{CH}_3$	methyl *n*-propyl ether
$\text{CH}_3\text{CH}_2-\text{O}-\text{CH}_2\text{CH}_3$	diethyl ether or ethyl ether

$-\text{O}-\text{CH}_3$ methyl phenyl ether (anisole)

$-\text{O}-$ diphenyl ether or phenyl ether

Check Test Number 6

Give the common name for each of the ethers given below:

a. $\text{CH}_3\text{CH}_2\text{CH}_2-\text{O}-\text{CH}_2\text{CH}_3$

b. $\text{CH}_3\text{CH}_2-\text{O}-$

c. $\text{CH}_3\text{CH}_2-\text{O}-\text{CH}_2\text{CH}_3$

Answers: (a) ethyl *n*-propyl ether, (b) ethyl phenyl ether, (c) diethyl ether or ethyl ether.

The Cyclic Ethers

Cyclic ethers contain an ether functional group as part of a ring system. They are heterocyclic compounds. The simplest cyclic ether is ethylene oxide, a colorless gas. Ethylene oxide has excellent germicidal properties, and it is used in special autoclaves to sterilize surgical instruments. It is also used as an agricultural fungicide to destroy fungus on seeds and grains. Much of the ethylene glycol produced in the U.S. is made from ethylene oxide.

Ethylene oxide

Three other common cyclic ethers are tetrahydrofuran, tetrahydropyran, and dioxane. They are all colorless liquids that are widely used as solvents.

Tetrahydrofuran   Tetrahydropyran   Dioxane

Many biological compounds contain cyclic ether structures. You will find one in the structure of tetrahydrocannabinol (THC), which was given earlier in this chapter. Also, when you study carbohydrates, you will find that they too contain cyclic ether structures.

Properties of Ethers

Most of the common ethers are colorless, highly volatile liquids. Dimethyl ether and ethyl methyl ether are gases at room temperature. Ethers have boiling points that are considerably lower than those of alcohols of similar molecular weight, as shown by the comparisons in Table 17.2. Because ethers cannot hydrogen bond to each other, as alcohols can, the weaker forces of attraction result in lower boiling points than alcohols of similar molecular weight. Ethers have boiling points more like those of the comparable alkanes, which cannot participate in hydrogen bonding either.

Because of the high volatility of many ethers, they should only be used in an area with very good ventilation. Ether vapors are more dense than air, and they can gather in layers on floors and table tops instead of diffusing away. These vapors can form explosive mixtures with air that can be ignited by static electricity. Ethers should never be used around an open flame or sparking electrical device. There have been occasions in the past when ether vapors exploded in an operating room causing severe injuries and even death. Whenever flammable anesthetics are used, precautions must be taken to avoid such accidents.

There is an additional hazard associated with the aliphatic ethers. When exposed to air for prolonged periods, aliphatic ethers slowly form peroxides. These peroxides are unstable, and even in small quantities can detonate, causing a violent explosion. The formation of peroxides can be prevented by the addition of certain antioxidants.

Ethers are not very reactive chemically. For this reason they are used as solvents for many organic compounds in reactions. Ethers are excellent "fat solvents." If a liquid ether is brought in contact with skin, it not only dissolves and removes oils from the surface, but fatty substances in cells can also be removed. Prolonged contact with ethers can result in a "chemical burn." Ethers are not very soluble in water, but they are somewhat more soluble than alkanes. Ether molecules are somewhat polar by virtue of their bent shape. Also, water can form a hydrogen bond with the oxygen atom of an ether.

$$
\begin{array}{c}
R \\
\diagdown \\
O\text{---}H \diagup \overset{O}{} \diagdown H \\
\diagup \\
R \\
\uparrow \\
|
\end{array}
$$

Hydrogen bond

The lower-molecular-weight ethers, methyl ether and ethyl methyl ether, are moderately soluble in water, but as the organic groups get larger the solubility becomes much less. The higher-molecular-weight ethers are more alkanelike in their solubility behavior. This is seen in Table 17.2.

Ethers as Anesthetics

In 1846 William Morton, a Boston dentist, was the first to publicly use diethyl ether as an anesthetic during the extraction of a tooth. Before that time, a stiff slug of whisky or brandy provided only a minimal relief from the pain of surgery, and patients often suffered from shock brought on by the pain. The discovery of the anesthetic property of ether was one of the greatest breakthroughs in medicine. Even today when most people think of ether, they associate it with anesthesia.

Diethyl ether is administered by inhaling the vapors of the volatile liquid. It quickly relaxes muscles and renders the patient unconscious and insensitive to pain. It achieves these results by depressing the central nervous system. Its effect on blood pressure, pulse rate, and respiration is slight. Unfortunately, diethyl ether is very flammable, and you have already been made aware of the consequences of this property. Also, the vapors are irritating to the respiratory tract, and postoperative nausea frequently follows its use. Because diethyl ether has these disadvantages, other ethers have been sought out and employed as anesthetics. Methyl *n*-propyl ether, also known as neothyl, and divinyl ether (vinethene) have fewer adverse side effects, though they both are flammable.

$$CH_3—O—CH_2CH_2CH_3 \qquad H_2C=CH—O—CH=CH_2$$

<table>
<tr><td>Methyl *n*-propyl ether</td><td>Divinyl ether</td></tr>
<tr><td>(neothyl)</td><td>(vinethene)</td></tr>
</table>

Recent years have seen the development of several new inhalation anesthetics that avoid the problems associated with ethers, yet render the patient unconscious quickly. One of these is halothane. Though not an ether, halothane is an extremely volatile liquid that is not flammable, and its use is not accompanied by serious side effects.

$$F—\overset{\displaystyle F}{\underset{\displaystyle F}{C}}—\overset{\displaystyle H}{\underset{\displaystyle Br}{C}}—Cl$$

Halothane

Terms

Several important terms appeared in Chapter 17. You should know the meaning of each of the following:

alcohol	diol
hydroxyl group	triol
primary alcohol	denatured alcohol
secondary alcohol	thiol
tertiary alcohol	mercaptan
1°, 2°, 3°	phenols

hydration

dehydration

inorganic ester

ether

ether functional group

cyclic ether

Questions

Answers to starred questions appear at the end of the text.

1. Classify each of the following compounds as either an alcohol, phenol, or ether:

    *a. $CH_3CH_2CH_2OH$

    *b. $CH_3-O-CH_2CH_3$

    c. $HOCH_2CH_2OH$

    * d. 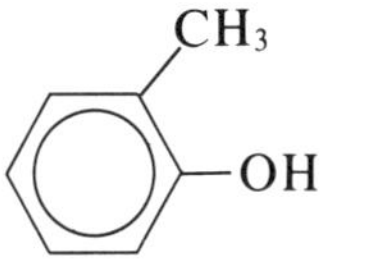 (benzene ring with CH_3) $-OH$

    e. 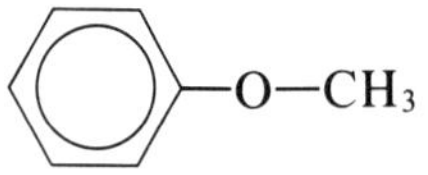 (benzene ring) $-O-CH_3$

    * f. 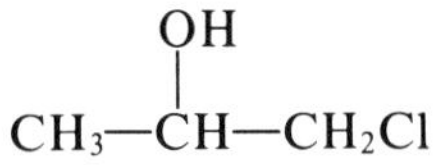 $CH_3-\overset{\overset{\displaystyle OH}{|}}{CH}-CH_2Cl$

    g. 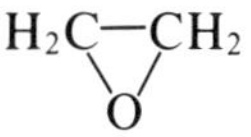 H_2C-CH_2 with O bridging (cyclic ether)

2. Why is ethyl alcohol, CH_3CH_2OH, a liquid at room temperature, while dimethyl ether, CH_3-O-CH_3, a compound with the same molecular formula and molecular weight, is a gas?

3. Give the correct IUPAC name for each of the following alcohols:

    * a. $CH_3CH_2CH_2CH_2CH_2OH$

    b. 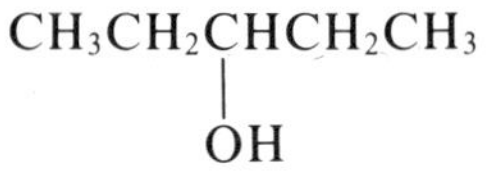 $CH_3CH_2\overset{\overset{\displaystyle }{}}{C}HCH_2CH_3$ with OH below

    * c. $CH_3CH_2CH_2-\overset{\overset{\displaystyle CH_3}{|}}{\underset{\underset{\displaystyle OH}{|}}{C}}-CH_3$

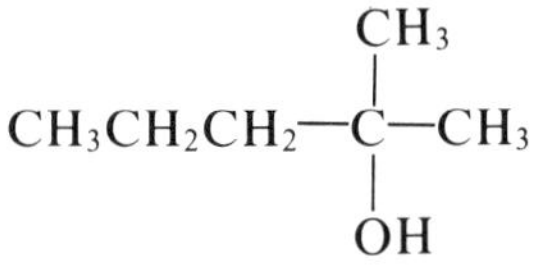

    d. 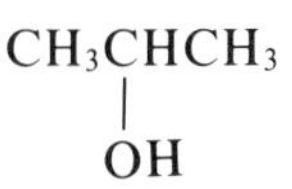 $CH_3\overset{}{C}HCH_3$ with OH below

    *e. $ClCH_2CH_2CH_2OH$

    f. $HOCH_2CH_2CH_2OH$

    * g. 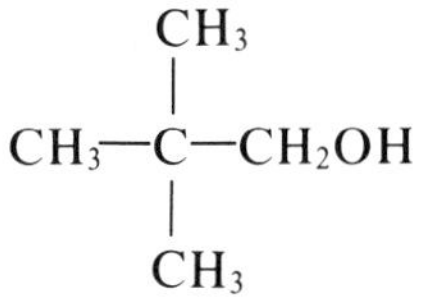 $CH_3-\overset{\overset{\displaystyle CH_3}{|}}{\underset{\underset{\displaystyle CH_3}{|}}{C}}-CH_2OH$

    h. 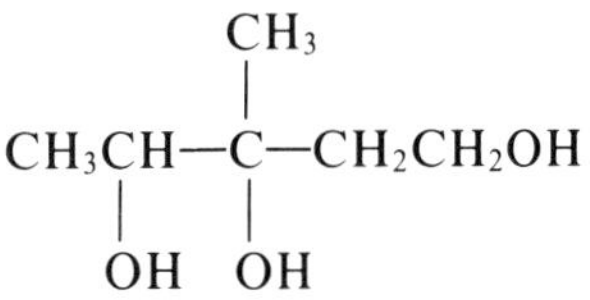 $CH_3CH-\overset{\overset{\displaystyle CH_3}{|}}{\underset{\underset{\displaystyle OH}{|}}{C}}-CH_2CH_2OH$ with OH under CH

4. Classify each of the following alcohols as either primary, secondary, or tertiary:

*a.
$$CH_3-\underset{\underset{CH_3}{|}}{\overset{\overset{H}{|}}{C}}-OH$$

d. $CH_3CH_2CHCH_3$
$\quad\ \ \overset{|}{CH_2OH}$

b. CH_3OH

e. ⟨benzene ring⟩$-CH_2CH_2OH$

*c.
$$CH_3-\underset{\underset{CH_3}{|}}{\overset{\overset{OH}{|}}{C}}-CH_2CH_3$$

*f. ⟨cyclopentane ring with H and OH⟩

5. Write the formula for the alcohol produced in each of the following hydration reactions:

a. $CH_3-\overset{\overset{H}{|}}{C}=CH_2 + HOH \xrightarrow{H_2SO_4}$

c. $H_2C=CH_2 + HOH \xrightarrow{H_2SO_4}$

*b. $CH_3-\underset{\underset{CH_3}{|}}{\overset{\overset{H}{|}}{C}}=C-CH_3 + HOH \xrightarrow{H_2SO_4}$

d. $CH_3-\overset{\overset{H}{|}}{C}=\overset{\overset{H}{|}}{C}-CH_3 + HOH \xrightarrow{H_2SO_4}$

6. Give the formula of the alcohol formed in each of the following reduction reactions:

*a. $CH_3-\overset{\overset{O}{||}}{C}-CH_3 + H_2 \xrightarrow{catalyst}$

c. $CH_3CH_2-\overset{\overset{O}{||}}{C}-CH_3 + H_2 \xrightarrow{catalyst}$

b. $CH_3CH_2\underset{\underset{H}{|}}{C}=O + H_2 \xrightarrow{catalyst}$

7. Write the formula of the product formed in the controlled oxidation of each of the following alcohols. If there is no reaction, write N.R. as the answer.

*a. $CH_3CH_2CH_2OH \xrightarrow{[O]}$

*c. $CH_3\underset{\underset{OH}{|}}{CH}CH_2CH_3 \xrightarrow{[O]}$

b. $CH_3-\underset{\underset{CH_3}{|}}{\overset{\overset{OH}{|}}{C}}-CH_3 \xrightarrow{[O]}$

d. $CH_3\underset{\underset{H_2COH}{|}}{CH}CH_3 \xrightarrow{[O]}$

8. Give the formula for the *major* alkene product formed when each of the following alcohols are dehydrated at high temperatures:

*a. $CH_3\underset{\underset{OH}{|}}{CH}CH_3$

b. $CH_3CH_2CH_2CH_2OH$

*c. $CH_3\underset{\underset{OH}{|}}{\overset{\overset{CH_3}{|}}{C}}-CH_2CH_3$

d.
$$\underset{\underset{CH_2CH_3}{|}}{\overset{\overset{H\quad CH_3}{|\quad\,|}}{HOC-CH-CH_2CH_3}}$$

*e.
$$CH_3\underset{\underset{OH}{|}}{CHCH_2}-\underset{\underset{CH_3}{|}}{\overset{\overset{CH_3}{|}}{C}}-CH_3$$

9. Write the formula for the ether that would form as the following pairs of alcohols are dehydrated:

*a. $CH_3CH_2OH \;+\; HOCH_2CH_3 \xrightarrow{H_2SO_4}$

b. $CH_3\underset{\underset{CH_3}{|}}{\overset{\overset{H}{|}}{C}}-OH \;+\; HOCH_3 \xrightarrow{H_2SO_4}$

c. $CH_3CH_2\underset{\underset{CH_3}{|}}{CHOH} \;+\; HOCH_2CH_3 \xrightarrow{H_2SO_4}$

10. Give the correct IUPAC names for the following compounds which are identified here by their common names:(a) wood alcohol, (b) glycol, (c) grain alcohol, (d) glycerine, (e) rubbing alcohol.

11. What is the %(v/v) of ethyl alcohol in a 100 proof vodka?

12. Why is ethyl alcohol denatured?

*13. Why is methanol so very toxic to humans?

*14. Which compound in each group would you expect to be the most soluble in water? The least soluble?
   a. CH_3CH_2OH, CH_3-O-CH_3, $CH_3CH_2CH_3$
   b. $HOCH_2CH_2OH$, $CH_3CH_2CH_2OH$, $CH_3CH_2CH_2CH_2CH_2CH_2OH$
   c. CH_3CH_2OH, $CH_3CH_2CH_2CH_2OH$, $CH_3CH_2CH_2CH_2CH_2CH_2CH_2OH$

15. Draw the condensed formula for the following compounds from their common or IUPAC names:
   *a. 3-pentanol
   b. 1,4-hexanediol
   *c. 3-chloro-1-butanol
   d. phenol
   *e. ethyl methyl ether
   f. isopropyl alcohol
   *g. 3,4-dimethyl-3-hexanol
   h. ethylene glycol
   *i. 1,2,3-pentanetriol
   j. 3-methyl-1-butanol
   *k. methyl phenyl ether

16. What is a thiol?

17. What class of compound is nitroglycerine? What are two uses for nitroglycerine?

18. What precautions should be taken when working with a volatile organic liquid such as diethyl ether?

Aldehydes, Ketones, Carboxylic Acids, and Esters

In the previous chapter, you were told about organic compounds that contain oxygen bonded to two other atoms by single bonds, the alcohols, phenols, and ethers. As we continue the discussion of organic chemistry, we will now turn our attention to four classes of compounds that contain oxygen joined to carbon by a double bond. The $-\overset{\overset{\displaystyle O}{\|}}{C}-$ group is called the **carbonyl group,** and you will recognize its presence in the general formulas of aldehydes, ketones, carboxylic acids, and esters:

aldehyde   ketone   carboxylic acid   ester

In this chapter, each of these classes of organic compounds will be described in terms of their preparation, physical and chemical properties, and the ways they are named. Several important examples of each class will also be discussed.

Objectives

By the time you finish this chapter you should be able to do the following.

1. Write the general formula for an aldehyde, ketone, carboxylic acid, and an ester.
2. Give the common and IUPAC names for the first few members of the aldehyde and ketone families.
3. Write the structural formula for an aldehyde, ketone, carboxylic acid, or ester given its IUPAC name.
4. Describe one method that can be used to prepare an aldehyde, a ketone, a carboxylic acid, or an ester.
5. Compare the boiling points and solubilities in water of aldehydes, ketones, carboxylic acids, and esters of similar molecular weight.
6. Describe the addition of reagents to the carbonyl group in aldehydes and ketones.
7. Give the common and IUPAC names for the first few members of the carboxylic acid family.
8. Describe three ways to prepare salts of carboxylic acids, and how the salts are named.
9. Give the common and IUPAC names of an ester given its structural formula.
10. Describe three esters of salicylic acid that have medicinal value.

Aldehydes and Ketones

Aldehydes and ketones both contain the carbonyl group, $>C=O$. **Ketones** have two organic groups $(R-)$ bonded to the carbon atom of the carbonyl group. **Aldehydes** have both a hydrogen atom $(H-)$ and an organic group $(R-)$ bonded to the carbonyl carbon. The only exception to this is the simplest aldehyde, formaldehyde, that has both bonds to the carbonyl occupied by hydrogen.

$$\underset{\text{aldehyde}}{\overset{\displaystyle O}{\underset{R\diagup\,\diagdown H}{\parallel}{C}}} \qquad \underset{\text{ketone}}{\overset{\displaystyle O}{\underset{R\diagup\,\diagdown R}{\parallel}{C}}}$$

Just as we saw with the alkenes, which also have the two-single, one-double bond pattern about carbon, $\diagup$C=, all bond angles about the carbonyl group in aldehydes and ketones are 120°. The atoms joined to the carbonyl carbon all lie in the same plane. Formaldehyde is a planar molecule with 120° bond angles:

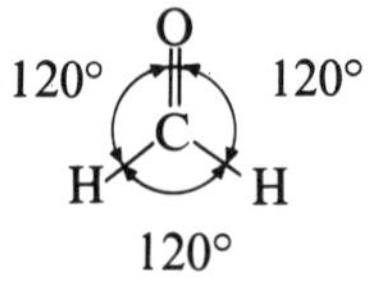

formaldehyde

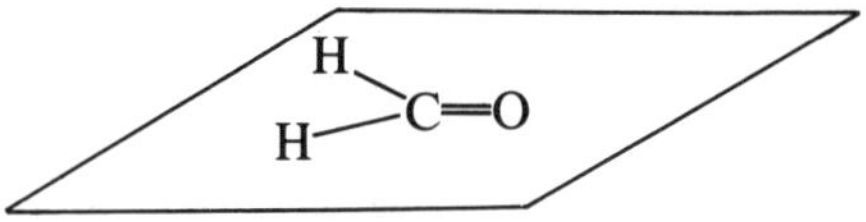

The bond angles in acetone, the simplest ketone, are also 120°. The carbon atoms of the two methyl groups lie in a plane with the carbonyl group:

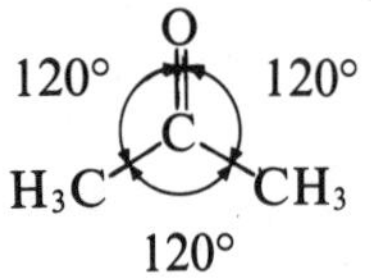

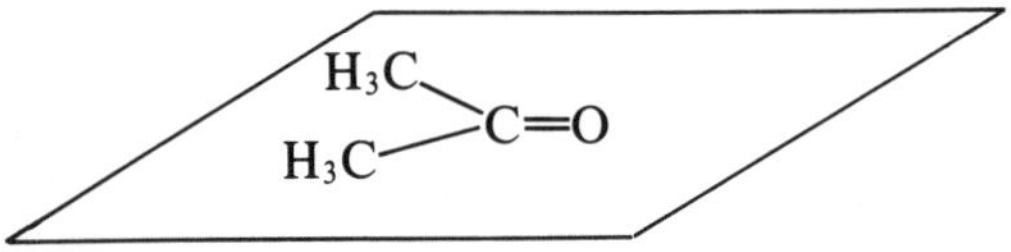

The aldehyde functional group is written in condensed form as –CHO. Notice the sequence of the symbols, with H placed between the C and the O. Be certain you do not confuse the aldehyde (–CHO) with an alcohol functional group ($-\overset{|}{\underset{|}{C}}OH$). The two compounds are quite different. The condensed formulas for ketones are written RCOR.

Naming the Aldehydes

Since most of the common aldehydes were known long before the development of the IUPAC system, their common names are still used quite extensively. The common name of an aldehyde is derived from the common name of the carboxylic acid that contains the same number and arrangement of carbon atoms. The -*ic* or -*oic* ending of the name of the acid is dropped and replaced with "aldehyde." The common name is written as one word. Thus, formaldehyde, the common name of the simplest aldehyde, is derived from formic acid (form- + aldehyde = formaldehyde). The name of the next higher member of the aldehyde family, acetaldehyde, is derived from acetic acid (acet- + aldehyde = acetaldehyde).

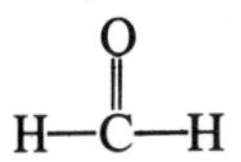

formaldehyde

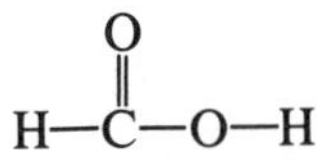

formic
acid

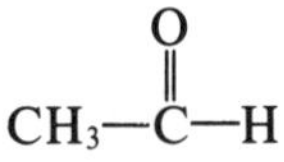

acetaldehyde

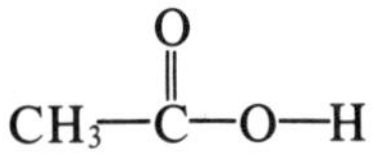

acetic
acid

The IUPAC rules for naming aldehydes are similar to those used earlier for other organic compounds. Let us go through the step-by-step process to develop the IUPAC name for the following aldehyde. Notice the location of the aldehyde functional group, $-\overset{\overset{\displaystyle O}{\displaystyle \|}}{C}-H$, on the right-hand end of the structural formula:

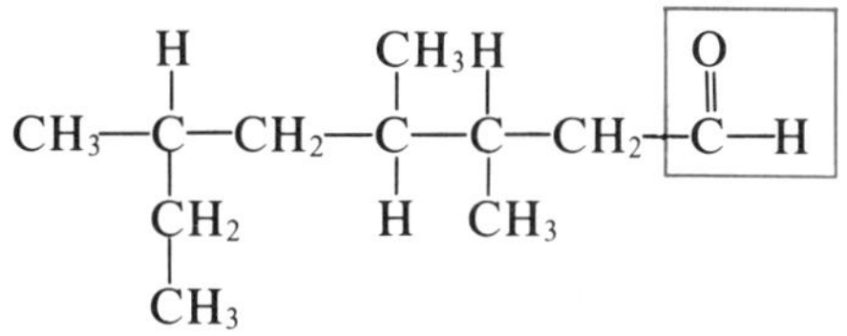

Step 1. Examine the structural formula and locate the longest chain of carbon atoms that contains the carbonyl group. This is the main chain of the compound:

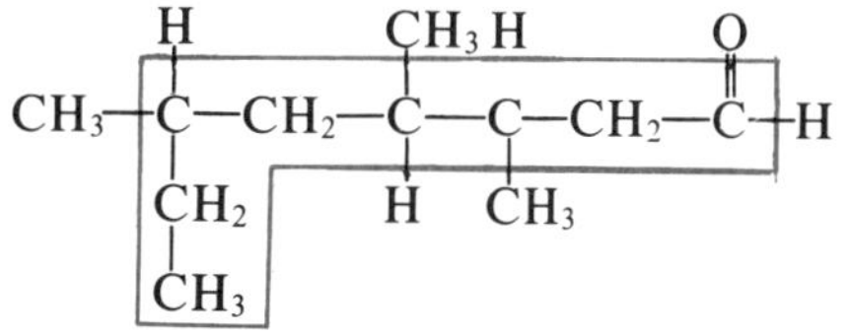

Step 2. The main chain is given the name of the alkane (Table 15.1) with the same number of carbon atoms, except the -e ending is replaced by -al, the characteristic ending used to identify a compound as an aldehyde. In the example, the main chain has eight carbon atoms (octane), so the name of the parent aldehyde is:

octanal

Step 3. The carbon atoms in the main chain are consecutively numbered beginning at the carbonyl group. The carbonyl carbon atom is always taken as carbon number one. However, this number will not appear in the IUPAC name, since the location of the aldehyde group is understood to be at carbon 1.

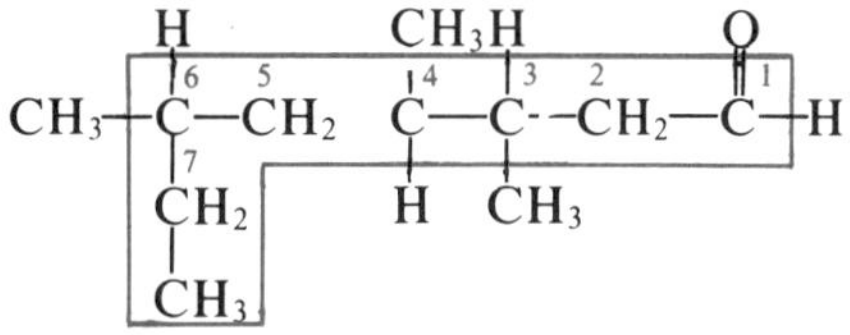

Step 4.   Alkyl groups bonded to the main chain are then located and numbered, and their names are written before the name of the parent aldehyde. The names and arrangement of groups attached to the main chain follow the same rules that were used for naming alkanes. In the example, there are three methyl groups that must be indicated in the name. They are attached to carbon atoms 3, 4, and 6 of the main chain.

$$CH_3-\overset{\overset{\displaystyle H}{|}}{\underset{\underset{\displaystyle CH_3}{\underset{|}{CH_2}}}{C}}-CH_2-\overset{\overset{\displaystyle CH_3}{|}}{\underset{\underset{\displaystyle H}{|}}{C}}-\overset{\overset{\displaystyle H}{|}}{\underset{\underset{\displaystyle CH_3}{|}}{C}}-CH_2-\overset{\overset{\displaystyle O}{||}}{C}-H$$

The IUPAC name is then:

3,4,6-trimethyloctanal

The IUPAC and common names of several other aldehydes appear in Table 18.1. Study these examples carefully, then try to name the aldehydes in Check Test Number 1.

TABLE 18.1 Common and IUPAC Names of Several Aldehydes		
Structural formula	Common name	IUPAC name
$H-\overset{\overset{\displaystyle O}{||}}{C}-H$	Formaldehyde	Methanal
$CH_3-\overset{\overset{\displaystyle O}{||}}{C}-H$	Acetaldehyde	Ethanal
$CH_3CH_2-\overset{\overset{\displaystyle O}{||}}{C}-H$	Propionaldehyde	Propanal
$CH_3CH_2CH_2-\overset{\overset{\displaystyle O}{||}}{C}-H$	Butyraldehyde	Butanal
$\begin{array}{l} H-C=O \\ H-\overset{|}{\underset{\underset{\displaystyle CH_2OH}{|}}{C}}-OH \end{array}$	Glyceraldehyde	2,3-Dihydroxypropanal

Structural formula	Common name	IUPAC name
TABLE 18.1 (cont.)		
Common and IUPAC Names of Several Aldehydes		

Structural formula	Common name	IUPAC name
(benzaldehyde structure)	Benzaldehyde	Benzaldehyde
(vanillin structure)	Vanillin	4-Hydroxy-3-methoxybenzaldehyde*
(cinnamaldehyde structure)	Cinnamaldehyde	—

*The methoxy group is CH_3O-.

Check Test Number 1

1. Give the common names for each of the following aldehydes. The common names of the related carboxylic acids are given as guides.

   a.   H—C=O

   (benzene ring structure)   (related to benzoic acid)

   b. $CH_3(CH_2)_3$—$\overset{\overset{\displaystyle O}{\|}}{C}$—H   (related to valeric acid)

2. Give the IUPAC name for each of the following aldehydes:

   a. $CH_3CH_2\overset{\overset{\displaystyle CH_3}{|}}{C}H$—$\overset{\overset{\displaystyle O}{\|}}{C}$—H

   b. $CH_3\overset{\underset{\displaystyle Br}{|}}{C}H$—$\overset{\overset{\displaystyle O}{\|}}{C}$—H

   c. $CH_3(CH_2)_3\overset{\overset{\displaystyle O}{\|}}{C}$—H

   d. $CH_3\overset{\underset{\displaystyle Br}{|}}{C}H\overset{\overset{\displaystyle CH_3}{|}}{C}HCH_2\overset{\overset{\displaystyle O}{\|}}{C}$—H

Answers:
1. a. Benzaldehyde   b. Valeraldehyde
2. a. 2-Methylbutanal   c. Pentanal
   b. 2-Bromopropanal   d. 3-Bromo-4-methylpentanal

Naming Ketones

The common names of ketones are derived by listing the names of the groups attached to the carbonyl carbon atom, usually in alphabetical order, followed by the word "ketone" written separately. This system is similar to that used to derive the common names of ethers. Acetone could also be called dimethyl ketone.

$$CH_3-\overset{\overset{\textstyle O}{\|}}{C}-CH_3$$

Acetone (dimethyl ketone)

The following ketone has the common name ethyl methyl ketone, though it is also called methyl ethyl ketone:

$$CH_3-\overset{\overset{\textstyle O}{\|}}{C}-CH_2CH_3$$

The common names of many aromatic ketones are named in the same way, but some have old established names that are used more often. For example,

$$CH_3-\overset{\overset{\textstyle O}{\|}}{C}-\bigcirc$$ could be called methyl phenyl ketone, but it is more often

called acetophenone by chemists. Benzophenone could also be named diphenyl ketone:

Benzophenone (diphenyl ketone)

Complex cyclic ketones are almost always identified with short, common names and not by IUPAC names. Progesterone and camphor are two such compounds:

Camphor

Progesterone

When naming ketones by the IUPAC method, we will follow the same basic procedures used for the aldehydes, with the following changes:

(1) The carbon atoms in the longest continuous chain that contains the carbonyl group (the main chain) will be numbered consecutively from the end that will give the carbonyl group the smaller number.

(2) The main chain will be given the name of the alkane (Table 15.1) with the same number of carbon atoms, except the -e ending will be replaced by -one, the characteristic ending used for ketones. For example, acetone would be called propanone. A 2- in front of the name would be redundant in this case, since the carbonyl group in this ketone could only be at the second position. Several ketones appear in Table 18.2 with their common and IUPAC names. Study them carefully, then try to name the ketones in the second check test.

TABLE   18.2
Common and IUPAC Names of Several Ketones

Structural formula	Common name	IUPAC name
$CH_3\!-\!\overset{\displaystyle O}{\overset{\|}{C}}\!-\!CH_3$	Acetone	Propanone
$CH_3CH_2CH_2\!-\!\overset{\displaystyle O}{\overset{\|}{C}}\!-\!CH_3$	Methyl n-propyl ketone	2-Pentanone
$CH_3CH_2\!-\!\overset{\displaystyle O}{\overset{\|}{C}}\!-\!CH_2CH_3$	Diethyl ketone	3-Pentanone
$CH_3\overset{\displaystyle CH_3}{\overset{\|}{CH}}\!-\!\overset{\displaystyle O}{\overset{\|}{C}}\!-\!CH_2CH_3$	Ethyl isopropyl ketone	2-Methyl-3-pentanone
$CH_3CH_2\!-\!\overset{\displaystyle CH_3}{\underset{\displaystyle CH_3}{\overset{\|}{C}}}\!-\!\overset{\displaystyle O}{\overset{\|}{C}}\!-\!\overset{\displaystyle }{\underset{\displaystyle CH_3}{CH}}CH_3$	—	2,4,4-trimethyl-3-hexanone

Check Test Number 2

Give the common and IUPAC names for each of the following:

a $\quad$ $CH_3CH_2CH_2CH_2$—$\overset{\overset{\displaystyle O}{\|}}{C}$—$CH_3$

b $\quad$ CH_3CH_2—$\overset{\overset{\displaystyle O}{\|}}{C}$—$CH_2CH_2CH_3$

c $\quad$ CH_3—$\overset{\overset{\displaystyle O}{\|}}{C}$—$CH_2CH_3$

Answers:
a. Common name: *n*-butyl methyl ketone
   IUPAC name: 2-hexanone
b. Common name: ethyl *n*-propyl ketone
   IUPAC name: 3-hexanone
c. Common name: Ethyl methyl ketone
   IUPAC name: Butanone (2- not needed)

Physical Properties of Aldehydes and Ketones

Many of the physical properties of aldehydes and ketones can be attributed to the polar nature of the carbonyl group. The electrons shared between the carbon and oxygen atoms of the carbonyl are not shared equally. The oxygen has a greater electronegativity (a greater attraction for the shared electrons) and draws the shared electrons somewhat away from carbon and toward itself. This causes the oxygen atom to bear a partial negative charge, $\delta-$, and the carbon atom to bear a partial positive charge, $\delta+$. The carbonyl group, then, acts as a small dipole within the molecule:

$$\overset{\overset{\displaystyle O}{\|}}{C}\overset{\delta-}{}\,\delta+$$

Because of the polarity of the carbonyl group, dipole–dipole intermolecular forces will attract the molecules to each other in the liquids and solids. Because of this dipole–dipole force, aldehydes and ketones have boiling points markedly higher than those of the alkanes of similar molecular weight. However, the aldehydes and ketones do not have boiling points as high as the corresponding alcohols, because unlike the alcohols, these carbonyl compounds are unable to engage in intermolecular hydrogen bonding.

dipole-dipole attraction between aldehyde or ketone molecules

hydrogen bonding between alcohol molecules

The polarity of the carbonyl group also increases the solubility of the smaller aldehydes and ketones in water. Water is able to hydrogen bond to the oxygen atom of the carbonyl, accommodating them in solution.

aldehyde-water hydrogen bonding

ketone-water hydrogen bonding

Generally, an aldehyde or ketone with fewer than five carbon atoms will be quite soluble in water, but as the size of the R– groups increases, the solubility decreases rapidly. With large R– groups, the compounds become increasingly hydrocarbonlike and cannot be readily accommodated by the polar water molecules.

A comparison of the boiling points and solubility in water of aldehydes, ketones, alcohols, and alkanes of similar molecular weight appears in Table 18.3.

TABLE   18.3
Comparison of Boiling Points and Solubilities of Several Aldehydes and Ketones with Alkanes and Alcohols of Similar Molecular Weight

Name	Structural formula	Molecular weight	Boiling point (°C)	Solubility g/100 ml H_2O
Ethane	CH_3CH_3	30	− 89	Insoluble
Formaldehyde	H–C=O (with H below)	30	− 21	All proportions
Methyl alcohol	CH_3OH	32	65	All proportions

TABLE   18.3
Comparison of Boiling Points and Solubilities of Several Aldehydes and
Ketones with Alkanes and Alcohols of Similar Molecular Weight (cont.)

Name	Structural formula	Molecular weight	Boiling point (°C)	Solubility g/100 ml H_2O
Propane	$CH_3CH_2CH_3$	44	− 42	Insoluble
Acetaldehyde	$CH_3-\overset{H}{\underset{}{C}}=O$	44	20	All proportions
Ethyl alcohol	CH_3CH_2OH	46	78	All proportions
n-Butane	$CH_3CH_2CH_2CH_3$	58	0	Insoluble
Propionaldehyde	$CH_3CH_2-\overset{H}{\underset{}{C}}=O$	58	49	16.0
Acetone	$CH_3-\overset{O}{\underset{}{C}}-CH_3$	58	56	All proportions
n-Propyl alcohol	$CH_3CH_2CH_2OH$	60	97	All proportions
n-Pentane	$CH_3CH_2CH_2CH_2CH_3$	72	36	Insoluble
Butyraldehyde	$CH_3CH_2CH_2-\overset{H}{\underset{}{C}}=O$	72	76	7.0
Methyl ethyl ketone	$CH_3-\overset{O}{\underset{}{C}}-CH_2CH_3$	72	80	26.0
n-Butyl alcohol	$CH_3CH_2CH_2CH_2OH$	74	118	7.9

liquid

Table 18.4 compares the boiling points and solubilities in water of several aldehydes and ketones. After reviewing the trends in these tables, try predicting trends in Check Test Number 3.

Check Test Number 3

1. Use a figure to show the hydrogen bonding that will exist between acetaldehyde and water in an aqueous solution of acetaldehyde.
2. Consider the following compounds of similar molecular weight:

n-hexane $\qquad\qquad CH_3CH_2CH_2CH_2CH_2CH_3$

3-pentanone $\qquad\qquad CH_3CH_2-\overset{O}{\underset{}{C}}-CH_2CH_3$

TABLE 18.4
Physical Properties of Some Aldehydes and Ketones

Aldehydes				
Common name	Structural formula	Molecular weight	Boiling point (°C)	Solubility g/100 ml H_2O
Formaldehyde	$H-\overset{H}{\underset{}{C}}=O$	40	−21	All proportions
Acetaldehyde	$CH_3-\overset{H}{\underset{}{C}}=O$	44	20	All proportions
Propionaldehyde	$CH_3CH_2-\overset{H}{\underset{}{C}}=O$	58	49	16
Benzaldehyde	(benzene ring)$-\overset{O}{\underset{}{C}}-H$	106	178	0.3

Molecular weight: Increases ↓
Boiling point: Increases ↓
Solubility: Decreases ↓

Ketones				
Common name	Structural formula	Molecular weight	Boiling point (°C)	Solubility g/100 ml H_2O
Acetone	$CH_3-\overset{O}{\underset{}{C}}-CH_3$	58	56	All proportions
Methyl ethyl ketone	$CH_3CH_2-\overset{O}{\underset{}{C}}-CH_3$	72	80	26
Diethyl ketone	$CH_3CH_2-\overset{O}{\underset{}{C}}-CH_2CH_3$	86	101	5
Cyclohexanone	(cyclohexane ring)$=O$	98	157	2
Acetophenone	$CH_3-\overset{O}{\underset{}{C}}-$(benzene ring)	120	202	Insoluble
Benzophenone	(benzene ring)$-\overset{O}{\underset{}{C}}-$(benzene ring)	182	306	Insoluble

Molecular weight: Increases ↓
Boiling point: Increases ↓
Solubility: Decreases ↓

pentanal

$$CH_3CH_2CH_2CH_2\overset{\overset{\displaystyle O}{\|}}{C}-H$$

1-pentanol

$$CH_3CH_2CH_2CH_2CH_2OH$$

a. List the compounds from left to right in order of decreasing boiling point.

b. Which compound of the four would have the lowest solubility in water?

Answers:

1. Something like this would be appropriate:

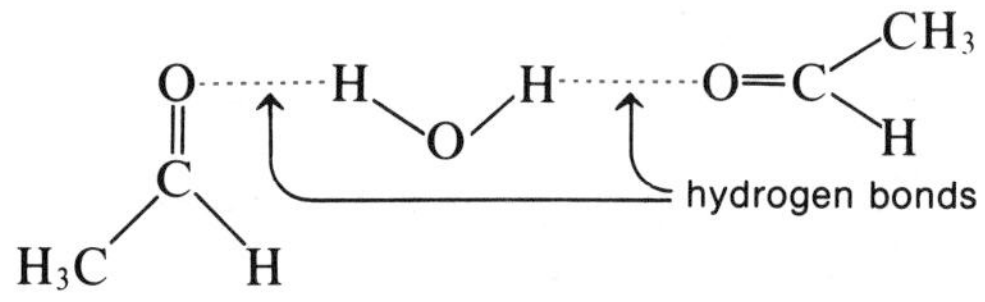

2. a. 1-pentanol (highest b.p.), 3-pentanone, pentanal, *n*-hexane (lowest b.p.)

  b. *n*-hexane

The Preparation of Aldehydes and Ketones

There are several methods that may be used to prepare aldehydes and ketones, but one of the more important means is the oxidation of alcohols. Aldehydes may be prepared by the controlled oxidation of a primary (1°) alcohol. The aldehyde must be removed from the reaction mixture as it is formed to prevent further oxidation to the carboxylic acid. Several oxidizing agents can be used to carry out the synthesis, but we will use the general symbol [O] to indicate oxidation by an appropriate agent, as was done in Chapter 17. Remember that oxidation can be regarded as the removal of two hydrogen atoms from a molecule. They most often combine with oxygen to form water.

$$RCH_2OH \xrightarrow{[o]} R-\overset{\overset{\displaystyle O}{\|}}{C}-H \ + \ H_2O$$

1° alcohol   aldehyde

$$CH_3CH_2OH \xrightarrow{[o]} CH_3\overset{\overset{\displaystyle O}{\|}}{C}-H \ + \ H_2O$$

ethyl alcohol   acetaldehyde

Ketones may be prepared by the controlled oxidation of secondary (2°) alcohols. Ketones are not oxidized further under usual laboratory conditions.

$$R-\underset{\underset{H}{|}}{\overset{\overset{OH}{|}}{C}}-R \xrightarrow{[o]} \underset{R}{\overset{O}{\underset{\diagdown}{\overset{||}{C}}}} {R} + H_2O$$

2° alcohol   ketone

$$CH_3-\underset{\underset{H}{|}}{\overset{\overset{OH}{|}}{C}}-CH_3 \xrightarrow{[o]} \underset{H_3C}{\overset{O}{\underset{\diagdown}{\overset{||}{C}}}} {CH_3} + H_2O$$

isopropyl alcohol   acetone

Chemical Properties of Aldehydes and Ketones

Oxidation of Aldehydes   Aldehydes are easily oxidized under mild conditions to carboxylic acids. In fact, simple exposure to air will slowly oxidize aldehydes to acids. The ease of the oxidation can be used to determine if an unknown compound is an aldehyde. There are three specific reagents that are used for this purpose: Tollen's reagent, Fehling's reagent, and Benedict's reagent.

Tollen's reagent is a mixture of silver nitrate, $AgNO_3$, in aqueous ammonia, $NH_{3(aq)}$. If a small amount of an aldehyde is added to Tollen's reagent in a test tube, the silver ion is reduced to metallic silver and plates out on the inner wall of the test tube forming a silver mirror. The silver mirror is an excellent visual indication of the presence of an aldehyde. The general equation is

$$R-\overset{\overset{O}{||}}{C}-H + 2[Ag(NH_3)_2]^+OH^- \longrightarrow$$

aldehyde   Tollen's reagent
(colorless)

$$R-\overset{\overset{O}{||}}{C}-O^\ominus, NH_4^+ \quad 2\,Ag\!\!\downarrow + 3NH_{3\,(aq)} + H_2O$$

ammonium salt   silver
of the car-   mirror
boxylic acid

Fehling's reagent and **Benedict's reagent** are both alkaline (basic) solutions of copper (II) sulfate, $CuSO_4$. The only difference between them is the reagent that is added to prevent the copper (II) ion from precipitating as $Cu(OH)_2$. Sodium tartrate is used in Fehling's solution for this purpose, and sodium citrate is used in Benedict's solution. The visual evidence of the oxidation of an aldehyde using these reagents is the formation of a brick-red precipitate of copper (I) oxide, Cu_2O. The general equation for either reagent is the same:

$$\underset{\text{aldehyde}}{R-\overset{\overset{\displaystyle O}{\|}}{C}-H} \; + \; \underset{\substack{\text{Fehling's or Bene-}\\\text{dict's reagent (blue)}}}{2Cu^{2+} \; + \; Na^+ \; + \; 5OH^-} \longrightarrow \underset{\substack{\text{sodium salt of the}\\\text{carboxylic acid}}}{R-\overset{\overset{\displaystyle O}{\|}}{C}-O^{\ominus}, Na^+} \; + \; \underset{\substack{\text{brick-red}\\\text{precipitate}}}{Cu_2O\downarrow} + \; H_2O$$

Both the Fehling and Benedict tests have been used for years to detect the presence of glucose in the urine of diabetic patients. Glucose contains a free aldehyde group and is oxidized to gluconic acid as Cu^{2+} is reduced to Cu^+ in Cu_2O. This test is discussed further in Chapter 21.

Ketones and alcohols are not oxidized by these three reagents. However, certain powerful oxidizing agents, such as hot, concentrated nitric acid, can break ketone molecules apart and oxidize the fragments to carboxylic acids.

Reduction of Aldehydes and Ketones In contrast to oxidation, reduction can be viewed as the addition of two hydrogen atoms to a molecule. Hydrogen gas, H_2, under high pressure can be used as the source of the hydrogen atoms, and the reactions require the use of catalysts, such as finely divided platinum metal. The reduction of an aldehyde produces a primary alcohol. Acetaldehyde can be reduced to ethyl alcohol.

$$\underset{\text{acetaldehyde}}{CH_3-\overset{\overset{\displaystyle O}{\|}}{C}-H} \; + \; H_2 \; \xrightarrow[\text{high pressure}]{Pt} \; \underset{\text{ethyl alcohol (1°)}}{CH_3CH_2OH}$$

The reduction of a ketone produces a secondary alcohol. Acetone can be reduced to isopropyl alcohol.

$$\underset{\text{acetone}}{CH_3-\overset{\overset{\displaystyle O}{\|}}{C}-CH_3} \; + \; H_2 \; \xrightarrow[\text{high pressure}]{Pt} \; \underset{\text{isopropyl alcohol (2°)}}{CH_3-\underset{\underset{\displaystyle H}{|}}{\overset{\overset{\displaystyle OH}{|}}{C}}-CH_3}$$

You will recognize that these reduction reactions are just the reverse of the oxidation reactions used to prepare aldehydes and ketones. In the body, special enzymes catalyze the addition of hydrogen atoms to aldehydes and ketones.

Addition of Reagents to the Carbonyl Group   The most characteristic type of reaction of both aldehydes and ketones is the addition of reagents across the carbon–oxygen double bond. The reactions are similar to those seen earlier in which reagents were added to the carbon–carbon double bonds in alkenes. There are differences, though, that arise principally from the polarity of the carbonyl group. The polarity of the carbonyl will influence the way in which a reagent will add to the double bond. In the examples that will be presented, a hydrogen atom from the adding molecule will bond to the oxygen of the carbonyl, and the remainder will join to the carbon atom of the carbonyl. The more electronegative oxygen joins to the hydrogen. The general equation for the addition to the carbonyl shows the specific way a reagent adds to the carbon–oxygen double bond:

$$
HX \; + \; \underset{\substack{R \quad R \\ (H)}}{\overset{\overset{O}{\|}}{C}} \longrightarrow \underset{\substack{| \\ X}}{\overset{\overset{OH}{|}}{R-C-R}} \; (H)
$$

| adding reagent | aldehyde or ketone | addition product |

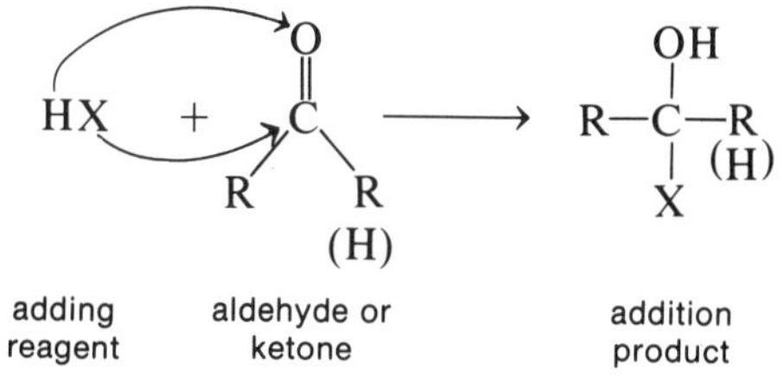

Let us look at some specific examples of addition reactions to the carbonyl group.

Addition of Hydrogen Cyanide   The addition of hydrogen cyanide (HCN) to a ketone produces a **cyanohydrin**. Acetone can be converted into a cyanohydrin by the following reaction:

$$
\underset{\text{acetone}}{CH_3-\overset{\overset{O}{\|}}{C}-CH_3} \; + \; HCN \longrightarrow \underset{\text{2-cyano-2-propanol}}{CH_3-\overset{\overset{OH}{|}}{\underset{\underset{CN}{|}}{C}}-CH_3} \qquad \text{(—CN is the cyano group)}
$$

Cyanohydrins are important intermediates in the synthesis of amino acids, hydroxy acids, and many carbohydrates in the laboratory.

Addition of Alcohols   The addition of alcohols to aldehydes and ketones is catalyzed by acids, and these reactions are readily reversible. The addition of an alcohol to an aldehyde produces a **hemiacetal**; whereas the ketones produce **hemiketals**. The hemiacetal and hemiketal structures are found in carbohydrates. The hydrogen of the alcohol joins to the oxygen of the carbonyl, and the remainder of the alcohol adds to the carbon of the carbonyl.

General Reactions:

$$\text{R'OH} \ + \ \underset{\text{aldehyde}}{\overset{\overset{\displaystyle O}{\overset{\|}{\text{R—C—H}}}}{}} \ \underset{\text{catalyst}}{\overset{\text{acid}}{\rightleftarrows}} \ \underset{\text{hemiacetal}}{\overset{\overset{\displaystyle OH}{|}}{\underset{\overset{|}{\text{O—R'}}}{\text{R—C—H}}}}$$

alcohol   aldehyde   hemiacetal

$$\text{R'OH} \ + \ \underset{\text{ketone}}{\overset{\overset{\displaystyle O}{\overset{\|}{\text{R—C—R}}}}{}} \ \underset{\text{catalyst}}{\overset{\text{acid}}{\rightleftarrows}} \ \underset{\text{hemiketal}}{\overset{\overset{\displaystyle OH}{|}}{\underset{\overset{|}{\text{O—R'}}}{\text{R—C—R}}}}$$

alcohol   ketone   hemiketal

Notice that the hemiacetal and the hemiketal have both alcohol and ether functional groups that involve the same carbon atom.

Specific Reactions:

(1) $\underset{\text{acetaldehyde}}{\overset{\overset{\displaystyle O}{\overset{\|}{\text{CH}_3\text{—C—H}}}}{}} + \underset{\text{methyl alcohol}}{\text{CH}_3\text{OH}} \overset{\text{HCl}}{\rightleftarrows} \underset{\text{a hemiacetal}}{\overset{\overset{\displaystyle OH}{|}}{\underset{\overset{|}{\text{O—CH}_3}}{\text{CH}_3\text{—C—H}}}}$

(2) $\underset{\text{acetone}}{\overset{\overset{\displaystyle O}{\overset{\|}{\text{CH}_3\text{—C—CH}_3}}}{}} + \underset{\text{methyl alcohol}}{\text{CH}_3\text{OH}} \overset{\text{HCl}}{\rightleftarrows} \underset{\text{a hemiketal}}{\overset{\overset{\displaystyle OH}{|}}{\underset{\overset{|}{\text{O—CH}_3}}{\text{CH}_3\text{—C—CH}_3}}}$

Hemiacetals and hemiketals readily react further with alcohol to form **acetals** and **ketals**, respectively. Because of the ease of this reaction, hemiacetals and hemiketals are usually not isolated in the laboratory, and they continue to react to form acetals and ketals. The reaction involves the removal of a molecule of water between the hydroxyl group of the hemiacetal or hemiketal and the alcohol functional group. Ketals and acetals are important in carbohydrate chemistry.

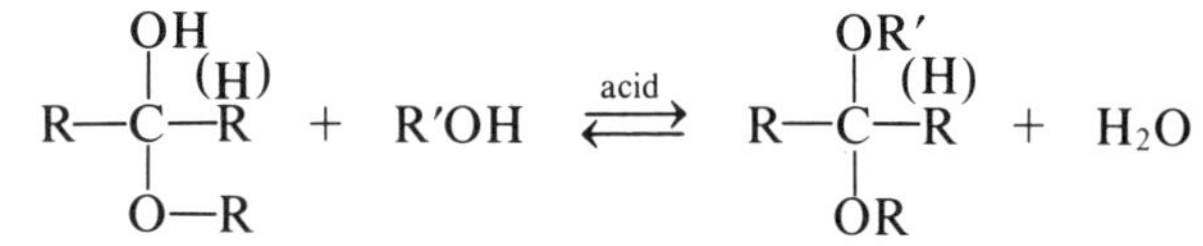

$$\underset{\substack{\text{hemiketal or} \\ \text{hemiacetal (H)}}}{R-\overset{\overset{\text{OH}}{|}\text{(H)}}{\underset{\underset{\text{O}-R}{|}}{C}}-R} \;+\; \underset{\text{alcohol}}{R'OH} \;\overset{\text{acid}}{\rightleftarrows}\; \underset{\substack{\text{ketal or} \\ \text{acetal (H)}}}{R-\overset{\overset{\text{OR}'}{|}\text{(H)}}{\underset{\underset{\text{OR}}{|}}{C}}-R} \;+\; H_2O$$

Addition of Ammonia
The reactions of aldehydes and ketones with ammonia, NH_3, occur in two steps: (1) the normal addition of ammonia to the carbonyl to produce an unstable compound, and (2) the rapid release of water from the addition product to produce an **imine**. This kind of reaction is used by living organisms in the synthesis of amino acids from compounds containing a ketone group. The general equation for the two-step sequence is the same for either an aldehyde or ketone.

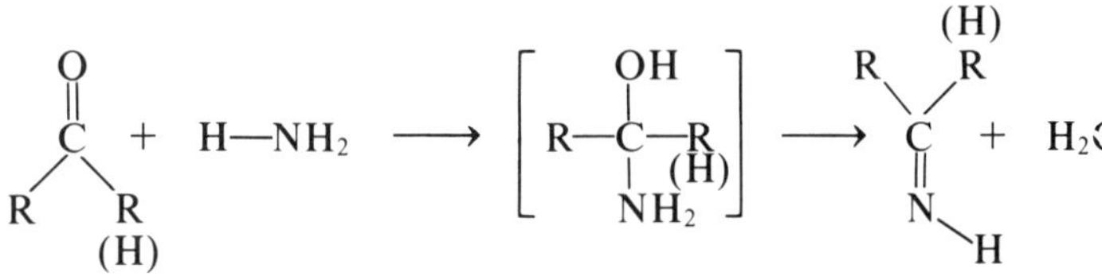

$$\underset{\substack{\text{ketone or} \\ \text{aldehyde (H)}}}{\overset{\overset{\text{O}}{||}}{\underset{R\quad R}{\underset{\text{(H)}}{C}}}} + \underset{\text{ammonia}}{H-NH_2} \longrightarrow \left[\underset{\substack{\text{unstable} \\ \text{addition} \\ \text{product}}}{R-\overset{\overset{\text{OH}}{|}}{\underset{\underset{NH_2}{|}}{C}}-\underset{\text{(H)}}{R}} \right] \longrightarrow \underset{\text{an imine}}{\overset{\overset{\text{(H)}}{R\diagdown\diagup R}}{\underset{\underset{H}{N}}{C}}} + H_2O$$

Check Test Number 4

Complete the following reactions involving aldehydes and ketones:
a. Acetone + Tollen's reagent $\longrightarrow$
b. Acetaldehyde + $2CH_3OH \overset{\text{acid}}{\longrightarrow}$
c. Acetone + $NH_3 \rightarrow$

Answers:
a. No reaction. Since acetone is a ketone, it will not be oxidized by Tollen's reagent.

$$\text{b.} \quad \underset{}{CH_3-\overset{\overset{\text{O}}{||}}{C}-H} + CH_3OH \overset{\text{acid}}{\rightleftarrows} CH_3-\overset{\overset{\text{OH}}{|}}{\underset{\underset{\text{O}-CH_3}{|}}{C}}-H \underset{\substack{\text{(second} \\ \text{molecule)} \\ + CH_3OH}}{\overset{}{\underset{\text{acid}}{\rightleftarrows}}} CH_3-\overset{\overset{\text{O}-CH_3}{|}}{\underset{\underset{\text{O}-CH_3}{|}}{C}}-H + H_2O$$

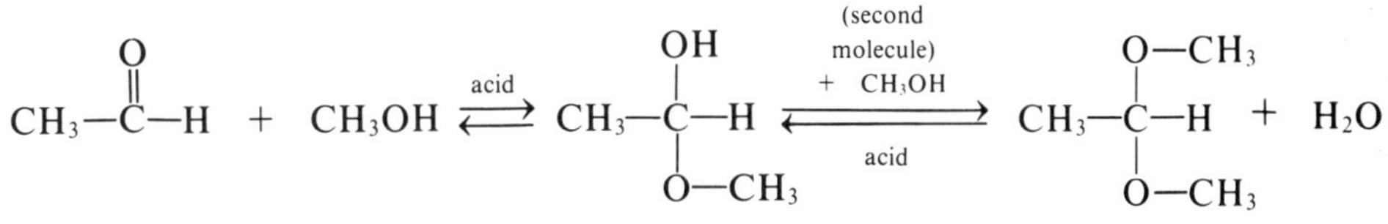

$$\qquad\qquad\qquad\qquad\qquad\qquad \underset{\text{hemiacetal}}{} \qquad\qquad\qquad \underset{\text{acetal}}{}$$

c.

$$CH_3-\overset{\overset{\displaystyle O}{\|}}{C}-CH_3 \ + \ NH_3 \ \longrightarrow \ \left[CH_3-\overset{\overset{\displaystyle OH}{|}}{\underset{\underset{\displaystyle NH_2}{|}}{C}}-CH_3 \right] \longrightarrow \ CH_3-\overset{\overset{\displaystyle N\diagdown_{\displaystyle H}}{\|}}{C}-CH_3 \ + \ H_2O$$

an imine

Some Important Aldehydes and Ketones

Formaldehyde— HCHO

Formaldehyde is a gas at room temperature, and when mixed with air is easily oxidized to formic acid, HCOOH. However, if dissolved in water, the ease of oxidation is greatly reduced, and the solution of formaldehyde is more convenient to handle than the gas. Commercially, formaldehyde is available as a 40% aqueous solution known as **formalin.** The solution can be used to sterilize surgical instruments and gloves. It is also used as an embalming fluid since it hardens tissue and retards its decomposition. Perhaps you are familiar with the characteristic pungent odor of many biology laboratories. It is the odor of formaldehyde from the solutions used to preserve tissue samples.

Under certain conditions of temperature and pressure, formaldehyde will polymerize to form a straight chain solid polymer called **paraformaldehyde:**

$$\left(-O-\overset{\overset{\displaystyle H}{|}}{\underset{\underset{\displaystyle H}{|}}{C}}-O-\overset{\overset{\displaystyle H}{|}}{\underset{\underset{\displaystyle H}{|}}{C}}-O-\overset{\overset{\displaystyle H}{|}}{\underset{\underset{\displaystyle H}{|}}{C}}- \right)$$

paraformaldehyde

When heated to around 190°C, paraformaldehyde decomposes to release formaldehyde gas. Formaldehyde is used in the manufacture of Bakelite and Melamine plastics. Also, formaldehyde is used with the Tollen's reagent to produce silver mirrors commercially.

Acetaldehyde— CH₃CHO

Acetaldehyde is a liquid at room temperature, and like formaldehyde, is readily oxidized upon exposure to air to acetic acid. Because of its high susceptibility to oxidation, acetaldehyde is difficult to store for long periods of time. For this reason, it is usually sold commercially as the solid, cyclic trimer called **paraldehyde.** One molecule of the trimer is formed as three molecules of the aldehyde join together to form a six-membered ring of alternating carbon and oxygen atoms. A strong acid is needed to catalyze the reaction. The formation of paraldehyde is reversible, and the acetaldehyde can be regenerated as needed.

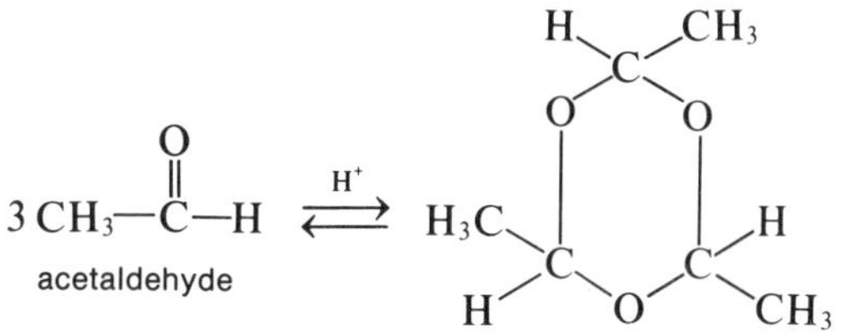

$$3 \text{ CH}_3-\overset{\overset{\displaystyle O}{\|}}{C}-H \; \underset{}{\overset{H^+}{\rightleftharpoons}} \;$$

acetaldehyde

paraldehyde

Paraldehyde has been used as a hypnotic drug to induce sleep. It has also been used as a sedative and an obstetric analgesic.

Acetaldehyde has a sharp, irritating odor, and like formaldehyde it can cause irritation of the mucous membranes if its vapors are inhaled for prolonged periods. Acetaldehyde is used in the manufacture of acetic acid, synthetic rubber, plastics, perfumes, flavors, dyes, and silvered mirrors.

Other Aldehydes

Many of the higher molecular weight aldehydes have pleasant odors and tastes and they are used in fragrances and artificial flavors. Butyraldehyde, $\text{CH}_3\text{CH}_2\text{CH}_2\text{CHO}$, is one of the chemicals responsible for the appetizing odor of fresh bread. Citral, shown below, has the odor and flavor of lemon and is used in foods, household products, and colognes. Cinnamaldehyde has the odor and flavor of cinnamon, and vanillin has the odor and flavor of vanilla. Benzaldehyde has the flavor of almonds.

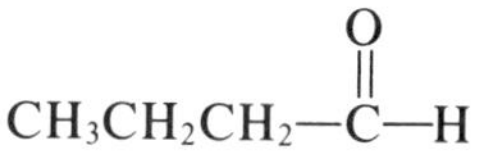

$$\text{CH}_3\text{CH}_2\text{CH}_2-\overset{\overset{\displaystyle O}{\|}}{C}-H$$

Butyraldehyde
(Fresh bread)

$$\text{CH}_3-\overset{\overset{\displaystyle CH_3}{|}}{C}=\overset{\overset{\displaystyle H}{|}}{C}-\text{CH}_2\text{CH}_2-\overset{\overset{\displaystyle CH_3}{|}}{C}=\overset{\overset{\displaystyle H}{|}}{C}-\overset{\overset{\displaystyle O}{\|}}{C}-H$$

Citral
(Lemon)

$$\text{---CH}=\text{CH}-\overset{\overset{\displaystyle O}{\|}}{C}-H$$

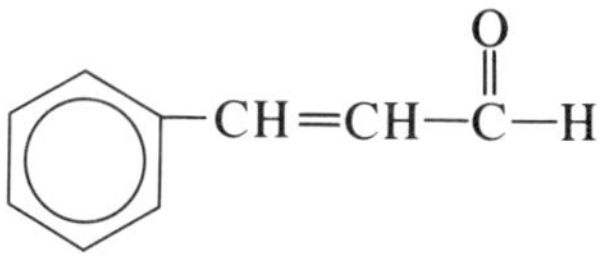

Cinnamaldehyde
(Cinnamon)

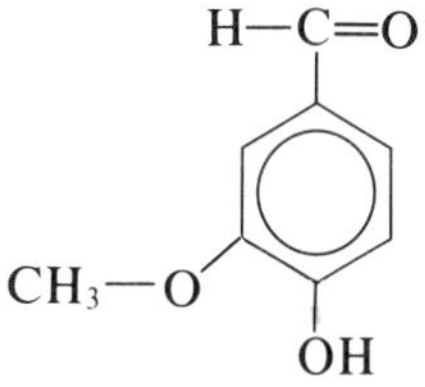

Vanillin
(Vanilla)

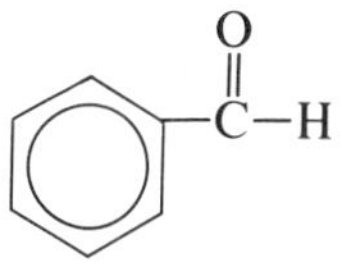

Benzaldehyde
(Almond)

Acetone—
(CH₃)₂C=O

Industrially, the most important ketone is acetone. It is a liquid at room temperature and is very soluble in water. Acetone is widely used as a solvent for many organic compounds, and it is the principal component of most fingernail polish removers. Acetone is used in the manufacture of many organic compounds. Inhalation of acetone vapor may produce headache, fatigue, excitement, bronchial irritation, and, in large amounts, narcosis.

Small amounts of acetone are normally produced by the body during the metabolism of fats. However, under conditions of starvation or severe diabetes mellitus, the amount of acetone and other ketones can increase to levels that, unless corrected, can bring on coma or even death. The sweet, fruity odor of acetone may be detected on the breath of a diabetic experiencing elevated levels of acetone in the blood.

Other Ketones

Methyl ethyl ketone is widely used in industry as a solvent, as is the cyclic ketone, cyclohexanone.

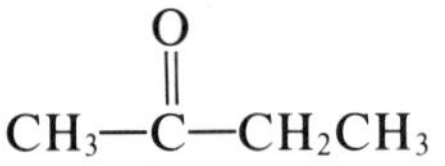

Methyl Ethyl Ketone

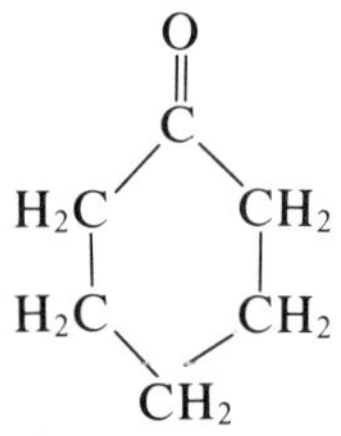

Cyclohexanone

The camphor tree produces a fragrant cyclic ketone that is used in cold medicines and liniments. Not surprisingly, the common name of this ketone is camphor. Muscone is another cyclic ketone that is used to enhance the fragrance of perfumes. It contains 15 carbon atoms in the ring, including the carbonyl carbon.

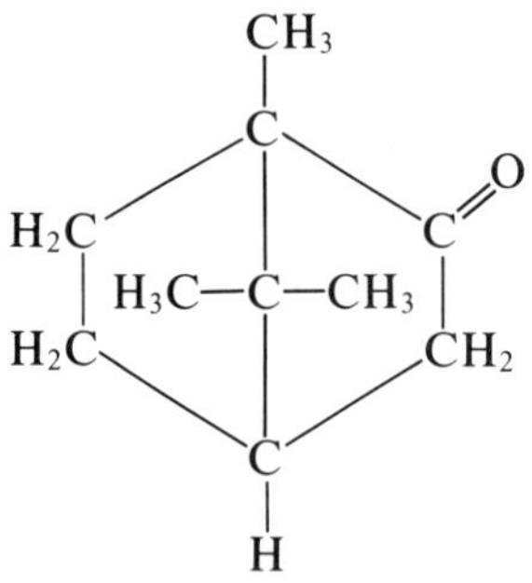

Camphor

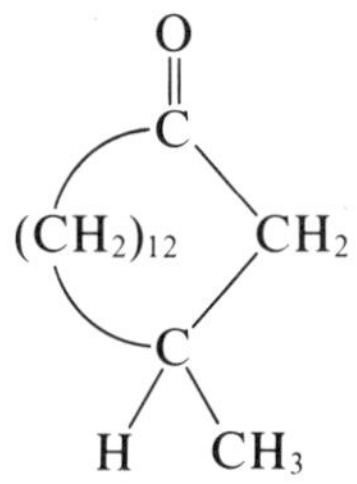

Muscone

The Carboxylic Acids

Carboxylic acids are organic compounds that contain the **carboxyl functional**

group, $-\overset{\displaystyle O}{\overset{\|}{C}}-OH$. Unlike aldehydes and ketones, the carbonyl group in a carboxylic acid does not generally enter into addition reactions. Rather, the properties of these compounds are largely the result of their weak acidity. Carboxylic acids are weak acids, and only a small fraction of the acid molecules exist as ions when dissolved in water. Acetic acid is a typical example, and its behavior as a weak acid was described in Chapter 13.

$$CH_3-C\underset{OH}{\overset{O}{\diagdown}} \quad + \quad H_2O \quad \rightleftharpoons \quad CH_3-C\underset{O^{\ominus}}{\overset{O}{\diagdown}} \quad + \quad H_3O^+$$

acetic acid   acetate ion   hydronium ion

The carboxyl group is found in many biological molecules. Most of these compounds contain other functional groups as well. Glycine, an amino acid, contains the carboxyl group and an amino group, $-NH_2$. Citric acid contains three carboxyl groups and the hydroxyl group, $-OH$. Oleic acid, a long chain carboxylic acid, also contains a carbon–carbon double bond.

$$H_2N-\overset{\displaystyle H}{\underset{\displaystyle H}{\overset{|}{\underset{|}{C}}}}-C\underset{OH}{\overset{O}{\diagdown}}$$

Glycine
(an amino acid)

$$H_2C-C\underset{OH}{\overset{O}{\diagdown}}$$
$$|$$
$$HO-C-C\underset{OH}{\overset{O}{\diagdown}}$$
$$|$$
$$H_2C-C\underset{OH}{\overset{O}{\diagdown}}$$

$$CH_3(CH_2)_7CH=CH-(CH_2)_7COOH$$

Oleic acid

Citric acid
(an important compound in metabolism
and in citrus fruits)

The carboxyl group can be written in condensed form as $-COOH$, and occasionally it is seen written as $-CO_2H$.

Naming the Carboxylic Acids

Because carboxylic acids and their derivatives occur widely in nature, they were some of the earliest organic compounds studied. The common names given the acids at that time are still widely used today, though the IUPAC names are frequently used for the less familiar members of the family. The common names for several carboxylic acids appear in Tables 18.5 and 18.6 (p. 478).

The IUPAC names of the aliphatic carboxylic acids can be derived by applying the same rules used to name aldehydes, with the following alterations:

1. The main chain of the molecule is the longest chain of carbon atoms that contains the carboxyl group.
2. The name of the main chain is taken from the alkane with the same number of carbon atoms (Table 15.1), but with the *-e* ending replaced by *-oic* and followed by the word "acid."
3. The carboxyl carbon is taken as carbon 1 when numbering the carbon chain.

Several carboxylic acids are given below with their IUPAC names. Study them carefully until you understand the derivation of each name.

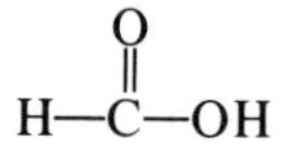

$$
\underset{\substack{\text{Methanoic acid}\\\text{(formic acid)}}}{\text{H}-\overset{\displaystyle O}{\overset{\|}{C}}-\text{OH}}
\qquad
\underset{\substack{\text{Ethanoic acid}\\\text{(acetic acid)}}}{\text{CH}_3-\overset{\displaystyle O}{\overset{\|}{C}}-\text{OH}}
\qquad
\underset{\text{3-Methylbutanoic acid}}{\text{CH}_3-\overset{\displaystyle \text{CH}_3}{\overset{|}{\text{CH}}}\text{CH}_2\overset{\displaystyle O}{\overset{\|}{C}}-\text{OH}}
$$

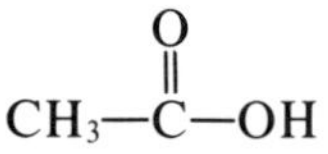

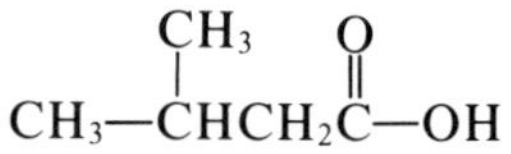

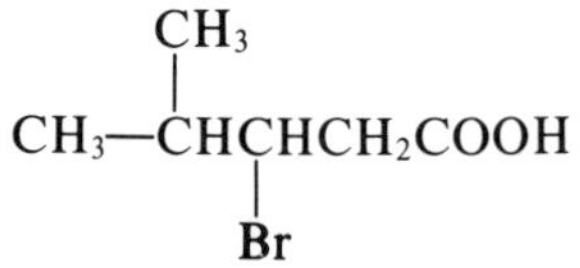

$$
\underset{\substack{\\ \text{3-Bromo-4-methylpentanoic acid}}}{\text{CH}_3-\overset{\displaystyle \text{CH}_3}{\overset{|}{\text{CH}}}\underset{\displaystyle \underset{\text{Br}}{|}}{\text{CH}}\text{CH}_2\text{COOH}}
\qquad
\underset{\text{4-chloroheptanoic acid}}{\text{CH}_3\text{CH}_2\text{CH}_2\overset{\displaystyle \text{Cl}}{\overset{|}{\text{CH}}}\text{CH}_2\text{CH}_2\text{COOH}}
$$

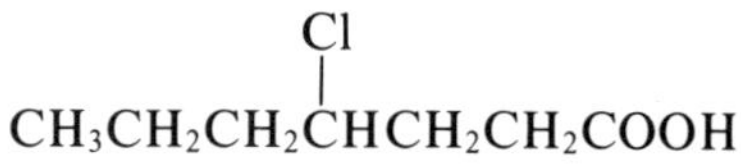

$$
\underset{\text{3-hydroxypropanoic acid}}{\text{HOCH}_2\text{CH}_2\text{COOH}}
\qquad
\underset{\text{3-ethyl-5,5-dimethylhexanoic acid}}{\text{CH}_3-\overset{\displaystyle \text{CH}_3}{\underset{\displaystyle \underset{\text{CH}_3}{|}}{\overset{|}{\text{C}}}}-\text{CH}_2-\underset{\displaystyle \underset{\underset{\displaystyle \underset{\text{CH}_3}{|}}{\text{CH}_2}}{|}}{\text{CH}}-\text{CH}_2-\text{COOH}}
$$

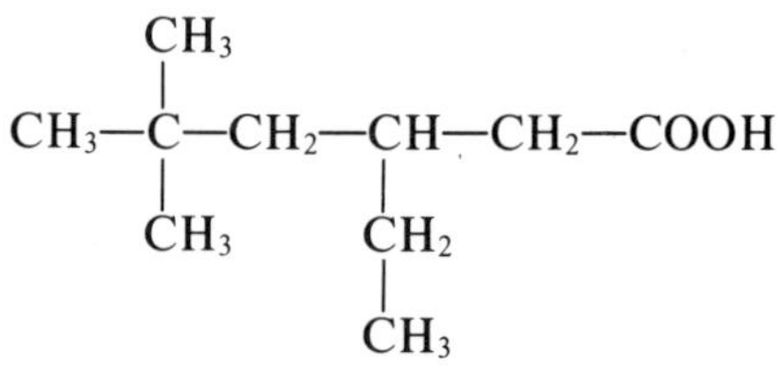

The aromatic carboxylic acids are known primarily by common names.

Benzoic acid

Salicylic acid

Several of the more familiar aliphatic acids are listed in Table 18.5 along with their sources, common names, and IUPAC names. Table 18.6 lists several of the more important aromatic and aliphatic carboxylic acids. Several contain two or more carboxyl groups and other functional groups as well.

TABLE 18.5
Common Aliphatic Carboxylic Acids

Number of carbon atoms	Structural formula	Common name	IUPAC name	Source
Saturated carboxylic acid				
1	HCOOH	Formic acid	Methanoic acid	Ants and bees
2	CH_3COOH	Acetic acid	Ethanoic acid	Vinegar
3	CH_3CH_2COOH	Propionic acid	Propanoic acid	Fat
4	$CH_3(CH_2)_2COOH$	Butyric acid	Butanoic acid	Butter
5	$CH_3(CH_2)_3COOH$	Valeric acid	Pentanoic acid	Valerian roots
6	$CH_3(CH_2)_4COOH$	Caproic acid	Hexanoic acid	Goat
*12	$CH_3(CH_2)_{10}COOH$	Lauric acid	Dodecanoic acid	Laurel tree
*14	$CH_3(CH_2)_{12}COOH$	Myristic acid	Tetradecanoic acid	Nutmeg
*16	$CH_3(CH_2)_{14}COOH$	Palmitic acid	Hexadecanoic acid	Palm oil
*18	$CH_3(CH_2)_{16}COOH$	Stearic acid	Octadecanoic acid	Tallow
*20	$CH_3(CH_2)_{18}COOH$	Arachidic acid	Eicosanoic acid	Peanuts
Unsaturated carboxylic acid				
*18	$C_{17}H_{33}COOH$(one double bond)	Oleic acid	usually not used	Most fats and oils
*18	$C_{17}H_{31}COOH$(two double bonds)	Linolenic acid	usually not used	Most fats and oils
*18	$C_{17}H_{29}COOH$(three double bonds)	Linolenoic acid	usually not used	Most fats and oils
*20	$C_{19}H_{31}COOH$(four double bonds)	Arachidonic acid	usually not used	Most fats and oils

*These acids are known as fatty acids.

TABLE 18.6
Common Aromatic and Aliphatic Carboxylic Acids with More Than One Functional Group

Common Name	Structure	
Benzoic acid	$\bigcirc$—COOH	Used as a food preservative
Terephthalic acid	COOH / $\bigcirc$ / COOH (a dicarboxylic acid)	Used in the production of Dacron fibers
Phthalic acid	COOH / $\bigcirc$—COOH (a dicarboxylic acid)	Used in perfumes
Salicylic acid	COOH / $\bigcirc$—OH	Used in the synthesis of aspirin
Lactic acid	OH / CH_3—C—COOH / H	Produced in muscles, sour milk
Citric acid	H_2C—COOH / HO—C—COOH (tricarboxylic acid) / H_2C—COOH	Metabolism intermediate; added to soft drinks, and found in citrus fruits
Pyruvic acid	O / CH_3—C—COOH	Produced in metabolism
Malic acid	OH / HOOC—CH_2—CH—COOH (a dicarboxylic acid)	Metabolism intermediate
Tartaric acid	OH OH / HOOC—C—C—COOH / H H (a dicarboxylic acid)	Found in grapes, and used in soft drinks. Its potassium salt is known as cream of tartar.

TABLE 18.6 Common Aromatic and Aliphatic Carboxylic Acids with More Than One Functional Group (cont.)		
Common Name	**Structure**	
Oxalic acid	HOOC-COOH (dicarboxylic acid)	Tart taste in rhubarb
Malonic acid	HOOC-CH$_2$-COOH (dicarboxylic acid)	Industrial uses
Succinic acid	HOOC-(CH$_2$)$_2$COOH (dicarboxylic acid)	Metabolism intermediate
Glutaric acid	HOOC-(CH$_2$)$_3$COOH (dicarboxylic acid)	Metabolism intermediate

Check Test Number 5

Give the IUPAC names for the following:

a. 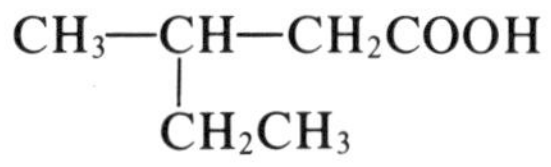   CH$_3$—CH—CH$_2$COOH
                  |
                  CH$_2$CH$_3$

b. 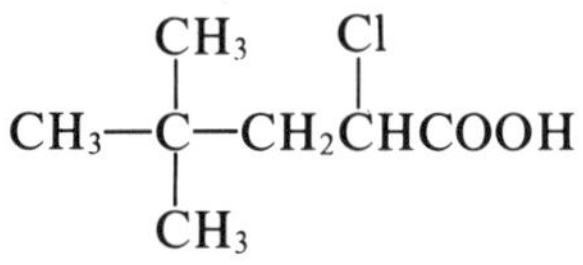   CH$_3$ Cl
                     | |
        CH$_3$—C—CH$_2$CHCOOH
                     |
                  CH$_3$

c.   CH$_3$(CH$_2$)$_3$COOH

Answers:

a. 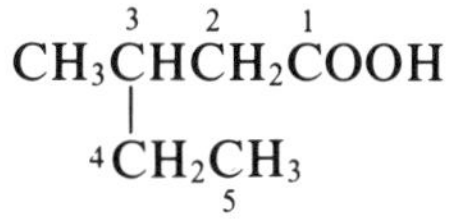   CH$_3$CHCH$_2$COOH   3-Methylpentanoic Acid
                  |
              CH$_2$CH$_3$

b. 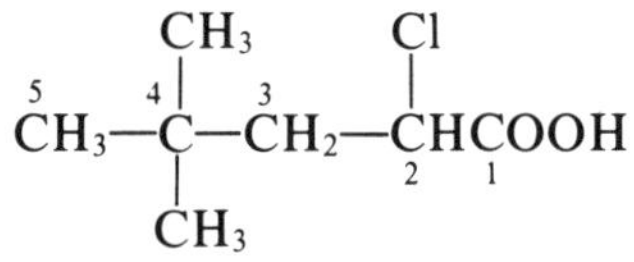   CH$_3$ Cl
                     | |
        CH$_3$—C—CH$_2$—CHCOOH
                     |
                  CH$_3$

       2-Chloro-4,4-dimethylpentanoic Acid

c.   Pentanoic Acid

The Preparation of Carboxylic Acids

Carboxylic acids are commonly prepared by the oxidation of aldehydes, or the two-step oxidation of primary alcohols. These reactions have been discussed previously in this chapter and in Chapter 17, so the preparations will be shown here with a simple general equation:

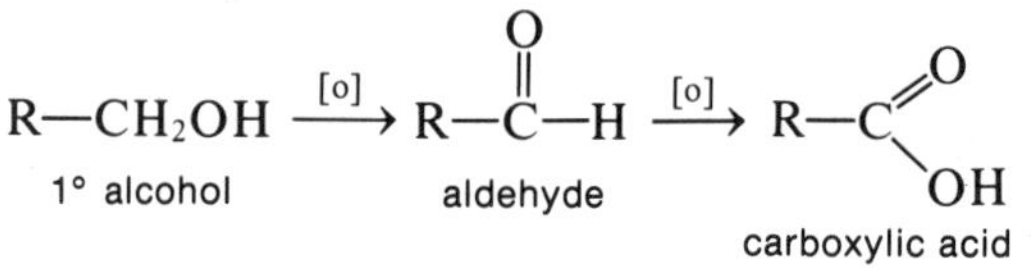

$$R—CH_2OH \xrightarrow{[o]} R—\overset{\displaystyle O}{\underset{\displaystyle \|}{C}}—H \xrightarrow{[o]} R—C\overset{\diagup O}{\diagdown OH}$$

$$\text{1° alcohol} \qquad \text{aldehyde} \qquad \text{carboxylic acid}$$

The Physical Properties of Carboxylic Acids

Carboxylic acids can strongly hydrogen bond to each other and with water. In pure samples of carboxylic acids, and in some of their aqueous solutions, the hydrogen bonding between acid molecules results in the formation of dimers, two acid molecules quite firmly held together as a unit. Two hydrogen bonds hold the dimer together:

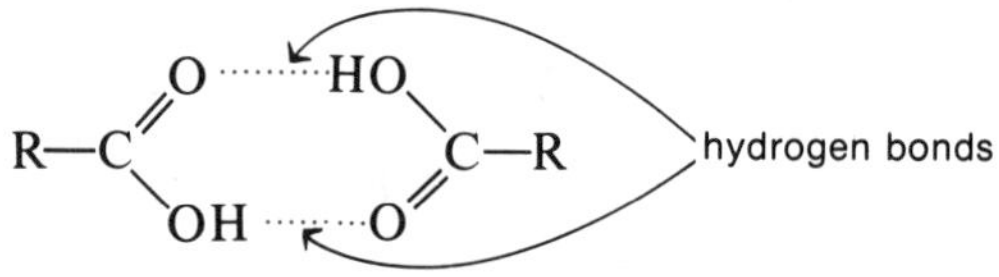

a carboxylic acid dimer

The ability of carboxylic acids to form dimers explains why they have higher boiling points than alcohols of comparable molecular weight.

Carboxylic acids are also more soluble in water than the corresponding alcohols, because of their ability to engage in extensive and stronger hydrogen bonding with the solvent.

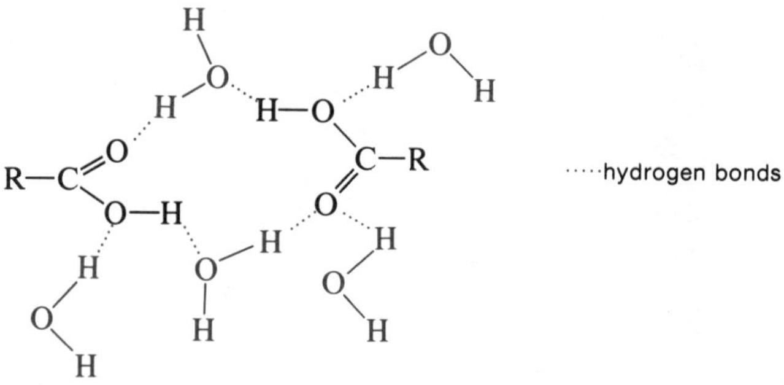

As the size of the carboxylic acids increases, their boiling points increase also, but their solubility in water decreases. These same trends were also seen with the aldehydes and ketones. The solubility in water and the boiling points of several carboxylic acids and alcohols of similar molecular weight are compared in Table 18.7.

TABLE 18.7
Comparison of the Boiling Points and Solubility in Water of Several Carboxylic Acids and Alcohols

Name	Structural formula	Molecular weight	Boiling point (°C)	Solubility in grams per 100 ml H_2O
Formic acid	HCOOH	46	100	All proportions
Ethyl alcohol	CH_3CH_2OH	46	78	All proportions
Acetic acid	CH_3COOH	60	118	All proportions
n-Propyl alcohol	$CH_3CH_2CH_2OH$	60	97	All proportions
Propionic acid	CH_3CH_2COOH	74	141	All proportions
n-Butyl alcohol	$CH_3CH_2CH_2CH_2OH$	74	118	7.9
Butyric acid	$CH_3CH_2CH_2COOH$	88	163	All proportions
n-Pentyl alcohol	$CH_3(CH_2)_3CH_2OH$	88	138	2.3
Valeric acid	$CH_3CH_2CH_2CH_2COOH$	102	187	4.0
n-Hexyl alcohol	$CH_3(CH_2)_4CH_2OH$	102	156	0.6

(Molecular weight: Increases ↓; Boiling point: Increases ↓; Solubility: Decreases ↓)

The first nine members (C_1 through C_9) of the saturated carboxylic acid family are colorless liquids with sharp irritating odors. The odor of vinegar is that of acetic acid, and butyric acid has the unpleasant odor of rancid butter. Carboxylic acids containing more than nine carbon atoms are solids at room temperature, and because of their lower volatility, they have little noticeable odor.

Chemical Properties of Carboxylic Acids

Much of the chemistry of carboxylic acids is centered around their acidic behavior. Of course, carboxylic acids are weak acids and dissociate only to a small extent in solution. The molecular acid exists in equilibrium with the ionized species. In a $0.1M$ solution of formic acid, about 42 of every 1000 molecules of the acid exist as formate and hydronium ions in equilibrium with 958 molecules of formic acid:

$$HCOOH + H_2O \rightleftharpoons HCOO^{\ominus} + H_3O^+$$

Formic acid   Formate ion   Hydronium ion

Solutions of carboxylic acids can turn blue litmus red, neutralize bases, react with active metals to produce hydrogen, and react with carbonate salts to produce carbon dioxide. You may recall these characteristics of acids from the discussion in Chapter 13.

The Formation of Carboxylate Salts

Carboxylic acids can react with bases, active metals (Ca, Mg, Na, K, Al, etc.), sodium carbonate, and sodium bicarbonate to produce salts of the acids. Salts of carboxylic acids are named by replacing the -ic ending of the common or IUPAC name of the acid with -ate. This provides the name of the anion of the salt, which is then written after the name of the cation (the positive ion) of the salt. The salt $CH_3COO^{\ominus} Na^+$ would be named sodium acetate or, in the IUPAC system, sodium ethanoate.

Several reactions of carboxylic acids that produce salts are given below. The common and IUPAC names of each acid and salt are also given. The formulas of the salts will be written to emphasize the existence of the ions, as was done for sodium acetate above. The charge on the carboxylate ion is circled to avoid confusion with the dash symbol used for bonds.

Reactions with Bases

A carboxylic acid will react with base to produce a salt of the acid and water. This is the typical acid–base neutralization reaction:

$$CH_3COOH + KOH \rightarrow CH_3COO^{\ominus}K^+ + H_2O$$

Common name: Acetic acid   Potassium acetate
IUPAC name: Ethanoic acid   Potassium ethanoate

Reactions with Active Metals

Carboxylic acids react with active metals to produce a salt of the acid and hydrogen gas:

$$2HCOOH + Mg \rightarrow (HCOO^{\ominus})_2Mg^{2+} + H_2$$

Common name: Formic acid   Magnesium formate
IUPAC name: Methanoic acid   Magnesium methanoate

Reaction with Sodium Carbonate and Sodium Bicarbonate

Carboxylic acids react with sodium carbonate, Na_2CO_3, and sodium bicarbonate, $NaHCO_3$, to produce the sodium salt of the acid, carbon dioxide gas, and water:

$$2CH_3CH_2COOH + Na_2CO_3 \rightarrow 2CH_3CH_2COO^{\ominus}Na^+ + CO_2 + H_2O$$

Common name: Propionic acid   Sodium propionate
IUPAC name: Propanoic acid   Sodium propanoate

$$\text{C}_6\text{H}_5-\text{COOH} + \text{NaHCO}_3 \longrightarrow \text{C}_6\text{H}_5-\text{COO}^{\ominus}\text{Na}^+ + \text{CO}_2 + \text{H}_2\text{O}$$

Common name: Benzoic acid   Sodium benzoate
IUPAC name: Benzoic acid   Sodium benzoate

    The sodium and potassium salts of all the carboxylic acids have at least some solubility in water. Those of the lower molecular weight acids are quite soluble. The salts of the large fatty acids, sodium stearate, sodium palmitate, and sodium oleate are the principal components of bar soap. The cleansing action of soap is described in Chapter 20.

$$\text{CH}_3(\text{CH}_2)_{16}\text{COO}^{\ominus}\text{Na}^+ \qquad \text{CH}_3(\text{CH}_2)_{14}\text{COO}^{\ominus}\text{Na}^+$$

Sodium stearate   Sodium palmitate

$$\text{CH}_3(\text{CH}_2)_7\text{CH}=\text{CH}(\text{CH}_2)_7\text{COO}^{\ominus}\text{Na}^+$$

Sodium oleate

The Formation of Esters

    The reactions of carboxylic acids with alcohols to produce esters was described in Chapter 17, so the formation of esters will be described only briefly here. The reactions are catalyzed by mineral acids, such as H_2SO_4 or H_3PO_4, and water is eliminated as the ester is formed. The way esters are named, and their importance, will be discussed later on in this chapter. Acetic acid can react with ethyl alcohol to produce ethyl acetate, a sweet-smelling ester with excellent solvent properties:

$$\text{CH}_3-\underset{\text{OH}}{\overset{\text{O}}{\text{C}}} + \text{CH}_3\text{CH}_2\text{OH} \overset{\text{H}^+}{\rightleftharpoons} \text{CH}_3-\underset{\text{OCH}_2\text{CH}_3}{\overset{\text{O}}{\text{C}}} + \text{H}_2\text{O}$$

acetic acid   ethyl alcohol   ethyl acetate (ester)

    The formation of esters is an important reaction in body chemistry. In the body, enzymes catalyze the reactions.

Check Test Number 6

    1. Complete the following equations that describe reactions of carboxylic acids.
      a. $\text{HCOOH} + \text{NaHCO}_3 \rightarrow$
      b. $\text{CH}_3\text{COOH} + \text{Ca} \rightarrow$

      c. $\text{C}_6\text{H}_5-\text{COOH} + \text{NaOH} \longrightarrow$

      d. $\text{CH}_3\text{CH}_2\text{COOH} + \text{CH}_3\text{OH} \overset{\text{H}^+}{\longrightarrow}$

2. Give the common names for the following carboxylate salts. You may need to consult Tables 18.5 and 18.6 for the names of carboxylic acids.
   a. $CH_3CH_2CH_2COO^{\ominus}Na^+$

   b.

3. Give the IUPAC name for the following salts.
   a. $CH_3CH_2CH_2COO^{\ominus}Na^+$
   b. $CH_3(CH_2)_3COO^{\ominus}Na^+$

Answers:

1. a. $HCOOH + NaHCO_3 \rightarrow HCOO^{\ominus}Na^+ + CO_2 + H_2O$
   b. $2CH_3COOH + Ca \rightarrow (CH_3COO^{\ominus})_2Ca^{2+} + H_2$

   c.

   d.

2. a. Sodium butyrate   b. Sodium salicylate
3. a. Sodium butanoate   b. Sodium pentanoate

Some Important Carboxylic Acids

Formic Acid—
HCOOH

Formic acid is a colorless liquid at room temperature with a sharp, penetrating odor. Though a weak acid, it dissociates to a greater extent in solution than the other aliphatic carboxylic acids. Formic acid is excreted by ants and bees, and is responsible for much of the discomfort of bites and stings by these creatures. Formic acid is dangerously corrosive to skin and will produce blisters and severe rash. Commercially, formic acid is used in tanning leather, dyeing wool, and in the manufacture of plastics.

Acetic Acid—
CH3COOH

Acetic acid is a colorless liquid at room temperature. Pure acetic acid freezes at 16°C forming a solid that resembles ice. For this reason, pure acetic acid is often called glacial acetic acid. The solid resembles a glacier resting at the bottom of a partially frozen bottle of the acid. Vinegar is a 5% solution of acetic acid in water, and the sharp odor and sour taste of acetic acid are easily recognized. Acetic acid is used in the manufacture of various acetate fibers and films, dyes, plastics, and many organic compounds. The lead salt of acetic acid,

$Pb(CH_3COO)_2$, is used as an astringent in the treatment of poison oak and poison ivy.

Oxalic Acid—
HOOC–COOH

Oxalic acid is a white, solid dicarboxylic acid. It is most often obtained as the dihydrate, in which one molecule of acid is associated with two molecules of water in the crystal, $HOOC–COOH \cdot 2H_2O$. Oxalic acid is one of the strongest naturally occurring acids, and ingestion of even moderate amounts of the acid can cause severe gastroenteritis with vomiting and diarrhea. The tart taste of rhubarb is due to the presence of oxalic acid. People who have eaten large quantities of rhubarb have developed the symptoms of oxalic acid poisoning described above. Solutions of oxalic acid are used to remove rust stains from porcelain sinks and tubs, and blood stains from fabric. Calcium oxalate, CaC_2O_4, can form in the kidneys, and because of its very low solubility, may result in the development of kidney stones.

Lactic Acid —

$$\underset{\displaystyle CH_3CHCOOH}{\overset{\displaystyle OH}{|}}$$

Lactic acid is produced in milk by the action of lactobacillus bacteria on lactose, the sugar in milk. It is primarily responsible for the sour taste of "sour milk." Lactic acid is produced in muscle tissue during vigorous exercise as a product of the glycolysis of glucose, and it causes much of the muscle soreness after hard physical work. Lactic acid is added to certain foods to impart a sour taste.

Citric Acid —

$$\begin{array}{l} H_2C–COOH \\ | \\ HO–C–COOH \\ | \\ H_2C–COOH \end{array}$$

Citric acid is a solid, tricarboxylic acid found in citrus fruits, and it is largely responsible for their tart taste. In fact, citric acid is widely used as a food additive in candies and beverages to provide some of the same tartness as citrus fruits. Citric acid is a vital compound in the metabolism of food to provide energy for the body. Sodium citrate, $Na_2C_6H_6O_7$, is used as an anticoagulant for blood samples in medical laboratories. Solutions of magnesium citrate, $MgC_6H_6O_7$, are often used as laxatives.

Esters

Esters can be considered derivatives of carboxylic acids in which an organic group replaces the acidic hydrogen of the carboxyl group. The general formula of an ester is

$$R–C\overset{\displaystyle O}{\underset{\displaystyle O–R'}{\diagdown}}$$

where R may be an organic group or hydrogen if the ester is a derivative of formic acid, HCOOH. Esters may be written in condensed form as RCOOR′

or RCO_2R'. The ester functional group is $-\overset{\overset{O}{\|}}{C}-O-\overset{|}{\underset{|}{C}}-$.

The synthesis of esters from carboxylic acids and alcohols has been described earlier in this chapter and in Chapter 17 and will not be reproduced here. You may wish to review the discussion in Chapter 17.

Naming Esters

Any ester can be viewed as the product of a reaction between a carboxylic acid and an alcohol. When deriving the name of an ester, the name of the group bonded to the oxygen atom (the group from the alcohol) appears first in the name, followed by the name of the carboxylic acid that corresponds to the remainder of the ester. The *-ic* ending of the name of the acid is replaced by *-ate,* just as was done when naming salts of these acids. Both the common and IUPAC names are derived in the same way. The only difference is that the IUPAC names of the carboxylic acids are used for the IUPAC names of the esters.

Notice how the common name of the following ester is derived:

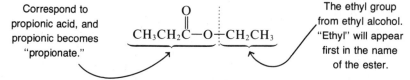

Correspond to propionic acid, and propionic becomes "propionate."

$$CH_3CH_2\overset{\overset{O}{\|}}{C}-O\dashv CH_2CH_3$$

The ethyl group from ethyl alcohol. "Ethyl" will appear first in the name of the ester.

Common name: ethyl propionate
The IUPAC name of the acid is propanoic acid, and the IUPAC name would then be ethyl propanoate.

The name of the following ester is derived in the same manner:

$$\underset{\text{butyrate} \atop \text{or butanoate}}{\underline{CH_3CH_2CH_2\overset{\overset{O}{\|}}{C}-O}}-\underset{\text{methyl (from methyl alcohol)}}{\underline{CH_3}}$$

Common name: methyl butyrate
IUPAC name: methyl butanoate

Several esters with their common and IUPAC names are given below. Examine each name carefully until you can see how it is derived from the formula of the ester. You may wish to refer to the names of the carboxylic acids in Tables 18.5 and 18.6.

	Common Name	IUPAC Name
(benzene ring)—$\overset{\overset{\displaystyle O}{\|\|}}{C}$—O—CH$_2CH_3$	ethyl benzoate	ethyl benzoate
CH$_3\overset{\overset{\displaystyle O}{\|\|}}{C}$—O—CH$_2CH_2CH_3$	n-propyl acetate	n-propyl ethanoate
H—$\overset{\overset{\displaystyle O}{\|\|}}{C}$—O—CH$_2CH_3$	ethyl formate	ethyl methanoate
CH$_3\overset{\overset{\displaystyle O}{\|\|}}{C}$—O—(benzene ring)	phenyl acetate	phenyl ethanoate

—(benzene ring)   is the phenyl group from phenol.

It is not difficult to write the structural formula of an ester given its name. The first part of the name indicates the group attached to the oxygen atom of the ester functional group, and the last part corresponds to the carboxylic acid. The structure of methyl acetate can be derived this way:

$$\underset{\text{—CH}_3 \quad \text{CH}_3\overset{\overset{\displaystyle O}{\|\|}}{C}\text{—O—}}{\overset{\text{methyl}\qquad\qquad\text{acetate}}{}}$$

The structural formula is: CH$_3\overset{\overset{\displaystyle O}{\|\|}}{C}$—O—CH$_3$

Now try naming esters and writing their structural formulas in Check Test Number 7.

Check Test Number 7

1. Give the common names for the following esters.

    a. $CH_3CH_2\overset{\displaystyle O}{\overset{\|}{C}}-OCH_3$ b. 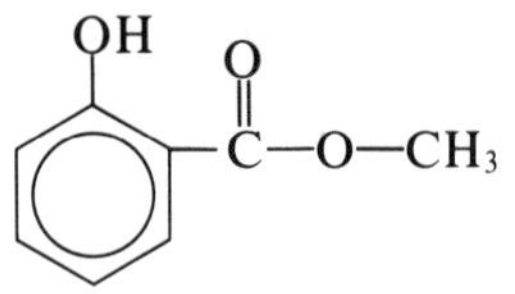 [structure: benzene ring with OH group and $\overset{\displaystyle O}{\overset{\|}{C}}-O-CH_3$ group]

2. Give the IUPAC names for the following esters.

    a. $CH_3-O-\overset{\displaystyle O}{\overset{\|}{C}}-CH_2CH_2CH_3$ b $CH_3CH_2-\overset{\displaystyle O}{\overset{\|}{C}}-O-CH_2CH_3$

3. Write the structural formulas of the following esters.
    a. *n*-Octyl acetate   b. Ethyl ethanoate

Answers:
1. a. Methyl propionate   b. Methyl salicylate
2. a. Methyl butanoate   b. Ethyl propanoate

3. a. $CH_3-\overset{\displaystyle O}{\overset{\|}{C}}-O-CH_2(CH_2)_6CH_3$ b. $CH_3\overset{\displaystyle O}{\overset{\|}{C}}-O-CH_2CH_3$

Physical Properties of Esters

Esters are not as soluble in water as the carboxylic acids of similar molecular weight. About 8 g of ethyl acetate will dissolve in 100 g of water, but butyric acid, which has the same molecular weight, is soluble in water in all proportions. The polar carbonyl group in an ester does not instill enough polarity in the molecule to allow adequate solvation by water. The boiling points of esters are lower than those of both carboxylic acids and alcohols of similar size. In fact, esters generally boil at temperatures closer to those of ethers or alkanes of similar size. A comparison of the boiling point of ethyl acetate with those of similar molecular weight compounds appears below.

Name	Structural formula	Molecular weight	Boiling point (°C)
Ethyl acetate	$CH_3CH_2-O-\overset{\displaystyle O}{\overset{\|}{C}}-CH_3$	88	77
Ethyl *n*-propyl ether	$CH_3CH_2-O-CH_2CH_2CH_3$	88	64
n-Hexane	$CH_3(CH_2)_4CH_3$	86	69
n-Pentyl alcohol	$CH_3(CH_2)_3CH_2OH$	88	138
Butyric acid	$CH_3CH_2CH_2COOH$	88	164

TABLE   18.8
Esters That Are Used as Flavoring Agents

Name	Structure	Flavor
Methyl butyrate	$CH_3{-}O{-}\overset{\displaystyle O}{\overset{\|}{C}}{-}CH_2CH_2CH_3$	Apple
n-Pentyl butyrate	$CH_3(CH_2)_4{-}O{-}\overset{\displaystyle O}{\overset{\|}{C}}{-}CH_2CH_2CH_3$	Apricot
n-Pentyl acetate	$CH_3(CH_2)_4{-}O{-}\overset{\displaystyle O}{\overset{\|}{C}}{-}CH_3$	Banana
n-Octyl acetate	$CH_3(CH_2)_7{-}O{-}\overset{\displaystyle O}{\overset{\|}{C}}{-}CH_3$	Orange
Ethyl butyrate	$CH_3CH_2{-}O{-}\overset{\displaystyle O}{\overset{\|}{C}}{-}CH_2CH_2CH_3$	Pineapple
Isobutyl formate	$CH_3\overset{\displaystyle CH_3}{\overset{\|}{CH}}{-}CH_2{-}O{-}\overset{\displaystyle O}{\overset{\|}{C}}{-}H$	Raspberry
Ethyl formate	$CH_3CH_2{-}O{-}\overset{\displaystyle O}{\overset{\|}{C}}{-}H$	Rum
Methyl salicylate	benzene ring with $\overset{\displaystyle O}{\overset{\|}{C}}{-}O{-}CH_3$ and OH	Wintergreen

Perhaps the most familiar physical property of esters is their pleasant odor. The aromas of many flowers, fruits, and perfumes are due to mixtures of esters. Esters are widely used as food additives to impart specific flavors and aromas to processed foods. A list of several esters that are used in processed foods as flavoring agents appears in Table 18.8.

Chemical Properties of Esters

Hydrolysis of Esters   The most important chemical reaction of esters is their hydrolysis to a carboxylic acid and an alcohol. Hydrolysis, then, is the reverse of their preparation. The reactions are catalyzed by strong acids. Methyl salicylate, the ester with the aroma of wintergreen, can be hydrolyzed to salicylic acid and methyl alcohol:

methyl salicylate   salicylic acid   methyl alcohol

Under laboratory conditions, the acid catalyzed hydrolysis of an ester usually does not go to completion, that is, only part of the ester is converted into the carboxylic acid and the alcohol. The ester, acid, and alcohol reach a state of equilibrium, symbolized by the double arrow in the equation. In living organisms, enzymes catalyze the hydrolysis of esters.

Saponification of Esters   **Saponification** is the base catalyzed hydrolysis of an ester. Usually, enough base is present so that the salt of the carboxylic acid is produced along with the alcohol. Unlike the acid catalyzed hydrolysis reactions, those catalyzed with strong bases (NaOH, KOH) convert all the ester into products. The reactions are not reversible because the salt cannot react with the alcohol to reform the ester. Methyl acetate is converted to sodium acetate and methyl alcohol when heated in the presence of an aqueous solution of sodium hydroxide:

methyl acetate   sodium hydroxide   sodium acetate

An important application of saponification is the hydrolysis of animal fats and vegetable oils to produce soap. In fact, the term saponification comes from the Latin terms, *sapo* (soap) and *facare* (to make), which shows the importance of base hydrolysis in soap making. A fat (triglyceride) is a triester and it can be viewed as the product of a reaction between glycerol and three large carboxylic acids, commonly called fatty acids. Base hydrolysis of the triester reforms the alcohol and produces the salts of the fatty acids. The saponification is carried out by adding the fat to a heated aqueous solution of base.

a typical fat
(a triester)

glycerol

sodium salt of the
fatty acid (a soap)

The Esters of Salicylic Acid

Salicylic acid is a remarkable compound. When taken internally, it has both antipyretic (fever-reducing) and analgesic (pain-killing) properties.

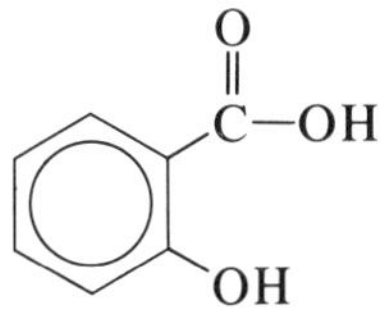

salicylic acid

Unfortunately, the compound is quite irritating to the lining of the stomach, and prolonged use could lead to the development of ulcers. In the late 1800's, ways were sought to modify the structure of salicylic acid to reduce its acidity while retaining its medicinal properties. An extremely successful modification in the 1890's produced the acetate ester of salicylic acid, acetylsalicylic acid. You are surely familiar with this compound by its generic name, aspirin, the most widely used drug in the world:

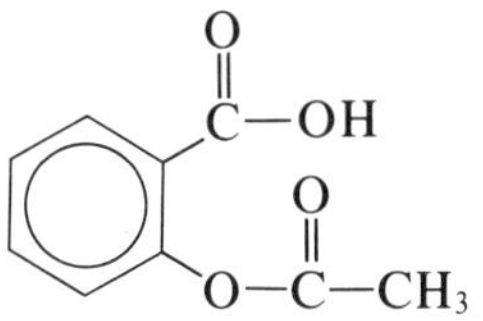

acetylsalicylic acid
(aspirin)

Though aspirin can irritate the lining of the stomach, the effect is much less than that of salicylic acid alone. Aspirin passes through the stomach into the small intestine, where it is hydrolyzed to salicylic acid and absorbed by the body. Notice, the hydroxyl group (–OH) on salicylic acid reacts to form the acetate ester; the carboxylic acid group remains intact.

Methyl salicylate, the methyl ester of salicylic acid, is also used for its analgesic properties. You may recognize that this compound is also used for its wintergreen flavor and aroma. When added to liniments, it is absorbed into the skin and eventually hydrolyzes releasing salicylic acid. The pleasant wintergreen fragrance of many over-the-counter liniment creams is really due to the pain killer it contains.

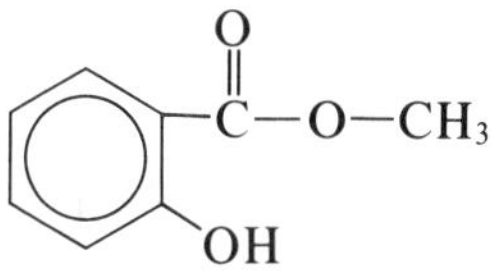

methyl salicylate

Another ester of salicylic acid that is used in medicine is phenyl salicylate, commonly called Salol. It is used as an intestinal antiseptic in veterinary medicine (it hydrolyzes into salicylic acid and phenol in the intestines) and in humans as an analgesic and antipyretic. It is also used to coat pills that are designed to dissolve in the small intestine and not in the stomach. Phenyl salicylate is not affected by the acidic environment of the stomach, but hydrolyzes in the mildly basic condition of the intestine.

phenyl salicylate
(Salol)

Terms

Several important terms appeared in Chapter 18. You should know the meaning of each of the following.

carbonyl group	hemiacetal
aldehyde	hemiketal
ketone	acetal
carboxylic acid	ketal
ester	imine
Tollen's reagent	saponification
Fehling's reagent	cyanohydrin
Benedict's reagent	formalin

Questions

Answers to the starred questions appear at the end of the text.

1. Write the general formulas for an aldehyde, ketone, carboxylic acid, and an ester.
2. Give the common and IUPAC names for the following aldehydes and ketones.

*a. $H-\overset{\overset{\displaystyle O}{\|}}{C}-H$ b. $\underset{H_3C}{\overset{\overset{\displaystyle O}{\|}}{C}}\overset{}{\diagdown}CH_3$

*c. $CH_3\overset{\displaystyle O}{\overset{\|}{C}}-H$ *f $CH_3-\overset{\displaystyle O}{\overset{\|}{C}}-CH_2CH_2CH_3$

*d. $CH_3CH_2-\overset{\displaystyle O}{\overset{\|}{C}}-CH_2CH_3$ g $CH_3CH_2\overset{\displaystyle O}{\overset{\|}{C}}-H$

e   （benzene ring）$-\overset{\displaystyle O}{\overset{\|}{C}}-H$ h $CH_3CH_2CH_2CH_2\overset{\displaystyle O}{\overset{\|}{C}}-H$

3. Give the IUPAC name for the following compounds.

*a. $CH_3\overset{\displaystyle CH_3}{\overset{|}{C}}HCH_2\overset{\displaystyle O}{\overset{\|}{C}}-H$ d $CH_3CH_2-\underset{\underset{\displaystyle CH_3}{|}}{\overset{\overset{\displaystyle H}{|}}{C}}-\overset{\displaystyle O}{\overset{\|}{C}}-H$

*b. $CH_3-\underset{\underset{\displaystyle CH_2}{\underset{\underset{\displaystyle CH_3}{|}}{|}}}{\overset{\overset{\displaystyle H}{|}}{C}}-CH_2-\overset{\displaystyle O}{\overset{\|}{C}}-CH_3$

e $CH_3-\underset{\underset{\displaystyle CH_3}{|}}{\overset{\overset{\displaystyle CH_3}{|}}{C}}-CH_2-\overset{\displaystyle O}{\overset{\|}{C}}-CH_3$

*c. $CH_3\underset{\underset{\displaystyle Br}{|}}{C}H-\overset{\displaystyle O}{\overset{\|}{C}}-CH_2CH_3$ f $CH_3(CH_2)_4\overset{\displaystyle O}{\overset{\|}{C}}-H$

*4. Arrange the following ketones in order of increasing solubility in water.

（benzene ring）$-\overset{\displaystyle O}{\overset{\|}{C}}\diagdown_{CH_3}$ $H_3C-\overset{\displaystyle O}{\overset{\|}{C}}-CH_3$ $CH_3CH_2-\overset{\displaystyle O}{\overset{\|}{C}}-CH_2CH_3$

5. Arrange the following aldehydes in order of increasing boiling points;

$CH_3CH_2CH_2CH_2\overset{\displaystyle O}{\overset{\|}{C}}-H$ $CH_3CH_2\overset{\displaystyle O}{\overset{\|}{C}}H$ $H-\overset{\displaystyle O}{\overset{\|}{C}}-H$

6. Complete the following equations that describe the reactions of aldehydes and ketones.

*a. $CH_3\overset{\displaystyle O}{\overset{\|}{C}}-H \xrightarrow{[o]}$

b. $CH_3CH_2-\overset{\displaystyle O}{\overset{\|}{C}}-CH_3 \; + \; HCN \longrightarrow$

*c. $HCHO \; + \; CH_3OH \xrightarrow{H^+}$

d. $CH_3{-}\underset{\underset{OCH_3}{|}}{\overset{\overset{OH}{|}}{C}}{-}CH_3$ + CH_3OH $\xrightarrow{H^+}$

*e. $CH_3\overset{\overset{O}{\|}}{C}{-}H$ + $2[Ag(NH_3)_2]^+OH^-$ $\longrightarrow$

*f. $CH_3CH_2{-}\overset{\overset{O}{\|}}{C}{-}CH_3$ + H_2 $\xrightarrow[\text{high pressure}]{Pt}$

g. $CH_3\overset{\overset{O}{\|}}{C}{-}H$ + NH_3 $\longrightarrow$

7.   What are two uses of formalin?

*8.   What chemical property of an aldehyde is demonstrated in the reduction of silver ion to silver metal in the Tollen's test?

9.   Complete the following oxidation equations.

*a. $CH_3CH_2CH_2OH$ $\xrightarrow[\text{(one step)}]{[O]}$

*b. $CH_3{-}\underset{\underset{}{}}{\overset{\overset{OH}{|}}{C}}H{-}CH_3$ $\xrightarrow{[o]}$

c. $CH_3\overset{\overset{OH}{|}}{C}HCH_2CH_2CH_3$ $\xrightarrow{[o]}$

d. CH_3OH $\xrightarrow{[o]}$

10.   Give the common and IUPAC names for the following carboxylic acids and esters.

*a. CH_3COOH

b. CH_3CH_2COOH

* c. $CH_3CH_2\overset{\overset{O}{\|}}{C}{-}O{-}CH_2CH_3$

d. 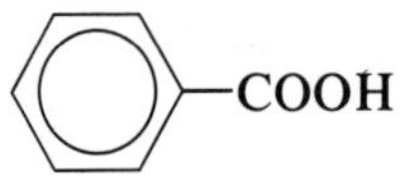 $\bigcirc{-}COOH$

*e. $H{-}\overset{\overset{O}{\|}}{C}{-}O{-}CH_2CH_2CH_2CH_3$

f. $CH_3{-}\underset{\underset{CH_3}{|}}{\overset{\overset{H}{|}}{C}}{-}O{-}\overset{\overset{O}{\|}}{C}{-}CH_3$

11. Give the common and IUPAC names for the following salts of carboxylic acids.

    *a. $CH_3CH_2CH_2CH_2\overset{\displaystyle O}{\overset{\|}{C}}-O^{\ominus}K^+$

    b. $\left(\bigcirc\!\!\!\!-\overset{\displaystyle O}{\overset{\|}{C}}-O^{\ominus}\right)_2 Ca^{2+}$

    c. $H-\overset{\displaystyle O}{\overset{\|}{C}}-O^{\ominus}Na^+$

    d. $\left(CH_3CH_2\overset{\displaystyle O}{\overset{\|}{C}}-O^{\ominus}\right)_2 Mg^{2+}$

12. Complete the following equations describing reactions of the carboxylic acids.
    *a. $CH_3COOH + KOH \rightarrow$
    b. $CH_3CH_2COOH + NaHCO_3 \rightarrow$
    *c. $2HCOOH + Na_2CO_3 \rightarrow$

    d. 2 $\bigcirc\!\!\!\!-COOH + Ca \longrightarrow$

13. Write the structural formulas for the carboxylic acid and the alcohol that must react to prepare the following esters.

    a. $\bigcirc\!\!\!\!-\overset{\displaystyle O}{\overset{\|}{C}}-O-CH_2CH_3$

    *b. $CH_3CH_2CH_2-O-\overset{\displaystyle O}{\overset{\|}{C}}-CH_3$

    c. $CH_3CH_2CH_2CH_2\overset{\displaystyle O}{\overset{\|}{C}}-O-CH_2CH_2CH_2CH_3$

    *d. $\bigcirc\!\!\!\!-\overset{\displaystyle O}{\overset{\|}{C}}-O-\bigcirc$

*14. Why would aspirin tablets that have been exposed to moist air begin to smell like vinegar?

15. What is saponification?

*16. $CH_3-\overset{\displaystyle O}{\overset{\|}{C}}-O-CH_2CH_3 + NaOH \xrightarrow{\Delta}$

17. Write structural formulas for the following compounds.
    a. oxalic acid
    ***b.** 2-methylbutanoic acid
    ***c.** 3,3-dichloropropionaldehyde
    d. 3-hydroxypropanal
    e. citric acid
    f. salicylic acid
    ***g.** *n*-butyl formate
    h. methyl salicylate
    ***i.** 2-hexanone
    ***j.** decanoic acid
    k. sodium stearate

Organic Nitrogen Compounds

Carbon, hydrogen, and oxygen are the most abundant elements in organic compounds, and the previous chapters presented several important classes of compounds that arise from the different ways these elements combine with each other. Nitrogen is the next most abundant element in organic compounds, and many nitrogen-containing substances are of vital importance in the chemistry of the body. Proteins, nucleic acids, and most hormones contain nitrogen, and the properties of these compounds are in large part due to the presence of this element.

In this chapter we will deal with three types of organic compounds that contain nitrogen: the **amines**, the **amides**, and the **nitrogen heterocyclics**. The amines contain alkyl or aryl groups joined to nitrogen by single covalent bonds. The amides contain nitrogen bonded to a carbonyl group. The heterocyclic compounds include nitrogen in rings of atoms.

$$CH_3-\overset{\displaystyle H}{\underset{\displaystyle H}{N}}-H \qquad CH_3-\overset{\displaystyle O}{C}\diagdown_{NH_2}$$

An amine
(methylamine)

An amide
(acetamide)

A nitrogen
heterocyclic
(pyridine)

The discussion of organic nitrogen compounds will begin with the amines, compounds that are derivatives of ammonia. Their preparation and some of their physical and chemical properties will be described, along with the way they are named. Amides can be considered derivatives of the carboxylic acids, and their names reflect this relationship. After describing their properties, several drugs that contain the amide structure will be introduced. Heterocyclic compounds that include nitrogen in rings of atoms are important in body chemistry and in the structures of many drugs. Several five- and six-membered heterocyclic compounds will be described along with biologically important compounds that include these rings in their structures. The chapter will end with a brief description of four classes of nitrogen-containing drugs: amphetamines, barbiturates, narcotics, and the hallucinogens.

Objectives

By the time you finish this chapter you should be able to do the following.

1. Describe the classification of amines.
2. Derive the common and IUPAC names of an amine given its structural formula and vice versa.
3. Describe three ways in which amines can be prepared.
4. Describe the formation of amine salts and the way they are named.
5. Describe the amide functional group, and derive the common and IUPAC names of an amide given its structural formula and vice versa.
6. Describe the solubility characteristics of amines, amine salts, and amides in water.
7. Describe how amides are prepared.
8. Describe the hydrolysis of amides under basic and acidic conditions.
9. Draw the structural formulas of pyrrolidine, pyrrole, indole, imidazole, pyridine, quinoline, pyrimidine, and purine.
10. Describe the classes of drugs: amphetamines, barbiturates, narcotics, and hallucinogens.

Amines

Amines are organic, nitrogen-containing compounds that can be considered derivatives of ammonia, NH_3. Either one, two, or all three of the hydrogen atoms in ammonia can be replaced by an organic group to form an amine, and they may be classed as **primary** (1°), **secondary** (2°), or **tertiary** (3°) based on the number of alkyl or aryl groups (R–) directly bonded to nitrogen.

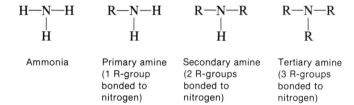

Ammonia	Primary amine (1 R-group bonded to nitrogen)	Secondary amine (2 R-groups bonded to nitrogen)	Tertiary amine (3 R-groups bonded to nitrogen)

Amines may also be classified as either aliphatic or aromatic based on the type of organic groups bonded to nitrogen. An aliphatic amine has only alkyl groups (methyl, ethyl, propyl, etc.) joined to nitrogen along with the required number of hydrogen atoms to fill the three bonds. Aromatic amines have at least one

aromatic (aryl) group (such as the phenyl group, ⬡) bonded to nitro-

gen. A single aromatic group will result in an aromatic amine, even though there may also be alkyl groups present. The following amines are classified as being 1°, 2°, or 3°, and also as aliphatic or aromatic.

$$H-N-CH_3 \qquad H-N-⬡ \qquad CH_3CH_2-N-CH_3 \qquad H-N-CH_2-⬡$$
$$\;|\qquad\qquad\quad\; | \qquad\qquad\qquad\qquad | \qquad\qquad\qquad | $$
$$H \qquad\qquad\quad CH_3 \qquad\qquad\qquad CH_3 \qquad\qquad\quad H$$

1°-aliphatic   2°-aromatic   3°-aliphatic   1°-aliphatic (the aromatic ring is not *directly* bonded to nitrogen)

Naming the Amines   The common names for the aliphatic amines are derived by simply listing the alkyl groups bonded to nitrogen, either alphabetically or in order of increasing size, followed by -**amine**. The names are written as one word, and the prefixes di- and tri- are used if a particular alkyl group appears more than once in the formula. Study the names of the following examples so that you can see how they describe the compound.

$$H-N-CH_3 \qquad H-N-CH_3 \qquad CH_3-N-CH_3$$
$$\;|\qquad\qquad\qquad | \qquad\qquad\qquad | $$
$$H \qquad\qquad\quad CH_3 \qquad\qquad\quad CH_3$$

Methylamine   Dimethylamine   Trimethylamine

$$\underset{\underset{\displaystyle CH_3}{|}}{H-N-CH_2CH_3} \qquad \underset{\underset{\displaystyle CH_3}{|}}{CH_3-N-CH_2CH_3} \qquad \underset{\underset{\displaystyle CH_3}{|}}{CH_3-\overset{\overset{\displaystyle H}{|}}{CH}-N-CH_2CH_3}$$

| Methylethylalmine | Dimethylethylamine | Ethylisopropylamine |

The larger, more complex amines are more conveniently named using the IUPAC system. The $-NH_2$ group is called the **amino group** in the IUPAC system, and it is treated as a substituent on the main chain of the molecule. Methylamine, CH_3NH_2, would be aminomethane in the IUPAC system, though it is most often called by its common name. The following amines are named according to the IUPAC rules. By now you should be sufficiently familiar with the IUPAC system to understand how the names are derived.

$$\underset{\underset{\displaystyle NH_2}{|}}{CH_3-CH-CH_2CH_3} \qquad \underset{\underset{\displaystyle NH_2}{|}}{CH_3-CH-CH_2OH} \qquad \underset{\underset{\displaystyle NH_2}{|}}{CH_3-CH-CH_2}-\underset{\underset{\displaystyle NH_2}{|}}{CH-CH_3}$$

| 2-aminobutane | 2-amino-1-propanol | 2,4-diaminopentane |

In the IUPAC system, $-\overset{\overset{\displaystyle H}{|}}{N}CH_3$ is called the methylamino group, and $-N(CH_3)_2$ is the dimethylamino group. Methylamino and dimethylamino are enclosed in parentheses when included in the name to indicate that both methyl- and amino- are part of the same group.

$$\underset{\underset{\displaystyle H}{\diagdown}\underset{\displaystyle CH_3}{\diagup}}{\overset{\overset{\displaystyle CH_3-CH-CH_3}{|}}{N}} \qquad \underset{\underset{\displaystyle N(CH_3)_2}{|}}{CH_3-CH_2-CH-CH_2OH}$$

2-(methylamino)propane · 2-(dimethylamino)-1-butanol

Aromatic amines are most often given common names as derivatives of aniline, the simplest aromatic amine.

| Aniline | 3-methylaniline or o-methylaniline (o-stands for ortho-) | 3,4-dimethylaniline |

Alkyl or aryl groups bonded to the nitrogen atom in aniline are given a N- or N,N- prefix to indicate that they are bonded to nitrogen.

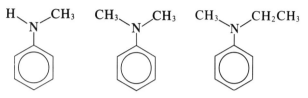

N-methylaniline   N,N-dimethylaniline   N-methyl-N-ethylaniline

Check Test Number 1

1. Give the common name for each of the following amines, and indicate whether it is a primary, secondary, or tertiary amine:

   a. $CH_3CH_2NH_2$

   b.   ⬡—N—CH_3 (with H above N)

   c. $CH_3CH_2CH_2CH_2N$—CH_3 (with CH_3 below N)

   d.   H—N bonded to two phenyl groups

   e. CH_3—N—CH_2CH_3 (with $CH_2CH_2CH_3$ below N)

2. Use the IUPAC system to name the following amines:

   a. CH_3CHCH_3 (with NH_2 below CH)

   b. CH_3—N—$CHCH_2CH_3$ (with CH_3 above N and CH_3 below CH)

   c. $CH_3CHCH_2CH_2OH$ (with N below CH, bonded to H and CH_3)

Answers:
1. a. Ethylamine (1°), b. N-methylaniline (2°), c. dimethyl-*n*-butylamine (3°),
   d. N-phenylaniline (2°), e. methylethyl-*n*-propylamine (3°).

2. a. 2-Aminopropane, b. 2-(dimethylamino)butane, c. 3-(methylamino)-1-butanol.

Physical Properties of Amines

Methylamine, dimethylamine, ethylamine, and trimethylamine are gases at room temperature. Aniline is a colorless liquid as are the other primary amines containing from 3 to 11 carbon atoms. The higher members of the amine family are solids at room temperature. Generally, amines have characteristic fishlike odors. Trimethylamine has a very "fishy" odor. Some amines have odors that are downright offensive. Part of the odor of decaying flesh is due to two amines produced in the breakdown of proteins. Their common names suggest their unusual odors: putrescine ($H_2N-CH_2CH_2CH_2CH_2-NH_2$) and cadaverine ($H_2N-CH_2CH_2CH_2CH_2CH_2-NH_2$).

Primary and secondary amines, because they contain hydrogen bonded to nitrogen, are able to engage in intermolecular hydrogen bonding, though their hydrogen bonds are not as strong as those in alcohols or carboxylic acids. These amines have boiling temperatures between those of the alkanes and alcohols of similar molecular weight. Tertiary amines cannot hydrogen bond, and they boil at temperatures nearer to those of ethers of similar size.

Amines containing fewer than five or six carbon atoms are reasonably soluble in water. Aromatic amines are less soluble than aliphatic amines. Amines are polar molecules and can hydrogen bond with water, but as the alkyl or aryl groups increase in size they become more hydrocarbonlike and their solubility in water decreases markedly.

The aromatic amines are generally toxic compounds and are readily absorbed through the skin. For this reason, they must be handled with caution. Some aromatic amines are known carcinogens (cancer-causing agents).

Preparation of Amines

Amines may be prepared by the reduction of amides, imines, or nitriles. Nitriles are compounds that contain the cyano group, $-C\equiv N$, bonded to carbon. Aniline can be prepared by the reduction of nitrobenzene. In each of the following examples, the reduction is carried out by the addition of hydrogen atoms to the reactant. Several reducing agents are capable of this, but for simplicity, the symbol [H] will be used to indicate the source of hydrogen in the reduction process.

Amines from Amides. Amides may be reduced to form either primary, secondary, or tertiary amines. The degree of substitution on the amide nitrogen determines the class of the amine produced.

$$CH_3-C\overset{O}{\underset{NH_2}{\diagup}} \xrightarrow{[H]} CH_3CH_2-\underset{H}{\overset{|}{N}}-H \ + \ H_2O$$

Acetamide   Ethylamine (1°)

$$CH_3-C\overset{\displaystyle O}{\underset{\underset{\displaystyle H}{\displaystyle N-CH_3}}{\Big\langle}} \xrightarrow{[H]} CH_3CH_2-\underset{\displaystyle H}{N}-CH_3 \ + \ H_2O$$

Methylethylamine (2°)

$$H-C\overset{\displaystyle O}{\underset{\underset{\displaystyle CH_3}{\displaystyle N-CH_3}}{\Big\langle}} \xrightarrow{[H]} CH_3-\underset{\displaystyle CH_3}{N}-CH_3 \ + \ H_2O$$

Trimethylamine (3°)

N,N-dimethylformamide

Amines from Imides. Imides may be reduced to form either primary or secondary amines.

$$CH_3-\overset{\displaystyle N\diagdown H}{\underset{}{\overset{\|}{C}}}-H \xrightarrow{[H]} CH_3-\overset{\displaystyle NH_2}{\underset{}{CH_2}}$$

Ethylamine (1°)

$$CH_3-\overset{\displaystyle N\diagdown CH_3}{\underset{}{\overset{\|}{C}}}-CH_3 \xrightarrow{[H]} CH_3-\overset{\displaystyle H\diagup N\diagdown CH_3}{\underset{}{CH}}-CH_3$$

Methylisopropylamine (2°)

Amines from Nitriles. The reduction of a nitrile produces only primary amines.

$$CH_3C\equiv N \xrightarrow{[H]} CH_3CH_2-\underset{\displaystyle H}{N}-H$$

Acetonitrile

Ethylamine (1°)

Aniline from Nitrobenzene. The reduction of nitrobenzene produces aniline.

$$\underset{\text{Nitrobenzene}}{\overset{\displaystyle NO_2}{\bigcirc}} \xrightarrow{[H]} \underset{\text{Aniline}}{\overset{\displaystyle NH_2}{\bigcirc}} \ + \ 2H_2O$$

Chemical Properties of Amines
Amines are bases since, when dissolved in water, they generate hydroxide ions, $OH^-_{(aq)}$. Like ammonia, they are weak bases, producing a hydroxide ion concentration much lower than that of the amine itself. Only a fraction of the amine molecules in solution react with water at any one time to form ions. As with ammonia, the amine exists in equilibrium with the ions.

$$NH_{3(aq)} + H_2O \rightleftharpoons NH^+_{4(aq)} + OH^-_{(aq)}$$
$$\text{Ammonia} \qquad\qquad \text{Ammonium ion}$$

$$CH_3NH_{2(aq)} + H_2O \rightleftharpoons CH_3NH^+_{3(aq)} + OH^-_{(aq)}$$
$$\text{Methylamine} \qquad\qquad \text{Methylammonium ion}$$

In a $0.1M$ solution of methylamine, about 65 out of every 1000 molecules of the amine exists as methylammonium ions. Compared to ammonia, which has about 13 of every 1000 molecules converted to ammonium ions, methylamine would be considered the stronger base. It is ionized to a greater extent, though it is still a weak base.

Formation of Amine Salts. Since amines are bases, they will react with acids to form salts. Even fairly large amines are soluble in solutions of acids because they can react to form soluble salts.

$$CH_3(CH_2)_9NH_2 + HCl_{(aq)} \rightarrow CH_3(CH_2)_9NH^+_3\,Cl^-_{(aq)}$$
$$\text{Decylamine} \qquad\qquad\qquad \text{Decylammonium chloride}$$

$$\text{(insoluble in water)} \qquad\qquad \text{(soluble salt of the amine)}$$

In living organisms, amines exist primarily as salts.

The formation of an amine salt involves the unshared pair of electrons on the nitrogen atom of the amine. Nitrogen is a Group V element, and it uses three of its five valence electrons to form the three bonds to hydrogen atoms or organic groups in amines. The two unused electrons, the unshared pair, are available to çovalently bond to the hydrogen ion. The positive charge on the hydrogen ion then becomes the positive charge on the organic ammonium ion.

$$
\begin{array}{ccc}
& \overset{\displaystyle \overset{\frown}{\quad}\!\!-H^+}{} & \qquad\qquad H \\
& & \qquad\quad\; \cdot\cdot \\
H_3C\!:\!\ddot{N}\!:\!CH_3 & \longrightarrow & H_3C\!:\!\overset{+}{\ddot{N}}\!:CH_3 \\
\overset{|}{C}H_3 & & \overset{|}{C}H_3
\end{array}
$$

Amine salts are named as derivatives of ammonium salts. The salt NH_4Cl (NH_4^+, Cl^-) is called ammonium chloride. In a similar way, then, CH_3NH_3Cl ($CH_3NH_3^+$, Cl^-) is named methylammonium chloride. This amine salt could be produced by bubbling methylamine into a solution of hydrochloric acid.

$$CH_3NH_{2(g)} + HCl_{(aq)} \longrightarrow CH_3NH_3^+, Cl^-_{(aq)}$$

Methylamine ·   Methylammonium chloride

Other amine salts are just as easily named.

$$(CH_3)_3NH^+, Cl^- \qquad\qquad CH_3CH_2-\overset{+}{N}H_2, Br^-$$

            |
            CH_3

trimethylammonium chloride

methylethylammonium bromide

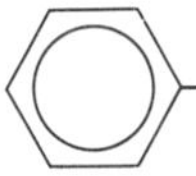

$\langle\bigcirc\rangle$—NH_3^+ is the anilinium ion, so $\langle\bigcirc\rangle$—NH_3^+, Br^- would be called anilinium bromide.

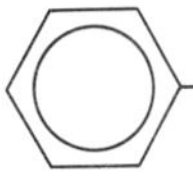

There is yet another kind of amine salt, that in which four organic groups are bonded to nitrogen. They are called **quaternary ammonium salts.** An example of a quaternary salt is tetramethylammonium chloride.

$$\left[CH_3-\overset{\overset{\displaystyle CH_3}{|}}{\underset{\underset{\displaystyle CH_3}{|}}{N^+}}-CH_3 \right] Cl^-$$

Tetramethylammonium chloride

Dilute solutions of certain quaternary ammonium salts are used as germicides for sterilizing medical instruments. They are also used in certain mouthwashes to reduce the surface tension of the solution and increase its ability to foam when used. The quaternary salt used in Scope® mouthwash is shown below. It is present at a 0.005% concentration, and even this very dilute solution exhibits some anti-infective properties.

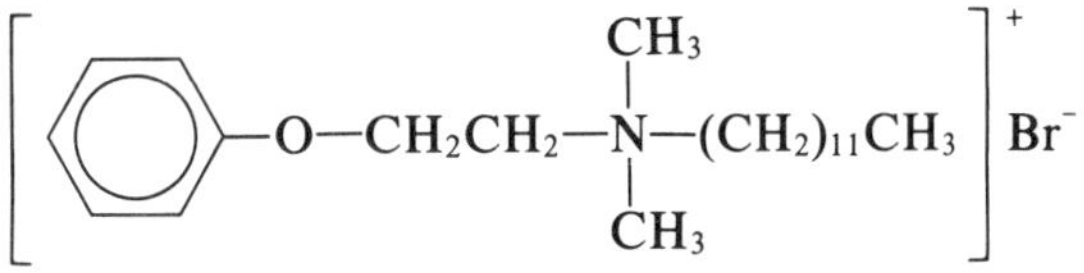

$$\left[\langle\bigcirc\rangle-O-CH_2CH_2-\overset{\overset{\displaystyle CH_3}{|}}{\underset{\underset{\displaystyle CH_3}{|}}{N}}-(CH_2)_{11}CH_3 \right]^+ Br^-$$

Two important quaternary compounds in the body are choline and acetylcholine, an ester of choline. Choline is required in the formation of cellular membranes, and acetylcholine is important in nerve transmission.

$$\left[HOCH_2CH_2-\overset{\overset{\displaystyle CH_3}{|}}{\underset{\underset{\displaystyle CH_3}{|}}{N}}-CH_3 \right]^+ OH^- \qquad \left[CH_3-\overset{\overset{\displaystyle O}{||}}{C}-O-CH_2CH_2-\overset{\overset{\displaystyle CH_3}{|}}{\underset{\underset{\displaystyle CH_3}{|}}{N}}-CH_3 \right]^+ OH^-$$

Choline   Acetylcholine

The quaternary amines and the amine salts have properties that are quite different from the neutral amines. They are ionic compounds, and are very soluble in water. They are solids at room temperature and have no odor. The custom of putting lemon juice or vinegar on fish simply converts the amines in fish (responsible for the fishy odor) into ammonium salts which have no odor. Both lemon juice and vinegar contain acids.

Check Test Number 2

1. Give the common name for the following:

   a. $\left[\begin{array}{c} CH_3CH_2NH_2 \\ | \\ CH_3 \end{array} \right]^+ I^-$

   c. $\left[\text{(benzene ring)} NH_3 \right]^+ Br^-$

   b. $\left[\begin{array}{c} CH_3 \\ | \\ CH_3CH_2-N-CH_3 \\ | \\ CH_3 \end{array} \right]^+ Cl^-$

2. Complete the following equations:

   a. $CH_3CH_2CH_2NH_2 + HBr \rightarrow$

   b. $CH_3-C\overset{\displaystyle O}{\underset{\displaystyle \underset{H}{\overset{|}{N}}-CH_3}{\diagdown}} \xrightarrow{[H]}$

   c. $CH_3CH_2NH_{2(aq)} + H_2O \rightleftharpoons$

Answers:

1. a. Methylethylammonium iodide, b. trimethylethylammonium chloride, c. anilinium bromide.

2. a. $\rightarrow CH_3CH_2CH_2NH_3^+ Br^-$, b. $\rightarrow CH_3CH_2\overset{\displaystyle H}{\overset{\displaystyle |}{N}}CH_3 + H_2O$,

   c. $\rightleftharpoons CH_3CH_2NH_{3(aq)}^+ + OH_{(aq)}^-$.

Amides

Amides are organic compounds with the functional group

$$-C\overset{\displaystyle O}{\underset{\displaystyle \underset{|}{N}-}{\diagdown}}$$

Amides can be considered derivatives of carboxylic acids in which the –OH group of the carboxyl is replaced by an amino group: –NH$_2$, –NHR, or –NR$_2$. The following compounds are amides. Those with organic groups (R–) bonded to the nitrogen atom are considered substituted amides.

The bond joining the nitrogen atom to the carbon atom of the carbonyl group is called the **amide linkage** or **amide bond.** It is a strong covalent bond, and is found in proteins and certain synthetic fibers such as nylon.

Naming the Amides

The amides are named as derivatives of carboxylic acids. The **-ic** or **-oic** ending of the name of the acid is dropped, and replaced by **-amide.** The name is written as one word. For example, the amide CH$_3$–C–NH$_2$ would be named as a derivative of acetic acid. It would be called acetamide (acet- + -amide = acetamide). This would be the common name for the amide, since the common name of the acid is used. The IUPAC name would be ethanamide, derived from the IUPAC name of the acid, ethanoic acid (ethan- + -amide = ethanamide). The common and IUPAC names of several carboxylic acids appear in Tables 18.5 and 18.6. You may wish to refer to these tables for names of carboxylic acids. Two other amides are named below.

CH$_3$CH$_2$CH$_2$C—NH$_2$

| Common name: | butyramide |
| IUPAC name: | butanamide |

C—NH$_2$

| Common name: | benzamide |
| IUPAC name: | benzamide |

The substituted amides are named in much the same way, as derivatives of the carboxylic acids. The organic groups bonded to the amide nitrogen must be included in the name and clearly shown to be bonded to nitrogen. This is accomplished by using the prefixes N- and N,N- before the name of the group or groups such as N-methyl, N,N-dimethyl, and N-methyl-N-ethyl. The group names with the N– prefixes are placed before the name of the amide. As before, the name is written as one word. Several substituted amides appear below with their common and IUPAC names. Study each one carefully then try to name the amides in Check Test Number 3.

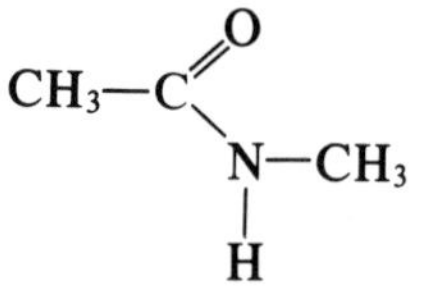

Common: N-methylacetamide
IUPAC: N-methylethanamide

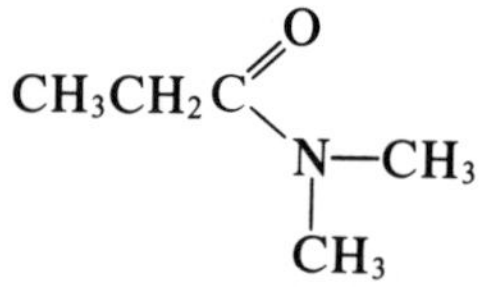

Common: N, N-dimethylpropionamide
IUPAC: N, N-dimethylpropanamide

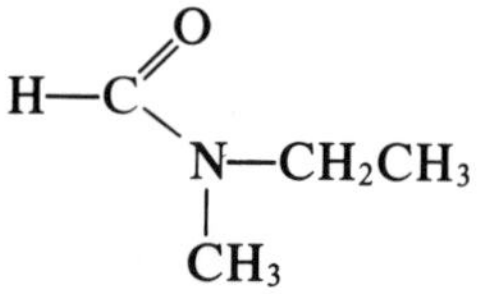

Common: N-methyl-N-ethylformamide
IUPAC: N-methyl-N-ethylmethanamide

In the following example, the importance of using the N- prefix to designate the group on the amide nitrogen is shown.

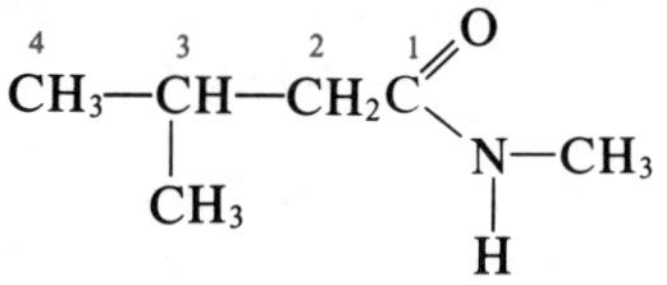

Common: N-methyl-3-methylbutyramide
IUPAC: N-methyl-3-methylbutanamide

Check Test Number 3

Derive the common and IUPAC names for the following amides:

a.

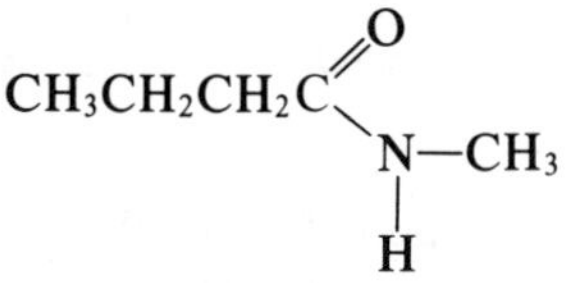

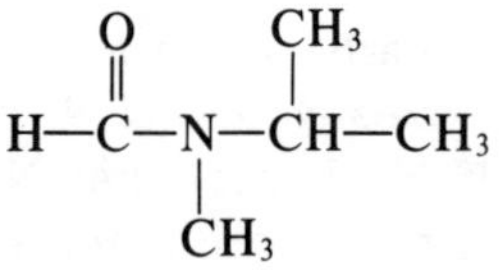

c.

b.

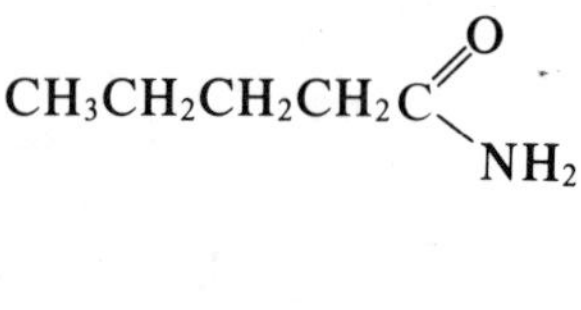

Answers:

	Common name	IUPAC name
a.	N–methylbutyramide	N–methylbutanamide
b.	N–methyl–N– isopropylformamide	N–methyl–N– isopropylmethanamide
c.	valeramide	pentanamide

Physical Properties of Amides

Amides are either liquids or solids at room temperature. Formamide is the only liquid, unsubstituted amide, the rest are solids. Most of the mono- and disubstituted amides are solids. Amides that contain fewer than about six carbon atoms are reasonably soluble in water. The smaller amides are very soluble in water, a property arising from the ability of amides to hydrogen bond with water.

Amides are also able to engage in intermolecular hydrogen bonding, which is largely responsible for their unusually high melting and boiling points compared to other organic compounds of similar molecular weight.

The disubstituted amides are unable to engage in intermolecular hydrogen bonding with themselves and have lower melting and boiling points than the non-substituted amides. A comparison of N,N-dimethylformamide and propionamide, two amides of the same molecular weight, shows the influence of hydrogen bonding on these physical properties.

$$H-C\overset{\displaystyle O}{\underset{\underset{\displaystyle CH_3}{|}}{\diagdown N-CH_3}} \qquad\qquad CH_3CH_2C\overset{\displaystyle O}{\diagdown NH_2}$$

Cannot engage in intermolecular hydrogen bonding	Can engage in intermolecular hydrogen bonding
MW = 73	MW = 73
m.p. = −60°C	m.p. = 81°C
b.p. = 153°C	b.p. = 213°C

The Preparation of Amides   Amides are most conveniently prepared by reacting ammonia or primary and secondary amines with derivatives of carboxylic acids. The reactions with the carboxylic acids themselves are usually quite slow and more difficult to carry out.

A carboxylic acid derivative that is frequently used is the **acid chloride,** in which the −OH group of the carboxyl group is replaced by chlorine.

$$R-C\overset{\displaystyle O}{\diagdown Cl}$$

An acid chloride

Acid chlorides are most often prepared by reacting a carboxylic acid with thionyl chloride, $SOCl_2$, in a suitable organic solvent that is free of any trace of water. Water must be avoided since it reacts with thionyl chloride.

$$R-C\overset{\displaystyle O}{\diagdown OH} + SOCl_2 \longrightarrow R-C\overset{\displaystyle O}{\diagdown Cl} + SO_2 + HCl$$

Carboxylic   Thionyl   Acid chloride
acid   chloride

The reaction of ammonia or an amine with the acid chloride takes place quickly, producing the amide and an ammonium salt. The formation of the amide produces hydrogen chloride, HCl, which then combines with the ammonia or amine to form the chloride salt. For this reason two moles of ammonia or amine are required for each mole of acid chloride.

$$R-C\overset{\displaystyle O}{\diagdown Cl} + 2NH_3 \longrightarrow R-C\overset{\displaystyle O}{\diagdown NH_2} + NH_4Cl$$

Acid chloride   Amide   Ammonium
chloride

Acetamide can be prepared by reacting ammonia with the acid chloride of acetic acid, acetyl chloride.

$$CH_3C\overset{\displaystyle O}{\underset{\displaystyle Cl}{\diagdown}} + 2NH_3 \longrightarrow CH_3-C\overset{\displaystyle O}{\underset{\displaystyle NH_2}{\diagdown}} + NH_4Cl$$

Acetyl chloride   Acetamide

N-methylacetamide can be prepared using methylamine, a primary amine. The salt, methylammonium chloride, is also formed.

$$CH_3C\overset{\displaystyle O}{\underset{\displaystyle Cl}{\diagdown}} + 2CH_3NH_2 \longrightarrow CH_3C\overset{\displaystyle O}{\underset{\displaystyle N-CH_3}{\diagdown}} + CH_3NH_3^+, Cl^-$$

Acetyl chloride   Methylamine   H   Methylammonium chloride

N-methylacetamide

N,N-dimethylacetamide requires the use of dimethylamine, a secondary amine. Dimethylammonium chloride is also formed.

$$CH_3C\overset{\displaystyle O}{\underset{\displaystyle Cl}{\diagdown}} + 2CH_3-\underset{\displaystyle CH_3}{\overset{\displaystyle H}{N}} \longrightarrow CH_3C\overset{\displaystyle O}{\underset{\displaystyle N-CH_3}{\diagdown}} + (CH_3)_2NH_2^+, Cl^-$$

Acetyl chloride   Dimethylamine   CH_3   Dimethylammonium chloride

N,N-dimethylacetamide

Chemical Properties of Amides

Aqueous solutions of amides are neutral, unlike solutions of amines, which are basic ($pH > 7$). Amides are stable in aqueous solution, and they do not hydrolyze (react with water) even at elevated temperatures. However, the addition of acid or base to the amide solution will cause hydrolysis to occur readily. Hydrolysis cleaves the carbonyl–nitrogen bond (the amide bond). The $-OH$ group from water bonds to the carbonyl to produce a carboxylic acid. The hydrogen from water, $-H$, bonds to the nitrogen atom to form either ammonia or an amine.

Water   Amide bond breaks   Carboxylic acid   Ammonia or amine

Hydrolysis under acidic or basic conditions will result in one of the two products appearing as a salt.

Acidic hydrolysis of an amide will produce a carboxylic acid and an ammonium salt. For example, the hydrolysis of acetamide in the presence of hydrochloric acid produces acetic acid and ammonium chloride. The ammonia produced in the hydrolysis immediately reacts with hydrochloric acid to produce the salt.

$$CH_3-C\underset{NH_2}{\overset{O}{\diagup}} + H_2O + HCl \xrightarrow{\Delta} CH_3-C\underset{OH}{\overset{O}{\diagup}} + NH_4Cl$$

Acetamide   Acetic acid   Ammonium chloride

The acid hydrolysis of N-methylformamide produces formic acid and the methylammonium salt.

$$H-C\underset{\underset{H}{N-CH_3}}{\overset{O}{\diagup}} + H_2O + HCl \xrightarrow{\Delta} H-C\underset{OH}{\overset{O}{\diagup}} + CH_3\overset{+}{N}H_3, Cl^-$$

N-methylformamide   Formic acid   Methylammonium chloride

Under basic conditions, hydrolysis of an amide produces a salt of the carboxylic acid, and the ammonia or amine is formed as a neutral compound. The hydrolysis of acetamide in the presence of aqueous sodium hydroxide produces sodium acetate and ammonia.

$$CH_3C\underset{NH_2}{\overset{O}{\diagup}} + NaOH_{(aq)} \xrightarrow{\Delta} CH_3-C\underset{O^\ominus, Na^+}{\overset{O}{\diagup}} + NH_3$$

Acetamide   Sodium acetate   Ammonia

Under the same conditions, N-methylformamide produces the salt of the acid, sodium formate, and methylamine.

$$H-C\underset{\underset{H}{N-CH_3}}{\overset{O}{\diagup}} + NaOH_{(aq)} \xrightarrow{\Delta} H-C\underset{O^\ominus, Na^+}{\overset{O}{\diagup}} + CH_3NH_2$$

N-methylformamide   Sodium formate   Methylamine

Proteins are polymers of amino acids joined together through amide bonds. The digestion of a protein in the body involves the enzyme-catalyzed hydrolysis of the amide bonds to produce the amino acids, which are then absorbed. Notice the amide bonds in the short section of a protein shown below. Each is hydrolyzed in digestion. Proteins and the digestion of proteins are discussed in Chapter 22.

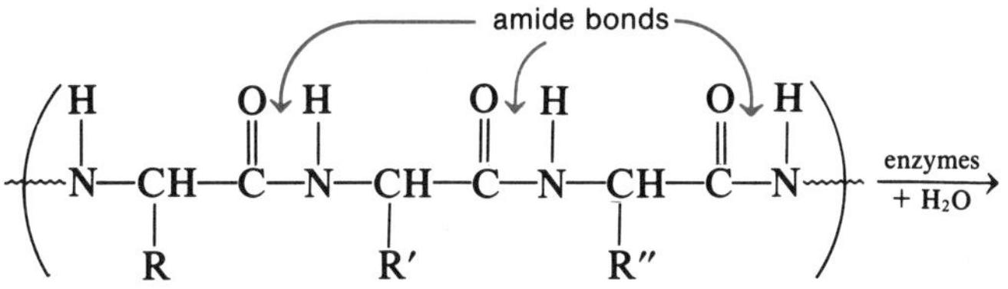

Section of protein

$$H_2N{-}\underset{\displaystyle R}{\underset{|}{CH}}{-}COOH + H_2N{-}\underset{\displaystyle R'}{\underset{|}{CH}}{-}COOH + H_2N{-}\underset{\displaystyle R''}{\underset{|}{CH}}{-}COOH$$

Amino acids

Check Test Number 4

1. Write the equations showing the preparation of N-methylacetamide starting with acetic acid. You will first need to produce the acid chloride.
2. Write the equations showing the hydrolysis of N,N-dimethylacetamide (a) in the presence of hydrochloric acid and (b) in the presence of sodium hydroxide.

Answers:

1. Acid chloride formation:

$$CH_3{-}C(=O){-}OH + SOCl_2 \longrightarrow CH_3{-}C(=O){-}Cl + SO_2 + HCl$$

Reaction of acid chloride with methylamine:

$$CH_3C(=O)Cl + 2CH_3NH_2 \longrightarrow CH_3C(=O){-}N(H){-}CH_3 + CH_3NH_3^+, Cl^-$$

2. Acidic hydrolysis:

$$CH_3-\overset{\displaystyle O}{\underset{\displaystyle \underset{\displaystyle CH_3}{|}}{\underset{\displaystyle N-CH_3}{\big\backslash}}}C \quad + \quad H_2O \quad + \quad HCl \quad \xrightarrow{\Delta} \quad CH_3-\overset{\displaystyle O}{C}\overset{\displaystyle \diagup}{\underset{\displaystyle OH}{\diagdown}} \quad + \quad (CH_3)_2NH_2^+, Cl^-$$

Basic hydrolysis:

$$CH_3-\overset{\displaystyle O}{\underset{\displaystyle \underset{\displaystyle CH_3}{|}}{\underset{\displaystyle N-CH_3}{\big\backslash}}}C \quad + \quad NaOH_{(aq)} \quad \xrightarrow{\Delta} \quad CH_3-\overset{\displaystyle O}{C}\overset{\displaystyle \diagup}{\underset{\displaystyle O^{\ominus}, Na^+}{\diagdown}} \quad + \quad (CH_3)_2NH$$

Some Important Amides

Urea. Urea is a diamide, having two amino groups bonded to a carbonyl group.

$$\underset{\text{Urea}}{H-\overset{\displaystyle \overset{\displaystyle H}{|}}{N}-\overset{\displaystyle \overset{\displaystyle O}{\|}}{C}-\overset{\displaystyle \overset{\displaystyle H}{|}}{N}-H}$$

Urea is a white solid at room temperature and it is very soluble in water. It is a waste product of protein metabolism, and a healthy adult will excrete about 30 g of urea in the urine each day.

Commercially, urea is used widely as a fertilizer since it can provide nitrogen to the soil. It is also used in the synthesis of drugs and plastics. Urea is prepared in industry by reacting carbon dioxide with ammonia under high pressure and in the presence of a catalyst.

$$CO_2 \quad + \quad 2NH_3 \quad \xrightarrow[\Delta, \text{ high pressure}]{\text{catalyst}} \quad H_2N-\overset{\displaystyle \overset{\displaystyle O}{\|}}{C}-NH_2 \quad + \quad H_2O$$

Nylon. Nylon is a synthetic polymer that contains many amide linkages. It is formed by reacting a diamine (two amino groups per molecule) with a diacid chloride (two acid chloride units per molecule). Nylon 66, the polymer used in many nylon fabrics, is formed by combining a diamide with six carbons with a diacid chloride with six carbons (6 + 6 = Nylon 66).

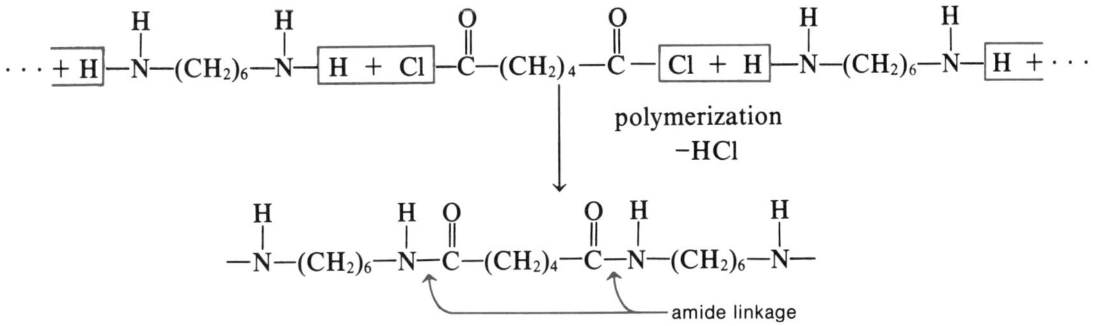

$$\cdots + \boxed{H} - N - (CH_2)_6 - N - \boxed{H + Cl} - \overset{O}{\underset{}{C}} - (CH_2)_4 - \overset{O}{\underset{}{C}} - \boxed{Cl + H} - N - (CH_2)_6 - N - \boxed{H} + \cdots$$

polymerization
−HCl

$$- N - (CH_2)_6 - N - \overset{O}{\underset{}{C}} - (CH_2)_4 - \overset{O}{\underset{}{C}} - N - (CH_2)_6 - N -$$

amide linkage

Nylon 66

Nylon is a **polyamide**, so called because it contains repeating amide units.
Xylocaine. Xylocaine is a local anesthetic commonly used by dentists and physicians to block pain. It is also known as lidocaine. Notice the amide bond in the structural formula.

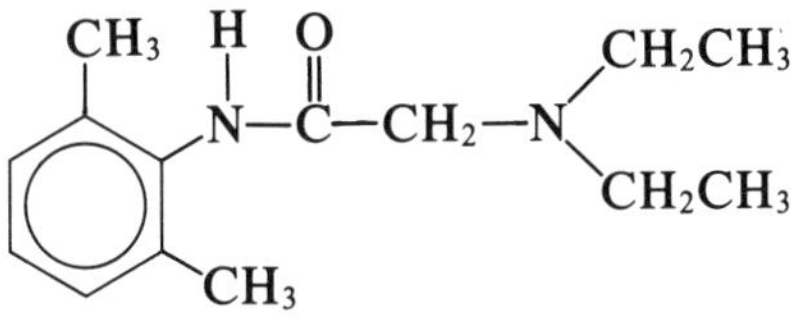

Xylocaine

The injectable form of xylocaine is an aqueous solution of the chloride salt. Xylocaine is not soluble in water, but the salt is very much so.

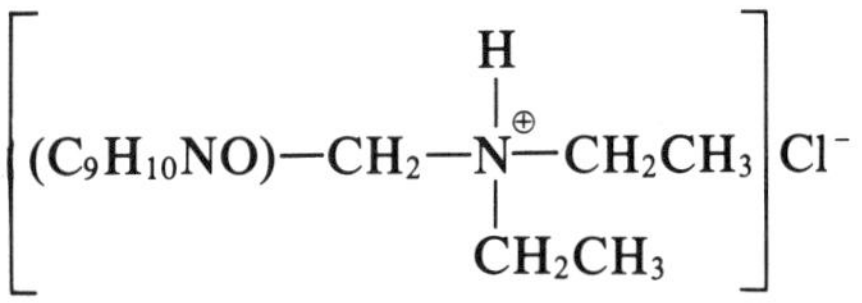

$$\left[(C_9H_{10}NO) - CH_2 - \overset{H}{\underset{CH_2CH_3}{\overset{|}{N}}} - CH_2CH_3 \right] Cl^-$$

Acetanilide and its Derivatives. Acetanilide is an aromatic amide that, at one time, was used as an antipyretic and an analgesic. However, because of its toxicity, its use has been supplanted by other compounds such as aspirin or derivatives of acetanilide.

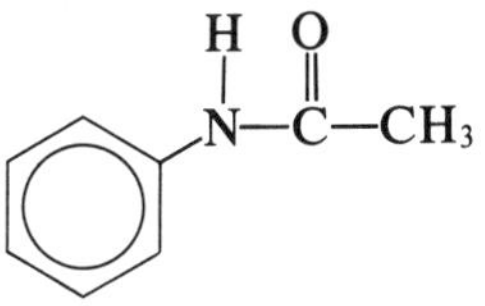

Acetanilide

Phenacetin and acetaminophen are derivatives of acetanilide, and both have been used as analgesics and antipyretics for years in many over-the-counter medicines.

Phenacetin   Acetaminophen

In recent years, evidence has suggested that phenacetin may cause anemia as well as liver and kidney damage, and its use in over-the-counter medicines has been greatly reduced. For many years phenacetin was combined with aspirin and caffeine and sold as a pain reliever commonly known as APC, though it was available under various trade names as well. Phenacetin has been replaced by acetaminophen in the last few years in several of these combination analgesics. Supposedly, caffeine acts as a mild stimulant in these preparations, but recent findings suggest that caffeine counteracts the antipyretic property of aspirin. It would be better to use simple aspirin to reduce a fever.

Acetaminophen is the active ingredient in Tylenol® and Datril®, and it is often used by people who cannot take aspirin. It is roughly equivalent to aspirin in its ability to reduce pain and fever, though it does not have the anti-inflammatory property of aspirin.

Heterocyclic Nitrogen Compounds

Nitrogen atoms are commonly included in ring systems. You may recall that a heterocyclic compound contains atoms of two or more different elements joined together to form a ring of atoms. Heterocyclic compounds with carbon and oxygen atoms in rings were seen in Chapter 17. They were the cyclic ethers. Now, we are going to look at some heterocyclic compounds that contain carbon and nitrogen atoms in rings. These kinds of ring systems are very common in living organisms, and they appear in many drugs as well. Most often, nitrogen-containing heterocyclics exist as five- and six-membered rings, so these will be the ones we will be concerned with here. They are all identified by common names.

The Five-Membered Rings

The important five-membered heterocyclic compounds we will describe are: pyrrolidine, pyrrole, indole, and imidazole.

Pyrrolidine. Pyrrolidine is a cyclic, secondary amine. The pyrrolidine structure is found in many compounds, notably in the amino acid proline and in nicotine.

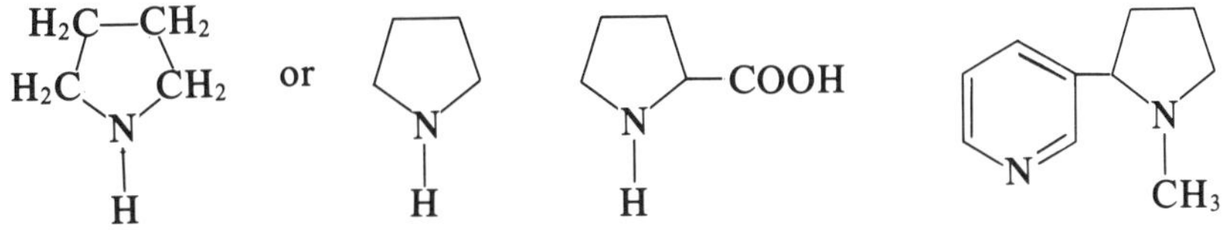

Pyrrolidine   Proline   Nicotine

Pyrrole. Pyrrole is an unsaturated, cyclic secondary amine. The pyrrole structure is found in several compounds of biological importance such as chlorophyll, hemoglobin, vitamin B_{12}, and the cytochromes. In each of these compounds there are structures composed of four pyrrole units joined together which surround a metal ion. The heme group in hemoglobin reveals the arrangement of the four pyrrole units in a plane surrounding an Fe^{2+} ion.

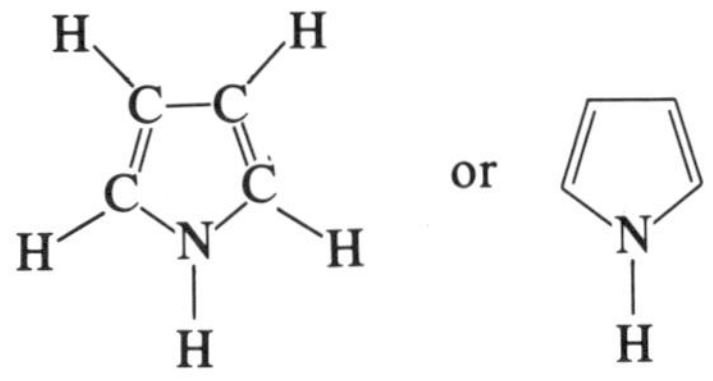

Pyrrole

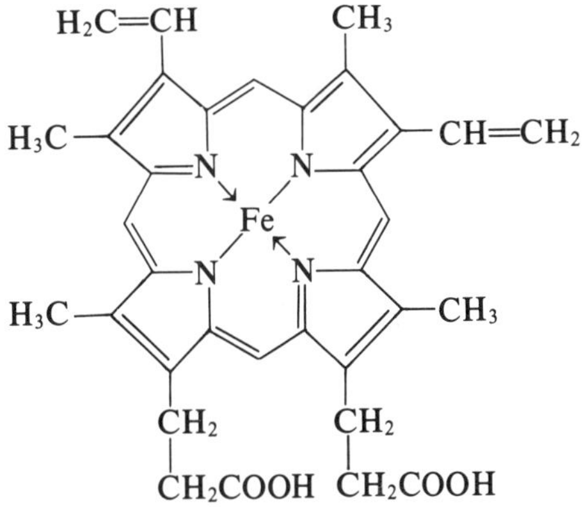

The heme group of hemoglobin

Indole. Indole contains a five-membered heterocyclic ring fused to a benzene ring. It is part of tryptophan, an amino acid, and is incorporated in the structures of many naturally occurring drugs. Serotonin, a hormone, is a derivative of indole. It is believed to be important in brain chemistry.

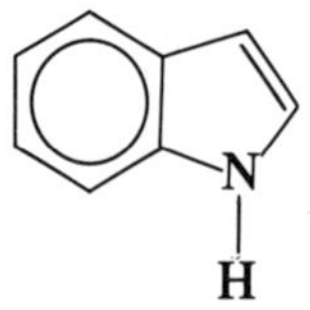

Indole

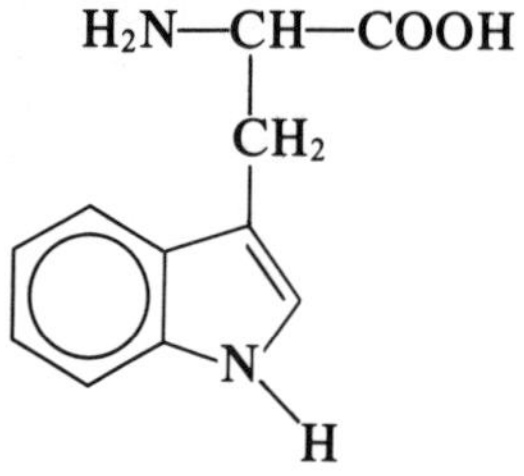

Tryptophan

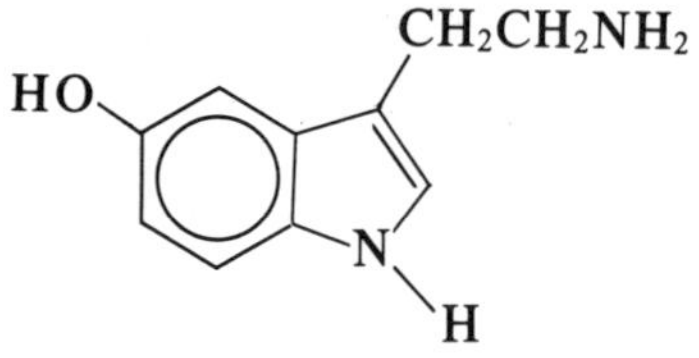

Serotonin

Imidazole.   Imidazole is a five-membered heterocyclic compound containing two nitrogen atoms in the ring. The imidazole structure is found in histidine, an amino acid.

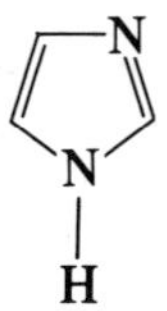

Imidazole

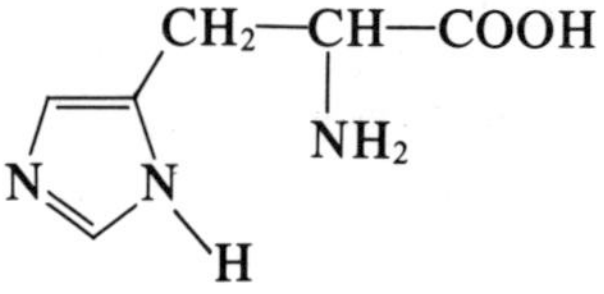

Histidine

The Six-Membered Rings

There are four important heterocyclic structures that involve six-membered rings: pyridine, quinoline, pyrimidine, and purine.

Pyridine.   The structural formula of pyridine resembles that of benzene except, of course, pyridine contains one nitrogen atom in the ring. Pyridine is obtained from coal tar, a liquid mixture that is baked out of coal as it is heated in the absence of oxygen. It is widely used in industry as a solvent and reactant in the synthesis of plastics, drugs, dyes, and other materials. It is also found incorporated into many biological compounds such as nicotine, seen earlier, and the vitamins B_3 (niacin) and B_6.

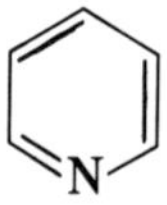

Pyridine

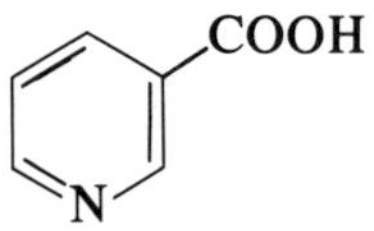

Nicotinic
acid

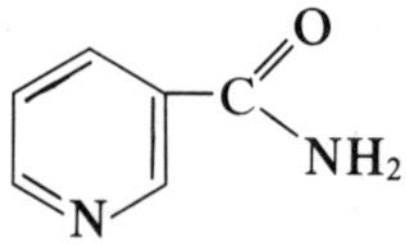

Nicotinamide

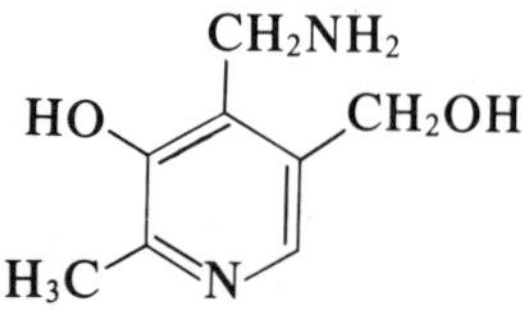

Pyridoxamine
one of the B_6
vitamins

The B_3 vitamins

Quinoline. Quinoline contains two rings fused together, one a benzene ring and the other a pyridine ring. The quinoline structure is found in some naturally occurring drugs such as quinine. Quinine, which was first isolated from the bark of the cinchona tree, is used in the treatment of malaria.

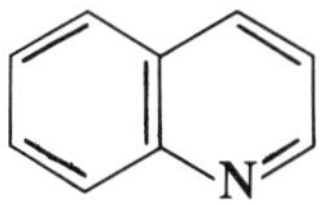

Quinoline

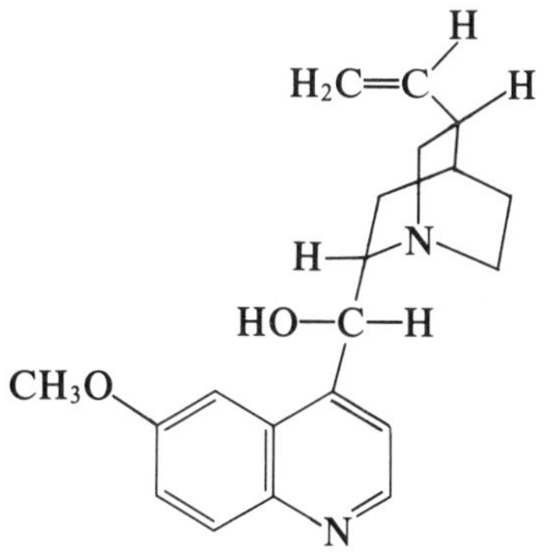

Quinine

Pyrimidine. Pyrimidine is a six-membered ring that includes two nitrogen atoms. It is part of the structure of thiamine (vitamin B_1), and is incorporated in the structures of the nucleic acids, DNA and RNA.

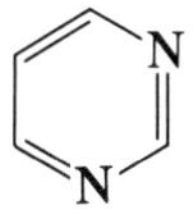

Pyrimidine

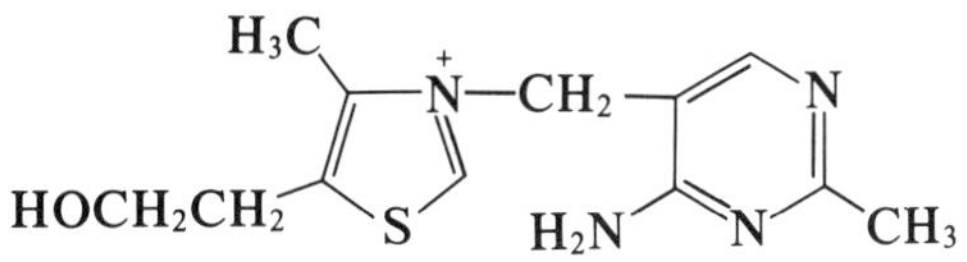

Thiamine (vitamin B_1)

Purine. Purine contains two heterocyclic rings fused together. Derivatives of purine are found in the nucleic acids, DNA and RNA, and in adenosine triphosphate (ATP), a compound used in cells to store and transfer energy for cellular operation. Caffeine is also a derivative of purine.

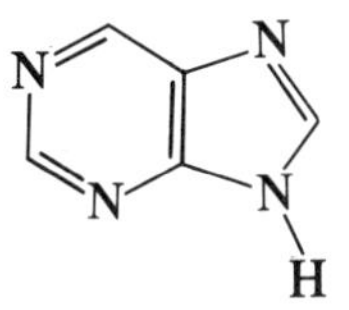

Purine

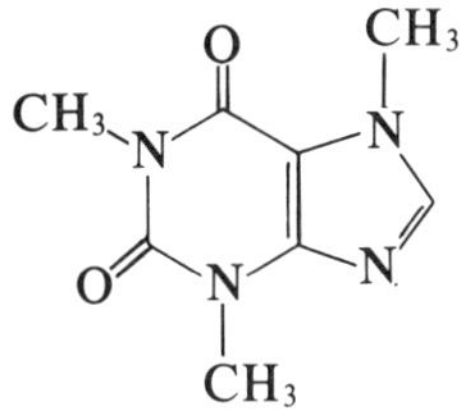

Caffeine

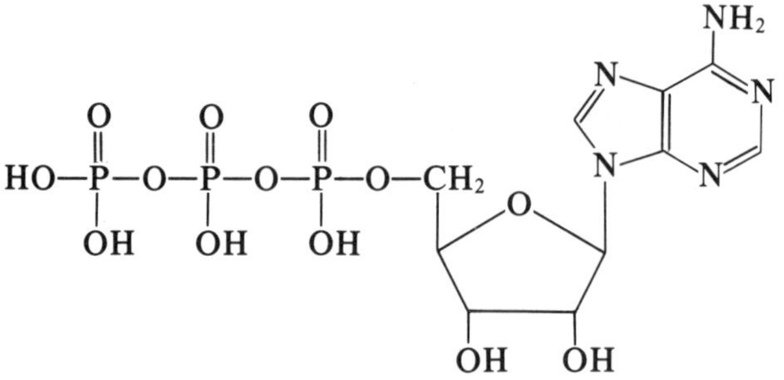

Adenosine triphosphate (ATP)

Some Nitrogen-Containing Drugs

For centuries drugs have been obtained from mushrooms, herbs, bark, roots, flowers, and leaves to induce sleep, relieve pain, poison game, or to see visions. Many of these drugs have been isolated and found to be complex heterocyclic amines. Because of their mild alkaline properties and plant origin, they are all classed as **alkaloids.** Many of these alkaloids have been synthesized in the laboratory or modified to enhance their properties. Nicotine, quinine, and caffeine are alkaloids, and their structures appeared in the previous section. Many other drugs are synthetic, such as the amphetamines and barbiturates. Let us look at these two classes of drugs first, then some narcotic and hallucinogenic drugs will be described.

Amphetamines   The **amphetamines** are nonalkaloid drugs synthesized to mimic the stimulant action of epinephrine (adrenalin) and norepinephrine on the central nervous system. Amphetamines apparently act by interfering with the use of certain compounds involved in the transmission of nerve impulses. The structural formula of amphetamine is shown below, along with those of epinephrine and norepinephrine so that you can see the similarity in their structures. Amphetamine has a rather simple structure compared to many other drugs. Amphetamine is marketed under the name Benzedrine® (Smith, Kline & French).

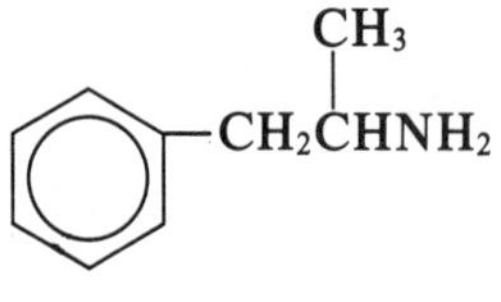

Amphetamine

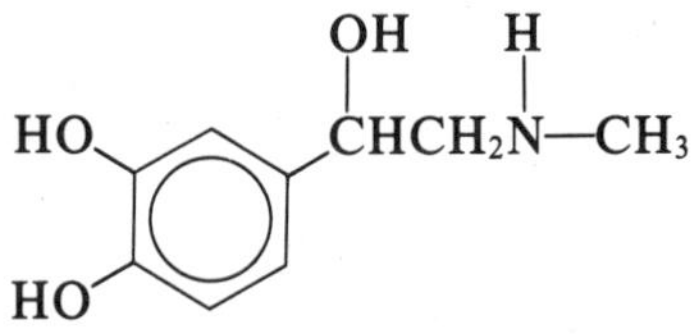

Epinephrine

Norepinephrine

Amphetamine was the first member of the family of amphetamine drugs. Meth-amphetamine and methoxyamphetamine are two other members of the family.

Methamphetamine

Methoxyamphetamine

The amphetamines are moderate appetite depressants, and because they can combat fatigue, the have gained a variety of nicknames: pep pills, uppers, bennies, speed, and others. The use of these nicknames points to a serious problem associated with these drugs, that is, their illicit use solely for their stimulant action. Methamphetamine is the "speed" that is injected into the bloodstream to produce euphoria. Methoxyamphetamine has the street name STP, which stands for serenity, tranquility, peace. Only too often, amphet-amines do not produce these effects. When abused, they can induce hyper-activity, palpitations, dizziness, tremor, headache, and, in extreme cases, amphetamine psychosis, a condition clinically indistinguishable from schizo-phrenia. Once the amphetamine-induced state wears off, a period of fatigue and depression often sets in. It is the continued use of amphetamines to avoid this depressive state that may ultimately lead to the "crash," a state of complete mental and physical exhaustion.

After reading about some of the problems associated with the abuse of am-phetamines, you may wonder what medical use they have. During the 1970's the prescribing of amphetamines by the medical profession was severely re-duced in response to the abuse of the drugs. They are used, though, in the treatment of narcolepsy, a condition characterized by an uncontrollable ten-dency to fall asleep at any time. Though quite controversial, they have also been used to reduce the activity of hyperkinetic children. Paradoxically, am-phetamines have a calming rather than a stimulating effect on some of these children. Amphetamines are no longer recommended for use in weight-reduction programs.

Barbiturates   Unlike the amphetamines, which have a stimulant effect, the **barbiturates** act to depress the central nervous system, and are referred to commonly as "downers." In large doses, barbiturates can also depress a wide range of bodily functions, especially respiration and heartbeat, and the results can be fatal. For this reason barbiturates must be used under the strict management of a physician.

Barbiturates are used as sedatives and to induce sleep. They also work in combination with general anesthetics allowing a milder application of anesthetic to achieve the same degree of anesthesia. The synergistic (mutual enhancement) effect of barbiturates with alcohol can be extremely dangerous. In combination the total depressive effect is many times greater than the effect of either when used alone. Many deaths occur every year when barbiturate "sleeping pills" are taken after an evening of social drinking. Barbiturates are also used in the treatment of epilepsy and as part of a program of psychotherapy.

Barbiturates are derivatives of barbituric acid, a heterocyclic amide.

Barbituric acid

Well over 2000 derivatives of barbituric acid have been prepared, but only a few have any medical use. Barbiturates can become habit forming, and prolonged use can develop a tolerance for the drug. Larger doses are then required to produced the same effect.

Several barbiturates are listed in Table 19.1, along with their structures and uses. Nicknames, arising from their use in the illegal drug scene, are given for a few of the most abused members of the family.

Narcotics   **Narcotics** are powerful pain killers, and they are also some of the most addicting drugs known. Because of their habit-forming nature, narcotics are used only in cases of severe pain such as that associated with cancer, surgery, injury, or heart attack. The fact that narcotics are addictive causes many physicians to avoid or greatly minimize their use. However, it has been cited that fewer than 1% of hospitalized patients given high doses of narcotics for 10 days develop true addiction.

The oriental poppy is the source of most naturally occurring narcotic drugs. **Opium,** the milky syrup obtained from the poppy, contains more than 20 different alkaloids, the most abundant of which is **morphine.** Morphine is the most potent pain-relieving drug currently available. It reduces the perception of pain and the emotional response to it. Morphine produces its analgesic action at

TABLE 19.1
Some Barbiturate Drugs

Name (commercial name)	Structural formula	Use
Barbital (Veronal®)		sedative and anesthetic
Phenobarbital (Luminol®)		anticonvulsant, sedative
Secobarbital sodium (Seconal®) "red devils" (sold in red capsules)		sedative, sleep aid, very addictive
Amobarbital (Amytal®) "blue heaven" (sold in blue capsules)		sleep-aid, very addictive

dose levels of about 10 mg given intramuscularly without causing severe alterations of consciousness. The analgesic action lasts from 3 to 4 hours.

the heavy line represents part of a ring coming out of the page

Morphine

Heroin is a derivative of morphine and has an even greater ability to reduce pain. Unfortunately, it is extremely addictive, and even brief use can quickly bring on dependence. Because of its addictive powers, heroin is not legally available in the United States, though it is illegally brought into the country and distributed throughout the illicit drug market. Heroin can produce an extreme sense of euphoria.

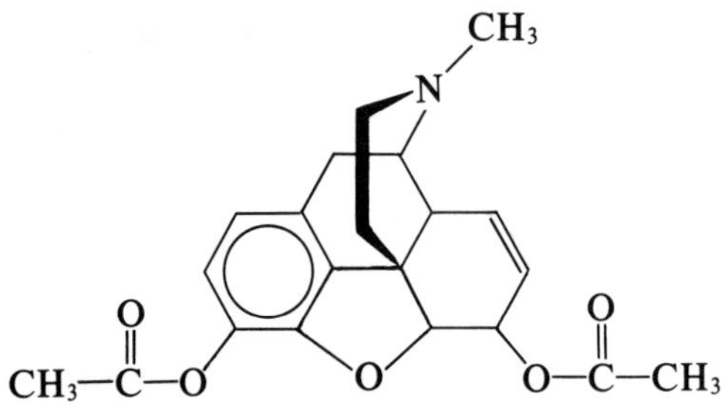

Heroin

Codeine is also obtained from opium but is less effective at reducing pain than morphine, as well as being much less addictive. It seems that the ability to relieve pain goes hand-in-hand with the ability to addict. Research has not yet been able to produce a drug with the pain-killing power of morphine without the side effect of addiction. It seems that the two properties go together. Codeine is a commonly used narcotic, and is available in several prescription pain relief drugs. About 65 mg of codeine has the same analgesic effect of two 5-grain aspirin tablets (650 mg of aspirin). It also suppresses the cough reflex, and is included in several prescription cough medicines.

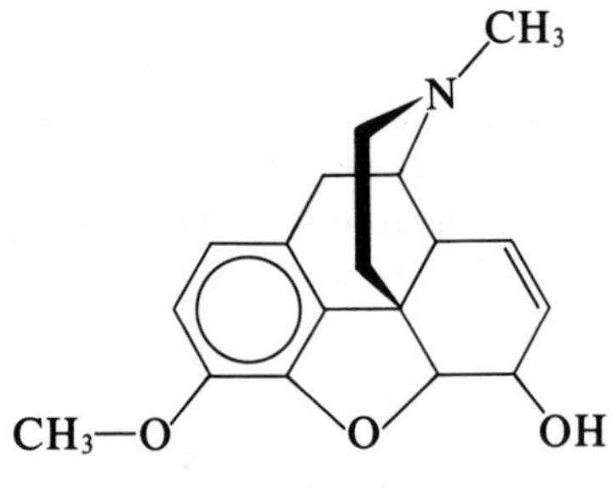

Codeine

Meperidine is a widely used synthetic narcotic analgesic. Its commercial name is Demerol®. Its pain-relieving power is between that of codeine and morphine.

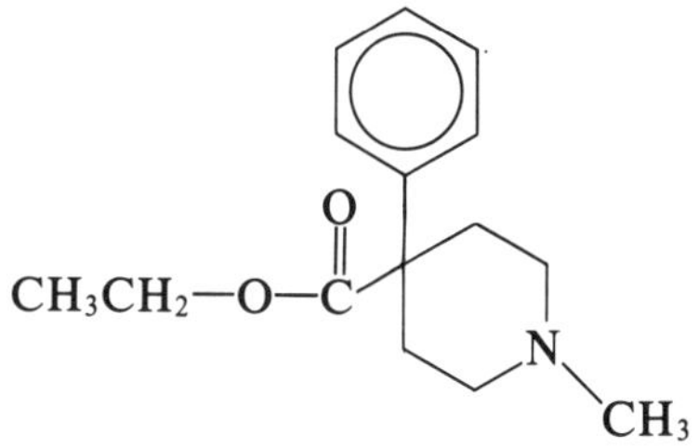

$$CH_3CH_2-O-\overset{\overset{\displaystyle O}{\|}}{C}$$

Meperidine
(Demerol)

Hallucinogens

Hallucinogens are drugs that affect the perception of mood (sometimes depressive, usually euphoric), time, space, the senses, and the self. They are also called psychedelic drugs.

The most powerful hallucinogen is **lysergic acid diethylamide** (LSD). It is derived from lysergic acid, an alkaloid obtained from a mold that grows on rye and other cereal grains. Oral doses of less than 0.05 mg can bring on hallucinations. The impairment of judgment that can accompany the use of LSD can lead to dangerous decisions that may bring harm to the user or others.

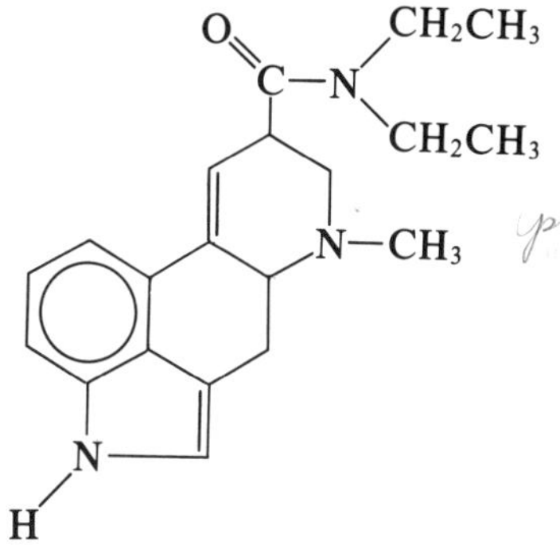

piperidine

Lysergic acid diethylamide (LSD)

Another hallucinogen is **mescaline,** obtained from the peyote cactus. Its effects are similar to those of LSD. Mescaline, or peyote as it is sometimes called, has been used by certain American Indian tribes in rituals. Though an illegal drug in the United States, its use is still allowed in certain instances by one of these groups.

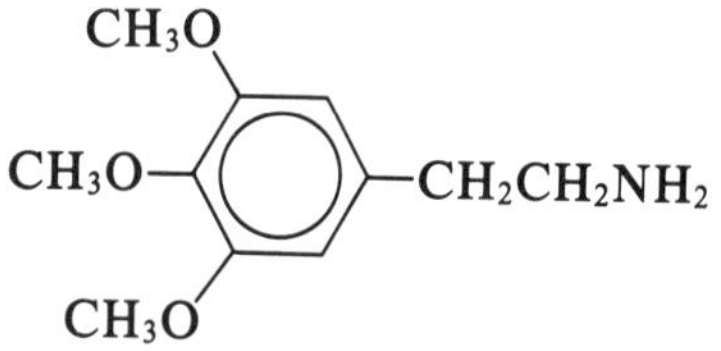

$$CH_3O-\langle\rangle-CH_2CH_2NH_2$$

Mescaline

Psilocybin (silent p) is a potent hallucinogen obtained from the Mexican hallucinogenic fungus. Notice the indole heterocyclic structure in the compound. The indole ring system is also present in LSD.

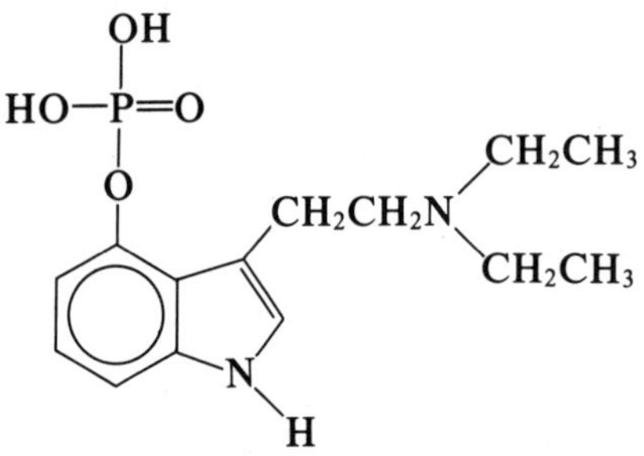

Psilocybin

The hallucinogenic action of these drugs is thought to result from an interference with the role of serotonin, an indole-based hormone in the brain.

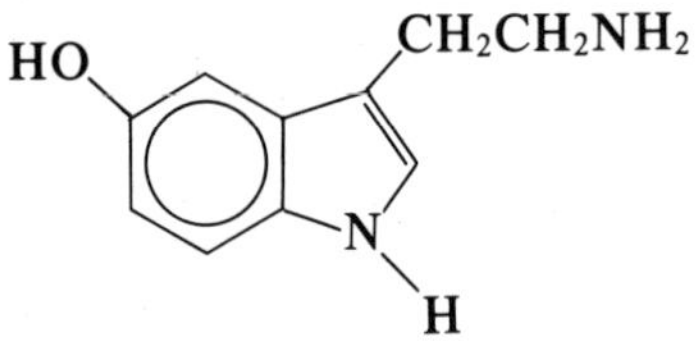

Serotonin

Perhaps the most widely used hallucinogen today is **phencyclidine,** more commonly known by its street name, PCP. Designed to be used as an anesthetic, the drug when misused can produce hallucinogenic and depressive states. It is reported to be used primarily by juveniles.

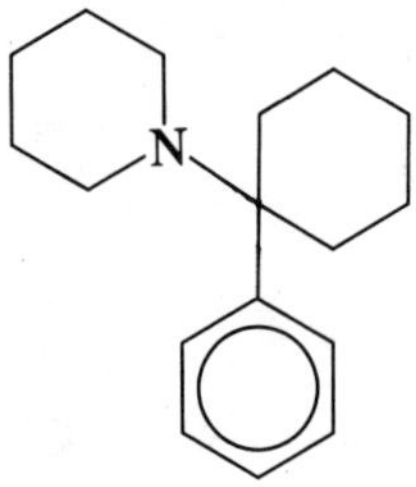

phencyclidine (PCP)

Terms

Several important terms appeared in Chapter 19. You should know the meaning of each of the following:

amine   acid chloride
amide   alkaloid
amide bond   opium
primary amine   narcotic
secondary amine   hallucinogen
tertiary amine   amphetamine
quaternary amine   barbiturate
amine salt

Questions

Answers to the starred questions appear at the end of the text.

*1. Classify each of the following amines as either primary, secondary, or tertiary:

   a. CH_3—N—H
               |
              CH_3

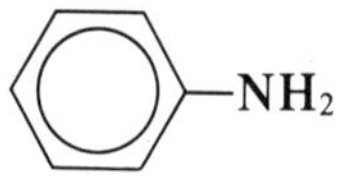

   c. CH_3CH_2—N—CH_3
                   |
                  CH_3

   b. ⬡—NH_2

   d. CH_3CH_2—N—CH_3
                   |
                   H

2. Give the common name for each of the following amines:

   a. CH_3—N—CH_3
               |
              CH_3

   *b. $CH_3CH_2CH_2NH_2$

   *c. 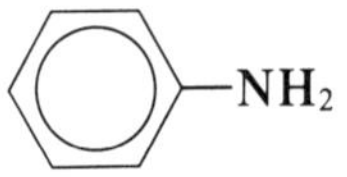 ⬡—NH_2

   d. CH_3CH_2—N—CH_3
                   |
                  CH_3

   *e. CH_3—⬡—N—H
                    |
                    H

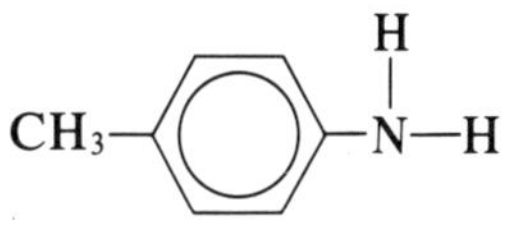

   f. CH_3CH—N—CH_3
          | |
         CH_3 CH_3

   g. CH_3CH_2—N—CH_2CH_3
                   |
                   H

* h. 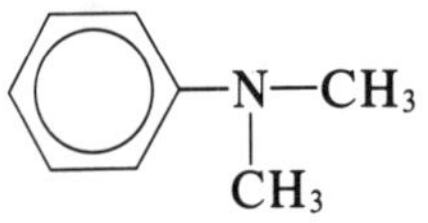 C_6H_5—N—CH$_3$ | CH$_3$ j. 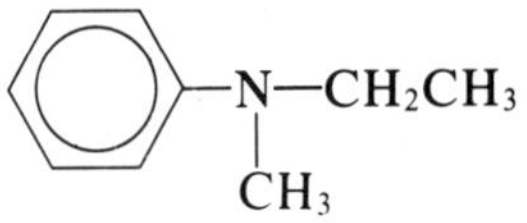 C_6H_5—N—CH$_2$CH$_3$ | CH$_3$

* i. 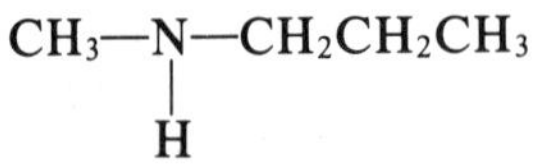   CH$_3$—N—CH$_2$CH$_2$CH$_3$ | H

3.   Draw the structural formulas for the following amines:

*a. methylethylisopropylamine   f. ethylamine
 b. 2-aminopentane   *g. 2-methylaniline
*c. N-ethylaniline   h. 2-(methylamino)butane
*d. 2,3-diaminopentane   i. trimethylamine
 e. N,N-diethylaniline   *j. N-methyl-N-ethylaniline

4.   Give the IUPAC name for the following amines:

                NH$_2$
                |
* a. 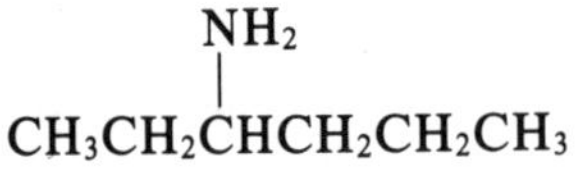   CH$_3$CH$_2$CHCH$_2$CH$_2$CH$_3$ c. 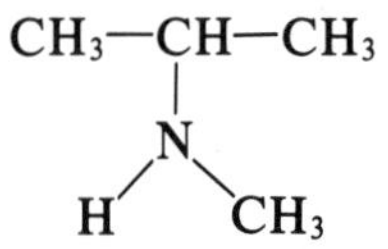   CH$_3$—CH—CH$_3$
                                                          |
                                                          N
                                                         / \
                                                        H   CH$_3$

* b.   CH$_3$—CH—CH$_2$CH$_3$ d. 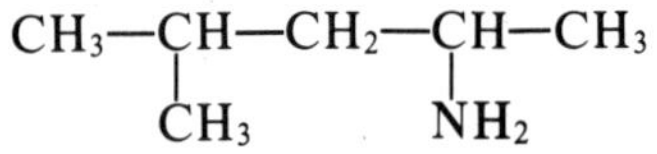   CH$_3$—CH—CH$_2$—CH—CH$_3$
              | | |
              N   CH$_3$ NH$_2$
             / \
          CH$_3$ CH$_3$

5.   Complete the following equations that describe the preparation of amines:

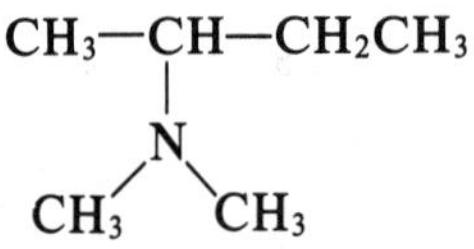

         O
         ‖
a. 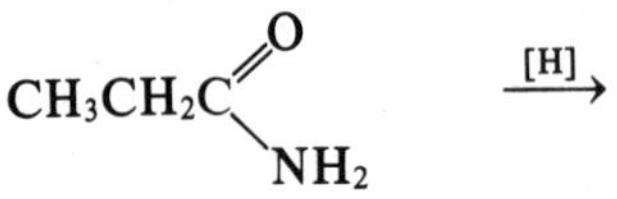   CH$_3$CH$_2$C $\xrightarrow{[H]}$
         \
          NH$_2$

         O
         ‖
* b.   CH$_3$CH$_2$C $\xrightarrow{[H]}$
             \
              N—CH$_3$
              |
              H

              H
             /
            N
            ‖
c.   CH$_3$CH$_2$C—H $\xrightarrow{[H]}$

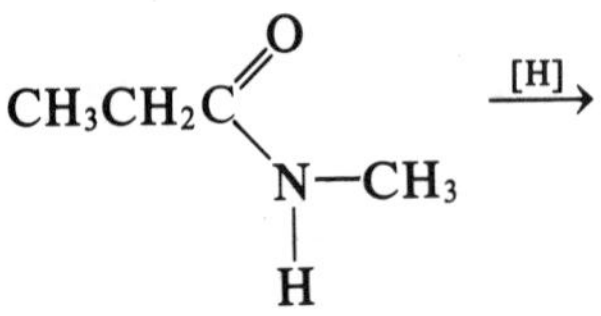

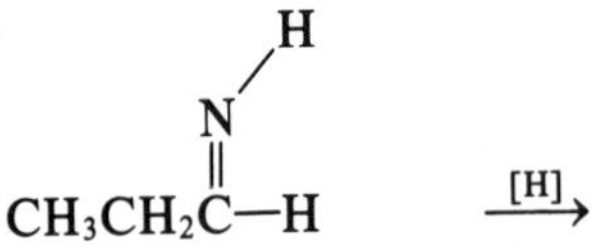

*d. 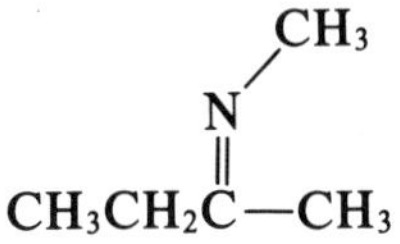   CH₃CH₂C—CH₃ $\xrightarrow{[H]}$

(with N—CH₃ group double bonded above the C)

e. 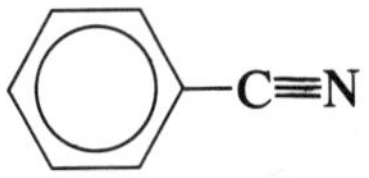   ⬡—C≡N $\xrightarrow{[H]}$

f. 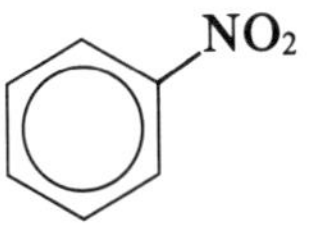   ⬡—NO₂ $\xrightarrow{[H]}$

6. Amines are considered to be weak bases. Use an equation to show the basic character of dimethylamine in water.

7. Complete the following equations, and name the product:

* a.   CH₃CH₂—NH₂ + HCl₍ₐq₎ ⟶

  b. 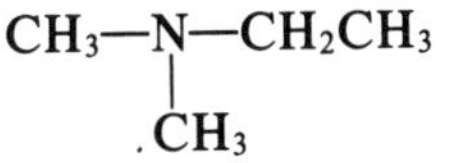   CH₃—N—CH₂CH₃ + HBr₍ₐq₎ ⟶
           |
          CH₃

8. What kind of ammonium salt is choline?

9. Give the common name for the following amides:

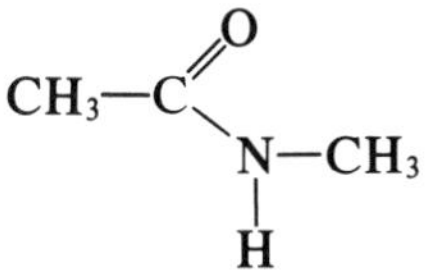

* a. 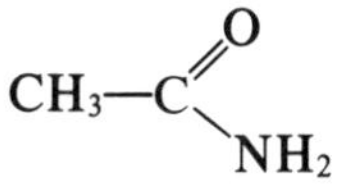   CH₃—C(=O)—NH₂

  c.   CH₃CH₂C(=O)—N(—CH₃)—CH₃

  b.   CH₃—C(=O)—N(—CH₃)—H

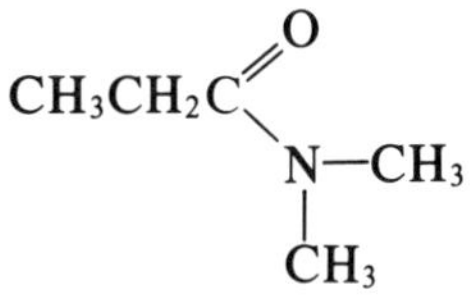

  d. 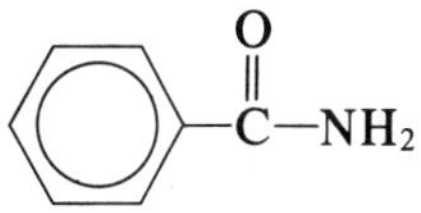   ⬡—C(=O)—NH₂

10. Draw structural formulas of the following amides:

   *a. N-isopropylacetamide   *d. methanamide
    b. butyramide   e. N-ethyl-N-methylacetamide
    c. N,N-dimethylbenzamide   f. propanamide

11. Draw a figure showing how acetamide can hydrogen bond with water in an aqueous solution of the amide.

*12. Can N,N-dimethylacetamide hydrogen bond with its neighbors in a sample of the liquid?

*13. Give the equation showing the preparation of acetyl chloride.

14. Starting with the appropriate acid chloride, give the equation for the preparation of the following amides:

* **a.** 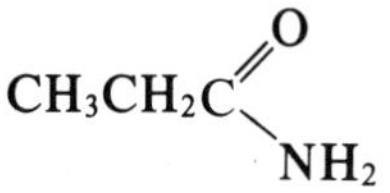 $CH_3CH_2C\underset{NH_2}{\overset{O}{\diagdown}}$

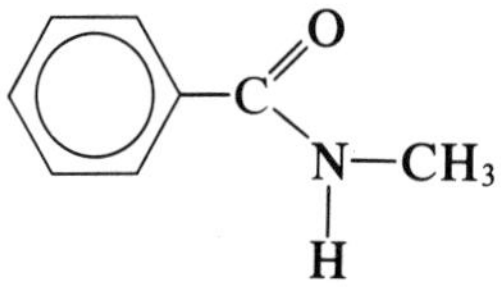

c. 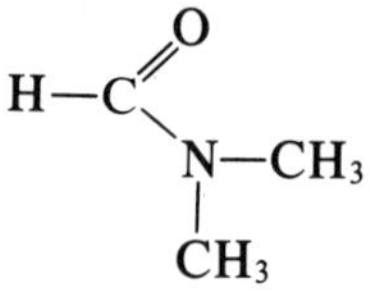 $H-C\underset{\underset{CH_3}{\overset{|}{N}}-CH_3}{\overset{O}{\diagup}}$

* **b.**   (benzene ring)$-C\underset{\underset{H}{\overset{|}{N}-CH_3}}{\overset{O}{\diagup}}$

15. Give the products of the following hydrolysis reactions:

* **a.** 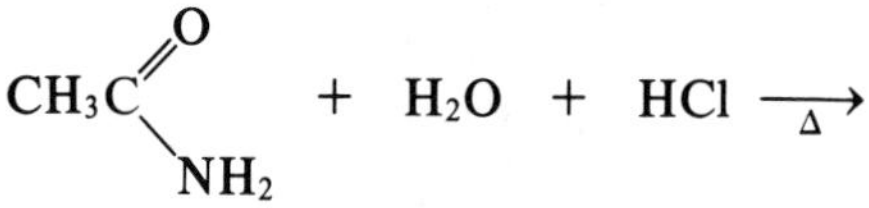 $CH_3C\underset{NH_2}{\overset{O}{\diagup}}$ + H_2O + $HCl \xrightarrow{\Delta}$

* **b.** 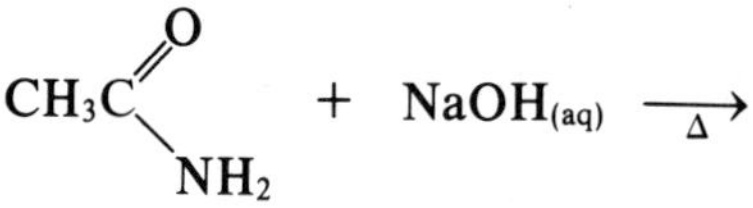 $CH_3C\underset{NH_2}{\overset{O}{\diagup}}$ + $NaOH_{(aq)} \xrightarrow{\Delta}$

c. 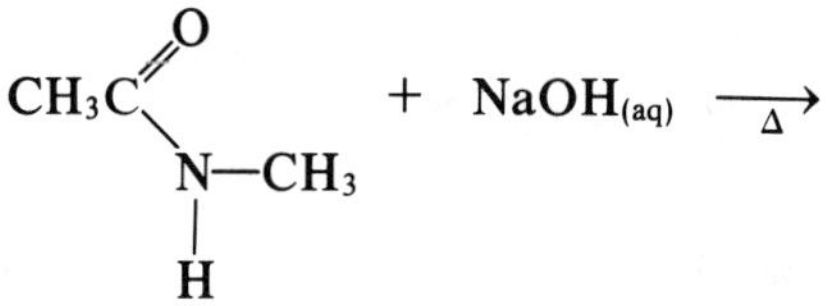 $CH_3C\underset{\underset{H}{\overset{|}{N}-CH_3}}{\overset{O}{\diagdown}}$ + $NaOH_{(aq)} \xrightarrow{\Delta}$

d. 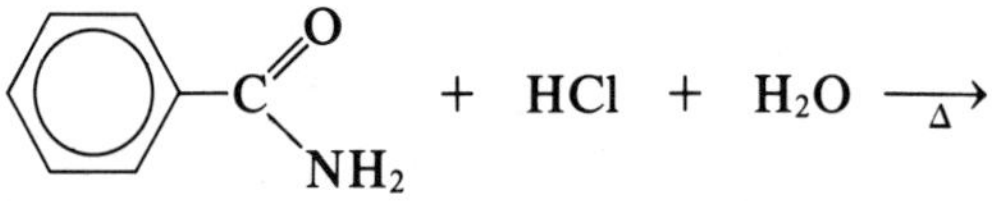   (benzene ring)$-C\underset{NH_2}{\overset{O}{\diagup}}$ + HCl + $H_2O \xrightarrow{\Delta}$

***16.** Why can proteins and nylon be called polyamides?

17. Xylocaine is a water-insoluble solid. How can it be made water soluble for use in injection solutions?

18. Why would an ulcer patient prefer to use acetaminophen instead of aspirin for the relief of minor pain?

19. Identify the nitrogen heterocyclic unit(s) in each of the following compounds:

* **a.**

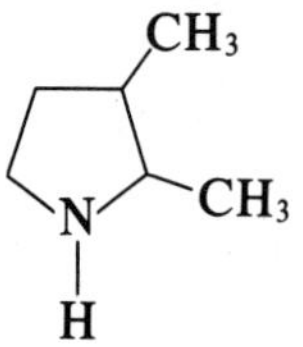

b.

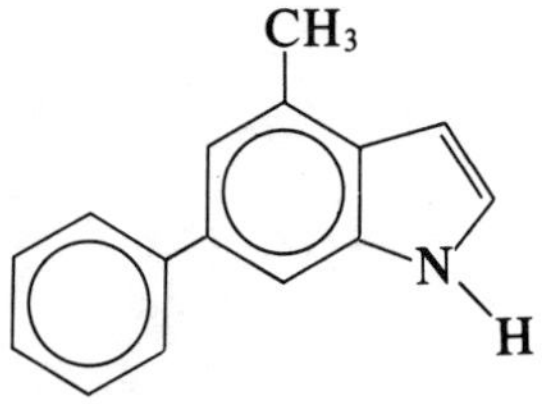

* c.

d.

* e. in LSD

20. Would amphetamine be classed as a stimulant or a depressant?
*21. What is a principal medical use of the barbiturates?
22. What has proven to be a rich source for the alkaloid narcotics?
*23. In what way is heroin related to morphine?

Lipids and the Fat-Soluble Vitamins

Lipids are a biological class of compounds that serve as one of the structural materials forming cell membranes and a source of stored energy for the cell. Lipids are not composed of repeating units as are proteins (repeating amino acid units) and carbohydrates (repeating monosaccharide units).

When plant or animal tissue is finely ground in one of the so-called fat solvents (diethyl ether, benzene, chloroform, and others), the material that dissolves is called the **lipid fraction**. Though soluble in these solvents, the fraction is not soluble in water. We will define a **lipid** as a biological compound that is insoluble in water, but very soluble in the fat solvents. This definition, though useful, is more a matter of convenience than one based on the chemical nature of lipids.

Fewer than 50 different kinds of lipids are found in the body. From a nutritional point of view, the compounds that form the most significant subclass of lipids are the triglycerides, commonly called fats, and so a major portion of this chapter will deal with the properties and importance of triglycerides.

532

Objectives

By the time you finish this chapter you should be able to do the following:

1. Classify a lipid into one of the five lipid subclasses.
2. Name and draw the structures of several triglycerides.
3. Describe the physical and chemical properties of fatty acids and triglycerides.
4. Explain how soaps and detergents function as cleansing agents.
5. Describe the functions of the phospholipids.
6. Draw the basic structure found in all steroids.
7. Describe how cholesterol and the triglycerides are related to varying stages of atherosclerosis.
8. List and describe the important steroids.
9. Describe the functions of the fat-soluble vitamins.
10. Describe the digestion and absorption of lipids in the body.

Classification of Lipids

Lipids are a heterogenerous class of biological compounds grouped together by solubility, but they can be separated into subclasses based on differences in their chemical composition and hydrolysis (reaction with water) products. There are five subclasses of lipids.

1. Fatty acids: Long-chain monocarboxylic acids.
2. Simple lipids: Yield an alcohol and fatty acids upon hydrolysis. There are two categories of simple lipids.
   a. Waxes: Yield a long-chain alcohol and a fatty acid on hydrolysis.
   b. Triglycerides: Yield glycerol and fatty acids on hydrolysis.
3. Compound lipids: Yield an alcohol, fatty acids, and other usually complex compounds upon hydrolysis. There are two categories of compound lipids.
   a. Phospholipids: Yield glycerol, fatty acids, phosphoric acid, and a nitrogen-containing alcohol on hydrolysis.
   b. Sphingolipids: Yield a fatty acid, a nitrogen-containing alcohol, or a monosaccharide (usually galactose) sphingosine, and sometimes phosphoric acid on hydrolysis.

4. Steroids: All steroids contain a substituted four-member fused ring system. This subclass will be divided into five categories based on a similarity in structure and/or function.
   a. Cholesterol
   b. Bile salts
   c. Adrenocortical hormones
   d. Sex hormones
   e. Vitamin D

5. Terpenes: Each compound in this subclass can be considered as being composed of units of isoprene, C_5H_8.

The lipids will be discussed in the same sequence as they appear in the outline above.

The Fatty Acids

Fatty acids are long-chain monocarboxylic acids. In living systems, fatty acids exist either as free species or in combination with other substances, though the amount of free fatty acid is usually small. Most is combined with alcohol or amine groups in the form of esters ($-\overset{\overset{O}{\|}}{C}-O-R$) or amides ($-\overset{\overset{O}{\|}}{C}-\overset{\overset{|}{}}{N}-$).

The term "fatty acid" is used to describe (1) the free fatty acid, (2) the fatty acid portion of an ester or amide, or (3) the long hydrocarbon chain of the acid. For example, part of an ester might be described as a fatty acid, though only that part of the ester that came from the fatty acid is being indicated:

$$H_3C-CH_2-CH_2-CH_2-CH_2-CH_2-CH_2-CH_2-CH_2-CH_2-CH_2-\overset{\overset{O}{\|}}{C}-OH$$

A fatty acid (long-chain monocarboxylic)

$$H_3C-CH_2-CH_2-CH_2-CH_2-CH_2-CH_2-CH_2-CH_2-CH_2-CH_2-\overset{\overset{O}{\|}}{C}\diagdown$$
$$O-CH_2-CH_2-CH_2-CH_2-CH_3$$

This portion of an ester is from the fatty acid. It may be referred to as a fatty acid or a fatty acid group.

Fatty acids whether found free on in the combined state have the following characteristics:

1. They are long-chain monocarboxylic acids symbolized $R-COOH$.
2. There is usually an even number of carbon atoms in the acid.

3. The alkyl group (R–) of the carboxylic acid is a straight-chain carbon unit, rather than being branched or cyclic.

4. The alkyl group of the carboxylic acid may be either saturated (single bonds only) or unsaturated (one or more carbon-to-carbon double bonds).

The names and formulas of the most common saturated and unsaturated fatty acids are found in Tables 20.1 and 20.2.

Palmitic and stearic acids are the most prevalent saturated fatty acids in vegetable and animal tissue, since they compose a major portion of the fatty acids that make up triglycerides.

The unsaturated fatty acids are found primarily in vegetable oils, with oleic acid being most prevalent. In fact, considering all vegetable and animal triglycerides, oleic acid is found to be the most abundant fatty acid in nature.

From Tables 20.1 and 20.2, one can see that oleic, linoleic, and linolenic acids all contain 18 carbon atoms, the same number found in stearic acid. Yet the physical and chemical properties of stearic acid vary considerably from those of the other three, and these differences stem completely from the fact that stearic acid is a saturated acid while the other three are unsaturated, containing one or more carbon-to-carbon double bonds.

The melting points of the saturated fatty acids increase as the number of carbon atoms increases, a trend that is also seen in the alkanes (see Figure 20.1, p. 537). The same general trend is seen in the unsaturated fatty acids, but a more important trend here is that the melting points decrease substantially as the number of carbon-to-carbon double bonds increases. The effect is so marked that as a general rule, the greater the degree of unsaturation (number of double bonds), the lower will be the melting point of an unsaturated fatty acid regardless of the number of carbon atoms it contains.

The larger fatty acids, because of their long-chain alkyl groups, are not soluble in water. They act more like long-chain hydrocarbons in that respect, but they

TABLE 20.1
Common Saturated Fatty Acids

Name	Melting point (°C)	RCOOH formula*	Condensed formula	Source
1. Butyric	– 4	C_3H_7COOH	$CH_3(CH_2)_2COOH$	butter fat
2. Lauric	44	$C_{11}H_{23}COOH$	$CH_3(CH_2)_{10}COOH$	coconut oil
3. Myristic	54	$C_{13}H_{27}COOH$	$CH_3(CH_2)_{12}COOH$	coconut oil; nutmeg oil
4. Palmitic	63	$C_{15}H_{31}COOH$	$CH_3(CH_2)_{14}COOH$	fats and oils
5. Stearic	70	$C_{17}H_{35}COOH$	$CH_3(CH_2)_{16}COOH$	fats and oils
6. Arachidic	77	$C_{19}H_{39}COOH$	$CH_3(CH_2)_{18}COOH$	peanut oil

*Saturated fatty acids have the general formula (C_nH_{2n+1})–COOH. The alkyl group (R–) is (C_nH_{2n+1}).

TABLE 20.2
Common Unsaturated Fatty Acids

Name	Melting point (°C)	RCOOH formula*	Number of double bonds	Position of double bonds‡	Source
1. Oleic	14	$C_{17}H_{33}COOH$	1	9	olive oil, butter
2. Linoleic	−5	$C_{17}H_{31}COOH$	2	9, 12	corn oil, soybean oil
3. Linolenic	−11	$C_{17}H_{29}COOH$	3	9, 12, 15	linseed oil
4. Arachidonic	−50	$C_{19}H_{31}COOH$	4	5, 8, 11, 14	lecithin

Condensed formula†

1. Oleic

$$\overset{9}{C}H_3(CH_2)_7CH=CH(CH_2)_7\overset{1}{C}OOH$$

2. Linoleic

$$\overset{12}{C}H_3(CH_2)_4CH=CH-CH_2-\overset{9}{C}H=CH(CH_2)_7\overset{1}{C}OOH$$

3. Linolenic

$$\overset{15}{C}H_3CH_2CH=CH-CH_2-\overset{12}{C}H=CH-CH_2-\overset{9}{C}H=CH(CH_2)_7\overset{1}{C}OOH$$

4. Arachidonic

$$\overset{14}{C}H_3(CH_2)_4CH=CH-CH_2-\overset{11}{C}H=CH-CH_2-\overset{8}{C}H=CH-CH_2-\overset{5}{C}H=CH-(CH_2)_3\overset{1}{C}OOH$$

*Unsaturated fatty acids have the general formulas $C_nH_{2n-1}COOH$; $C_nH_{2n-3}COOH$; $C_nH_{2n-5}COOH$; etc.

†Since there are double bonds in these compounds, there is a possibility for cis-trans isomerism. In nature, the unsaturated fatty acids are found in the cis configuration.

‡Carbon atoms in carboxylic acids are numbered from the carboxylate carbon (#1) and increasing to the end of the chain. The position of the double bond is given by the lowest-numbered carbon atom involved in the double bond.

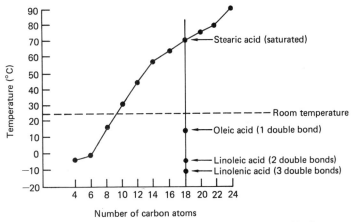

Fig. 20.1. As the carbon chain in saturated fatty acids increases in the number of carbon atoms, the melting point of the acids increase. As the degree of unsaturation increases, the melting points decrease. Three unsaturated fatty acids, each containing 18 carbon atoms, are shown for comparison.

are soluble in many organic, nonpolar solvents like benzene and carbon tetrachloride.

Unsaturated fatty acids that contain more than one double bond cannot be synthesized by the body and must be supplied by the diet. These **essential fatty acids** are linolenic, linoleic, and arachidonic acids.

Check Test Number 1

1. Draw the structural formulas of butyric and lauric acid.
2. Why is pelargonic acid, $CH_3(CH_2)_7COOH$, not found in many lipids?
3. Would palmitoleic acid, $C_{15}H_{29}COOH$, be saturated or unsaturated?

Answers:

1. Butyric Acid:

$$H-\overset{\overset{\displaystyle H}{|}}{\underset{\underset{\displaystyle H}{|}}{C}}-\overset{\overset{\displaystyle H}{|}}{\underset{\underset{\displaystyle H}{|}}{C}}-\overset{\overset{\displaystyle H}{|}}{\underset{\underset{\displaystyle H}{|}}{C}}-\overset{\overset{\displaystyle O}{||}}{C}-OH$$

Lauric Acid:

$$H-\overset{\overset{\displaystyle H}{|}}{\underset{\underset{\displaystyle H}{|}}{C}}-\overset{\overset{\displaystyle H}{|}}{\underset{\underset{\displaystyle H}{|}}{C}}-\overset{\overset{\displaystyle H}{|}}{\underset{\underset{\displaystyle H}{|}}{C}}-\overset{\overset{\displaystyle H}{|}}{\underset{\underset{\displaystyle H}{|}}{C}}-\overset{\overset{\displaystyle H}{|}}{\underset{\underset{\displaystyle H}{|}}{C}}-\overset{\overset{\displaystyle H}{|}}{\underset{\underset{\displaystyle H}{|}}{C}}-\overset{\overset{\displaystyle H}{|}}{\underset{\underset{\displaystyle H}{|}}{C}}-\overset{\overset{\displaystyle H}{|}}{\underset{\underset{\displaystyle H}{|}}{C}}-\overset{\overset{\displaystyle H}{|}}{\underset{\underset{\displaystyle H}{|}}{C}}-\overset{\overset{\displaystyle H}{|}}{\underset{\underset{\displaystyle H}{|}}{C}}-\overset{\overset{\displaystyle H}{|}}{\underset{\underset{\displaystyle H}{|}}{C}}-\overset{\overset{\displaystyle O}{||}}{C}-OH$$

2. Pelargonic acid contains *nine* carbon atoms, an odd number. Usually only fatty acids with an even number of carbon atoms are found in nature since fatty acids are biosynthesized two carbon atoms at a time.
3. Palmitoleic acid fits the general formula of an unsaturated fatty acid with *one* double bond. All other bonds in the alkyl group are single bonds.

Recently, there have been hints of a "morning-after" birth control pill. The new hormonelike substances being tested are known as **prostaglandins**, compounds that were first isolated from semen. Twenty or so of these compounds have been found in small amounts in many different tissues and biological fluids. Prostaglandins comprise a family of unsaturated fatty acids that are very different from the common unsaturated fatty acids seen in Table 20.2. The prostaglandins contain 20 carbon atoms and are derived from arachidonic acid. Two examples are given below. A small structure change in these molecules, as in most biological molecules, produces a drastic change in their biological effect.

Prostaglandin E$_2$ Prostaglandin F$_{2\alpha}$

Present research shows that prostaglandins can prevent conception, induce labor, lower blood pressure, and induce healing of ulcers. Aspirin and other antiinflammation drugs seem to operate by inhibiting some types of prostaglandin synthesis. A great deal of research is currently underway with the hope that improvements can be made in the isolation, understanding, and use of natural and synthetic prostaglandins.

Chemical Properties of Fatty Acids

Chemically, the fatty acids exhibit the same properties as short-chain carboxylic acids.

1. They react with alcohols to form esters and water:

$$R-\overset{O}{\overset{\|}{C}}-OH + HO-R' \xrightarrow{catalyst} R-\overset{O}{\overset{\|}{C}}-O-R' + H_2O$$

Acid   Alcohol   Ester   Water

In the laboratory, the catalyst is a mineral acid; in biological systems, it is an enzyme. The alkyl group attached to the oxygen atom in the ester

(−R) always comes from the alcohol. Remember, when naming an ester, give the name of the alcohol first, followed by the name of the acid with its *-ic* ending replaced by *-ate*, the characteristic ending of all esters. For example, the ester formed from ethyl alcohol and acetic acid, $CH_3COOCH_2CH_3$, would be named ethyl acetate.

2. The reverse of esterification also occurs in nature. The reaction is called **hydrolysis**, and the products are a carboxylic acid and an alcohol:

$$R\overset{\overset{\displaystyle O}{\|}}{-C}-O-R' \;+\; H_2O \xrightarrow{\text{catalyst}} R\overset{\overset{\displaystyle O}{\|}}{-C}-OH \;+\; R'OH$$

| Ester | Water | | Acid | Alcohol |

3. The reaction of a fatty acid with a base produces a salt and water:

$$R\overset{\overset{\displaystyle O}{\|}}{-C}-O-H \;+\; \underset{\text{(or KOH)}}{NaOH} \longrightarrow R\overset{\overset{\displaystyle O}{\|}}{-C}-O^{\ominus}\,Na^{\oplus} + H_2O$$

| Acid | Base | Salt | Water |

If the acid contains more than eight carbon atoms, the salt that forms is called a soap. Common hand soaps are nothing more than sodium salts of the larger fatty acids. Soaps, being salts of carboxylic acids, are named accordingly. The name of the metal ion is given first, followed by the name of the acid, with its *-ic* ending replaced by *-ate*. The sodium salt of oleic acid, $C_{17}H_{33}COONa$, would be named sodium oleate. Soaps and detergents are not exactly the same thing, and both will be discussed later on in this chapter.

4. The unsaturated fatty acids contain double bonds, and therefore they undergo addition reactions similar to the alkenes, adding hydrogen for example. Any unsaturated fatty acid can be converted to a saturated acid by **hydrogenation**:

$$CH_3(CH_2)_4CH=CHCH_2CH=CH(CH_2)_7COOH + 2H_2 \xrightarrow{Ni} CH_3(CH_2)_{16}COOH$$

| Linoleic acid | Hydrogen gas | Stearic acid |

The carbon-to-carbon double bonds in unsaturated fatty acids can also react with molecular iodine (I_2), adding one molecule of iodine per double bond. The degree of unsaturation of a fatty acid, or compound containing fatty acid groups, can be expressed as the number of grams of iodine consumed by 100 g of the compound. This value is commonly called the **iodine number** of the compound.

Simple Lipids

Simple lipids are esters formed from fatty acids and alcohols. There are two categories of simple lipids: waxes and triglycerides.

Waxes   **Waxes** are large monoesters formed by the combination of a long-chain fatty acid and a long-chain monohydroxy alcohol (contains only one -OH group). Waxes are among the simplest members of the lipid family since they undergo few chemical reactions. An example of a wax is myricyl palmitate, the major component of beeswax. It is an ester formed from palmitic acid and myricyl alcohol, $C_{30}H_{61}OH$:

$$C_{15}H_{31}-\overset{\overset{\displaystyle O}{\|}}{C}-O-C_{30}H_{61}$$

Myricyl palmitate

In animals, waxes act as protective coatings for hair, skin, and feathers, and they protect plants from dehydration and insects. Some naturally occurring waxes are lanolin, beeswax, carnauba wax, and whale oil. Lanolin is a coating on wool, and it is used commercially in ointments and creams. Recently, a mixture of lanolin and female sex hormones (steroids) has shown promise in preventing the loss of hair in aging men. Whale oil, which is really a low-melting wax, has been used in cosmetics, candles, transmission oil, and ointments. These uses have been greatly reduced, though, because of the recent ban on the killing of whales. Artificial waxes have replaced whale oil in some uses, but as yet, not very well. Beeswax is secreted by glands of the honey bee, and it serves as the supporting medium for the honeycomb. Beeswax is used in pharmaceutical products, expensive candles, and in dentistry for teeth impressions. Carnauba wax is obtained from the carnauba palm and is used in car and floor waxes.

Check Test Number 2

1. Give an equation showing how myricyl palmitate in beeswax would be formed from an alcohol and a fatty acid.
2. Write equations describing the reactions of palmitic acid with sodium hydroxide, and oleic acid with calcium hydroxide.

3. How many hydrogen molecules would be needed to convert one molecule of prostaglandin E_2 into a saturated fatty acid?

Answers:

1. The general reaction for the formation of an ester $(R-\overset{\overset{O}{\|}}{C}-OR')$ from an alcohol and carboxylic acid shows that R' always comes from the alcohol. Thus, the alcohol is myricyl alcohol and the fatty acid is palmitic acid:

$$C_{15}H_{31}COOH \;+\; C_{30}H_{61}OH \;\xrightarrow{\text{enzyme}}\; C_{15}H_{31}-\overset{\overset{O}{\|}}{C}-O-C_{30}H_{61} \;+\; H_2O$$

Palmitic acid　　　Myricyl alcohol　　　　　Myricyl palmitate　　　　Water

2. $$C_{15}H_{31}COOH \;+\; KOH \;\longrightarrow\; C_{15}H_{31}COO^{\ominus}K^{+} \;+\; H_2O$$

Palmitic acid　　Potassium　　　　Potassium　　　　Water
　　　　　　　hydroxide　　　　palmitate

$$2\,C_{17}H_{33}COOH \;+\; Ca(OH)_2 \;\longrightarrow\; (C_{17}H_{33}COO^{\ominus})_2Ca \downarrow \;+\; 2\,H_2O$$

Oleic acid　　　Calcium hydroxide　　　Calcium oleate　　　Water

(Remember that carboxylate ions, $RCOO^{\ominus}$, bear a single negative charge. Since the charge on the calcium ion is $2+$, two carboxylate ions must be combined with one calcium ion to form a neutral compound.) Group IIA salts of fatty acids are not soluble in water, so calcium oleate is written as a precipitate in the equation above.

3. Prostaglandin E_2 contains two double bonds and would require two molecules of hydrogen per molecule of the acid to convert it to a saturated fatty acid.

Triglycerides

The triglycerides comprise the most abundant group of lipids in nature and in our diet. **Triglycerides** are triesters of glycerol and three fatty acids:

$$\begin{array}{c} H_2C-OH \\ | \\ H-C-OH \\ | \\ H_2C-OH \end{array}$$

Glycerol

One, two, or all three of the hydroxyl groups of glycerol can react with fatty acids to produce mono-, di-, or triglycerides, respectively. Though mono- and diglycerides are present in nature and are important in lipid metabolism, the principal species is the triglyceride.

The general structure of a triglyceride is

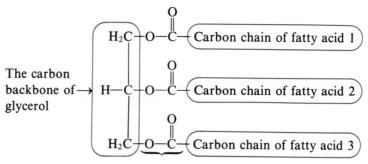

Ester linkages join the glycerol backbone to
the carbon chains of the fatty acids

The triglyceride structure can also be written as

$$\begin{array}{ccc}
CH_2OOCR & & RCOOCH_2 \\
| & & | \\
CHOOCR' & or & R'COOCH \\
| & & | \\
CH_2OOCR'' & & R''COOCH_2
\end{array}$$

If the three fatty acids in a triglyceride are the same, then the lipid is known as a **simple triglyceride**. However, the triglycerides usually found in nature contain carbon chains from two or three different fatty acids and so are called **mixed triglycerides**. The following are examples of simple and mixed triglycerides. Can you name the fatty acids in each? The names of the compounds will give you clues:

$$\begin{array}{ccc}
\underset{\substack{\\ \text{Glyceryl tristearate} \\ \text{M.P. } 71°C \\ \text{(simple triglyceride)}}}{\begin{array}{l} H_2C-O-\overset{\overset{\displaystyle O}{\|}}{C}-C_{17}H_{35} \\[1em] H-\overset{\overset{\displaystyle O}{\|}}{C}-O-\overset{\overset{\displaystyle O}{\|}}{C}-C_{17}H_{35} \\[1em] H_2C-O-\overset{\overset{\displaystyle O}{\|}}{C}-C_{17}H_{35} \end{array}}
\end{array}$$

Glyceryl tristearate
M.P. 71°C
(simple triglyceride)

Glyceryl trioleate
M. P. −17°C
(simple triglyceride)

Glyceryl oleolaurostearate
(mixed triglyceride)

The names of the glycerides must indicate the alcohol (always glycerol) and the acids that combined to form the compound. As with all esters, the name of the alcohol is given first, and its -ol ending is changed to -yl (glycer*ol* to glycer*yl*). The names of the acids then follow, and are written as one long, separate word. If two or three acids are the same, then the prefixes di- or tri- are used before the name of the acid. The ending of the acid name is changed from -ic to -o (stear*ic* to stear*o*) unless it is the only or last acid in the name, in which case the -ic ending is changed to -ate, to indicate the compound is an ester. In mixed triglycerides, the acids are named in sequence starting from one end of the glycerol carbon chain. Go back and review the names of the three previous examples.

The three carbon atoms in glycerol can be numbered 1, 2, and 3, starting at one end. These numbers can be used to indicate to which carbon atom a particular fatty acid is joined in mono- and diglycerides. Number the carbon atoms so that the smallest number will appear in the name. This naming scheme is used to identify glyceryl 1-monooleate, a monoglyceride. The prefix mono- indicates a monoglyceride. It is written in the expanded form to show the position of the double bond in the fatty acid chain:

$$H-\overset{\overset{\displaystyle H}{|}}{\underset{\underset{\displaystyle H}{|}}{\underset{|}{\overset{|}{C}}}}-O-\overset{\overset{\displaystyle O}{\|}}{C}-CH_2CH_2CH_2CH_2CH_2CH_2CH_2CH=CHCH_2CH_2CH_2CH_2CH_2CH_2CH_2CH_3$$

$$H-\overset{2}{C}-OH$$

$$H-\overset{3}{C}-OH$$

Glyceryl 1-monooleate
(a monoglyceride, M.P. 35°C)

In living systems, triglycerides are formed in a stepwise sequence, with one fatty acid joining each time:

$$\begin{array}{ccc} CH_2O\boxed{H} \;+\; H-O\boxed{-} & \overset{O}{\overset{\|}{C}}-R & \\ CH OH & \xrightarrow{enzyme} & CH_2-O-\overset{O}{\overset{\|}{C}}-R \\ CH_2OH & & CH-OH \quad + H_2O \\ & & CH_2-OH \end{array}$$

Glycerol   First fatty acid   Monoglyceride

$$\begin{array}{ccc} CH_2OOCR & & CH_2OOCR \\ H-\overset{|}{C}-O\boxed{H} \;+\; H-O\boxed{-}\overset{O}{\overset{\|}{C}}-R' & \xrightarrow{enzyme} & CHOOCR' \;+\; H_2O \\ CH_2OH & & CH_2OH \end{array}$$

Monoglyceride   Second fatty acid   Diglyceride

$$
\begin{array}{l}
CH_2OOCR \\
CHOOCR' \\
CH_2O\boxed{H\ +\ HO}\!-\!\overset{\displaystyle O}{\overset{\|}{C}}\!-\!R'' \xrightarrow{\ enzyme\ }
\end{array}
\qquad
\begin{array}{l}
CH_2OOCR \\
CHOOCR' \\
CH_2OOCR'' + H_2O
\end{array}
$$

Diglyceride Third fatty acid   Triglyceride

It is more convenient to write the synthesis as a single equation:

$$
\begin{array}{l}
CH_2OH \\
CHOH \\
CH_2OH
\end{array}
+\ RCOOH\ +\ R'COOH\ +\ R''COOH\ \xrightarrow{\ enzyme\ }
\begin{array}{l}
CH_2OOCR \\
CHOOCR' \\
CH_2OOCR''
\end{array}
+\ 3H_2O
$$

Glycerol   The three fatty acids   Triglyceride

Triglycerides are normally classified into one of two subdivisions based on their physical and chemical properties. Sometimes it is not perfectly clear into which subdivision a particular triglyceride should be placed, but all will be considered as either a fat or an oil. **Fats** are those triglycerides that contain mainly saturated fatty acid groups, and they are sometimes called saturated triglycerides. **Oils** contain mostly unsaturated fatty acid groups, and they are often called unsaturated triglycerides or polyunsaturated fats.

Earlier it was noted that as the alkyl groups in a series of saturated fatty acids become larger the melting points become higher. This trend is also seen in the saturated triglycerides. Those with the larger fatty acid groups have the higher melting points. Also, like the fatty acids, as the degree of unsaturation (number of carbon-to-carbon double bonds) increases, the melting points become lower, so low in fact that most unsaturated triglycerides are liquids at room temperature. This is why unsaturated triglycerides are often called oils.* The physical state of a triglyceride at room temperature can often be used to distinguish an unsaturated triglyceride from a saturated one. If it is a liquid, chances are that it is unsaturated; if it is a solid, then it is probably saturated. There are exceptions to this, but as a rule it works in most cases.

It is not possible to write a single formula that describes a triglyceride sample as it is found in nature, since natural samples are always mixtures of several different triglycerides. When samples from natural sources are subjected to chemical analysis, they are broken down into glycerol and fatty acids. The fatty acids that are produced are then identified. The result allows chemists to say only which fatty acids are present and in what amounts. They do not have enough information to reconstruct the triglyceride mixture that made up the sample. But the results do show that the kinds and amounts of fatty acids produced will depend on the source of the sample (plant of animal) as well as the climate and diet of the source. This is shown in Table 20.3.

---

*Do not confuse the unsaturated triglyceride "oil" with the common name for petroleum. Petroleum oil is a mixture of hydrocarbons, with properties much different from those of triglycerides.

TABLE   20.3
Composition of Some Natural Fats and Oils

| | Range of fatty acid composition (% by weight) | | | | | |
| | Saturated acids | | | Unsaturated acids | | Melting |
	Myristic	Palmitic	Stearic	Oleic	Linoleic	point (°C)
Animal fats						
Butter*	7–10	23–29	9–13	27–35	4–5	32
Lard	1–2	28–30	12–18	40–50	6–7	31
Human Fat	2–3	22–24	7–9	45–48	8–10	15
Vegetable oils						
Olive	0–1	9–12	2–4	60–84	4–10	– 6
Peanut	0	6–8	2–5	50–65	13–26	3
Corn	0–2	7–10	3–4	43–50	34–40	–20
Soybean	0–2	6–10	2–4	21–29	50–58	–16
Cottonseed	0–2	19–23	1–2	23–30	40–48	– 1
Linseed†	0	4–8	2–4	10–37	4–40	–24

*3–5% butyric acid.
†25–60% linolenic acid.

Physical Properties of Triglycerides

Some of the physical properties of triglycerides have already been mentioned, such as their physical state at room temperature, solid (fats) or liquid (oils). Fats and oils are both less dense than water and will float on the surface as a film or as particles. They both are poor conductors of heat and electricity, and thus serve as heat insulators for the body. Many Americans are well insulated. Though fats and oils are not soluble in water, they are very soluble in many organic liquids as well as in other lipids. Though pure triglycerides are odorless, tasteless, and colorless, they dissolve many flavorful substances of varying colors, and in that way become a good-tasting part of our diets.

Check Test Number 3

1. What is the chemical difference between a fat and an oil?
2. Draw the structure of glyceryl palmitodiarachidonate.
3. Name the following:

$$H_2COOC-C_3H_7$$
$$H-COOC-(CH_2)_{16}CH_3$$
$$H_2COOC-C_{17}H_{31}$$

Answers:

1. The alkyl groups of fats are saturated, while those of oils are mostly unsaturated.

2. $\quad$ $\underset{\displaystyle \text{from palmitic acid}}{H_2C-O-\overset{\displaystyle O}{\overset{\displaystyle \|}{C}}-C_{15}H_{31}} \longleftarrow$ from palmitic acid

$$H-\underset{|}{\overset{|}{C}}-O-\overset{\displaystyle O}{\overset{\displaystyle \|}{C}}-C_{19}H_{31}$$

from arachidonic acid

$$H_2C-O-\overset{\displaystyle O}{\overset{\displaystyle \|}{C}}-C_{19}H_{31}$$

3. Glyceryl   butyro   stearo   linoleate ⟵ The compound

   from   from   from   from   is an ester
   glycerol   butric   stearic   linoleic   so that the
               acid   acid   acid   last acid ends
                                                 in   *ate*

(The acids are named in sequence as they appear along the glycerol backbone.)

Chemical Properties of Triglycerides

1. Hydrolysis. This is the only reaction that is involved during the digestion process of fats and oils. In the body, the dietary triglycerides are broken down in the small intestines by hydrolysis, first into diglycerides, then monoglycerides, and finally ending up as three fatty acids and glycerol. The enzyme, lipase, catalyzes the reaction in the body:

$$\text{triglyceride} + 3H_2O \xrightarrow{\text{catalyst}} 3 \text{ fatty acids} + \text{glycerol}$$

Commercially, fats and oils are the least expensive starting materials for the production of glycerol and fatty acids. Mineral acids are used for catalysts in industry.

2. Hydrogenation. The carbon-to-carbon double bonds in unsaturated triglycerides can add hydrogen to form more saturated ones, just as was seen with fatty acids. By controlling this reaction carefully, oily and semisolid triglycerides can be converted to the familiar shortenings, margarines, and peanut butter spreads. Oleomargarine, the first synthetic food, is a blend of hydrogenated vegetable oils, unsaturated vegetable oils, emulsifying agents, preservatives, flavorings, and carotene for the yellow color. By varying the amount of hydrogenation, margarines can be of the near liquid type sold in squeeze bottles, the soft spreadable-when-cold type, or the usual hard form packaged in sticks.

3. Oxidation. Triglycerides with unsaturated fatty acid groups are susceptible to oxidation, a reaction in which oxygen from the air reacts with the carbon-to-carbon double bonds in the compound. The products of the reactions are mixtures of short-chain carboxylic acids and aldehydes. Both of these kinds of compounds have disagreeable odors and tastes (especially butyric acid); consequently, oxidation reactions play a major role in food spoilage by turning food rancid.

$$\underset{\substack{\text{Part of a fatty} \\ \text{acid chain}}}{\text{~~C=C~~}} \ + \ O_2 \ \longrightarrow \ \underset{}{\text{~~C=O}} \ + \ \underset{\text{Aldehydes}}{O\text{=C~~}}$$

$$2 \text{~~C=O} \ + \ O_2 \ \longrightarrow \ 2 \underset{\text{Acids}}{\text{~~C}} \underset{}{\overset{O}{\overset{\|}{\diagdown}}} \text{OH}$$

To prevent the oxidation of triglycerides in foods, small amounts of **antioxidants** are added. The antioxidants have a higher affinity for reaction with oxygen than the unsaturated lipids. Vitamin C and vitamin E are natural antioxidants. Butylated hydroxyanisole (BHA) and butylated hydroxytoluene (BHT) are two synthetic antioxidants that are added to many of the foods we buy:

Butylated hydroxyanisole (BHA)   Butylated hydroxytoluene (BHT)

Hydrolysis reactions can also be responsible for turning foods rancid, since the products of the reactions can also be unpleasant short-chain carboxylic acids. Hygroscopic (moisture absorbing) agents are added to processed food to inhibit the hydrolysis of triglycerides.

4. Saponification. Triglycerides, when hydrolyzed in the presence of a base like sodium hydroxide, produce **soaps** and glycerol. The sodium salts of fatty acids are generally solid soaps and are formed into bars. Potassium salts, which form if the base is potassium hydroxide, are called soft soaps, and can be liquids at room temperature. The general equation describing the saponification reaction is given below:

$$
\begin{array}{cccc}
\underset{\text{Triglyceride}}{
\begin{array}{c}
H_2C-O-\overset{\overset{\displaystyle O}{\|}}{C}-R \\[2mm]
H-\overset{}{\underset{}{C}}-O-\overset{\overset{\displaystyle O}{\|}}{C}-R' \\[2mm]
H_2C-O-\overset{\overset{\displaystyle O}{\|}}{C}-R''
\end{array}}
& \underset{\text{Base}}{+\ 3\,NaOH_{(aq)}} \xrightarrow{\text{heat}} &
\underset{\text{Glycerol}}{\begin{array}{c} H_2COH \\[2mm] HCOH \\[2mm] H_2COH \end{array}}
& \underset{\text{Soaps}}{\begin{array}{c} R-\overset{\overset{\displaystyle O}{\|}}{C}-O^{\ominus}Na^+ \\[2mm] R'-\overset{\overset{\displaystyle O}{\|}}{C}-O^{\ominus}Na^+ \\[2mm] R''-\overset{\overset{\displaystyle O}{\|}}{C}-O^{\ominus}Na^+ \end{array}}
\end{array}
$$

or (KOH)

A specific example of saponification is that of glyceryl stearooleolaurate:

$$H_2C-O-\overset{\overset{O}{\|}}{C}-(CH_2)_{16}CH_3$$

$$H-\overset{|}{\underset{|}{C}}-O-\overset{\overset{O}{\|}}{C}-(CH_2)_7CH=CH(CH_2)_7CH_3 \ + \ 3 \ KOH_{(aq)} \ \xrightarrow{\text{heat}} $$

$$H_2\overset{|}{C}-O-\overset{\overset{O}{\|}}{C}-(CH_2)_{10}CH_3$$

$$\begin{array}{c} H_2COH \\ | \\ H-C-OH \\ | \\ H_2COH \end{array}$$

Glyceryl stearooleolaurate   Potassium hydroxide   Glycerol

$$+ \ CH_3(CH_2)_{16}-\overset{\overset{O}{\|}}{C}-O^{\ominus}K^+ \ + \ CH_3(CH_2)_7CH=CH(CH_2)_7-\overset{\overset{O}{\|}}{C}-O^{\ominus}K^+$$

Potassium stearate   Potassium oleate

$$+ \ CH_3(CH_2)_{10}-\overset{\overset{O}{\|}}{C}-O^{\ominus}K^+$$

Potassium laurate

The sodium and potassium salts of the long-chain fatty acids are soluble in water, but their salts with many other metals are insoluble. This is responsible for one of the major drawbacks of soap since the insoluble salts can precipitate on clothing and washbasins as a dull, dirty scum. The problem is especially serious in areas with hard water. Hard water contains many dissolved salts, the most common ones being salts of Mg^{2+}, Ca^{2+}, and Fe^{3+}. When soap is mixed with hard water, the negatively charged fatty acid ions combine with these metal ions forming the insoluble scum that floats on the surface and can form the common bathtub ring. The bathtub ring (soap scum) reaction is given below:

$$2 \ CH_3(CH_2)_{14}-\overset{\overset{O}{\|}}{C}-ONa \ + \ CaCl_2 \ \longrightarrow$$

Sodium palmitate   Calcium chloride
(soluble in water)   (from hard water)

$$\left[CH_3(CH_2)_{14}-\overset{\overset{O}{\|}}{C}-O\right]_2Ca\downarrow + \ 2 \ NaCl$$

Calcium palmitate   Sodium chloride
(insoluble in water—   (soluble in water)
"the bathtub ring")

Synthetic detergents do not form precipitates in hard water and because of this they have largely replaced soaps in laundry use.

The Cleansing Action of Soaps and Detergents

Both soaps and detergents operate in a similar manner as cleansing agents. Both are composed of a large alkyl group with a polar acid group on one end that is in the form of a sodium or potassium salt. The alkyl end is nonpolar and hydrophobic ("water hating") and is not readily soluble in polar materials like water. It is very soluble, though, in nonpolar substances like greasy oils. The polar end is hydrophylic ("water loving") and is very soluble in polar solvents like water but not in oils. Soaps contain the carboxylate group on the polar end. Sodium stearate, an important part of many commercial bar soaps, is shown below:

$$CH_3CH_2CH_2CH_2CH_2CH_2CH_2CH_2CH_2CH_2CH_2CH_2CH_2CH_2CH_2CH_2CH_2 \overset{\overset{\displaystyle O}{\parallel}}{-C} -O^{\ominus}Na^+$$

Hydrophobic end
(soluble in oils and grease)

Hydrophylic end
(soluble in water)

Most of the soil on our clothing and skin is combined with the natural oils from the body and oils from other sources. When soap is mixed with water in the presence of this greasy dirt, the hydrophobic end of the soap (the long carbon chain) dissolves in the oil leaving the negatively charged hydrophylic end on the outside dissolved in the water. This dual solubility behavior is responsible for the cleaning action of soap (see Figure 20.2). As more soap dissolves in the oily dirt, it is released into the water forming an emulsion (a dispersion of one liquid in another). The scrubbing action that accompanies the cleansing process breaks up this emulsion into much smaller bodies known as **micelles**. Once formed, micelles will not recombine since their surfaces are coated with the negatively charged carboxylate ends of the soap, and bodies with a like charge repel each other. With rinsing, the micelles containing the oil and dirt are washed away. Soap also aids the cleansing process in another way by lowering the surface tension of water. This makes it easier for the soap–water mixture to get into and surround the oily dirt.

Detergents can be made from petroleum as well as from fats, and they have structures that are similar to soaps. One end of the structure is a long string of carbon atoms that is soluble in greases and oils, while the other end is polar and capable of forming ions in water. This dual solubility allows a detergent to act much like a soap, but with one major difference. The polar end of the detergent molecule is not a carboxylate group as is found in soaps, but rather may be a sulfonate group ($-SO_3^-$), sulfate group ($-SO_4^-$), or other acid group that will not form water-insoluble precipitates with the metal ions usually found in hard water. Consequently, detergents will not produce the dulling scum that is so common when soaps are used.

The first synthetic detergents, though satisfactory in the laundry room, caused major problems in municipal sewage plants and lakes and rivers, since they were not **biodegradable**. The carbon chain on the early detergents contained branches and rings of carbon atoms which resisted degradation by bacteria. The result was an abnormal buildup of detergents in water that resulted in foam-

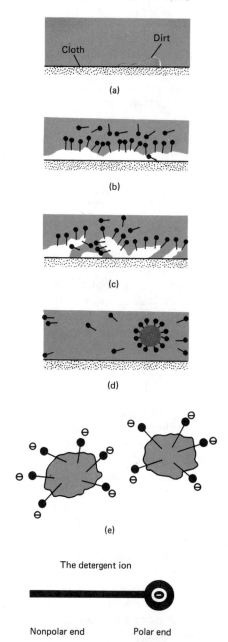

The detergent ion

Nonpolar end   Polar end

Fig. 20.2.   The dual solubility behavior of a soap or detergent is responsible for its cleansing action. (a) Greasy dirt on a piece of cloth. (b) When placed in a solution of detergent in water, the hydrophobic ends of the detergent ions become dissolved in the greasy dirt. (c) The greasy dirt is broken into smaller particles and pulled away from the cloth, and (d) brought into the water forming micelles and is held in suspension. (e) The micelles do not group back together because their surfaces bear the negative charge of the polar ends of the detergent ions and are repelled by nearby micelles.

clogged sewage plants and water systems. The problem was solved by changing the carbon chain to one that looks more like those found in the original soaps, which were known to be biodegradable:

$$CH_3(CH_2)_{10}CH_2-O-\overset{\overset{O}{\|}}{\underset{\underset{O}{\|}}{S}}-O^{\ominus}Na^+$$

A biodegradable detergent

$$CH_3-\overset{\overset{CH_3}{|}}{\underset{\underset{CH_3}{|}}{C}}-CH_2-\overset{\overset{CH_3}{|}}{\underset{\underset{CH_3}{|}}{C}}-CH_2-\overset{\overset{CH_3}{|}}{CH}-\langle\bigcirc\rangle-\overset{\overset{O}{\|}}{\underset{\underset{O}{\|}}{S}}-O^{\ominus}Na^+$$

A nonbiodegradable detergent

Compound Lipids

Compound lipids are of physiological rather than nutritional importance. There are three principal kinds of compound lipids: phospholipids, sphingolipids, and glycolipids. The phospholipids produce glycerol when hydrolyzed, but sphingolipids and glycolipids do not.

Phospholipids
The phospholipids are the second most abundant kind of naturally occurring lipid. (Remember, the triglycerides are the most abundant.) The phospholipids are involved in the transport of fatty acids in the blood system, but their main function is in the formation of cell membranes.

Phospholipids contain two fatty acid units and a phosphate group attached to glycerol by ester linkages, and thus are considered as derivatives of **glyceryl phosphate**. A **phosphatidic acid** can be considered an ester of glyceryl phosphate formed by its reaction with two fatty acids. Several different fatty acids can be present in phosphatidic acids, but usually one is saturated and the other one is unsaturated.

$$\begin{array}{l}
H_2C-OH \\
\ | \\
H-C-OH \\
\ | \quad\quad\overset{O}{\|} \\
H_2C-O-P-OH \\
\quad\quad\quad | \\
\quad\quad\quad OH
\end{array}$$

Glyceryl phosphate

$$\begin{array}{l}
\quad\quad\quad\quad\overset{O}{\|} \\
H_2C-O-C-R \\
\ | \quad\quad\overset{O}{\|} \\
H-C-O-C-R' \\
\ | \quad\quad\overset{O}{\|} \\
H_2C-O-P-OH \\
\quad\quad\quad | \\
\quad\quad\quad OH
\end{array}$$

A phosphatidic acid

Phospholipids are esters of the phosphatidic acids, formed by linking various nitrogen-containing groups to the phosphate group. The **phosphatides** are the largest group of phospholipids and are formed when phosphatidic acids react with nitrogen-containing alcohols. The major nitrogen-containing alcohols are shown below. The structures are drawn with a positive charge on the nitrogen atom of each, since they exist in this form in the phosphatides.

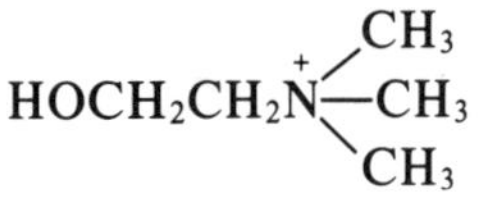

$$HOCH_2CH_2\overset{+}{N}\!\!\overset{\displaystyle CH_3}{\underset{\displaystyle CH_3}{\diagup\!\!\diagdown}}\!\!CH_3 \qquad HOCH_2CH_2\overset{+}{N}H_3 \qquad HOCH_2\overset{+}{\underset{\displaystyle COOH}{C}}HNH_3$$

| Choline | Ethanolamine | Serine (an amino acid) |

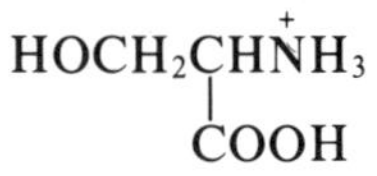

The general reaction for phosphatide synthesis is

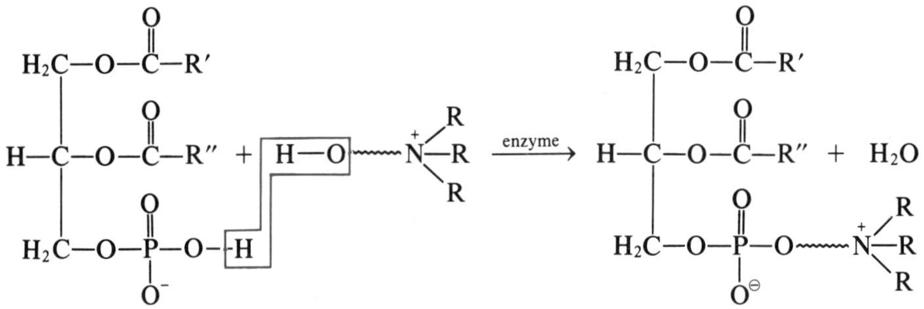

| Phosphatidic acid | A nitrogen-containing alcohol (R = —H or —CH₃) | A phosphatide (a phospholipid) |

Three important phosphatides are phosphatidyl choline, more commonly known as **lecithin**, and phosphatidyl ethanolamine and phosphatidyl serine, both known as **cephalins**:

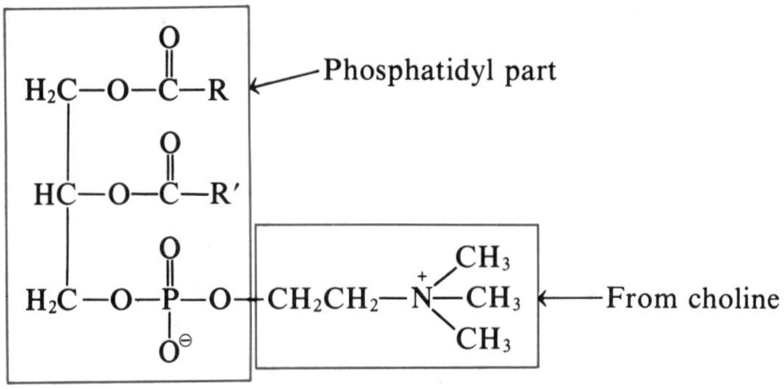

Phosphatidyl choline
(commonly called lecithin)

Phosphatidyl ethanolamine .... Phosphatidyl serine

(Both are known as cephalins)

The phospholipids contain a polar and a nonpolar end; thus they act as good emulsifying agents in biological systems. These phospholipids are found in layers in cell membranes and play a definite role in determining what enters and leaves the cell.

The lecithins are involved in the transport of triglycerides from one part of the body to another. The protoplasm of each cell contains small amounts of lecithins but their function within the cell is as yet unknown. Commercially, lecithins are extracted from soybeans and are used as emulsifying agents in food and candies.

The venom of some poisonous snakes and spiders contains enzymes that convert blood lecithins into lysolecithin, a compound that destroys blood cells:

Lecithin .... Lysolecithin

The cephalins are complex mixtures of phosphatidyl serine and phosphatidyl ethanolamine. The cephalins are found in brain and spinal tissue and are known to function in blood clotting.

Members of a third class of phosphatides are the **plasmalogens**. The only difference between plasmalogen and the other phosphatides is that the first carbon of glycerol is bonded to an unsaturated fatty acid alkyl group by an **ether**

linkage. The plasmalogens can have either ethanolamine, serine, or choline bonded to the phosphate group, as in the other phosphatides.

$$H_2C-O{\sim}CH{=}CH{\sim}R$$

$$H-\overset{\displaystyle |}{C}-O-\overset{\displaystyle \overset{O}{\|}}{C}-R'$$

$$H_2\overset{\displaystyle |}{C}-O-\overset{\displaystyle \overset{O}{\|}}{\underset{\underset{O^{\ominus}}{|}}{P}}-O-\boxed{\begin{array}{l}\text{Nitrogen*}\\ \text{containing}\\ \text{group}\end{array}}$$

A plasmalogen   *Ethanolamine, serine, or choline.

The plasmalogens are found in heart and brain tissue and are not found in plants. The functions of the plasmalogens are unclear.

Cell Membranes   The cell is surrounded on all sides by the **cell membrane**, which serves as a structural boundary, separating the cell's contents from its environment. The most important function of the cell membrane, however, is that it controls the flow of substances into and out of the cell. Cell membranes in plants and animals are composed of phospholipids and proteins. The phospholipids can be viewed as long, stringlike molecules with two nonpolar fatty acid chains pointing in one direction, and a shorter, polar chain pointing in the opposite direction. This polar–nonpolar arrangement of the phosphatide gives it good membrane-forming characteristics.

Nonpolar end
insoluble in water, but soluble
in nonpolar substances

Polar end
soluble in water, but insoluble
in nonpolar substances

It will be more convenient to use a simple picture to represent the phospholipid molecule. A circle will be used to symbolize the polar end of the molecule, and two wavy tails will represent the nonpolar fatty acid chains:

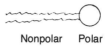

Nonpolar   Polar

The current model for the cell membrane places the phospholipids in a double layer, with their polar ends on the outside and their nonpolar tails together on the inside. It is something like a sandwich, with the tails between two polar layers. This **bilayer**, as it is called, is about 0.00001 mm thick and is composed of a mixture of phospholipids (see Figure 20.3). The properties of the membranes will be determined by the kinds of fatty acids and the kinds of polar groups in the phospholipids.

Protein molecules are anchored to the bilayer in several different ways. The various arrangements of protein in the cell membrane are listed below and illustrated in Figure 20.4.

1. Some protein molecules are on the outer surface of the cell exposed to the cell's environment.

2. Some form channels from the outer surface to the inner surface.

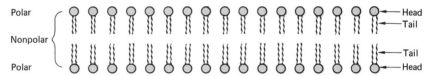

Fig. 20.3. The arrangement of phospholipids in the bilayer. The polar ends form the outer surface of the bilayer, and the nonpolar fatty acid chains are together on the inside. The polar–nonpolar property of the phospholipids gives them good membrane-forming characteristics.

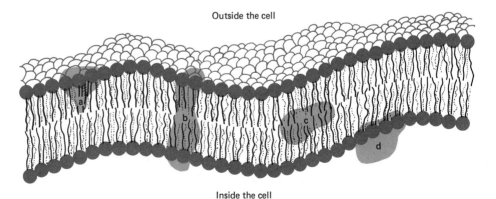

Fig. 20.4. The cell membrane contains protein molecules distributed throughout the phospholipid bilayer. (a) Protein exposed to the outer surface of the cell; (b) protein passing through the bilayer forming a channel across the membrane; (c) protein imbedded within the nonpolar region of the bilayer; and (d) protein on the inner surface of the bilayer facing the cell contents.

3. A few protein molecules are imbedded within the bilayer.

4. Some protein molecules are on the inner surface of the bilayer exposed to the contents of the cell.

Lipid bilayers are highly resistant to the transfer of polar substances across their nonpolar inner regions. An ion cannot easily pass through this region, though neutral molecules, like water, can with little difficulty. Ionic and highly polar molecules can enter the cell, though, through the proteins associated with the bilayer. The proteins, then, perform special functions that the phospholipids cannot do, and vice versa. In this way the cell can have a stable membrane that can pass both neutral and ionic materials in both directions.

Sphingolipids ꞏ The sphingolipids are compound lipids that may be subclassified into the **sphingomyelins** and the **glycolipids**. Though the sphingolipids are not derivatives of glycerol, they are derivatives of a long-chain amino alcohol, **sphingosine**:

$$CH_3(CH_2)_{12}-CH{=}CH-\overset{\overset{\displaystyle H}{|}}{\underset{3}{C}}-OH$$
$$H-\overset{}{\underset{2}{C}}-NH_2$$
$$\underset{1}{CH_2OH}$$

Sphingosine

Note the numbering of the carbon atoms in the structure of sphingosine. The sphingomyelins contain a fatty acid alkyl group bonded to the second carbon of sphingosine through an **amide linkage** $-\overset{\overset{\displaystyle H}{|}}{N}-\overset{\overset{\displaystyle O}{||}}{C}-$ and a phosphocholine bonded to the first position:

$$CH_3(CH_2)_{12}-CH{=}CH-\overset{\overset{\displaystyle H}{|}}{C}-OH$$
$$H-\overset{}{C}-\overset{\overset{\displaystyle H}{|}}{N}-\overset{\overset{\displaystyle O}{||}}{C}-R$$
$$CH_2-O-\overset{\overset{\displaystyle O}{||}}{\underset{\underset{\displaystyle O^{\ominus}}{|}}{P}}-O-CH_2CH_2-\overset{+}{N}{\Big\langle}^{CH_3}_{CH_3}{-}CH_3$$

A sphingomyelin

Appreciable amounts of sphingomyelins are found in brain and nerve tissue. In Niemann–Pick disease, sphingomyelins that are constantly being produced by the body are not broken down as normal, leading to mental retardation. The disease is caused by a missing enzyme that is required for the correct breakdown of these compounds. As in almost all diseases that are the result of missing enzymes, there is no known treatment.

The **glycolipids** (cerebrosides) have structures similar to the sphingomyelins, except that a carbohydrate unit is bonded to the first carbon of sphingosine through an ether linkage instead of the phosphocholine structure. The monosaccharide (sugar) unit is usually galactose. Milk is needed by newborn infants not only for the energy it contains but also for the galactose which is used in the formation of glycolipids. Tay–Sachs disease is the result of an abnormal accumulation of glycolipids in brain tissue, again due to a missing enzyme required to break them down correctly.

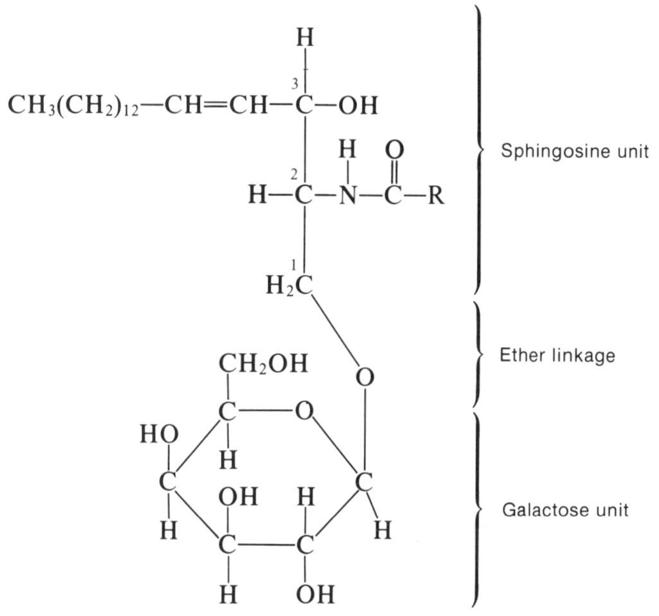

$$CH_3(CH_2)_{12}-CH{=}CH-\overset{\overset{\displaystyle H}{|}}{\underset{3}{C}}-OH$$

Sphingosine unit

Ether linkage

Galactose unit

A glycolipid

Steroids

The **steroids** are lipids which contain no ester functional unit in their main structure. Many different plant and animal steroids have been isolated, but none have been found in bacteria. All steroid structures are based on the **steroid**

nucleus, which is a four-member fused ring system. The steroid nucleus is shown below:

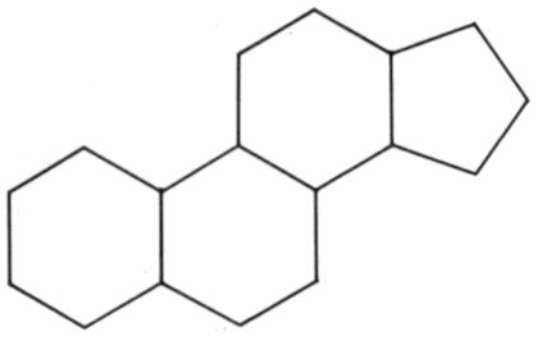

Steroid nucleus

Many modifications of the steroid nucleus are found in nature. One or more double bonds may exist within the ring framework, and various alkyl groups can be attached at several different places. If hydroxyl groups (–OH) are attached, the steroid is often called a sterol, which indicates its alcohol nature.

Cholesterol

Cholesterol is the most abundant sterol found in the human body, and bile salts, sex hormones, other hormones, and vitamin D are synthesized from it. Cholesterol seems to aid the absorption of fatty acids from the small intestine, and about two-thirds of the cholesterol in the blood is combined with unsaturated fatty acids through an ester linkage. The rest is in a free, uncombined form.

The overabundance of cholesterol and triglycerides in the bloodstream may be linked to **atherosclerosis**, a condition that results as lipids are deposited on the walls of arteries and vessels. (See Figure 20.5.) The lipid becomes associated with connective tissue in the artery, forming a plaque along the inner lining. This "roughening" of the vessel or artery wall may cause blood clots to form, which may partially or completely block the flow of blood. Occasionally, one of these clots will loosen and be carried by the blood to block a vessel or artery elsewhere in the body. If the blockage affects an artery feeding the heart, the result is a heart attack. If it blocks an artery in the area of the brain, a stroke results.

A diet that contains no cholesterol from animal fat, is low in saturated triglycerides, and has a reasonable number of calories has been shown to lower the blood levels of cholesterol and triglycerides. It should be pointed out that a diet free of cholesterol will not necessarily lower the blood cholesterol level, since it can be synthesized in the body very quickly from excess carbohydrates, proteins, and other fats.

Bile Salts

The **bile salts** are salts of certain steroids called the **bile acids**. These salts are very good emulsifying agents and aid in the digestion and absorption of fats. Bile is made in the liver, and besides the bile salts it contains cholesterol and bile pigments. After leaving the liver, the bile mixture is stored in the gallbladder and then empties into the small intestines when triglycerides enter. Gallstones are rocklike formations containing large amounts of hardened cholesterol which forms in the gallbladder. They can cause inflammation of the gallbladder and can obstruct the flow of bile to the small intestines.

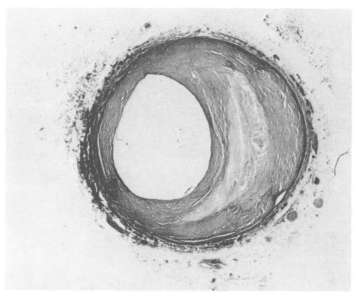

Fig. 20.5. Partially closed artery. (Earl P. Benditt, M.D., University of Washington, Seattle, Washington.)

Cholic acid
(a major bile acid)

Sodium salt of
glycocholic acid
(one of the bile salts)

Steroid Hormones of the Adrenal Cortex

The steroids produced by the adrenal cortex are important in controlling the metabolism of carbohydrates and proteins and maintaining the electrolyte and water balance of the body. If the gland does not produce sufficient amounts of these hormones, a condition known as Addison's disease results. In the past, Addison's disease was fatal, but now patients may be treated by receiving daily doses of these hormones.

The adrenocortical hormones are used in the treatment of rheumatic fever and arthritis. Examples of two of these hormones are given below, cortisone being the more familiar:

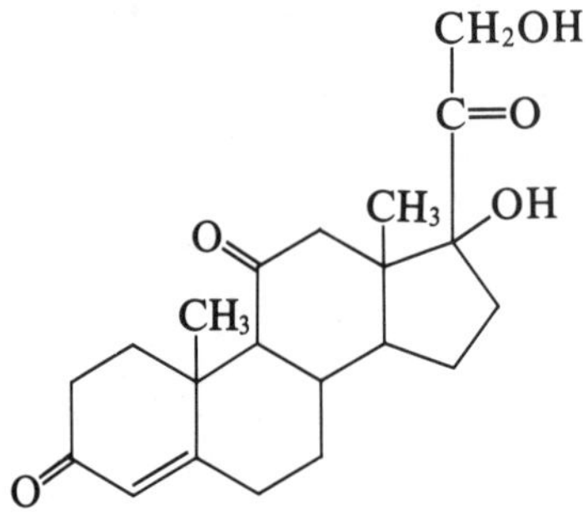

Cortisone

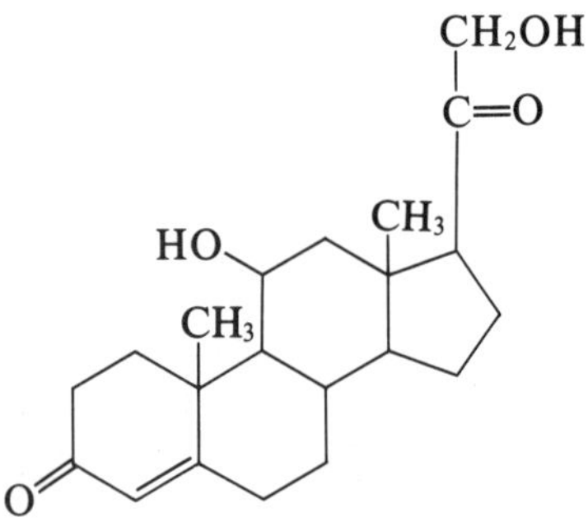

Corticosterone

Sex Hormones

Testosterone is one of the principal male hormones responsible for the development of the secondary sex characteristics. It is produced in the testes. Testosterone is converted into other hormones in the body, notably androsterone.

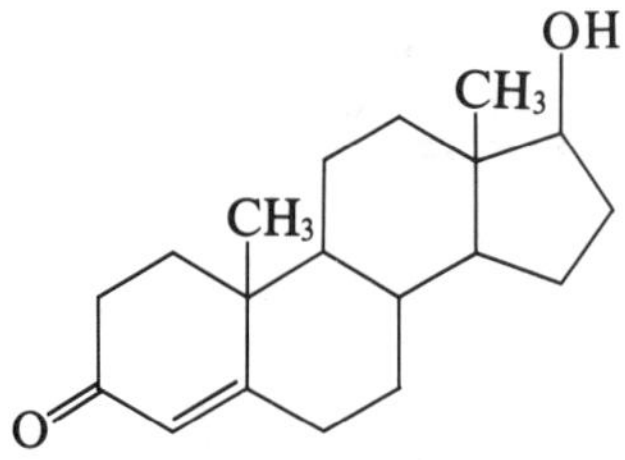

Testosterone

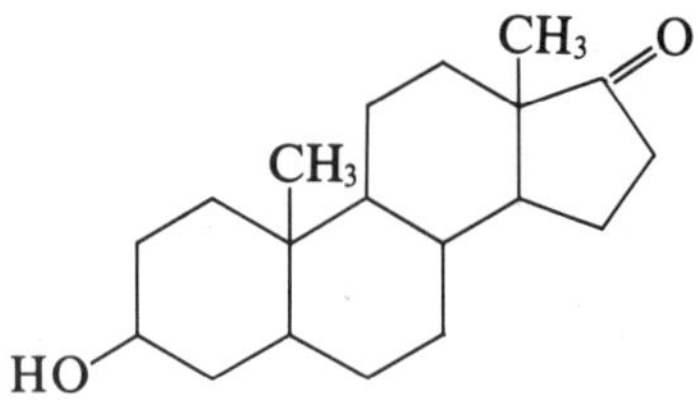

Androsterone

In the female, sex hormones are formed in the ovaries and they regulate the menstrual cycle and development of sexual characteristics. The menstrual cycle is controlled by two different kinds of hormones, **estrogens** and **progesterone**. The estrogens are responsible for the ovulation cycle and progesterone for the maintaining of pregnancy.

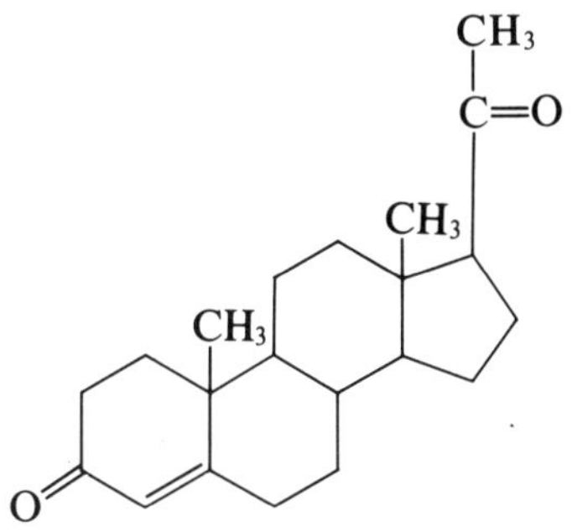

Progesterone

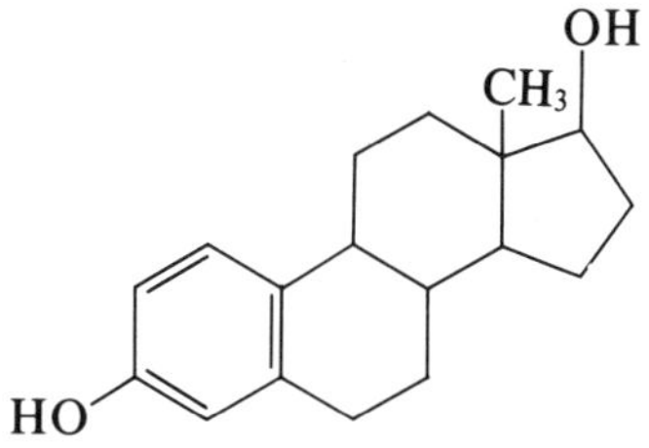

Estradiol (an estrogen)

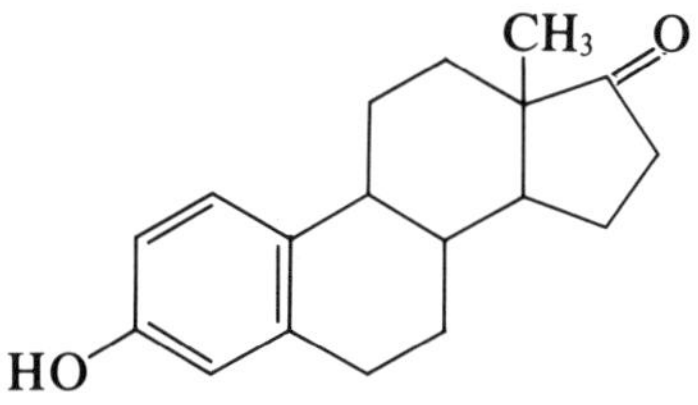

Estrone (an estrogen)

Ovulation in the female may be stopped by injections of progesterone–estrogen combinations; but the combination cannot be taken by mouth since they break down in the digestive system. This inconvenience was overcome when derivatives of the female hormones were developed that could be taken orally on a daily basis to prevent the release of the ovum. The first birth control pills were mixtures of synthetic estrogen–progesterone-like compounds. They prevented ovulation while allowing a normal menstrual cycle. Birth control pills manufactured after 1970 contain no estrogens but they do contain the more effective progesteronelike steroids. They prevent the migration of sperm into the oviduct.

Three synthetic derivatives of female sex hormones are shown below:

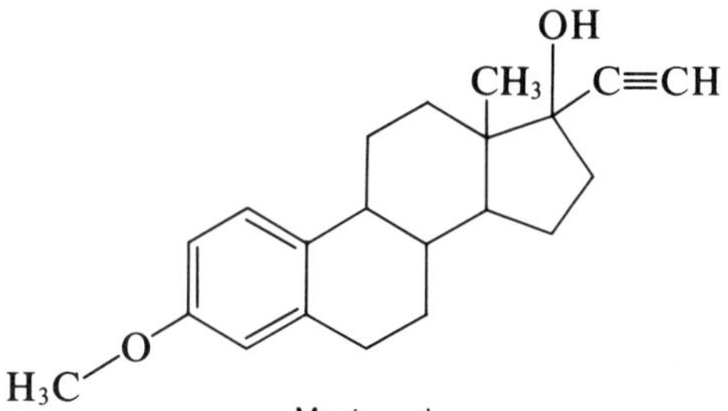

Mestranol
(estrogen-like)

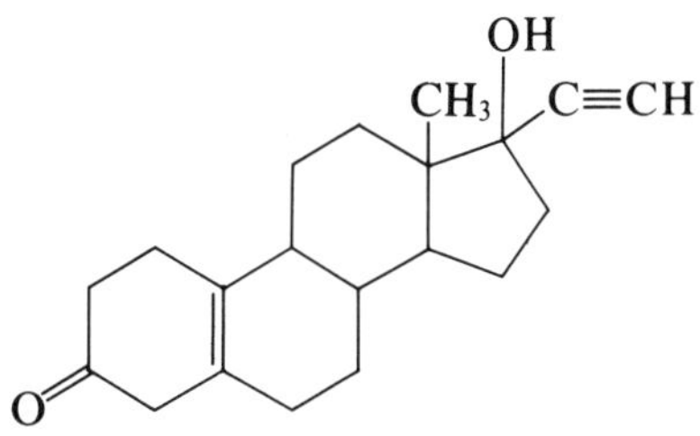

Norethynodrel
(progesterone-like)

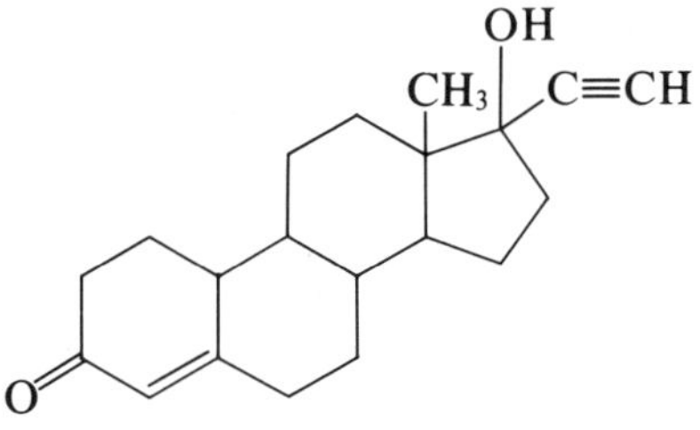

$$CH_3 \quad \overset{OH}{\underset{|}{C}} \equiv CH$$

Norethinodrone
(progesterone-like)

The Fat-Soluble Vitamins

Vitamins are organic compounds that are not synthesized by the body but must be present for good health. Many vitamins are actually mixtures of two or more similar compounds that perform the same basic function. These similar structures are known as **vitamers**. The structure given for a vitamin in this text will normally be of only one of its vitamers. The vitamins that will be discussed in this section are those found dissolved in plant and animal lipids, and as such are incorporated into our diets. The fat-soluble vitamins are vitamins D, E, A, and K. Vitamin D is a steroid, while the others are terpenes.

Vitamin D   Vitamin D is the only vitamin that is a steroid compound, and there are two principal vitamers, D_2 and D_3. The compound 7-dehydrocholesterol is formed in the body from cholesterol and is found in the skin. When the skin is exposed to ultraviolet radiation from the sun, 7-dehydrocholesterol is converted into vitamin D_3.

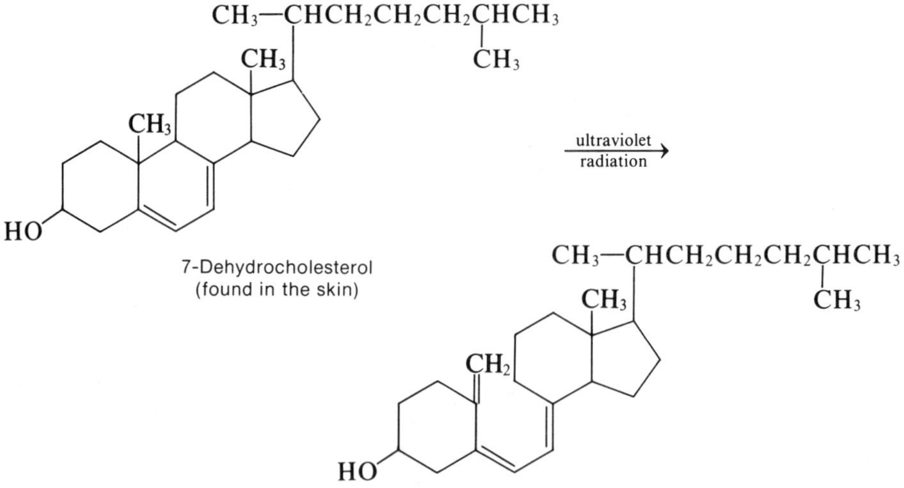

$$CH_3-CHCH_2CH_2CH_2CHCH_3$$

7-Dehydrocholesterol
(found in the skin)

$$\xrightarrow[\text{radiation}]{\text{ultraviolet}}$$

$$CH_3-CHCH_2CH_2CH_2CHCH_3$$

Vitamin D_3

The other vitamer, D_2, also known as calciferol, is obtained from yeast that has been irradiated with ultraviolet light. Notice that D_1 has not been mentioned. This is because what was originally believed to be vitamin D_1 turned out to be a mixture of D_2 and other compounds.

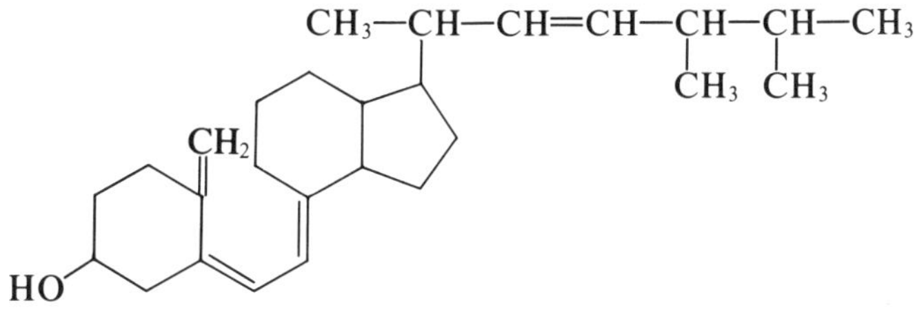

$$CH_3-CH-CH=CH-CH-CH-CH_3$$

Vitamin D_2 (calciferol)

Vitamin D is necessary for the absorption of calcium ions, Ca^{2+}, from food in the small intestines, and it is required for proper use of calcium and the phosphates in bone formation. A deficiency of this vitamin results in rickets, a disease that causes skeletal deformities such as bowlegs and knock knees. Vitamin D is found in fish liver oils and animal livers, but not in many other foods. Since milk is the main source of calcium for children, vitamin D_2 is added to it to provide a source that will ensure normal bone development.

Vitamin D is toxic in very large amounts and can cause softening of the bones due to calcium and phosphate loss. Bones of children and adults that have been affected are easily broken. Less severe effects of a large excess of vitamin D are nausea, diarrhea, and general weakness.

Terpenes   Vitamins A, E, and K are fat-soluble vitamins that are related to a subclass of lipid compounds known as terpenes. The terpenes are derivatives of **isoprene**, an unsaturated hydrocarbon that can join together with other isoprene molecules to form a wide variety of terpene compounds.

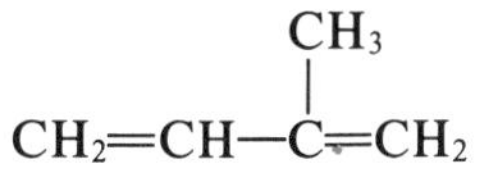

$$CH_2=CH-\underset{\underset{\displaystyle CH_3}{|}}{C}=CH_2$$

Isoprene (2-methylbutadiene)

Terpenes are important components of many aromatic oils (camphor, lemon, pine, mint) and plant pigments, like carotene, which is responsible for much of the orange-yellow color of carrots. The structure of β-carotene is shown below, and it is marked off to show several of the five carbon isoprene units. The structure of vitamin A is also shown. Notice how the structure of vitamin A is related to the structure of β-carotene. Carrots are considered an excellent source for vitamin A since the body can synthesize it from β-carotene.

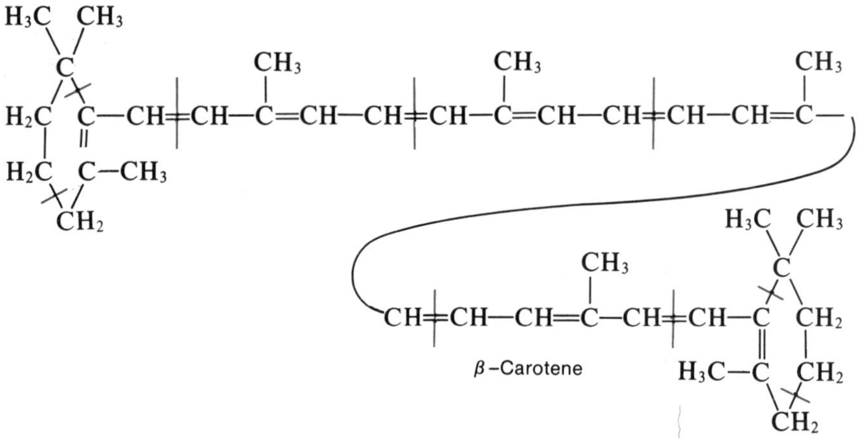

β-Carotene

Vitamin A An early sign of vitamin A deficiency is the inability to observe objects in dim light, a condition known as night blindness. A deficiency also increases the chances of respiratory tract infections and causes an impairment of kidney and reproductive functions. The main use of vitamin A in the body is its conversion, in the retina of the eye, into a compound required in the visual cycle. Low-level toxicity of this vitamin produces tender bones and some weakness; however, larger amounts cause headache, nosebleeds, and often dermatitis.

Leafy green and yellow vegetables and fruits supply most of the vitamin A in our diets. One of the most concentrated sources of this vitamin is fish liver oil.

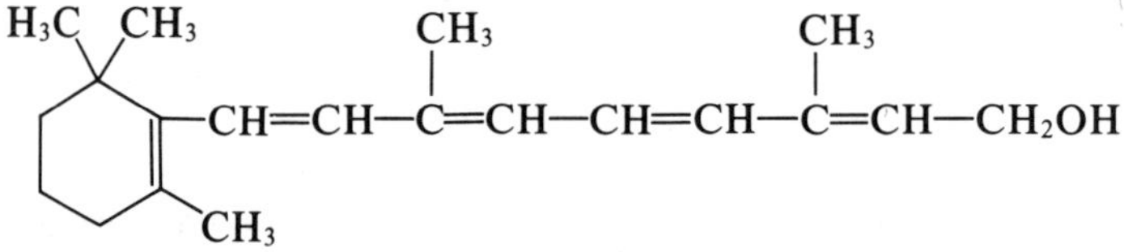

Vitamin A

Vitamin E Vitamin E contains an aromatic ring structure and a long chain of isoprenelike units. Vitamin E is found in abundance in green vegetables and vegetable oils. The exact function of vitamin E in the body is not known, but it may prevent the oxidation of unsaturated triglycerides in the body since it is known to be a good antioxidant.

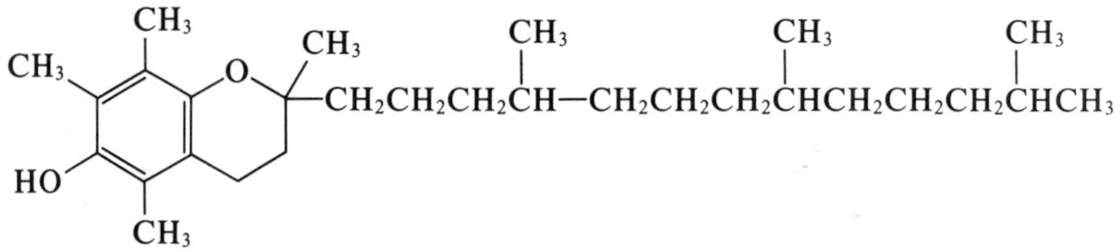

Vitamin E

Vitamin K Vitamin K is required for the rapid clotting of blood. Since a form of vitamin K can be synthesized by bacteria in the small intestines, conditions of vitamin deficiency are rarely encountered. Ingestion of large amounts of sulfa drugs can cause destruction of the intestinal bacteria, and a deficiency in vitamin K can result. Newborn infants may also show signs of deficiency until the bacteria are established in their intestines. Green leafy vegetables and tomatoes are good sources of vitamin K.

Vitamin K

Digestion and Absorption of Fats

The principal lipid present in the diet is the triglyceride. Triglycerides (fats and oils) pass unchanged through the digestive tract until they reach the small intestines. The presence of triglycerides triggers the gallbladder to pass bile into the small intestines. The bile, being slightly basic, can neutralize the acidic mixture of partially digested food from the stomach, but its main function is to emulsify the triglycerides into small droplets. The bile salts also aid the absorption of the fat-soluble vitamins from the small intestines.

The emulsified triglycerides are acted upon by **lipases**, enzymes supplied by pancreatic secretions. These hydrolytic enzymes catalyze the hydrolysis of triglycerides into glycerol and fatty acids, though some of the triglycerides may be reduced only to diglycerides or monoglycerides. The hydrolysis products are absorbed through the intestinal membrane and are then resynthesized into triglycerides or phospholipids. The resynthesized lipids are combined with protein in a lipoprotein complex and are transported in the lymph system until they enter the venous blood. The lipoprotein is then carried to the liver, where a portion of the triglycerides may be used for energy. The remaining lipoprotein is transported to the cells, where the protein is removed and the triglycerides are used for energy. If more triglycerides are present in the diet than are needed, then they are stored in the body as **adipose tissue**, the fatty layer of tissue just below the skin.

Terms

Several important terms appeared in Chapter 20. You should understand the meaning of each.

fat solvents	terpene	sphingosine
lipid	fatty acid	cholesterol
simple lipid	fats and oils	bile salts
wax	polyunsaturated fat	sex hormones
triglyceride	saponification	fat-soluble vitamins
compound lipid	soap	vitamers
phospholipid	detergent	isoprene
sphingolipid	lecithin	lipase
steroid	cephalins	lipoprotein

Questions

Answers to the starred questions appear at the end of the text.

*1.   What property of the lipids allows them to be classed together as a family?

2.   What are the five subclasses of lipids?

*3.   What is a fatty acid?

4.   Give general structural formulas for the following.
   *a. a saturated fatty acid
   *b. an unsaturated fatty acid
   c. a triglyceride

5.   Which lipid is found in both plants and animals? How does this lipid differ in each?

6.   As the number of carbon atoms in the alkyl group of saturated fatty acids increases, what trend is seen in their melting points? How are the melting points affected by an increasing number of carbon-to-carbon double bonds?

*7.   What reactions cause a fat or oil to turn rancid? How can this be prevented?

8.   Draw the structure of the following species.
   *a. a triglyceride that contains oleic, stearic, and lauric acids arranged in the order given
   *b. the general structure of a mono- and a diglyceride (mono- the first position, di- the first and second positions)
   c. glyceryl myristostearolaurate

9.   What is meant by the term hydrogenation? How is hydrogenation used commercially? Give an equation that describes the hydrogenation of a triglyceride that contains oleic, stearic, and linolenic acids.

*10. Give a general equation for the saponification reaction. Describe the products of the saponification reaction and give their uses.

11. What causes the "bathtub ring"?

12. In what way are detergents better than soaps? How do both work in the cleansing process?

*13. Give the general reaction describing the hydrolysis of a triglyceride. How is this reaction important in digestion?

*14. Draw the general structure of a wax. From what compounds are waxes made?

*15. What is a prostaglandin? Name some uses of prostaglandins.

16. What is a phospholipid? Is lecithin a phospholipid?

*17. What are the two categories of phospholipids? How is one category different from the other?

18. What are glycolipids? Where are they found in the body?

*19. Sphingosine is found in what kind of lipid?

20. Draw the basic four-member fused-ring structure found in all steroids. Which steroid is most abundant in the body?

*21. List some of the steroids found in the body. Give the uses of each.

*22. What are the fat-soluble vitamins? Which ones are steroids?

23. For each of the fat-soluble vitamins, list the problems that would result if a deficiency of each existed in the body.

24. Describe the digestion and absorption process for lipids.

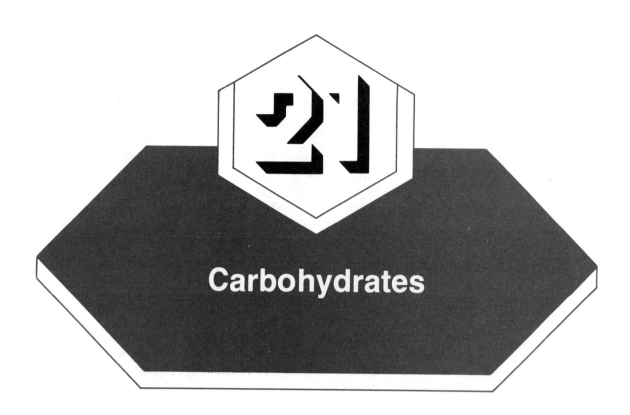

Carbohydrates

Carbohydrates are widely distributed in nature. Plants synthesize carbohydrates from carbon dioxide and water in the presence of sunlight. The process is called photosynthesis. Between 60% and 90% of the dry weight of plants is carbohydrate. They use it as a source of energy (glucose), as a reserve food supply (starch), and as structural material (cellulose). Animals are unable to synthesize carbohydrates and so must obtain them from plants in the diet. Less than 1% of the dry weight of animal tissue is carbohydrate, but this small amount is of critical importance. Carbohydrates are the primary source of energy for the body. Glucose circulates in the blood and is delivered to all areas of the body where it is used as a fuel in cellular respiration.

The discussion of carbohydrates will include descriptions of several important members of the carbohydrate family, from the simplest small-molecule sugars to the large polymeric substances like starch and cellulose. The role of glucose in diabetes and the digestion and absorption of carbohydrates will also be included. The chapter will end with a brief discussion of optical isomerism, a subtle type of isomerism exhibited by many carbohydrates.

Objectives

By the time you finish this chapter you should be able to do the following:

1. Describe the functional groups found in carbohydrates.
2. Define monosaccharide, disaccharide, polysaccharide, ketose, aldose, pentose, and hexose.
3. Write the equation describing the photosynthesis reaction, and describe how energy is stored in plants.
4. Draw the open-chain structures of an aldopentose, and an aldohexose.
5. Draw Haworth structures and Fischer structures for ribose, glucose, and fructose and identify the anomeric carbon atom.
6. Name three important disaccharides and name the two monosaccharide units of each.
7. Name some of the important polysaccharides in the body and discuss their function.
8. Describe three tests used to detect or measure glucose in body fluids.
9. Describe how carbohydrates are digested and absorbed in the body.
10. Discuss the relationship between blood sugar levels and diabetes.
11. Explain what is meant by optical activity, and describe what kinds of molecules will display optical activity.
12. Place a monosaccharide in either the D or L family given its Fischer structure.

Photosynthesis

Carbohydrates are produced in chlorophyll-containing green plants by a process called **photosynthesis**. In this complex process, carbon dioxide and water are combined in the presence of sunlight to form carbohydrates and oxygen. The process requires an input of energy. Chlorophyll, which is able to absorb light energy from the sun, is able to convert this energy into chemical energy as the carbohydrates are synthesized. The energy is then stored as chemical energy in the "higher-energy" carbohydrate molecules. Photosynthesis is one of the most important chemical reactions on the earth, and though the process proceeds through many steps, it can be summed up in a rather simple equation:

$$n\,CO_2 + n\,H_2O \xrightarrow[\text{chlorophyll}]{\text{sunlight}} (CH_2O)_n + n\,O_2$$
$$\text{Carbohydrate}$$

The photosynthesis of glucose, a monosaccharide, would be described in the following equation:

$$\text{energy} + 6CO_2 + 6H_2O \xrightarrow[\text{chlorophyll}]{\text{sunlight}} C_6H_{12}O_6 + 6O_2$$

The glucose units and those of other monosaccharides can then be joined together in plants to form disaccharides and polysaccharides, notably starch and cellulose. Various enzymes in plants are needed to catalyze these reactions.

$$2C_6H_{12}O_6 \xrightarrow{\text{enzymes}} C_{12}H_{22}O_{11} + H_2O$$

Monosaccharides $\qquad$ Disaccharide

$$nC_6H_{12}O_6 \xrightarrow{\text{enzymes}} (C_6H_{10}O_5)_n + nH_2O$$

Monosaccharides $\qquad$ Polysaccharide

It should also be noted that plants are also able to prepare amino acids (for protein synthesis) and a variety of lipids such as fats, waxes, and steroids during photosynthesis.

The chemical energy stored in carbohydrates can be released as they are metabolized in the body to form carbon dioxide and water. The equation that sums up the metabolism of glucose is the reverse of that describing photosynthesis except that the energy component is not light energy.

$$C_6H_{12}O_6 + 6O_2 \xrightarrow{\text{enzymes}} 6CO_2 + 6H_2O + \text{energy}$$

At this point, you can perhaps see how photosynthesis and metabolism depend on each other. Metabolism in animals consumes oxygen and releases carbon dioxide and water, both of which are in turn consumed by plants in photosynthesis, which in turn produce oxygen. Plants and animals are important parts of the oxygen–carbon dioxide cycle in nature.

Check Test Number 1

1. Briefly describe how light energy is stored in plants.
2. Concerning the oxygen–carbon dioxide cycle, how are plants and animals dependent on each other?

Answers:

1. Light energy is absorbed by chlorophyll, which aids in its conversion to chemical energy during carbohydrate synthesis. The high-energy carbohydrate molecules store the light energy as chemical energy.

2. Animals require O_2 and produce CO_2. Plants require CO_2 for photosynthesis, a process that produces O_2.

The Classes of Carbohydrates

Carbohydrates are formed from carbon, hydrogen, and oxygen, and they can be represented by the general formula $C_x(H_2O)_y$. Both x and y are small whole numbers that may be equal or unequal with x being larger than y. Early chemists believed these compounds were "hydrates of carbon" and so they were named carbohydrates. But as time passed, it became clear that carbohydrates were not hydrates at all, and a more appropriate definition was developed based on the functional groups they contained, usually several hydroxyl groups, $-OH$,

and a carbonyl group $-\overset{\overset{\displaystyle O}{\|}}{C}-$ which was either part of an aldehyde or ketone. Thus, carbohydrates can be described as either polyhydroxyaldehydes or polyhydroxyketones, or substances that yield these upon hydrolysis.

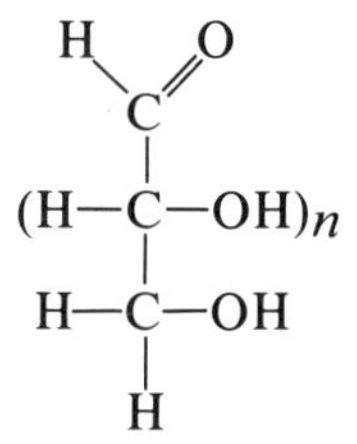

General formula of a
polyhydroxyaldehyde

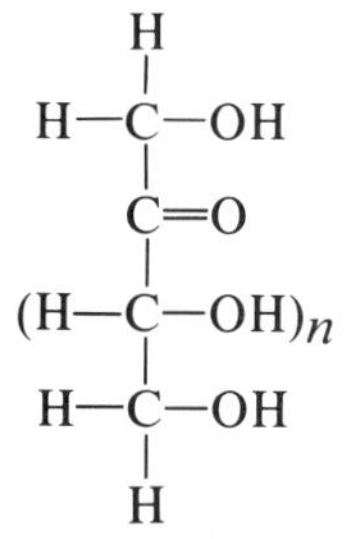

General formula of a
polyhydroxyketone

Perhaps the easiest way to classify carbohydrates is by their size, though they can also be classified by their carbonyl functional group as you will see later.

The simplest carbohydrates are the **monosaccharides**. The word saccharide comes from Greek *sakchar* meaning sugar. Monosaccharides are also called simple sugars, since they are not able to be broken down by acid hydrolysis (reaction with water in the presence of an acid) into smaller sugar molecules. Monosaccharides usually have the simple formula $(CH_2O)_n$ and they can contain three or more carbon atoms. The class names for monosaccharides are formed from a numerical prefix that signifies the number of carbon atoms they contain, joined to the *-ose* ending, as shown below.

Number of carbon atoms n	General name	General formula	Examples
3	triose	$C_3H_6O_3$	dihydroxyacetone
4	tetrose	$C_4H_8O_4$	erythrose, threose
5	pentose	$C_5H_{10}O_5$	xylose, ribose
6	hexose	$C_6H_{12}O_6$	glucose, fructose

Notice that different monosaccharides with the same formula are known, such as glucose and fructose (both $C_6H_{12}O_6$). These compounds differ in structure as we will see later, and can be considered isomers of one another. Larger saccharides, that is, those with larger molecules, are formed from the linking together of two or more monosaccharide units.

Oligosaccharides are composed of two to ten monosaccharide units joined together. When hydrolyzed, they are broken down into the monosaccharides. The most important oligosaccharides are those that contain just two saccharide units, and they are called **disaccharides**. Sucrose, lactose, and maltose are common disaccharides. Oligosaccharides containing three or more saccharide units are much less common than disaccharide.

Starch and cellulose are examples of **polysaccharides** which contain hundreds of simple sugar units joined together. These saccharide polymers can be broken apart by hydrolysis to form oligo- and monosaccharides. The term **sugar** is generally used to denote the sweet tasting crystalline, water-soluble monosaccharides and disaccharides. The larger oligo- and polysaccharides are water insoluble and tasteless.

monosaccharides $\xrightarrow{\text{hydrolysis}}$ no reaction

disaccharides $\xrightarrow{\text{hydrolysis}}$ 2 monosaccharides

oligosaccharides $\xrightarrow{\text{hydrolysis}}$ 2 to 10 monosaccharides

polysaccharides $\xrightarrow{\text{hydrolysis}}$ many monosaccharides (usually glucose)

A further step in the classification of **monosaccharides** takes into account the nature of the carbonyl group in the molecule. If the carbonyl is in an aldehyde group $-C\!\!\begin{array}{c}\nearrow O\\[-2pt]\searrow H\end{array}$, the monosaccharide is called an **aldose**, and if it is a ketone $-\overset{|}{\underset{|}{C}}-\overset{O}{\overset{\|}{C}}-\overset{|}{\underset{|}{C}}-$ the monosaccharide is a **ketose**. Glucose is an aldohexose, a six-carbon-atom saccharide with an aldehyde group. Fructose is a ketohexose, a six-carbon-atom saccharide with a ketone group.

Check Test Number 2

1. Indicate which of the following monosaccharides would be classed as aldose, ketose, ketopentose, hexose, polyhydroxyketone, aldohexose, carbohydrate:

a.

$$
\begin{array}{c}
H\diagdown \\
\quad C^{\diagup\!\!\diagup O} \\
| \\
H\!-\!C\!-\!OH \\
| \\
HO\!-\!C\!-\!H \\
| \\
H\!-\!C\!-\!OH \\
| \\
H\!-\!C\!-\!OH \\
| \\
H\!-\!C\!-\!OH \\
| \\
H
\end{array}
$$

b.

$$
\begin{array}{c}
H_2C\!-\!OH \\
| \\
C\!=\!O \\
| \\
H\!-\!C\!-\!OH \\
| \\
H\!-\!C\!-\!OH \\
| \\
H\!-\!C\!-\!OH \\
| \\
H
\end{array}
$$

2. Define hydrolysis.

Answers:
1. aldose, (a); ketose, (b); ketopentose, (b); hexose, (a); polyhydroxyketone, (b); aldohexose, (a); carbohydrate, (a) and (b).
2. Hydrolysis is a chemical reaction in which water is a reactant.

Monosaccharides

Monosaccharides are simple sugars that are white, water-soluble crystalline compounds, and most of them have a sweet taste. The **hexoses** (six-carbon-atom sugars) are the most important monosaccharides, and most of the carbohydrates in our diet contain hexoses either as free molecules or combined with one another in disaccharides and polysaccharides. The **trioses** (three-carbon-atom sugars) are important metabolic intermediates in muscle metabolism. Tetroses (four-carbon-atom sugars) are not of great biological importance. Two **pentoses** (five-carbon-atom sugars) are important though, since they are essential parts of the ribose nucleic acids (RNA) and deoxyribose nucleic acids (DNA).

Trioses:
$C_3H_6O_3$
The simplest monosaccharides are the trioses. They arise from the degradation of glucose in metabolism. Glyceraldehyde is an aldotriose, and dihydroxyacetone is a ketotriose.

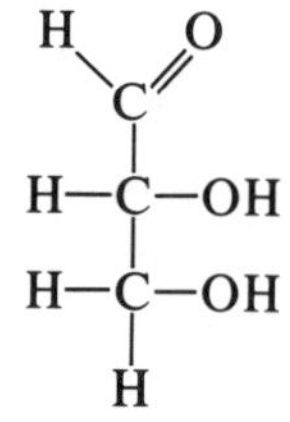

Glyceraldehyde

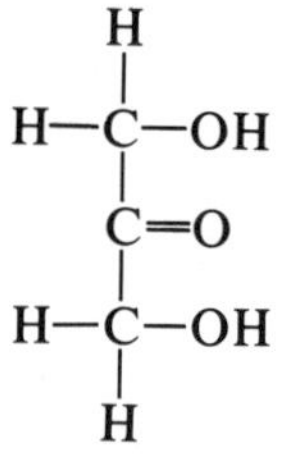

Dihydroxyacetone

Dihydroxyacetone is also used in some suntan lotions to produce an "instant synthetic suntan." Both trioses appear as phosphate derivatives of the above structures in muscle metabolism.

Pentoses: $C_5H_{10}O_5$ (and $C_5H_{10}O_4$)

The most important pentoses are ribose and deoxyribose (it has one less oxygen atom than ribose). They are components of the nucleic acids, ribonucleic acid (RNA) and deoxyribonucleic acid (DNA), which are in the nucleus of every cell and, acting together, control heredity and protein synthesis. Ribose is also found as an intermediate in carbohydrate metabolism, and is incorporated in other important biological molecules as well. Both ribose and deoxyribose are aldopentoses. Though similar in structure to ribose, deoxyribose lacks an oxygen atom at carbon number 2, that is, the carbon atom adjacent to the aldehyde group carbon atom.

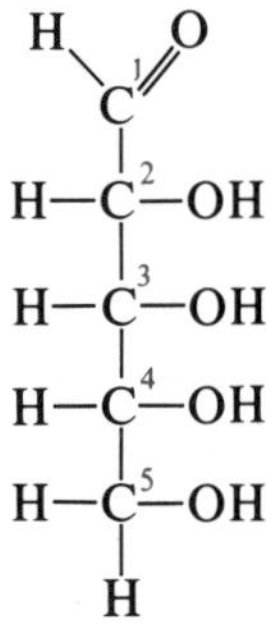

Ribose
$C_5H_{10}O_5$

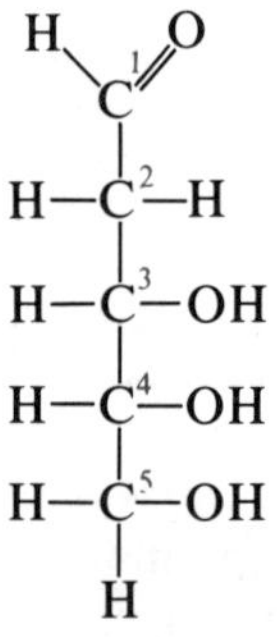

Deoxyribose
$C_5H_{10}O_4$

The above structures for ribose and deoxyribose are the **open-chain formulas** for these monosaccharides. Though the open-chain structures show the correct functional groups on each carbon atom, they do not correctly represent the structure of the molecules as they most often exist. In solution, both ribose and deoxyribose exist predominately as cyclic molecules (closed-chain structures). The cyclic structures are formed when the flexible, open-chain molecule bends around to bring the aldehyde group up to the hydroxyl group on carbon atom 4, allowing them to react. The aldehyde group and the hydroxyl group can undergo an additional reaction forming a **cyclic hemiacetal**, as shown below:

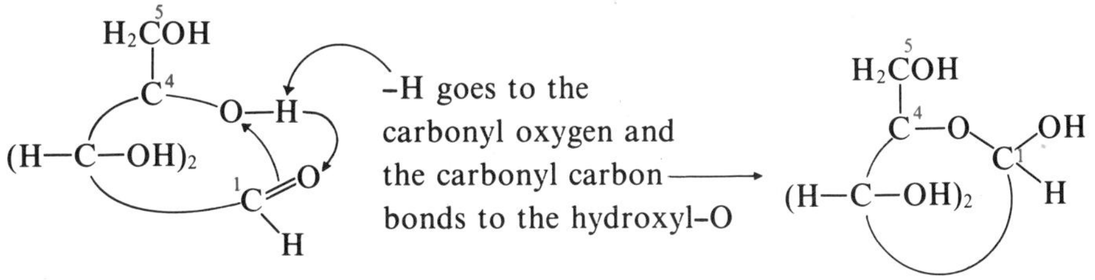

$$H_2\overset{5}{C}OH$$

$$\overset{4}{C}\!-\!O\!-\!H$$

$$(H\!-\!C\!-\!OH)_2 \quad \overset{1}{C}\!=\!O$$

$$H$$

−H goes to the
carbonyl oxygen and
the carbonyl carbon ⟶
bonds to the hydroxyl–O

$$H_2\overset{5}{C}OH$$

$$\overset{4}{C}\!-\!O\quad OH$$

$$(H\!-\!C\!-\!OH)_2 \quad \overset{1}{C}$$

$$H$$

Open-chain form   Cyclic form

The cyclic hemiacetal has an ether link and a hydroxyl group bonded to carbon atom 1, originally the aldehyde carbon. Because of the importance of this carbon atom in the ring, as you will see later, it is called the **anomeric carbon atom.**

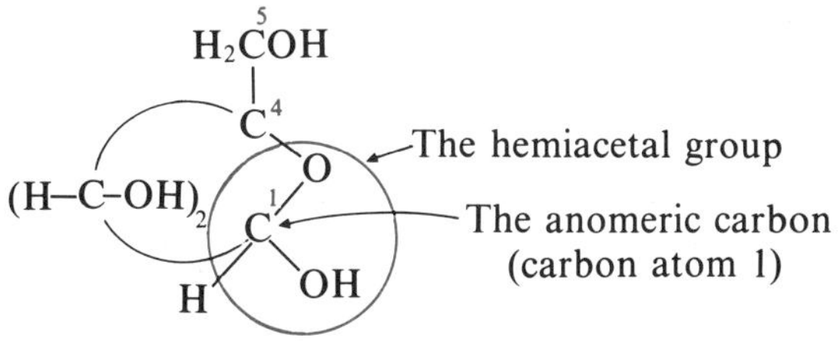

$$H_2\overset{5}{C}OH$$

$$(H\!-\!C\!-\!OH)_2 \quad \overset{4}{C}\;O$$

$$\overset{1}{C}$$

$$H \quad OH$$

The hemiacetal group

The anomeric carbon
(carbon atom 1)

The cyclic structures of ribose and deoxyribose are five-membered rings that are nearly planar. The convention is to draw the ring with the ring oxygen pointing to the rear and into the page and with the anomeric carbon on the right. Cyclic structures drawn in this way are called **Haworth structures** or **Haworth projections.** The five-membered ring is to be viewed as if it is going into and coming out of the page with the hydroxyl groups pointing above or below the plane of the ring. There are two cyclic forms of ribose:

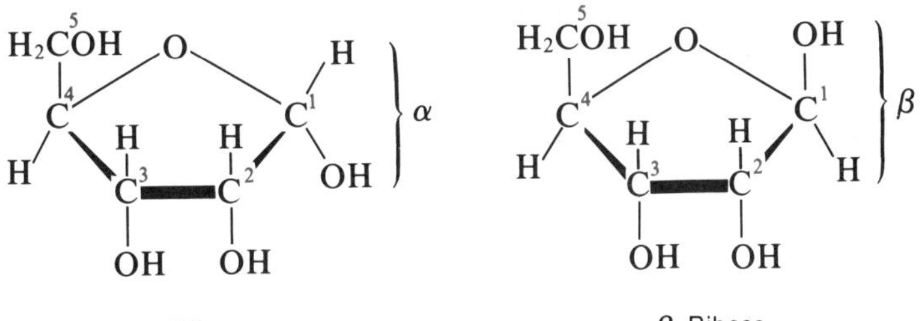

α–Ribose   β–Ribose

Abbreviated Haworth structures for α- and β-ribose show the orientation of the hydroxyl groups about a simplified five-membered ring:

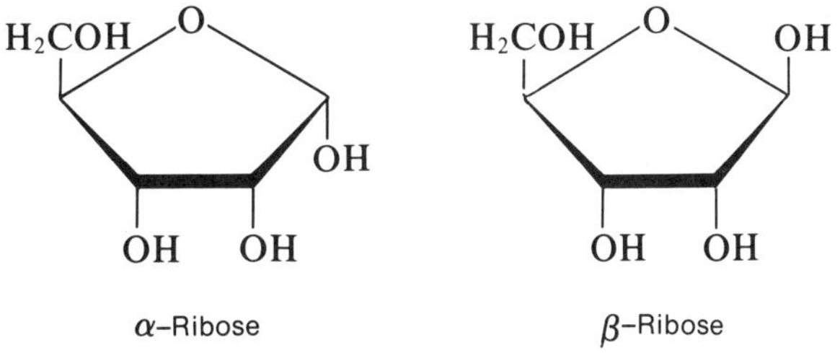

α–Ribose   β–Ribose

The **alpha** (α) and **beta** (β) forms of ribose differ only in the orientation of the hydroxyl group on the anomeric carbon. If the anomeric hydroxyl is below the ring, it is the alpha form of ribose; if above the ring, it is the beta form of ribose. When the ring closes and the cyclic hemiacetal forms, the hydroxyl group will be above the ring in some cases and below the ring in the others, depending on the orientation of the aldehyde group (up or down) when ring closure occurs. Notice that there is no difference in the orientations of the other hydroxyl groups on the other carbon atoms in either structure. The α and β forms are determined *only* by the hydroxyl orientation about the anomeric carbon.

The difference between the alpha and beta forms of ribose may seem small at this point, but it turns out to have an important effect on the properties of polysaccharides, as you will see later.

The Haworth structures of deoxyribose are

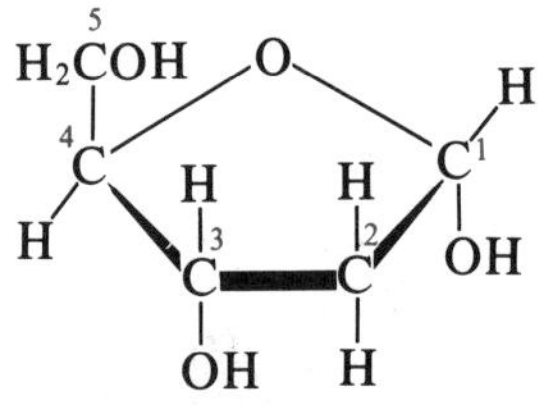

α–Deoxyribose 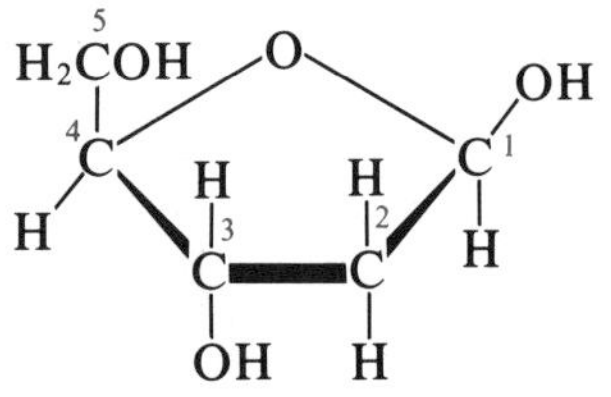 β–Deoxyribose

The abbreviated structures are

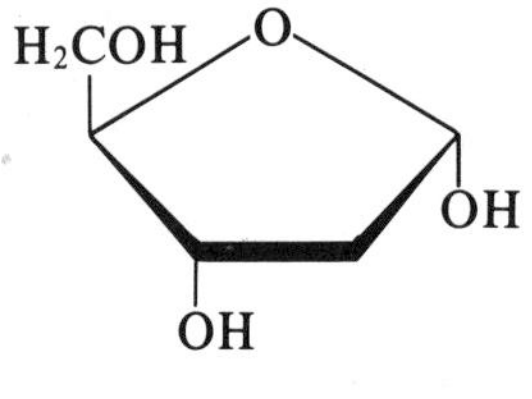

α–Deoxyribose 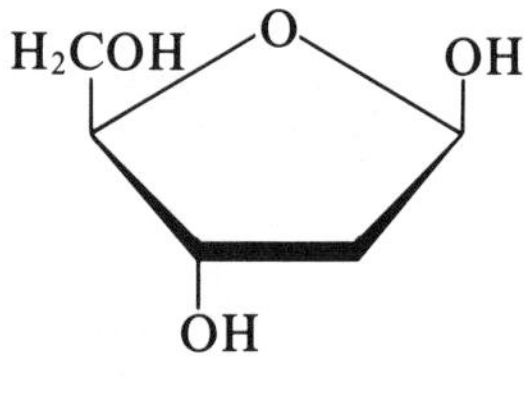 β–Deoxyribose

An alternate method that can be used to draw the structures of monosaccharides like ribose and deoxyribose uses **Fischer structures** or **Fischer projections**. The carbon chain is drawn vertically with the carbonyl group (in open-chain structures) on the anomeric carbon (in cyclic structures) at the top. The Fischer structure of the open-chain form of ribose is shown below:

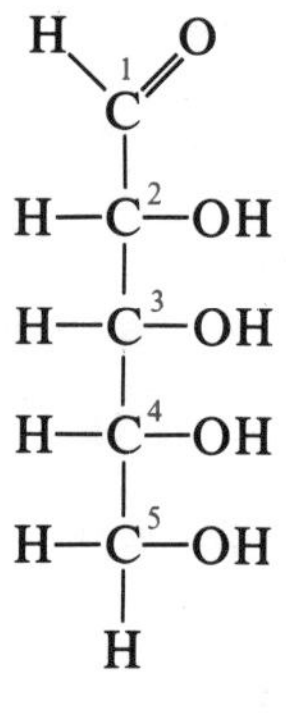

To represent the alpha and beta cyclic forms, the ether linkage is drawn to the *right side* of the vertical chain with bonds connecting the oxygen atom to carbon 1 and carbon 4. The alpha form of ribose, which has the anomeric hydroxyl below the ring on the Haworth structures, is drawn with the hydroxyl group to the *right*. The beta form has it drawn to the *left* of the anomeric carbon.

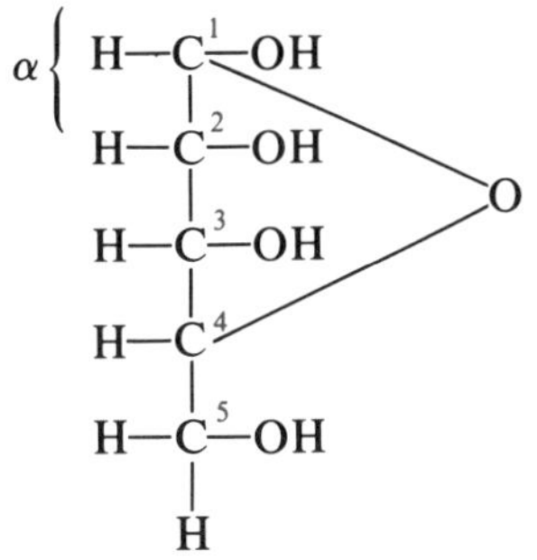

α–Ribose 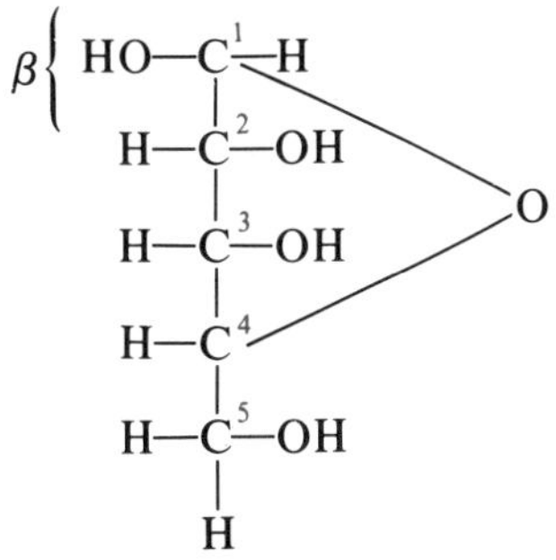   β–Ribose

Fischer structures are relatively awkward compared to Haworth structures, but both are widely used to represent structures of monosaccharides. Haworth structures can be changed to Fischer structures and vice versa by following two general rules governing orientation of the hydroxyl groups:

Fischer	**Haworth**
1. The –OH is placed to the *right* of the vertical chain	If –OH is ~~above~~ *below* the plane of the ring
2. The –OH is placed to the *left* of the vertical chain	If –OH is ~~below~~ *above* the plane of the ring

For comparison, the Haworth and Fischer structures for α-deoxyribose are given below. Note how the above rules are followed for the hydroxyl groups. The orientation of the hydroxyl group on carbon 5 which is not in the ring can be drawn either to the right or left, since it is not locked into any fixed orientation.

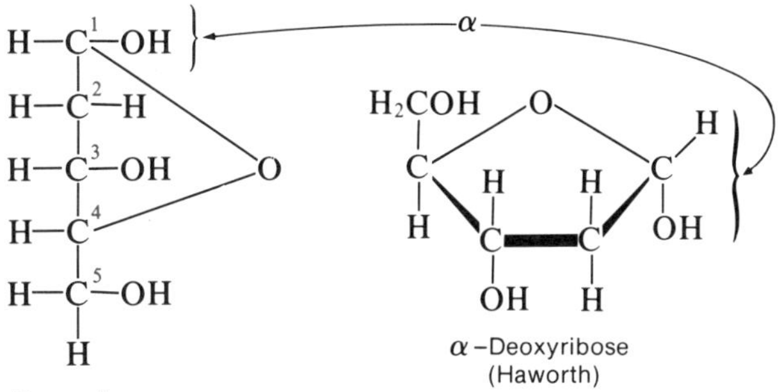

α–Deoxyribose
(Fischer)

α–Deoxyribose
(Haworth)

The cyclic hemiacetal structures of ribose and other monosaccharides are not very stable, and in solution they can open up to form open-chain molecules.

The open-chain molecules, which are present only in small amounts, can reclose forming the cyclic structure once again. But an interesting thing can happen in this process. An α-ribose cyclic molecule can open to form the open-chain species, which will then reclose to form a β-ribose molecule. If you prepared a solution of α-ribose in water, after a time the solution would contain both α-ribose and β-ribose in equilibrium with a small amount of the open-chain form of ribose. This conversion of alpha to beta and vice versa is termed **mutarotation**, and it is known to occur with many monosaccharides and disaccharides.

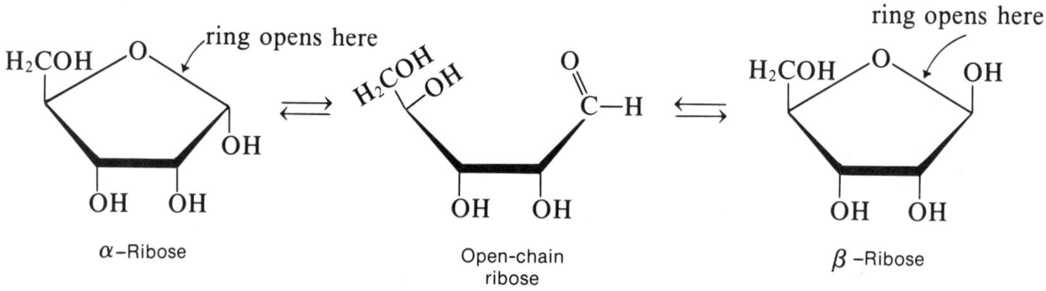

Chemical tests for monosaccharides that rely on the presence of a free aldehyde group can successfully be used with sugars that are over 99.9% in the cyclic form in solution. As the small amount of the open-chain structure (the only form with a free aldehyde group) is consumed by the test reagent, it is replenished as some of the cyclic forms open up to re-establish the equilibrium.

Hexoses:
$C_6H_{12}O_6$

There are many possible isomers for hexose monosaccharides, but only three of them are commonly encountered: glucose, fructose, and galactose. Glucose and galactose are aldohexoses and fructose is a ketohexose, as shown below in their open-chain Fischer structures:

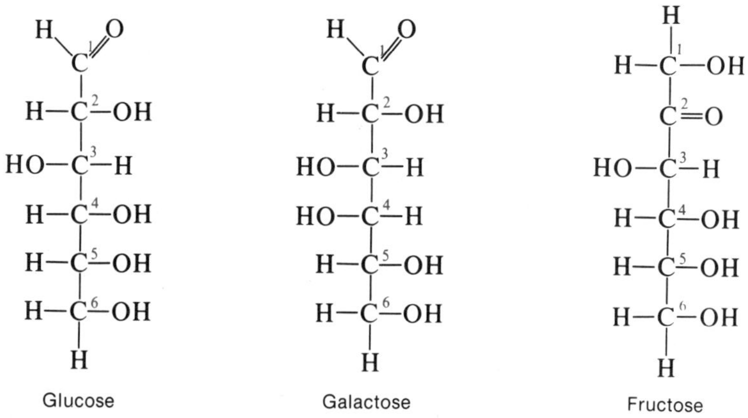

Glucose and galactose differ only in the orientation of the hydroxyl group on carbon atom number 4. Fructose has the same orientation of hydroxyl groups as glucose on carbon atoms 3, 4, and 5, but since fructose is a ketohexose, its first and second carbon atoms are unlike those in glucose, an aldose. Enzymes are present in animals that can convert one hexose into another.

Galactose is most commonly found combined with glucose in lactose (milk sugar), a disaccharide found in milk. Galactose is formed from glucose by enzymes in the mammary glands during the synthesis of lactose. It also appears in glycolipids, constituents of nerve and brain tissue. During pregnancy and lactation, galactose often appears in urine, which can lead to a false-positive "glucose test," though tests are available that can distinguish between these two sugars in urine. A rare congenital condition called **galactosemia** appears in infants who are unable to convert galactose to glucose because of an enzyme deficiency in the liver. Though these infants may appear normal at birth, the buildup of a galactose derivative in the tissue can cause irreparable brain damage and cataracts. Removal of milk from the child's diet (the source of lactose) until about the sixth year of life can considerably reduce the effects of the disease.

Galactose exists in alpha and beta cyclic hemiacetal structures formed by addition of the hydroxyl on carbon atom 5 to the aldehyde group, forming a six-membered ring. Carbon atom 1 is the anomeric carbon. In solution, the two cyclic structures are present in equilibrium with the open-chain structures seen earlier.

α –Galactose β –Galactose

The abbreviated Haworth structures for α- and β-galactose are

α –Galactose β –Galactose

Fructose is a very important ketohexose. It is combined with glucose to form the important disaccharide, sucrose (table sugar). Fructose is also known as fruit sugar or levulose, and it has a very sweet taste. A rare hereditary condition of fructose intolerance does exist which, if untreated, can lead to cirrhosis of the liver and mental deterioration. Treatment requires removal of all fructose and fructose-containing sugars from the diet. Fructose exists in both alpha and beta cyclic structures. The hydroxyl on carbon atom 5 adds to the ketone on carbon atom 2, forming a five-membered **cyclic hemiketal**. Hemiketal formation is similar to that of the hemiacetal except that a ketone rather than an aldehyde is used in reaction with the hydroxyl group. The anomeric carbon is carbon atom 2, and the alpha and beta forms are determined by the orientation of the –OH group on this carbon atom.

Open-chain fructose

α –Fructose

β –Fructose

The abbreviated structures are

α –Fructose

β –Fructose

In solution, fructose exists in both cyclic forms in equilibrium with the open-chain structure.

Glucose is the most abundant monosaccharide found in nature. It appears in fruits and fruit juices, and it is the fundamental building block of starch and cellulose, two polysaccharides found in plants. Glucose is the primary monosaccharide in the blood and, accordingly, it is also called blood sugar. Another name for glucose is dextrose. Since glucose does not require digestion to be used by the body, it is often given intravenously as a 5% solution to patients as a source of energy. The 5% solution is isotonic with blood, and it provides about 200 kcal (200 nutritional Calories) of energy per liter. Under normal conditions, the glucose level in blood serum will not exceed about 100 mg per 100 ml of serum. This normal level will frequently be exceeded by people with diabetes mellitus and glucose may appear in the urine if levels exceed 160 mg per 100 ml, the renal threshold.

Alcoholic beverages are commonly prepared by the fermentation of glucose or fructose solutions prepared from fruits or grains. These monosaccharides in the presence of yeast (a source of a particular group of enzymes called zymases) are converted to ethyl alcohol and carbon dioxide. Fermentation can produce solutions that are from 6% to 20% ethyl alcohol. Higher alcohol content is obtained by distilling the alcohol from the fermentation mixture.

$$C_6H_{12}O_6 \xrightarrow{\text{zymases}} 2C_2H_5OH + 2CO_2$$

Glucose or fructose   Ethyl alcohol

Glucose also exists in alpha and beta cyclic forms. The hydroxyl group on carbon atom 5 adds to the aldehyde, forming a six-membered hemiacetal ring. In solution, a very small amount of the open-chain form is in equilibrium with the cyclic structures. Carbon atom 1 is the anomeric carbon.

α –Glucose β –Glucose

The abbreviated Haworth structures for α- and β-glucose are

α –Glucose β –Glucose

Actually the six-membered ring of glucose as well as that of galactose are not planar as suggested by the Haworth structures. Rather, the ring is bent, forming a less-strained chairlike structure:

α –Glucose β –Glucose

For the sake of consistency, the planar Haworth structures will continue to be used to describe the six-membered rings, though you should be aware that they only approximate the true ring structure.

Chemical Tests for Glucose

Because of the high incidence of diabetes mellitus in the population, several procedures have been developed to detect and measure the concentration of glucose in urine and blood serum or blood plasma. Since urine normally does not contain glucose, it is usually sufficient in a urine test to determine if it is simply present or absent. But in blood serum or plasma a quantitative measurement is preferred.

Glucose is able to reduce copper(II) ion, Cu^{++}, to copper(I) ion, Cu^{+}, in basic solution.* This fact is the basis for the **Benedict's test**, which is used to detect glucose in urine. In the basic solution, Cu^{+} precipitates as Cu_2O, a brick-red solid that signals the presence of glucose. Benedict's test is also discussed in Chapter 18.

The success of the Benedict test is due to the presence of a small amount of glucose in the open-chain form which provides the free aldehyde group. The aldehyde is oxidized to the carboxylic acid as Cu^{++} is reduced to Cu^{+}.

$$
\begin{array}{c}
\text{H}\quad\text{O} \\
\diagdown\diagup \\
\text{C} \\
| \\
\text{H—C—OH} \\
| \\
\text{HO—C—H} \\
| \\
\text{H—C—OH} \\
| \\
\text{H—C—OH} \\
| \\
\text{H—C—OH} \\
| \\
\text{H} \\
\text{Glucose}
\end{array}
\;+\; 2Cu^{++} \;+\; 4OH^{-} \;\longrightarrow\;
\begin{array}{c}
\text{HO}\quad\text{O} \\
\diagdown\diagup \\
\text{C} \\
| \\
\text{H—C—OH} \\
| \\
\text{HO—C—H} \\
| \\
\text{H—C—OH} \\
| \\
\text{H—C—OH} \\
| \\
\text{H—C—OH} \\
| \\
\text{H} \\
\text{Gluconic acid}
\end{array}
\;+\; 2H_2O \;+\; Cu_2O
$$

Brick-red precipitate

Because glucose is capable of reducing copper (II) ion, it is classified as a **reducing sugar**. All monosaccharides are reducing sugars as it turns out, and are therefore able to interfere with the Benedict test for glucose. Fortunately, in most cases, other reducing sugars are absent or present in such low concentrations in urine that they will not interfere with the detection of the more prevalent glucose. Certain commercially available testing materials for glucose, such as Clinitest® tablets (Ames Laboratories), operate on the same principle

*A small amount of sodium citrate or sodium tartrate is also present to prevent Cu^{++} from precipitating as $Cu(OH)_2$ in the basic solution.

as Benedict's test, except certain modifications are made that generate precipitates of different colors for different glucose levels in urine: green (trace), yellow (up to 0.5%), orange (0.5–1.5%), and red (over 1.5%).

Interference by other reducing sugars is avoidable if the reactivity of the aldehyde group can be used in an alternate way (not as a reducing agent).

Glucose will react with *ortho*-toluidine in pure acetic acid at 100°C to form a soluble blue-green product. The intensity of the blue-green color in the acetic acid solution is directly related to the amount of glucose in the sample being tested. This procedure is widely used to measure the concentrations of glucose in blood serum or plasma and cerebrospinal fluid. Galactose can interfere with the *o*-toluidine test for glucose, but it is generally not a problem since it is normally present in very low concentrations in serum.

$$
\underset{\text{Glucose}}{\overset{\text{H}}{\underset{|}{\overset{|}{\underset{|}{\text{C}}}}}=\text{O}} \quad + \quad \underset{\text{o-Toluidine}}{\text{H}_2\text{N}-\underset{\text{H}_3\text{C}}{\bigcirc}} \quad \xrightarrow[\text{100°C}]{\overset{\text{acetic}}{\text{acid}}} \quad \underset{\text{Blue-green}\atop\text{product}}{\overset{\text{H}}{\underset{|}{\overset{|}{\underset{|}{\text{C}}}}}=\text{N}-\underset{\text{H}_3\text{C}}{\bigcirc}} \quad + \quad \text{H}_2\text{O}
$$

A very specific qualitative test for glucose has been developed that involves enzymes. Glucose oxidase is an enzyme that will catalyze the oxidation of glucose to gluconic acid, and it has no effect on other sugars that might also be present in the test sample. Hydrogen peroxide (H_2O_2) is also produced in the reaction, and a second enzyme catalyzes the oxidation of *o*-toluidine by H_2O_2 to form a colored product. The appearance of the color serves as a visual

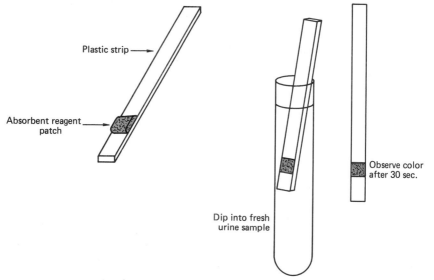

Fig. 21.1. The Clinistix® is a convenient device to detect glucose in urine. The absorbent paper patch contains subtances that react with glucose to form colored products.

indication that glucose is present in the sample. If no color appears, glucose is absent.

A number of commercially available test sticks use the enzyme method to detect glucose in urine. Clinistix® (Ames Laboratories) uses a specially treated patch of absorbent paper attached to a flat plastic strip. The paper is impregnated with *o*-toluidine and the required enzymes. The stick is simply dipped into a urine sample and the color observed.

Check Test Number 3

1. What is the significance of the "anomeric" carbon in a monosaccharide?
2. Draw the Haworth structure of α-glucose from the Fischer structure given below.

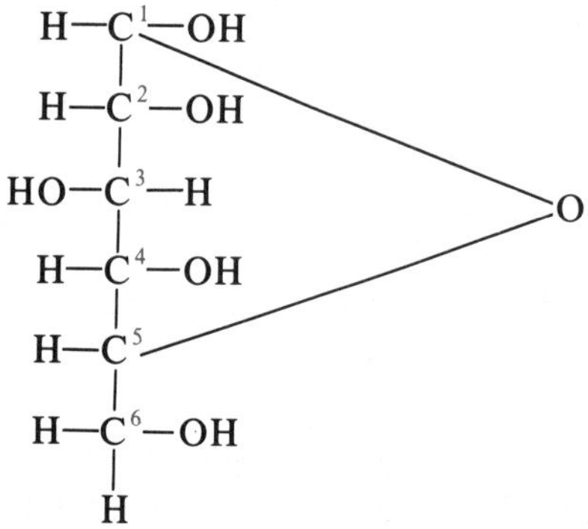

$$
\begin{array}{c}
\text{H—C}^1\text{—OH} \\
\text{H—C}^2\text{—OH} \\
\text{HO—C}^3\text{—H} \\
\text{H—C}^4\text{—OH} \\
\text{H—C}^5 \\
\text{H—C}^6\text{—OH} \\
\text{H}
\end{array}
\quad \text{O}
$$

3. Does the glucose test based on enzymes have any advantages over the Benedict test?

Answers:

1. The anomeric carbon in a closed-chain monosaccharide is the hemiacetal carbon atom. It reverts to the aldehyde or ketone carbon upon ring opening, and the orientation of the hydroxyl group on the anomeric carbon determines whether it is an α or β structure.

2. H_2COH

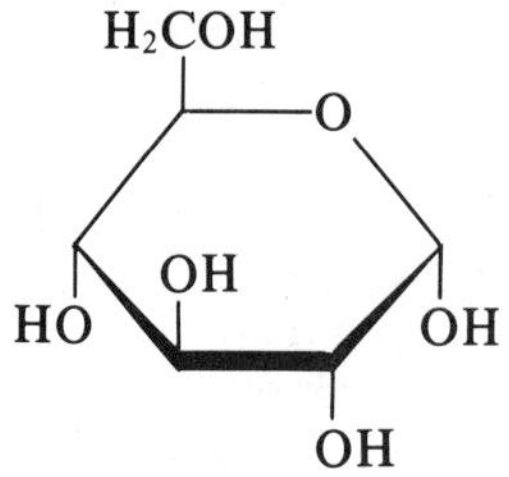

3. Yes, the enzyme-based test is specific for glucose, whereas any monosaccharide can interfere with the Benedict test.

Disaccharides

The general molecular formula for the disaccharides is $C_{12}H_{22}O_{11}$, and they are formed by the combination of two hexose monosaccharides. The hexose units are joined together through an ether "bridge" which forms as two hydroxyl groups (one on either hexose unit) react and eliminate water.

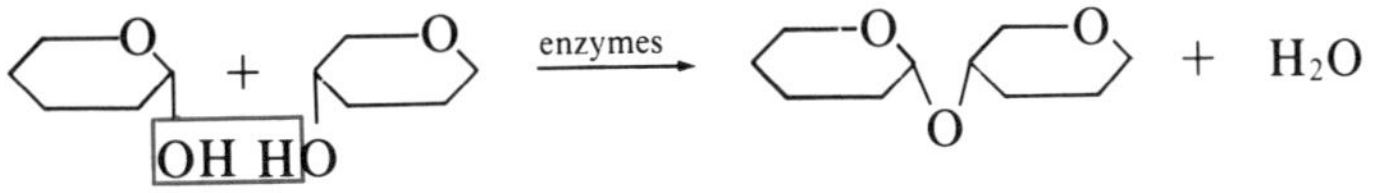

Two hexose monosaccharide units

Disaccharide

The three most common dissaccharides are **sucrose, maltose,** and **lactose.** Sucrose and maltose are found in plants, and lactose is found in milk. Disaccharides can be split by hydrolysis into the two monosaccharides. The hydrolysis reactions are catalyzed by acids or the enzymes: sucrase, maltase, and lactase.

$$\text{sucrose} \ + \ H_2O \ \xrightarrow{\text{sucrase}} \ \text{glucose} \ + \ \text{fructose}$$
$$\text{maltose} \ + \ H_2O \ \xrightarrow{\text{maltase}} \ 2 \ \text{glucose}$$
$$\text{lactose} \ + \ H_2O \ \xrightarrow{\text{lactase}} \ \text{glucose} \ + \ \text{galactose}$$

The three disaccharides are white, crystalline, sweet-tasting sugars. Sucrose is very soluble in water, maltose less so, and lactose is only slightly soluble. Each requires digestion (hydrolysis) before they can be absorbed by the body.

Maltose   **Maltose,** or malt sugar as it is commonly called, is found in germinating grains. It is not widely found in nature though it is the principal product formed in the partial hydrolysis of starch. Maltose consists of two glucose units joined by an ether linkage between carbon 1 of one glucose unit (labeled A below) and carbon 4 of the second (labeled B).

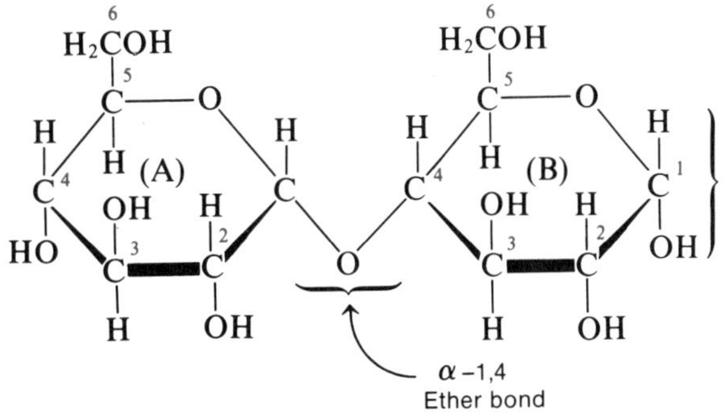

The alpha form is shown, but the beta form can also exist.

α−1,4
Ether bond

The abbreviated structure of maltose would be

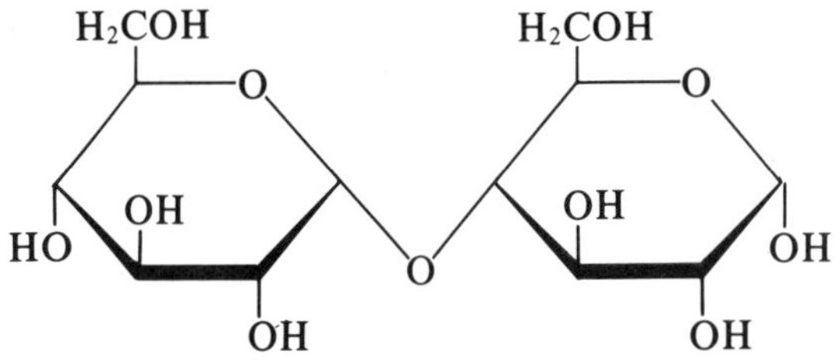

Alpha form shown

The ether bond is identified as an α-1,4 bond, since it is formed from the alpha hydroxyl group on carbon 1 (the anomeric carbon) of the first glucose unit and the hydroxyl group on carbon 4 of the second unit. The oxygen linkage between monosaccharide units is sometimes called a **glycoside bond**, and in maltose it would be described as an α-1,4 glycoside bond. Maltose is a reducing sugar, and it will give a positive Benedict test (formation of the brick-red Cu_2O precipitate). The glucose molecule labeled B, unlike the A glucose molecule, retains the anomeric carbon (carbon 1) which allows the ring to open up, forming the free aldehyde group as it undergoes mutarotation. The free aldehyde group, though present in small amounts, is responsible for the reducing property of maltose.

Lactose   Lactose is the disaccharide in milk, and it is commonly called milk sugar. Human milk contains from 5 to 8% lactose, higher than what is found in cow's milk. Lactose is commercially obtained as a byproduct of cheese production, and it is used in the preparation of liquid or powdered milk for infants or for those on high-calcium diets.

Lactose is composed of galactose and glucose joined by a β-1,4 ether (glycoside) bond. The beta hydroxyl group ($-OH$ above the ring) on carbon 1 of β-galactose combines with the hydroxyl group on carbon 4 of glucose, eliminating water thus joining the monosaccharides.

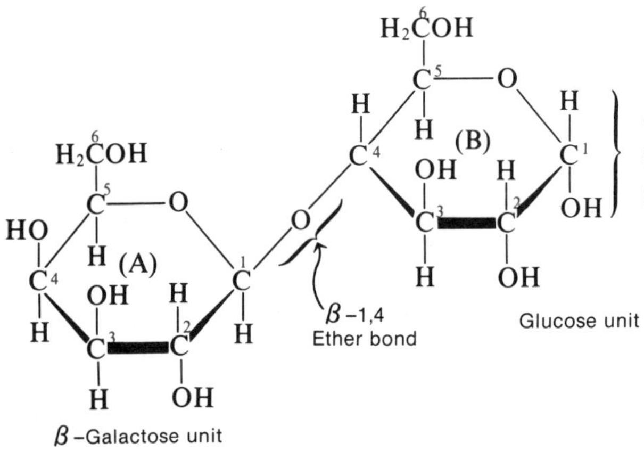

Could be an alpha or beta glucose

β-Galactose unit

Lactose may appear in the urine and blood during pregnancy and following childbirth during the period of lactation. It is a reducing sugar and will give a positive Benedict test just as maltose will. What may appear as a positive test for glucose in a routine urinalysis for a healthy woman may actually be a symptom of pregnancy.

Many people, especially those over 50, have a lactose enzyme deficiency in their digestive tracts. This can lead to diarrhea since lactose cannot be hydrolyzed into glucose and galactose. An infant with a lactose enzyme deficiency must be fed a milk-free diet. If the condition is not recognized, dehydration and malnutrition can result because of chronic diarrhea.

Sucrose

Sucrose, common table sugar, is the most important disaccharide. It is obtained principally from sugar cane and sugar beets, and it is also found in many fruits and fruit juices. Sucrose is composed of an α-glucose unit and a β-fructose unit joined together with an α-1,2 ether linkage.

The body has no particular need for sucrose since it can readily use other carbohydrates as sources of energy. But the sweet taste of sucrose is very desirable to most people and in the United States the average person consumes about *130 pounds* of sucrose and other sugars each year. This high rate of consumption of sucrose is a point of concern for many medical specialists since it can lead to obesity and also increase the likelihood of heart disease and diabetes. Sucrose is difficult to avoid in the diet since it appears in many processed foods to increase acceptance in the market place.

Sucrose is not a reducing sugar. This may seem surprising since both glucose and fructose are reducing sugars. But in sucrose, both the glucose and the fructose anomeric carbons are involved in the formation of the α-1,2 ether

TABLE 21.1 Approximate Sugar Content of Several Processed Foods	
Food	Percent by weight
Peanut butter	9
Coffee creamer	60
Frozen whipped topping	21
Chocolate bar	51
Ice cream	21
Catsup	28
Cola beverage	9

Source: *Consumer Reports*, March 1978, p. 136.

linkage, which eliminates the possibility of ring opening with the formation of reducing groups.

Sucrose is readily hydrolyzed to glucose and fructose by reaction with water in the presence of an acid or the enzyme sucrase. Even heating an aqueous solution of sucrose will cause hydrolysis of some of the sucrose, which results in a solution that is sweeter than the original sucrose solution. This practice is used in the manufacture of candy to obtain a sweeter product. A mixture of equal amounts of glucose and fructose is called **invert sugar**.

The sweet taste of mono- and disaccharides involves physiological effects that are, as yet, not well understood. However, sweetness is not a property of carbohydrates alone. Several organic compounds have been synthesized in laboratories or found in nature that are even sweeter than sucrose. Some of these have little or no caloric value and have been used to sweeten foods, allowing people to reduce their caloric intake while enjoying a more normal diet. They are also used by diabetics who must control their intake of carbohydrates.

One of the most widely used "artificial" sweeteners up until the early 1970's was cyclamate. Cyclamate is the sodium or calcium salt of cyclamic acid. Since it is much sweeter than an equal amount of sucrose, only a small amount is required to achieve a desired level of sweetness in beverages and foods.

$$\text{H}\diagdown\underset{\text{N}}{}\diagup\text{SO}_3^{\ominus}\text{ , Na}^+$$

Sodium cyclamate

Cyclamates were banned from foods in the early 1970's after tests with laboratory animals suggested that they may be capable of inducing cancer in humans.

Saccharin was used as an artificial sweetener long before cyclamates came on the scene. It is about 300 times sweeter than sucrose, but it leaves a mildly bitter aftertaste. In 1975, animal studies suggested that saccharin may be a very weak cancer-causing agent in humans, a fact that may bring its use in foods to an end.

Saccharin (not a saccharide)

Sorbitol, mannitol, and xylitol are polyalcohols that have a sweet taste which has led to their use as sweeteners. All three are found in nature in fruits and plant tissue, and they can also be prepared from sugars in the laboratory. They break down very slowly in the mouth, a property which has led to their use in "sugarless" gums.

Sorbitol	Mannitol	Xylitol
H	H	
H—C—OH	H—C—OH	H
H—C—OH	HO—C—H	H—C—OH
HO—C—H	HO—C—H	H—C—OH
H—C—OH	H—C—OH	HO—C—H
H—C—OH	H—C—OH	H—C—OH
H—C—OH	H—C—OH	H—C—OH
H	H	H

The relative sweetness levels of various natural and artificial sweetening agents are found in Table 21.2. Sucrose is used as the reference and is assigned a value of 1. A sweetener with a value of 5 is judged to be five times sweeter than sucrose.

TABLE 21.2
Relative Sweetness of Natural and Artificial Substances

Substance	Relative sweetness
Sucrose	1
Glucose	0.7
Fructose	1.7
Calcium cyclamate	30
Saccharin	300
Xylitol	1

Check Test Number 4

1. Identify the two monosaccharides that are combined in sucrose, maltose, and lactose.

Answers:
1. Sucrose: glucose and fructose; maltose: 2 glucose; lactose: glucose and galactose

Polysaccharides

Polysaccharides are high-molecular-weight polymers of monosaccharides. Some, such as starch and glycogen, are used as storage houses for simple sugars, and others, such as cellulose, are used as structural material in plants. Starch, glycogen, and cellulose are the most important polysaccharides and each is a polymer made from a single monosaccharide, glucose.

Starch **Starch** is the most important carbohydrate in our diet. About 25% of the food we eat is starch, which is primarily obtained from cereal grains (corn, wheat, and rice), potatoes, and vegetables (beans and peas). Plants store glucose in the cytoplasm of cells as starch granules, which are insoluble and unable to pass through cell membranes.

Starch is a mixture of two polysaccharides, amylose and amylopectin. Between 10 and 20% is amylose, a polymer containing from 60 to 300 glucose units joined by α-1,4 ether linkages. The amylose polymer which is a long strand of glucose units winds around to form a helical structure.

Amylose (n=60–300)

Amylose helix

The remaining 80 to 90% of starch is amylopectin, a polymer formed from 300 to 6000 glucose units joined by α-1,4 ether bonds. But unlike amylose, there are several short chains of from 25 to 30 glucose units that branch out from the longer chain. The branching points connect through formation of α-1,6 ether linkages. Even with branching, each chain of polymerized glucose units exists in a helical form.

Amylopectin

Amylopectin branching

In the presence of certain enzymes, starch can be hydrolyzed and broken down into smaller fragments of polymerized glucose units. Initially, in the hydrolysis reactions, fragments made up of several glucose units are formed which are called **dextrins**. Dextrins, in turn, are further broken down to form maltose, a disaccharide described earlier. Eventually maltose is hydrolyzed to glucose. If the hydrolysis is carried out in a laboratory, the various stages of the hydrolysis can be followed by observing how the color of an iodine(I_2)–starch solution changes as it is gently heated. Starch, in the presence of iodine, forms a blue-black color; dextrins in the presence of iodine form a red color; maltose and glucose do not cause a color change with iodine. As the hydrolysis proceeds the solution changes from blue-black to red and eventually to a pale brown, the usual color of iodine in water.

$$\text{starch} \xrightarrow{\text{hydrolysis}} \text{dextrins} \xrightarrow{\text{hydrolysis}} \text{maltose} \xrightarrow{\text{hydrolysis}} \text{glucose}$$

$$\text{blue-black} \longrightarrow \text{red} \longrightarrow \text{pale brown} \longrightarrow$$

Dextrins are formed on the surface of baking bread or toast by heat, and they are responsible for the golden color of the crust. Because dextrin solutions are viscous and sticky, dextrin is commonly used as a nontoxic adhesive for stamps, labels, and children's glue. A mixture of dextrins and maltose is readily digestable, and for that reason it is frequently used in special diets and infant formulas as a source of energy.

Starch forms a colloid in water due to the large size of the polysaccharide molecules. These colloids are used to thicken sauces, gravies, pie fillings, and ice cream. Chemically modified starches are frequently used to improve the appearance and texture of bread, and a mixture of low-molecular-weight dextrins, maltose, and glucose is commonly called corn syrup.

Glycogen   **Glycogen** is the storage form of carbohydrates in animals and for that reason it is sometimes called animal starch. It is found to some extent in all the tissues of the body, but primarily in muscle tissue (about 1% glycogen) and the liver (1.5–4.0% glycogen). Glycogen plays an important role in maintaining the correct glucose level in the blood and muscle tissue.

Glycogen has basically the same branched structure as amylopectin, and it is formed from glucose units joined through α-1,4 and α-1,6 ether links. But unlike amylopectin branching occurs at shorter intervals, about every 12 glucose units along each chain. Because of the large size, glycogen cannot pass through cell walls, but it is somewhat more soluble than starch. Whenever the amount of glucose in the blood is in excess of the needs of the body (such as after a meal), some of the excess is converted into glycogen and stored. The conversion of glucose to glycogen in the body is called **glycogenesis**. When glucose is again required by the body, glycogen is hydrolyzed back to glucose, a process called **glycogenolysis**. The average adult has enough glycogen to provide normal energy needs for about 16 hr.

Cellulose   **Cellulose** is a linear polymer of glucose, but unlike starch and glycogen, the oxygen linkages joining the glucose units are β-1,4 linkages (not α-1,4). This single difference makes cellulose undigestable in humans and other carnivorous animals, and it passes through the digestive system essentially unchanged.

Cellulose

Cellulose is the structural component of the cells of higher plants. Cotton fibers and paper are over 90% cellulose; wood is nearly 50% cellulose. The linear cellulose molecules lie along side each other and are held together by hydrogen bonding. This kind of interaction between the strands is largely responsible for the great strength of cellulose fibers.

Cellulose can be hydrolyzed in acid solution or in the presence of an enzyme, cellulase. The final product is glucose. Our digestive system lacks the cellulase enzyme needed to cleave the β-1,4 links. However, ruminants, such as cows, horses, and sheep, can survive on grass due to microorganisms in their digestive systems that produce this enzyme. Termites do a good job of digesting wood because they contain the proper enzyme-producing microorganisms in their digestive system.

Cellulose provides necessary bulk in our diets to prevent constipation and aid elimination. Medical specialists have recommended an increase in the fiber (cellulose) content of our diet to reduce the risk of digestive diseases. Unfortunately, modern food processing frequently removes much of the fiber content from food, and people need to make a conscious effort to include it in their diet. An abnormally large amount of fiber in the diet can reduce the absorption of certain minerals by the body.

Chitin   Chitin (kī'-tin) is found in the cell walls of fungi, and in the exoskeletons of anthropods (lobsters, crabs) and insects. It is a linear polymer of β-1,4 bonded glucose units, similar to cellulose, but it also has an acetamide $\overset{\displaystyle O \quad H}{\underset{\displaystyle \|\quad\;\; |}{H_3C-C-N-}}$ group bonded to the second carbon atom in place of the hydroxyl group.

Chitin (β-1,4)

Hydrocolloids   Seaweed contains polysaccharides that form gelatinous or near solid colloids in water. They are called **hydrocolloids**. One of these polysaccharides, alginate, is used as an impression medium in dentistry.

Dextrans (not Dextrins)   **Dextrans** are linear polymers of glucose joined through α-1,6 ether bridges. Partially hydrolyzed dextrans are used to increase blood volume in patients suffering from an acute loss of blood. The dextrans expand in the bloodstream and "extend" the volume of the blood in the body. Bacteria in the mouth produce dextrans which end up as a major part of dental plaque.

Heparin   **Heparin** is a natural anticoagulant in blood. It inhibits the ability of thrombin to break apart fibrinogen in the clot-forming mechanism. Heparin is made in lung and liver tissue and released into the bloodstream as a complex polysaccharide of sulfate- and amino-sulfate-substituted glucose units. Heparin is the most powerful natural anticoagulant known and is widely used for that purpose in medicine.

Heparin

Mucopoly-saccharides

Mucopolysaccharides are polysaccharides that (with the exception of hyaluronic acid) are associated with proteins. They are complex polymers containing hexoses, sulfate- and amide-substituted sugars, and acid sugars. Hyaluronic acid forms viscous solutions that are found in the eyeball (vitreous humor), in cell walls, and in synovial fluid, the slippery liquid that lubricates our joints. It is also an essential component of the cement that holds cells together in connective tissue.

Keratin sulfate and chondroitin sulfate are mucopolysaccharides that are found in cartilage, the cornea of the eye, and skin. They lubricate and add flexibility to these tissues while providing toughness and strength. Other mucoproteins in association with blood proteins are responsible for the differences between Type A, B, and O blood and the Rh factor.

Digestion and Absorption of Carbohydrates

Digestion of carbohydrates is simply the enzyme-controlled hydrolysis of oligo- and polysaccharides to simple sugars, primarily glucose, galactose, and fructose. Starch and glycogen are partially digested in the mouth by the action of **ptyalin**, a carbohydrase secreted in the saliva. Once in the stomach, the highly acid environment quenches the effectiveness of ptyalin and carbohydrate digestion ceases until passed into the basic environment of the small intestine. The pancreatic juice secreted into the small intestine contains several enzymes important in carbohydrate digestion. Pancreatic **amylase** converts the partially digested polysaccharide to maltose, and **maltase** splits maltose into two glucose units. The mucosal cells of the intestines also produce enzymes which are needed to hydrolyze maltose, sucrose, and lactose to the simple sugars.

The body can digest only those polysaccharides with alpha links between the sugar units. Cellulose with beta links is not hydrolyzed in humans since the proper enzymes are absent.

The simple sugars are absorbed through the intestinal wall directly into the bloodstream and carried by the portal vein to the liver. Here galactose and fructose are converted to glucose. Most of the glucose is sent to cells throughout the body and used for energy, while excess glucose is converted to glycogen and stored in liver and muscle tissue. If a large excess of carbohydrate is digested, it will ultimately be converted in a multistepped process to triglycerides (fats) and stored in adipose tissue.

Diabetes and the Blood Sugar Level

There are approximately 4.5 million diabetics in the United States, and it is estimated that nearly 300,000 people die each year as a direct result of this disease. There has been a 300% increase in the number of diabetics since 1950, and much of this increase is attributed to the increased use of sucrose in the diet. In laboratory experiments with strains of diabetes-prone rats, those fed a high sugar diet developed diabetes while those fed a sugar-free diet remained healthy.

Diabetes mellitus is characterized by the inability of the body to regulate the glucose level in the blood, which is usually at a much higher level than normal. Typical fasting glucose levels (after 8–10 hr without food of any kind) are between 70 and 100 mg of glucose per 100 ml of blood serum. Consistent glucose levels below 50 mg/100 ml indicate **hypoglycemia**, a condition nearly opposite that of diabetes. Hypoglycemia can be brought on by fasting, excessive physical exercise, and hormonal irregularities. In extreme cases, it can be fatal. A consistent glucose level above about 130 mg/100 ml indicates a condition known as **hyperglycemia**, characteristic of diabetes.

Diabetes is usually determined using a 3- or 5-hr **glucose tolerance test** (G.T.T.). Fasting blood and urine samples are first obtained from the patient, who is then given a solution containing about 100 g of glucose. Blood and urine samples continue to be taken every half-hour for 2 hr, then every hour. Typical results of a G.T.T. are shown in Figure 21.2 for normal individuals, and those with moderate and severe diabetes. When glucose levels in serum exceed 160–170 mg/100 ml, the **renal threshold**, glucose will pass through the kidneys and enter the urine, a condition known as **glucosuria**.

The most important processes that affect the level of glucose in blood are the formation and breakdown of glycogen. Insulin, a hormone secreted by the pancreas, reduces the glucose level in the blood by stimulating the formation of glycogen (glycogenesis). Insulin also enhances utilization of glucose by cells. Diabetes is caused by an insufficient supply of insulin, which results in increased blood glucose levels and deficient glucose use by cells. Tragically, the paradox of the diabetic condition is cell starvation in the presence of a markedly high glucose level. Only the brain remains unaffected, since its use of glucose is not insulin dependent. Cells starved for energy begin to use fats and to a lesser degree proteins as sources of energy, raising the fatty acid and ketone body

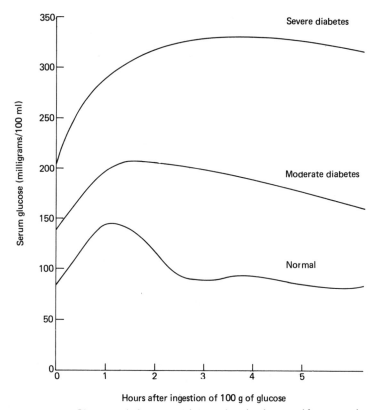

Fig. 21.2. Characteristic serum glucose levels observed for normal and diabetic patients during a 5-hour glucose tolerance test.

content of the blood. This destructive metabolism (catabolism) is responsible for the "acetone" breath of an untreated diabetic.

Treatment of diabetes centers around strict control of the diet. Frequently the use of drugs to stimulate the production of insulin, or the direct administration of carefully measured doses of insulin on a daily basis is necessary. The goal of any treatment is to maintain a relatively normal serum glucose level. With proper treatment, a diabetic can lead a near normal life.

Optical Isomerism*

Carbohydrate chemistry is in many ways the study of isomers. The hexoses, for example, have the same formula, $C_6H_{12}O_6$, but because the atoms can be arranged in space differently, a given hexose could be glucose, mannose, gal-

*This section may be omitted without loss of continuity.

actose, or another monosaccharide. The branch of chemistry that deals with the spatial arrangement of atoms in molecules is called **stereochemistry**. The area of stereochemistry that deals with differences in properties of isomers is called **stereoisomerism**. In a pair of stereoisomers, such as *cis-* and *trans-*2-butene, each molecule has the *same sequence* of bonded atoms, but they differ in the spatial arrangement of the atoms and groups bonded to the $C=C$ center of the molecule. *Cis-* and *trans-* isomerism, then, is one kind of stereoisomerism, commonly called **geometric isomerism**.

$$H_3C\diagdown \quad \diagup CH_3$$
$$C=C$$
$$H \diagup \quad \diagdown H$$

cis-2-Butene

$$H_3C\diagdown \quad \diagup H$$
$$C=C$$
$$H \diagup \quad \diagdown CH_3$$

trans-2-Butene

A second kind of stereoisomerism, one we will be concerned with here, is **optical isomerism**. Optical isomers differ in the way they affect plane-polarized light that is passed through solutions of them. A brief discussion of plane-polarized light would be in order before proceeding further.

Ordinary light vibrates in all directions at right angles to the direction of travel of the light beam. The direction of the vibrations can be indicated with small arrows drawn at right angles to the light beam as shown in Figure 21.3. If a polarizing filter is placed in the beam, only those vibrations in a single plane (the vertical plane in Figure 21.4) will pass through, and the others are held back. The light passed through the filter is called plane-polarized light. The polarizing filter acts something like a picket fence, allowing only those vibrations that line up with the long narrow "slots" between the pickets to pass through.

Plane-polarized light can interact with certain kinds of molecule in a way that changes the direction of the plane of vibration of the light. For example, the light entering the sample may be polarized in the vertical plane, but when it passes out of the sample, the plane may be *rotated* to the right or to the left by several degrees (α). If it was rotated to the right by 90°, the plane polarization

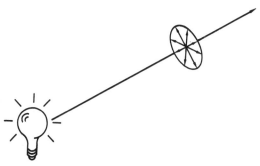

Fig. 21.3. Ordinary light vibrates in all planes at right angles to the direction of travel of the beam.

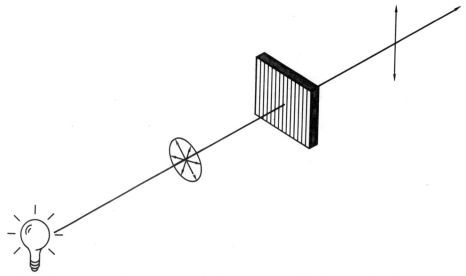

Fig. 21.4. A polarizing filter will only allow light vibrating in a single plane to pass through. The light is then said to be plane polarized.

of the emitted light would be horizontal. Substances that can rotate plane-polarized light are termed **optically active.** (See Figure 21.5.) The device used to determine optical activity is called a **polarimeter**, and is shown in Figure 21.6. A solution of an optically active compound that rotates the light plane to the right is termed **dextrorotatory**, and is symbolized with a (+) before the name of the compound, such as (+)-glucose. If a compound rotates light to the left, it is termed **levorotatory**, and is symbolized with a (−), such as (−)-fructose. For every dextrorotatory compound, such as (+)-glucose, there will exist a levorotatory isomer, (−)-glucose. These are optical isomers which have identical physical properties except that the (+) isomer rotates plane-polarized light to the right a certain number of degrees, and the (−) isomer rotates plane-polarized light to the left by the identical number of degrees. It is interesting to note that a solution containing a mixture of equal amounts of the (+) and (−) isomers of an optically active compound will not rotate plane-polarized light. The rotating power of each isomer seems to cancel out, and such a mixture is called a **racemic** mixture.

It is possible to predict whether or not a given compound is likely to display the property of optical activity based on its structure. For a molecule to be optically active, it must be **chiral**, that is, *the molecule must not be superimposable on its mirror image*. Perhaps an example will explain the meaning of chiral a bit better. Consider your right hand as the structure of some molecule. When you hold your right hand in front of a mirror, the image you see looks like your left hand (you can overlook fingerprints and wrinkles). Now, imagine taking the image out of the mirror (or use your left hand to act as the image)

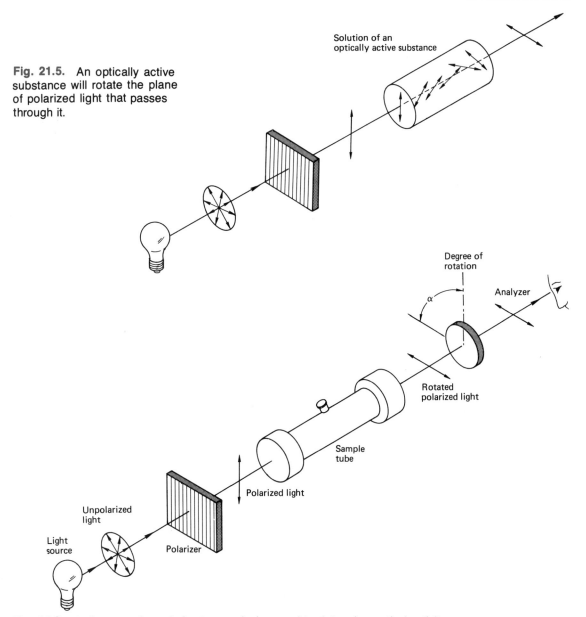

Fig. 21.5. An optically active substance will rotate the plane of polarized light that passes through it.

Solution of an optically active substance

Degree of rotation

Analyzer

Rotated polarized light

Sample tube

Polarized light

Unpolarized light

Light source

Polarizer

α

Fig. 21.6. A diagram of a polarimeter—a device used to determine optical activity.

and place it on top of your right hand with both palms facing the same direction. The two hands, the right hand and its mirror image are *not* superimposable (Figure 21.7). The thumb of one hand will be over the little finger of the other

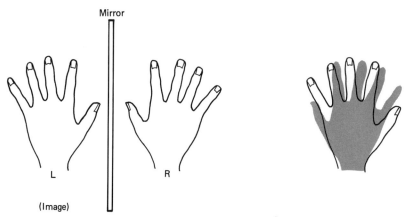

Fig. 21.7. The image of your right hand when held before a mirror will look like your left hand. But if you could remove the mirror image and place it on top of your right hand, the two would not be superimposable. Your right hand is therefore chiral. Imagine doing the same thing with the letter "H." H is not chiral.

and vice versa. Therefore, your right hand is chiral, and so is your left hand. If we think of them as molecules, one would be dextrorotatory and the other would be levorotatory. As you will learn later, it is not possible to predict which hand would be dextrorotatory, or which levorotatory, but since they are non-superimposable mirror images of each other, they would affect polarized light in equal but opposite ways. Pairs of molecules that are nonsuperimposable mirror images of one another are termed **enantiomers**. Enantiomers are also optical isomers.

It is not an easy job to imagine mirror images of molecules in your mind to see if they are chiral. But fortunately, there is an easier method. If an organic molecule contains a carbon atom to which *four different* atoms or groups are bonded, the molecule is chiral. The carbon atom bonded to the four different groups is the **chiral center** of the molecule. The middle carbon atom of glycer-aldehyde is a chiral center and so the compound can exist in two optically active forms, (+) and (−). Notice how the optical isomers are nonsuperim-posable mirror images of one another. One of the structures can be regarded as being right-handed, the other left-handed. Perspective drawings are used to emphasize the three-dimensional structure of the molecule.

The chiral center in each of the following molecules is indicated with an asterisk (*). Each compound can exist in two optically active forms (+) and (−), that is, in two molecular forms that are enantiomers.

Chlorobromoiodomethane

$$Cl-\overset{\overset{\displaystyle Br}{|}}{\underset{\underset{\displaystyle I}{|}}{C^*}}-H$$

Amphetamine

$$\underset{}{\text{〈O〉}}-CH_2-\overset{\overset{\displaystyle H}{|}}{\underset{\underset{\displaystyle NH_2}{|}}{C^*}}-CH_3$$

2-Butanol

$$H_3C-\overset{\overset{\displaystyle OH}{|}}{\underset{\underset{\displaystyle H}{|}}{C^*}}-CH_2-CH_3$$

Lactic acid

$$H_3C-\overset{\overset{\displaystyle OH}{|}}{\underset{\underset{\displaystyle H}{|}}{C^*}}-C\overset{\displaystyle O}{\underset{\displaystyle OH}{}}$$

Many compounds have more than one chiral center, which can lead to a large number of stereoisomers. Glucose has four chiral centers.

$$
\begin{array}{c}
H\diagdown{}\diagup O \\
C \\
H-C^*-OH \\
HO-C^*-H \\
H-C^*-OH \\
H-C^*-OH \\
H-C-OH \\
H
\end{array}
$$
(+)−Glucose

Earlier it was said that there was no way of predicting which enantiomer, the right-hand structure or the left-hand structure, causes dextrorotatory or levorotatory behavior. But there is a way of relating the spatial arrangement of atoms or groups about a chiral center to a given optical behavior.

In 1891, Emil Fischer, a German chemist, quite arbitrarily assigned a particular spatial arrangement of atoms to (+)-glyceraldehyde and said it had a D configuration. Likewise, (−)-glyceraldehyde was said to have the L configuration. The D **configuration** had the −OH group on the chiral center to the right, and the L **configuration** had it to the left. Fischer simply picked which isomer he thought was dextrorotatory and called that configuration "D," the other was "L." The two enantiomers were named D-(+)-glyceraldehyde and L-(−)-glyceraldehyde.

$$
\begin{array}{c}
H\diagdown{}\diagup O \\
C \\
H\blacktriangleright C \blacktriangleleft \boxed{OH} \\
H_2COH
\end{array}
$$
→
right
D configuration
assigned to
(+)−glyceraldehyde

$$
\begin{array}{c}
O\diagdown{}\diagup H \\
C \\
\boxed{HO}\blacktriangleright C \blacktriangleleft H \\
HOCH_2
\end{array}
$$
←
left
L configuration
assigned to
(−)−glyceraldehyde

Many years later, in 1951, it was shown that the spatial configurations proposed by Fischer were correct. The spatial arrangement of atoms chosen for the D configuration was found to be as he said they were, and the L configuration was therefore also correct.

Other molecules can be placed in D- or L-configurational families by comparing their structures with the reference glyceraldehyde molecules. Monosaccharides especially can be classified as D or L simply by comparing the placement of the –OH group on the chiral center on the Fischer structure furthest away from the carbonyl group with the glyceraldehyde models. If the –OH is on the right, it is in the D family. If it is on the left, it is in the L family.

Just because a monosaccharide is in the D family does not mean that it is also dextrorotatory, or if in the L family that it is levorotatory. It is that way with glyceraldehyde [D-(+)-glyceraldehyde, and L-(−)-glyceraldehyde] but that is only a coincidence. D-glucose happens to be dextrorotatory, so it can be named D-(+)-glucose, but D-ribose is levorotatory, and named D-(−)-ribose.

Most monosaccharides found in nature belong to the D family. We lack the proper enzymes in our bodies to metabolize monosaccharides in the L family. Yeast can ferment D glucose to produce alcohol, but it cannot do so with L-glucose. L-dopamine is an important drug in fighting Parkinson's disease, but D-dopamine has no effect at all. The spatial arrangement of atoms in chiral molecules is very important in biological systems. In all cases, the body will use only one enantiomer of a compound, either D or L, in the same way that only one of our hands can use a right-hand glove comfortably.

Check Test Number 5

1. How would a solution of (+)-glucose affect plane-polarized light passed through it? How would a solution of (−)-fructose affect plane-polarized light?
2. Which of the following compounds would you expect to exhibit optical isomerism (look for the chiral center)?

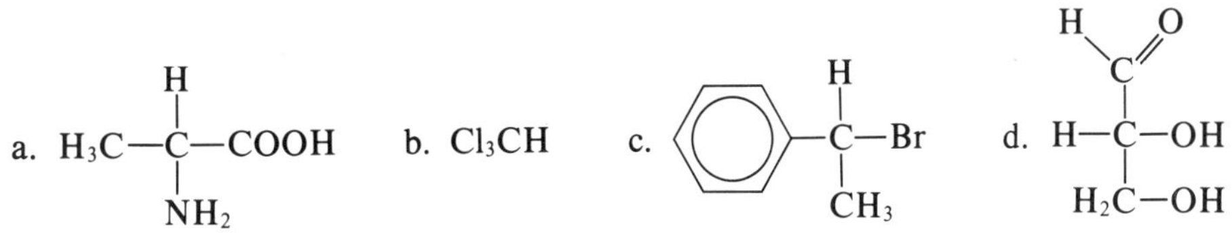

a. $H_3C-\overset{\overset{\displaystyle H}{|}}{\underset{\underset{\displaystyle NH_2}{|}}{C}}-COOH$ b. Cl_3CH c. ⟨benzene ring⟩$-\overset{\overset{\displaystyle H}{|}}{\underset{\underset{\displaystyle CH_3}{|}}{C}}-Br$ d. $H-\overset{\overset{\displaystyle H\diagdown_{C}\diagup^{O}}{}}{\underset{\underset{\displaystyle H_2C-OH}{}}{C}}-OH$

3. Would the sugar shown below be in the D or L family?

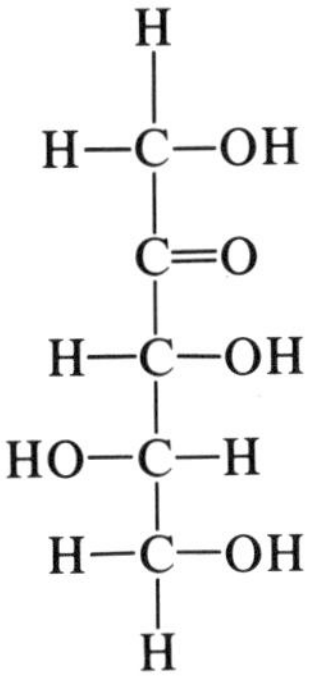

$$\overset{\displaystyle H}{\underset{\displaystyle H}{\overset{|}{\underset{|}{
\begin{array}{c}
H-C-OH \\
C=O \\
H-C-OH \\
HO-C-H \\
H-C-OH
\end{array}
}}}}$$

Answers:

1. (+)-glucose would rotate the polarized light plane to the right and (−)-fructose to the left.
2. Compounds (a), (c), and (d) have chiral centers and could exhibit optical isomerism.
3. L family. The hydroxyl group on the chiral carbon furthest from the carbonyl group is to the left on the Fischer structure.

Terms

Several important terms appeared in Chapter 21. You should understand the meaning of each.

carbohydrate
photosynthesis
monosaccharide
disaccharide
oligosaccharide
polysaccharide
simple sugar
aldose
ketose
triose
pentose

glycogenolysis
starch
cellulose
hydrocolloids
dextrans
mucopolysaccharide
hypoglycemia
hyperglycemia
G.T.T.
glucosuria
†optical isomerism

hexose   †plane-polarized light
anomeric carbon   †polarimeter
mutarotation   †chiral center
Fischer structure   †enantiomer
Haworth structure   †racemic mixture
cyclic hemiacetal   †dextrorotatory
reducing sugar   †levorotatory
Benedict's test   †D configuration
dextrins   †L configuration
glycogen
glycogenesis   †Optional terms.

Questions

Answers to the starred questions appear at the end of the text.

  *1. What two functional groups are found in all carbohydrates?

  2. Compare the old definition of a carbohydrate with the new definition.

  3. *a. Name and give the structure of the simplest organic compound that would fit the old definition of carbohydrates but not the new definition.

    b. Which monosaccharide would fit the new definition but not the old definition?

  4. a. Briefly outline the three principal subdivisions of carbohydrates.

    *b. What are the two classes of monsaccharides?

  5. Why is glycerol not considered a carbohydrate?

  *6. Give the name and structure of the simplest aldose and the simplest ketose.

  *7. Give the balanced equation describing the photosynthesis of glucose, $C_6H_{12}O_6$.

  8. Pentoses and higher monosaccharides can exist in different structural forms in solution. Draw the structure of these forms for glucose.

  9. Why can a glucose solution be administered directly into the bloodstream of a patient as an energy source?

  10. *a. Draw the Fischer and Haworth structures for the cyclic forms of ribose.

    *b. How does ribose differ from deoxyribose?

  11. What is mutarotation?

  *12. Name the three most important dietary hexoses, and give the open-chain structural formula for each one.

  13. Identify each of the following sugars with the proper name: **(a)** dextrose, **(b)** blood sugar, **(c)** cane sugar, **(d)** fruit sugar, **(e)** milk sugar, **(f)** levulose, **(g)** malt sugar, **(h)** table sugar.

  *14. Why is glucose considered a reducing sugar?

  *15. Of what use is the Benedict's test? Give the equation describing the Benedict test using glucose as the carbohydrate.

16. What advantage does the *ortho*-toluidine test for glucose have over the Benedict test? Give an equation describing the reaction of *ortho*-toluidine with glucose.

17. Name and give the structures for the three dietary disaccharides, and give the correct identification for the oxygen link that joins the monosaccharide units.

*18. Give word equations for the hydrolysis of maltose, sucrose, and lactose.

19. Why is sucrose not a reducing sugar?

*20. Compare the solubility in water of monosaccharides, disaccharides, and polysaccharides.

21. Comparing starch and cellulose, what structural difference allows one to be digested and not the other?

*22. What is the function of glycogen in the body?

23. Give at least one important function of the following polysaccharides:
(a) heparin, (b) chitin, (c) dextrans, (d) cellulose, (e) starch.

*24. What is the principal chemical reaction in the digestion of carbohydrates?

25. What is the "renal threshold" in reference to the glucose level in the blood?

26. Compare hypoglycemia and hyperglycemia in terms of blood glucose levels.

27. What general kind of result would you expect for a glucose tolerance test administered to a severe diabetic?

*†28. What is plane-polarized light? How would the plane of polarization be affected by a solution of dextrorotatory glucose?

*†29. What is the requirement that must be met for a substance to be chiral?

*†30. Which of the following have a chiral center?

a. $H_3C-\underset{\underset{H}{|}}{\overset{\overset{Br}{|}}{C}}-CH_2-CH_3$

c. $H_2N-\underset{\underset{CH_2SH}{|}}{\overset{\overset{H}{|}}{C}}-COOH$

b. $\langle\!\!\bigcirc\!\!\rangle-C\overset{\nearrow O}{\underset{\searrow OH}{}}$

d. $\underset{H}{\overset{H}{\diagdown}}C=C\underset{CH_3}{\overset{H}{\diagup}}$

*†31. Describe the affect that identical solutions of (+)-ribose and (−)-ribose have on plane-polarized light.

*†32. Draw the structural formulas of D- and L- glucose. Which configuration is found in nature?

†Optional questions.

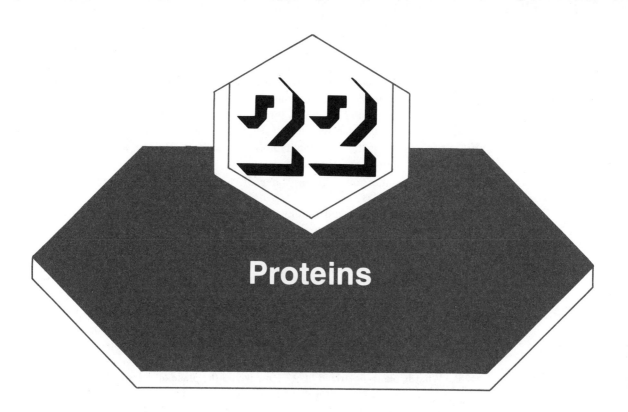

Proteins

The term "protein" was first used by Gerardus Mulder in 1838 to describe the complex nitrogen-containing organic compounds that are found in all living cells. It is derived from the Greek word *proteios*, which means "of first importance." This is an appropriate description of these important compounds since proteins are involved in essentially all biochemical processes in the body.

Living cells in both plants and animals are approximately 70% water and 30% solid material. About half of the solid material is protein. In plants, proteins are synthesized from carbon dioxide, water, nitrates, sulfates, and smaller amounts of several other compounds. Animals, on the other hand, are not able to synthesize proteins in this way, so proteinaceous material must be obtained in the diet.

Most of the protein material in the body is used in body building and repair, while the rest is involved in other vitally important biological functions. Protein is used little as a source of energy. While lipids and, to a lesser extent, carbohydrates are stored in the body as reserve sources of energy, the storage of protein is almost nonexistent. Consequently, for good health, it is necessary to have a regular intake of protein through the diet. An animal can survive for a limited time on a diet that contains only vitamins, minerals, and proteins (no carbohydrates or lipids). But if the animal is fed a diet containing everything but protein, premature death will follow.

After a brief discussion of the composition, function, and classes of proteins, we will examine the individual amino acids $H_2N-\overset{\overset{\displaystyle H}{|}}{\underset{\underset{\displaystyle R}{|}}{C}}-COOH$ and how they join together to form the giant protein molecules. Then, several chemical and physical properties of proteins will be discussed.

Objectives

By the time you finish this chapter, you should be able to do the following:

1. List and describe the main functions of proteins.
2. Describe the way in which proteins can be classified.
3. Name and give the structure of the 20 most important amino acids.
4. Describe the properties of amino acids.
5. Describe how electrophoresis and chromatography can effect separations of mixtures of amino acids.
6. Draw a peptide structure given its name.
7. Describe the causes of secondary, tertiary, and quaternary structures in proteins.
8. Explain how collagen, bone, and teeth are related.
9. Describe several ways proteins can be denatured.
10. Describe some of the more important protein tests.
11. Explain the basic digestion and absorption process of proteins.

The Composition and Molecular Weights of Proteins

Proteins are polymers of amino acids. They are extremely large molecules which have very high molecular weights ranging from about 5000 to over 40,000,000. As a matter of comparison, several familiar compounds with their molecular weights are listed in Table 22.1, along with four different proteins.

The elements that are principally found in proteins are carbon, oxygen, nitrogen, and hydrogen. Sulfur is the next most abundant, followed by several others which are found in relatively small amounts. Some proteins require the

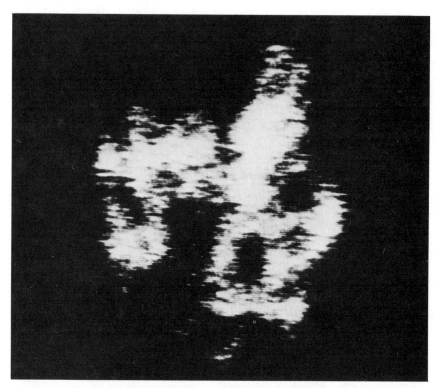

Fig. 22.1. An electronmicrograph of a single molecule of hemoglobin, the protein responsible for oxygen transport in the blood. Though the details are not clearly evident, the molecule resembles the folded and twisted structure of hemoglobin depicted in Figure 22.13. An electronmicrograph of a single molecule is a remarkable achievement. (From: A. V. Crewe, Scientific American, Vol. 224, p. 26, April 1971.)

TABLE 22.1 Molecular Weights of Several Compounds	
Compound	Molecular weight (amu)
Water	18.0
Ethyl alcohol	46.0
Glucose (blood sugar)	180.
Glyceryl tristearate (a fat)	891.
Proteins { Insulin	11,466
Egg albumin	48,000
Hemoglobin	64,500
Protein coat of a single virus	40,000,000

presence of a few atoms of certain elements in order to function, such as iodine, iron, or phosphorus. An average elemental composition of proteins is given in Table 22.2.

TABLE 22.2 Elemental Composition of Proteins	
Element	% by weight
C	50–55
O	20–24
N	15–18
H	6–8
S	0.0–3.5
P	0.0–1.5
Metals (Fe, Zn, Cu, and others)	trace

The Function and Classification of Proteins

Because proteins are very complex molecules, it has not been possible to classify them according to their structural similarities. One of the first attempts classified proteins as being either **fibrous** or **globular**. Though these names imply structural features, the classification is actually based on solubility. Fibrous proteins are insoluble in water. They are long, stringlike molecules that can wrap around each other to form fibers of great strength, such as collagen, elastin, and keratin. Globular proteins are soluble in water. They are compact, spherelike molecules. Albumins, globulins, and enzymes are examples of globular proteins.

A newer system of protein classification is based on the function served by the particular protein in the body.

1. Structural. Over 50% of the solid weight of the body (after water is removed) is composed of collagen, a major part of bone and connective tissue. Other proteins in this class include the keratins (hair, skin, nails) and elastin (blood vessels).

2. Contractile. Actin and myosin are proteins that stretch and contract as muscles work.

3. Enzymes.   Several proteins affect the speed of chemical reactions, such as the proteolytic enzymes that catalyze the breakdown (hydrolysis) of proteins into amino acids. Without enzymes, chemical reactions in the body would be so slow that life could not exist.

4. Hormones.   Many of the hormones in the body are proteins which regulate body processes. Insulin is a hormone that regulates the blood glucose level.

5. Antibodies.   These are proteins that combine with foreign species in the body to prevent injury to cells. Gamma-globulin is an antibody.

6. Blood proteins.   These proteins are circulated in the blood, such as the albumins, globulins, prothrombin, and fibrinogen. The last two are important in blood clotting.

Another protein classification system considers proteins as being either simple or conjugated. A **simple protein**, upon hydrolysis, produces only amino acids, while a **conjugated protein** produces amino acids plus a nonprotein unit, called a **prosthetic group**.

$$\text{simple protein} + H_2O \xrightarrow{\text{catalyst}} \text{amino acids}$$

$$\text{conjugated protein} + H_2O \xrightarrow{\text{catalyst}} \text{amino acids} + \underset{\text{(prosthetic group)}}{\text{nonprotein unit}}$$

Table 22.3 describes several of the more common classes of conjugated proteins.

TABLE 22.3 Conjugated Proteins and their Prosthetic Group		
Class	Prosthetic group	Examples
Lipoprotein	lipid (cholesterol, phospholipid)	blood lipoprotein (thought to have a role in heart disease)
Chromoprotein	pigmented group	hemoglobin
Glycoprotein	carbohydrate	mucin (found in saliva, it acts as a lubricant)
Phosphoprotein	phosphate or phosphoric acid	casein (found in milk)
Nucleoprotein	nucleic acids, RNA and DNA	some viruses and part of nucleus of the cell

Amino Acids

Like many large biological molecules, proteins can be broken down into smaller molecules by hydrolysis, the reaction with water. In the laboratory, hydrolysis reactions are usually carried out in the presence of strong acids (HCl or H_2SO_4), strong bases (NaOH or KOH), or proteolytic enzymes called proteases. The products of these reactions are amino acids. As the name implies, amino acids are carboxylic acids (R—COOH) which contain an amino group (—NH_2). Theoretically, the number of possible amino acids is unlimited, but only 20 are found in most proteins. With the exception of proline, which is an α-imino acid, all the others are α-amino acids. (α is the Greek letter alpha. "α-amino acid" is read "alpha-amino acid.")

In a carboxylic acid, the α position is the carbon atom adjacent to the carboxylic acid group.

The α position

An α-amino acid would be one with an amino group attached to the α-carbon atom.

An α—amino acid

All α-amino acids can be represented by the same general formula. The only difference between one amino acid and another is the nature of the R group that is also bonded to the α-carbon atom.

The α—carbon atom

Different for each kind of amino acid

The R group of an amino acid can be hydrogen, a simple alkyl group, a group containing an aromatic ring, a heterocyclic group, or other substituted group. Each different amino acid has a unique set of properties due to the composition of the different R groups. Often, the R group is referred to as the "side chain" of an amino acid.

Check Test Number 1

1. Most of the protein material in the body fills what important function?
2. What elements are found in all amino acids?
3. Draw the general structure of all amino acids, and point out the α position, the amino group, the side chain (R group), and the carboxyl group.

Answers:
1. Body building and repair.
2. C, H, O, and N.

3.

$$H_2N-\underset{\underset{R}{|}}{\overset{\overset{H}{|}}{C}}-COOH$$

α–Position

Amino group

Carboxyl group

Side chain

Amino acids are usually classified according to the number of carboxyl and amino groups in the molecule. The **neutral amino acids** contain only one carboxyl group and one amino group. Water solutions of the neutral amino acids have a neutral or near neutral pH. The **basic amino acids** contain one carboxyl group but more than one amino or aminolike group. Water solutions of the basic amino acids will have a basic pH. The **acidic amino acids** contain one amino group and more than one carboxyl group. Water solutions of the acidic amino acids will have an acidic pH.

A description of each of the 20 amino acids found in most proteins follows. They are grouped according to their neutral, acidic, or basic properties. The three-letter abbreviation used to symbolize each amino acid appears beneath the full name. The side chain of each amino acid is reproduced in color to emphasize its composition.

The Neutral Amino Acids.   Those which form solutions of near neutral pH and contain a single carboxyl and amino group.

Aliphatic amino acids

1. Glycine
   Gly

$$H_2N-\underset{\underset{\boxed{H}}{|}}{\overset{\overset{H}{|}}{C}}-COOH$$

The simplest amino acid, one of the first isolated from proteins.

2. Alanine
   Ala

$$H_2N-\underset{\underset{\boxed{CH_3}}{|}}{\overset{\overset{H}{|}}{C}}-COOH$$

3. Valine
   Val

$$H_2N-\underset{\underset{\underset{H_3C \quad CH_3}{}}{C-H}}{\overset{\overset{H}{|}}{C}}-COOH$$

It is easy to remember the structure of valine since the side chain is "V" shaped, like the first letter in the name.

4. Leucine
   Leu

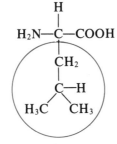

5. Isoleucine
   Ile

This is an isomer of
leucine

Hydroxy amino acids

6. Serine
   Ser

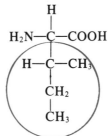

Serine can be remembered
as a hydroxy-substituted
alanine, since an –OH
group is substituted for
a –H on the methyl
group. Serine is found
in silk protein.

7. Threonine
   Thr

Aromatic amino acids

8. Phenylalanine
   Phe

This is another substituted
alanine.

9. Tyrosine
   Tyr

$$H_2N—\underset{\underset{CH_2}{|}}{\overset{\overset{H}{|}}{C}}—COOH$$

Notice the similarity between tyrosine and phenylalanine.

10. Tryptophan
    Trp

$$H_2N—\underset{\underset{CH_2}{|}}{\overset{\overset{H}{|}}{C}}—COOH$$

Though this is a heterocyclic amino acid, it is usually classified as aromatic because of the aromatic benzene ring.

Sulfur-containing amino acids

11. Cysteine
    Cys

$$H_2N—\underset{\underset{\underset{SH}{|}}{CH_2}}{\overset{\overset{H}{|}}{C}}—COOH$$

Another substituted alanine. Cysteine is present in many proteins, such as insulin.

12. Methionine
    Met

$$H_2N—\underset{\underset{\underset{\underset{\underset{CH_3}{|}}{S}}{CH_2}}{CH_2}}{\overset{\overset{H}{|}}{C}}—COOH$$

α-Imino acid

13. Proline
    Pro

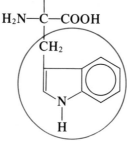

Proline, unlike the other amino acids, has the nitrogen in a heterocyclic system, forming an imine.

The Acidic Amino Acids and their Amide Derivatives. Acidic amino acids contain two carboxyl groups and a single amino group. Solutions of these amino acids are acidic (pH<7). Asparagine and glutamine (amide derivatives) are included in this category to show their relationship to aspartic and glutamic acid, though they form solutions of near neutral pH.

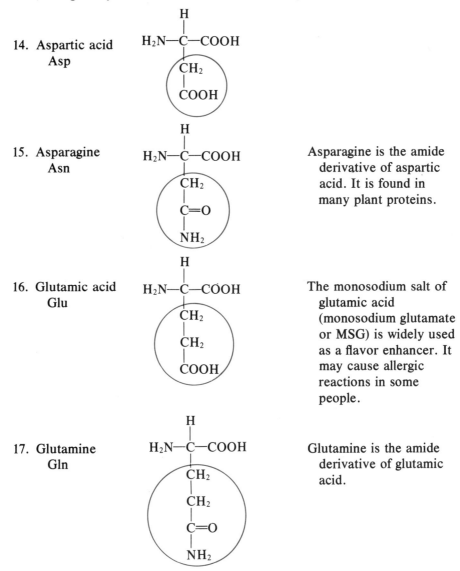

14. Aspartic acid
Asp

15. Asparagine
Asn

Asparagine is the amide derivative of aspartic acid. It is found in many plant proteins.

16. Glutamic acid
Glu

The monosodium salt of glutamic acid (monosodium glutamate or MSG) is widely used as a flavor enhancer. It may cause allergic reactions in some people.

17. Glutamine
Gln

Glutamine is the amide derivative of glutamic acid.

The Basic Amino Acids. The basic amino acids have one carboxyl group and more than one amino or aminolike group. Solutions of these amino acids are basic (pH>7).

18. Histidine
    His

$$H_2N-\underset{\underset{\displaystyle CH_2}{|}}{\overset{\displaystyle \overset{H}{|}}{C}}-COOH$$

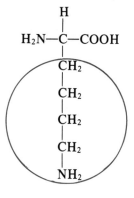

Histidine can be changed in the body to a compound that is responsible for many of the allergic reactions in the

body. The H–N– in the ring acts much like the usual amino group.

19. Arginine
    Arg

$$H_2N-\underset{\underset{\displaystyle CH_2}{|}}{\overset{\displaystyle \overset{H}{|}}{C}}-COOH$$

CH$_2$
CH$_2$
CH$_2$
C=NH
NH$_2$

20. Lysine
    Lys

$$H_2N-\underset{\underset{\displaystyle CH_2}{|}}{\overset{\displaystyle \overset{H}{|}}{C}}-COOH$$

CH$_2$
CH$_2$
CH$_2$
CH$_2$
NH$_2$

Lysine is found primarily in animal protein

Check Test Number 2

1. What general part of a particular α-amino acid molecule causes it to have properties different from those of other α-amino acids?
2. Does proline have a side chain?

Answers:
1. The α-amino acids differ from one another only in the composition of the side chains R–.
2. Proline is an α-imino acid, and as such does not have a side chain in the same sense that α-amino acids do.

$$H_2C \text{——} CH_2$$

Proline

Essential Amino Acids

Of the 20 common amino acids, 12 can be synthesized from carbohydrates or lipids in the body, provided a nitrogen source for the amino group is also present. The remaining eight amino acids are either not synthesized in the body, or if so, synthesized in an amount that is not sufficient at any time to meet the needs of the body. These are known as the **essential amino acids** and they must be supplied in the diet. They are listed in Table 22.4. Really, all 20 amino acids are essential for proper growth and maintenance, but the term "essential amino acids" will be reserved for those amino acids that must be provided by the diet.

TABLE 22.4 The Essential Amino Acids*	
valine	phenylalanine
leucine	tryptophan
isoleucine	methionine
threonine	lysine

* In addition, arginine and histidine are essential amino acids for infants.

Often, foods will be described as complete or incomplete protein sources. A **complete protein** contains all the essential amino acids, whereas an **incomplete protein** is lacking in at least one of the essential acids. Meats contain proteins that are normally complete, but plant proteins usually lack some of the essential amino acids, notably lysine. In underdeveloped countries where red meat or a variety of cereal grains are difficult to obtain, many people are in poor health due to deficiencies in the essential amino acids. Science is striving to produce plants that will be able to supply a larger number of the essential acids. They have successfully bred a lysine-rich variety of corn.

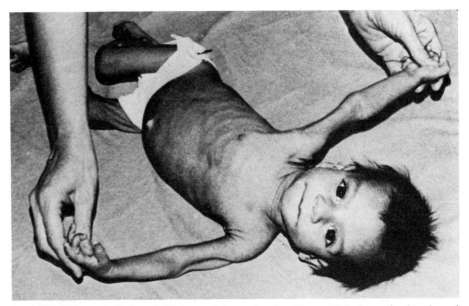

Fig. 22.2. Severe protein deficiency is a major health problem in underdeveloped countries. The lack of essential amino acids in the diet can produce anemia, edema, diminished growth, apathy, and reduced brain activity. After a child is deprived of the protein in the mother's milk, perhaps by a younger child, an improper diet can bring on the signs of protein deficiency. (Board of Global Ministries of the United Methodist Church)

Optical Activity of Amino Acids

With the exception of glycine, all amino acids contain an asymmetric carbon atom (the α-carbon) and thus will exist in either the D or L configuration. The configurations of amino acids are related to those of glyceraldehyde as shown below using the simplest amino acid that has an asymmetric carbon, alanine.

$$
\begin{array}{cccc}
\text{H—C=O} & \text{COOH} & \text{H—C=O} & \text{COOH} \\
| & | & | & | \\
\text{H—C—}\boxed{\text{OH}} & \text{H—C—}\boxed{\text{NH}_2} & \boxed{\text{HO}}\text{—C—H} & \boxed{\text{H}_2\text{N}}\text{—C—H} \\
| & | & | & | \\
\text{H}_2\text{C—OH} & \text{CH}_3 & \text{H}_2\text{C—OH} & \text{CH}_3 \\
\end{array}
$$

D–Glyceraldehyde   D–Alanine   L–Glyceraldehyde   L–Alanine

When drawing structures of amino acids to show their D or L configuration, place the carboxyl group on the top and the side chain (R group) on the bottom. The α-carbon is in the middle. Since the α-amino group is used as the reference group for amino acids, if you wish to show the D isomer, place it on the right-hand side of the α-carbon. If you wish to show the L isomer, place the amino group on the left-hand side. The structures above show this clearly.

Nearly all naturally occurring proteins contain only L-amino acids. However, the protein of some bacteria contains some of the D-amino acids. This may be responsible for their resistance to various antibiotics. The antibiotic actinomycin also contains some D-amino acids in its structure.

Check Test Number 3

1. Why are wheat and corn considered to be incomplete protein sources?
2. Draw the structures of D- and L-lysine.

Answers:

1. The proteins in wheat and corn do not contain all of the essential amino acids needed by the human body.

2.

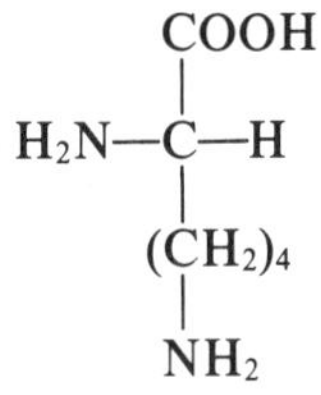

$$
\begin{array}{cc}
\text{COOH} & \text{COOH} \\
| & | \\
\text{H}_2\text{N—C—H} & \text{H—C—NH}_2 \\
| & | \\
(\text{CH}_2)_4 & (\text{CH}_2)_4 \\
| & | \\
\text{NH}_2 & \text{NH}_2
\end{array}
$$

L–Lysine 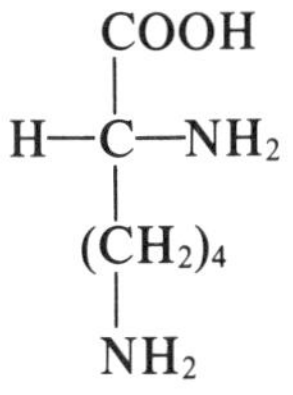   D–Lysine

The Dipolar Nature of Amino Acids

Amino acids are white crystalline solids melting at temperatures above 200°C. Most are very soluble in water but insoluble in nonpolar organic solvents like benzene, hexane, or carbon tetrachloride. Their solubility properties are not unlike those of typical inorganic salts like sodium chloride or potassium nitrate. Indeed, it is known that amino acids can exist in saltlike forms, the result of a transfer of a proton from the carboxyl group to the amino group.

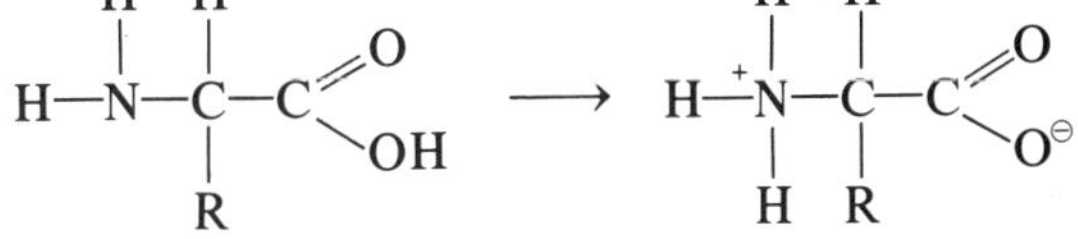

$$
\begin{array}{ccc}
\overset{\text{H}}{\underset{|}{}} \; \overset{\text{H}}{\underset{|}{}} & & \overset{\text{H}}{\underset{|}{}} \; \overset{\text{H}}{\underset{|}{}} \\
\text{H—N—C—C} \overset{\displaystyle O}{\underset{\displaystyle \text{OH}}{\Big\langle}} & \longrightarrow & \text{H—}\overset{+}{\text{N}}\text{—C—C}\overset{\displaystyle O}{\underset{\displaystyle O^{\ominus}}{\Big\langle}} \\
| & & | \quad | \\
\text{R} & & \text{H} \quad \text{R}
\end{array}
$$

Undissociated form   An internal salt (a zwitterion)

The **internal salt** is called a **zwitterion**, a dipolar ion with an equal number of positive and negative charges within the covalent structure. Up to this time amino acids have been shown in their undissociated forms only, but in biological systems, amino acids nearly always bear charges, either as zwitterions or as anions or cations. The zwitterion forms of three amino acids are shown below:

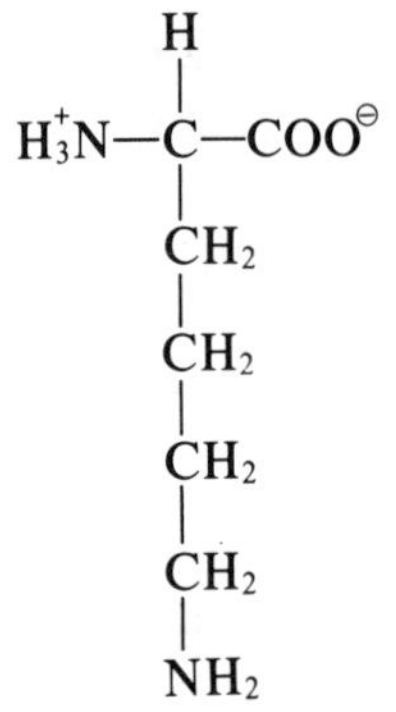

$$
\begin{array}{c}
H \\
| \\
H_3\overset{+}{N}-C-COO^{\ominus} \\
| \\
CH_2 \\
| \\
CH_2 \\
| \\
CH_2 \\
| \\
CH_2 \\
| \\
NH_2
\end{array}
$$

Lysine

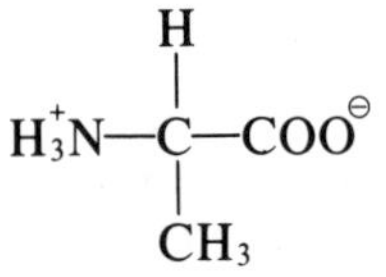

$$
\begin{array}{c}
H \\
| \\
H_3\overset{+}{N}-C-COO^{\ominus} \\
| \\
CH_3
\end{array}
$$

Alanine

$$
\begin{array}{c}
H \\
| \\
H_3\overset{+}{N}-C-COO^{\ominus} \\
| \\
CH_2
\end{array}
$$

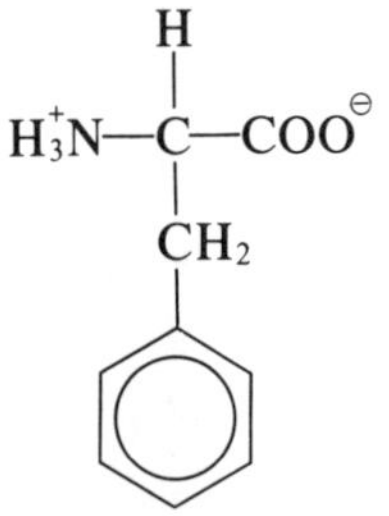

Tyrosine

For each different amino acid, there will be a certain pH at which it will exist in the zwitterion form. At this pH, the amino acid is electrically neutral and will not migrate toward the + or − poles of an electric field. The pH at which this behavior is observed is called the **isoelectric pH** or **isoelectric point** for that amino acid. The isoelectric pH will be different for each different amino acid.

Since amino acids are capable of donating or accepting protons, that is, can act as an acid or a base, they are considered to be **amphoteric** substances. When a small amount of strong acid is added to a solution of an amino acid at its isoelectric point, the added protons will react with the carboxylate ion, thereby reducing the H^+ concentration in the solution.

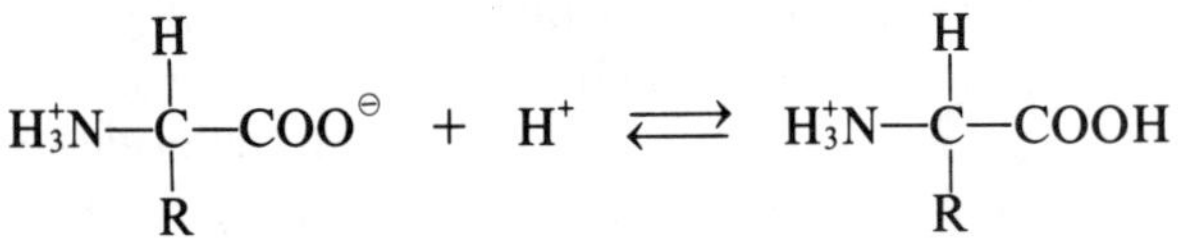

$$
\begin{array}{c}
H \\
| \\
H_3\overset{+}{N}-C-COO^{\ominus} \\
| \\
R
\end{array}
\quad + \quad H^+ \quad \rightleftharpoons \quad
\begin{array}{c}
H \\
| \\
H_3\overset{+}{N}-C-COOH \\
| \\
R
\end{array}
$$

zwitterion form   + acid $\longrightarrow$ positive ion

If a small amount of base is added to the amino acid solution, the mildly acidic $-\overset{+}{N}H_3$ group can give a proton to the hydroxide ion to form water, reducing the OH^- concentration.

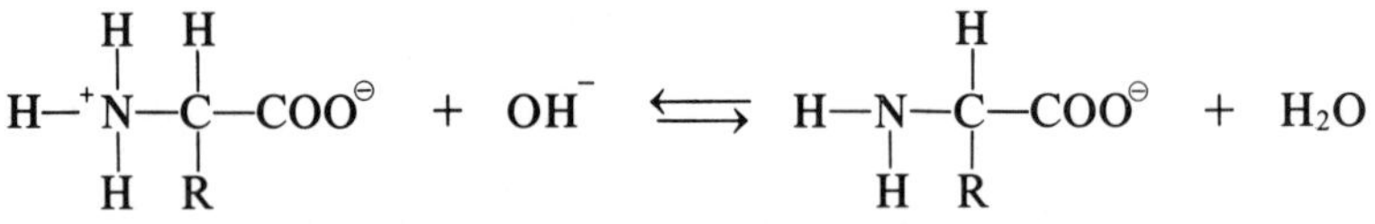

$$
\begin{array}{c}
H\ \ H \\
|\ \ | \\
H-\overset{+}{N}-C-COO^{\ominus} \\
|\ \ | \\
H\ \ R
\end{array}
\quad + \quad OH^- \quad \rightleftharpoons \quad
\begin{array}{c}
H \\
| \\
H-N-C-COO^{\ominus} \\
|\ \ | \\
H\ \ R
\end{array}
\quad + \quad H_2O
$$

zwitterion form   + base $\longrightarrow$ negative ion   + water

Thus, free amino acids and even those contained in proteins with free carboxyl or amino groups on their side chains are capable of neutralizing small amounts of acid *or* base. By doing so, they act as **buffers**. Proteins represent one of the major buffering components in blood and other body fluids.

It is possible to separate a mixture of amino acids based on the differences in their isoelectric pH's. If an amino acid mixture is placed in the center of a paper saturated with a buffer solution, those with isoelectric pH's less than the pH of the buffer will lose a proton and become negative ions. Those with greater

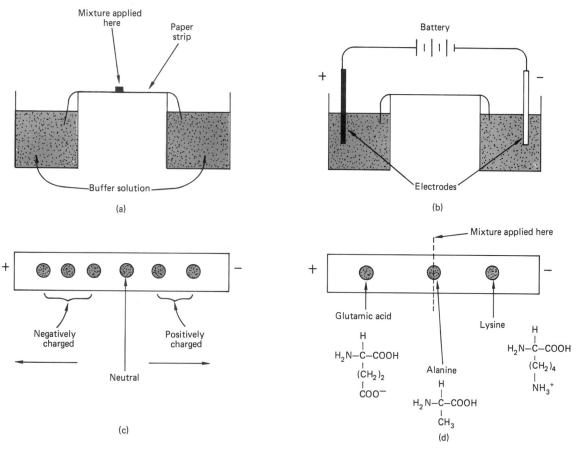

Fig. 22.3. Electrophoresis is used to separate mixtures of amino acids. (a) A small amount of the mixture is placed on a paper saturated with a buffer solution. Some amino acids will become charged + , some − , while some will remain neutral at the pH of the buffer. (b) Positive and negative electrodes are connected to opposite ends creating an electric field across the paper strip. (c) As time passes, the amino acids separate as they migrate at different rates to the electrode of opposite charge. (d) A mixture of alanine, lysine, and glutamic acid can be separated by using a buffer pH of 6.0. Glutamic acid forms a negative ion, lysine a positive ion, and alanine remains neutral at that pH.

isoelectric pH's will gain a proton and become positive ions. Then the (+) electrode of a high-voltage source is connected to one end of the paper, and the (−) electrode is connected to the other end, forming an electric field. The positive ions of the different amino acids will migrate at different speeds towards the negative electrode, and the negative ions of the different amino acids will migrate at different speeds toward the positive electrode. If an amino acid remains neutral, it will not move at all. The result is a separation of the mixture. This method of separation is known as **amino acid electrophoresis**.

Peptides

Complete hydrolysis of a protein will yield a mixture of amino acids. But if the hydrolysis reaction is stopped before the protein is completely broken down, small fragments made of several amino acids are obtained. In these fragments, amino acids are linked to each other by amide structures, $-\overset{\overset{\displaystyle O}{\|}}{C}-\overset{\overset{\displaystyle H}{|}}{N}-$ The fragments of linked amino acids are called **peptides**. Peptides can also be prepared from amino acids.

When two amino acids combine, the amino group of one acid reacts with the carboxyl group of the second to form water and a dipeptide (a peptide containing two amino acid units). By convention, when amino acids are written in an equation, the amino groups are placed to the left and the carboxyl groups to the right. The general equation for dipeptide formation is

$$H_2N-\overset{\overset{\displaystyle H}{|}}{\underset{\underset{\displaystyle R}{|}}{C}}-\overset{\overset{\displaystyle O}{\|}}{C}\!\!\boxed{-OH}\ +\ H\!\!\boxed{-}\!N-\overset{\overset{\displaystyle H}{|}}{\underset{\underset{\displaystyle R'}{|}}{C}}-\overset{\overset{\displaystyle O}{\|}}{C}-OH\ \xrightarrow{\text{catalyst}}$$

Amino acid 1   Amino acid 2

$$H_2N-\overset{\overset{\displaystyle H}{|}}{\underset{\underset{\displaystyle R}{|}}{C}}\!\!\boxed{-\overset{\overset{\displaystyle O}{\|}}{C}-N\!\!-}\!\overset{\overset{\displaystyle H}{|}}{\underset{\underset{\displaystyle R'}{|}}{C}}-\overset{\overset{\displaystyle O}{\|}}{C}-OH\ +\ H_2O$$

Amide structure   Dipeptide–A
(peptide bond)

To emphasize the significance of the amide structure in peptides, it is often called a **peptide bond** or **peptide linkage**. If amino acid 1 and amino acid 2 react

in the opposite sequence, a different dipeptide would result which would be
an isomer of the one above. The two dipeptides are labeled A and B to show
that they are not the same.

$$H_2N-\underset{\underset{R'}{|}}{\overset{\overset{H}{|}}{C}}-\overset{\overset{O}{||}}{C}-\underset{}{\overset{\overset{H}{|}}{N}}-\underset{\underset{R}{|}}{\overset{\overset{H}{|}}{C}}-\overset{\overset{O}{||}}{C}-OH$$

Dipeptide–B

A third amino acid could join with either dipeptide to form a tripeptide, and
the peptide could continue to grow in this way.

$$H_2N-\underset{\underset{R}{|}}{\overset{\overset{H}{|}}{C}}-\overset{\overset{O}{||}}{C}-\overset{\overset{H}{|}}{N}-\underset{\underset{R'}{|}}{\overset{\overset{H}{|}}{C}}-\overset{\overset{O}{||}}{C}-OH \quad + \quad H-\overset{\overset{H}{|}}{N}-\underset{\underset{R''}{|}}{\overset{\overset{H}{|}}{C}}-\overset{\overset{O}{||}}{C}-OH \quad \xrightarrow{\text{catalyst}}$$

Dipeptide–A   Amino acid 3

$$H_2N-\underset{\underset{R}{|}}{\overset{\overset{H}{|}}{C}}-\overset{\overset{O}{||}}{C}-\overset{\overset{H}{|}}{N}-\underset{\underset{R'}{|}}{\overset{\overset{H}{|}}{C}}-\overset{\overset{O}{||}}{C}-\overset{\overset{H}{|}}{N}-\underset{\underset{R''}{|}}{\overset{\overset{H}{|}}{C}}-\overset{\overset{O}{||}}{C}-OH \quad + \quad H_2O$$

A tripeptide

To form a specific tripeptide, let us react glutamine, valine, and cysteine in
that order. We will start by writing each structure in an equation with the amino
group on the left and carboxyl group on the right:

$$H-\overset{\overset{H}{|}}{N}-\underset{\underset{\underset{O=C-NH_2}{|}}{\underset{CH_2}{\underset{|}{CH_2}}}}{\overset{\overset{H}{|}}{C}}-\overset{\overset{O}{||}}{C}-O-H \quad + \quad H-\overset{\overset{H}{|}}{N}-\underset{CH_3 \quad CH_3}{\underset{\underset{CH}{}}{\overset{\overset{H}{|}}{C}}}-\overset{\overset{O}{||}}{C}-O-H \quad + \quad H-\overset{\overset{H}{|}}{N}-\underset{\underset{SH}{\underset{|}{CH_2}}}{\overset{\overset{H}{|}}{C}}-\overset{\overset{O}{||}}{C}-O-H \quad \xrightarrow{\text{catalyst}}$$

Glutamine   Valine   Cysteine

(*Continued on p. 624.*)

$$\longrightarrow \quad \underset{\substack{\displaystyle | \\ CH_2 \\ | \\ CH_2 \\ | \\ O=C-NH_2}}{H-N-C-C} - \underset{\substack{| \\ CH \\ \diagup \diagdown \\ CH_3 \quad CH_3}}{N-C-C} - \underset{\substack{| \\ CH_2 \\ | \\ SH}}{N-C-C} -O-H \quad + \quad 2\ H_2O$$

Notice that the tripeptide contains a free amino group on the left and a free carboxyl group on the right.

Biochemists have developed a shorthand method for describing peptides and peptide formation using the abbreviations for the amino acids. The amino groups are understood to be on the left and the carboxyl groups on the right. The formation of the tripeptide from glutamine, valine, and cysteine would be written this way:

$$Glu + Val + Cys \xrightarrow{\text{catalyst}} Glu-Val-Cys$$

Tripeptide

One can quickly realize that when a large number of amino acids react, an enormous number of different peptides could form. However, in living systems there is strict control over which amino acids will react and in what order.

The general term "peptide" is used to indicate a molecule containing a relatively small number of amino acids (usually less than 50) linked together by peptide bonds. Polypeptide generally refers to a larger molecule containing in excess of 50 amino acids, and "protein" is used to describe a very long chain of amino acids joined by peptide bonds, or many polypeptide units joined together.

Check Test Number 4

1. (a) Draw the structure of the tetrapeptide that forms when glycine, phenylalanine, histidine, and aspartic acid react. Combine them in the order given keeping all amino groups on the left. (b) Give the shorthand notation for the tetrapeptide.
2. Draw the structures of the two dipeptide isomers that can form when lysine and cysteine react. Remember, the side chains do not react.

Answers:

1. (a)

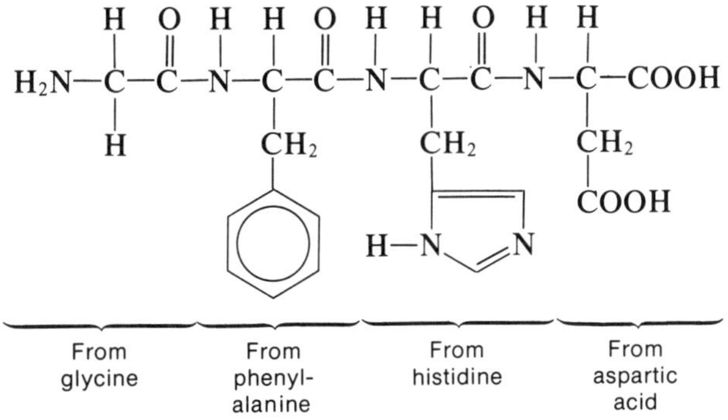

| From glycine | From phenyl-alanine | From histidine | From aspartic acid |

(b) Gly–Phe–His–Asp

2.

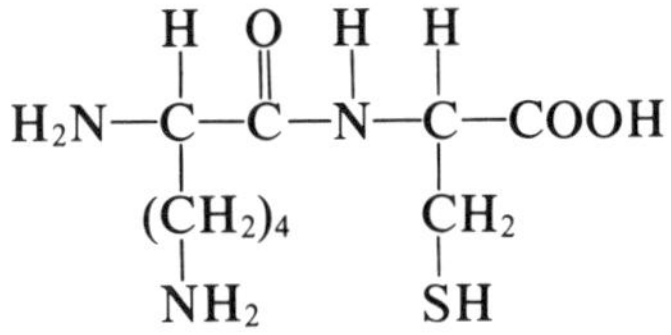

Lys–Cys

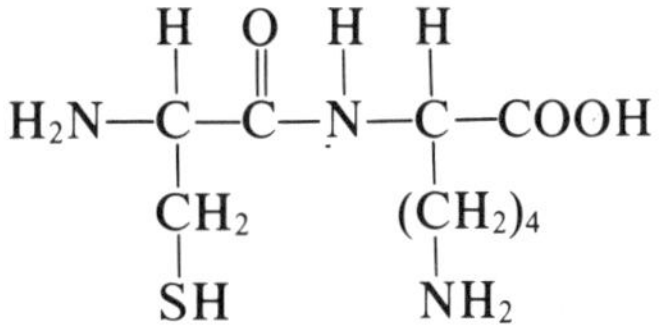

Cys–Lys

Separating Biological Mixtures

If we are to understand the myriad operations of a single cell, we must know the composition of the biological compounds that make it up. Before 1950, the separation and identification of these compounds was a tedious and difficult task, but since then, several methods have been developed that are capable of separating extremely complex mixtures into the individual components. One of these methods is electrophoresis. Another is **chromatography,** an area comprised of several methods which are all based on the same general principles. Chromatography has been instrumental in identifying many of the proteins, peptides, and nucleic acids found in the cell. Because of its importance in the understanding of the living system, and its use in medicine, you should be aware of the value and the operation of some of the chromatographic methods.

All chromatographic methods make use of the different degrees to which a molecule is attracted to different substances. Usually two substances are used, one an insoluble solid, such as paper, starch, minerals, or synthetic resins. The other is a liquid or gas which is made to flow over the solid. In some chromatographic methods, the solid is packed in a glass or metal tube so it cannot move, and is termed the **stationary phase**. The flowing liquid or gas passes through the tube and is called the **moving phase**. If a biological molecule is very soluble (attraction is high) in the moving phase, and not strongly attracted to the stationary phase, it will be quickly carried along as the moving phase flows over the solid. But, if a molecule is not as strongly attracted to the moving phase as it is to the stationary phase, it will be carried along at a slower rate. If two compounds in a mixture have different attractions for the stationary phase but are equally soluble in the moving phase, they would separate since the one with the lower attraction for the solid phase would be carried along at a faster rate, leaving the other one behind. At some point, samples of the moving phase can be collected to obtain the purified compounds.

The way that chromatography operates may sound complicated, but perhaps an analogy will help. Think of two different kinds of leaves (compounds) floating down a rocky stream (the moving phase). The leaf with the large, pointed edges will be held up by the rocks (the stationary phase) and move along at a slower rate. The other leaf, with smooth edges, can flow around a greater number of rocks and will travel at a higher rate. Downstream, the smooth leaf will be farther ahead of the pointed leaf, and you can think of them as being separated. Several different kinds of chromatography will be briefly described.

Column Chromatography

A finely divided solid is packed in a vertical glass tube which has a stopcock at the lower end. This serves as the stationary phase. A small volume of the mixture is poured onto the solid. Then the developing solvent, the moving phase, is passed through the column. The components of the mixture are separated as they pass over the solid *at different rates* and are collected in individual containers as the solution passes out the bottom of the tube.

Paper Chromatography

The stationary phase is a strip of porous paper, much like filter paper. The developing solvent is a liquid which covers the bottom of a large, covered glass jar, called the developing chamber. A small drop of the mixture is placed within an inch of one end of the paper and is allowed to dry. The paper is then mounted vertically in the developing chamber so just the end nearest the spot dips into the liquid. The spot is not immersed. As time passes, the liquid ascends the paper by capillary action, passing through the sample and carrying the components of the mixture *at different rates* up the paper. Those substances with a greater attraction for the solvent than for the cellulose fibers of the paper will move at the higher rates. Once the paper is dried, the individual components can be extracted from small pieces of paper cut out of the strip.

Thin-Layer Chromatography

Thin-layer chromatography (TLC) is very similar to paper chromatography except that the stationary phase can be a wide variety of finely divided solids held in a thin layer on a glass or plastic sheet. A droplet of the mixture is placed on the sheet and dried. The sheet is mounted vertically in the developing

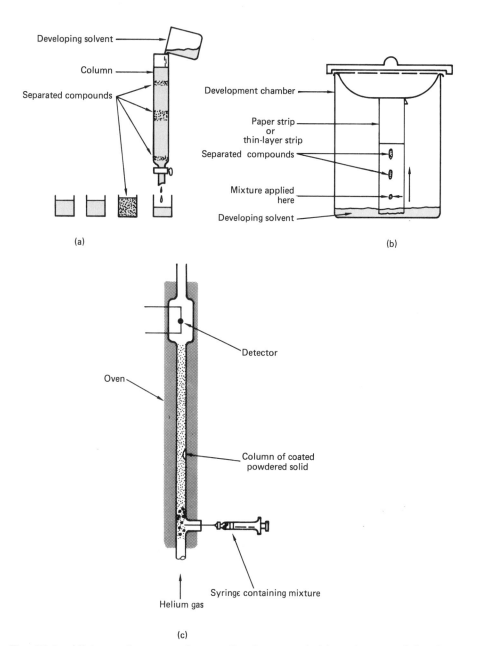

Fig. 22.4. Mixtures of compounds can often be separated by using one of the chromatographic methods. (a) Column chromatography—Separation occurs as a solution of the mixture passes over the stationary phase by the developing solvent. A solution of each component of the mixture can be collected as it passes out the bottom of the column. (b) Paper and thin-layer chromatography—The developing solvent ascends the stationary phase by capillary action allowing separation of the mixture as the solvent advances. (c) Gas chromatography—The vaporized mixture is carried by a stream of gas, usually helium, through a heated tube containing the stationary phase. An electrical device called a detector is used to sense (detect) each component of the mixture as it passes out the tube.

chamber, and the solvent is allowed to ascend through the sample spot. Very good separations of biological compounds are possible with TLC.

Gas Chromatography

Gas chromatography (GC) is used to separate gaseous mixtures, usually at temperatures well above 25°C. Liquid mixtures can be separated after being vaporized in the GC instrument. The stationary phase is a finely divided solid or an oil-coated solid packed in a thin metal tube that is housed in a heated chamber. The moving phase is a gas, usually helium. The GC method is easy to use, but the cost of the equipment is considerably higher than that used in the other methods. Because of the higher temperatures used in GC, many biological mixtures undergo thermal decomposition and cannot be separated. The technique is mentioned here because it is used in many hospitals to detect small amounts of alcohol, drugs, or toxic substances in the blood.

A Close-Up Look at Two Important Proteins

Insulin and hemoglobin are both of critical importance in the body. Insulin is involved in the regulation of blood sugar (glucose). Some individuals produce little or no insulin, and thus have an abnormally high glucose level, a condition known as **diabetes mellitus**. The tendency toward diabetes is thought to be hereditary. If diabetes is allowed to run its course, death can result. Fortunately, insulin can be injected at regular intervals to meet the needs of the body.

Hemoglobin is one of the most abundant proteins in the body. There are roughly 3×10^8 molecules of hemoglobin in each red blood cell, and about 2×10^{12} red cells in each pint of blood. Hemoglobin picks up oxygen in the lungs and transports it to other parts of the body for use in metabolism. A deficiency of hemoglobin results in **anemia,** a condition characterized by a listless feeling, fatigue, reduced resistance to disease, and increased heart and respiratory rates.

The detailed structures of insulin and hemoglobin are known. The structures of proteins can be viewed as a composite of four stages of organization. Each stage will be discussed with reference to insulin and hemoglobin later on in this chapter.

Insulin

The first protein to have its complete amino acid sequence determined was insulin. Frederick Sanger completed this work in 1953 and in 1958 received the Nobel Prize for this accomplishment. Insulin contains 51 amino acid units arranged in two polypeptide chains (labeled A and B in Figure 22.5). It has a molecular weight of 11,466. The chains are connected to each other by disulfide linkages formed between one cysteine in one chain and one cysteine in the other. There is also another disulfide linkage within the A chain.

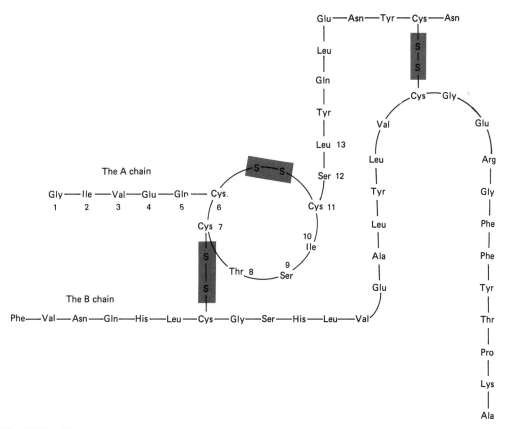

Fig. 22.5. The complete amino acid sequence of human insulin. The two amino acid chains (labeled A and B) are joined together in two places by disulfide links. A third disulfide link forms a loop in the A chain.

Even before Sanger's work, it was known that if insulin from either human or animal sources was injected into the body of a diabetic, regulation of the blood sugar level would return for several hours. The similarity shown by the insulins from the different sources was understood once the amino acid sequence of each was known. As it turns out, the differences were small in number. The sequence of amino acid units for human insulin is shown in Figure 22.5. Table 22.5 lists the differences that exist between human insulin and that from other sources. The only variations occur in the A chain at amino acid positions 8, 9, and 10.

Apparently, the amino acids in positions 8, 9, and 10 in the A chain are not actively involved in the physiological behavior of insulin, since changes in these amino acids do not alter its effectiveness in humans. But such is not always the case. In hemoglobin, the change of one amino acid unit at a particular site in the polypeptide chain causes a dramatic change in its properties, with tragic effects for the individual.

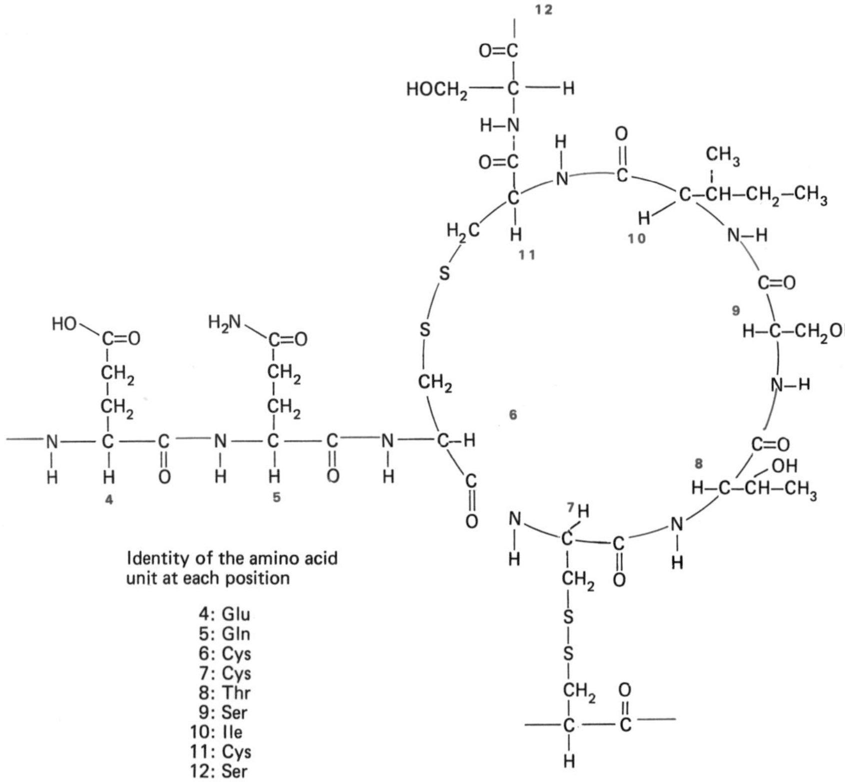

Fig. 22.6. The looped section of the A chain in human insulin showing the disulfide bond between cysteine units at positions 6 and 11. The peptide bonds are indicated in color. The cysteine unit at position 7 participates in a disulfide link connecting the A and B chains (see Figure 22.5).

Identity of the amino acid unit at each position

4: Glu
5: Gln
6: Cys
7: Cys
8: Thr
9: Ser
10: Ile
11: Cys
12: Ser

TABLE 22.5

Source	Amino acid at position		
	8	9	10
Human	Thr	Ser	Ile
Pig	Thr	Ser	Ile
Whale	Thr	Ser	Ile
Horse	Thr	Gly	Ile
Sheep	Ala	Gly	Val
Cow	Ala	Ser	Val

The amino acid sequence is identical in each except for positions 8, 9, and 10 in the A chain.

Hemoglobin   Hemoglobin is a large protein with a molecular weight of 64,500. It is composed of a globin unit and four heme units, complex planar cyclic units with an iron ion (Fe^{2+}) located in the center. The oxygen-carrying capacity of hemoglobin is centered in the heme groups. The globin unit is composed of four polypeptide chains. Two are identical α chains, each composed of 141 amino acid units, and the other two are identical β chains, each composed of 146 amino acid units. One heme group is bonded to each chain by coordination of the iron ion of the heme to a histidine side chain protruding from the polypeptide chain.

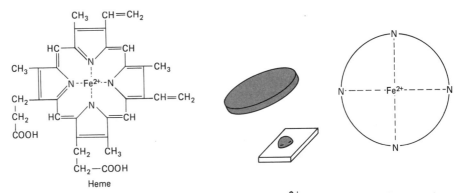

Fig. 22.7. The heme group contains an iron ion (Fe^{2+}) in the center of a complex, planar ring containing four nitrogen atoms. For convenience, the heme group is often represented as a flat disk, a square, or a circle of four nitrogen atoms with the iron ion in the center.

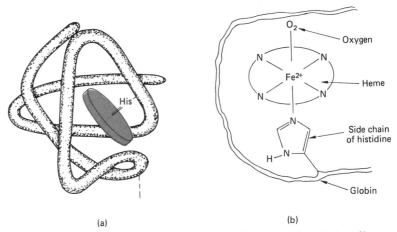

(a)   (b)

Fig. 22.8. (a) One heme group is bonded to each of the four globin chains of hemoglobin by coordination of the iron ion to a histidine unit in the protein chain. (b) Oxygen from the lungs bonds to the iron ion in the heme unit and is carried to various tissues in the body where it is needed.

The complete amino acid sequence in the four polypeptide chains that form the globin unit of hemoglobin has been determined. Since a large number of amino acid units are involved, it will not be reproduced here. But knowledge of the amino acid sequences found in normal adult hemoglobin, HbA, and those in the abnormal hemoglobin of sickle-cell patients, HbS, revealed the cause of sickle-cell anemia at the molecular level. The abnormal hemoglobin has a single amino acid alteration in each β chain in which valine is substituted for the normal glutamic acid unit. Linus Pauling was the first to confirm that the abnormal hemoglobin could carry oxygen from the lungs as readily as the normal hemoglobin. However, once the oxygen was released, the abnormal hemoglobin altered its shape resulting in the collapse of the red blood cells into the familiar sickle shape. The sickled cells are relatively weak and can rupture easily.

The sickle-cell syndrome can appear as sickle-cell trait or sickle-cell anemia. **Sickle-cell trait** is less serious since only a few of the cells are sickled. There are no harmful effects noticed in individuals with the trait. Those with **sickle-cell anemia** have a large number of sickle-shaped cells. Because these cells are destroyed at an abnormally high rate, the body is hard put to replace them as needed. Besides suffering the usual symptoms of severe anemia, these individuals have an abnormally high incidence of kidney and brain damage. Unfortunately, there is no known cure for this heredity-controlled disease.

The Structure of Proteins

The unique biological activity of insulin and hemoglobin is intimately related to the three-dimensional shape or structure of the molecule as it exists in the body. This shape is called the **native state** of the protein. In most cases, a minor alteration of the structure is sufficient to either greatly change or completely eliminate its biological function.

Both insulin and hemoglobin are large and complex chains of amino acids. But the sequence of amino acids only partly describes their structure, for the way in which the chains interact with themselves and with other chains must also be considered.

Protein structure can be described in a four-step sequence beginning with the primary structure of the chain.

The Primary Structure of Proteins   The sequence of amino acid units in a polypeptide chain is called the **primary structure** of the protein. The primary structure of bovine vasopressin, a hormone involved in the regulation of blood pressure in cattle, is

Cys–Tyr–Phe–Gln–Asn–Cys–Pro–Arg–Gln

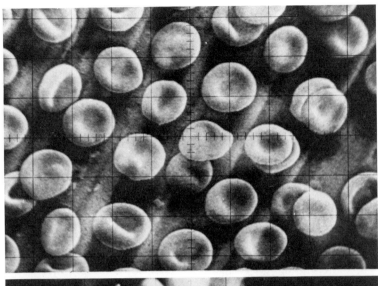

Fig. 22.9. (a) Normal red blood cells. (b) The deoxygenated red blood cells of an individual with sickle-cell anemia. These cells collapse to form a variety of abnormal shapes, most notably the sickle shape. These abnormal cells are subject to a much faster rate of destruction than normal red blood cells. (Philips Electronic Instruments, Inc., Mahwah, New Jersey)

Since the primary structure of bovine vasopressin only gives the amino acid sequence, it will not show the disulfide linkage which is present connecting the two cysteine units. This is actually part of the tertiary structure of the protein which will be described later.

Cys–Tyr–Phe–Gln–Asn–Cys–Pro–Arg–Gln

⌞——— S ——— S ———⌟

Bovine vasopressin

The primary structure of insulin is the sequence of amino acids in both the A and B chains, but again, without the disulfide linkages.

The Secondary Structure of Proteins

The long polypeptide chains described by the primary structure almost always exist in definite shapes. In most proteins, the peptide chain (the backbone of the molecule) coils around on itself forming a helix, a result of **hydrogen bonding** that exists within the molecule. The chain can also form other structures. The way in which a peptide chain is arranged in space due to hydrogen bonding is called the **secondary structure** of the protein.

Two principal kinds of secondary structures are found. The more common is the α **helix** and the other is the **pleated sheet**. The α helix is a right-handed coil structure held together by hydrogen bonds. The hydrogen bonds are formed between amide hydrogens on one loop and carbonyl oxygens on an adjacent loop of the helix.

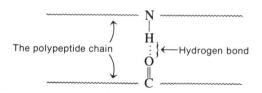

Several turns of an α-helix structure are shown in Figure 22.10 with the hydrogen bonds indicated by dotted lines. Each turn of the helix is spaced 1.5 Å from the turn above and the turn below. The helix makes one complete turn for every 3.6 amino acid units in the chain. The side chains of each amino acid protrude outward from the helix. Because they project away from the helix, they are in a position to affect the way the helical structure will fold over on itself generating the tertiary structure which will be discussed later.

The fibrous keratin proteins found in hair, skin, and nails are cables made of several α-helical structures. Notice in Figure 22.11 how they wind around each other, much like the fibers of a rope. The polypeptide chains in hemoglobin exist as helical structures along several lengths of the chain, while certain sections are in a more random arrangement.

The polypeptide chains of fibroin, a fibrous protein found in silk, are arranged in the pleated sheet secondary structure. Instead of coiling, several chains are arranged alongside each other with every other chain running in opposite directions. This allows the amide hydrogens on one chain to be properly positioned to engage in hydrogen bonding with carbonyl groups on the adjacent chains. The pleats or folds in the sheet result from the normal zig–zag arrangement of the atoms in the backbone of each chain. A pleated sheet structure involving

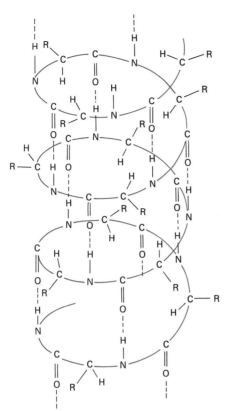

Fig.22.10. The α-helix secondary structure of a polypeptide chain. Note the hydrogen bonds (dotted lines) that hold the helix together and the side chains (R groups) that protrude outward from the structure.

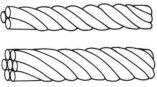

Fig. 22.11. Keratin has a ropelike secondary structure of coiled polypeptide chains.

only two polypeptide chains is shown in Figure 22.12. Notice the side chains of the amino acid protruding above and below the sheet.

Yet another secondary structure is found in the protein, collagen. It will be discussed later on in the chapter.

The Tertiary Structure of Proteins   Most globular proteins have a compact, rigid shape that results from a specific pattern of folding and bending of the primary and secondary structures. This specific three-dimensional structure is known as the **tertiary structure** of the

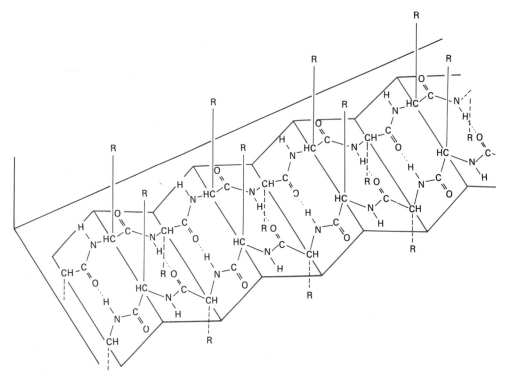

Fig. 22.12. The pleated sheet secondary structure of silk. The polypeptide chains travel in opposite directions. (From: Karlson, Introduction to Modern Biochemistry, 3rd Ed., Academic Press, New York, 1968.)

protein. The tertiary structure determines much of the biological function of a protein.

The forces responsible for the tertiary structure arise from the nature of the side chains on the amino acid units which protrude from the secondary structures. There are four principal types of forces that cause the folding and bending of the secondary structure to produce the tertiary structure of a protein:

1. the disulfide bond
2. ionic attraction
3. hydrogen bonding
4. hydrophobic attraction

The **disulfide bond** links together two cysteine units through reaction of the sulfhydryl functional groups (–SH) on both side chains. The covalent disulfide bond (–S–S–) will connect two regions of a polypeptide chain, or two individual chains, imparting a rigid structure to that part of the molecule. Two disulfide bonds connect the A and B polypeptide chains in insulin (Figure 22.5), though none exist in hemoglobin.

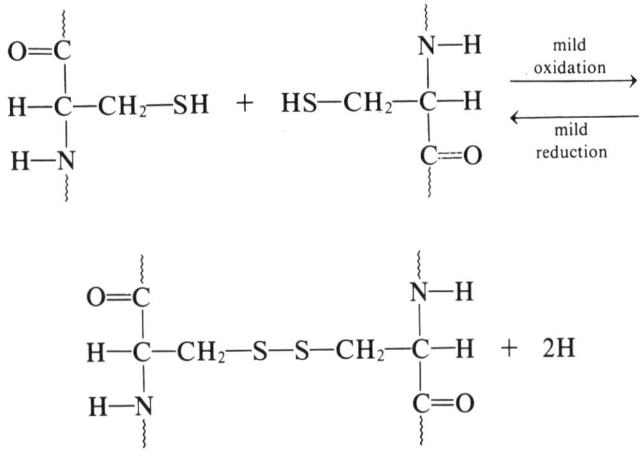

$$O=C \quad\quad N-H \quad\quad \xrightarrow[\text{reduction}]{\substack{\text{mild} \\ \text{oxidation}}}$$

$$H-C-CH_2-SH \; + \; HS-CH_2-C-H$$

$$H-N \quad\quad\quad\quad\quad\quad C=O$$

$$O=C \quad\quad\quad\quad N-H$$

$$H-C-CH_2-S-S-CH_2-C-H \; + \; 2H$$

$$H-N \quad\quad\quad\quad\quad\quad C=O$$

Disulfide linkages are stable and will not break with moderate changes in pH, or concentrations of dissolved salts.

The **ionic attractions** that exist between various sites along a polypeptide backbone result from interactions between oppositely charged, side-chain amino and carboxylate groups which extend from the secondary structures. The side chains of aspartic acid and glutamic acid can exist as charged carboxyl groups, $-\overset{\overset{\displaystyle O}{\|}}{C}-O^{\ominus}$. Those of lysine and arginine can exist as charged amino groups $(-\overset{+}{N}H_3)$. If these are brought together as the chain folds, attractions will result that hold those parts of the chain together. These attractions are also termed **salt bridges,** and their stability is quite pH dependent.

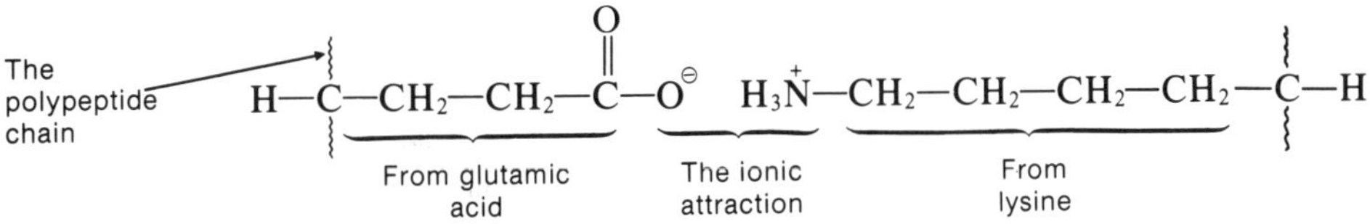

The polypeptide chain

$$H-C-CH_2-CH_2-\overset{\overset{\displaystyle O}{\|}}{C}-O^{\ominus} \quad H_3\overset{+}{N}-CH_2-CH_2-CH_2-CH_2-C-H$$

From glutamic acid   The ionic attraction   From lysine

Earlier you saw the importance of **hydrogen bonding** in the secondary structure of proteins. It is also of major importance in the formation of the tertiary structure. Those amino acid units with side chains that possess groups capable of engaging in hydrogen bonding will do so when properly matched up in a folded polypeptide chain. Two examples are shown below.

Several amino acid side chains cannot participate in any of the preceding kinds of attractions that are found in tertiary structures since they lack the required functional groups. The fourth type of attractive force represents the weak attraction of one nonpolar side chain for another due to their mutual insolubility in the water that surrounds them. This is termed a **hydrophobic attraction** to emphasize the important role of the solvent, water. Because of the hydrophobic (water-repelling) nature of the nonpolar side chains, many proteins fold into shapes that shield them from water, while projecting polar side chains outward. Even though each hydrophobic attraction is weak, the total of all the individual attractions is significant. The aromatic benzene rings and the alkyl groups both contribute to the hydrophobic attraction.

Check Test Number 5

1. (a) What kind of bonding is responsible for the secondary structure of proteins? (b) Draw adjacent sections of a polypeptide chain and indicate the places where this kind of bonding can occur.
2. If the side chains of the following pairs of amino acids are brought together in a polypeptide, how will they interact? Indicate each interaction in a drawing. (a) lysine and aspartic acid; (b) alanine and valine; (c) tyrosine and glutamic acid.

Answers:

1. (a) hydrogen bonding (b)

$$\begin{array}{c} \quad\quad\quad O \\ \quad\quad\quad \| \\ \sim\!\!\sim N - C \sim\!\!\sim \\ \quad | \\ \quad H \\ \quad \vdots \\ \quad O \\ \quad \| \\ \sim\!\!\sim C - N \sim\!\!\sim \\ \quad\quad\quad | \\ \quad\quad\quad H \end{array}$$

2. (a) ionic attraction

$$H - \overset{\displaystyle |}{\underset{\displaystyle |}{C}} - (CH_2)_4 - \overset{+}{\underline{NH_3}} \quad \underline{\overset{\ominus}{O}} - \overset{O}{\overset{\|}{C}} - CH_2 - \overset{\displaystyle |}{\underset{\displaystyle |}{C}} - H$$

(b) hydrophobic attraction

$$H - \overset{\displaystyle |}{\underset{\displaystyle |}{C}} - CH_3 \quad \underline{\begin{array}{c} H_3C \quad\quad H \\ \quad\diagdown\!\diagup\;\; \\ C - \overset{\displaystyle |}{\underset{\displaystyle |}{C}} - H \\ \quad\diagup \\ H_3C \end{array}}$$

(c) hydrogen bonding

$$H - \overset{\displaystyle |}{\underset{\displaystyle |}{C}} - CH_2 - \hexagon - \underline{OH \cdots O} = \overset{\displaystyle |}{\underset{\displaystyle OH}{C}} - CH_2 - CH_2 - \overset{\displaystyle |}{\underset{\displaystyle |}{C}} - H$$

The Quaternary Structure of Proteins

Many proteins are composed of two or more tertiary structures grouped together to form a single unit. The arrangement of the tertiary subunits in the larger protein is called the **quaternary structure** of the protein. Each chain, or subunit, has its own primary, secondary, and tertiary structure, and they are held together in the quaternary structure by ionic, hydrogen, and hydrophobic attractions. Both insulin and hemoglobin are such proteins. Earlier you learned

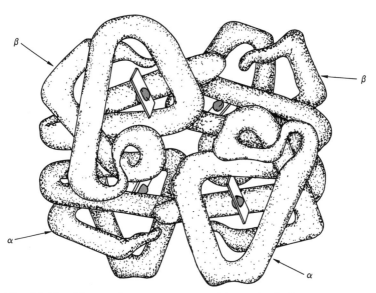

β

β

α

α

Fig. 22.13. The quaternary structure of hemoglobin formed from two alpha and two beta polypeptide units. Note the location of the four heme groups.

that hemoglobin is composed of two α chains and two β chains, each with a heme group. Hemoglobin performs its life-sustaining functions only when the four chains are positioned together as shown in Figure 22.13. Insulin is composed of two tertiary structures arranged in a single unit. Most proteins with a molecular weight over 50,000 have quaternary structure.

Ultimately, the shape of a protein is determined by its primary structure, the sequence of amino acids. The unique shape of each different protein is a result of the kinds of attractive forces that can be established along and between polypeptide chains. This, of course, depends completely on the kinds and order of amino acids forming the chain. The few proteins that have been synthesized in the laboratory fold into tertiary structures identical to those of the **native proteins** obtained from the body, when compared under the same conditions.

Collagen, Bone, and Teeth

Collagen is the most abundant protein in the human body. Collagen is the principal component of supportive tissue and connective tissue. A third kind of secondary structure is found in collagen. It involves three polypeptide chains twisted about each other forming a **triple helix** which is held together by hydrogen bonds. The triple helix units then line up alongside each other to form fibers of very high strength.

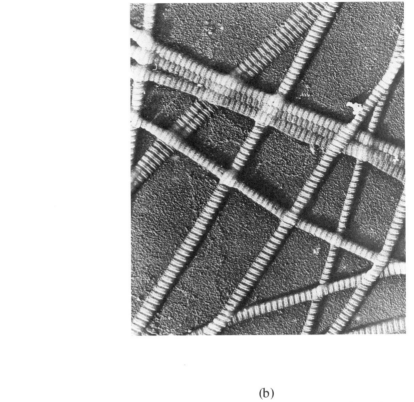

(a) (b)

Fig. 22.14.   (a) The secondary structure of collagen is a triple helix formed from three polypeptide chains held to one another by hydrogen bonds. (b) Many of these triple helix molecules overlap each other to form a collagen fibril. The electron micrograph shows collagen fibrils that were carefully pulled away from human skin. (From: J. Gross, *Scientific American*, Vol. 204, p. 120, May 1961).

Collagen fibers are an integral part of bone and teeth, and they are responsible for the springiness and much of the strength of both. Crystals of hydroxyapatite, a complex calcium phosphate mineral of formula $Ca_5(PO_4)_3OH$, are dispersed among the collagen fibers to add hardness and firmness. The minerals are deposited in a process known as **calcification.** Vitamins A, C, and D are necessary in this process. Bones and teeth are really a mixture of protein and mineral, and the differences that exist between different kinds of bone are due to the different relative amounts of protein and mineral they contain.

The inner structure of a tooth is shown in Figure 22.15. The largest component of a tooth is **dentine,** a substance similar to bone, but much higher in mineral content causing it to be harder than normal bone. The enamel, which covers the exposed area of the tooth, is nearly all mineral, so it too is very hard. **Dental caries** are the result of **decalcification,** the removal of minerals from the protein structure. When carbohydrates (too often from sweets) are acted upon by

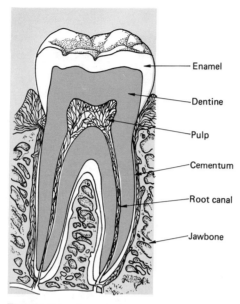

Fig. 22.15. The structure of a tooth.

bacteria in the mouth, lactic acid is produced which can remove minerals from the enamel surface. The end result is tooth decay. The demineralization process can be reduced considerably by treating teeth with fluoride ion, either through the drinking water or by direct application. The fluoride ion F^- gets into the enamel and takes the place of some of the hydroxide ions OH^- in hydroxy apatite, forming a very hard structure with increased resistance to decay. Repeated application of fluoride ion is necessary every five to seven months to maintain its effect. Stannous fluoride, SnF_2, is often used as a source of fluoride ion in fluoride treatments and in toothpaste, and sodium fluoride, NaF, is used in drinking water at a concentration of one part per million.

The Properties of Proteins

Because proteins are such large molecules, they form **colloidal** mixtures as opposed to true solutions, and thus show the light dispersion of the Tyndall effect (see Chapter 12). The large size of proteins prevents their movement through cell membranes. A damaged membrane, something that can occur in certain kidney diseases, will allow protein to pass into the urine and be discharged. A routine urinalysis will always include a check for protein. Because proteins will not pass through healthy cell membranes, they contribute to the total osmotic pressure of body fluids. Several other properties of proteins are

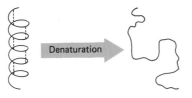

α Helix   Denatured protein strand

Fig. 22.16. Denaturation destroys the native structure of a protein by breaking the attractive forces responsible for the secondary, tertiary, and quaternary organization.

revealed in the denaturing process, the digestion of proteins, and in the chemical tests that are used to detect the presence of proteins.

Denaturation of Proteins

When either the quaternary, tertiary, or secondary structure of a protein is altered, the protein is said to have been **denatured**. The denaturation process can disrupt the hydrogen bonds, the disulfide linkages, or any of the other attractive interactions that exist, though the peptide bonds are not broken. Once the native structure is changed, the protein can no longer perform its biological function. There are several ways by which a protein can be denatured:

1. Use of heat or ultraviolet radiation. Both heat and energetic ultraviolet light can rupture the attractive forces responsible for protein structure. This results in coagulation and a complete loss of activity. When an egg is fried, an irreversible coagulation of the protein occurs. Few proteins can retain their biological activity above 50°C. For this reason, high temperatures and ultraviolet light are used to sterilize dental and surgical tools. Bacteria are destroyed as the protein they contain is denatured.

2. Use of acids and bases. Several of the amino acid side chains that protrude from the surface of a protein carry a positive charge, $-\overset{+}{N}H_3$, or a negative charge, $-\overset{\displaystyle O}{\overset{\|}{C}}-O^-$. These can participate in salt bridge formation only under certain conditions of pH. If the pH is changed by adding an acid or base, these salt bridges can break and the protein can coagulate. Hydrogen bonding can also be disrupted. The very acidic conditions of the stomach begin the digestion of proteins by this process.

3. Use of organic solvents. Acetone, ethyl alcohol, and isopropyl alcohol can disrupt secondary, tertiary, and quaternary protein structure. For this reason, these solvents can be used as disinfectants.

4. Use of heavy-metal salts. Solutions of mercury (Hg^{2+}), silver (Ag^+), and lead (Pb^{2+}) salts provide heavy-metal ions that form strong bonds with the side chains of aspartic and glutamic acids. Several ionic attractive forces within the protein structure are thus lost, and the protein precipitates from solution as a metal complex. Mercurochrome and merthiolate are common antiseptics that contain mercury.

5. Use of mechanical agitation. Most globular proteins readily denature if stirred or shaken vigorously. Egg whites are denatured as they are whipped into meringue.

Tests for Proteins

Several chemical tests are used to determine if peptide bonds, amino acids, or proteins are present. The three described below all produce characteristic colors if the test is positive.

1. The biuret test. A few drops of dilute copper(II) sulfate ($CuSO_4$) is added to a strongly basic solution of the sample to be checked. A violet color indicates the presence of peptide bonds.

2. The ninhydrin test. The sample to be checked is heated in the presence of ninhydrin. The appearance of a blue color indicates the presence of amino acids. The more intense the color, the greater the amino acid concentration.

Ninhydrin

3. The xanthoproteic test. Addition of nitric acid to a sample produces a yellow color if proteins are present. The color arises from the reaction of nitric acid with the aromatic rings of tryptophan, tyrosine, and phenylalanine. Nitric acid will produce a yellow stain if it contacts the skin.

The Digestion and Absorption of Proteins

Protein digestion begins in the stomach as the hydrochloric acid brings about denaturation. The denatured protein exposes several peptide bonds that are then hydrolyzed in the presence of the enzyme **pepsin**. About 10% of the peptide bonds are broken forming smaller, more soluble peptides. After entering the small intestines, the remaining peptide bonds are hydrolyzed in the presence of enzymes from the pancreas and intestinal cells. The hydrolysis of a peptide bond is the reverse of its formation.

$$\underset{\substack{|\\R}}{\overset{\substack{H\\|}}{C}}-\underset{}{\overset{\substack{O\\\|}}{C}}-\underset{}{\overset{\substack{H\\|}}{N}}-\underset{\substack{|\\R'}}{\overset{\substack{H\\|}}{C}} + H_2O \xrightarrow{\text{enzyme}} \underset{\substack{|\\R}}{\overset{\substack{H\\|}}{C}}-\overset{\substack{O\\\|}}{C}-OH + H-\underset{}{\overset{\substack{H\\|}}{N}}-\underset{\substack{|\\R'}}{\overset{}{C}}$$

The amino acids are absorbed through the intestinal wall directly into the bloodstream. They are then carried by the portal circulation to the liver, the principal organ responsible for the breakdown and synthesis of amino acids. Then they are sent to all parts of the body where they are synthesized into protein for repair of old tissue or generation of new tissue. Some of the amino acids are also used as a source of carbon and nitrogen for the synthesis of nonprotein biomolecules such as heme.

Unlike carbohydrates and fats, excess amino acids cannot be stored for later use. However, the amino groups can be removed, and the carbon skeletons can be oxidized to carbon dioxide and water for energy, or they may be converted to carbohydrates and fats and stored. The amino groups that were removed from the amino acids are converted into urea, $H_2N\text{-}\overset{\substack{O\\\|}}{C}NH_2$, and excreted in the urine.

Terms

Several important terms appeared in Chapter 22. You should understand the meaning of each.

protein

α-amino acid

peptide bond

peptide

polypeptide

amphoteric

zwitterion

isoelectric point

primary structure

secondary structure

tertiary structure

quaternary structure

chromatography

simple protein

conjugated protein

prosthetic group

sickle-cell anemia

α helix

denaturation

electrophoresis

complete protein

collagen

calcification

dental caries

disulfide bond

Questions

Answers to the starred questions appear at the end of the text.

*1. What is an α-amino acid?

2. Draw the structure of threonine at its isoelectric point. Using equations, show how it can act as an amphoteric substance.

*3. Show the zwitterion form of tyrosine.

4. List the main functions of proteins in the body.

5. Give an example of each of the following.
   *a. hydroxy amino acid
   *b. acidic amino acid
   c. aromatic amino acid
   d. heterocyclic amino acid

6. List the essential amino acids. What do the aliphatic essential amino acids have in common?

7. Use equations to show how solutions of amino acids can act as buffers.

*8. List the dicarboxylic amino acids.

9. Into what subclasses are the conjugated proteins subdivided?

*10. What chemical reagents will hydrolyze a protein?

11. In what principal way does gas chromatography differ from column chromatography?

*12. Of the four levels of structure for proteins, which is found in all proteins?

13. Why can proteins act as buffers?

14. Draw the structural formula for the following peptide:

Phe–Ile–Gln–Pro–His

*15. In terms of the approximate number of amino acid units each contains, what is a peptide, a polypeptide, a protein.

*16. What functional groups other than –COOH and –NH_2 are present in the side chains of proteins?

17. Show how the peptides Gln–Cys–Ala and Thr–Cys–Pro could form a covalent bond linking them together.

*18. Draw the structural formula of serine, showing the optically active form that would be found in the body.

19. Describe the four levels of protein structure: primary, secondary, tertiary, and quaternary.

*20. Do proteins form true solutions or colloids? Why?

21. Denaturation will affect which of the four levels of protein structure?

*22. Show how the tetrapeptide Gly–Asp–Gly–Ala would be coagulated by lead ion (Pb^{2+}).

*23. In what way does the heat of an autoclave destroy bacteria?

24. When a child eats lead paint, why is milk or egg given as an antidote until the stomach may be pumped?

*25. Describe the ninhydrin and biuret tests.

26. Describe the composition of tooth enamel in terms of both protein and mineral. What is responsible for the positive effects of fluoride ion in fighting tooth decay?

*27. Oxytocin, a peptide hormone of the pituitary gland which causes contraction of the smooth muscles, is often used in obstetrics to initiate labor. The formula of oxytocin is given below. Draw its structural formula, indicating each peptide bond.

$$Cys-Trp-Ile-Gln-Asn-Cys-Pro-Lev-Gly$$
$$\lfloor \underline{\quad} S \underline{\quad} S \underline{\quad} \rfloor$$

28. Describe each of the following interactions that are important in determining the tertiary structure of proteins.
a. salt bridge
b. hydrophobic interaction
c. side-chain hydrogen bonding

*29. What kind of bonding holds the protein chains of the pleated sheet structure together?

*30. The protein in hair (keratin) contains a large number of disulfide bonds. What would happen to the disulfide bonds if (a) the hair was rolled into curls, (b) treated with a mild oxidizing solution, and then (c) a mild reducing solution?

31. Describe the process by which dietary proteins are digested and absorbed.

32. Use an equation to describe the hydrolysis of Gly–His–Asp.

*33. Why must insulin be injected, and not taken by mouth?

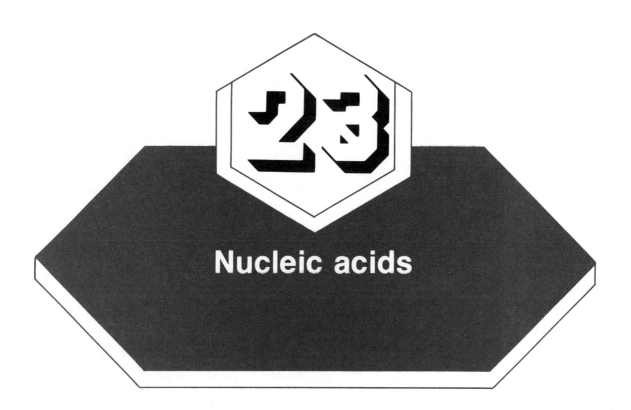

Nucleic acids

Thus far you have learned that lipids, carbohydrates, and proteins are required for the normal growth and energy requirements of an organism. But, what is it in the cells of our bodies that determines why we are human beings and not a plant or another animal? Why do we have physical characteristics like those of our parents? Nucleic acids provide the answers to these questions.

The nucleic acids were first isolated from the nuclei of cells, thus the name, **nucleic acids**. There are two types of nucleic acids: the **ribonucleic acids (RNA)** and the **deoxyribonucleic acids (DNA)**. DNA serves as the master blueprint, in coded form, of an organism. The ribonucleic acids read the code on DNA and participate in protein synthesis.

All the information that determines the identity, and is required for the growth of an organism is contained within the structure of DNA. The DNA code determines the complexity of an organism and, to a large extent, its general life span. Also, in humans, features such as eye color, height, and general physical appearance are determined by the information contained in DNA.

The chapter will begin with a discussion of the compounds that form the nucleic acids, and how they combine to form DNA and RNA. Then the structure of DNA will be discussed along with the process by which it can duplicate itself during cell division. The role of DNA and RNA in protein synthesis will then be described, followed by the problems that can develop if there is a mutation of the DNA structure. The mode of action of certain antibiotics is then described, followed by a description of viral infections and the problems and possible solutions associated with combating viral disease.

Objectives

By the time you finish this chapter, you should be able to do the following:

1. Name and draw the structural formulas for the sugars and bases that are present in DNA and RNA, and describe how they are combined in nucleosides and nucleotides.
2. Name the compounds represented by the symbols A, T, C, G, U, and interpret the meaning of symbols such as AMP, TTP, dGMP, CDP.
3. Describe base pairing in DNA and RNA, and state which bases pair with each other in these nucleic acids.
4. Describe the primary and secondary structures of nucleic acids.
5. Describe the double-helix structure of DNA, and show how it is held in that structure.
6. Describe in general terms how the sequence of bases in DNA can represent the genetic code of an organism.
7. Describe the process of DNA replication.
8. Name and describe the function of the three types of ribonucleic acids.
9. Describe the significance of the codon and anticodon in protein synthesis.
10. Describe transcription, the process of RNA synthesis.
11. Describe translation, the process of protein synthesis in cells.
12. Describe the kinds of mutations that can alter a DNA molecule and how a mutation can affect protein synthesis.
13. Describe the cause of a genetic disease.

The Composition of Nucleic Acids

Nucleic acids in cells are usually in the form of a protein–nucleic acid complex called a **nucleoprotein**. The protein can be released from the nucleic acid by mild **hydrolysis,** a reaction of the complex with water in the presence of a catalyst. The nucleic acid can then be broken down to produce the substances that make it up. Hydrolysis of a nucleic acid will first produce a mixture of **nucleotides,** smaller molecules that are composed of three distinct units. The nucleotides are the repeating units found in all nucleic acids. The hydrolysis of the nucleotides yields phosphoric acid, H_3PO_4, plus a smaller two-membered

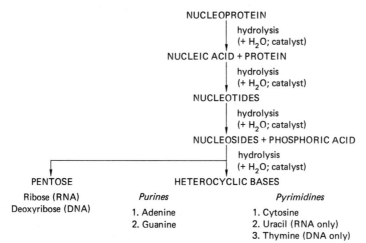

NUCLEOPROTEIN
| hydrolysis
| (+ H_2O; catalyst)
NUCLEIC ACID + PROTEIN
| hydrolysis
| (+ H_2O; catalyst)
NUCLEOTIDES
| hydrolysis
| (+ H_2O; catalyst)
NUCLEOSIDES + PHOSPHORIC ACID
| hydrolysis
| (+ H_2O; catalyst)

PENTOSE	HETEROCYCLIC BASES	
Ribose (RNA)	*Purines*	*Pyrimidines*
Deoxyribose (DNA)	1. Adenine	1. Cytosine
	2. Guanine	2. Uracil (RNA only)
		3. Thymine (DNA only)

Fig. 23.1. Stages of the controlled hydrolysis of nucleic acids.

unit called a **nucleoside.** Finally, the nucleosides can be broken down to produce a **pentose sugar** (either ribose or deoxyribose) and a mixture of **heterocyclic bases,** compounds with nitrogen atoms included in rings. There are two kinds of bases found in nucleosides, the **purines** and the **pyrimidines.** Each will be described later.

The step-by-step hydrolysis of a nucleoprotein is shown in Figure 23.1.

The Sugars in Nucleic Acids—Ribose and Deoxyribose

A given nucleic acid will contain either ribose or deoxyribose as the pentose sugar, never both. The nucleic acids that contain ribose are called **ribonucleic acids,** abbreviated **RNA.** Those containing deoxyribose are called **deoxyribonucleic acids, DNA.** Both sugars are present in the β (beta) form, with the –OH group on the anomeric carbon (carbon number 1') pointing up in the Haworth structures. When included in a nucleic acid, the number of each carbon atom in the sugar molecule is accompanied by a prime symbol ('), as 1', 2', 3', etc. This convention is used in the Haworth structures of β-ribose and β-deoxyribose shown below.

β–ribose β–deoxyribose

The Bases in Nucleic Acids—Pyrimidines and Purines

The nitrogen-containing heterocyclic compounds found in nucleic acids are called bases because aqueous solutions of these compounds have pH's on the basic side of 7. There are three pyrimidine bases found in nucleic acids: **cytosine (C), uracil (U),** and **thymine (T).** The capital letters are abbreviations for the bases. DNA contains only cytosine and thymine. RNA contains only cytosine and uracil. The structure of each pyrimidine base is given below. Notice how the atoms in the ring are numbered. The atom in each compound that joins to the sugar will be referred to by number as we go along. The prime symbol is not used with numbers in the base structures. This will help us distinguish the atoms in bases from those in the sugars.

The Pyrimidine Bases

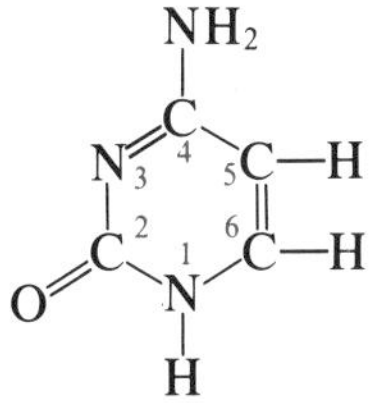

Cytosine (C) 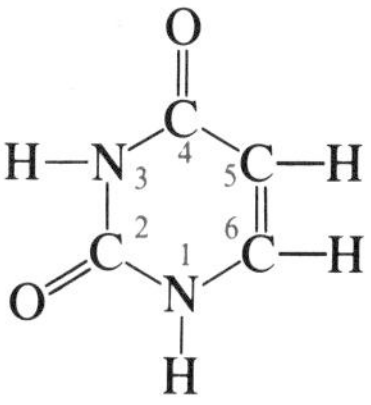   Uracil (U) 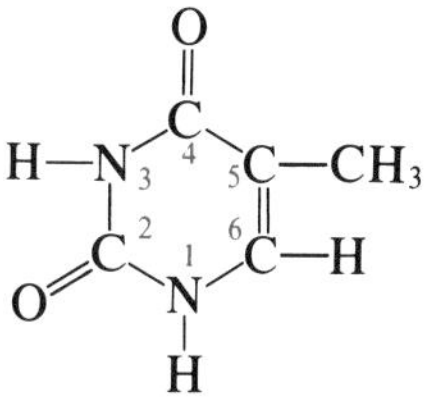   Thymine (T)

There are two purine bases found in nucleic acids: **adenine (A)** and **guanine (G).** Both occur in DNA and RNA. The structures of both are shown below.

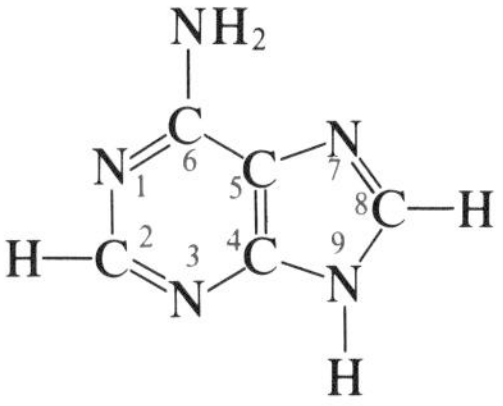

Adenine (A) 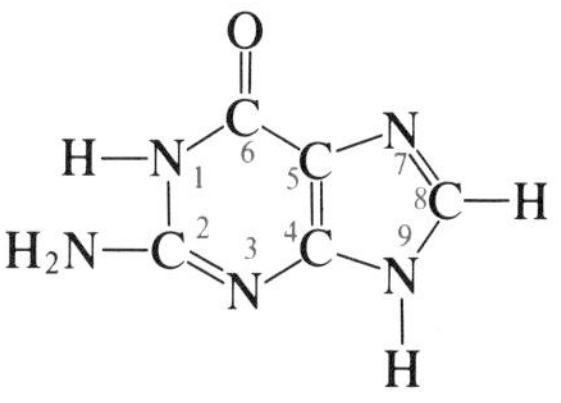   Guanine (G)

The compounds that come together to form nucleic acids are determined by studying the hydrolysis products of DNA and RNA. The composition of each is summarized in Table 23.1. Notice that four of the six compounds are identical in each nucleic acid.

Now, let us use the compounds listed in Table 23.1 to show how a nucleic acid is formed from its parts. We will start with a pentose sugar and a base to form a nucleoside. Then phosphoric acid will be combined with the nucleoside

	TABLE 23.1 The Complete Hydrolysis Products of DNA and RNA		
		DNA	RNA
Purine bases		adenine guanine	adenine guanine
Pyrimidine bases		cytosine thymine	cytosine uracil
Pentose sugar		deoxyribose	ribose
Inorganic acid		phosphoric acid	phosphoric acid

to form a nucleotide. A little later, after the nucleotides are modified slightly, we will show how they combine to form a nucleic acid.

The Nucleosides

When any one of the heterocyclic bases reacts with either β-ribose or β-deoxyribose, a **nucleoside** is formed. These reactions are catalyzed by enzymes:

$$\left.\begin{array}{c} \text{purine base} \\ \text{or} \\ \text{pyrimidine base} \end{array}\right\} + \beta\text{-pentose} \xrightarrow{\text{enzyme}} \text{nucleoside} + H_2O$$

If the pentose is ribose, the nucleosides that result are named *adenosine* and *guanosine* for the purine bases, and *cytidine, uridine*, and *thymidine* for the pyrimidine bases. Notice that those with purine bases end in *-osine*, while those with pyrimidine bases end in *-idine*. If the sugar is deoxyribose, the prefix *deoxy-* is added to the name given previously, as in deoxycytidine or deoxyadenosine. The names of the nucleosides for each base–sugar combination appear in Table 23.2.

The bond joining the pyrimidine bases to the pentose always forms between carbon 1' of the pentose and nitrogen 1 of the base. This is shown in the formation of thymidine below. A molecule of water is eliminated between the molecules as they join.

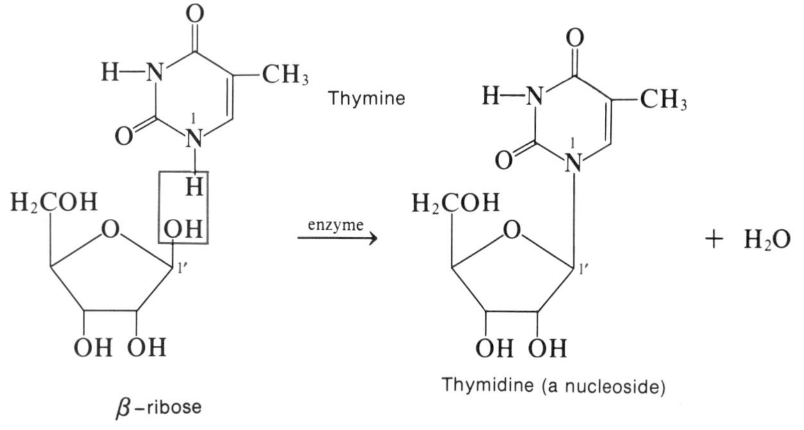

β-ribose

Thymine

Thymidine (a nucleoside)

+ H₂O

The bond between the purine bases and the pentose always forms between carbon 1' on the pentose and nitrogen 9 of the base. This is shown below in the formation of deoxyadenosine.

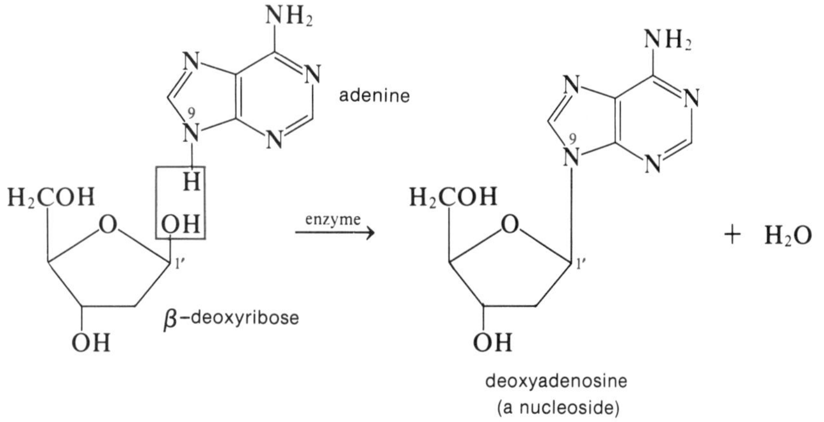

adenine

β-deoxyribose

deoxyadenosine
(a nucleoside)

+ H₂O

Check Test Number 1

1. Which pentose and which base combine to form (a) cytidine, (b) deoxyguanosine?
2. What is the name of the nucleoside formed from β-ribose and uracil?

Answers:
1. (a) β-ribose + cytosine; (b) β-deoxyribose + guanine.
2. Uridine.

The Nucleotides

A **nucleotide** is formed when a nucleoside combines with phosphoric acid with the enzyme-catalyzed elimination of water. The nucleotides are then phosphate esters of the nucleosides. As a memory device, the word nucleotide contains the letter **t**, which can remind you that three units are present in nucleotides, a sugar, a base, and a phosphate group.

The nucleotides that are present in nucleic acids all contain the phosphate group joined to carbon 5′ of the pentose. Addition of the phosphate group to adenosine, a nucleoside, produces adenosine-5′-monophosphate **(AMP)**, a nucleotide.

adenosine
(a nucleoside)

adenosine–5′–monophosphate (AMP)
(a nucleotide)

As you can see in the formal name for AMP, nucleotides are named by the addition of "monophosphate" after the name of the nucleoside. A primed number, such as 5′, is placed before "monophosphate" to indicate the particular pentose carbon to which the phosphate group has bonded. If a primed number is absent from the name of a nucleotide, 5′ is to be understood. The names of the nucleosides and the nucleotides are summarized in Table 23.2. Note the abbreviations used for the nucleotides. They will be used later on in the chapter, so you may need to refer back to Table 23.2 occasionally to determine their meaning. In the abbreviation **GMP**, G stands for guanosine, M for mono-, and P for phosphate. A "d" stands for deoxy-.

Check Test Number 2

1. Draw the structure of deoxythymidine-5′-monophosphate, dTMP. Refer to Table 23.2 if necessary.
2. What is the difference between dGMP and GMP?

Answers:

1.

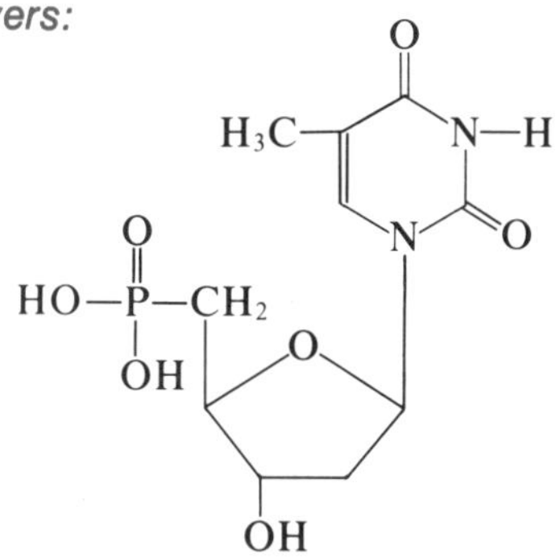

dTMP

2. GMP contains ribose, dGMP, deoxyribose. Otherwise they are the same.

TABLE 23.2
Name of Nucleosides and Nucleotides

The nucleosides		
Base	Ribose present	Deoxyribose present
Adenine	adenosine	deoxyadenosine
Guanine	guanosine	deoxyguanosine
Cytosine	cytidine	deoxycytidine
Uracil	uridine	deoxyuridine
Thymine	thymidine	deoxythymidine*

The nucleotides				
Base	Ribose present	Abbreviation	Deoxyribose present	Abbreviation
Adenine	adenosine-5′-monophosphate	AMP	deoxyadenosine-5′-monophosphate	dAMP
Guanine	guanosine-5′-monophosphate	GMP	deoxyguanosine-5′-monophosphate	dGMP
Cytosine	cytidine-5′-monophosphate	CMP	deoxycytidine-5′-monophosphate	dCMP
Uracil	uridine-5′-monophosphate	UMP	deoxyuridine-5′-monophosphate	dUMP
Thymine	thymidine-5′-monophosphate	TMP	deoxythymidine-5′-monophosphate*	dTMP

*The prefix deoxy- means the pentose sugar is deoxyribose rather than ribose. However, since thymine is usually found only in DNA, with deoxyribose, the prefix is often omitted from the name. For clarity, the deoxy- prefix will be used in this textbook.

Nucleotides are the building blocks of nucleic acids in the same way that amino acids are the building blocks of proteins. Nucleotides are found in all cells, most often with the phosphate group joined to the 5′ carbon of the pentose. A smaller number of nucleotides may also be present that will have the phosphate group joined to the 3′ or 2′ carbon of the pentose. They are all monophosphates and differ only in the location of the phosphate group. However, free nucleotides in the cell can also exist as diphosphates and triphosphates. Adenosine monophosphate (AMP) can react with phosphoric acid to produce adenosine diphosphate (**ADP**). ADP can combine with still a third unit of phosphoric acid to produce adenosine triphosphate (**ATP**).

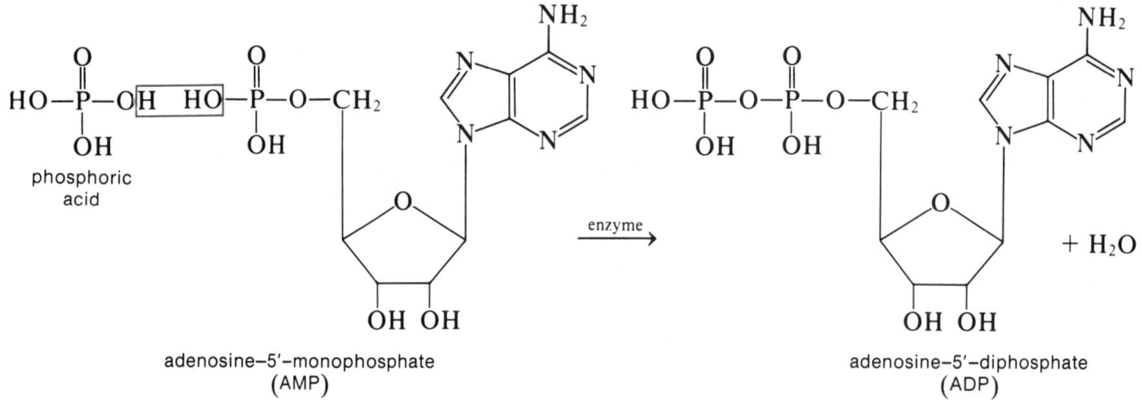

phosphoric
acid

adenosine–5′–monophosphate
(AMP)

enzyme

adenosine–5′–diphosphate
(ADP)

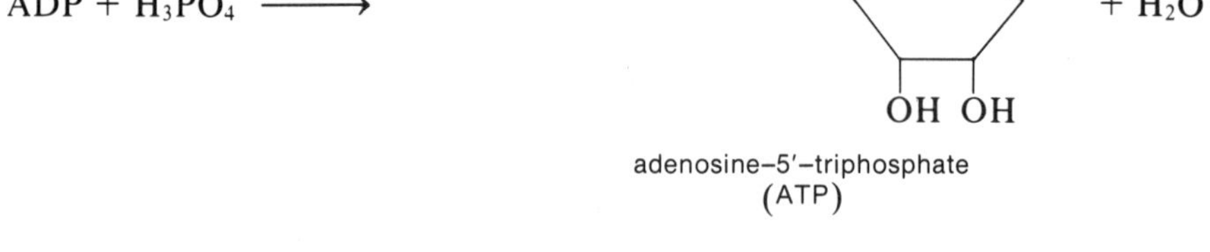

$$\text{ADP} + \text{H}_3\text{PO}_4 \xrightarrow{\text{enzyme}}$$

adenosine–5′–triphosphate
(ATP)

The other monophosphate nucleotides also form di- and triphosphates in the cell. Though the di- and triphosphates are present as free molecules in the cell, they do not appear as such in DNA and RNA. The triphosphates are necessary

TABLE 23.3
Abbreviations for the Nucleotide-5′-phosphates

Base	Ribonucleotides			Deoxyribonucleotides		
	mono P	di P	tri P	mono P	di P	tri P
Adenine	AMP	ADP	ATP	dAMP	dADP	dATP
Guanine	GMP	GDP	GTP	dGMP	dGDP	dGTP
Cytosine	CMP	CDP	CTP	dCMP	dCDP	dCTP
Uracil	UMP	UDP	UTP	dUMP	dUDP	dUTP
Thymine	TMP	TDP	TTP	dTMP	dTDP	dTTP

☐ : These nucleotide phosphates are used to synthesize DNA and RNA.

for the formation of the nucleic acids, as you shall see. The three- and four-letter abbreviations for the di- and triphosphate nucleotides appear in Table 23.3. Those nucleotides that are used in nucleic acid synthesis appear in boxes. The remaining nucleotides exist in cells and many perform functions not directly related to DNA or RNA synthesis.

Check Test Number 3

Write the names for the nucleotide-5′-phosphates symbolized by the following: (a) UTP, (b) dGTP, (c) GDP.

Answers:

(a) uridine-5′-triphosphate, (b) deoxyguanosine-5′-triphosphate, (c) guanosine-5′-diphosphate.

Combining the Nucleotides—The Primary Structure of Nucleic Acids

Nucleic acids are polymers of nucleotides. The **primary structure** of a nucleic acid is the sequence of nucleotides in the polymer chain. Each nucleotide in the chain is joined to two other nucleotides by phosphate groups connecting carbon 5′ of one nucleotide to carbon 3′ of the next. The phosphate group that joins two nucleotides together (through the sugar in each nucleotide) is a **phos-**

phodiester, and the link that joins the two nucleotides is a **3′, 5′-phosphodiester link**. A portion of the primary structure of a nucleic acid can be represented this way:

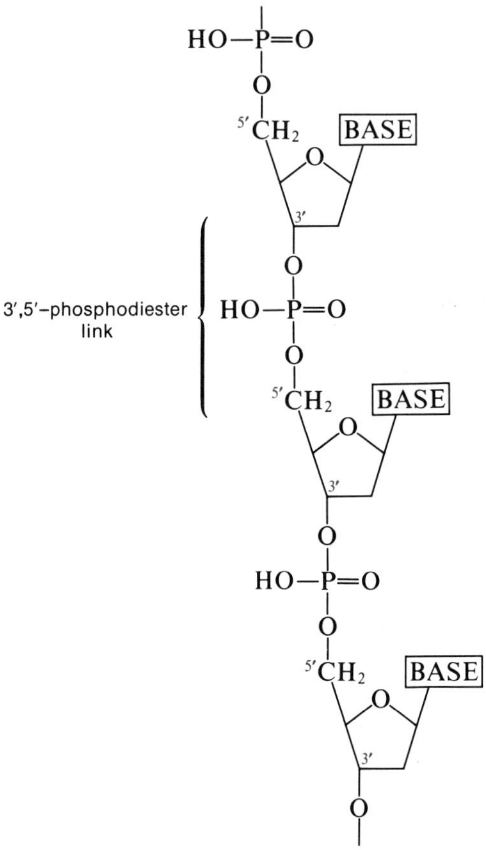

In cells, nucleic acids are formed from the nucleotide-5′-triphosphates. As two nucleotides are joined together, a pyrophosphate unit is eliminated between them.

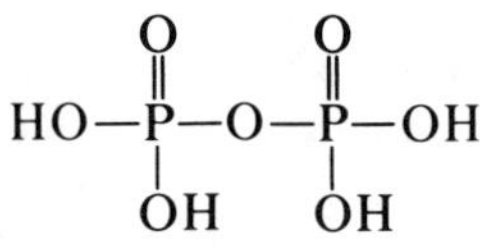

pyrophosphate
(pyrophosphoric acid)

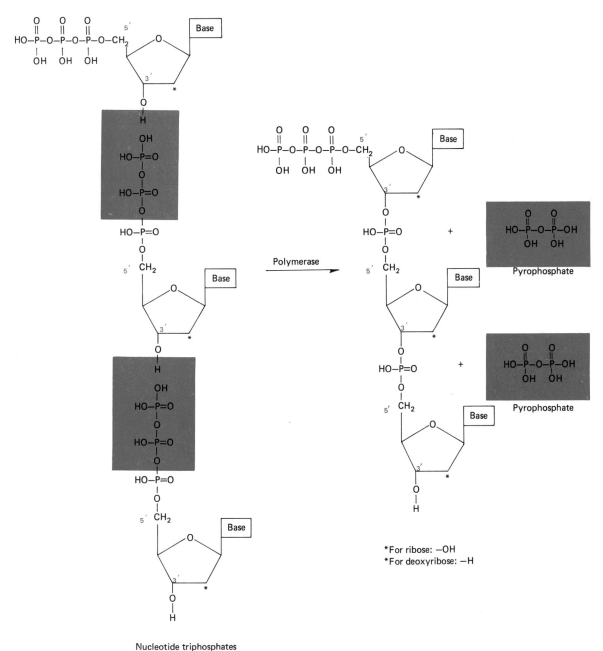

Nucleotide triphosphates

Fig. 23.2. The synthesis of a portion of a nucleic acid from nucleotide triphosphates showing the elimination of pyrophosphate.

The joining of the nucleotide triphosphates to form the nucleic acid polymer is catalyzed by enzymes, collectively called **polymerases**. The joining of three nucleotide triphosphate units to form a portion of a nucleic acid is shown in Figure 23.2.

Nucleic acids, then, are polymers of repeating sugar–phosphate units. The bases, which are joined to the pentose sugars, lie off to the side. A nucleic acid can be simply represented this way. The repeating sugar–phosphate units can be regarded as the backbone of the molecule.

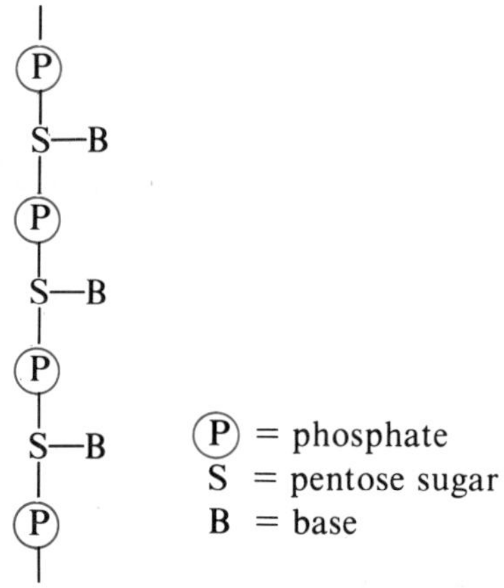

Ⓟ = phosphate
S = pentose sugar
B = base

If the sugar is ribose, the nucleic acid is an RNA. If deoxyribose, it is a DNA. The sequence of bases that lie along the polymer chain represents an immense amount of information in coded form, but more about this later.

Check Test Number 4

Draw a section of a DNA molecule which would contain the bases adenine (A), cytosine (C), and guanine (G) in that order. Being DNA, the sugar is deoxyribose. You may need to refer to structures presented earlier. Draw the section vertically, with adenine at the top.

Answer:

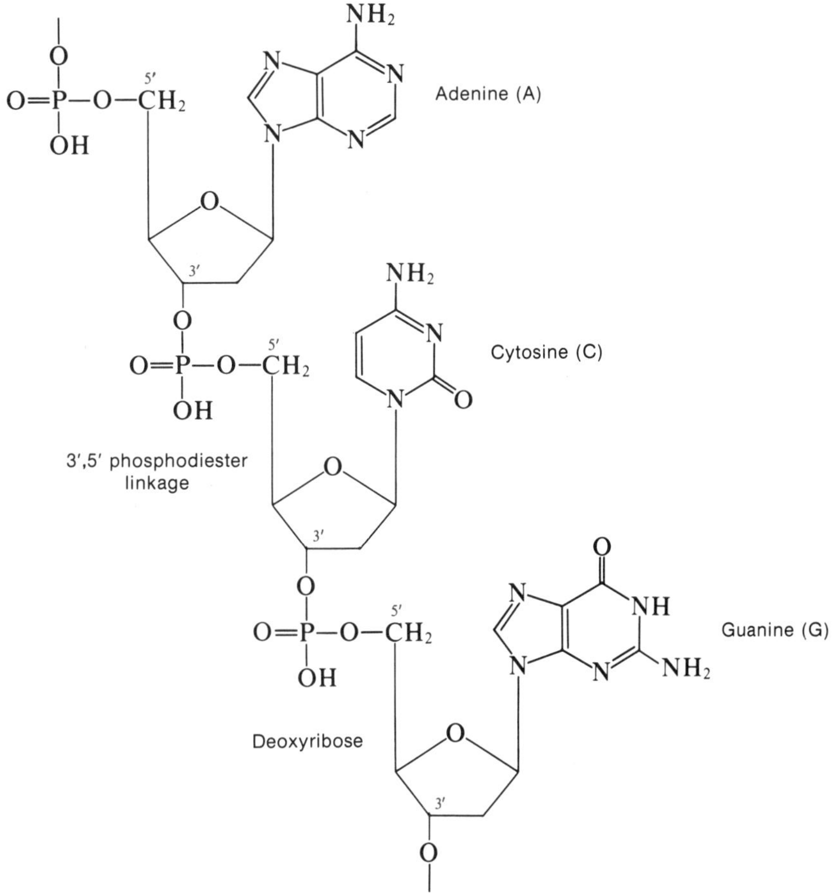

The Secondary Structure of Nucleic Acids—DNA and the Double Helix

One of the greatest discoveries of modern biology, if not *the* greatest discovery, was the determination of the structure of DNA in 1953. Much was known about DNA before that time, but fitting the pieces of knowledge together to give a complete picture of the molecule was an exceedingly difficult task. Finally, in 1953, it all came together. Using information obtained from x-ray studies of DNA, chemical analysis of the molecule, and molecular models, James D. Watson and Francis Crick concluded that DNA must exist in the form of a double helix of two DNA chains. Furthermore, the two helical DNA chains that wind together to form the double helix must be held together by hydrogen bonding between pairs of bases—one base of each pair coming from each chain. The theory was ingenious, and in 1962, Crick and Watson shared the Nobel Prize for their discovery. Let us examine the structure of DNA in more detail.

The complete DNA molecule exists as two strands of DNA wound together to form a **double helix**. It is called a double helix simply because each DNA strand is in helical form. A space-filling model of a section of the DNA double helix appears in Figure 23.3. The repeating sugar–phosphate backbone of each DNA molecule forms the outside of the double helix. The bases which are bonded to the sugars point inward with the plane of each molecule parallel with the other. Each base in one helix, either adenine (A), guanine (G), cytosine (C), or thymine (T), is lined up with another base which points toward it from the other helix. It is the hydrogen bonding between these bases that holds the double helix together. But in DNA, the bases do not line up in just any way. There is a definite pairing that occurs. Cytosine always pairs with guanine (C to G), and adenine always pairs with thymine (A to T). Only these two combinations of the four bases allow the strongest sets of hydrogen bonds to form. The base pairs, C–G and A–T, are shown in Figure 23.4.

There are several features to note about each base pair. Two hydrogen bonds exist in the A–T pair, and three in the G–C pair. The bases must lie parallel

Fig. 23.3. A space-filling model of a portion of the DNA double helix. (Science Related Materials, Inc., Janesville, Wisconsin)

with each other for these multiple hydrogen bonds to form. The hydrogen bonds between each pair of bases are symbolized: –A :::: T– and –G ::::: C–. One base of each pair is a purine, A and G, and the other is a pyrimidine, C and T. The purine–pyrimidine combination allows a constant separation between the two DNA strands along the length of the double helix. Figure 23.5 shows an expanded view of the double helix, revealing the pairing of bases and yet another fact about the double helix. The two DNA strands run in opposite directions, that is, their directions are antiparallel. In a sense, one strand begins at one end of the molecule and travels to the other end. It is here that the second strand begins and continues in the opposite direction to the point where the other began.

The DNA strands are not identical in terms of their base sequence; rather they are complementary. The complementary nature of the two DNA strands arises from the specific, G–C and A–T, base pairing in the double helix. Notice also, in Figure 23.5, that the phosphate groups are ionized. The phosphate groups are mildly acidic, and lose their protons to proteins that are closely associated with DNA in the nucleus of the cell. DNA and RNA are called nucleic acids because of their mild acidity.

The double helix of DNA is a **secondary structure** of nucleic acids, analogous to the helical structure (a single, not a double helix) of proteins. The DNA

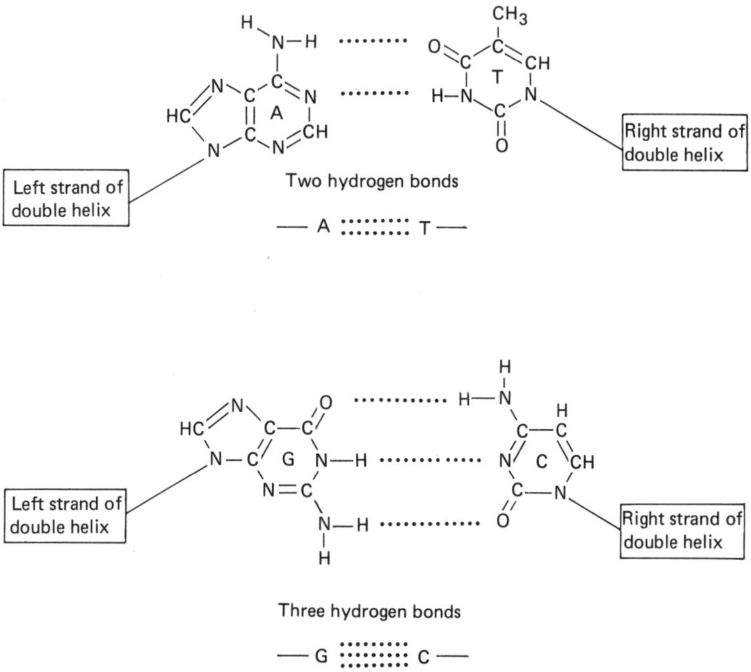

Fig. 23.4. Hydrogen-bond formation in base pairing. In DNA, adenine (A) always pairs with thymine (T), and guanine (G) always pairs with cytosine (C).

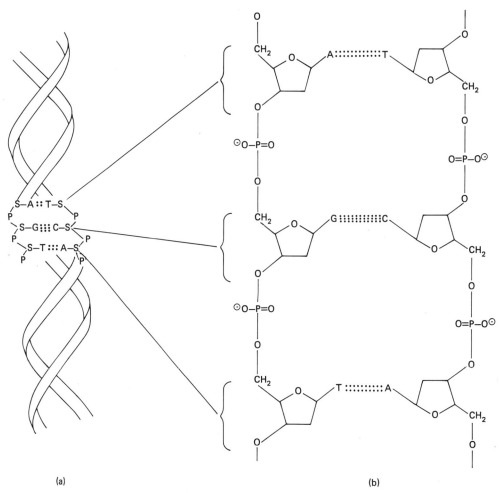

(a)

(b)

Fig. 23.5.   An expanded view of the DNA double helix. In (A), P symbolizes phosphate and S symbolizes the deoxyribose sugar. In (B), complementary base pairing holds the two DNA strands together.

double helix can be viewed as an elegant spiral staircase with chains of sugar–phosphate units forming railings on either side. Stairsteps composed of pairs of bases span from one railing to the other at regular intervals as it turns. There are about 10 steps, or base pairs, in each complete turn of the double helix.

RNA also exhibits some helical secondary structure, though RNA molecules do not pair up as DNA molecules do. The nucleic acid chain of an RNA molecule can fold over onto itself and form a double helix structure in places, but generally the secondary structure of RNA is not a single, fixed, definite structure as is found in DNA. The importance of secondary structure in RNA will appear when we discuss RNA later on.

Heredity, Chromosomes, and DNA

The transfer of genetic characteristics from parent to child, generation after generation, is known as **heredity**. If you compare your physical appearance with those of your parents, you will notice certain similarities; perhaps in physical stature, skin, hair or eye color, or facial shape. But beyond these minor similarities, you and they are human. That fact may sound humorous at first, but why is it that humans only produce humans or that eagles only produce eagles, or that tarantulas only produce tarantulas? What kind of information transfer is involved during reproduction that causes each species to reproduce its own kind? The study of heredity is called **genetics,** and we introduce the topic here because DNA is deeply involved.

The fundamental heredity-bearing material is the **chromosome,** which is located in the nucleus in cells of higher organisms. Bacterial cells contain one chromosome, human cells contain 46, 23 coming from each parent during sexual reproduction. The 46 human chromosomes appear in Figure 23.6. Each chro-

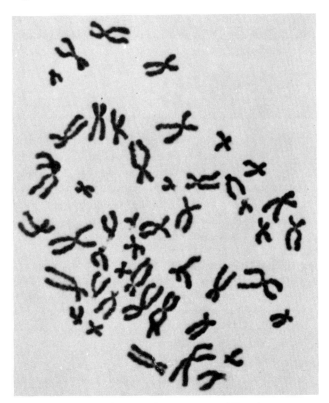

Fig. 23.6. The 46 human chromosomes. The chromosomes contain the DNA of the cell in combination with protein. Photograph by L. J. Sciorra as reproduced in C. Avers, *Basic Cell Biology,* D. Van Nostrand Co., New York, 1978.

mosome contains a single, double-helical DNA molecule associated with protein. The DNA present in the cells of higher organisms has the same deoxyribose–phosphate backbone, the same double-helical structure, and the same four bases (A, G, C, T) attached individually to the sugars. But that is where the structural similarity ends. The DNA of every different organism is different in terms of its length and the sequence of bases along the DNA strand. In human sexual reproduction, the chromosomes that combine to form the fertilized egg contain the characteristic DNA of humans. The fertilized egg can only grow to form a human. The particular base sequence in the DNA strand of an organism is its **genetic code,** and it is different for every organism. The **genes** are short lengths of the DNA molecule that represent, in coded form, the heredity of the organism. The number of genes in the single DNA molecule of a bacterium may be as high as 4000, while for humans, who have much larger DNA molecules, the number of genes exceeds an average of about 1,000,000 per molecule. Each gene codes for a specific amino acid sequence in a particular protein. The proteins are needed for maintaining the life processes in an organism. When a cell divides, it must transfer its heredity message in the form of identical DNA molecules to the new cell, while retaining the same message for itself. This is accomplished by the synthesis of a new DNA molecule. The process of duplicating DNA in the nucleus of a cell is called **DNA replication,** so let us see how this is done.

DNA Replication

The duplication of a DNA molecule must be exact if the correct genetic information is to be transferred during cell division. This is ensured by the specific way in which the four bases pair through hydrogen bonding. You will recall that cytosine (C) always pairs with guanine (G), and thymine (T) always pairs with adenine (A). This pairing between complementary bases is evident in the section of DNA shown in Figure 23.7 (a). The P and S letters along each backbone of the DNA double helix symbolize phosphate and the sugar, deoxyribose.

Replication begins as the DNA double helix slowly unwinds exposing two DNA strands. Each strand will serve as a template for the synthesis of two new strands of DNA, Figure 23.7 (b). The fluid surrounding DNA in the nucleus is rich in dATP, dCTP, dGTP, and dTTP, the nucleotide triphosphates. They will move into position along the exposed, unwound strands of DNA, pairing A to T or T to A and C to G or G to C. This specific pairing of bases results in an exact complementary copy of each strand. Once the nucleotide triphosphates are in place, the enzyme, DNA polymerase, catalyzes the elimination of pyrophosphate between adjacent triphosphates, joining the nucleotides into a strand, Figures 23.7 (c), (d), pp. 668–669. As the original DNA continues to

(Continued, p. 671.)

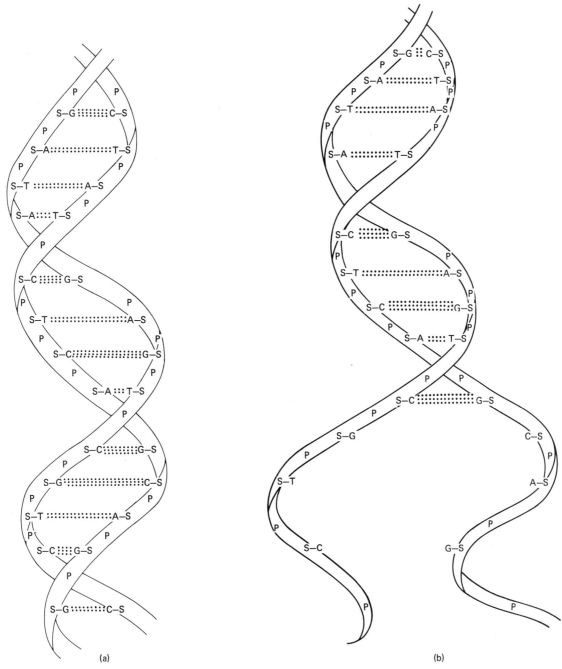

Fig. 23.7. DNA replication. (a) The DNA double helix showing base pairing. (b) Replication begins as the double helix partially unwinds exposing two DNA strands.

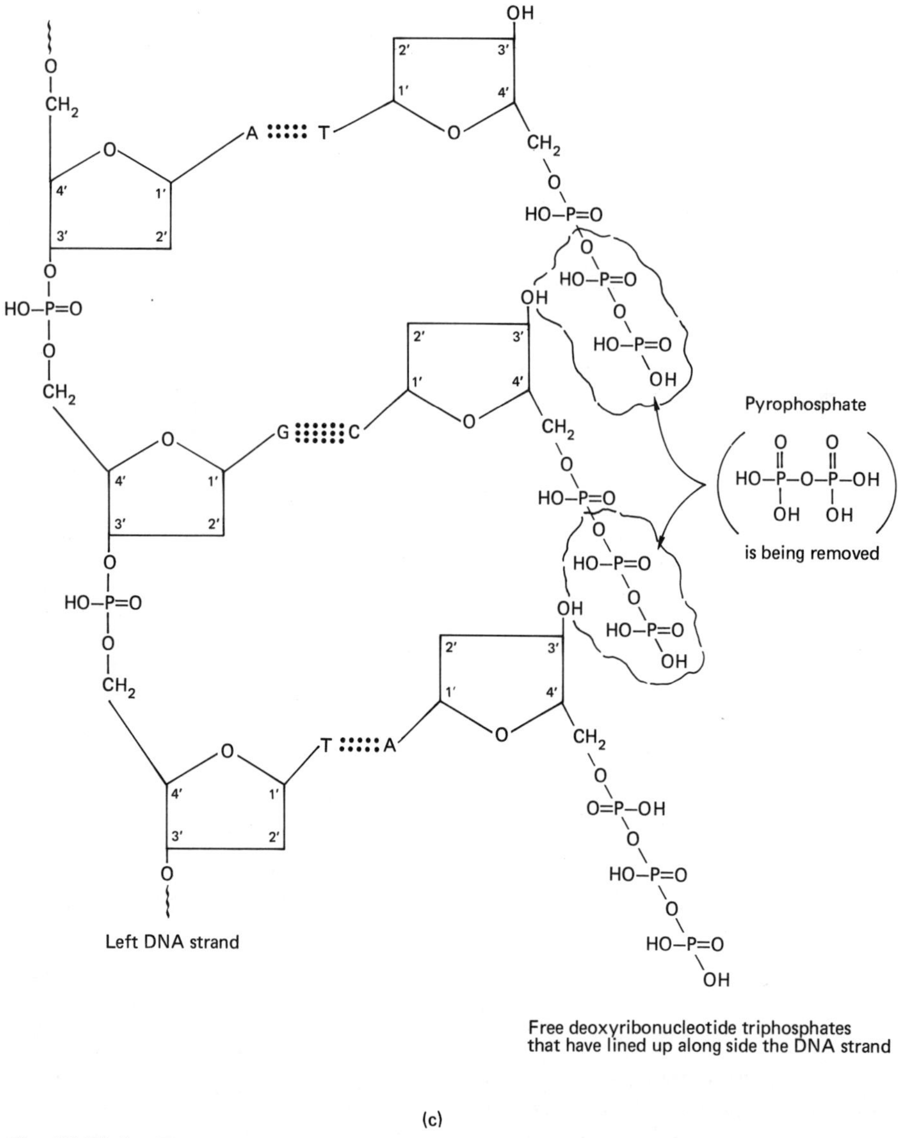

Fig. 23.7(c). The complementary nucleotide triphosphates line up alongside the DNA strand and are joined together with the elimination of pyrophosphate to form the new, complementary DNA strands.

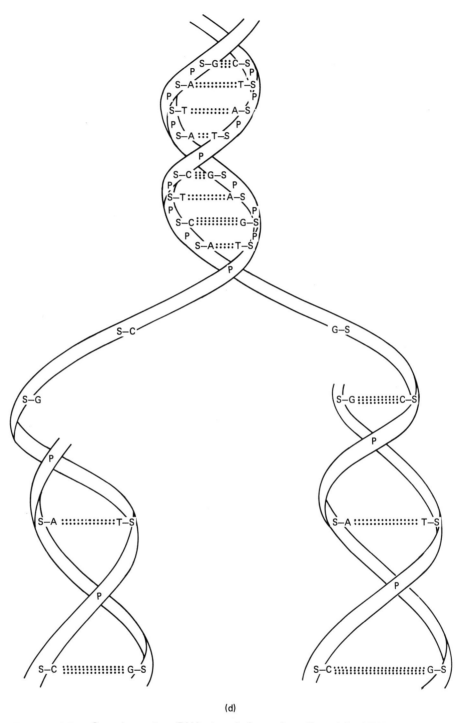

(d)

Fig. 23.7(d). Complementary DNA strands form along the original DNA strands, and when they are complete. . . . .

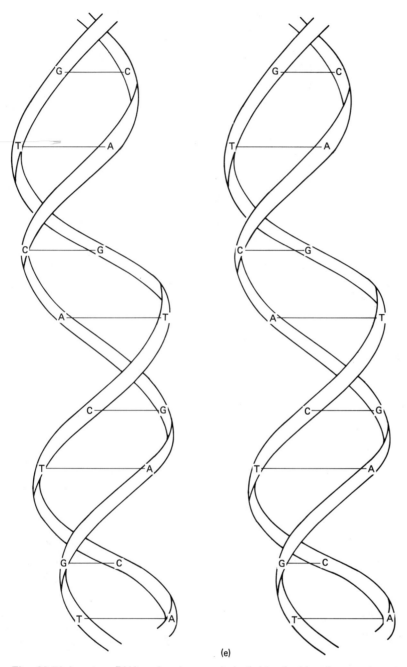

(e)

Fig. 23.7(e).   two DNA molecules result, both identical in all respects.

unwind, new nucleotide triphosphates pair with complementary bases along the exposed strand, to be soon joined together lengthening the new DNA strand. Eventually, a new complementary copy is made of each strand of the original DNA, and they separate as two complete DNA molecules. Each contains one "old" DNA strand, and one "new" strand, Figure 23.7 (e). One DNA double helix will be passed to the daughter cell during cell division, bringing with it the genetic information needed to survive. The other remains with the parent cell.

In 1967 Arthur Kornberg of Stanford University successfully carried out the duplication of a DNA molecule in a test tube. He extracted the DNA from a virus that is able to infect *E. coli* bacteria. The virus DNA was added to a special solution containing dCTP, dATP, dTTP, and dGTP, along with Mg^{2+} and DNA polymerase. The virus DNA replicated, producing a copy that was as equally able to infect *E. coli* as the natural DNA.

The Ribonucleic Acids

Ribonucleic acids are present in the nucleus and the cytoplasm of cells. The four bases in all ribonucleic acids are cytosine (C), guanine (G), adenine (A), and uracil (U). They are joined to ribose sugars which, along with phosphate, form the repeating ribose–phosphate backbone of the molecule. The ribonucleic acids do not exhibit the degree of secondary structure found in DNA. They exist principally as single strands that may fold over on themselves to generate a short double-helix region. If a double helix does form, there must certainly be base pairing holding it together. The bases in RNA pair C to G (–C :::: G–) and A to U (–A ::: U–).

There are three different ribonucleic acids found in cells, and each has a particular function. They are called **ribosomal RNA, messenger RNA,** and **transfer RNA**. Each kind of RNA is synthesized in the nucleus from the codes present in the genes on DNA. RNA synthesis is described as part of protein synthesis later on (see Figure 23.9).

The largest of the three ribonucleic acids are the **ribosomal RNA's (rRNA)**, which have molecular weights of a few million. Ribosomal RNA is combined with protein to form the ribosomes, the sites of protein synthesis in cells. Between 60% and 80% of the total RNA in cells is rRNA.

The molecules of **messenger RNA (mRNA)** carry the genetic message instructing the synthesis of a particular protein from the genes on DNA to the ribosomes. At the ribosomes they direct protein synthesis. Messenger RNA molecules can be relatively small or quite large with molecular weights up to 2,000,000. About 10% of the RNA in cells is mRNA.

The smallest RNA molecules are **transfer RNA's (tRNA)**. They transfer specific amino acids from the cytoplasm to the ribosomes for use in protein synthesis. The structures of many tRNA's have been determined, and are all

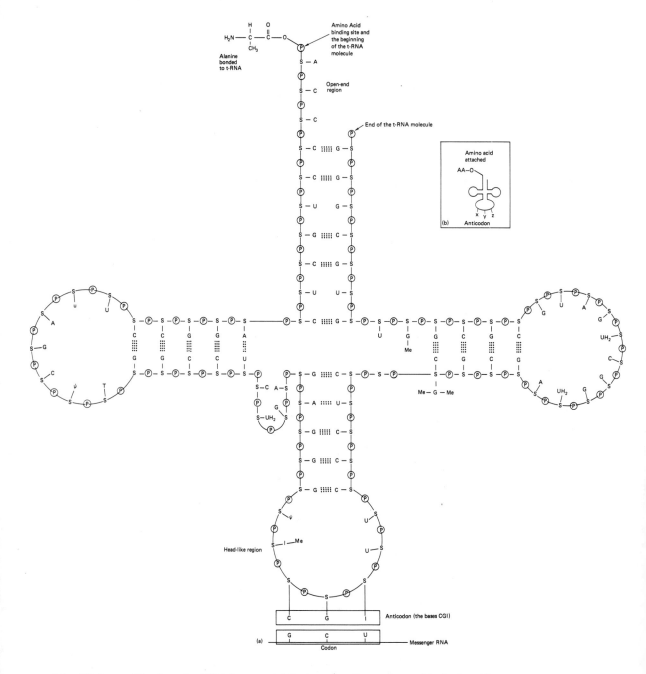

Fig. 23.8. The transfer RNA from yeast that carries the amino acid alanine. The amino acid is carried at the beginning of the RNA chain located at the top of the figure. The rounded region on the opposite end of the tRNA molecule carries a sequence of three bases forming the anticodon which identifies the amino acid. The anticodon matches with a codon on mRNA. This particular tRNA contains some bases not usually found in nucleic acid: I = inosine, ψ = pseudouridine, Me = methyl derivative of base, UH$_2$ = dihydrouridine. Note the hydrogen bonding that results in some secondary structure in tRNA. In RNA bases pair C to G, A to U. (b) A symbolic representation of a tRNA.

composed of a single RNA strand that is folded into a structure that when reproduced on paper is not unlike a four-leaf clover. The structure of the transfer RNA that carries alanine in yeast cells, alanine-tRNA, is shown in Figure 23.8. The shape of the molecule is maintained by four regions in which some base pairing exists. There are two distinct regions of every tRNA molecule that have important biological function. The first is the site at which the amino acid is joined to the tRNA molecule. This is at the beginning of the RNA chain located at the open end of the molecule. Opposite the open end, across the molecule, is the second region of special importance. This is at the bottom of the tRNA molecule in Figure 23.8. Across the lower rounded end of the tRNA molecule are three bases that represent a specific code for the amino acid carried by the tRNA. This three-base region is called the **anticodon** of the tRNA molecule. Notice, the anticodon for alanine in Figure 23.8 is CGI. C stands for cytosine, G for guanine, and I for the base inosine. Inosine is not one of the four bases that are supposed to appear in a RNA, but its presence reveals another fact about tRNA molecules. Occasionally, bases other than C, G, A, and U will be included in tRNA molecules. Sometimes they are derivatives of the normal RNA bases and other times they may be altogether different. Nonetheless, the CGI anticodon identifies a tRNA as a carrier of alanine.

If transfer RNA's have anticodons, how are they used in protein synthesis? There are corresponding three-base regions along the length of all mRNA molecules. These three-base regions are called **codons.** Each codon is specific for a particular amino acid that is carried by a tRNA. It is the attraction for the codon by the specific anticodon, in the form of hydrogen bonding, that causes amino acids to be delivered to the ribosomes in the proper sequence during protein synthesis.

There is usually more than one codon that signals for a given amino acid. Alanine has four: GCC, GCA, GCG, and GCU. The anticodon CGI can match with any one of them, but it will not match with any other. The codons for the 20 amino acids (actually for the tRNA's that carry them) are universal; that is, they are used by all organisms in protein synthesis. The codons for each of the 20 amino acids are listed in Table 23.4. p. 676.

Protein Synthesis

When a specific protein is needed by a cell, a definite gene (region) of the organism's DNA is copied to form a complimentary RNA molecule known as messenger RNA (mRNA). The process of obtaining a specific code from DNA through the synthesis of a mRNA molecule is called **transcription,** which means ''to copy.'' In transcription, a specific region of a DNA molecule acts as a template for the synthesis of the mRNA molecule. The process also requires the enzyme RNA polymerase, Mg^{2+} ions, and the ribonucleotide triphosphates ATP, CTP, GTP, and UTP.

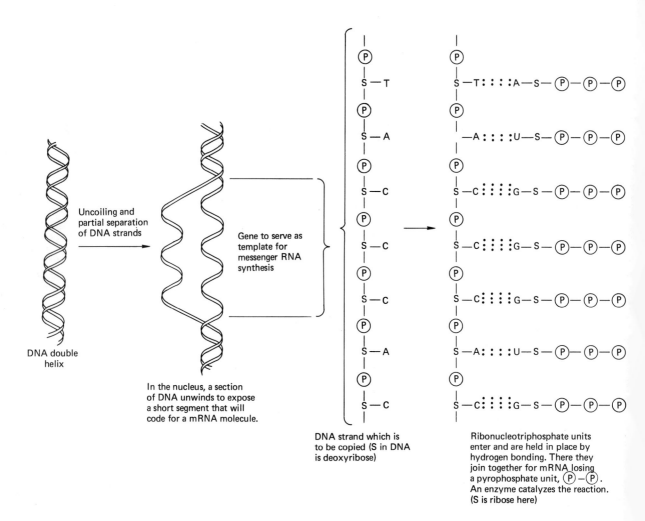

In the nucleus, a section
of DNA unwinds to expose
a short segment that will
code for a mRNA molecule.

DNA strand which is
to be copied (S in DNA
is deoxyribose)

Ribonucleotriphosphate units
enter and are held in place by
hydrogen bonding. There they
join together for mRNA losing
a pyrophosphate unit, (P)–(P).
An enzyme catalyzes the reaction.
(S is ribose here)

Fig. 23.9.   The transcription process. Here a particular mRNA is synthesized from a
gene on DNA. The mRNA codes for a specific protein.

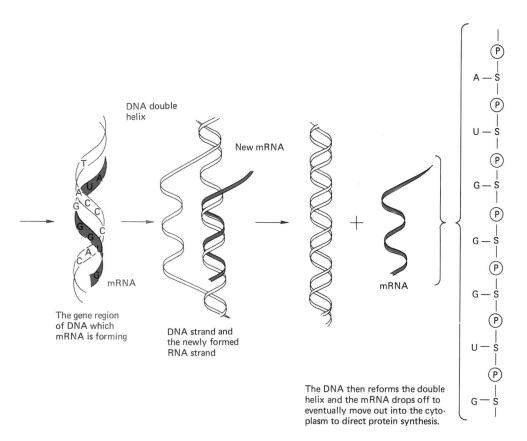

DNA double helix

New mRNA

mRNA

The gene region of DNA which mRNA is forming

DNA strand and the newly formed RNA strand

mRNA

+

The DNA then reforms the double helix and the mRNA drops off to eventually move out into the cytoplasm to direct protein synthesis.

TABLE 23.4 The Genetic Codons	
Amino acid	Codons
Alanine	GCA, GCC, GCG, GCU
Arginine	AGA, AGG, CGA, CGC, CGG, CGU
Asparagine	AAC, AAU
Aspartic acid	GAC, GAU
Cysteine	UGC, UGU
Glutamic acid	GAA, CAG
Glutamine	CAA, CAG
Glycine	GGA, GGC, GGG, GGU
Histadine	CAU, CAC
Isoleucine	AUA, AUC, AUU
Leucine	UUA, UUG, CUA, CUC, CUG, CUU
Lysine	AAA, AAG
Methionine	AUG (also chain initiation signal)
Phenylalanine	UUC, UUU
Proline	CCA, CCC, CCG, CCU
Serine	AGC, AGU, UCG, UCU, UCA, UCC
Threonine	ACA, ACC, ACG, ACU
Tryptophan	UGG
Tyrosine	UAC, UAU
Valine	GUG, GUA, GUC, GUU (GUU is sometimes used as a chain initiation signal)
Termination signals	UAA, UAG, UGA

Transcription begins in the nucleus of the cell when the specific section of the DNA helix that is to be copied separates, or unwinds, from the helix. The ribonucleotide triphosphates, which are present in the nuclear fluid, then line up along the DNA strand and are held in place by hydrogen bonding to the complimentary bases in DNA. RNA polymerase catalyzes the elimination of pyrophosphate units between neighboring ribonucleotide triphosphates, thus producing a mRNA molecule that is a complementary copy of that specific gene of DNA. Once complete, the mRNA drops off the DNA strand and enters the nuclear fluid. The DNA strand, having served its task, then returns to the normal helical form. The newly synthesized mRNA migrates out of the nucleus into the cytoplasm of the cell to make its way to the ribosomes. There, the mRNA will direct the synthesis of proteins. The process of transcription, the DNA-directed synthesis of mRNA, is shown in Figure 23.9.

The directing of protein synthesis at the ribosomes by mRNA is called **translation**. The ribosomes are knobby attachments that cover portions of the endoplasmic reticulum in a cell. They are also found in the cytoplasm. Each ribosome is composed of two distinct, near-spherical subunits, one smaller than the other. As the mRNA contacts the ribosome, one end of the mRNA strand becomes attached to the smaller subunit. This end of mRNA contains the codon AUG, the initiator codon which signals for protein synthesis. The AUG codon matches the anticodon UAC of a transfer RNA (tRNA) which carries the amino acid, methionine. The first amino acid in the protein to be synthesized will always be methionine. Once the protein is complete, the methionine residue will be removed. (In bacteria, the tRNA with the UAC anticodon carries N-formyl-methionine, a derivative of methionine.) The smaller ribosome subunit with the attached mRNA, along with the methionine-tRNA, then combines with the larger ribosome subunit to form an active ribosome unit. Other species, collectively called initiation factors, are also associated with the active ribosome unit.

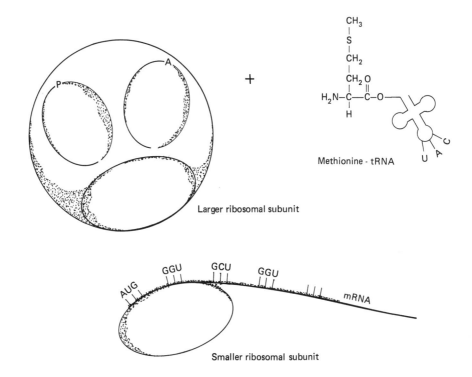

Methionine - tRNA

Larger ribosomal subunit

Smaller ribosomal subunit

The larger subunit has two sites on its surface that are actively involved in protein synthesis. One is the amino-acid–tRNA site (the *A* site) that accepts incoming amino-acid–tRNA units. The adjacent site serves as a holding place for the tRNA that carries the growing protein chain (the *P* site). The actual process of building the protein chain occurs at and between these two sites. Once methionine–tRNA (Met–tRNA) is seated in the *P* site, the ribosome unit is ready to begin protein synthesis.

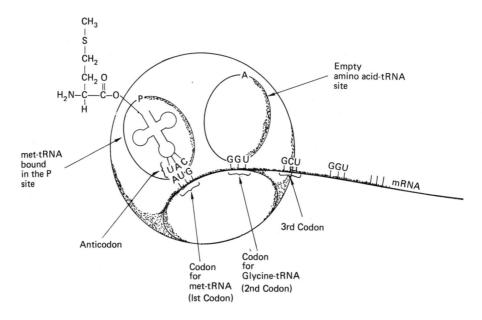

Note on the figure above how the second codon of mRNA is positioned just below the empty *A* site. An amino-acid–tRNA with the specific anticodon to match this second codon on the mRNA strand will enter the *A* site. The second codon is GGU, which pairs to the anticodon CCA of a glycine–tRNA unit. A glycine–tRNA unit migrates toward the ribosome from the cytoplasm and is attracted to the *A* site. As the codon–anticodon pairing is established, the glycine–tRNA unit becomes positioned in the *A* site. The first step in building the proteins can now take place.

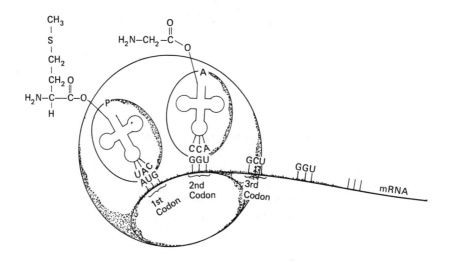

The enzyme peptidyl synthetase catalyzes the reaction that joins the methionine amino acid to glycine. Specifically, the reaction is between the carboxyl carbon of met–tRNA (at the *P* site) and the free amino group of Gly–tRNA (at the *A* site), resulting in the formation of the first peptide bond of the protein. Notice that methionine is transferred from its tRNA in the *P* site to the amino acid associated with the *A* site. This kind of *P* to *A* site handoff will occur over and over as the protein grows.

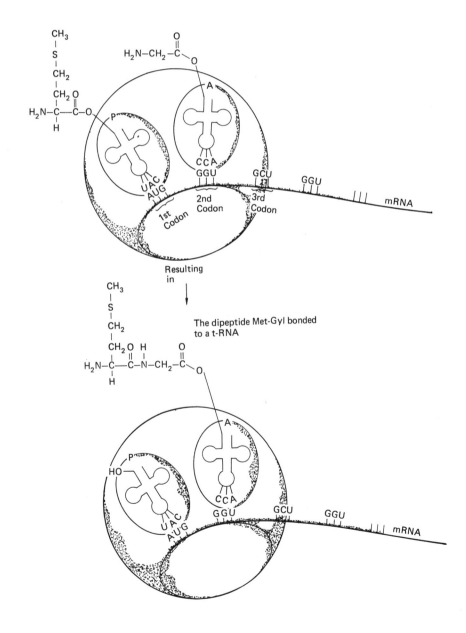

The tRNA remaining in the *P* site, having lost its amino acid (methionine), drops off, leaving the *P* site empty. The tRNA then moves back out into the cytoplasm to find another molecule of methionine.

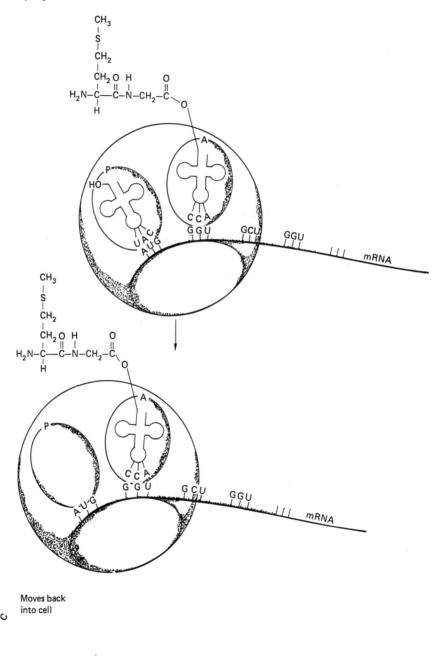

Moves back into cell

The dipeptide–tRNA shifts from the *A* site to the *P* site as the entire ribosome unit moves down the mRNA strand just enough to place the third codon, GCU, under the empty *A* site.

An amino-acid–tRNA with the proper anticodon (CGA) then enters the *A* site. This anticodon corresponds to an alanine–tRNA unit.

Peptidyl synthetase catalyzes the formation of the next peptide bond, this time between the carboxyl carbon of the glycine residue (at *P*) and the amino group of alanine (at *A*). The dipeptide moves from the *P* to the *A* site and a tripeptide is formed, met–gly–ala.

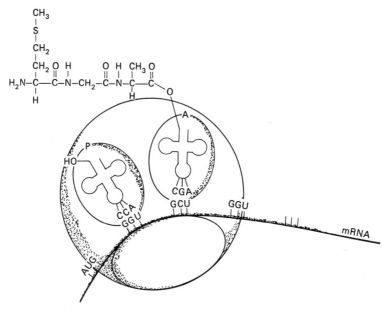

The tRNA remaining in the *P* site, having lost its amino acid, leaves the ribosome and returns to the cytoplasm to find another glycine molecule. The tripeptide–tRNA shifts from *A* to *P* as the ribosome again moves down the mRNA strand to the next codon.

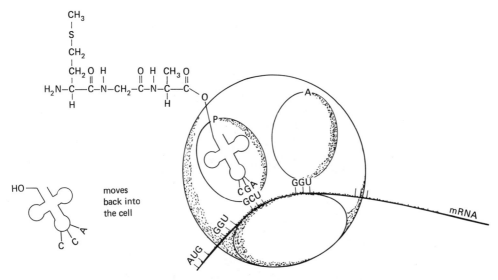

The process is repeated again and again until a specific codon on mRNA is reached that signals the protein synthesis to stop. UAA is one such stop signal. Since there is no amino-acid–tRNA anticodon that can match with UAA, the protein can no longer grow and the protein–tRNA unit drops off the ribosome. After leaving the ribosome, the tRNA is removed, and the protein is set free. The methionine residue, the first amino acid in the protein chain, is then removed by enzyme action. This leaves the required protein, which in the environment of the cell folds into its proper tertiary structure. The protein is then transferred out of the cell.

Once the protein–tRNA unit leaves the ribosome, the two subunits come apart, releasing the mRNA, which is then broken down by enzymes in the cell to the component ribonucleotides, sugars and bases. They will be used again to synthesize other mRNA molecules.

In the previous discussion of protein synthesis, we saw what happens when just one ribosome moves down the mRNA strand. In actual practice, several ribosomes move down a single mRNA strand at the same time. Each ribosome is at a different stage in the synthesis of its molecule of the protein. The involvement of many ribosomes, collectively called **polyribosomes,** preparing many molecules of the same protein at the same time allows the cell to be an efficient protein producer. It takes only a few seconds for a protein molecule to be prepared.

Figure 23.10 is an electron micrograph showing both transcription and translation occuring in *Escherichia coli.* In bacteria, transcription and translation occur in the same region of the cell and at the same time. In higher organisms, transcription takes place in the nucleus and translation outside the nucleus. A summation of the events involved in protein synthesis appears in Figure 23.11.

The Regulation of Protein Synthesis

Every human cell, no matter what its function in the body, contains the same amount of DNA. Each cell contains the same information needed to synthesize each of the specific proteins needed by the body. Yet, a given cell will only produce certain proteins. For instance, insulin is only produced by certain cells in the pancreas, and nowhere else. The reasons why this is so in higher organisms is not completely understood. The best information currently available has been obtained from the study of bacteria. In bacteria, there is a repression–activation mechanism that turns specific genes off and on for mRNA synthesis. Whether this same mechanism also operates in humans is not known. Further research in this area offers the possibility of better control of bacterial and viral infection and perhaps even the controlled destruction of cancer cells.

Mutations

It is clear that the base sequence of DNA ultimately determines the structure of every protein molecule in an organism. Any replacement, modification, loss, or gain of a base in the DNA of an organism will usually result in the production

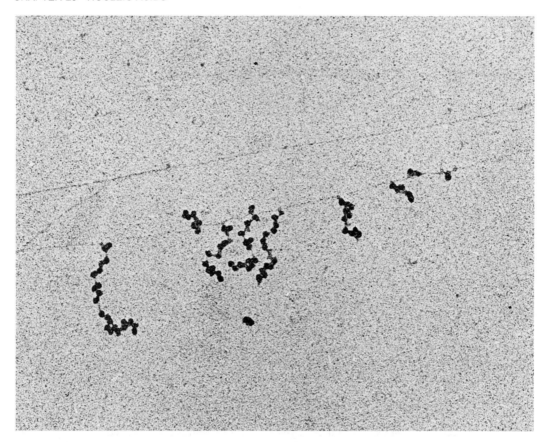

Fig. 23.10. An electron photomicrograph of transcription–translation complexes from *Escherichia coli*. Messenger RNA is being synthesized at the sites indicated by arrows. Molecules of RNA polymerase can be seen at these sites. The large dark bodies are ribosomes along the strands of mRNA. From O.L. Miller, Jr., *et al.,* 1970, *Science,* Vol. 169, pp. 392–395, Fig. 3.

of abnormal proteins. Any change in the base sequence of DNA is known as a **mutation**.

The most serious kind of mutation involves the loss or gain of one or more bases. This would cause at least one, and possibly more, genes to be read incorrectly because the base sequence would be shifted out of order by one or more units. Any mRNA synthesized with the mutated gene would have the wrong codons and would therefore produce the wrong protein.

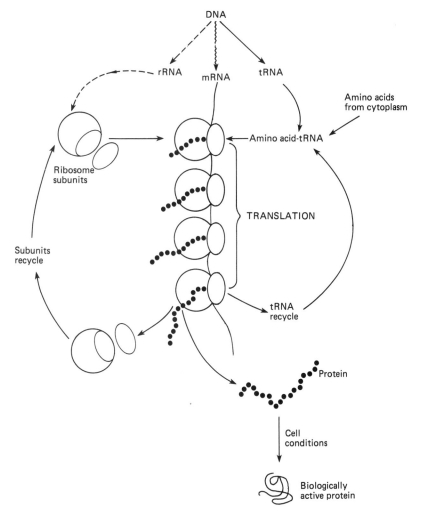

Fig. 23.11. A summary of protein synthesis in the cell. Adapted from Avers, *Basic Cell Biology,* D. Van Nostrand Co., New York, 1978.

Below, a section of one DNA strand with the complimentary mRNA is shown along with the amino acid sequence that would result in translation.

Segment of DNA → –AAG CCG ACA GAT AGC–

Complimentary mRNA → –UUC GGC UGU CUA UCG–

Resulting amino acid sequence → –Phe– Gly– Cys– Leu– Ser–

If the two bases CG were absent from DNA, the result of a mutation, an improper mRNA would be formed which would result in a completely different protein in translation.

Lost in a mutation

Segment of original DNA → –AAG C CG ACA GAT AGC–

Segment of mutated DNA → –AAG CAC AGA TAG C–

Complimentary mRNA → –UUC GUG UCU AUC G–

Resulting amino acid sequence →–Phe– Val– Ser– Ile–

If, as a result of the mutation of DNA, the abnormal protein causes the death of the organism, the alteration of the gene is called a **lethal mutation**. However, many if not most mutations involve only the replacement of one base by another. This is called a **point mutation,** and the usual result is the substitution of one amino acid for another in the primary structure of a protein. A given substitutional change in the amino acid sequence may or may not have a detrimental effect on the organism.

Mutations can occur simply by chance or by exposure to radiation or various chemicals. Some dyes, drugs, and compounds found in smoke have been shown to cause mutations in test animals. Visible examples of human mutations can be seen in some of the children born to Japanese citizens who were exposed to extremely high levels of radiation during the atomic bombing of Japan in World War II.

Not all mutations need be considered bad. In fact, many of the evolutionary changes that have occurred over the past millions of years of human development may have been the result of mutations.

Genetic Diseases

It has been estimated that there are over 1500 inherited diseases. Because they are due to defective or missing genes in the DNA of an individual, they are classed as **genetic diseases**. The genetic abnormalities are probably due to mutations. A genetic disease may be quite evident, such as a physical deformity or a chronic painful disease. On the other hand, the condition may be so well hidden the individual may be unaware that she even carries a genetic abnormality.

Higher organisms contain pairs of chromosomes. One member of the pair may be identical to the other, or it may be nonidentical. If both chromosomes of a pair are identical and have the same defective gene, the individual will have a specific genetic disease. This is because all of the protein made from the code on that gene will be defective, if it is made at all. If only one chromosome of the pair is abnormal, then at least half the protein coded by the specific gene region on each DNA will be normal, the other half coded by the faulty gene will be abnormal, or perhaps it will not be produced at all. If only one chromosome of each pair is faulty, at least some correct protein will be produced, and an individual may show no signs of the genetic abnormality.

In **sickle-cell anemia,** a genetic disease, the protein that is abnormal is β-hemoglobin. The amino acid valine is substituted for glutamic acid at one specific point in the protein chain. If an individual inherits the defective genes from both parents, she will develop the disease. However, if only one chromosome in each pair is defective, the individual is said to carry the **sickle-cell trait.** Half of the β-hemoglobin units will be normal, and the individual usually suffers few clinical symptoms. (Refer to Chapter 22, "Proteins," for a further discussion of sickle-cell anemia.) More than 80 different single amino acid substitutions in hemoglobin have been detected in humans. Fortunately, only a few of these cause serious adverse effects.

Phenylketonuria (PKU) is another genetic disease. It is caused by a defective gene that codes for phenylalanine hydroxylase, an enzyme that catalyzes the conversion of phenylalanine to tyrosine in the body. About 2% of the population

phenylalanine   tyrosine

carries this defective gene and about one in 20,000 newborn infants have the disease. Because the proper enzyme is missing, phenylalanine accumulates in the body. Some of it is converted into phenylpyruvic acid and similar compounds, as the body attempts to reduce the accumulated phenylalanine through an alternate metabolic route. The buildup of phenylpyruvic acid and the other metabolites results in damage to the growing brain, and the individual will inevitably suffer from severe mental retardation.

phenylpyruvic acid

The disease gets its name, phenylketonuria, from phenylpyruvic acid, a phenyl ketone (and an acid) that appears in the urine of individuals with this disease.

TABLE 23.5 Some Human Genetic Diseases	
Disease	Defective or missing protein
Sickle-cell anemia	β-hemoglobin
PKU	phenylalanine hydroxylase (enzyme)
Albinism	tyrosinase (enzyme)
Tay–Sach's disease	hexosaminidase (enzyme)
Gaucher's disease	glycolipid enzyme
Hemophilia	antihemophilic factor (clotting factor)
Diabetes mellitus	insulin (hormone)
Dwarfism	growth hormone

Phenylalanine can be readily detected in the blood. Many states require that every newborn be tested within a few hours of birth for PKU. Any infant diagnosed to have PKU, because of abnormally high phenylalanine and low tyrosine levels in blood, is immediately placed on a special diet that limits the intake of phenylalanine to just that needed for normal development. If the child remains on this diet until the period of rapid brain growth is over, the symptoms of retardation will not occur. However, the individual still has the genetic disease.

Unfortunately, little can be done for individuals suffering from many genetic diseases, such as albinism, Tay–Sach's disease, Gaucher's disease, or sickle-cell anemia. The error that exists in their DNA cannot be corrected. Several genetic diseases appear in Table 23.5 with the missing or defective protein responsible for the condition.

For couples who have, within one of their families, a history of genetic disease, it is wise for them to have genetic counseling in order to understand the risk of having an abnormal child. If the woman becomes pregnant, she is often advised to have **amniocentesis** performed during the third month of pregnancy. This procedure requires withdrawing a small amount of the amniotic fluid which surrounds the growing fetus. Stray fetal cells are isolated and allowed to grow in the laboratory and their proteins and chromosomes are checked for possible genetic errors. If errors are found, the parents are informed of the possible effects to the child and a clinical abortion is a possibility, but the abortion issue is very controversial. This same procedure is also recommended for pregnant women over 35, since the risk of having an abnormal child increases as a woman grows older.

Genetic Engineering

During the past few years, techniques have been developed that allow researchers to isolate the genes of one organism, and incorporate these genes into the DNA of another organism. The procedure is known as **recombinant DNA**. The organism most often studied in recombinant DNA research is *Escherichia coli,* a bacterium found in the intestines of humans. However, the specific *E. coli* used in these studies is a mutant that can only survive under the carefully controlled conditions of the laboratory.

E. coli contains a plasmid that is a very small circular unit of DNA separate from the principal chromosome unit. The plasmids are first isolated from cultures of *E. coli,* and then cleaved at one point in the circle to produce a linear DNA unit. Another gene, either synthetic or obtained from another organism, is then fused to one end of the linear DNA strand. The strands carrying the added gene are then allowed to reclose, forming circular plasmids once again. The new plasmids are then put back into *E. coli* cells, where they act normally. Whenever the bacterium divides, new plasmids are synthesized which are copies of the plasmid containing the added gene. The new gene will direct the synthesis of protein in the same manner as those originally present in the plasmid.

In 1977 a synthesized gene coded for the human hormone somatostatin (14AA) was introduced into *E. coli* bacteria. The gene worked, and for the first time a specific human protein was produced by another organism. Somatostatin is involved in the release of insulin and glucagon in the body. Up to this time, only a few milligrams of somatostatin had been isolated, and that required the tedious extraction of the hormone from 500,000 sheep brains. Two gallons of recombinant *E. coli* produced 5 mg of the hormone in a relatively short period of time.

The potential benefits of recombinant DNA research can be enormous, but at the same time the risks can also be great. The possibility of producing a new and deadly bacteria or virus, or even a human mutation, has turned recombinant DNA research into a highly controversial issue, with strong social and political overtones. The National Institute of Health has provided guidelines by which this type of research can be carried out. With the proper controls in place, research may continue with a minimum risk of a DNA accident.

Viruses

A **virus** is a biological unit that exists at the borderline separating the living from the nonliving. They appear as crystals of either DNA or RNA surrounded by a tightly packed protein coat. The coat may also contain carbohydrate and lipid components. Viruses have no need of metabolism when they are not in a living cell, and they may be stored for long periods of time in laboratories. But when added to the cell of a host organism, they become active and take over the metabolism of the cell. Viruses are very specific in terms of which plant, animal, or bacterium they infect.

Since either DNA or RNA serves as the genetic material of a virus, they may be classed as either DNA viruses or RNA viruses. Chicken pox, mumps, and rabies are caused by DNA viruses. Influenza, polio, and encephalitis are caused by RNA viruses. The molecular weights of viruses vary from one million to 100 million, and they exist in different shapes, such as rods or spheres.

One type of virus, the **bacteriophage,** which infects bacteria, has been extensively studied. The bacteriophage attaches itself to the wall of a bacterium and injects its nucleic acid into the cell. The nucleic acid of the virus takes over the metabolism of the cell and directs the synthesis of additional virus particles. As the bacterium becomes filled with virus particles, it dies and releases them to neighboring cells. Some types of virus have been observed entering host cells and remaining there in an inactive state. Then, for some unknown reason, they later become active and take over the cell. Still other viruses are able to fuse with the DNA of the host, and be replicated by the host through many generations before the virus particles again reappear.

Evidence has been obtained that suggests some human cancers, such as breast cancer and some forms of leukemia, may be caused by viruses. The evidence is in the form of "viral footprints"—molecules such as mRNA, enzymes, and other proteins that appear in the cancer cells and not in normal cells. These foreign molecules have properties that suggest they were produced under the direction of a virus.

Drugs That Inhibit Nucleic Acid and Protein Synthesis

There are many natural and synthetic compounds that are known to fight bacterial infections. Several of these compounds, or drugs, act by inhibiting either nucleic acid or protein synthesis in bacterial cells. We know them as **antibiotics** or **antimetabolites**. The antibiotics are obtained from microorganisms, such as fungi or bacteria, while the antimetabolites are synthesized drugs. In most cases, both operate by inhibiting the activity of enzymes. Several drugs currently in use that operate by inhibiting nucleic acid and protein synthesis are listed in Table 23.6. The listed antibiotics selectively attack bacterial cells and do not interfere with the normal cells in the body. Some of the antimetabolites interfere with enzyme activity in both bacterial and normal body cells. Both the dosage and the length of time these drugs are administered must be carefully controlled to avoid harm to the patient. Certain antimetabolites have proven effective as anticancer drugs. Both 6-mercaptopurine and azaserine exhibit antineoplastic activity.

Perhaps the best-known antibiotic is penicillin, obtained from the fermentation of a certain mold. It destroys bacterial cells by interfering with the formation of the cell wall, preventing its completion. The bacteria, incompletely formed, then dies. Several chemical modifications of natural penicillin are currently used as antibiotics.

TABLE 23.6 Selected Drugs That Inhibit Nucleic Acid and Protein Synthesis	
Drug	Mode of action
Antibiotics	
Actinomycin D	Forms complexes with guanosine bases in DNA and in doing so interferes with DNA replication and mRNA synthesis.
Chloramphenicol	Binds to larger ribosomal particle and disrupts peptide synthesis.
Puromycin	Part of its structure is similar to that of an amino-acid–tRNA unit. The drug bonds to the growing peptide chain, causing it to be dropped by mRNA as an abnormal protein.
Streptomycin	Binds to the smaller ribosomal particle and causes errors in the reading of mRNA.
Tetracycline	Binds to the smaller ribosomal particle and inhibits the binding of amino-acid–tRNA's.
Antimetabolites	
Sulfanilamide (a sulfa drug)	Has a structure similar to p-aminobenzoic acid, which is needed by bacterial cells in the formation of purine bases and thus nucleic acids. Discussed in Chapter 24.
Azaserine	Disrupts synthesis of purine nucleotides and thus nucleic acid.
6-Mercaptopurine	Structure is similar to the purines and when used by cells in nucleic acid synthesis produces faulty nucleic acids.
5-Fluorodeoxyuridine	Structure is similar to pyrimidines, and also produces faulty nucleic acids.

Viral infections, up until the last few years, have not been readily combated with drugs. If an antiviral drug is to be effective, it must be able to enter the infected cell and destroy the virus without causing irreparable harm to the host. In late 1977, adenosine arabinoside (ara-A) was licensed by the Food and Drug Administration as the first effective drug for use against a viral infection. The drug is marketed under the trade name, Vira-A® (Parke-Davis), and it has proven effective in destroying the herpes encephalitis virus in humans. Before its use, 70% of the people who contracted this viral brain disease died, and most of the remaining 30% suffered some form of permanent neurological damage. Today, because of the drug, less than 10% die or suffer serious aftereffects. The drug is also effective in combating herpes zoster, commonly called shingles, and against herpes keratitis, a virus that infects the cornea of the eye sometimes causing scarring and loss of vision.

Even though ara-A is limited in its range of antivirus activity, scientists are confident that other antiviral drugs can be developed. It is interesting to note that cytosine arabinoside (ara-C), the most effective drug for combating one form of acute leukemia, differs from the structure of ara-A only in the base it contains.

There has been protection from viral diseases for many years, but it involved the production of antibodies in the body. Polio, mumps, and measles, all viral diseases, can be prevented through vaccines. The vaccines contain dead or weakened virus that, when placed in the body, will bring about the production of specific antibodies. These antibodies will attack that virus if it enters the body again at another time. The immunity thus obtained must be strengthened periodically with "booster shots."

Interferon—The Body's Defense Against Viral Infection

In 1957, Dr. Alick Issacs and Dr. Jean Lindemann discovered that fluid isolated from a culture of cells obtained from a virus-infected animal would protect other cells of that animal from the virus. Other researchers soon learned that every animal cell that is attacked by a virus produces a protein known as **interferon,** and that each species of animal produces its own kind of interferon that has the ability to combat any kind of virus infection it may contract. Human interferon has a molecular weight of about 26,000. The interferon produced in chickens has a higher molecular weight, around 38,000.

Interferon appears to fight viral infections by inhibiting, within the cell, the synthesis of proteins needed to make new virus molecules. The infected cell and the virus dies, but no new virus is formed. Also, interferon seems to be involved in the production of antibodies, which in turn combat specific viruses. The body begins producing interferon within a few hours following the onset of a viral infection. Antibodies do not begin to appear until quite a bit later. Thus, in humans and other animals, interferon is the first line of defense against a viral infection.

Interferon would seem to be the perfect antiviral agent, since it is produced naturally by the body. But the promise of a broad spectrum antiviral agent based on interferon has been slow to be realized. It is a very complex protein that is made in only small amounts by the body. This makes it difficult to obtain and purify. Partially purified samples of human interferon have been shown to destroy herpes virus, measles virus, and hepatitis-B virus in humans. The American Cancer Society has established a multimillion dollar program for evaluating the use of interferon in cancer therapy. Experimental evidence suggests that interferon may accelerate recovery from certain kinds of cancer.

By far the greatest news to those experimenting with human interferon was the announcement in December of 1978 that a quantity of pure human interferon was now available to allow scientists to determine the amino acid sequence of the protein. This will not be an easy task, but the benefits could be great. We will all have to wait to see if interferon is the answer to the ravages of viral infection.

Terms

Several important terms appeared in Chapter 23. You should understand the meaning of each of the following.

nucleic acid

nucleoside

nucleotide

DNA

tRNA

mRNA

rRNA

3', 5'-phosphodiester link

double helix

base pairing

secondary structure of DNA

chromosome

gene

DNA replication

genetic code

transcription

translation

codon

anticodon

ribosome

A site

P site

mutation

genetic disease

recombinant DNA

virus

interferon

Questions

Answers to the starred questions appear at the end of the text.

*1. What is a nucleoside, and how does it differ from a nucleotide?

*2. Draw the structures of the two sugars that are present in the nucleic acids. Which is present in DNA?

*3. Name the bases found in DNA, and those found in RNA. Which are purines and which are pyrimidines?

4. Draw the structures of the five bases found in the nucleic acids.

5. Which sugar and which base combine to form *(a) guanosine, (b) uridine, *(c) deoxyadenosine, (d) deoxythymidine?

6. Give the name of the nucleotide symbolized by each of the following abbreviations: *(a) CMP, (b) UTP, *(c) dADP, (d) dGMP.

7. Draw the structures of *(a) thymidine-5'-monophosphate, (b) adenosine-5'-diphosphate.

*8. What is the 3', 5'-phosphodiester link in a nucleic acid?

*9. Which bases pair in DNA? In RNA?

10. What type of bonding holds base pairs together?

*11. Describe the secondary structure of DNA. How is it held in this structure?

12. Of what importance is the secondary structure of DNA in terms of DNA replication?

*13. Briefly describe the process of DNA replication.

*14. Which nucleotide phosphates are used in DNA replication?

15. What guides the sequence of nucleotide phosphates in DNA replication and RNA synthesis?

*16. Name and briefly describe the function of each of the three ribonucleic acids.

17. What is the form of the genetic code in DNA?

*18. What is a codon? How is it related to the anticodon on tRNA?

19. Describe the process of transcription, the DNA-directed RNA synthesis.

*20. What is the particular codon that commands the start of protein synthesis? On which kind of RNA is it found, and at what point in the molecule?

21. Which amino acids are indicated by the following codons: *(a) UGC, (b) AAU, *(c) GGG, (d) UUU.

22. What is a mutation in terms of DNA?

*23. A section of mRNA has the following sequence of bases: –GGA AAA UGG GUA–. In a mutation, the sequence became –GGA AAU GGG UAC–. (a) What mutation occurred? (b) What protein sequence is represented by the codons prior to the mutation? (See Table 23.4.) (c) What protein sequence is represented after the mutation?

24. A codon sequence was altered by a mutation of DNA. In what way does the composition of the protein change? Before mutation: –GCU UGU GGU AGC A–; After mutation: –CUU GUG GUA GCA–?

*25. Why is sickle-cell anemia considered a genetic disease? How does the DNA of an individual with the sickle-cell disease differ from one with sickle-cell trait?

26. What biological pathway is interrupted in infants with PKU?

27. What is interferon?

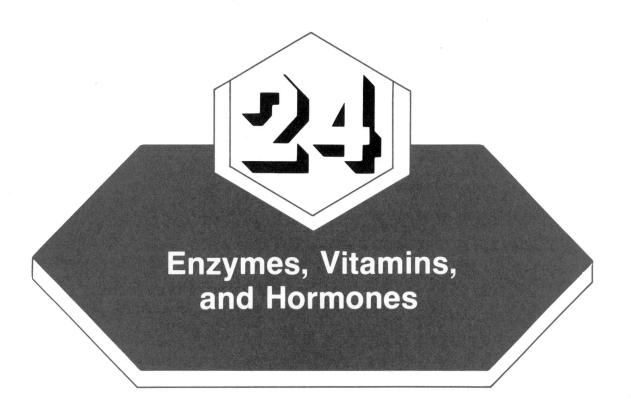

Enzymes, Vitamins, and Hormones

Thousands of chemical reactions take place within the cells of living organisms. Many of these reactions have also been carried out under laboratory conditions but, outside the cell, they were often very much slower. To increase the rates of the reactions so they would proceed about as fast as they do in cells, high temperatures or extreme pH's were required, conditions that simply could not exist in a living cell. How then is it possible for reactions to occur in cells at rates fast enough to meet the needs of the body? The answer is that cells produce special proteins known as enzymes that catalyze biological reactions, markedly increasing their rates.

A cell may contain as many as 3000 different enzymes, and each one has a specific purpose in the operation of the cell. Besides increasing the rates of reactions, enzymes are also involved in the control of the amounts and kinds of substances produced in cells. But the enzymes are not without control either. The activity of many enzymes is controlled by hormones secreted by the endocrine glands. Hormones are important communicators between cells.

Enzymes are globular proteins that may contain one or more peptide chains folded into a specific three-dimensional unit. Their molecular weights range from around 10,000 to over a million. In some cases, a nonprotein unit, called a cofactor, may be associated with an enzyme. For certain enzymes, the cofactor may be a substance derived from one of the water-soluble vitamins, for others it may be a specific metal ion.

The chapter will begin with a study of the six different classes of enzymes and how enzymes are commonly named. Since the water-soluble vitamins are important in the formation of certain cofactors, they will be discussed next. Then the way that enzymes are involved in catalyzing biological reactions will be described, along with the factors that influence the activity of an enzyme. The chapter will end with a discussion of hormones.

Objectives

By the time you finish this chapter, you should be able to do the following:

1. Describe the function of enzymes in the chemistry of the body.
2. List the six types of enzymes and, given an equation describing an enzyme-catalyzed reaction, name the type of enzyme involved.
3. Define: active site, coenzyme, cofactor, metal-ion activator, and substrate.
4. Name the water-soluble vitamins and give at least one important function of each.
5. Describe the lock and key theory and the induced-fit theory of enzyme action.
6. Outline the three-step mechanism that describes an enzyme-catalyzed conversion of a substrate to a product.
7. List the four categories of enzyme specificity.
8. Describe how temperature and pH can influence the activity of an enzyme.
9. Describe the four major types of enzyme inhibition.
10. Describe the function of hormones in the body.

The Names of Enzymes and Their Classification

When researchers began to isolate and study large numbers of enzymes, no definite scheme existed that could be used to assign each one a systematic name. As a result, many trivial names came into being that did not indicate the specific role of a particular enzyme nor the compound it acted on. The primary compound acted on by an enzyme is called the **substrate** of that enzyme. Eventually, it became accepted practice to name each enzyme after the substrate upon which it acted by simply adding "-ase" to the root of the name of the substrate. A lipase, then, would be an enzyme that acts on lipids, sucrase on sucrose, and so forth. These are now considered the common names for enzymes. The "-ase" ending indicates the compound is an enzyme. Three very common enzymes, pepsin, trypsin, and chymotrypsin, are still called by these older names that end in "-in." The "-in" ending was used to indicate they were proteins.

In 1961, the International Union of Biochemistry (the IUB) adopted a more complex, though more precise, method of naming enzymes. Each enzyme was to be named according to the type of reaction it catalyzed, its substrate, and the product of the reaction. Each enzyme ends up with a rather long name, and though they are useful to enzyme chemists, they will not be used here. We will use the common names (-ase endings) since they are easier for most people to use and remember.

Each enzyme can be classified into one of six different enzyme classes based on the kind of reaction it catalyzes. Let us examine each enzyme class along with several specific examples of enzyme-catalyzed reactions in each one. You may see some reactions and compounds that are unfamiliar, but each is important in the chemistry of the body. Those reactions that are reversible are indicated with opposing arrows ($\rightleftharpoons$) in the equations.

1. The Hydrolases

These enzymes catalyze hydrolysis reactions (reactions with water). The digestive enzymes are hydrolylases, and they are responsible for the digestion (hydrolysis breakdown) of large food molecules into smaller units that are able to be absorbed by the body.

General reaction:

$$\text{Substrate} + H_2O \xrightleftharpoons{\text{hydrolase}} \text{products}$$

Reactions that are catalyzed by some of the more familiar types of hydrolases are given below.

(a) Lipases. These catalyze the hydrolysis of lipids.

$$\text{triglyceride} + 3H_2O \xrightleftharpoons{\text{lipase}} \text{glycerol} + 3 \text{ fatty acids}$$

(b) Carbohydrate Hydrolases. These catalyze the hydrolysis of carbohydrates to monosaccharides.

$$\text{starch} + H_2O \xrightleftharpoons[\text{(mouth)}]{\text{amylase}} \text{dextrins}$$

$$\text{dextrins} + H_2O \xrightleftharpoons[\text{(small intestine)}]{\text{pancreatic amylase}} \text{maltose}$$

$$\text{maltose} + H_2O \overset{\underset{\text{maltase}}{\text{(small intestine)}}}{\rightleftharpoons} \text{2 glucose units}$$

$$\text{sucrose} + H_2O \overset{\underset{\text{sucrase}}{\text{(small intestine)}}}{\rightleftharpoons} \text{glucose} + \text{fructose}$$

$$\text{lactose} + H_2O \overset{\underset{\text{lactase}}{\text{(small intestine)}}}{\rightleftharpoons} \text{glucose} + \text{galactose}$$

(c) Proteases.   These enzymes act together to catalyze the hydrolysis of proteins and peptides into mixtures of amino acids. Examples are pepsin (stomach), trypsin, chrymotrypsin, and the carboxypeptidases (small intestines).

$$\text{protein} + H_2O \overset{\underset{\text{HCl; pepsin}}{\text{(stomach)}}}{\rightleftharpoons} \text{peptides}$$

$$\text{peptides} + H_2O \overset{\underset{\text{other enzymes}}{\text{(small intestines)}}}{\rightleftharpoons} \text{amino acids}$$

(d) Nucleases.   These catalyze the hydrolysis of nucleic acids.

$$\text{nucleic acid} + H_2O \overset{\text{nuclease}}{\rightleftharpoons} \text{a mixture composed of nucleic bases} + \text{pentose sugars} + \text{phosphoric acid}$$

2. The Oxido-reductases   These enzymes catalyze oxidation–reduction reactions that occur in the body that involve the simultaneous removal or addition of two electrons *and* two protons ($2e^-$, $2H^+$). Remember that oxidation and reduction *always* occur together.

General reaction:

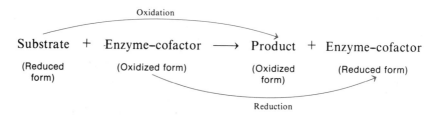

The most important type of oxidoreductases in the human body are the dehydrogenases. These enzymes are involved in the addition or removal of hydrogen (along with electrons) to or from their substrates. Two examples of oxidoreductase reactions are given below.

(a) The Oxidation and Phosphorlyaton (the Substitution of a Phosphate Group) of Glyceraldehyde-3-phosphate.

The enzyme-cofactor complex

$$
\begin{array}{c}
O{=}C{-}H \\
| \\
H{-}C{-}OH \quad\quad O \\
| \quad\quad\quad || \\
CH_2{-}O{-}P{-}OH \\
| \\
OH
\end{array}
+ H_3PO_4 + \text{glyceraldehyde-3-phosphate dehydrogenase-NAD}^{\oplus}
$$

Glyceraldehyde-3-
phosphate

(reduced form)

(oxidized form)

$$
\begin{array}{c}
O \\
|| \\
O{=}C{-}O{-}P{-}OH \\
| \quad\quad OH \\
H{-}C{-}OH \\
| \quad\quad O \\
\quad\quad || \\
H_2C{-}O{-}P{-}OH \\
| \\
OH
\end{array}
+ \text{glyceraldehyde-3-phosphate dehydrogenase-NADH + H}^+
$$

(reduced form)

1,3-diphosphoglyceric
acid

(oxidized form)

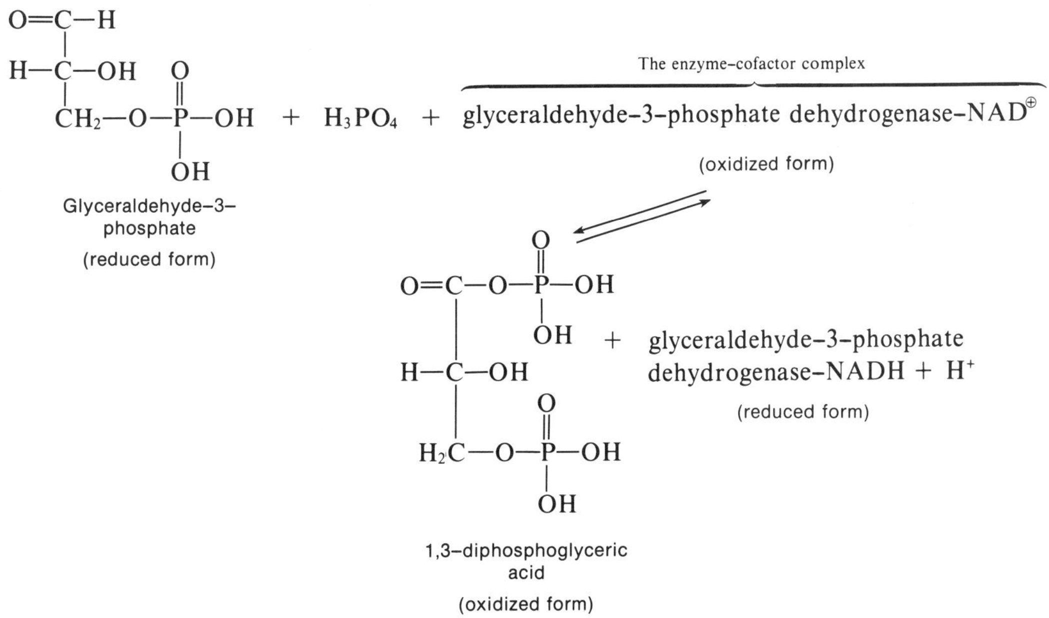

This same reaction can be written in an abbreviated form that shows only the structures of the substrate and product. The enzyme-cofactor complex is written simply as $NAD^{\oplus}$ and $NADH+H^+$. They, along with phosphoric acid, are placed above the double arrows, and whether they are consumed or produced in the reaction is indicated with curved arrows.

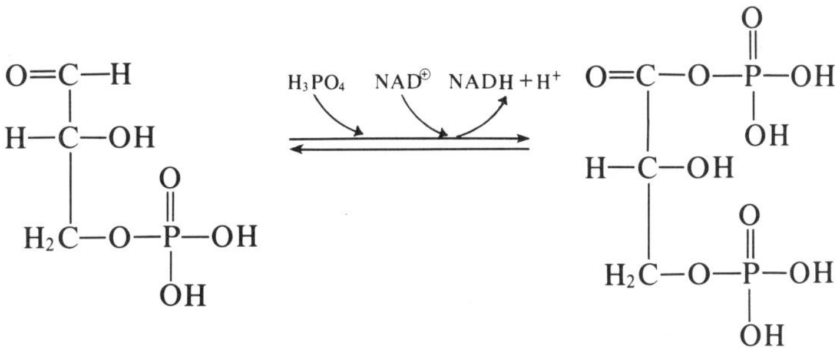

(b) The Reduction of Acetaldehyde in the Fermentation Process that Produces Ethyl Alcohol.

$$CH_3-\overset{\overset{\text{O}}{\|}}{C}-H \;+\; \text{alcohol dehydrogenase-NADH} + H^+ \;\rightleftharpoons\; CH_3CH_2OH$$

| Acetaldehyde (oxidized form) | The enzyme–cofactor complex (reduced form) | Ethyl alcohol (reduced form) |

+

alcohol dehydrogenase–NAD$^\oplus$
(oxidized form)

The structure of the NAD$^\oplus$ coenzyme appears later on in this chapter in the discussion of niacin, one of the water-soluble vitamins.

3. The Transferases These enzymes are involved in the transfer of groups of atoms from one compound to another. The transferred groups are fragments of substrate molecules, like amino, alkyl, or phosphate groups. Transferases are subdivided according to the type of group transferred.

(a) Transaminases. These enzymes catalyze the reversible transfer of an amino group from an amino acid to a keto acid.

$$R-\overset{\overset{\text{O}}{\|}}{C}-COOH \;+\; R'-\overset{\overset{\text{NH}_2}{|}}{\underset{\underset{\text{H}}{|}}{C}}-COOH \;\underset{\xrightarrow{\text{transaminase}}}{\rightleftharpoons}\; R-\overset{\overset{\text{NH}_2}{|}}{\underset{\underset{\text{H}}{|}}{C}}-COOH \;+\; R'-\overset{\overset{\text{O}}{\|}}{C}-COOH$$

α–Keto acid   Amino acid   New amino acid   New α–keto acid

(b) Transmethylases. Transfer methyl groups, —CH$_3$

(c) Transformylases. Transfer formyl groups, $-\overset{\overset{\text{O}}{\|}}{C}-H$

(d) Transacylases. Transfer acyl groups, $-\overset{\overset{\text{O}}{\|}}{C}-R$,

such as the acetyl group, $H_3C-\overset{\overset{\text{O}}{\|}}{C}-$.

(e) Kinases (Transphosphatases). Transfer phosphate groups,

$$-O-\overset{\overset{\text{O}}{\|}}{\underset{\underset{\text{OH}}{|}}{P}}-OH.$$

Glycolysis is one of the processes used by the cells to produce energy from glucose. In the first step of glycolysis, glucose is converted into glucose-6-phosphate by the action of the enzyme hexokinase, which transfers a phosphate group from ATP to glucose.

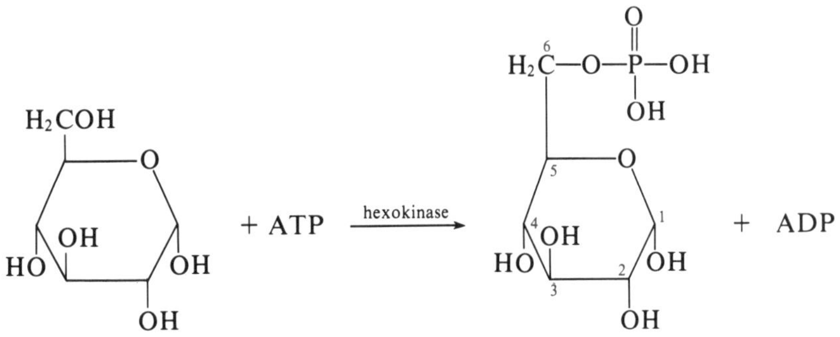

Glucose

Glucose–6–phosphate

4. The Lyases   These enzymes catalyze the removal or addition of groups from substrates, without the use of hydrolysis or oxidation–reduction reactions. Lyases differ from transferases since the group that is added or removed is in the form of a molecule, and not just a part of a molecule as was the case with the transferases. The most important type of lyases are the hydrases, dehydrases, and the decarboxylases.

(a) Addition (Hydrases) and Removal (Dehydrases) of Water.   These enzymes catalyze the addition of water to double bonds, or the removal of water to form a double bond. There are two different lyases that are important in the **Krebs cycle** (Chapter 25), an important set of reactions that releases energy for cells by the breakdown of food. The enzyme, aconitase, catalyzes the removal of water from citric acid to form aconitic acid. Then water is added to aconitic acid to form isocitric acid, an isomer of citric acid.

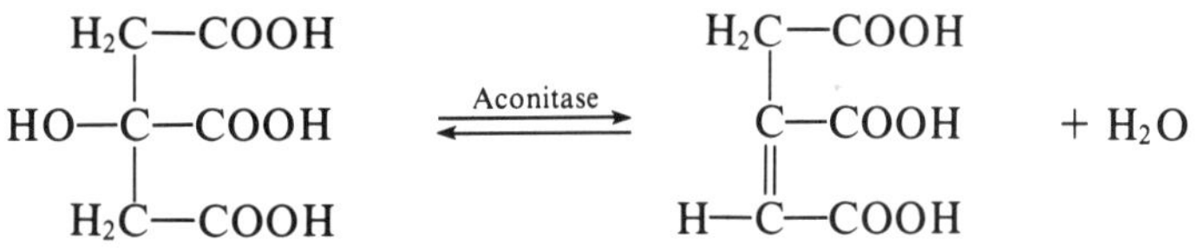

Citric acid

Aconitic acid

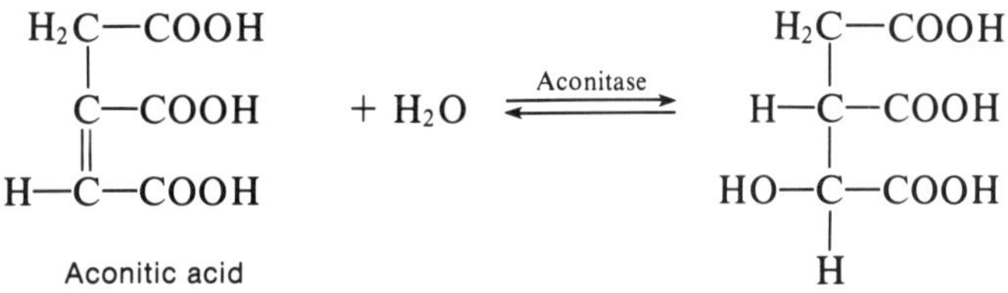

Aconitic acid

Isocitric acid

In the next to last step in the Krebs cycle, the enzyme, fumarase, catalyzes the addition of water to the double bond of fumaric acid to form malic acid:

$$\underset{\text{Fumaric acid}}{\underset{H}{\overset{HOOC}{}}C=C\underset{H}{\overset{COOH}{}}} + H_2O \; \underset{\text{fumarase}}{\rightleftharpoons} \; \underset{\text{Malic acid}}{HOOC-CH_2-\underset{H}{\overset{OH}{\underset{|}{\overset{|}{C}}}}-COOH}$$

(b) **Decarboxylases.** These lyases eliminate carboxyl groups from substrates by removing carbon dioxide (CO_2). The decarboxylation of pyruvic acid to form acetaldehyde in the glycolysis pathway (Chapter 25) prepares it for entry into the Krebs cycle. There are also two other decarboxylation steps in the Krebs cycle.

$$\underset{\text{Pyruvic acid}}{CH_3-\overset{O}{\overset{\|}{C}}-\overset{O}{\overset{\|}{C}}-OH} \; \xrightarrow[\text{decarboxylase}]{\text{Pyruvic acid}} \; \underset{\text{Acetyaldehyde}}{CH_3-\overset{O}{\overset{\|}{C}}-H} + CO_2$$

5. The Isomerases   These enzymes are involved in the intramolecular rearrangements of substrates to form different isomers.

(a) **Cis – Trans Isomerases.** These catalyze conversion of one geometrical isomer to the other about a carbon–carbon double bond. The conversion of *trans*-retinene to the *cis*-isomer is a necessary step in the chemistry of vision.

(b) **Sugar Isomerases.** These convert one sugar into another. The following are two important reactions in the glycolysis pathway.

$$\text{glucose-6-phosphate} \; \underset{\text{isomerase}}{\overset{\text{phosphohexose}}{\rightleftharpoons}} \; \text{fructose-6-phosphate}$$

$$\text{dihydroxyacetone phosphate} \; \underset{\text{isomerase}}{\overset{\text{triose phosphate}}{\rightleftharpoons}} \; \text{glyceraldehyde-3-phosphate}$$

6. The Ligases   These enzymes are often called synthetases, and they catalyze the joining of two molecules to form a larger one. Energy is required for this synthesis, and it is usually obtained by the conversion of ATP to ADP + Pi (inorganic phosphate). The body uses ligases in the synthesis of proteins, lipids, polysaccharides, and nucleic acids.

In the first step of the urea cycle, which involves the production of urea from ammonia, the required ligase is carbamyl phosphate synthetase. Ammonia is a waste product of metabolism.

$$NH_3 + CO_2 + 2ATP \xrightarrow[\text{Synthetase}]{\text{Carbamyl phosphate}} \underset{\underset{OH}{|}}{H_2N-\overset{\overset{O}{\|}}{C}-O-\overset{\overset{O}{\|}}{P}-OH} + 2ADP + Pi$$

The conversion of glutamic acid into glutamine, another amino acid, is catalyzed by the enzyme glutamic acid synthetase:

$$\underset{\text{Glutamic acid}}{\underset{\overset{|}{COOH}}{\overset{|}{\underset{CH_2}{\overset{|}{\underset{CH_2}{\overset{|}{H_2N-\overset{\overset{H}{|}}{C}-COOH}}}}}}} + NH_3 + ATP \underset{Mg^{2+}}{\xrightarrow[\text{synthetase}]{\text{glutamic acid}}} \underset{\text{Glutamine}}{\underset{\overset{|}{O=C-NH_2}}{\overset{|}{\underset{CH_2}{\overset{|}{\underset{CH_2}{\overset{|}{H_2N-\overset{\overset{H}{|}}{C}-COOH}}}}}}} + H_2O + ADP + Pi$$

Check Test Number 1

1. Give the common names for the enzymes that would catalyze reactions that use urea and maltose as substrates.
2. For each of the following, give the general class and the more specific subclass of the enzyme that would catalyze the reaction. (For example, the general class may be a hydrolase, while the specific subclass might be a lipase.)

a. $$\underset{\text{Dipeptide}}{H_2N-\underset{\overset{|}{CH_3}}{\overset{\overset{H}{|}}{C}}-\overset{\overset{O}{\|}}{C}-\underset{\overset{|}{H}}{\overset{\overset{H}{|}}{N}}-\overset{\overset{H}{|}}{C}-COOH} + H_2O \rightleftharpoons \overset{enzyme}{\longrightarrow}$$

$$\underset{\text{Alanine}}{H_2N-\underset{\overset{|}{CH_3}}{\overset{\overset{H}{|}}{C}}-COOH} + \underset{\text{Glycine}}{H_2N-\underset{\overset{|}{H}}{\overset{\overset{H}{|}}{C}}-COOH}$$

b. A reaction that is catalyzed by yeast enzymes:

$$\underset{\text{Pyruvic acid}}{CH_3-\overset{\overset{O}{\|}}{C}-COOH} \xrightarrow[Mg^{+2}, TPP]{enzyme} \underset{\text{Acetaldehyde}}{CH_3-\overset{\overset{O}{\|}}{C}-H} + CO_2$$

c. $HOOC-CH_2-\underset{\underset{\displaystyle COOH}{|}}{\overset{\overset{\displaystyle H}{|}}{C}}-NH_2$ + $HOOC-CH_2CH_2-\overset{\overset{\displaystyle O}{\|}}{C}-COOH$ $\xrightarrow{\text{enzyme}}$

Aspartic acid α-Ketoglutaric acid

$HOOC-CH_2-\overset{\overset{\displaystyle O}{\|}}{C}-COOH$ + $HOOC-CH_2CH_2-\overset{\overset{\displaystyle NH_2}{|}}{C}H-COOH$

Oxaloacetic acid   Glutamic acid

d. ADP + $\underset{\underset{\displaystyle\underset{\displaystyle COOH}{|}}{\underset{\displaystyle CH_2=C}{|}}}{\overset{\overset{\displaystyle OH}{|}}{\underset{\displaystyle O}{\overset{\displaystyle HO-P=O}{|}}}}$ $\underset{\xleftarrow{\hspace{1em}}}{\xrightarrow{\text{enzyme}}}$ ATP + $CH_3-\overset{\overset{\displaystyle O}{\|}}{C}-COOH$ + H_2O

Phosphoenolpyruvic   Pyruvic acid
acid

Answers:

1. Urease; maltase.
2. a. hydrolase—protease
   b. lyase—decarboxylase
   c. transferase—transaminase
   d. transferase—kinase (transphosphatase)

Enzymes as Functioning Units

Many enzymes, such as chymotrypsin, trypsin, and lysozyme, owe their reactivity only to their specific protein structure. However, other enzymes are conjugated proteins which require the presence of some specific nonprotein unit before they can become an active enzyme. The protein part of a conjugated enzyme is known as an **apoenzyme**, and its nonprotein part is called a **cofactor**. The active enzyme unit which results when the apoenzymes and the cofactor(s) are combined is called a **holoenzyme** (a complete enzyme). If the cofactor is firmly attached to the apoenzyme, it is also known as a **prosthetic group**.

There are two different kinds of cofactors. If the cofactor is an organic unit, it is commonly called a **coenzyme**. If the cofactor is a metal ion, it is called a **metal ion activator**. Some enzymes require both kinds of cofactors. Some of the metal ion activators are Na^+, K^+, Mg^{2+}, Ca^{2+}, Mn^{3+}, Co^{2+}, and Zn^{2+}. Many of the trace metal ions found in the human body are probably important in enzyme reactions. Some of the vitamins, or their derivatives, are coenzymes. Since vitamins and metal ions (minerals) are essential for proper enzyme function, it is easy to see why they are essential parts of the diet.

Some enzymes are secreted by cells in inactive forms known as **proenzymes** or **zymogens**. When the inactive form reaches its required destination, other enzymes remove a peptide section producing the active enzyme. The digestive enzymes trypsin and pepsin break down proteins into amino acids. Pepsinogen is secreted by cells in the stomach and is then converted into pepsin, the active enzyme, by the action of enzymes and hydrochloric acid in the stomach. Trypsinogen is produced by the pancreas and is converted into trypsin by enzymes in the small intestine. You can understand why these digestive enzymes must be synthesized in an inactive form, otherwise they would digest the proteins in the cells where they were produced.

Vitamins as Cofactors

We have observed so far that the human body requires carbohydrates, proteins, fats, and minerals for good health. However, if these were the only materials present, the body would soon stop functioning. What else, then, is needed? Vitamins are the last necessary nutrient. Vitamins occur in small amounts in nearly all foods. They are essential organic compounds that for the most part cannot be synthesized by the body. A lack of one or more vitamins can produce **vitamin deficiency diseases**. Most of these diseases are a direct result of poor nutrition, and as a result they are frequently seen in underdeveloped countries.

Vitamins can be classed as being either **water-soluble vitamins** or **fat-soluble vitamins**. The fat-soluble vitamins (A, D, E, and K) were discussed in Chapter 20. Because of their solubility in fatty tissue, they remain in the body for a considerable time once ingested. The fat-soluble vitamins do not act as enzyme cofactors. The water-soluble vitamins are vitamin C, biotin, folic acid, thiamine, pyridoxal, pantothenic acid, cyanocobalamin, and riboflavin. They are not retained by the body for extended periods and must be replenished through the diet on a regular basis. The water-soluble vitamins do serve as enzyme cofactors, so it is appropriate to study them as you learn about enzymes.

Vitamin C   Early seamen, while on long voyages, often developed bleeding gums and general muscular weakness, symptoms of the disease scurvy. In the middle 1700's, the British discovered that fresh fruit and vegetables in the diet could cure and prevent scurvy. Since fresh vegetables could not be carried on long voyages, barrels of lemons and limes became staples on British ships, and British sailors soon became known as "limeys." In the 1930's, it was discovered that fruits and vegetables contained ascorbic acid (vitamin C), the nutrient that prevented scurvy.

Most animals can synthesize vitamin C from carbohydrates in their diets. However, humans, guinea pigs, and monkeys cannot, so they must obtain the vitamin in their diets. Vitamin C can be destroyed by cooking vegetables in boiling water or baking them at high temperature. In the United States, many people obtain vitamin C from frozen or canned orange juice or powdered drink mixes that have vitamin C added. The way the body uses vitamin C is unclear, but it is thought to function as a coenzyme in the transfer of hydroxyl groups (−OH) during the formation of the structural protein, collagen.

Dr. Linus Pauling, winner of Nobel Prizes in Chemistry (1954) and in Peace (1962), announced in the early 1970's that ingesting large amounts of vitamin C would prevent or at least lessen the severity of the common cold. The sale of vitamin C tablets has skyrocketed since that time. However, little clinical evidence seems to support his claim, but since the public has learned that large amounts of this vitamin seem to have little harmful effect, they continue taking it, hoping for the best. One side effect accompanying the use of massive amounts of this vitamin is diarrhea.

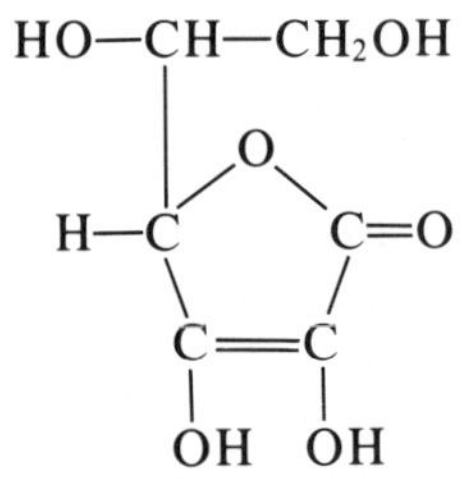

$$HO-CH-CH_2OH$$

Vitamic C (Ascorbic acid)

Biotin   Biotin bonds covalently to carboxylase enzymes and acts as a carrier for the removal or addition of carbon dioxide in processes such as the synthesis of fatty acids in the body. Evidence of a deficiency of this vitamin are skin disorders such as dermatitis. Good sources of biotin are eggs, nuts, and beef liver, but since it is also synthesized by intestinal bacteria, deficiencies are rare.

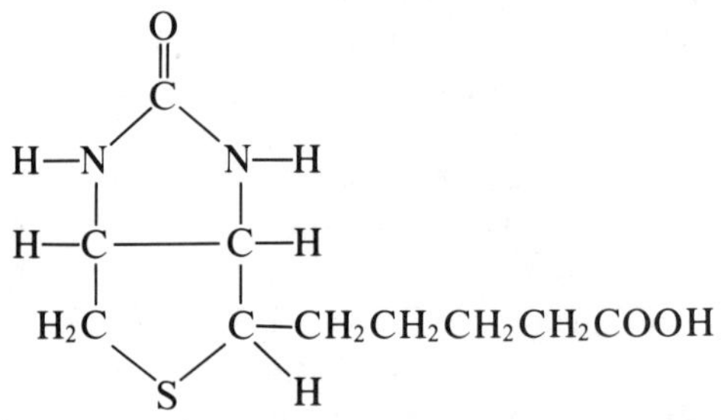

$$C-CH_2CH_2CH_2CH_2COOH$$

Biotin

Folic Acid   Folic acid is converted by the body into the coenzyme tetrahydrofolic acid. The transformation requires the addition of hydrogen to two double bonds in the vitamin. As a coenzyme, it acts with enzymes that catalyze the synthesis of some purines and amino acids by assisting in the transfer of CH_3-, $-\overset{\overset{O}{\|}}{C}-H$, and $-CH_2OH$ groups. Deficiency of this vitamin leads to anemia and other blood disorders, gastric disorders, and a general slowing down of cell growth and division. Good sources of this vitamin are soybeans, eggs, wheat germ, mushrooms, and liver.

Folic acid

Tetrahydrofolic acid (Coenzyme)

Thiamine (Vitamin B$_1$ or Thiamine Chloride)   This vitamin is the last of the water-soluble vitamins that requires only minor changes before it can act as a coenzyme. The coenzyme form is known as thiamine pyrophosphate (TPP). The coenzyme acts with certain enzymes that remove carbon dioxide from compounds. It functions with the decarboxylase enzyme used by cells to convert pyruvic acid, a breakdown product of glucose, to acetyl CoA and carbon dioxide. A deficiency of thiamine can cause beriberi and circulatory malfunctions. Enriched bread, beef, pork, eggs, beans, peas, and whole grain cereals are good sources of this vitamin. Cooking can destroy much of this vitamin in foods.

Thiamine chloride
(vitamin B$_1$)

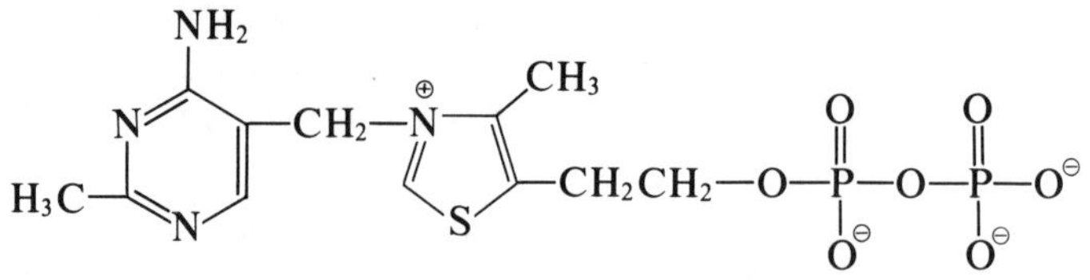

Thiamine pyrophosphate (TPP)
(Coenzyme)

Pyridoxine, Pyridoxamine, and Pyridoxal (the Vitamin B$_6$ Family)

In 1938, pyridoxine was isolated from plants and classified as vitamin B$_6$. However, in 1944, it was observed that pyridoxal and pyridoxamine isolated from animal tissues also served the same vitamin function, but were even more active in their effect. Since all three serve the same vitamin function, they are considered vitamers of the vitamin B$_6$ family.

Any of the three may be used in the body to form the two coenzymes pyridoxal phosphate and pyridoxalamine phosphate. These coenzymes serve with enzymes that either transfer amino groups or remove carbon dioxide from compounds. These enzymes are especially important in amino acid and fatty acid metabolism. A deficiency of this vitamin can produce dermatitis, anemia, weight loss, and convulsions in infants. Good sources of this vitamin are cereals, liver, beans, peas, and milk.

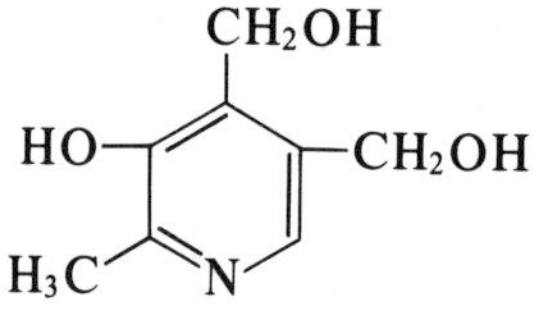

| Pyridoxine | Pyridoxal | Pyridoxamine |

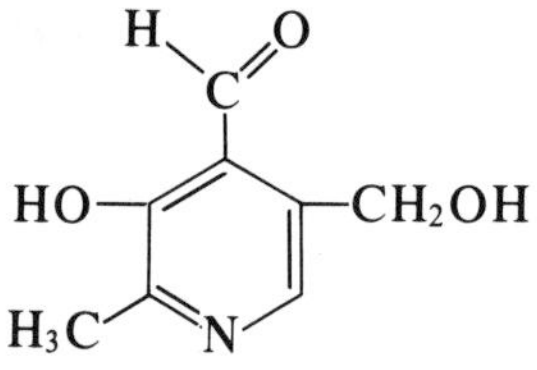

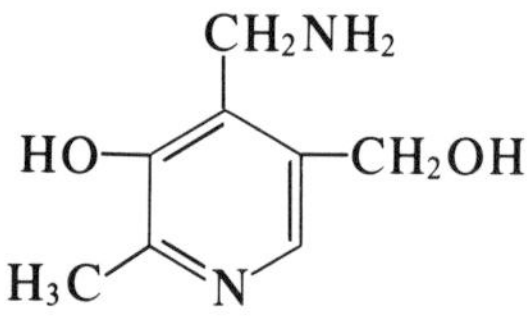

vitamin B$_6$ vitamers

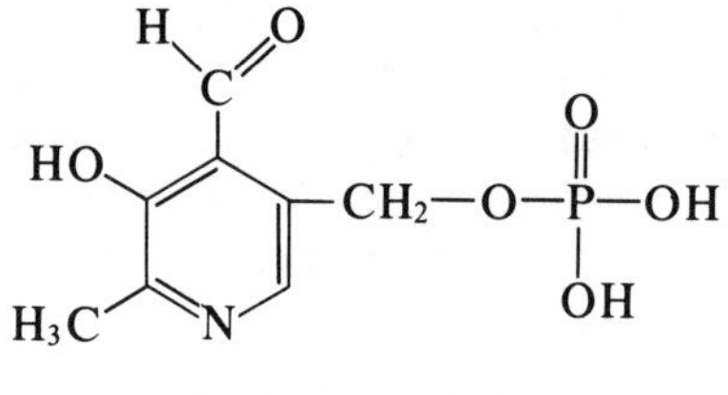

Pyridoxal phosphate
(Coenzyme)

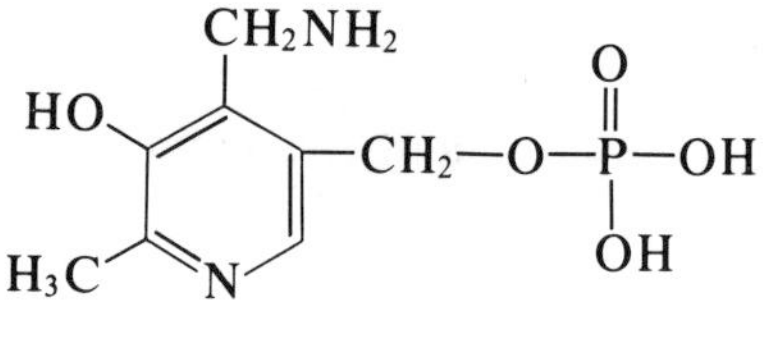

Pyridoxamine phosphate
(Coenzyme)

Pantothenic Acid

Pantothenic acid is required by the body in the synthesis of coenzyme A, one of the central compounds in the metabolism of carbohydrates, proteins, and fats into water, carbon dioxide, and energy. When coenzyme A is used by cells, a $CH_3-\overset{O}{\overset{\|}{C}}-$ group is substituted on the molecule and it is called acetyl coenzyme A (acetyl CoA). Acetyl CoA is also involved in the synthesis of fatty acids and steroids. Signs of pantothenic acid deficiency are digestive tract disorders, cardiovascular problems, general weakness, mental confusion, and a reduced resistance to infection. Good sources of this vitamin are milk, eggs, liver, and yeast.

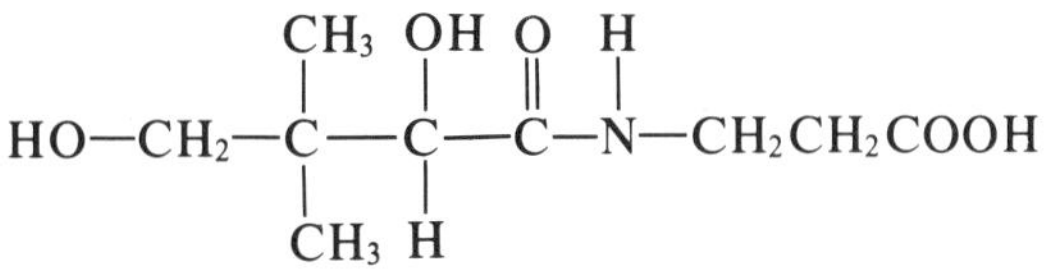

$$HO{-}CH_2{-}\underset{\underset{CH_3}{|}}{\overset{\overset{CH_3}{|}}{C}}{-}\underset{\underset{H}{|}}{\overset{\overset{OH}{|}}{C}}{-}\overset{\overset{O}{\|}}{C}{-}\underset{\underset{H}{|}}{N}{-}CH_2CH_2COOH$$

Pantothenic acid

The structure of coenzyme A is shown below. The modification that converts it to acetyl coenzyme A replaces the hydrogen on the –SH group with an acetyl group. Notice how pantothenic acid is incorporated into the structure of the coenzyme.

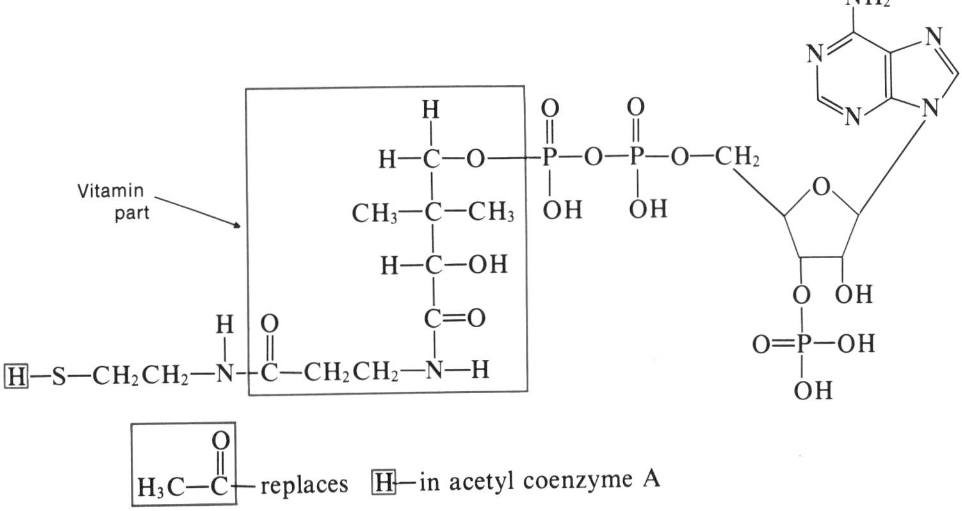

$$H_3C{-}\overset{\overset{O}{\|}}{C}{-} \text{ replaces } \boxed{H}{-} \text{ in acetyl coenzyme A}$$

Coenzyme A is also called coenzyme ASH (CoASH) since the –SH group is the one that functions in the transfer of acetyl groups.

Cyano-cobalamin (Vitamin B_{12})

Before 1926, pernicious anemia was often a fatal disease. It primarily affected the elderly, though on rare occasions it also affected children. Pernicious anemia is characterized by enlarged, easily ruptured red blood cells, considerable weight loss, and very poor muscle coordination. During the 1920's, it was learned that the anemic condition could be reversed if patients consumed large quantities of near raw liver, a food that is now known to be a good source of vitamin B_{12}. It was a deficiency of vitamin B_{12} that brought on the disease.

In 1948, vitamin B_{12} was isolated from beef liver. Tedious chemical analysis has shown it to be a large, complex molecule built around a central cobalt ion, as shown in Figure 24.1.

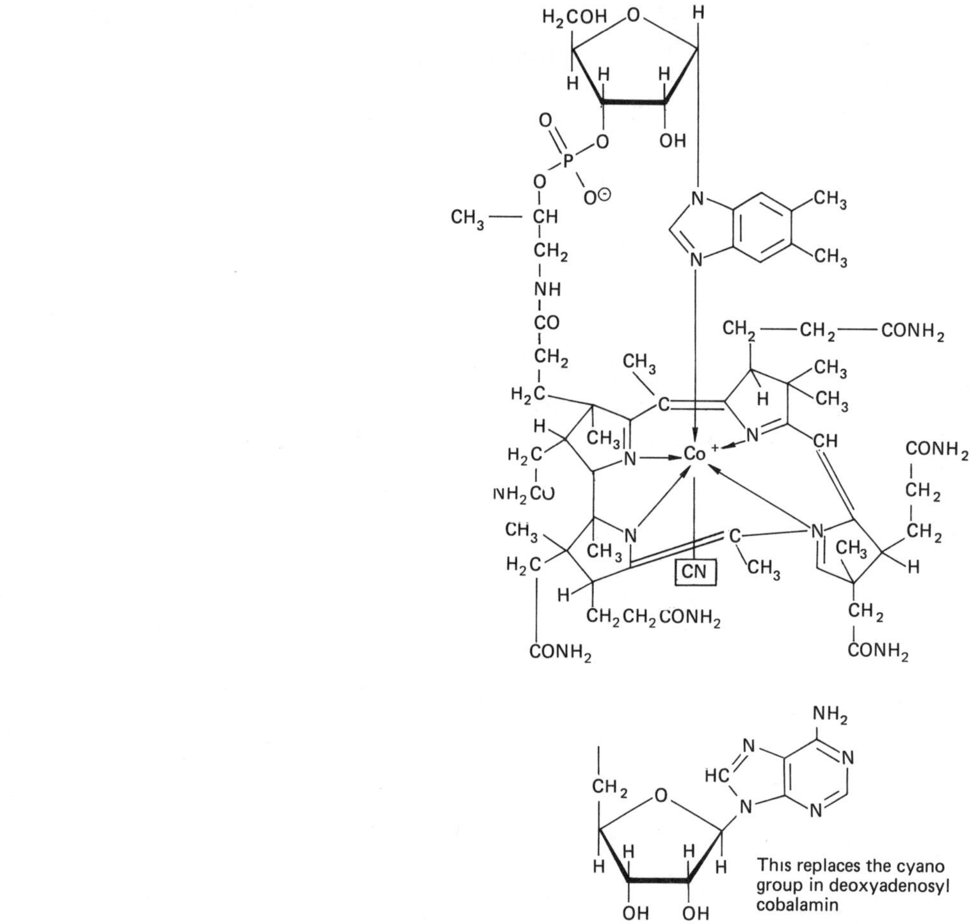

Fig. 24.1. Cyanocobalamin (vitamin B_{12}) and the modification that converts it to the coenzyme deoxyadenosyl cobalamin.

Vitamin B_{12} is used by the body to synthesize deoxyadenosyl cobalamin, a coenzyme required in carbohydrate and lipid metabolism and in the synthesis of DNA and RNA. The best sources of vitamin B_{12} are liver, lean meats, milk products, and eggs. However, about 40% of the body's requirement of this vitamin is obtained from bacteria in the intestines.

Today, pernicious anemia is relatively uncommon. Strict vegetarians are more susceptible to the disease because plants do not provide an adequate supply of vitamin B_{12}. Some elderly people suffer a deficiency of the vitamin, not because it is missing from their diets, but because it is not adequately absorbed by the body. Vitamin B_{12} requires the presence of a certain protein in the intestinal tract, called the "intrinsic factor," if it is to be absorbed by the body. An insufficient level of the "intrinsic factor" can bring on the symptoms of vitamin B_{12} deficiency. The vitamin can be administered by intramuscular injection to overcome the problem of reduced absorption. Vitamin B_{12} is stored in the liver in sufficient quantities to meet the needs of the body for three to five years.

Riboflavin (Vitamin B_2)   Riboflavin is used in the synthesis of two coenzymes, flavin mononucleotide (FMN) and flavin adenine dinucleotide (FAD). These coenzymes assist in the transfer of two electrons and two protons in certain enzyme-catalyzed oxidation–reduction reactions that are important in the metabolism of carbohydrates, amino acids, and fats. The symptoms of riboflavin deficiency are dermatitis, impaired vision, and sores within the mouth. Good sources of the vitamin are milk, liver, eggs, yeast, and green vegetables. The structure of riboflavin, FMN, and FAD are shown below.

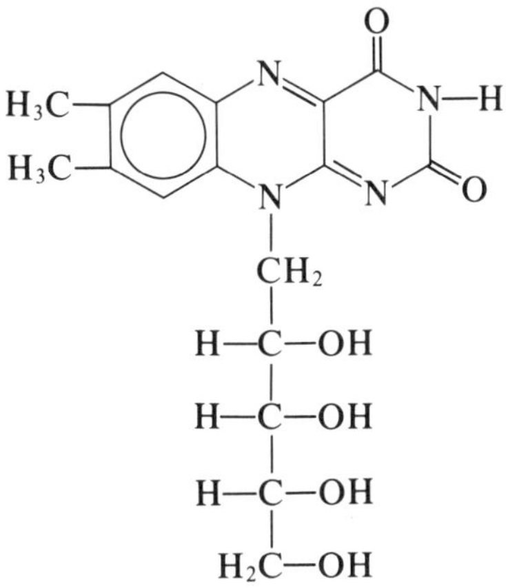

Riboflavin (vitamin)

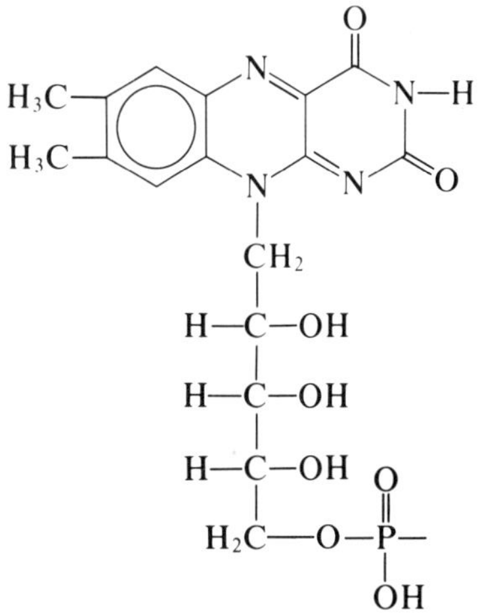

Flavin mononucleotide (FMN) (coenzyme)

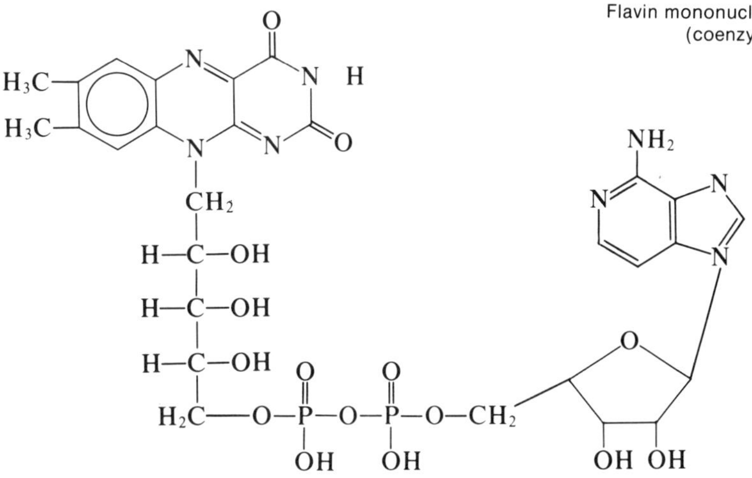

Flavin adenine dinucleotide (FAD) (coenzyme)

An example of an oxidation–reduction reaction using an enzyme–FAD coenzyme combination is the conversion (oxidation) of succinic acid to fumaric acid, an important step in the Krebs cycle. Note the transfer of two hydrogen atoms from succinic acid to the enzyme–coenzyme unit in the reaction.

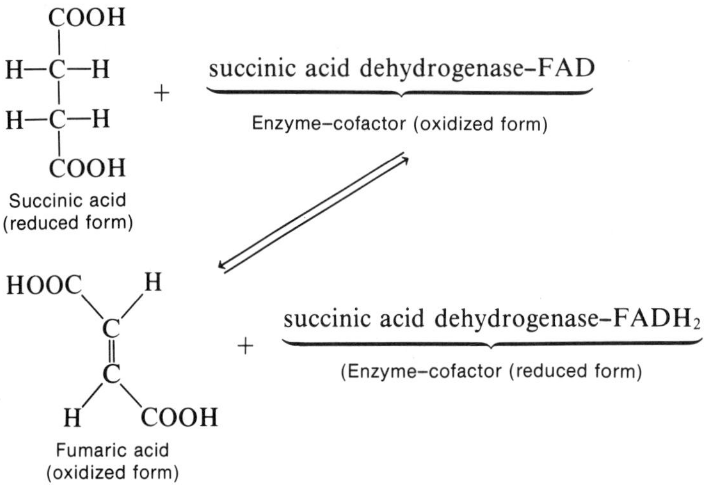

COOH
|
H—C—H   +   succinic acid dehydrogenase–FAD
|
H—C—H   Enzyme–cofactor (oxidized form)
|
COOH
Succinic acid
(reduced form)

HOOC   H
    \   /
     C
     ‖
     C
    /   \
  H   COOH   +   succinic acid dehydrogenase–FADH$_2$
                                (Enzyme–cofactor (reduced form)
Fumaric acid
(oxidized form)

The above reaction can also be written in an abbreviated form just as was done with the NAD$^\oplus$–NADH + H$^\oplus$ coenzyme complexes. Only the substrate and product are shown in the equation, and the FAD–FADH$_2$ coenzymes are shown above the double arrow.

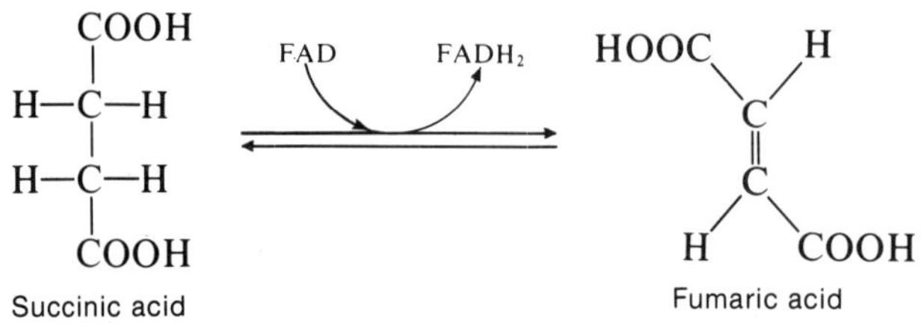

COOH   FAD   FADH$_2$ HOOC   H
| \   /
H—C—H   C
| ‖
H—C—H   C
| /   \
COOH   H   COOH
Succinic acid   Fumaric acid

Niacin   Niacin is composed of two compounds (vitamers) that display vitamin ac-
(Vitamin B$_3$)   tivity, nicotinic acid and nicotinamide.

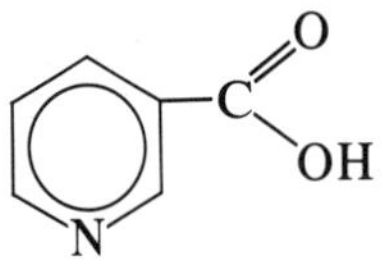

Nicotinic acid 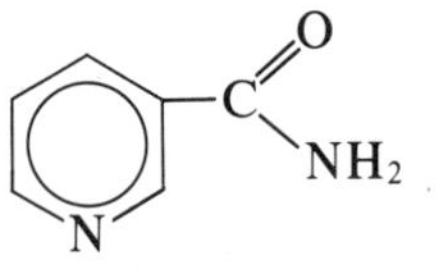   Nicotinamide

Severe niacin deficiency is a principal cause of pellagra, a disease characterized by general weakness, dermatitis, indigestion, diarrhea, and nervous disorders.

The deficiency usually occurs in areas where maize (Indian corn) forms a major part of the diet. Niacin in maize is tightly bound and cannot be absorbed in the intestinal tract. Also, maize contains only low levels of tryptophan, an amino acid that can be used by the body to synthesize small but inadequate amounts of niacin.

Niacin is required by the body to make nicotinamide adenine dinucleotide (NAD$^\oplus$) and its phosphate derivative (NADP$^\oplus$). Both serve as coenzymes in oxidation–reduction reactions. Good sources of this vitamin are yeast, lean red meat, liver, and beans.

The structure of NAD$^\oplus$, along with the modification that converts it to NADP$^\oplus$, is given below.

The NAD$^\oplus$ and NADP$^\oplus$ coenzymes function in a manner similar to FMN and FAD. Pyruvic acid is reduced by an enzyme–NADH + H$^+$ complex to lactic acid. Notice, the reduction adds two hydrogen atoms to lactic acid, which come from NADH + H$^+$.

$$CH_3-\overset{\overset{\displaystyle O}{\|}}{C}-COOH \ + \ \text{lactic acid dehydrogenase-NADH}_2 \ \underset{\xrightarrow{}}{\overset{\text{muscle tissue}}{\rightleftharpoons}}$$

$$CH_3-\overset{\overset{\displaystyle OH}{|}}{\underset{\underset{\displaystyle H}{|}}{C}}-COOH \ + \ \text{lactic acid dehydrogenase-NAD}^+$$

Lactic acid

In abbreviated form, the equation is

$$CH_3-\overset{\overset{\displaystyle O}{\|}}{C}-COOH \ \rightleftharpoons \ CH_3-\overset{\overset{\displaystyle OH}{|}}{\underset{\underset{\displaystyle H}{|}}{C}}-COOH$$

NADH₂ → NAD⁺

Check Test Number 2

1. The enzyme pyruvic acid decarboxylase requires the presence of magnesium ion, Mg^{2+}, to function. How would you classify Mg^{2+} as an enzyme cofactor?
2. Why are the proteolytic enzymes pepsin and trypsin secreted in an inactive form?
3. Give the name of the vitamin that is described by each of the following:
   a. Contains a cobalt ion.
   b. Deficiency of this vitamin causes beriberi.
   c. It is used by the body to make coenzyme A.
   d. The coenzyme FMN is made from this vitamin.

Answers:
1. Mg^{2+} would be classed as a metal-ion activator.
2. If they were initially prepared in an active form in the secretory cells, they would catalyze the digestion of protein within the cells.
3. a. cyanocobalamin (vitamin B_{12})
   b. thiamine (vitamin B_1)
   c. pantothenic acid
   d. riboflavin (vitamin B_2)

Enzyme Action

Enzymes are biological catalysts. They increase the rates of reactions by lowering the activation energies (Chapter 11). For example, the decomposition of hydrogen peroxide, H_2O_2, to form oxygen, O_2, and water has an activation energy of 18 kcal/mole of H_2O_2 in the absence of a catalyst. The same reaction occurs in cells, but in the presence of an enzyme called catalase. Catalase reduces the activation energy to around 7 kcal/mole, a substantial reduction which in turn allows the reaction to take place quickly at body temperature. The effect of catalase on the activation energy of this reaction is shown in Figure 24.2. Catalase increases the rate of the decomposition reaction about 100-million-fold compared to the uncatalyzed reaction carried out under similar conditions in the laboratory.

The efficiency of an enzyme is given by its **turnover number**, the number of substrate molecules (reactant molecules) that one enzyme unit can transform in one minute. At body temperature, the turnover number for catalase is 5.6 million. This means one catalase molecule can pick up, decompose, and release 5.6 million H_2O_2 molecules each minute. The highest turnover number measured to date is 36 million for carbonic anhydrase, the enzyme that catalyzes the conversion of carbon dioxide and water to carbonic acid, H_2CO_3, in the blood.

An enzyme catalyzes a reaction by changing the pathway that leads from reactants to products; that is, it changes the mechanism of the reaction. It does this by providing a special surface for the substrate molecule (or molecules) to fit into so that bonds can be broken and formed much easier. The special surface is called the **active site** of the enzyme. The active site is usually a small crevice or cavity formed in the tertiary structure of the protein–remember, enzymes are proteins. The active site on an enzyme has a specific geometry or shape that will accommodate only specific substrate molecules that can fit into it. The substrate molecule, therefore, must have a structure that comple-

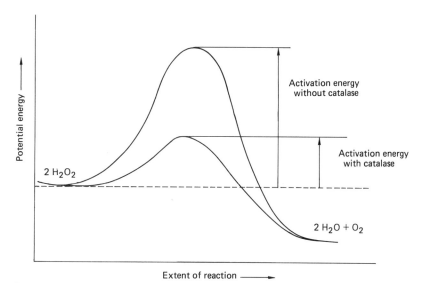

Fig. 24.2. Catalase increases the rate of decomposition of hydrogen peroxide, H_2O_2, by reducing the activation energy of the reaction. At 38°C, the reaction is 10^8 times faster with catalase than without it.

ments the structure of the active site, so the two fit together like a key in a lock. In fact, the idea that the enzyme and substrate molecules must fit together in a specific way is the basis for the **lock and key theory** of enzyme action. The enzyme is the key since it "unlocks" or changes the substrate molecule.

The cofactors (coenzymes or metal-ion activators) that are required by enzymes to function may either be part of the active sites or they may bond elsewhere to the surface of the enzymes and assist in a necessary but less direct way.

Let us examine a step-by-step sequence of events that would be typical for an enzyme-catalyzed reaction that breaks a substrate molecule into two parts. The mechanism of the reaction can be described by three equations that represent the three steps of the process. Let E stand for the enzyme and S–S stand for the substrate. Notice that each of the three steps is a reversible reaction.

Step 1. $E + S-S \rightleftharpoons E(S-S)$ Formation of the enzyme–substrate complex.

Step 2. $E(S-S) \rightleftharpoons E(S-S)^*$ Formation of the high-energy transition state complex, $E(S-S)^*$.

Step 3. $E(S-S)^* \rightleftharpoons E + S + S$ The high-energy transition state complex breaks apart forming free enzyme and product.

If all three steps are added together, the overall reaction is simply the dissociation of the substrate molecule. Since an enzyme catalyzes the reaction, "enzyme" is written over the arrows.

$$S-S \underset{\longleftarrow}{\overset{enzyme}{\rightleftharpoons}} S + S$$

Now let us look at each step in greater detail, paying close attention to the way the substrate and enzyme interact.

Step 1. The substrate becomes attached to the enzyme by fitting into the active site. Another region on the enzyme's surface, called the **holding site**, grips the substrate molecule in such a way that the substrate-active site fit is optimum for reaction.

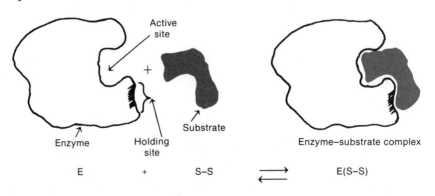

Active site			
Enzyme	Holding site	Substrate	Enzyme–substrate complex
E	+	S–S	E(S–S)

Step 2. Once the enzyme–substrate complex is formed, interactions between the substrate and the active site weaken a specific bond in the substrate molecule so it can break more easily, and the energy required to weaken the bond represents a major part of the activation energy for the reaction. The enzyme–substrate complex with the weakened substrate bond is the short-lived transition state complex in the reaction. It represents the point of highest energy in the dissociation of the substrate molecule.

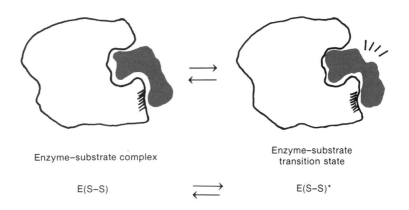

Enzyme–substrate complex   Enzyme–substrate transition state

E(S–S)   E(S–S)*

Step 3. The enzyme–substrate transition state complex then dissociates once the substrate molecule breaks apart. This leaves the enzyme free to catalyze the dissociation of yet another molecule.

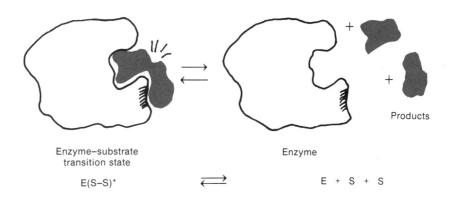

Enzyme–substrate transition state   Enzyme   Products

E(S–S)*   E + S + S

Though it takes a few minutes to read this description of an enzyme-catalyzed reaction, the entire process can take place in less than a ten-thousandth of a second in the body.

The lock and key theory of enzyme action provides a good description of many enzyme-catalyzed reactions. However, the action of some enzymes is better described by the **induced fit theory**, which is a modification of the lock and key idea. Some enzymes can catalyze reactions involving several different substrates. No one substrate represents a perfect fit for the active site on the enzyme. But with enzymes of this type, the active site is somewhat flexible, and as a substrate enters the active site, it "induces" the enzyme to change shape slightly so the active site "fits" the substrate properly. A different substrate would induce a different change in the shape of the active site so it too can achieve a proper fit.

Enzyme Specificity

Enzymes are very specific catalysts. At the extreme, some enzymes interact with only one substrate and catalyze only one reaction with that substrate. This high degree of specificity is due to the unique character of the active site on the enzyme. For this reason, many cells contain over 1000 different enzymes, each with a specific task in the total chemistry of the cell.

There are four general classes of enzyme specificity:

1. *Absolute Specificity:* There are a limited number of enzymes that catalyze only one reaction with only one substrate. An example is the enzyme sucrase, which only catalyzes the hydrolysis of sucrose to form glucose and fructose. It does nothing else.

2. *Stereochemical Specificity:* Some enzymes will catalyze reactions involving substrates with only a specific spatial geometry. In humans, there are enzymes that catalyze reactions involving only D–sugars or L–amino acids. L–sugars and D–amino acids cannot be used by the body. We do not have enzymes for these isomers. Other enzymes will interact only with *cis*-alkenes and not the *trans*-isomers.

3. *Group Specificity:* Certain enzymes only act on groups that have specific surrounding structures. For example, the digestive enzyme trypsin will only hydrolyze peptide bonds ($-\overset{\text{O}}{\overset{\|}{\text{C}}}-\overset{\text{H}}{\overset{|}{\text{N}}}-$) that are adjacent to lysine or arginine residues on a protein chain.

4. *Linkage Specificity:* There are enzymes that will catalyze reactions involving only one specific kind of chemical bond. Esterases will catalyze the hydrolysis of ester links ($-\overset{\text{O}}{\overset{\|}{\text{C}}}-\text{O}-\overset{|}{\underset{|}{\text{C}}}-$) in esters, lipids, and even in complex nonbiological esters.

Check Test Number 3

1. What is the principal difference between the lock and key theory of enzyme action and the induced fit theory?
2. Which class of enzyme specificity is described by each of the following?
   a. The enzyme will interact only with *cis*–unsaturated fatty acids.
   b. Chymotrypsin cleaves peptide bonds adjacent to aromatic amino acids.
   c. The enzyme only converts maltose into two glucose molecules.
   d. Lipase will catalyze the hydrolysis of many different esters.

Answers:

1. The active site in the lock and key theory is fixed in its size and shape, but in the induced-fit theory, the enzyme can be induced by the substrate to change shape slightly, allowing the substrate to achieve a proper fit in the active site.
2. a. stereochemical
   b. group
   c. absolute
   d. linkage

Factors That Influence Enzyme Action

Enzyme-catalyzed reactions proceed at rapid rates in most cases, a fact reflected in the large turnover numbers for these enzymes. Several factors can influence the rate of these reactions, factors that in turn can be interpreted as affecting the activity of an enzyme. The rates of enzyme-catalyzed reactions are affected by temperature, pH, and the concentrations of the substrate, enzyme, and the necessary cofactors. Let us examine each one of these factors in greater detail.

Temperature The rate of an enzyme-catalyzed reaction increases as the temperature of the reaction medium increases, but only up to a point. For every enzyme, there is one temperature at which the reaction rate will be at a maximum, and that temperature is called the **optimum temperature** for that enzyme; see Figure 24.3. The rate of reaction at temperatures above or below the optimum temperature will be slower. As it turns out, the optimum temperature for most of the enzymes in the body is approximately 37°C (98.6°F), normal body temperature.

At temperatures above 60°C, many enzymes will denature, destroying their secondary and tertiary structures, which in turn destroys the active sites. At these temperatures, the activity of the enzyme is reduced to zero. When milk

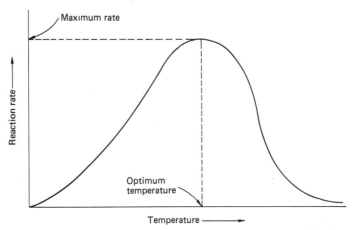

Fig. 24.3. The effect of temperature on the rate of an enzyme-catalyzed reaction. The temperature at which the rate is at a maximum is the optimum temperature for that enzyme.

is pasteurized, or when food is canned, the high temperatures used in these processes destroy not only bacteria, but also the enzymes. Low temperatures do not denature enzymes. Tissue samples, sperm, or other biological materials can be stored at subfreezing temperatures for years without markedly reducing the catalyzing power of the enzymes present in the samples.

pH Just as was seen with the effect of temperature on enzyme activity, there is an **optimum pH** at which an enzyme's activity is greatest. At pH's above and below the optimum pH, the activity of the enzyme is reduced, and reaction rates are slower, as shown in Figure 24.4. The optimum pH for pepsin, a proteolytic enzyme in the stomach, is around 2, close to that of the acid environment in the stomach. Trypsin, another proteolytic enzyme, has an optimum pH of around 8, close to that of the upper intestinal tract where it is found. If trypsin were placed in the highly acidic environment of the stomach, it would likely denature and lose its catalytic activity.

Substrate Concentration For any given enzyme, there will be some maximum number of substrate molecules that it can transform or turn over each minute. If this number is very large, the reaction will have a high maximum rate. Now let us consider a single enzyme molecule in a solution of substrate molecules. If the concentration of the substrate is very low, the number of substrate molecules that can be "caught" by the enzyme each minute will also be low—lower than the number of substrate molecules it is capable of handling during that time. The rate of reaction, as measured by the number of substrate molecules changed each minute, will also be low, lower than the maximum rate. If the concentration of substrate is increased, the rate of the reaction will also increase until a point

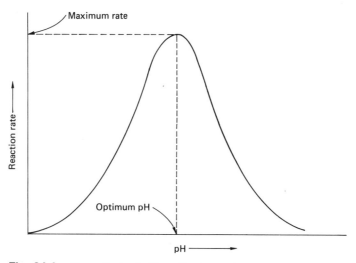

Fig. 24.4. The effect of pH on the rate of an enzyme-catalyzed reaction. The pH at which the rate is at a maximum is the optimum pH for that enzyme.

is reached where the enzyme is working as fast as it can; that is, it is transforming its maximum number of substrate molecules each minute. At this point, the enzyme is said to be saturated, and any further increase in the concentration of the substrate will not increase the rate of reaction. The enzyme can work no faster. The effect of substrate concentration on the rate of an enzyme-catalyzed reaction is shown in Figure 24.5.

Enzyme Concentration   As the concentration of an enzyme increases in a solution, the rate at which the substrate is changed increases also. Look at it this way: If one enzyme

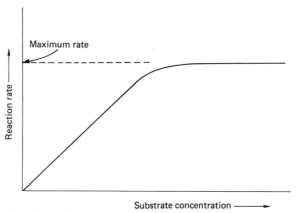

Fig. 24.5. The effect of substrate concentration on the rate of an enzyme-catalyzed reaction.

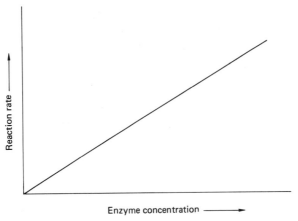

Fig. 24.6. The effect of enzyme concentration on the rate of an enzyme-catalyzed reaction when the substrate concentration is high.

molecule can transform a million substrate molecules each minute, then two enzyme molecules will handle twice that number of substrate molecules in the same time. The substrate will be consumed twice as fast, which doubles the rate of reaction. The way the rate of an enzyme-catalyzed reaction is affected by the concentration of enzyme is shown in Figure 24.6. As long as sufficient substrate is present, the reaction rate is proportional to the enzyme concentration.

Cofactor Concentration
Cofactors are essential parts of active enzymes. If the proper cofactor is absent, an enzyme will have little or no activity. For this reason, cofactors must be present in sufficient concentration to "activate" the enzyme molecules. If they are not, then only a fraction of the enzyme concentration will be effective in catalyzing reactions.

Enzyme Inhibition

Enzyme-catalyzed reactions can be slowed down or completely stopped by substances that reduce the ability of an enzyme to function. These substances are called **inhibitors**, since they inhibit enzyme action. There are four major types of inhibitors: competitive, noncompetitive, end-product, and irreversible. Let us look at each one to see how it affects an enzyme.

Competitive Inhibition
Competitive inhibitors are molecules that look very much like the substrate molecules that a given enzyme is designed to handle. They are able to enter the active site of an enzyme, just like the correct substrate would, but once there, the enzyme is unable to cause the proper reaction to occur so the inhibitor

molecule is not released quickly. While it stays in the active site, it inhibits the enzyme from catalyzing the reaction with the correct substrate. The reason this type of inhibition is called competitive is because both the substrate and the inhibitor compete for the same active site. As more active sites are occupied by inhibitors, the concentration of working enzyme decreases, and the rate of the reaction decreases. Fortunately, competitive inhibition is reversible, and increasing the concentration of substrate will reduce the effectiveness of the inhibitor. The substrate can compete better for the active site if its concentration is higher.

A good example of competitive inhibition is the enzyme-catalyzed oxidation of succinic acid to fumaric acid in the Krebs cycle. Succinic acid is the substrate, but it has a structure that is very similar to that of malonic acid, a substance that will inhibit the reaction.

Succinic acid
(substrate)

Malonic acid
(inhibitor)

If the malonic acid concentration is very high, it will tie up the active sites on the enzyme molecules, inhibiting the oxidation of succinic acid.

Another example of competitive inhibition describes the action of sulfa drugs in fighting a bacterial infection in the body. Some disease-causing bacteria require folic acid, a water-soluble vitamin, if they are to grow and mutiply. These bacteria can synthesize folic acid within themselves from *para*-aminobenzoic acid in a series of enzyme-catalyzed reactions. If this synthesis of folic acid can be slowed down or stopped, bacterial growth will be reduced so the normal defense mechanisms of the body will be more able to destroy those that remain. Sulfanilamide, a sulfa drug, has a structure that is very similar to *para*-aminobenzoic acid. It acts as a competitive inhibitor toward the enzyme that uses *para*-aminobenzoic acid as a substrate. Once this enzyme bonds to sulfanilamide, its ability to use *para*-aminobenzoic acid is inhibited, and the production of folic acid is interrupted. The administration of large quantities of sulfanilamide to patients will thus inhibit specific bacterial enzymes, and in this way fight the infection. Of course, patients require folic acid too, but it is obtained in their diets.

O=C—OH ... NH₂ ... O=S=O ... NH₂

Para–aminobenzoic   Sulfanilamide
acid

Noncompetitive
Inhibition

Noncompetitive inhibitors combine reversibly with some part of an enzyme other than the active site. This will cause a change in the three-dimensional structure of the enzyme and greatly reduce its ability to act as a catalyst. Since the activity of the enzyme is reduced, in a noncompetitive way, increasing the substrate concentration will not increase the rate of reaction as was seen in competitive inhibition.

End-Product
Inhibition
(Feedback
Inhibition)

Cells continually synthesize materials that are needed for proper function and growth. Most often each synthesis proceeds through a series of steps that will lead to the required product. Each step is catalyzed by an enzyme, and the product of one step is used as the substrate for the next. Often, the final product of a sequence of cellular reactions will inhibit the enzyme needed to catalyze the first step of the sequence. As the end-product concentration increases, more of the enzyme molecules used in the first step become inhibited and the rate at which the end product is formed slows down. In **end-product inhibition,** the rising concentration of the product of a series of reactions brings about an inhibition of its own synthesis.

The enzyme that is inhibited is called a **regulatory enzyme** or an **allosteric enzyme.** Allosteric enzymes are more complex than ordinary enzymes. They contain not only an active site for the substrate molecule, but also a second site for the inhibitor molecule, called the **allosteric site**; Figure 24.7. If the allosteric site is occupied by the inhibitor, the active site ceases to function. Once the end-product (inhibitor) concentration drops in the cell, the inhibitor can be released from the allosteric site, allowing the enzyme to again be active.

End-product inhibition allows a cell to control the concentration of a substance within reasonably narrow limits. An example of this kind of control is the synthesis of isoleucine, an amino acid, from threonine, another amino acid. The conversion of the one amino acid to the other involves a five-step process in cells. Isoleucine, the end-product of the reactions, inhibits the activity of the allosteric enzyme (E_1) used in the first step. If the concentration of isoleucine gets too high, it will be produced at a much slower rate as more and more E_1 is inhibited.

$$H_2N-\underset{\underset{CH_3}{|}}{\overset{\overset{H}{|}}{C}}-COOH \overset{E_1}{\longrightarrow} \overset{E_2}{\longrightarrow} \overset{E_3}{\longrightarrow} \overset{E_4}{\longrightarrow} \overset{E_5}{\longrightarrow} H_2N-\underset{\underset{\underset{CH_3}{|}}{\overset{}{CH_2}}}{\overset{\overset{H}{|}}{C}}-COOH$$

Isoleucine inhibits E_1 when its
concentration gets too high

Threonine   Isoleucine

Irreversible
Inhibition

The three modes of enzyme inhibition described thus far have one important feature in common. Each is reversible, allowing the enzyme to return to its active role as a catalyst once the inhibitor is released. **Irreversible inhibitors**, on the other hand, combine with an enzyme so firmly that its catalytic ability is destroyed permanently. Toxic heavy metal ions, like Ag^+, Hg^{2+}, and Pb^{2+}, are irreversible inhibitors. They bond very strongly to the protein structure of the enzyme, distorting its shape and eliminating its catalytic activity. You may recall how heavy-metal ions can denature proteins (Chapter 22).

The cyanide ion, CN^-, irreversibly inhibits enzymes that contain iron as a cofactor, such as catalase and cytochrome oxidase, an enzyme involved in the use of molecular oxygen in cells. You can see why cyanide salts are so highly toxic. They inhibit the use of oxygen in cells. Cyanide ions also block the oxygen-carrying capacity of hemoglobin in the blood. Military nerve gases and

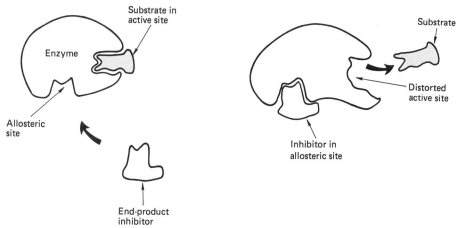

Fig. 24.7. End-product inhibition of an enzyme. The active site can catalyze substrate reactions if the allosteric site is unoccupied. When an end-product inhibitor occupies the allosteric site, the active site is deformed and rendered inactive. The inhibition is reversible.

certain organophosphate insecticides are irreversible inhibitors of enzymes involved in nerve transmission.

Check Test Number 4

1. How is the rate of an enzyme-catalyzed reaction affected by temperature?
2. As the substrate concentration increases in an enzyme-catalyzed reaction, how is the rate of substrate conversion affected?
3. Which types of enzyme inhibitors combine reversibly with enzymes?

Answers:
1. The rate increases as temperature increases until the optimum temperature is reached. Further increases in temperature reduce the rate.
2. As substrate concentration increases, the rate of substrate conversion increases, until the enzyme is saturated. Beyond that point of maximum rate, further increases in substrate concentration will not increase the rate.
3. Competitive, noncompetitive, and end-product inhibitors.

Uses of Enzymes

Clearly, enzymes serve many critical functions in the body, catalyzing nearly every reaction that takes place there. In recent years, researchers have learned to use enzymes to serve our needs in several ways. Enzymes can be powerful diagnostic tools in medicine, and it is not difficult to measure the level of activity of a specific enzyme in blood serum or urine quickly and accurately. Normally, enzymes appear in blood serum, or other extracellular fluids, at very low concentrations, but certain disease states can markedly increase the level of one or more of them. Enzymes principally are found within cells. If disease or injury damages the cell membrane, the enzymes will flow out into the extracellular fluid and eventually enter the bloodstream, where they can then be detected easily.

An elevated serum **acid phosphatase** level in men can indicate a cancer of the prostate that has metastasized. Serum **alkaline phosphatase** is elevated in diseases of the liver, pancreas, lung, and bone. Alkaline phosphatase is actually a group of enzymes that hydrolyze organic phosphate ester bonds under alkaline (basic) conditions. Acute inflammation of the pancreas can be detected through elevated levels of serum **amylase**. If the amylase level is elevated for an extended period of time, it may indicate cysts or cancer of the pancreas. Elevated levels of **gamma-glutamyl transpeptidase** (GGT) aid in the diagnosis of liver abnormalities. It becomes elevated under the same conditons that cause serum alkaline phosphatase to rise, though it is considered to be more sensitive to liver conditions. Two other enzymes that can also be used to monitor liver conditions are serum **glutamic pyruvic transaminase** (SGPT) and serum **lactic dehydrogenase**

(LDH). LDH is a mixture of five isomeric enzymes, called isoenzymes. They can be separated from one another by electrophoresis (Chapter 22). The patterns produced by the five isomers can be used to indicate specific diseases other than just liver abnormalities.

The use of serum enzyme levels in the diagnosis of cardiovascular disease may allow the nature of the disease to be determined as well as its severity. Within three to four hours after suffering a heart attack (myocardial infarction), the level of serum **creatine phosphokinase** (CPK) begins to rise, reaching a peak within 36 hr. It then drops sharply to normal values within two to four days, as shown in Figure 24.8. The extent of the rise reflects the amount of myocardial damage. CPK is the most specific serum enzyme in diagnosing heart attacks. Another enzyme, serum **glutamic oxaloacetic transaminase** (SGOT), also becomes elevated in nearly all patients following an acute heart attack. It begins to rise within about 12 hr, reaching a peak level in about 24 hr, then returns to normal in four to seven days. (SGOT is also greatly increased in acute liver disease and kidney disease and if there is serious muscle damage.) LDH also rises following an acute heart attack, reaching a peak value within four to five days after the attack.

Of course, the use of enzymes is not restricted to medicine. Commercially, proteolytic enzymes, such as papain, are used to tenderize meat. They catalyze the hydrolysis of connective tissue, reducing the toughness of meat. Food processors use enzymes to partially digest food for infants and others with digestive problems. In recent years, certain proteolytic enzymes have been

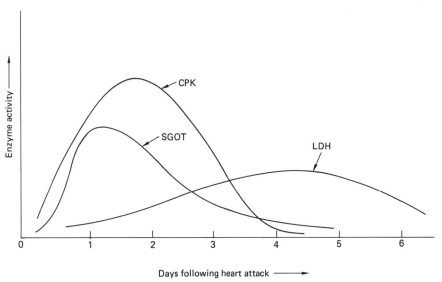

Fig. 24.8. Following a heart attack, the levels of the enzymes CPK, SGOT, and LDH in serum change in a characteristic way. The extent of damage to the heart muscle is related to the extent of elevation of the enzyme activity above normal values.

used to remove cataracts from the eye. Enzymes have been added to laundry detergents to aid the removal of stains, such as blood, from clothing. Eventually, many of these products had to be removed from the marketplace because of allergic reactions suffered by consumers. Enzymes are also involved in fermentation processes to produce ethyl alcohol from grain, sauerkraut, and other foods.

Hormones

Located throughout the body are several ductless glands that comprise the **endocrine system**; Figure 24.9. These glands produce **hormones**, substances that act as messengers to stimulate various changes in organs and tissue. The name "hormone" is derived from a Greek word, *hormon*, which means "to set into motion," an apt description of the way hormones function. Hormones are carried through the body in the bloodstream to produce a specific effect on

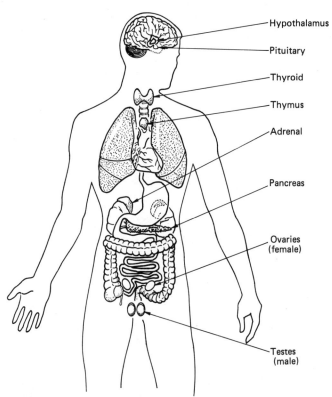

Fig. 24.9. The location of the endocrine glands.

certain cells. The cells activated by hormones are called **target cells**. Each target cell has a specific bonding site that is designed to interact with only one kind of hormone. All other hormones will be unable to stimulate the cell.

Most of the hormones in the body are required in only microgram quantities to produce the desired response. However, each endodrine gland has a relatively large hormone reserve that can be called upon in time of stress to ready the body to meet unusual demands. Hormones can be classified according to their structure as being either: (a) a steroid, (b) a peptide or protein, or (c) an amino-acid derivative.

The two major pathways for the transmission of information from one organ or tissue to another are the circulatory system and the nervous system. The endodrine system cross-links these two systems through the hypothalamus. The hypothalamus is an endocrine gland that lies at the base of the brain. It receives nerve impulses and then responds by secreting special types of hormones called **releasing factors**. The releasing factors pass directly into the anterior (forward) lobe of the pituitary gland (located just below the hypothalamus) causing the release of other specific hormones that are concerned with normal growth and metabolism. These hormones, along with a brief description of their importance, are listed below.

1. *Growth Hormone or Somatotropin.* A hormone that is concerned with the growth of bones, muscles, and other organs of the body. It increases the rate of protein synthesis, promotes the use of fats, and decreases the use of carbohydrates. Reduced secretion causes dwarfism; oversecretion before puberty causes giantism, or if after puberty, enlargement of hands, feet, face, or other parts of the body.

2. *Thyrotropin.* Controls the production of hormones by the thyroid gland, as well as its growth.

3. *Adrenocorticotrophic Hormone (ACTH).* Controls the growth and hormone production of the adrenal cortex.

4. *Prolactin.*

5. *Follicle-Stimulating Hormone.*

6. *Luteinizing Hormone.*

These last three hormones control the growth, development, and functioning of the ovaries and testes and the stimulation of milk by the mammary glands. The growth hormone is the only one of the six that does not stimulate the release of hormones by other glands of the endocrine system.

The hypothalamus also acts on the posterior (rearward) lobe of the pituitary gland causing it to release two important hormones, vasopressin and oxytocin. Both are polypeptides composed of nine amino acid units. Vasopressin, also known as the antidiuretic hormone, acts on the tubules and collecting ducts of the kidneys, controlling water reabsorption. An insufficient amount of this hormone can result in excess water loss by the body. Higher level of vasopressin can cause an increase in blood pressure by constricting the walls of veins and arteries.

Oxytocin promotes the release of milk in the breasts of new mothers. During childbirth, it causes contractions of the uterine muscles. Oxytocin is frequently administered to pregnant women to induce childbirth.

Several hormones that are not under direct control of the pituitary gland are insulin, glucagon, epinephrine (adrenalin), and norepinephrine. The discussion of insulin and glucagon will be set aside until the following chapter on metabolism (Chapter 25).

Epinephrine and norepinephrine are produced by the medulla (middle) of the adrenal gland. Unlike most of the other endocrine glands, the adrenal medulla is regulated by nerve impulses. During stressful situations, such as "fight or run" confrontations, epinephrine and norepinephrine are produced to ready the body for action. Epinephrine is predominant, and it affects many different target cells, causing an increased heart beat, while increasing blood sugar by accelerating the breakdown of glycogen in the liver. With an increased level of blood sugar, the muscles and brain are more able to cope with the problem. Norepinephrine has a lesser overall effect than epinephrine, but it is important in the conduction of nerve impulses, perhaps making the brain more efficient.

When hormones reach their target cells, they bind to a specific hormone receptor. The steroid-type hormones, which are lipid soluble, pass directly through the cell membrane combining with a receptor molecule inside. This combination then interacts with definite genes on DNA to increase the synthesis of a required protein. Some of the steroid hormones were discussed in Chapter 22.

Nonsteroid hormones combine with receptor sites on the surfaces of cell membranes. The binding stimulates the enzyme, adenylate cyclase, to convert ATP to cyclic AMP. The cyclic AMP, which is inside the cell, initiates a specific set of reactions to produce some required protein. Cyclic AMP is often called the hormone's second messenger.

The overall regulation of hormone production is a complex series of interactions between the nervous system and the endocrine glands. Several feedback relationships exist to maintain correct hormone levels.

Table 24.1 summarizes the characteristics of the major hormones in the body.

Check Test Number 5

1. Which gland in the endocrine system relates the nervous system to the circulatory system?
2. Name two endocrine glands that are stimulated by nerve impulses.
3. Which type of hormone can pass through cell membranes?
4. Which hormone is used to induce labor?

(*Answers on p. 733.*)

TABLE 24.1			
The Sites and Functions of the Major Human Hormones			
Endocrine gland and its hormones	Hormone type*	Site of action	Effects
I. Hypothalamus releasing factors	peptides	pituitary	activates the synthesis of hormones or inhibits their synthesis
II. Anterior lobe of the pituitary gland (the adenohypophysis)			
A. Growth hormone (somatotropin)	protein (191AA)	all tissues	general body and bone growth; also involves carbohydrate and fat metabolism
B. Thyrotropin	protein (220AA)	thyroid gland	synthesis of thyroid hormones
		adipose tissue	activates fat use
C. Adrenocorticotropic hormone (ACTH)	peptide (39AA)	adrenal cortex	synthesis of adrenal cortical steroid hormones
D. Prolactin	protein (191AA)	mammary glands	starts flow of milk
		corpus luteum	maintains estrogen and progesterone levels
E. Follicle-stimulating hormone	protein (200AA)	ovaries	development of follicles, estrogen, and ovulation
F. Luteinizing hormone	protein (200AA)	ovaries	estrogens and progesterone synthesis
		testes	production of testosterone
III. Posterior lobe of the pituitary gland (the neurohypophysis)			
A. Vasopressin	peptide (9AA)	kidneys	reabsorption of water
		arteries	increases blood pressure
B. Oxytocin	peptide (9AA)	uterus	uterine contractions
		mammary glands	milk secretion

TABLE 24.1 (cont.) The Sites and Functions of the Major Human Hormones			
Endocrine gland and its hormones	Hormone type*	Site of action	Effects
IV. Adrenal cortex (outside portion of adrenal gland)			
A. Cortisone	steroid	most tissues	interaction of carbohydrate, protein, and fat metabolism
B. Aldosterone	steroid	kidneys	reabsorption of Na^+
V. Adrenal medulla (middle region of adrenal gland)			
A. Epinephrine	amino acid derivative	most tissues	prepares body for emergency action: increases pulse rate; blood pressure rise; conversion of glycogen to glucose
B. Norepinephrine	amino acid derivative	most tissues	similar to epinephrine; also stimulates nervous system
VI. Pancreas			
A. Insulin (from beta cells)	protein (51AA)	most tissues	control of glucose metabolism; synthesis of glycogen
B. Glucagon (from alpha cells)	peptide (29AA)	liver	conversion of glycogen to glucose
		adipose tissue	use of fat
C. Somatostatin (from D cells)	peptide (14AA)	pancreas	inhibits release of glucagon
		anterior pituitary	inhibits release o growth hormones
VII. Thyroid			
A. Thyroxine	amino acid derivative	most tissues	increases rate of cellular metabolism
B. Calcitonin	peptide (32AA)	bones and kidneys	control of Ca^{+2} and phosphate levels

TABLE 24.1 (cont.) The Sites and Functions of the Major Human Hormones			
Endocrine gland and its hormones	Hormone type*	Site of action	Effects
VIII. Parathyroid			
Parathyrin	protein (84AA)	bones and kidneys	control of Ca^{+2} and phosphate levels
IX. Testes			
Testosterone	steroid	most tissues and accessory sex organs	development of male sex characteristics
X. Ovaries			
A. Estradiol†	steroid	accessory sex organs	development and normal cycle function
		mammary glands	formation of duct system
B. Estrone†	steroid	most tissues	female secondary sex characteristics

*AA = Amino acid.

†These act together and are collectively called the estrogens.

Answers:
1. The hypothalamus.
2. The hypothalamus and the adrenal medulla.
3. Steroid hormones.
4. Oxytocin.

Terms

Several important terms appear in Chapter 24. You should understand the meaning of each.

enzyme
substrate
cofactor
coenzyme
metal ion activator
apoenzyme

isomerase
ligase
enzyme inhibition
competitive inhibition
noncompetitive inhibition
end-product inhibition

holoenzyme

active site

holding site

lock and key theory

induced-fit theory

turnover number

hydrolase

oxidoreductase

transferase

lyase

allosteric enzyme

irreversible inhibition

optimum pH

optimum temperature

endocrine system

hormone

releasing factors

target cell

proenzyme

enzyme–substrate complex

Questions

Answers to starred questions appear at the end of the text.

*1. Enzymes were important to humans long before we knew their function. What were two processes used for years that were catalyzed by enzymes?

2. What is the characteristic ending used for enzyme names?

*3. In what ways are enzymes similar to ordinary chemical catalysts? In what ways are they different?

4. Give the general *type* of enzyme that would catalyze each of the following reactions.

*a. 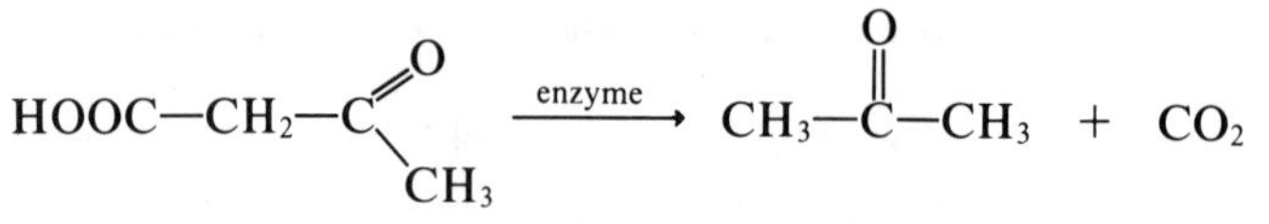 $HOOC-CH_2-C\overset{\displaystyle O}{\underset{\displaystyle CH_3}{\big<}} \xrightarrow{\text{enzyme}} CH_3-\overset{\displaystyle O}{\overset{\|}{C}}-CH_3 \; + \; CO_2$

*b. glucose $\underset{\text{enzyme}}{\rightleftharpoons}$ fructose

*c. 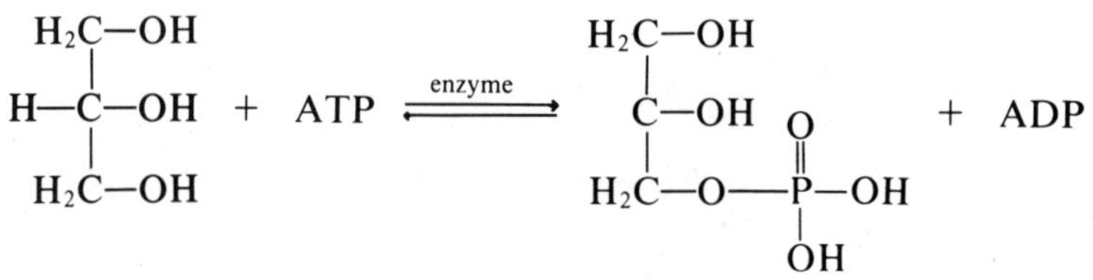
$$\begin{array}{l} H_2C-OH \\ | \\ H-C-OH \\ | \\ H_2C-OH \end{array} \; + \; ATP \; \underset{\text{enzyme}}{\rightleftharpoons} \; \begin{array}{l} H_2C-OH \\ | \\ C-OH \quad O \\ | \qquad \| \\ H_2C-O-P-OH \\ \qquad\quad | \\ \qquad\quad OH \end{array} \; + \; ADP$$

5. If an excess of vitamins is taken into the body, what happens to the water-soluble vitamins? The fat-soluble vitamins?

*6. What is the principal function of biotin in the body?

7. What is the principal function of folic acid in the body?

*8. What are the two theories of enzyme action?

9. Using E to symbolize the enzyme, S to symbolize the substrate, and P the product, give the general equations describing an enzyme-catalyzed reaction that converts the substrate to product.

10. For each of the following, state which kind of enzyme specificity is indicated.

 *a. Carboxypeptidase is specific for terminal amino acids that have their amino groups free on the peptide chain.

 *b. Thrombin cleaves only the bond between arginine and glycine.

 c. Carbonic anhydrase is involved only in the formation of carbonic acid.

 d. $R-O-\overset{\overset{O}{\|}}{C}-R' + H_2O \xrightleftharpoons{\text{esterase}} ROH + R'COOH$

11. Why does pepsin function in the stomach, but is inactive in the intestinal tract?

*12. At what point will an increased substrate concentration not increase the rate of reaction catalyzed by an uninhibited enzyme?

13. In what way would the lead in lead paint affect body enzymes if ingested by children?

*14. Which type of enzyme inhibition is most damaging to enzymes?

15. How are the serum levels of CPK, SGOT, and LDH affected by damage to heart tissue as usually occurs in heart attacks?

*16. What single food is a rich source for vitamins?

17. What is a hormone?

18. What condition can result if an excess of the growth hormone is secreted prior to puberty in a human?

*19. In what way is the hypothalamus important as a link between the two transmission systems of the body?

20. Give the name of the water-soluble vitamin(s) that is (are) described by each of the following statements.

 *a. A cure for pellagra.

 *b. Used to make $NAD^{\oplus}$.

 *c. Addition of a pyrophosphate unit converts it to a coenzyme.

 *d. Contains cobalt.

 *e. Serves in the transfer of methyl groups.

 *f. Used in collagen formation.

 g. Use in coenzymes involved in CO_2 transfer.

 h. Used to make FAD.

 i. Can be deficient in strict vegetarians.

 j. Found in fruit juice.

 k. Acetyl CoA needs this one for its synthesis.

 l. Used to produce coenzymes involved in oxidation–reduction reactions.

 m. Tryptophan may be used to make it.

 n. Can cure beriberi.

 o. Nicotinic acid is one of its vitamers.

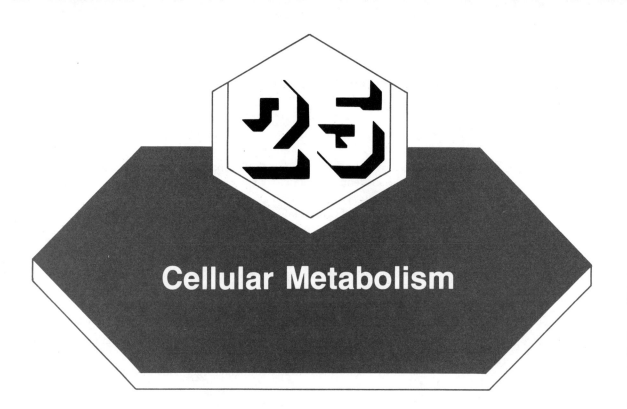

Cellular Metabolism

In the previous chapters you have been told that the principal foods we ingest are fats, carbohydrates, and proteins. Upon digestion, they yield glycerol and fatty acids, monosaccharides, and amino acids, respectively. Once absorbed by the body, these digestion products are transported to cells and used as fuel for energy or as building material for the growth and repair of cells. The complex series of reactions that transform absorbed food into cellular compounds and energy is called **metabolism**.

There are two broad categories of metabolic activity: **catabolism**, which involves those processes that break down absorbed food into smaller units; and **anabolism**, which involves those processes that synthesize larger molecules from smaller ones. Catabolism releases energy and anabolism consumes energy. The cell couples these processes so the synthesis of required materials can be accomplished.

In this chapter you will learn how energy is transferred in cellular processes, and the pathways followed as fats, carbohydrates, and, to a lesser extent, proteins are metabolized. You will also see how each of these metabolic pathways come together in the Krebs cycle, a series of reactions that convert much of the chemical energy stored in food to chemical energy for use in cellular processes.

Objectives

By the time you finish this chapter you should be able to do the following:

1. Describe the role of ATP as an energy carrier and how it is regenerated in cells.
2. Describe the function of the electron transport system in cellular metabolism.
3. Describe the function of the Krebs cycle, and the reactions that make it up.
4. Describe the digestion of carbohydrates, fats, and proteins.
5. Describe the Embden–Meyerhof pathway of glycolysis, and the conditions that lead to pyruvic acid as opposed to lactic acid formation.
6. Describe the central role of acetyl coenzyme A in metabolism.
7. Describe the fatty acid cycle, and the process that is required to bring a fatty acid into the cycle.
8. Compare the metabolism of a fat with that of glucose in terms of ATP production and the percent of stored chemical energy that is converted to ATP.
9. Describe the conditions that may lead to elevated levels of ketone bodies in the blood.
10. Describe the deamination reactions in amino acid metabolism, and the ways that different amino acids may enter the Krebs cycle.
11. Describe the function of the urea cycle, and how it operates in conjunction with amino acid metabolism.

Cellular Energy—The Role of ATP

Catabolism of foods releases energy to the body in the form of heat and chemical energy. The heat is used to maintain body temperature, but because it flows out of the body to the surroundings, it is not available to provide energy for use in anabolism reactions. Chemical energy, on the other hand, is retained by the body. It is stored in high-energy compounds that can be used quickly when energy is needed. These compounds must be able to interact with chemical processes to deliver energy when and where it is needed.

There are several high-energy compounds produced in cells, but one serves as the principal energy-storage compound, **adenosine triphosphate (ATP)**. All cells contain from 0.5–2.5 mg of ATP per ml of cellular volume. ATP is the most important carrier of energy to and from cellular activities. It provides energy for the synthesis of proteins, nucleic acids, for nerve transmission, muscle contraction, and the active transport of ions across membranes. Essentially all energy-producing and energy-consuming reactions are linked by this compound. ATP is a relatively unstable compound, that is, it exists at a high potential energy. Upon hydrolysis (reaction with water) it yields lower-energy compounds. The potential energy difference between ATP plus water and the hydrolysis products equals the energy released for use in cellular reactions. The structure of ATP is shown below. You may recognize this compound as one used by cells in the synthesis of DNA and RNA (Chapter 23). In cells, ATP exists as an anion, having lost one or more of the ionizable hydrogens on the triphosphate group. The symbol ATP really represents this molecule in any one ionized form or as a mixture of the ionized forms. The neutral compound is shown below with all hydrogens in place:

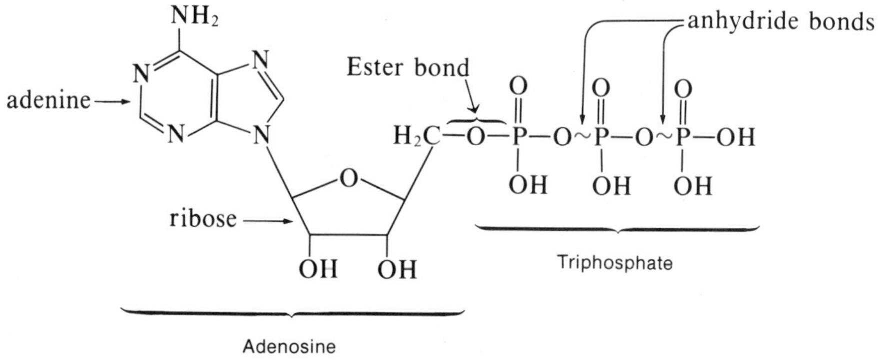

The bonds that are able to be cleaved in hydrolysis with a substantial release of energy are indicated with wavy lines ($\sim$). These are the anhydride linkages in ATP. The ester linkage that joins the phosphate to the ribose sugar does not release a large amount of energy when hydrolyzed, so it is indicated with the usual dash (–). The wavy line will also be used to indicate bonds in other high-energy compounds that are cleaved with the release of substantial amounts of energy when hydrolyzed.

Be certain you realize that simply breaking a chemical bond does not release energy. The wavy-line bonds in ATP only indicate the points in the molecule where cleavage occurs in hydrolysis. The release of energy results from the production of lower-energy, more stable compounds, and the released energy is equal to the difference between the potential energies of the reactants and the potential energies of the products. The anhydride linkages are sometimes described as being high-energy bonds, a somewhat deceptive term at best.

The principal way in which ATP is hydrolyzed cleaves the anhydride bond that connects the terminal phosphate group to the molecule. The products of the reaction are adenosine diphosphate (ADP) and inorganic phosphate, symbolized Pi. It is common practice to call the phosphate product simply "inorganic phosphate," and it represents the ionized forms of phosphoric acid.

In cellular conditions it exists as a mixture of the dihydrogen phosphate ion $H_2PO_4^-$, and the hydrogen phosphate ion, HPO_4^{2-}. For convenience, we will symbolize the removed phosphate group as Pi. The structure of ATP will be abbreviated to emphasize the triphosphate group.

$$
\text{adenosine}-O-\overset{\overset{\displaystyle O}{\|}}{\underset{\underset{\displaystyle OH}{|}}{P}}-O\sim\overset{\overset{\displaystyle O}{\|}}{\underset{\underset{\displaystyle OH}{|}}{P}}-O\sim\overset{\overset{\displaystyle O}{\|}}{\underset{\underset{\displaystyle OH}{|}}{P}}-OH \;+\; H_2O \;\underset{\xleftarrow{\hspace{1cm}}}{\overset{enzyme}{\xrightarrow{\hspace{1cm}}}}
$$

ATP

$$
\text{adenosine}-O-\overset{\overset{\displaystyle O}{\|}}{\underset{\underset{\displaystyle OH}{|}}{P}}-O\sim\overset{\overset{\displaystyle O}{\|}}{\underset{\underset{\displaystyle OH}{|}}{P}}-OH \;+\; Pi \;+\; 7.3 \text{ kcal}
$$

ADP

The hydrolysis of ATP → ADP + Pi releases 7.3 kcal of energy per mole of ATP. Adenosine diphosphate (ADP) can also be hydrolyzed to produce energy, producing adenosine monophosphate and Pi and 7.3 kcal of energy per mole of ADP:

$$
\text{adenosine}-O-\overset{\overset{\displaystyle O}{\|}}{\underset{\underset{\displaystyle OH}{|}}{P}}-O\sim\overset{\overset{\displaystyle O}{\|}}{\underset{\underset{\displaystyle OH}{|}}{P}}-OH \;+\; H_2O \;\underset{\xleftarrow{\hspace{1cm}}}{\overset{enzyme}{\xrightarrow{\hspace{1cm}}}}
$$

ADP

$$
\text{adenosine}-O-\overset{\overset{\displaystyle O}{\|}}{\underset{\underset{\displaystyle OH}{|}}{P}}-OH \;+\; Pi \;+\; 7.3 \text{ kcal}
$$

AMP

Adenosine monophosphate (AMP) can also undergo hydrolysis, but cleavage of the ester linkage only produces 3.4 kcal of energy per mole of AMP. Because the energy produced is only about half of that obtained in the hydrolysis of ATP and ADP, cells rarely use AMP as an energy source.

$$
\text{adenosine}-O-\overset{\overset{\displaystyle O}{\|}}{\underset{\underset{\displaystyle OH}{|}}{P}}-OH \;+\; H_2O \;\underset{\xleftarrow{\hspace{1cm}}}{\overset{enzyme}{\xrightarrow{\hspace{1cm}}}}\; \text{adenosine} + Pi \;+\; 3.4 \text{ kcal}
$$

AMP

Occasionally, ATP will be hydrolyzed to produce AMP and inorganic pyrophosphate, symbolized PPi. This releases 10.0 kcal of energy per mole of ATP. You will see this reaction only once during the discussion of metabolism. ATP most often is converted to ADP and Pi.

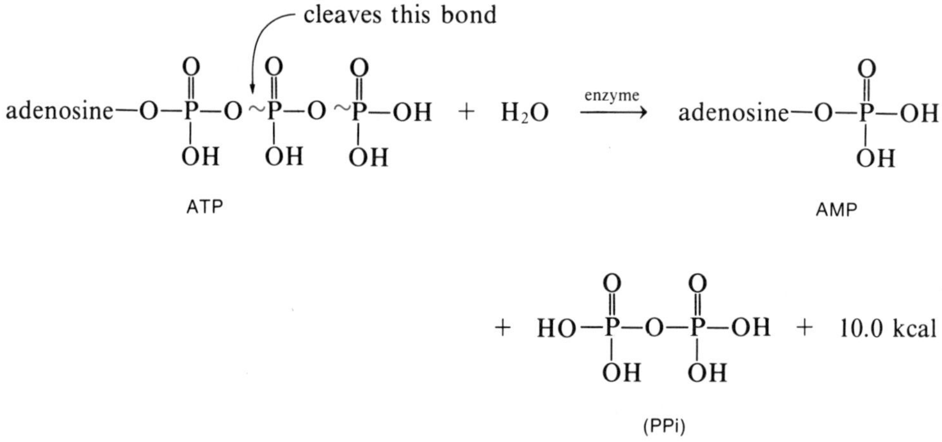

cleaves this bond

$$\text{adenosine—O—} \overset{\overset{O}{\|}}{\underset{\underset{OH}{|}}{P}} \text{—O} \sim \overset{\overset{O}{\|}}{\underset{\underset{OH}{|}}{P}} \text{—O} \sim \overset{\overset{O}{\|}}{\underset{\underset{OH}{|}}{P}} \text{—OH} \; + \; H_2O \; \xrightarrow{enzyme} \; \text{adenosine—O—} \overset{\overset{O}{\|}}{\underset{\underset{OH}{|}}{P}} \text{—OH}$$

ATP   AMP

$$+ \; HO \text{—} \overset{\overset{O}{\|}}{\underset{\underset{OH}{|}}{P}} \text{—O—} \overset{\overset{O}{\|}}{\underset{\underset{OH}{|}}{P}} \text{—OH} \; + \; 10.0 \text{ kcal}$$

(PPi)

Pyrophosphate actually exists as an ion in cells, but it is represented as the neutral species above.

The energy-producing hydrolysis reactions of ATP, ADP, and AMP are summarized below:

$$\text{ATP} + H_2O \; \underset{\longleftarrow}{\overset{enzyme}{\longrightarrow}} \; \text{ADP} + \text{Pi} + 7.3 \text{ kcal/mole}$$

$$\text{ADP} + H_2O \; \underset{\longleftarrow}{\overset{enzyme}{\longrightarrow}} \; \text{AMP} + \text{Pi} + 7.3 \text{ kcal/mole}$$

$$\text{AMP} + H_2O \; \underset{\longleftarrow}{\overset{enzyme}{\longrightarrow}} \; \text{adenosine} + \text{Pi} + 3.4 \text{ kcal/mole}$$

$$\text{ATP} + H_2O \; \xrightarrow{enzyme} \; \text{AMP} + \text{PPi} + 10.0 \text{ kcal/mole}$$

An important feature of three of these reactions is that they are reversible. Once ATP is converted to ADP + Pi with the release of energy, it can be reformed in a process called **phosphorylation**. If it were not possible to readily reform ATP, cells would quickly become filled with ADP, AMP, and Pi and cease to function.

The body couples the energy-producing conversion of ATP to ADP + Pi with other reactions that provide the energy needed to drive ADP + Pi back to ATP again. The interrelationship of these coupled reactions is shown below:

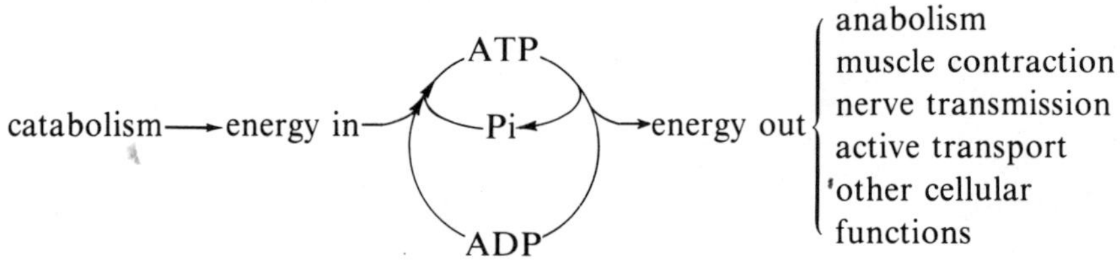

catabolism⟶energy in⟶ (cycle: ATP, Pi, ADP) ⟶energy out { anabolism
muscle contraction
nerve transmission
active transport
'other cellular
functions

Phosphorylation of ADP can proceed through two different routes in animals. Let us look at each of them to see how it is done.

Substrate Level Phosphorylation

Substrate level phosphorylation is the transfer of a phosphate group from one compound of higher energy to another of lower energy. The transfer of the phosphate group is made directly from one compound to the other, and the process is catalyzed by an enzyme.

Several compounds containing phosphate groups that are found in cells are listed in Table 25.1. The energy released as each is hydrolyzed is also given. Generally, any compound that releases a larger amount of energy upon hydrolysis than ATP can transfer its phosphate to ADP and convert it to ATP. For example, 1,3-diphosphoglyceric acid releases 11.8 kcal of energy upon hydrolysis. This is greater than the 7.3 kcal of energy released by ATP. In glycolysis, the transfer of phosphate from the higher-energy 1,3-diphosphoglyceric acid to ADP does take place, regenerating ATP:

1,3–Diphosphoglyceric acid   + ADP → (phosphoglycerokinase) → 3–Phosphoglyceric acid   + ATP

The compounds listed in Table 25.1 that release substantial amounts of energy (> 7 kcal/mole) upon hydrolysis can serve as energy sources in cells, but each ultimately interacts with the conversion of ADP to ATP.

Oxidative Phosphorylation—The Electron Transport System

Most of the conversion of ADP + Pi to ATP occurs in small membrane-bound granula known as the **mitochondria**. The mitochondria are the power-houses of a cell, and the greater the energy needs of a cell, the more mitochondria

TABLE 25.1
Some High- and Low-Energy Cellular Phosphate Compounds

Compound	Structure*	Energy released when hydrolyzed (kcal/mole)	Use
Phosphoenolpyruvic acid		14.8	glucose metabolism (Embden–Meyerhof pathway)
1,3-diphosphoglyceric acid		11.8	glucose metabolism (Embden–Meyerhof pathway)
Creatine phosphate		10.3	energy storage in mammalian muscle tissue
Adenosine triphosphate (ATP) as: $ATP + H_2O \rightleftharpoons AMP + PPi$	Structures in text	10.0	fatty acid activation, leading to fatty acid cycle
as: $ATP + H_2O \rightleftharpoons ADP + Pi$	Structures in text	7.3	principal source of chemical energy for cellular needs
Adenosine diphosphate (ADP) $ADP + H_2O \rightleftharpoons AMP + Pi$		7.3	cellular energy
Adenosine monophosphate (AMP) $AMP + H_2O \rightleftharpoons$ adenosine $+ Pi$	Structures in text	3.4	—
Glucose-1-phosphate		5.0	glucose metabolism

		Energy released when hydrolyzed	
Compound	Structure*	(kcal/mole)	Use
Fructose-6-phosphate		3.8	glucose metabolism

$$\underset{\substack{\text{HO}-\overset{\displaystyle O}{\underset{\displaystyle OH}{\overset{\displaystyle \|}{P}}}-O-CH_2}}{}$$

TABLE   25.1 (Cont.)
Some High- and Low-Energy Cellular Posphate Compounds

*~ indicates a bond that upon hydrolysis will yield greater than 5 kcal of energy per mole.

it will contain. The "ATP factory" in mitochondria is a series of enzymes grouped together and attached to the inner membrane surface. The purpose of these enzymes is to oxidize the coenzymes NADH + H$^+$ and FADH$_2$ back to NAD$^\oplus$ and FAD for use in the oxidation of food. Before going further, it would be good to refresh your memory about these coenzymes, which were introduced in Chapter 24.

NAD represents nicotinamide adenine dinucleotide. It exists in two forms, an oxidized form NAD$^\oplus$ and a reduced form NADH. NAD$^\oplus$ can accept two electrons, $2e^-$, and a proton, H$^+$, to become NADH, but in practice, two electrons and *two* protons are removed simultaneously from the molecule it is oxidizing, so in equations the coenzyme in reduced form is usually written as NADH + H$^+$:

$$NAD^\oplus + 2e^- + 2H^+ \rightarrow NADH + H^+$$

Oxidized form   Reduced form

FAD represents flavin adenine dinucleotide, the oxidized form is FAD, and it can accept two electrons and two protons and be reduced to FADH$_2$:

$$FAD + 2e^- + 2H^+ \rightarrow FADH_2$$

Oxidized form   Reduced form

Both NAD$^\oplus$ and FAD are important oxidizing agents in metabolism. Once reduced to NADH + H$^+$ and FADH$_2$, they must be oxidized back to NAD$^\oplus$ and FAD for use again. As was said earlier, this is the function of the enzyme groups in the mitochondria, and they comprise the **electron transport system** (ETS) which is also called the **respiratory chain**.

The electron transport system is connected with the conversion of ADP + Pi to ATP. The details of this connection will be put off for a moment, but if one molecule of NADH is oxidized to $NAD^{\oplus}$ by the ETS, three molecules of ATP are produced. The equation describing this reaction shows that oxygen is consumed, and thus this means of converting ADP to ATP is termed **oxidative phosphorylation**. The consumption of oxygen is also the reason the ETS is freqently called the respiratory chain.

$$NADH + H^+ + 3ADP + 3Pi + \tfrac{1}{2}O_2 \rightarrow NAD^{\oplus} + 3ATP + H_2O$$

The oxidation of $FADH_2$ to FAD converts two molecules of ADP to ATP:

$$FADH_2 + 2ADP + 2Pi + \tfrac{1}{2}O_2 \rightarrow FAD + 2ATP + H_2O$$

In the ETS, then,

$$\text{each } (NADH + H^+ \rightarrow NAD^{\oplus}) \text{ produces 3ATP}$$

$$\text{and each } (FADH_2 \rightarrow FAD) \qquad \text{produces 2ATP}$$

The electron transport system is diagramed in Figure 25.1. It consists of a series of enzymes and coenzymes that remove two electrons and two protons $(2e^-, 2H^+)$ from either $NADH + H^+$ or $FADH_2$ and combines them with an oxygen atom (from O_2) to produce water. The overall equation is

$$2H^+ + 2e^- + \tfrac{1}{2}O_2 \rightarrow H_2O + \text{energy}$$

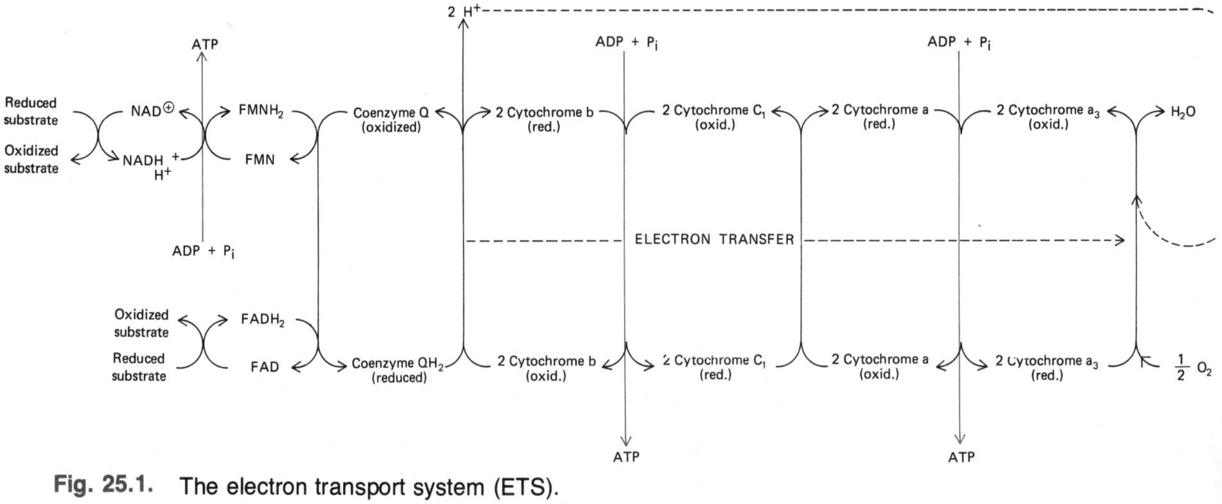

Fig. 25.1.   The electron transport system (ETS).

This is the reaction that produces the energy needed to convert ADP + Pi back to ATP.

There are two points at which electrons and protons can enter the ETS. One is used by NADH + H^+ and the other by $FADH_2$. Let us begin with NADH + H^+ in the upper left-hand corner of Figure 25.1. NADH + H^+ transfers $2e^-$ and $2H^+$ to the coenzyme FMN (flavin mononucleotide), reducing it to $FMNH_2$. Enough energy is released here to convert one ADP + Pi to ATP. $NAD^\oplus$ is formed and will then return to oxidize another molecule. $FMNH_2$ transfers the $2e^-$ and $2H^+$ to coenzyme Q, reducing it to coenzyme QH_2. Coenzyme QH_2 then releases two H^+ to the surrounding medium and transfers the $2e^-$ into a series of connecting cytochrome enzymes. Each cytochrome (labeled b, C_1, a, and a_3 in the figure) is associated with an iron-containing hemelike coenzyme. The iron rapidly changes from Fe^{2+} to Fe^{3+} as it transfers an electron to the next cytochrome. By changing back and forth, $Fe^{2+} \rightarrow Fe^{3+} \rightarrow Fe^{2+} \rightarrow$ . . ., the iron ions can transport an electron quickly down the cytochrome chain to molecular oxygen. Since $2e^-$ are transferred into the cytochrome chain, and since each iron ion can only transfer one electron at a time, it is necessary that the cytochrome enzymes occur in pairs. One cytochrome of the pair handles one e^-, and the second e^- is handled by the other. In this way both electrons reach the end of the cytochrome chain simultaneously. ATP is produced at two points in the cytochrome system using the energy released by the electron transport.

Let us return to $FADH_2$ for a moment before concluding the electron transport operation. $FADH_2$ can transfer $2e^-$ and $2H^+$ directly to coenzyme Q, bypassing the FMN–$FMNH_2$ step. Because $FADH_2$ can do this, it produces one less ATP than NADH + H^+, for a total of two ATP.

Returning to the electron transport down the cytochrome system, the final step in the operation is the simultaneous transfer of the $2e^-$ to an oxygen molecule, which, in the presence of $2H^+$, is reduced to water. Nearly all the oxygen used by the body in metabolism is consumed in this last step of the ETS.

The steps involved in the electron transport system are irreversible, and if any one is inhibited, the entire process stops. Certain drugs and poisons are known to block specific steps. Rotenone, an insecticide, and amytal, a barbiturate, prevent the transfer of electrons and protons from $FMNH_2$ to coenzyme Q. If either is taken in large dose, the operation of the ETS in cells will be reduced and death may result from the inability of the body to use oxygen. Cyanide ion ($C \equiv N^-$) and carbon monoxide (CO) both bond strongly to the iron ions in the cytochromes. Both can inhibit the ETS, and if taken into the body in sufficient quantity, the results are fatal.

When you think of cellular respiration, the convenient notion that oxygen combines with absorbed food in cells to produce carbon dioxide, water, and energy usually comes to mind. Basically this is right, but from what you have just seen, it is not quite that straightforward. Oxygen is involved in the oxidation of food most certainly, but in an indirect way. Its connection is through the coenzymes NAD$^\oplus$ and FAD and the electron transport system. Now you need to find out how NAD$^\oplus$ and FAD get involved in the oxidation of food. This will bring us to the central oxidizing system in metabolism, the Krebs cycle.

Check Test Number 1

1. What species are transferred into the electron transport system as NADH + H$^+$ is oxidized to NAD$^\oplus$?
2. Write a single equation that describes the overall reaction that takes place in the ETS.
3. If one mole of FADH$_2$ is oxidized by the ETS, how many moles of ATP are produced?
4. Could fructose-6-phosphate convert ADP to ATP by substrate level phosphorylation?

Answers:

1. $2e^-$ and $2H^+$
2. $2H^+ + 2e^- + \frac{1}{2}O_2 \rightarrow H_2O + \text{energy}$
3. 2 moles of ATP
4. No. The energy released when fructose-6-phosphate is hydrolyzed is 3.8 kcal/mole, less than that for ATP (see Table 25.1).

The Krebs Cycle

The **Krebs cycle** is a series of chemical reactions that produces the electrons and protons (e^-, H$^+$) needed by the electron transport system to produce ATP. The Krebs cycle, in conjunction with the ETS, is then an "energy producer," producing energy in the form of ATP. The enzyme-catalyzed reactions that comprise the Krebs cycle take place in the mitochondria, the same location as the ETS. The reactions comprise a cycle because the product of the last reaction in the series is the reactant for the first reaction. In a sense, it is a sophisticated chemical merry-go-round.

The fuel for the Krebs cycle is the acetyl group, $CH_3\overset{\overset{\displaystyle O}{\|}}{C}-$, which is carried into the cycle by coenzyme A in the form of acetyl coenzyme A:

$$CH_3-\overset{\overset{\displaystyle O}{\|}}{C}-SCoA$$

Acetyl coenzyme A

The structural formula of coenzyme A appeared in Chapter 24, but for simplicity it is usually abbreviated as CoASH or HSCoA. In acetyl coenzyme A, the acetyl group replaces the sulfhydryl hydrogen ($-SH$) and is bonded to sulfur.

Virtually all absorbed monosaccharides, fatty acids, and nearly half the absorbed amino acids can be converted in the body into acetyl groups. For example, the aerobic catabolism of glucose in the glycolysis pathway (to be discussed later) results in the formation of pyruvic acid:

$$CH_3\overset{\overset{\displaystyle O}{\|}}{C}-COOH$$

Pyruvic acid

A large part of the energy that glucose provides to the body is extracted in the Krebs cycle–ETS combination. If this is to happen, pyruvic acid must be converted to acetyl coenzyme A. This important step leading into the Krebs cycle is accomplished through a series of steps. Pyruvic acid is first oxidized by the $NAD^{\oplus}$ coenzyme associated with pyruvic acid dehydrogenase (the enzyme), followed by the removal of carbon dioxide (decarboxylation). Coenzyme A bonds to the resulting acetyl group to form acetyl coenzyme A:

$$CH_3\overset{\overset{\displaystyle O}{\|}}{C}-COOH \;+\; CoASH \;+\; NAD^{\oplus} \;\xrightarrow{\text{pyruvic acid dehydrogenase}}$$

Pyruvic acid

$$CH_3\overset{\overset{\displaystyle O}{\|}}{C}-SCoA \;+\; NADH \;+\; H^+ \;+\; CO_2$$

The oxidation of pyruvic acid reduces $NAD^{\oplus}$ to $NADH + H^+$. NADH is then oxidized in the ETS back to $NAD^{\oplus}$ producing three molecules of ATP.

The conversion of pyruvic acid to acetyl coenzyme A along with the involvement of the ETS can be symbolized this way:

$$CH_3\overset{\overset{\displaystyle O}{\|}}{C}-COOH\xrightarrow{\hspace{2cm}}CH_3\overset{\overset{\displaystyle O}{\|}}{C}-SCoA$$

CoASH   CO_2

NAD^{\oplus}   NADH + H^+

3ATP   ETS

CoASH and $NAD^\oplus$ enter the conversion; CO_2 and $NADH + H^+$ are produced as pyruvic acid is converted to acetyl-CoA.

Once in the form of acetyl coenzyme A, the two carbon fragment from pyruvic acid can be oxidized to carbon dioxide and water in the Krebs cycle. The reactions within the cycle do not require oxygen—they are *anaerobic* reactions. But the reactions in the ETS do require oxygen for the production of water— they are *aerobic* reactions. The Krebs cycle–ETS combination then must operate under aerobic conditions so the required oxidizing agents ($NAD^\oplus$ and FAD) can be regenerated.

Now, let us follow the ten steps of the Krebs cycle. Some of the compounds appearing in the cycle may be new to you, and their structures might seem complicated at first glance. Look each one over as it appears, and look for what is added or removed as it becomes the next compound in the cycle. This approach will help you get a handle on the changes that take place. The involvement of the coenzymes $NAD^\oplus$ and FAD will be symbolized using curved arrows on the larger "yield" arrows of the equations. Each reaction in the cycle must be catalyzed by enzymes. You will notice that the names of the enzymes describe the reactions they catalyze. A dehydrogenase catalyzes the removal of hydrogen atoms (actually an e^- and a H^+ which adds up to the composition of a hydrogen atom) in an oxidation.

The Krebs cycle appears in Figure 25.2. A more complete Krebs cycle showing formulas of all compounds, involvement of the ETS, and sources of acetyl coenzyme A appears in Figure 25.3.

Step 1: Acetyl-CoA transfers the acetyl group to *oxaloacetic acid* to form *citric acid*. Coenzyme-ASH is re-formed to be used again. In the conditions that exist in cells, oxaloacetic acid and citric acid, along with the other acids

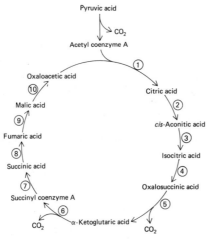

Fig. 25.2. An abbreviated version of the Krebs cycle.

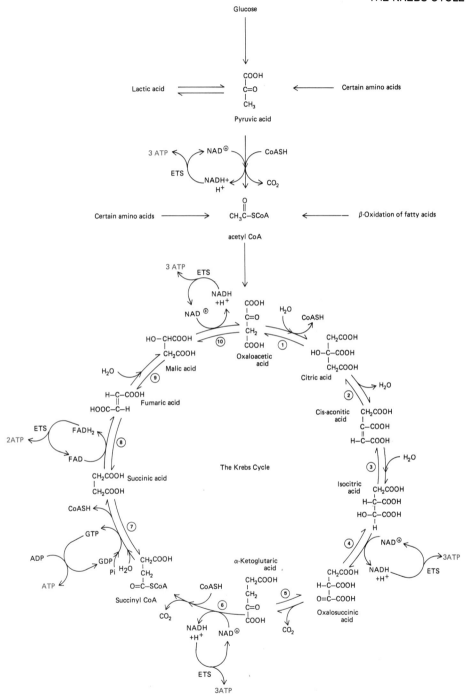

Fig. 25.3 A more complete picture of the Krebs cycle showing the involvement of the electron transport system and sources of acetyl coenzyme A.

in the cycle, exist as anions; but, for simplicity, they will be written as neutral acids with the ionizable hydrogens in place:

$$\text{Oxaloacetic acid} + H—\overset{\overset{\displaystyle H}{|}}{\underset{\underset{\displaystyle H}{|}}{C}}—\overset{\overset{\displaystyle O}{\|}}{C}—SCoA + H_2O \underset{\text{synthetase}}{\overset{\text{citric acid}}{\rightleftharpoons}} \text{Citric acid} + CoASH$$

Oxaloacetic acid:
$$\begin{array}{c} COOH \\ | \\ C{=}O \\ | \\ CH_2 \\ | \\ COOH \end{array}$$

Citric acid:
$$\begin{array}{c} COOH \\ | \\ CH_2 \\ | \\ HO—C—COOH \\ | \\ CH_2 \\ | \\ COOH \end{array}$$

Steps 2 and 3: The enzyme, aconitase, first catalyzes the removal of water from *citric acid* to form *cis-aconitic acid*. Then, in Step 3, the enzyme catalyzes the addition of water to aconitic acid to form *isocitric acid*, an isomer of citric acid.

Citric acid $\underset{\text{aconitase}}{\overset{H_2O}{\rightleftharpoons}}$ *cis*-aconitic acid $\underset{\text{aconitase}}{\overset{H_2O}{\rightleftharpoons}}$ Isocitric acid

Citric acid:
$$\begin{array}{c} COOH \\ | \\ CH_2 \\ | \\ HO—C—COOH \\ | \\ H—C—H \\ | \\ COOH \end{array}$$

cis-aconitic acid:
$$\begin{array}{c} HOOC \quad CH_2 \quad COOH \\ \diagdown \quad | \quad / \\ C \\ \| \\ C \\ / \quad \diagdown \\ H \quad COOH \end{array}$$

Isocitric acid:
$$\begin{array}{c} COOH \\ | \\ CH_2 \\ | \\ H—C—COOH \\ | \\ HO—C—H \\ | \\ COOH \end{array}$$

Steps 4 and 5: *Isocitric acid* is converted to *oxalosuccinic acid*. The conversion is catalyzed by the isocitric acid dehydrogenase complex (a complex containing more than one enzyme). The oxidation is carried out by $NAD^{\oplus}$, and the conversion of NADH back to $NAD^{\oplus}$ is accompanied by the formation of three molecules of ATP. The same enzyme complex catalyzes the conversion of oxalosuccinic acid to *α-ketoglutaric acid* as CO_2 is removed.

Isocitric acid $\underset{\text{isocitric acid dehydrogenase complex}}{\rightleftharpoons}$ Oxalosuccinic acid $\underset{\text{isocitric acid dehydrogenase complex}}{\overset{CO_2}{\rightleftharpoons}}$ α-ketoglutaric acid

(3ATP, ETS, $NAD^{\oplus}$, NADH+H⁺)

Isocitric acid:
$$\begin{array}{c} COOH \\ | \\ CH_2 \\ | \\ H—C—COOH \\ | \\ HO—C—H \\ | \\ COOH \end{array}$$

Oxalosuccinic acid:
$$\begin{array}{c} COOH \\ | \\ CH_2 \\ | \\ H—C—COOH \\ | \\ C{=}O \\ | \\ COOH \end{array}$$

α-ketoglutaric acid:
$$\begin{array}{c} COOH \\ | \\ CH_2 \\ | \\ CH_2 \\ | \\ C{=}O \\ | \\ COOH \end{array}$$

Step 6: *α-Ketoglutaric acid* is converted to *succinyl coenzyme A*. α-Ketoglutaric acid is oxidized by $NAD^{\oplus}$, a coenzyme in the α-ketoglutaric acid dehydrogenase complex (again, more than one enzyme). Carbon dioxide is removed and CoASH is introduced. This is the only nonreversible reaction in the Krebs cycle, and it ensures that the cycle only goes in one direction:

$$
\begin{array}{c}
COOH \\
| \\
CH_2 \\
| \\
CH_2 \\
| \\
C{=}O \\
| \\
COOH
\end{array}
\qquad
\begin{array}{c}
COOH \\
| \\
CH_2 \\
| \\
CH_2 \\
| \\
C{=}O \\
| \\
SCoA
\end{array}
$$

α-ketoglutaric acid dehydrogenase complex

$NAD^{\ominus}$ $NADH + H^+$

CoASH CO_2

ETS

3ATP

α–ketoglutaric acid   Succinyl CoA

Step 7: *Succinyl-CoA* is released by the coenzyme as *succinic acid* is formed. The hydrolysis of succinyl CoA to form succinic acid provides the energy for the formation of GTP from GDP and Pi. GTP then transfers its phosphate group to ADP to form ATP. This is a good example of substrate phosphorylation. The reaction is catalyzed by succinic acid thiokinase.

ATP   ADP

$$
\begin{array}{c}
COOH \\
| \\
CH_2 \\
| \\
CH_2 \\
| \\
C{=}O \\
| \\
SCoA
\end{array}
\qquad
\begin{array}{c}
COOH \\
| \\
CH_2 \\
| \\
CH_2 \\
| \\
COOH
\end{array}
$$

H_2O GDP   GTP   CoASH

Pi→

succinic acid thiokinase

Succinyl CoA   Succinic acid

Step 8: *Succinic acid* is oxidized to *fumaric acid*, the coenzyme, FAD, is the oxidizing agent, and the recycling of $FADH_2$ to FAD in the ETS is accomplished by the production of two molecules of ATP. This is the only time FAD is used in the Krebs cycle.

2ATP

ETS

$$
\begin{array}{c}
COOH \\
| \\
H{-}C{-}H \\
| \\
H{-}C{-}H \\
| \\
COOH
\end{array}
$$

FAD $FADH_2$

succinic acid dehydrogenase

$$
\begin{array}{c}
HOOC \qquad H \\
\diagdown \quad \diagup \\
C \\
\| \\
C \\
\diagup \quad \diagdown \\
H \qquad COOH
\end{array}
$$

Succinic acid   Fumaric acid

Step 9: *Fumaric acid* is hydrolyzed to *malic acid*. The enzyme is fumarase.

Fumaric acid   Malic acid

Step 10: In this final step of the Krebs cycle, *malic acid* is oxidized by $NAD^{\oplus}$ to *oxaloacetic acid*. This completes the cycle and makes available oxaloacetic acid for the start of another cycle. The NADH is oxidized back to $NAD^{\oplus}$ by the ETS and produces three more molecules of ATP for cellular use. Malic acid dehydrogenase is required for this step.

Malic acid   Oxaloacetic acid

The overall conversion of the acetyl group in acetyl coenzyme A to carbon dioxide and water in the Krebs cycle–ETS combination can be summed up in a single equation. At this point you can appreciate what a single equation does not reveal about the steps involved in converting reactants to products.

$$\overset{\overset{\text{O}}{\underset{\|}{}}}{\text{CH}_3\text{C-SCoA}} + 2O_2 + 12ADP + 12Pi \rightarrow 2CO_2 + H_2O + 12ATP + HSCoA$$

It is not difficult to remember the series of ten compounds that make up the Krebs cycle if you use the following memory device: "Cindy Ann is our kitten. She sure finds mice offensive." The first letter of each word corresponds to the first letter of the name of each compound. It begins with *c*itric acid (*C*indy) and continues around the cycle.

Compound	Memory device
*c*itric acid	*C*indy
*a*conitic acid (*cis* form)	*A*nn
*i*socitric acid	*i*s
*o*xalosuccinic acid	*o*ur

Compound	Memory device
α-*k*etoglutaric acid	*k*itten.*
*s*uccinyl CoA	*S*he
*s*uccinic acid	*s*ure
*f*umaric acid	*f*inds
*m*alic acid	*m*ice
*o*xaloacetic acid	*o*ffensive.

The Krebs cycle is named in honor of Hans Krebs, who proposed the series of reactions in 1937. In 1953 he received the Nobel Prize for his work. The Krebs cycle is also called the citric acid cycle, because citric acid is the first species formed in the cycle; and the tricarboxylic acid cycle, because four of the compounds are carboxylic acids that have three –COOH groups in the molecule.

The Krebs cycle complete with the structures of all the compounds involved, the interactions of coenzymes in oxidation steps, and the several sources of acetyl coenzyme A appears in Figure 25.3.

The two carbon atoms that enter the Krebs cycle in the acetyl group are not the same carbon atoms lost in the two carbon dioxide molecules, but since two carbons entered, we can think of the two that left as being equivalent to those in the acetyl group. If we start with acetyl coenzyme A and move through the cycle, a total of 12 molecules of ATP are formed in the Krebs cycle–ETS combination. If we start with pyruvic acid, a total of 15 molecules of ATP are formed, the additional three coming from the ETS as pyruvic acid is oxidized by $NAD^{\oplus}$ in the step leading to its conversion to acetyl coenzyme A. A summary of the generation of ATP in the Krebs cycle–ETS combination appears in Table 25.2.

Check Test Number 2

1. What is the principal function of the Krebs cycle?
2. There are several oxidation steps in the Krebs cycle. What are the oxidizing agents used in them?
3. Which is the nonreversible step in the Krebs cycle?

Answers:

1. The oxidation of the acetyl group with the production of electrons and protons to drive the ETS, which in turn produces ATP.
2. The coenzymes $NAD^{\oplus}$ and FAD.
3. Step 6, the conversion of α-ketoglutaric acid to succinyl CoA.

---

*The transition from the end of the first sentence to the beginning of the second is the only irreversible step in the Krebs cycle: α-ketoglutaric acid $\rightarrow$ succinyl-CoA. Also, the two places where carbon dioxide is lost in the cycle are those that form α-ketoglutaric acid (*k*itten.) and the next step, the irreversible one that forms succinyl-CoA (*S*he), the beginning of the second sentence.

TABLE   25.2
ATP Production in the Krebs Cycle–ETS Combination

Step	Reaction	Source of ATP	Number of ATP Molecules
	pyruvic acid → acetyl-CoA	NADH → NAD$^{\oplus}$ in ETS	3
4	isocitric acid → oxalosuccinic acid	NADH → NAD$^{\oplus}$ in ETS	3
6	α-ketoglutaric acid → succinyl-CoA	NADH → NAD$^{\oplus}$ in ETS	3
7	succinyl-CoA → succinic acid	ADP $\xrightarrow{\text{GTP} \quad \text{GDP}}$ ATP (substrate phosphorylation)	1
8	succinic acid → fumaric acid	FADH$_2$ → FAD in ETS	2
10	malic acid → oxaloacetic acid	NADH → NAD$^{\oplus}$ in ETS	3

Grand Totals: Starting with pyruvic acid = 15 ATP
Starting with acetyl-CoA = 12 ATP

Carbohydrate Metabolism

Many of the carbohydrates discussed in this section were described in Chapter 21, and you may wish to refer to that chapter to review structures of compounds or the meaning of terms.

The principal dietary carbohydrates are starch, cellulose, and sucrose (table sugar). Small amounts of lactose are obtained from dairy products, and still smaller amounts of fructose, glucose, and other sugars enter the diet from a variety of sources. The cellulose in the diet cannot be digested, so it acts as bulk (fiber) for the development of feces in the large intestine. Starch is digested (hydrolyzed) and converted to glucose. Sucrose yields glucose and fructose when digested, and lactose yields glucose and galactose. Thus, the digestion of carbohydrates results principally in a mixture of three monosaccharides: glucose, fructose, and galactose. This monosaccharide mixture is transported across the membranes of the small intestine into the blood, which carries them by portal circulation to the liver. At the liver, fructose and galactose are either converted to glucose or into other compounds that can enter the same metabolic pathway as glucose. Thus, the metabolism of carbohydrates is essentially the metabolism of glucose. Carbohydrate digestion is summarized in Table 25.3.

		TABLE 25.3		
		A Summary of Carbohydrate Digestion		
Carbohydrate	Typical sources	Site of digestion	Required enzymes	Digestion products
Cellulose	fruits and vegetables	none	none	none
Starch	potatoes, beans, peas, breads	mouth ↓ small intestines ↓ small intestines	salivary amylase pancreatic amylase maltase	dextrins + maltose ↓ maltose ↓ glucose
Sucrose	table sugar; added to many processed foods	small intestines	sucrase	fructose + glucose
Lactose	milk and dairy products	small intestines	lactase	galactose + glucose
Maltose	small amounts in grain	small intestines	maltase	glucose
Glucose (free)	small amounts in fruit, vegetables	none required	none required	glucose
Fructose	equal amounts of glucose and fructose in honey, small amounts in fruits and vegetables	none required	none required	fructose

Glucose Metabolism

Let us begin this discussion of the metabolism of glucose at the point where it enters the cell.

Glucose can easily enter cells through the cell membranes, but to prevent it from leaving once inside, it is converted into glucose-6-phosphate. The reaction uses ATP as the source of phosphate, and is catalyzed by the hexokinase enzyme:

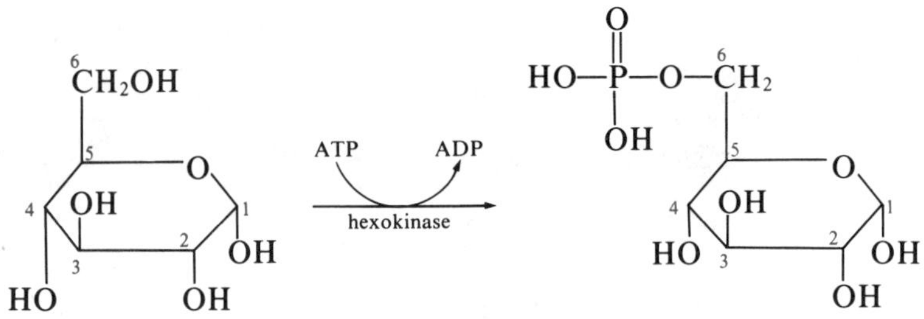

Glucose Glucose–6–phosphate

Glucose-6-phosphate is either used by the cell for energy (that is, metabolism continues) or it may be converted into glycogen and stored for further use. The generation of glycogen occurs in all cells, but glycogen storage is especially important in liver and muscle cells. Normally, there are about 100 g of glycogen in the liver, but it can be much higher. Glycogen is a large, highly branched polymer, composed of glucose units joined together through α-1,4 and α-1,6 ether linkages. It was described in Chapter 21.

Glycogen Formation—Glycogenesis

The conversion of glucose to glycogen is called **glycogenesis**. The first step in the formation of glycogen is the conversion of glucose to glucose-6-phosphate that was shown above. Glucose-6-phosphate is then converted to glucose-1-phosphate. The enzyme required in this conversion is phosphoglucomutase:

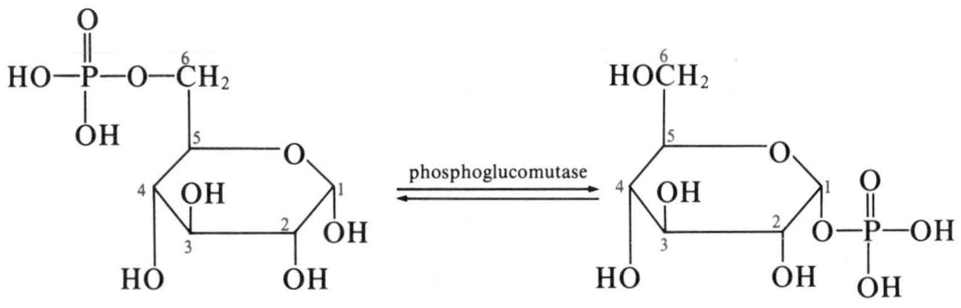

Glucose–6–phosphate

Glycogenesis is an anabolic process, and energy is required to build the polymer. The energy is obtained from the high-energy nucleotide uridine triphosphate (UTP), which reacts with glucose-1-phosphate to produce a uridine diphosphate-glucose molecule and pyrophosphate (PPi). For simplicity, the equation will be expressed in words:

$$\text{glucose-1-phosphate} + \text{UTP} \xrightleftharpoons{\text{pyrophosphorylase}} \text{UDP-glucose} + \text{PPi}$$

The UDP-glucose molecule carries glucose to the end of a growing glycogen chain. There, under the direction of the enzyme, glycogen synthesase, the glucose unit is transferred to the glycogen chain and becomes part of the glycogen polymer.

$$\text{UDP-glucose} + (\text{glucose})_n \xrightarrow{\text{glycogen synthetase}} (\text{glucose})_{n+1} + \text{UDP}$$

Glycogen — Glycogen (now with
one more glucose)

Glycogen Breakdown— Glycogenolysis

Glycogenolysis is the breaking down of glycogen to form glucose. It may be caused by a drop in the blood sugar level or by the release of the hormones epinephrine and glucagon. Though glycogen is principally stored in muscle and liver tissue, glycogenolysis can *only* occur in the liver where the necessary enzymes are present. The glycogen stored in muscle can be used as a source of energy once converted to glucose-6-phosphate, as you will see.

Glycogenolysis begins with the enzyme catalyzed cleavage of the α-1,4 ether links that hold the glucose units in the polymer chain. Glucose-1-phosphate is formed:

$$(\text{glucose})_n + \text{Pi} \xrightarrow{\text{phosphorylase}} (\text{glucose})_{n-1} + \text{glucose-1-phosphate}$$

Glycogen — Glycogen (now with
one less glucose unit)

Phosphoglucomutase then catalyzes the conversion of glucose-1-phosphate to glucose-6-phosphate, just the reverse of the reaction seen in glycogenesis:

$$\text{glucose-1-phosphate} \underset{\text{phosphoglucomutase}}{\rightleftharpoons} \text{glucose-6-phosphate}$$

Glucose-6-phosphate is then converted into glucose. The enzyme required for this process is in the liver, but it is the absent enzyme in muscle tissue. This is why glycogenolysis cannot occur in muscle tissue.

$$\text{glucose-6-phosphate} \xrightarrow{\text{glucose-6-phosphatase}} \text{glucose} + \text{Pi}$$

The glucose can then be released to the blood as required.

The Embden–Meyerhof Pathway

The first major step in the use of glucose for energy production in cells follows a series of reactions collectively called the **Embden–Meyerhof pathway**. The reactions take place in the cytoplasm of those cells that have a source of glucose, such as glycogen. The pathway of reactions can take place with or without the presence of oxygen. If oxygen is present, the end product is pyruvic acid, which can then enter the Krebs cycle via acetyl-CoA. If oxygen is absent, a typical occurence in vigorously exercised muscle tissue, the product of the pathway is lactic acid. Only a small part of the stored chemical energy in glucose is released in the form of ATP in the Embden–Meyerhof (EM) pathway. The remainder of the available energy is obtained as ATP in the formation of acetyl-CoA and the oxidation of the acetyl group in the Krebs cycle–ETS combination. Yet, the energy made available in the EM pathway fills an important energy requirement for the body, especially under conditions of oxygen depletion in cells. Since the EM pathway can operate anaerobically (in the absence of oxygen), it can continue to supply energy to cells after the aerobic processes have shut down. As was mentioned earlier, vigorous use of muscle tissue quickly exhausts its ability to carry out the aerobic oxidation of glucose (via the EM pathway and the Krebs cycle). The blood supply may not be able to replenish oxygen as fast as it is needed. The anaerobic EM pathway then becomes the principal energy source. At other times, when there is an adequate supply of oxygen, the EM pathway produces the pyruvic acid required for the formation of acetyl-CoA which enters the Krebs cycle.

The conversion of glucose through the Embden–Meyerhof pathway to either pyruvic acid (under aerobic conditions) or lactic acid (under anaerobic conditions) is called **glycolysis**. At one time glycolysis via the EM pathway referred only to the anaerobic sequence leading to lactic acid, but it may be used for either.

Let us go through the series of reactions that constitute the Embden–Meyerhof pathway. The compounds involved in the pathway are arranged in sequence in Figure 25.4. Some of the compounds may be new to you. But if you examine each one and note how it changes as it goes to the next step, the reactions can be understood. There are ten steps in the pathway that leads to pyruvic acid. The eleventh step, which reduces pyruvic acid to lactic acid, takes place in the absence of oxygen. The sixth step will ultimately determine if the eleventh step takes place. The two steps (6 and 11) interact.

Before looking at each step individually, it may be constructive to give an overview of the process. As the pathway begins, glucose is converted to glucose-6-phosphate, which in turn is converted into fructose-6-phosphate. A phosphate is then added, forming fructose-1,6-diphosphate. At this point in the pathway, the hexose is cleaved, forming dihydroxyacetone phosphate and glyceraldehyde-3-phosphate. The dihydroxyacetone phosphate isomerizes to glyceraldehyde-3-phosphate, which is then converted to 1,3-diphosphoglyceric acid, then into 3-phosphoglyceric acid, and further to 2-phosphoglyceric acid, which then is converted to pyruvic acid. Depending on the oxygen supply in the cell, the pyruvic acid may be reduced to lactic acid, or it may enter the Krebs cycle via acetyl-CoA.

Overall, the oxidation of glucose via the EM pathway in anaerobic glycolysis leading to lactic acid can be expressed in a single equation:

$$\text{glucose} + 2ADP + 2Pi \rightarrow 2 \text{ lactic acid} + 2ATP$$

Let us see what steps are involved that, when added together, produce such a straightforward equation.

Step 1. *Glucose* is converted to *glucose-6-phosphate*. This requires the use of one molecule of ATP and the enzyme hexokinase, and the reaction is irreversible:

$$\text{glucose} \xrightarrow[\text{hexokinase}]{\text{ATP} \quad \text{ADP}} \text{glucose-6-phosphate}$$

Step 2. *Glucose-6-phosphate* is converted to *fructose-6-phosphate*. The two compounds are isomers, and the enzyme phosphoglucoisomerase is required:

$$\text{glucose-6-phosphate} \xrightleftharpoons{\text{phosphoglucoisomerase}} \text{fructose-6-phosphate}$$

Step 3. ATP is required for the conversion of *fructose-6-phosphate* to *fructose-1,6-diphosphate*. This is an irreversible reaction and the last reaction in the pathway that consumes ATP. The enzyme phosphofructokinase is required.

$$\text{fructose-6-phosphate} \xrightarrow[\text{phosphofructokinase}]{\text{ATP} \quad \text{ADP}} \text{fructose-1,6-diphosphate}$$

Some fructose and galactose will be converted to fructose-1,6-diphosphate in the body and enter the EM pathway at this point. Most, though, is converted to glucose in the liver and enters the pathway at the beginning.

Step 4. The hexose *fructose-1,6-diphosphate* is split into two trioses, *dihydroxyacetone phosphate* and *glyceraldehyde-3-phosphate*. Aldolase catalyzes the breakdown.

$$\text{fructose-1,6-diphosphate} \xrightleftharpoons{\text{aldolase}} \begin{cases} \text{dihydroxyacetone phosphate} \\ \text{glyceraldehyde-3-phosphate} \end{cases}$$

Step 5. Dihydroxyacetone phosphate cannot be directly used by cells for the production of energy, but it can be converted by the enzyme triose phosphate isomerase into glyceraldehyde-3-phosphate, which can continue in the pathway. At this point, one molecule of glucose has resulted in the formation of two molecules of glyceraldehyde-3-phosphate. As the production of ATP occurs in the following steps, we need to keep in mind that two reactant molecules will be equivalent to one glucose molecule.

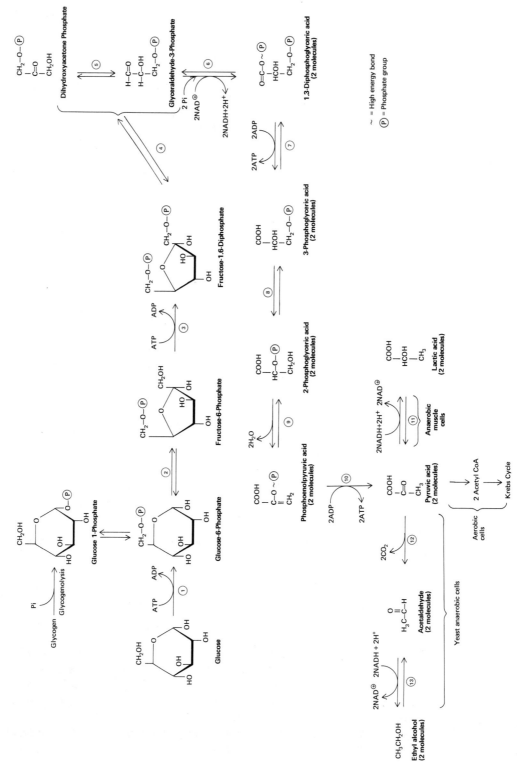

Fig. 25.4. The Embden–Meyerhof pathway of glycolysis.

This fifth step in the EM pathway is also important in the metabolism of triglycerides (fats). The glycerol that is formed when a triglyceride is hydrolyzed ends up as glyceraldehyde-6-phosphate. It can then enter the EM pathway at this point.

$$\text{dihydroxyacetone phosphate} \underset{}{\overset{\text{triose phosphate}}{\underset{\text{isomerase}}{\rightleftharpoons}}} \text{glyceraldehyde-3-phosphate}$$

Step 6. *Glyceraldehyde-3-phosphate* is oxidized and phosphorylated (addition of a phosphate group) to form *1,3-diphosphoglyceric acid*, a high-energy compound listed in Table 25.1. The oxidation is brought about by the coenzyme $NAD^{\oplus}$ associated with glyceraldehyde-3-phosphate dehydrogenase. We will use structural formulas here to show the high-energy phosphate bond. Two molecules of reactant and product are shown in the equation:

Glyceraldehyde–3–
phosphate

1,3–diphosphoglyceric
acid

This sixth step in the pathway is especially important in determining whether the final product of glycolysis is lactic acid or pyruvic acid. If the passage of compounds through the pathway is to continue, the NADH formed in step 6 must be oxidized back to $NAD^{\oplus}$, or the consumed $NAD^{\oplus}$ must be replaced from another source. If oxygen is present in the cell, NADH can be oxidized to $NAD^{\oplus}$ in the ETS. On the other hand, if oxygen is absent, the only way $NAD^{\oplus}$ can be obtained is in step 11, which converts pyruvic acid to lactic acid. This is what happens in anaerobic glycolysis. Step 6 and step 11 trade NADH and $NAD^{\oplus}$ back and forth so the pathway is complete. We will have more to say about this when we get to step 11.

Step 7. The high-energy phosphate bond ($\sim$) in *1,3-diphosphoglyceric acid* will convert ADP to ATP at this point, forming *3-phosphoglyceric acid*. Since we are now considering the results of two diphosphate molecules, two molecules of ATP will be produced. The cycle is now even in its use of ATP: 2ATP are consumed (steps 1 and 3) and 2ATP are produced in step 7. You may recall this process is described as substrate-level phosphorylation.

$$2(1,3\text{-diphosphoglyceric acid}) \xrightleftharpoons[\text{phosphoglycero-kinase}]{\overset{\text{2ADP} \quad \text{2ATP}}{}} 2 \text{ (3-phosphoglyceric acid)}$$

Step 8. *3-phosphoglyceric acid* is catalyzed by the enzyme, phosphoglycermutase, into its isomer, 2-phosphoglyceric acid:

$$2 \text{ (3-phosphoglyceric acid)} \xrightleftharpoons{\text{phosphoglycermutase}} 2(2\text{-phosphoglyceric acid})$$

Step 9. Water is removed from *2-phosphoglyceric acid* forming *phosphoenolpyruvic acid*. The phosphate group becomes a high-energy phosphate in the process. Enolase catalyzes the reaction.

2–phosphoglyceric acid   Phosphoenolpyruvic acid

Step 10. The high energy *phosphoenolpyruvic acid* transfers its phosphate to ADP producing ATP and *pyruvic acid*. The irreversible reaction is catalyzed by pyruvic acid kinase. Since we need to consider the energy production of two molecules in this step, a total of two molecules of ATP are produced.

Phosphoenolpyruvic acid   Pyruvic acid

Once pyruvic acid is formed, its fate will be determined by the availability of oxygen in the cell, as was described in step 6. If oxygen is deficient, pyruvic acid will be reduced to lactic acid by NADH, as shown in step 11.

Step 11. In anaerobic glycolysis, the pyruvic acid produced in step 10 is reduced to lactic acid by the NADH coenzyme. The reduction of pyruvic acid is accompanied by the simultaneous oxidation of NADH to $NAD^{\oplus}$, which is returned to step 6 to replace that used in the oxidation of glyceraldehyde-3-phosphate.

$$2 \quad CH_3-\overset{\overset{\displaystyle O}{\|}}{C}-COOH \underset{\text{lactic dehydrogenase}}{\overset{2(NADH+H^+) \quad 2NAD^{\oplus}}{\rightleftharpoons}} 2 \quad CH_3-\overset{\overset{\displaystyle OH}{|}}{\underset{\underset{\displaystyle H}{|}}{C}}-COOH$$

Pyruvic acid

Lactic acid

Anaerobic glycolysis is the principal energy source in muscle cells during vigorous exercise. The accumulation of lactic acid is largely responsible for the muscle soreness experienced by all of us during and after strenuous physical labor. About 80% of the lactic acid eventually diffuses into the bloodstream and is carried to the liver where it is either oxidized to pyruvic acid for eventual entry into the Krebs cycle (via acetyl-CoA), or it will be converted into glycogen and stored. The lactic acid remaining in muscle tissue will be later oxidized to pyruvic acid and eventually enter the Krebs cycle once the oxygen supply in the cell is restored.

Anaerobic glycolysis cannot continue indefinitely. The buildup of lactic acid, and its eventual entry into the blood, will bring about mild metabolic acidosis (Chapter 13). Also, the cells develop what is called an **oxygen debt**, indicating the eventual need for oxygen to return to aerobic operation. The oxygen debt must be repaid eventually so the cell can oxidize the accumulated lactic acid and return to the more efficient aerobic production of ATP.

The body responds to acidosis by increasing the depth and rate of breathing. The human limitation imposed by the maximum rate and depth at which an individual can breathe in response to acidosis limits vigorous exercise to only brief periods. We need periods of rest during physical exercise so the body can catch up with the oxygen demand.

Fermentation

Anaerobic glycolysis via the EM pathway is used as the principal energy source in yeast. You are aware that yeast is necessary for the fermentation of glucose to ethyl alcohol. The alcohol is produced by the conversion of pyruvic acid first to acetaldehyde, followed by the reduction of acetaldehyde by NADH to ethyl alcohol.

$$\underset{\text{Pyruvic acid}}{\overset{\displaystyle COOH}{\underset{\displaystyle CH_3}{\overset{\displaystyle |}{\underset{\displaystyle |}{C=O}}}}} \quad \xrightarrow[\text{enzyme}]{CO_2} \quad \underset{\text{Acetaldehyde}}{\overset{\displaystyle H\;\;O}{\underset{\displaystyle CH_3}{\underset{\displaystyle |}{C}}}} \quad \underset{\text{Ethyl alcohol}}{\overset{\displaystyle H_2C-OH}{\underset{\displaystyle CH_3}{\underset{\displaystyle |}{CH_3}}}}$$

The $NAD^{\oplus}$ is sent to step 6 in the pathway, so the series of reactions can continue. These conversions appear in Figure 25.4 as step 12 and step 13. The fermentation of grains to produce ethyl alcohol is anaerobic glycolysis.

Check Test Number 3

1. Which two steps in the Embden–Meyerhof pathway are connected in terms of $NAD^{\oplus}$ and NADH?
2. Under what conditions do muscle cells convert glucose to lactic acid?
3. Why is it necessary to consider two molecules of pyruvic acid being produced in the EM pathway for each glucose molecule that entered?
4. What is the oxidizing agent in anaerobic glycolysis?

Answers:

1. Steps 6 and 11.
2. Anaerobic conditions.
3. The 6-carbon hexose is split into two 3-carbon units, glyceraldehyde-6-phosphate and dihydroxyacetone phosphate, in step 4. One molecule → two molecules.
4. $NAD^{\oplus}$

The ATP Yield From the Embden–Meyerhof Pathway

Let us summarize the production of ATP in the Embden–Meyerhof pathway for both aerobic and anaerobic glycolysis.

One molecule of glucose undergoing anaerobic glycolysis will produce a total of two ATP molecules. The result is the same whether the product is lactic acid or ethyl alcohol (in fermentation).

Anaerobic glycolysis in muscle:

$$C_6H_{12}O_6 \;+\; 2ADP \;+\; 2Pi \;\longrightarrow\; \underset{\text{Lactic acid}}{2\;\overset{\displaystyle OH}{\underset{\displaystyle H}{\underset{\displaystyle |}{CH_3-\overset{|}{C}-COOH}}}} \;+\; 2ATP$$

Fermentation:

$$C_6H_{12}O_6 + 2ADP + 2Pi \rightarrow 2\ CH_3CH_2OH + 2CO_2 + 2ATP$$

Glucose   Ethyl alcohol

The use and production of ATP in both anaerobic oxidations are summarized below:

	Reactions involving ATP in anaerobic glycolysis	**Number of ATP molecules per molecule of glucose**
Step 1.	Glucose → glucose-6-phosphate	−1
Step 3.	Fructose-6-phosphate → fructose-1,6-diphosphate	−1
Step 7.	2 (1,3-diphosphoglyceric acid) → 2 (3-phosphoglyceric acid)	2
Step 10.	2 (phosphoenolpyruvic acid) → 2 (pyruvic acid)	2
	Net gain:	2

Under aerobic conditions, glycolysis through the EM pathway will produce an additional six molecules of ATP through the ETS in step 6. Because oxygen is present, the oxidation of NADH to $NAD^\oplus$ is able to be carried out in the ETS. Since two molecules of glyceraldehyde-3-phosphate arise from one molecule of glucose, a total of six ATP will be formed, three from each of the two glyceraldehyde-3-phosphate molecules oxidized. The equation summing up the aerobic oxidation of glucose in the EM pathway is

$$C_6H_{12}O_6 + 8ADP + 8Pi \longrightarrow 2\ CH_3\overset{\overset{\displaystyle O}{\|}}{C}-COOH + 8ATP$$

Pyruvic acid

The use and production of ATP in aerobic glycolysis is summarized below:

	Reactions involving ATP in aerobic glycolysis	**Number of ATP molecules per molecule of glucose**
Step 1.	Glucose → glucose-6-phosphate	−1
Step 3.	Fructose-6-phosphate → fructose-1,6-diphosphate	−1
Step 6.	The conversion of 2NADH to $2NAD^\oplus$ by the ETS	6
Step 7.	2 (1,3-diphosphoglyceric acid) → 2 (3-phosphoglyceric acid)	2
Step 10.	2 (phosphoenolypyruvic acid) → 2 (pyruvic acid)	2
	Net gain:	8

Continuing on with the aerobic oxidation of glucose, with the conversion of pyruvic acid to acetyl coenzyme A and its oxidation in the Krebs cycle, we can determine the total number of ATP molecules produced by one molecule of glucose as it is oxidized to CO_2 and water in cellular respiration. The oxidation of one molecule of glucose to two molecules of pyruvic acid yields eight ATP. Earlier it was shown that the oxidation required to convert one pyruvic acid molecule to the acetyl group on coenzyme A produced three ATP. The glucose equivalent in this conversion (two pyruvic acid) would yield a total of six ATP. Passage of acetyl-CoA through the Krebs cycle–ETS combination produces 12 ATP, as was shown earlier, so the two acetyl-CoA glucose equivalent would produce 24 ATP. Once through the Krebs cycle, glucose is completely oxidized to CO_2 and H_2O, and the oxidation is complete. Let us summarize the results.

The net reaction for the aerobic oxidation of glucose to CO_2 and water is

$$C_6H_{12}O_6 + 6O_2 + 38ADP + 38Pi \rightarrow 6H_2O + 6CO_2 + 38ATP$$

$\uparrow$ Used in ETS $\qquad\qquad\qquad\qquad\qquad$ $\uparrow$ Available for cellular use

A summary of the sources of ATP appears below for the aerobic oxidation of glucose:

glucose → 2 pyruvic acid	8 ATP
2(pyruvic acid) → 2 (acetyl-CoA)	6 ATP
2 (acetyl-CoA) → $4CO_2$ + $4H_2O$ in Krebs cycle–ETS combination	24 ATP
Grand total:	38 ATP

So far, our calculations have dealt with molecules of glucose and ATP, but we could just as easily speak in terms of moles. One mole of glucose produces 38 moles of ATP in complete aerobic oxidation. We could make a calculation, albeit crude, of the efficiency with which the body extracts the potential energy stored in glucose for storage as ATP. Earlier it was stated that one mole of ATP produced 7.3 kcal of energy as it is converted to ADP. Thus, 38 moles of ATP would produce (38 moles) × (7.3 kcal/mole) = 277 kcal of energy for cellular use. How does this compare with the oxidation of glucose in the laboratory, that is, the conversion of one mole of glucose to CO_2 and water? One mole of glucose when burned in an oxygen atmosphere to CO_2 and water produces 686 kcal of energy that can be used for useful work. The efficiency of aerobic oxidation of glucose in the body is then

$$\% \text{ efficiency} = \frac{277 \text{ kcal}}{686 \text{ kcal}} \times 100\% = 40.4\%$$

The body can obtain about 40% of the potential energy stored in glucose using its most productive route, aerobic oxidation. The energy not stored as ATP is largely lost as heat.

Lipid Metabolism

Lipids were discussed in Chapter 20, and you may wish to read over the introductory section of that chapter to refresh your memory concerning this class of compounds.

We obtain lipids in our diets from both plant and animal tissue. Most of this lipid material is fat, the common term used for triglycerides. The digestion of lipids takes place in the small intestine. The enzyme, pancreatic lipase, catalyzes the hydrolysis of the triglycerides to form free fatty acids, salts of fatty acids, glycerol, monoglycerides, and diglycerides:

$$
\begin{array}{c}
\underset{\text{Triglyceride}}{
\begin{array}{l}
H_2C-O-\overset{\displaystyle O}{\overset{\|}{C}}-R \\[4pt]
H-\underset{}{\overset{}{C}}-O-\overset{\displaystyle O}{\overset{\|}{C}}-R \\[4pt]
H_2C-O-\overset{\displaystyle O}{\overset{\|}{C}}-R
\end{array}}
\xrightarrow[\text{(+ H}_2\text{O)}]{\text{pancreatic lipase}}
\end{array}
$$

R—COOH
Fatty acid

$R-COO^{\ominus}, M^{+}$
Fatty acid salt

H_2C-OH
$H-C-OH$
H_2C-OH
Glycerol

$H_2C-O-\overset{O}{\overset{\|}{C}}-R$
$H-C-O-\overset{O}{\overset{\|}{C}}-R$
H_2C-OH
Diglyceride

$H_2C-O-\overset{O}{\overset{\|}{C}}-R$
$H-C-OH$
H_2C-OH
Monoglyceride

These hydrolysis products are able to pass through the intestinal wall into the lymph system. There, they re-form triglycerides and briefly exist as small droplets of fat suspended in the lymphatic fluid. The small amount of phospholipid material that appears in the diet is hydrolyzed, carried into the lymph system, and re-formed in much the same way as the triglycerides. Cholesterol requires no digestion, and is absorbed directly into the lymph system. Some of the cholesterol in lymph will combine with free fatty-acid-forming esters. In the lymph, the triglycerides, phospholipids, cholesterol, and the small amount of free fatty acid that is present all combine with protein to form *lipoprotein* complexes. Lipoprotein acts as a carrier for lipids as they are transported through the lymph system to eventually enter the bloodstream at the left subclavian vein.

At this point it would be useful to say something about lipid levels in the blood, since they are frequently measured as part of a thorough physical examination, and in the evaluation of patients with heart disease. Lipid levels in the blood may rise sharply following a meal rich in fats. In fact, the serum may become cloudy due to the presence of large amounts of lipoprotein. Much of this lipid-containing material will be removed from the blood within a few hours, and enter the liver and adipose tissue. The blood serum then will return to its normal, clear state. Because lipid levels rise and fall in response to the diet, they should be measured only after a period of fasting, usually overnight. Typical fasting levels for total lipid content for adults will be between 450 and 1000 mg/100 ml of blood serum. Between 150 and 280 mg of the total will be due to cholesterol, about 70% of which will be in the form of cholesterol esters. Between 40 and 150 mg of the total will be due to triglycerides. Abnormally high fasting serum levels of triglycerides and cholesterol are believed to contribute to hardening of the arteries, resulting in high blood pressure, strokes, and heart attacks, though recent evidence suggests that some forms of cholesterol, the high-density fraction, may actually retard the buildup of plaque on arterial walls.

Cholesterol and the phospholipids are not used as such by the body as sources of energy. They are sent to the liver, broken down, and used to synthesize other lipids needed by the body. The triglycerides, fatty acids, and glycerol are used to produce energy. If the triglycerides that enter the bloodstream are not needed right away as an energy source, they are stored in the liver and adipose tissue as body fat. On a per gram basis, the triglycerides represent the most efficient energy storage system in the body. One gram of fat can liberate about 9 kcal of energy in metabolism. Carbohydrates and proteins (when used in metabolism) both yield about 4 kcal/g.

The adipose tissue is in a steady state of activity, with the storage of newly arrived triglycerides and the hydrolysis of older ones to produce glycerol and fatty acids for use by cells. When the intake of food is in excess of that required to meet daily needs, carbohydrates and proteins can be converted to triglycerides through various pathways and also stored in adipose tissue. It is estimated that 30% of our carbohydrate intake ends up as adipose tissue and stored for later use.

The glycogen supply in cells can provide the energy needed by the body for only a few hours in the absence of an intake of food. Once the glycogen supply is depleted, the body calls out the reserves, the adipose tissue, to supply fuel for the body. Once the triglycerides stored in adipose tissue are hydrolyzed by enzyme action, the resulting fatty acids and glycerol are carried to the liver by the blood, the principal site of fatty acid metabolism. Fatty acids can also be metabolized in the heart and skeletal muscles. In fact, the majority of energy produced in resting muscle tissue is obtained from fatty acid metabolism, and not from the metabolism of glucose. Glycerol is converted to dihydroxyacetone phosphate, which will then enter the Embden–Meyerhof pathway:

Glycerol   Dihydroxyacetone phosphate (to EM pathway)

The subsequent oxidation of the fatty acids occurs in the mitochondria in a process known as **beta-oxidation** as shown in the reactions of the **fatty acid cycle**; Figure 25.5. As the fatty acid (in combination with coenzyme A) is carried around the cycle, two carbon atoms are removed each time, in a four-step sequence, to produce acetyl-CoA. The acetyl-CoA may then enter the Krebs cycle or it may be used in various synthesis (anabolism) reactions.

Let us go through the fatty acid cycle step by step to see how the fatty acid is converted two atoms at a time to acetyl-CoA. Refer to Figure 25.5 as you go along to follow the action.

Before a fatty acid can enter the cycle, it must be joined to coenzyme A. The resulting unit is called an **acyl coenzyme A**. Formation of acyl-CoA "activates" the fatty acid and readies it for oxidation in the cycle. Two carbon atoms in the fatty acid are especially important, and they are labeled α (alpha) and β (beta). As the acyl-CoA moves through the cycle, the fatty acid will be cleaved between the α and β carbons.

The "activation" of the fatty acid requires energy, which is supplied by ATP, and an enzyme, acyl coenzyme A synthetase. Notice that ATP is converted to ADP + PPi (pyrophosphate):

Fatty acid   acyl CoA

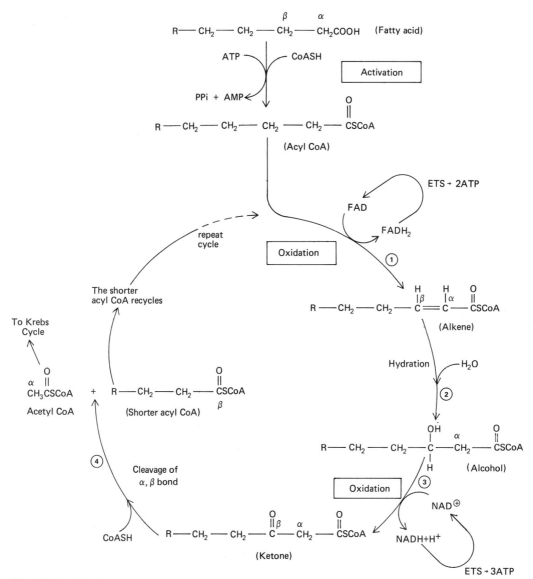

Fig. 25.5. The fatty acid cycle showing the β-oxidation of fatty acids.

Now, let us take the activated fatty acid in the form of an acyl-CoA through the fatty acid cycle. Each step is catalyzed by an enzyme though they will not be specifically identified.

Step 1. The coenzyme FAD, associated with one of several dehydrogenase enzymes, catalyzes the removal of two hydrogen atoms, one from the α and one from the β carbon. A double bond is produced between the α and β carbon atoms. The $FADH_2$ is oxidized in the electron transport system back to FAD, producing two ATP molecules.

$$R-CH_2CH_2-\overset{\underset{|}{H}}{\underset{H}{\overset{\overset{H}{|}}{\underset{\beta}{C}}}}-\overset{\underset{|}{H}}{\underset{H}{\overset{\overset{H}{|}}{\underset{\alpha}{C}}}}-\overset{\overset{O}{||}}{C}-SCoA \xrightarrow[\text{dehydrogenase}]{\overset{\text{2ATP}\quad\text{ETS}}{\text{FAD}\quad\text{FADH}_2}} R-CH_2CH_2-\overset{\underset{\beta}{}}{C}=\overset{\underset{\alpha}{}}{\underset{H}{\overset{}{C}}}-\overset{\overset{O}{||}}{C}-SCoA$$

Step 2. Water is added to the double bond (hydration). The hydroxyl (–OH) joins to the β-carbon, the hydrogen (–H) to the α-carbon. A secondary (2°) alcohol group is formed.

$$R-CH_2CH_2-C=\underset{H}{C}-\overset{\overset{O}{||}}{C}-SCoA \xrightarrow[\text{enzyme}]{H_2O} R-CH_2CH_2-\overset{\underset{|}{H}}{\underset{H}{\overset{\overset{OH}{|}}{\underset{\beta}{C}}}}-\overset{\underset{|}{H}}{\underset{H}{\overset{\overset{H}{|}}{\underset{\alpha}{C}}}}-\overset{\overset{O}{||}}{C}-SCoA$$

Step 3. The secondary alcohol group on the β-carbon is oxidized by $NAD^{\oplus}$ to a carbonyl group. A ketone is formed. The NADH is oxidized back to $NAD^{\oplus}$ in the ETS and three ATP molecules are produced.

$$R-CH_2CH_2-\overset{\underset{|}{H}}{\overset{\overset{OH}{|}}{C}}-CH_2-\overset{\overset{O}{||}}{C}-SCoA \xrightarrow[\quad]{\overset{\text{ETS}}{\underset{NAD^{\oplus}\quad NADH+H^+}{\text{3ATP}\leftarrow}}} R-CH_2CH_2-\overset{\overset{O}{||}}{\underset{\beta}{C}}-\overset{\alpha}{CH_2}-\overset{\overset{O}{||}}{C}-SCoA$$

The α–β carbon–carbon bond is weakened at this point.

Step 4. The α–β carbon–carbon bond is cleaved. A new molecule of CoASH enters and bonds to the β-carbon of the remaining fatty acid fragment forming another (two-carbon atoms shorter) acyl-CoA. The two carbons cleaved from the acid remain with the coenzyme A forming acetyl-coenzyme A. The new acyl-CoA takes another turn and around the cycle, and loses two more carbon atoms:

$$R-CH_2-CH_2-\overset{\overset{O}{||}}{\underset{\beta}{C}}-\overset{\alpha}{CH_2}-\overset{\overset{O}{||}}{C}-SCoA \xrightarrow[\text{enzyme}]{\text{CoASH}} \underset{\substack{\uparrow \\ \text{continues around} \\ \text{the cycle, back} \\ \text{to Step 1}}}{R-\overset{\text{new }\beta}{CH_2}\overset{\text{new }\alpha}{CH_2}-\overset{\overset{O}{||}}{C}-SCoA} + \underset{\substack{\uparrow \\ \text{may enter} \\ \text{Krebs} \\ \text{cycle}}}{CH_3\overset{\overset{O}{||}}{C}-SCoA}$$

The path around the fatty acid cycle will be followed again and again until the fatty acid is completely converted into acetyl-CoA. Since the β-carbon is oxidized in the cycle (to a ketone), you can see why the term β-oxidation is used to describe this series of reactions.

The ATP Yield from Fatty Acid Metabolism

Let us start with one mole of a typical fatty acid, palmitic acid, $CH_3-(CH_2)_{14}-COOH$, and determine the number of moles of ATP that will form as it becomes activated and is carried through the fatty acid cycle. Then, we will add to this the number of moles of ATP that will form as the eight moles of acetyl-CoA pass through the Krebs cycle–ETS combination. The sum of all this will be the total ATP production.

Activation of one mole of the fatty acid consumes the equivalent of two moles of ATP. Regenerating one mole of ATP from one mole of AMP is the energy equivalent of regenerating two moles of ATP from two moles of ADP. Both require the addition of two Pi. The fatty acid would need to pass around the cycle *seven* times, and this would produce eight moles of acetyl-CoA. (Unlike the first six passes, the seventh pass produces two moles of acetyl-CoA.) Seven turns around the fatty acid cycle would produce seven moles of $FADH_2$, and seven moles of $NADH + H^+$. The oxidation of the seven moles of $FADH_2$ back to FAD in the ETS produces (7 moles $\times$ 2 moles ATP/mole) 14 moles of ATP. The oxidation of seven moles of NADH back to NAD^+ in the ETS produces (7 moles $\times$ 3 moles ATP/mole) 21 moles of ATP. Adding this together produces quite a yield of ATP per mole of palmitic acid.

Reaction involving ATP in fatty acid metabolism	Moles of ATP per mole of palmitic acid
Activation of palmitic acid (ATP $\rightarrow$ AMP + PPi)	-2
8 acetyl CoA formed (12 ATP/acetyl CoA in Krebs cycle)	96
7 $FADH_2 \rightarrow$ 7 FAD (ETS=2 ATP/oxidation)	14
7 NADH $\rightarrow$ 7 $NAD^{\oplus}$ (ETS=3 ATP/oxidation)	21
Grand total:	129 moles

One mole of palmitic acid, when completely oxidized to CO_2 and water in the laboratory, produces 2,240 kcal of energy. The percentage of this energy that ends up being stored as ATP can be estimated. Assuming that each mole of ATP will be converted to one mole of ADP + Pi with the release of 7.3 kcal, the total amount of energy available from 129 moles of ATP is (129 mole × 7.3 kcal/mole) 942 kcal. The percentage of the energy in palmitic acid converted to ATP in its aerobic oxidation to CO_2 and water is a measure of the efficiency of the process:

$$\text{efficiency} = \frac{942 \text{ kcal}}{2240 \text{ kcal}} \times 100\% = 42.1\%$$

Thus, about 42% of the energy available in the complete oxidation of a typical fat ends up as ATP for cellular use. The rest of the energy, that not stored in ATP, is lost primarily as heat.

Check Test Number 4

1. What is the principal kind of lipid in the diet?
2. In what organ are the fatty acids converted to acetyl CoA?
3. Beginning at the activation step, how many moles of ATP would be produced if one mole of butyric acid is first converted to two moles of acetyl CoA, which are then oxidized to CO_2 and water in the Krebs cycle? Butyric acid is $CH_3CH_2CH_2COOH$.

Answers:
1. Triglycerides (fats).
2. The liver.
3. 32 moles of ATP

	ATP
activation	−2
fatty acid cycle (2 turns)	
2 FADH$_2$ → 2 FAD (in ETS)	4
2 NADH → 2 NAD$^{\oplus}$ (in ETS)	6
2 acetyl CoA in Krebs–ETS combination	24
Total =	32 moles of ATP

Acetyl Coenzyme A—The Supreme Intermediate

By this time you are certainly aware of the importance of acetyl coenzyme A. In carbohydrate metabolism, before the product of glycolysis can enter the Krebs cycle, it must first be converted into acetyl CoA. The breakdown of fatty acids in the fatty acid cycle produces acetyl CoA for eventual use in the Krebs cycle. Many of the amino acids produced in the digestion of protein end up as acetyl CoA, as you will see in the discussion of protein metabolism.

Acetyl coenzyme A is more than the carrier of fuel to the Krebs cycle; it is an important intermediate in the synthesis of many compounds in the body. It plays a vital role in the synthesis of fatty acids, triglycerides, cholesterol, and other steroids. It is also used to synthesize glucose, which leads to glycogen formation, and the synthesis of amino acids, which leads to protein formation.

The multiple roles played by acetyl coenzyme A in the chemistry of the body are summarized in Figure 25.6. It is the supreme intermediate in both anabolic and catabolic metabolism.

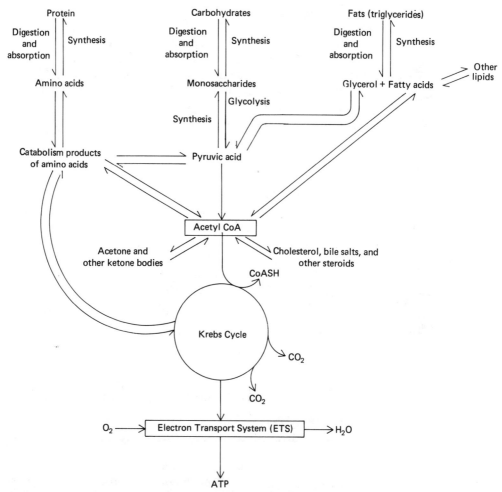

Fig. 25.6. The central role of acetyl coenzyme A in metabolism.

Metabolism Out of Balance—Ketosis

Under conditions of starvation or diabetes mellitus, carbohydrate metabolism is greatly reduced either because of a lack of carbohydrate intake or the inability of blood glucose to enter cells. The production of pyruvic acid is greatly reduced, which in turn, by means we will not discuss, reduces the availability of certain intermediates required in the Krebs cycle. Coupled with this, oxaloacetic acid is channeled away from its vital role in the Krebs cycle to be used in the synthesis of glucose in the liver. The weakened Krebs cycle is then unable to drive the ETS to produce enough ATP.

The absence of adequate carbohydrate metabolism naturally causes the body to increase fatty acid metabolism, resulting in the rapid production of acetyl-CoA. The acetyl-CoA level rises far above the level that can be handled by the now weakened Krebs cycle. The body responds to this flood of acetyl-CoA by converting the excess to acetoacetyl CoA. Two molecules of acetyl CoA produce one molecule of acetoacetyl CoA.

$$2 \quad CH_3-\overset{\overset{\displaystyle O}{\|}}{C}-SCoA \quad \xrightarrow[\text{ketothiolase}]{\text{CoASH}} \quad CH_3-\overset{\overset{\displaystyle O}{\|}}{C}-CH_2-\overset{\overset{\displaystyle O}{\|}}{C}-SCoA$$

Acetoacetyl CoA

Acetoacetyl CoA is converted in the liver to three compounds that are collectively known as **ketone bodies**: acetone, acetoacetic acid, and β-hydroxybutyric acid. Acetoacetic acid and β-hydroxybutyric acid exist as anions at physiological pH, but we will show them as the neutral compounds with the ionizable hydrogens in place.

$$CH_3-\overset{\overset{\displaystyle O}{\|}}{C}-CH_2-\overset{\overset{\displaystyle O}{\|}}{C}-SCoA$$

deacylase $\searrow$ CoASH

$$CH_3-\overset{\overset{\displaystyle O}{\|}}{C}-CH_3 \quad \xleftarrow[\text{decarboxylase}]{CO_2} \quad CH_3C-CH_2-\overset{\overset{\displaystyle O}{\|}}{C}-OH \quad \xrightarrow[\text{dehydrogenase}]{NADH+H^+ \quad NAD^\oplus} \quad CH_3-\overset{\overset{\displaystyle OH}{|}}{\underset{\underset{\displaystyle H}{|}}{C}}-CH_2-\overset{\overset{\displaystyle O}{\|}}{C}-OH$$

Acetone

Acetoacetic acid

β-hydroxybutyric acid
(not a ketone but classed as a ketone body)

Ketone bodies enter the bloodstream from the liver and normally appear in blood at levels close to 3 mg of ketone bodies per 100 ml of serum. Acetoacetic acid and β-hydroxybutyric acid can be used by heart and skeletal muscle as sources of energy. But if the disorders in carbohydrate metabolism that brought about the weakening of the Krebs cycle are not corrected, ketone body levels in blood may go as high as 90 mg/100 ml of serum. Abnormally high levels of ketone bodies in blood is a condition known as **ketoanemia**. The kidneys normally excrete about 100 mg of ketone bodies in a 24-hr day, and this may increase to 5000 mg/day under these conditions. The appearance of high levels of ketone bodies in urine is a condition known as **ketonuria**. High levels of ketone bodies in the blood will increase the diffusion of acetone into the lungs, resulting in the fruity-sweet "acetone breath" that is a symptom of the severe diabetic condition. Ketonuria and ketoanemia together characterize the condition of **ketosis**.

The buildup of acetoacetic acid and β-hydroxybutyric acid in the blood can result in acidosis by reducing the concentration of bicarbonate ion. This shifts the control region of the carbonic acid–bicarbonate buffer to lower pH. If uncorrected, oxygen transport by the blood decreases, rapid dehydration takes place, and an individual may fall into a coma, which may eventually lead to death.

By taking insulin at regular intervals, diabetics can maintain an adequate level of carbohydrate metabolism, thus reducing the risk of ketosis.

Protein Metabolism

Proteins were described in Chapter 22, and you may wish to refer to that chapter occasionally to refresh your memory if you are uncertain about terms or the names of compounds.

The digestion of proteins begins in the stomach with the combined action of hydrochloric acid and the enzyme pepsin. There, proteins are broken down into smaller fragments of several amino acids known as polypeptides. Once in the small intestine, proteolytic enzymes from the pancreatic and intestinal juices hydrolyze the polypeptides into amino acids. The amino acids pass through the intestinal wall and enter the bloodstream and become part of the **amino acid pool**. The amino acid pool is simply the total of all the amino acids circulating in blood and other fluids throughout the body. The composition of the pool is constantly changing as new amino acids enter and others leave for use in the synthesis of protein and other compounds.

The protein in the body is constantly being degraded and rebuilt. The time it takes for half of a specific protein to be degraded and replaced is the **half-life** or **turnover rate** for that protein. The turnover rate for liver and plasma proteins

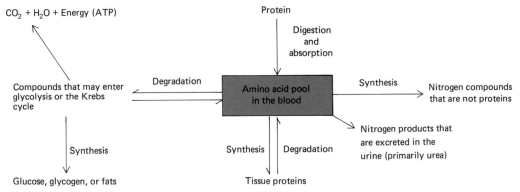

Fig. 25.7 The amino acid pool and how it is used in the body.

is about six days. Half of the protein molecules that make up your blood plasma today will be replaced by new protein in less than one week. The turnover rate for muscle tissue is about 180 days, and for collagen, the protein in structural tissue, it is about 1000 days.

Amino acids represent the principal dietary source of nitrogen for the body. They provide the nitrogen required for the synthesis of nonprotein, nitrogen-containing compounds, such as heme and the bases in DNA and RNA. If the nitrogen intake equals the nitrogen excreted, a condition of **nitrogen balance** exists. Growing children should have a positive nitrogen balance, taking in more nitrogen than they excrete. A negative nitrogen balance will occur in dissipating diseases and starvation.

The involvement of the amino acid pool in bodily processes is shown in Figure 25.7.

The Catabolism of Amino Acids

Protein metabolism is essentially the metabolism of the amino acids. Most of the amino acids in the amino acid pool are used for protein synthesis; the surplus is either used to make other nitrogen-containing compounds or converted to compounds that can be used in the Krebs cycle.

Catabolism of the amino acid surplus takes place in the liver. Each different amino acid is degraded in a specific pathway, though many of these pathways are interrelated. Because each amino acid has its own unique route for entry into the Krebs cycle, just those steps that are common to all will be described here, namely, the removal of the $-NH_2$ group and its conversion to urea.

The first step in the catabolism of an amino acid is the removal of the amino group. This can be accomplished by a process known as **transamination**. The amino group is removed from the amino acid as it becomes an α-keto acid. The amino group is transferred, usually to α-ketoglutaric acid (other α-keto acids are also sometimes used), to form glutamic acid:

$$
\begin{array}{c}
\begin{array}{c}
\text{COOH} \\
|\\
\text{H--C--NH}_2 \\
|\\
\text{R} \\
\text{Amino acid}
\end{array}
+
\begin{array}{c}
\text{COOH} \\
|\\
\text{C=O} \\
|\\
\text{CH}_2 \\
|\\
\text{CH}_2 \\
|\\
\text{COOH} \\
\alpha\text{--ketoglutaric} \\
\text{acid}
\end{array}
\xrightleftharpoons{\substack{\text{glutamate} \\ \text{transaminase,} \\ \text{Pyridoxal} \\ \text{phosphate}}}
\begin{array}{c}
\text{COOH} \\
|\\
\text{C=O} \\
|\\
\text{R} \\
\alpha\text{--keto} \\
\text{acid}
\end{array}
+
\begin{array}{c}
\text{COOH} \\
|\\
\text{H--C--NH}_2 \\
|\\
\text{CH}_2 \\
|\\
\text{CH}_2 \\
|\\
\text{COOH} \\
\text{Glutamic} \\
\text{acid}
\end{array}
\end{array}
$$

The new α-keto acids are converted in a number of steps to either pyruvic acid, acetyl coenzyme A, or one of the intermediates in the Krebs cycle. The different points of entry into the Krebs cycle for the amino acids are shown in Figure 25.8. Understand, though, that each amino acid must go through several intermediate steps before becoming one of the compounds in the Krebs cycle or one of those leading into it.

The glutamic acid formed in the transamination reaction can be converted back to α-keto glutaric acid by a process known as **oxidative deamination**:

$$
\begin{array}{c}
\begin{array}{c}
\text{COOH} \\
|\\
\text{H--C--NH}_2 \\
|\\
\text{CH}_2 \\
|\\
\text{CH}_2 \\
|\\
\text{COOH} \\
\text{Glutamic acid}
\end{array}
\xrightleftharpoons[\text{enzyme}]{\text{NAD}^{\oplus} \quad \text{NADH+H}^+ \quad \text{H}_2\text{O}}
\begin{array}{c}
\text{COOH} \\
|\\
\text{C=O} \\
|\\
\text{CH}_2 \quad + \quad \text{NH}_3 \\
|\\
\text{CH}_2 \\
|\\
\text{COOH} \\
\alpha\text{--ketoglutaric} \\
\text{acid}
\end{array}
\end{array}
$$

The reverse of the oxidative deamination reaction, reductive amination, and the transamination reaction are used by the body to synthesize needed amino acids. The essential amino acids, Table 22.4, cannot be prepared by the body because the required α-keto acids are not able to be synthesized, and these amino acids must be obtained in the diet.

Notice that ammonia, NH_3, is produced in the oxidative deamination reaction. The ammonia must be removed quickly from the body, since it is toxic. This is done by converting it to urea in the liver for eventual elimination in the urine. This is an important part of amino acid metabolism, cleaning up the waste, and it involves the last cycle we will discuss in this chapter, the urea cycle.

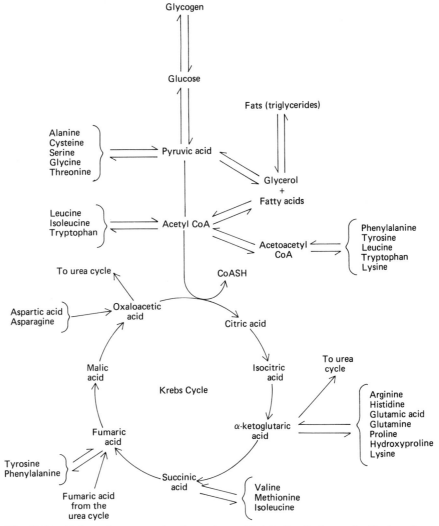

Fig. 25.8. Points of entrance for the amino acids into the Krebs cycle. Each amino acid must be converted to a suitable compound that can enter the cycle.

The Urea Cycle

The **urea cycle** is a series of reactions that convert ammonia to urea in the liver. The great majority of the nitrogen obtained in amino acid catabolism will end up being excreted as urea. The urea cycle is also called the **ornithine cycle,** and it too was proposed by Hans Krebs. The several steps that occur in the urea cycle can be summarized in a single equation:

$$2NH_3 \ + \ CO_2 \ \longrightarrow \ \underset{\text{Urea}}{H_2N-\overset{\overset{\textstyle O}{\|}}{C}-NH_2} \ + \ H_2O$$

Like most equations, this one does not begin to reveal the complex processes that are involved in going from reactants to products. Let us see what these processes are, one step at a time. Refer to Figure 25.9 as we go along so you can see how each step fits into the overall cycle. As has been done previously, compounds will be shown as neutral species though at physiological pH they may exist as ions.

Step 1. *Ammonia* and *carbon dioxide* combine in the presence of ATP to form *carbamyl phosphate*. Ammonia nitrogen enters the urea cycle as carbamyl phosphate. The synthesis requires the enzyme carbamyl phosphate synthetase. The passage of the nitrogen from ammonia, and carbon from carbon dioxide, through the cycle can be followed in the boxed portions of the structures:

Carbamyl phosphate

Step 2. *Carbamyl phosphate* reacts with ornithine to form *citrulline*. Ornithine will be regenerated at the end of the cycle to be used again. The reaction is catalyzed by ornithine transcarbamylase.

Step 3. *Citrulline* combines with *aspartic acid* to form *argininosuccinic acid*. This reaction brings another amino group into the sequence, and results in all the necessary atoms being in place for the formation of urea in step 5.

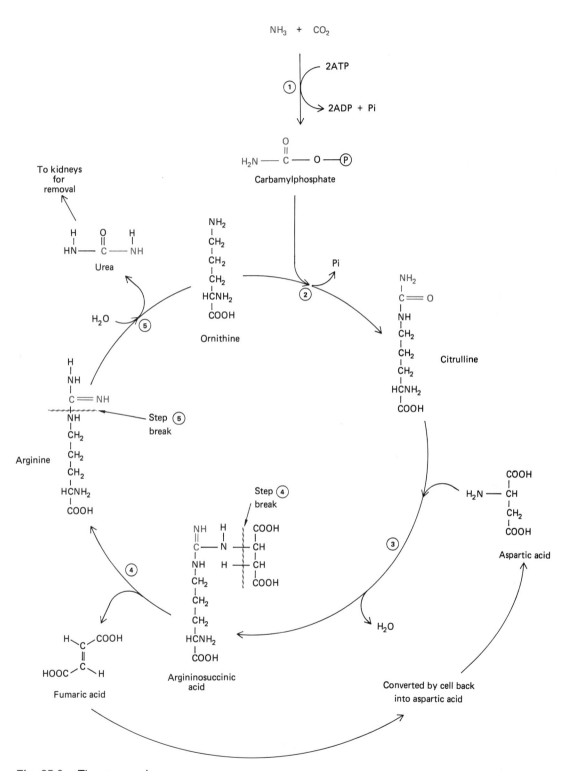

Fig. 25.9. The urea cycle.

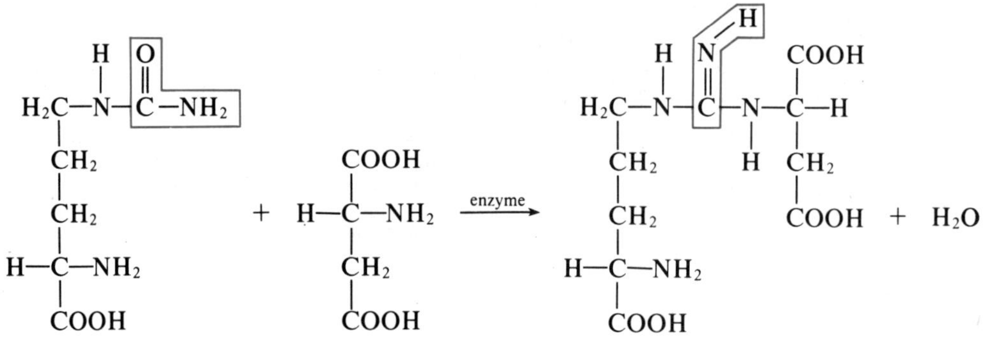

$$H_2C-N-C-NH_2 \quad + \quad H-C-NH_2 \xrightarrow{\text{enzyme}} H_2C-N-C-N-C-H \quad + \quad H_2O$$

Citrulline   Aspartic acid   Argininosuccinic acid

Step 4. The same enzyme required for step 3, argininosuccinic synthetase, splits the molecule produced in the third step into *fumaric acid* and *arginine*. Fumaric acid returns to the Krebs cycle, to be converted to oxaloacetic acid, which, by transamination, is converted back to aspartic acid. Arginine continues on in the cycle.

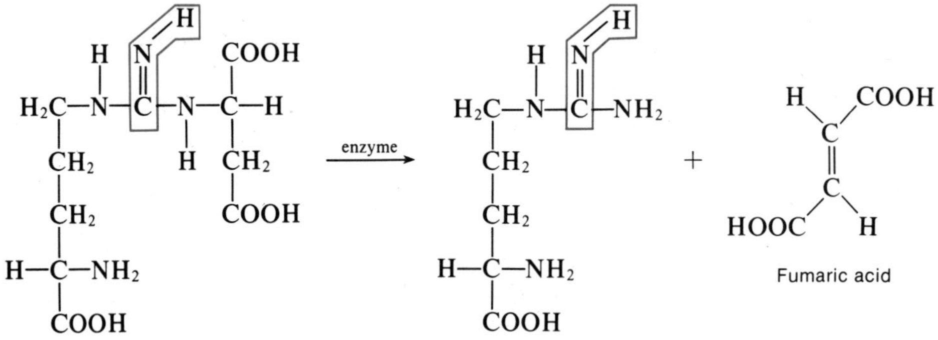

$$\xrightarrow{\text{enzyme}}$$

Argininosuccinic acid   Arginine   Fumaric acid

Step 5. *Arginine* is hydrolyzed to produce *ornithine* and *urea*:

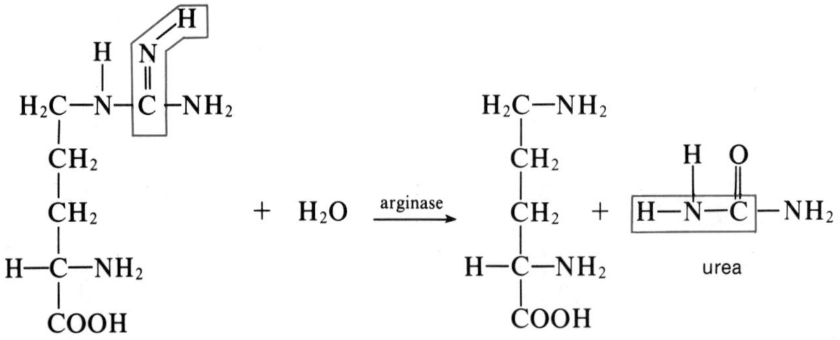

$$+ \quad H_2O \xrightarrow{\text{arginase}} \quad + \quad H-N-C-NH_2$$

urea

Arginine   Ornithine

Ornithine continues on in the cycle reacting with another molecule of carbamyl phosphate in step 2.

The urea is released into the bloodstream and carried to the kidneys for removal in urine. Any condition that prevents the release of urea by the kidneys can lead to **uremia**. If the retention of urea is not corrected, with the accompanying buildup of ammonia, the individual could fall into a coma, which may lead to death.

Normally, blood plasma contains about 0.5 mg of ammonia per liter. Levels only 2–3 times this value produce toxic symptoms. Hemodialysis can be used to cleanse the blood of these wastes, but it is only a temporary solution to a serious problem.

Check Test Number 5

1. What two kinds of compounds react in a transamination reaction?
2. In what form is the nitrogen in amino acids excreted from the body?
3. In which organ of the body do the reactions that comprise the urea cycle take place?

Answers:

1. Amino acid + an α-keto acid, most often α-ketoglutaric acid.

2. Urea, $H_2N-\overset{\overset{\textstyle O}{\|}}{C}-NH_2$

3. The liver.

Terms

Several important terms were introduced in Chapter 25. You should know the meaning of each of the following.

metabolism	acyl CoA
catabolism	ketoanemia
anabolism	ketonuria
ETS	ketosis
respiratory chain	nitrogen balance
Krebs cycle	transamination
glycogenesis	oxidative deamination
glycogenolysis	oxygen debt
glycolysis	urea cycle
fermentation	uremia

aerobic   substrate-level phosphorylation

anaerobic   oxidative phosphorylation

beta-oxidation   Embden–Meyerhof pathway

Questions

Answers to the starred questions appear at the end of the text.

*1. What are the two general categories of metabolism?

*2. What special function does ATP serve in cellular metabolism?

*3. How much energy is released when the following hydrolysis reactions take place?

a. $ATP + H_2O \xrightleftharpoons{\text{enzyme}} ADP + Pi$

b. $ATP + H_2O \xrightarrow{\text{enzyme}} AMP + PPi$

c. $ADP + H_2O \xrightleftharpoons{\text{enzyme}} AMP + Pi$

4. What does the symbol Pi represent?

5. What two ways can ADP be converted to ATP in animals?

*6. Referring to Table 25.1, would creatine phosphate be able to phosphorylate ADP to produce ATP?

7. Referring to Table 25.1, would glucose-1-phosphate be able to convert ADT to ATP by substrate level phosphorylation?

*8. Write the equations that describe the following.
a. The reduction of $NAD^\oplus$ to $NADH + H^+$.
b. The reduction of FAD to $FADH_2$.
c. The overall reaction in the electron transport system.

9. How many moles of ATP are produced in the ETS if one mole of NADH is oxidized to $NAD^\oplus$? If one mole of $FADH_2$ is oxidized to FAD?

*10. Why does the oxidation of $FADH_2$ by the ETS produce a lesser amount of ATP than does the oxidation of NADH?

*11. If the Krebs cycle itself does not require oxygen, how are oxidations carried out in this series of reactions?

12. Of what importance is acetyl coenzyme A to the Krebs cycle?

*13. How many moles of acetyl-CoA are produced in the oxidation and decarboxylation of five moles of pyruvic acid?

*14. Recite the memory device that may be used to remember the names of the compounds in the Krebs cycle.

*15. How many moles of ATP are produced if one mole of pyruvic acid is oxidized (via acetyl CoA) to CO_2 and H_2O in the Krebs cycle–ETS combination?

16. In what part of the cell do the reactions that comprise the Krebs cycle take place?

*17. Write the equation describing the conversion of pyruvic acid to acetyl coenzyme A, showing the involvement of the ETS.

*18. Beginning with one mole of glucose, how many moles of ATP are produced as it is oxidized to CO_2 and H_2O?

19. What is unique about the step in the Krebs cycle, compared to the others, in which α-ketoglutaric acid is converted to succinyl-CoA?

20. List the steps in the Krebs cycle–ETS combination that generate ATP.

*21. How many molecules of ATP would be generated by the ETS as two molecules of pyruvic acid are oxidized in the Krebs cycle to CO_2 and H_2O?

22. What are the three principal monosaccharides that are produced in carbohydrate digestion?

23. Why is it possible to regard the metabolism of carbohydrates as essentially the metabolism of glucose?

*24. What occurs in cells to trap glucose inside once it enters?

25. Why does glycogenolysis not occur in muscle tissue, though it does in the liver?

26. In what principal way does an aerobic process differ from an anaerobic process?

27. What are the end products of anaerobic glycolysis in yeast?

*28. What is the end product of the Embden–Meyerhof pathway when it occurs in muscle tissue under anaerobic conditions?

29. Why does the Embden–Meyerhof pathway produce pyruvic acid and not lactic acid when operating under aerobic conditions?

*30. Why cannot vigorous physical activity be carried out indefinitely by people?

31. If one mole of glucose passes through the EM pathway anaerobically to form lactic acid, how many moles of ATP are produced? If it passes through aerobically to produce pyruvic acid, how many moles of ATP are produced?

*32. Outline the production of ATP as one mole of glucose is oxidized to CO_2 and H_2O under aerobic conditions.

33. Describe the digestion of fats and the ways in which they are transmitted to and through the blood.

34. What changes must glycerol undergo before it can be metabolized in the Krebs cycle?

35. In which organ does β-oxidation occur?

*36. If one mole of lauric acid, $CH_3(CH_2)_{10}COOH$, is passed through the fatty acid cycle, how many moles of acetyl CoA are produced? Including the activation step forming acyl CoA, how many moles of ATP are produced in the FA cycle?

37. How many moles of ATP are produced if one mole of lauric acid, $CH_3(CH_2)_{10}COOH$, is oxidized to CO_2 and H_2O?

*38. The blocking of which metabolic pathway may lead to ketosis?

39. List the three compounds that are called the ketone bodies.

40. Describe how ketone bodies are formed from acetyl CoA.

41. What is the function of the amino acid pool?

42. What is the turnover rate for a particular protein?

43. What is transamination?

44. Why must the ammonia produced in oxidative deamination be removed from the body?

45. What is the function of the urea cycle?

A Review of Mathematics for the Health Sciences

A. Fractions

Fractions are commonly used to indicate the division of one number by another and they are written as

$$\frac{1}{10} \quad \text{or} \quad 1/10 \qquad \frac{4}{12} \quad \text{or} \quad 4/12$$

The number above the line is the **numerator,** which is divided by the number below the line, called the **denominator.** A whole number can also be considered a fraction in which the denominator equals 1:

$$3 = \frac{3}{1} \qquad 180 = \frac{180}{1}$$

If the numerator and denominator of a fraction are both multiplied by the same number, a fraction equal to the first is formed. The fraction 2/3 is equivalent to 4/6 and is also equivalent to 10/15:

$$\frac{2}{3} \xrightarrow{\text{numerator} \times 2} \frac{4}{6} \qquad \frac{2}{3} = \frac{4}{6}$$

$$\frac{2}{3} \xrightarrow{\text{numerator} \times 5} \frac{10}{15} \qquad \frac{2}{3} = \frac{10}{15}$$

Fractions can be converted into **decimals** by dividing the numerator by the denominator:

$$\frac{1}{10} = 0.1 \qquad \frac{4}{12} = 0.33 \qquad \frac{7}{8} = 0.875 \qquad \frac{4}{3} = 1.33$$

A decimal can be changed to a **percentage** term by multiplying it by 100%. Numbers expressed as **percent** are commonly used to report that part of a sample that is of a certain type. For example, if a nurse is in charge of 6 male and 8 female patients (sample equals 14 patients), the number of female patients can be expressed as a percentage of the group:

$$\text{percent female} = \frac{\text{number of female patients}}{\text{total number of patients in sample}} \times 100\%$$

$$= \frac{8}{14} \times 100\%$$

$$= 0.57 \times 100\% = 57\%$$

Dividing 8 by 14 generated a decimal which was then converted to percent.

Addition of fractions is possible if both fractions have the same denominator:

$$\frac{2}{5} + \frac{1}{5} = \frac{2 + 1}{5} = \frac{3}{5}$$

If the denominators are not the same, then one or both of the fractions must be changed to develop a common denominator:

$$\frac{3}{4} + \frac{1}{2} =$$

Change $\frac{1}{2}$ into $\frac{2}{4}$ $\left(\text{a fraction equal to } \frac{1}{2}\right)$

$$\frac{3}{4} + \frac{2}{4} = \frac{3 + 2}{4} = \frac{5}{4}$$

When both denominators must be changed to a common denominator, it is easily done by multiplying the denominators of both fractions to give the common denominator:

$$\frac{3}{7} + \frac{2}{3} =$$

common denominator $= 7 \times 3 = 21$

To convert the fraction 3/7 to the equivalent fraction with a denominator of 21, realize that if the common denominator is *three* times the old denominator, the new numerator will have to be *three* times the old numerator.

$$\frac{9}{21} + \frac{14}{21} = \frac{9 + 14}{21} = \frac{23}{21}$$

Here are some others:

$$\frac{8}{9} + \frac{1}{2} = \frac{16 + 9}{18} = \frac{25}{18}$$

$$\frac{3}{10} + \frac{2}{3} = \frac{9 + 20}{30} = \frac{29}{30}$$

$$\frac{2}{5} + 3 = \frac{2 + 15}{5} = \frac{17}{5} \quad \left(\text{remember 3 is actually } \frac{3}{1}\right)$$

Subtraction of fractions is done similarly to addition in that a common denominator is necessary. The following examples will demonstrate subtraction of fractions:

$$\frac{4}{5} - \frac{2}{5} = \frac{4 - 2}{5} = \frac{2}{5}$$

$$\frac{9}{10} - \frac{4}{5} = \frac{9 - 8}{10} = \frac{1}{10}$$

$$\frac{2}{3} - \frac{4}{7} = \frac{14 - 12}{21} = \frac{2}{21}$$

Multiplication of fractions is quite straightforward. The numerators of each fraction are multiplied to give the numerator of the product, and the denominators are multiplied to give the denominator of the product:

$$\frac{2}{3} \times \frac{4}{5} = \frac{2 \times 4}{3 \times 5} = \frac{8}{15}$$

$$\frac{1}{2} \times \frac{3}{5} \times \frac{4}{3} = \frac{1 \times 3 \times 4}{2 \times 5 \times 3} = \frac{12}{30}$$

Division of numbers that involve fractions is done by first inverting the divisor, then multiplying. The divisor is the number that is divided into the first number, the dividend.

$$\frac{2}{3} \div \frac{3}{5} = \frac{2}{3} \times \frac{5}{3} = \frac{10}{9}$$

Dividend   Divisor   Inverted

A fraction composed of other fractions indicates they are to be divided. The divisor is the denominator of the large fraction:

$$\text{Divisor} \rightarrow \left\{ \frac{\frac{10}{19}}{\frac{1}{2}} = \frac{10}{19} \div \frac{1}{2} = \frac{10}{19} \times \frac{2}{1} = \frac{20}{19} \right.$$

$$\text{Divisor} \longrightarrow \frac{\frac{4}{3}}{2} = \frac{4}{3} \div \frac{2}{1} = \frac{4}{3} \times \frac{1}{2} = \frac{4}{6}$$

Fractions can be reduced to simpler quantities in two ways. First, if *both* the numerator and denominator are divisible by the same whole number (greater than one), doing so will reduce or simplify the fraction:

Both can be divided by 100 $\longrightarrow$ $\dfrac{100}{200} = \dfrac{1}{2}$ $\dfrac{9}{27} = \dfrac{1}{3}$ $\dfrac{18}{48} = \dfrac{3}{8}$

If the fraction is what is called an **improper fraction**, that is, has a numerator that is larger than the denominator, then it is simplified by dividing the numerator by the denominator and stating the remainder as a fraction:

$$\frac{16}{5} = 3\frac{1}{5} \qquad \frac{29}{6} = 4\frac{5}{6} \qquad \frac{36}{8} = 4\frac{4}{8} = 4\frac{1}{2} \qquad \frac{5}{5} = 1$$

B. Scientific Notation (Exponential Notation)

Often numbers are encountered in science that are very large, such as the number of atoms in one gram of uranium (2 530 000 000 000 000 000 000), or numbers that are very small, such as the distance between two carbon atoms in a molecule (0.000 000 000 154 meters). Conventional numbers like these are cumbersome to write out and use in calculations, but they can be expressed more conveniently using scientific notation. In this method, numbers are expressed as the product of two terms written as

$$A \times 10^n$$

where A is a number between one and ten (1.000 to 9.999) and n, the exponent, is either a positive or negative whole number, such as 3 or -5.

Before going into the details of writing numbers in scientific notation, it would be instructive to say a few words about exponents.

In the term

$$\text{Base} \rightarrow 5^2 \leftarrow \text{Exponent}$$

the **exponent,** 2, is written as a superscript on the **base,** the number 5 in this example. Being a positive exponent, it indicates that 5 is to be multiplied by itself *two* times:

$$5^2 = 5 \times 5 = 25$$

An exponent is also called the **power** of the number with which it is used. Here we would say that five is raised to the second power. The following examples will demonstrate other positive exponents:

$$2^5 = 2 \times 2 \times 2 \times 2 \times 2 = 32$$

$$10^3 = 10 \times 10 \times 10 = 1000$$

$$10^1 = 10$$

It is also possible to have an exponent that equals zero. Any number raised to the zero power is equal to 1:

$$10^0 = 1$$

$$5^0 = 1$$

Exponents can also be negative (negative powers), as in the term 10^{-2}. The negative exponent (-2) indicates that 10 raised to the second power is to be written in the *denominator* of a fraction and then divided into 1:

$$10^{-2} = \frac{1}{10^2}$$

and

$$\frac{1}{10^2} = \frac{1}{10 \times 10} = \frac{1}{100} = 0.01$$

Notice that a negative exponent will generate small numbers, 1 or less, and positive exponents generate large numbers. Other examples of negative exponents are

$$5^{-2} = \frac{1}{5^2} = \frac{1}{5 \times 5} = \frac{1}{25} = 0.04$$

$$10^{-3} = \frac{1}{10^3} = \frac{1}{1000} = 0.001$$

Now, let us return to writing numbers in scientific notation. The number 5200 would be written as 5.2×10^3. This symbolizes that the original number written in conventional form could be obtained by multiplying 5.2 (a number between 1 and 10) by 10^3, that is, by 1000:

$$5.2 \times 10^3 = 5.2 \times (10 \times 10 \times 10)$$

$$= 5.2 \times (1000)$$

$$= 5200$$

An easier way to accomplish the same thing is to move the decimal point *three* places to the *right* for a positive exponent of 3:

$$5.2 \times 10^3$$

move decimal three places *to the right*:

$$5.200 = 5200$$

To correctly express the number 0.00665 in scientific notation we would write

$$6.65 \times 10^{-3}$$

To obtain the original number written in conventional form, we would have to multiply 6.65 by 0.001 ($10^{-3} = 0.001$):

$$6.65 \times 10^{-3} = 6.65 \times \left(\frac{1}{10 \times 10 \times 10} \right)$$

$$= 6.65 \times (0.001)$$

$$= 0.00665$$

Alternatively, the negative exponent tells you the number of places the decimal should be moved *to the left* to generate the original number.

$$6.65 \times 10^{-3}$$

move decimal three places *to the left*:

$$= 006.65$$

$$= 0.00665$$

The following examples will show other numbers written in scientific (exponential) notation:

$$0.52 = 5.2 \times 10^{-1}$$

$$0.0000075 = 7.5 \times 10^{-6}$$

$$1\ 000\ 000 = 1 \times 10^{6} \text{ or just } 10^{6}$$

$$0.000\ 000\ 000\ 154 = 1.54 \times 10^{-10}$$

$$139 = 1.39 \times 10^{2}$$

$$0.00001 = 1 \times 10^{-5} \text{ or just } 10^{-5}$$

Multiplication of numbers written in scientific notation is not difficult. First multiply the numerical part of each number, then *add* the exponents algebraically. When you add two numbers algebraically, you must pay careful attention to the sign (+ or −) of the numbers.

1. If both exponents are positive, like +3 and +5, the sum would be simply +8 or 8. The + sign is usually not written with the number for positive exponents.

$$(+3) + (+2) = +5 \qquad (+1) + (+2) = +3$$

2. If both exponents are negative, like -2 and -3, then the sum would be -5. The two negative numbers added together would equal a more negative number.

$$(-2) + (-6) = -8 \qquad (-4) + (-11) = -15$$

3. If one exponent is positive and the other negative, like $+4$ and -7, the smaller number is subtracted from the larger $(7 - 4 = 3)$ and the answer is given the sign of the larger digit. In this case the result would be -3.

$$(-9) + (+4) = -5 \quad (+6) + (-8) = -2 \quad (-1) + (+1) = 0$$

Here are some examples:

Add exponents

$$(2 \times 10^2) \times (3 \times 10^3) =$$

Multiply

$$= (2 \times 3) \times 10^{2+3}$$

$$= 6 \times 10^{2+3} = 6 \times 10^5$$

$$(3.5 \times 10^4) \times (4.0 \times 10^{-2}) = (3.5 \times 4.0) \times 10^{4+(-2)}$$

$$= 14 \times 10^2$$

$$= 1.4 \times 10^3$$

$$(6.10 \times 10^4) \times (5.20 \times 10^{-5}) = (6.1 \times 5.2) \times 10^{4+(-5)}$$

$$= 31.7 \times 10^{-1}$$

$$= 3.17$$

When dividing numbers written in scientific notation, the numerical parts are divided in the usual way and the sign of the exponent of the denominator is changed and added algebraically to the exponent of the numerator.

$$\frac{4 \times 10^4}{2 \times 10^2} = \frac{4}{2} \times 10^{4+(-2)}$$

Exponent of the numerator

Exponent of the denominator
(note, the sign is changed)

$$= 2 \times 10^2$$

$$\frac{6.6 \times 10^5}{2.0 \times 10^{-2}} = \frac{6.6}{2.0} \times 10^{5 + (+2)}$$

$$= 3.3 \times 10^{5 + (+2)} = 3.3 \times 10^7$$

$$\frac{4.85 \times 10^{-10}}{1.75 \times 10^4} = \frac{4.85}{1.75} \times 10^{-10 + (-4)}$$

$$= 2.77 \times 10^{-14}$$

C. Algebraic Operations

In any chemistry course you will find equations that relate two or more quantities. An equation relating °C and °F is such an equation. To determine the value of one of the quantities the equations often must be rearranged algebraically to isolate the quantity being sought on one side of the equation. We will be concerned with two basic operations in rearranging equations. The first is to bring terms that are added or subtracted on one side of the equation over to the other side. Suppose the following equation is to be solved for X, that is, rearranged so that X is the only term on one side:

$$X + 10 = Y$$

To isolate X on one side of the equation we can subtract 10 from both sides. Remember that *adding or subtracting the same term from both sides of an equation does not change the equality*.

$$(-10) \qquad X + 10 - 10 = Y - 10$$

$$X + \cancel{10} - \cancel{10} = Y - 10$$

$$X = Y - 10$$

To solve this next equation for X, you would need to add the fraction Y/Z to each side:

$$X - \frac{Y}{Z} = A$$

$$\left(+\frac{Y}{Z}\right) \qquad X - \frac{\cancel{Y}}{\cancel{Z}} + \frac{\cancel{Y}}{\cancel{Z}} = A + \frac{Y}{Z}$$

$$X = A + \frac{Y}{Z}$$

To solve the following equation for X, both Y and Z must be moved to the opposite side. You would first add Y to each side and then subtract Z from each side:

$$X - Y + Z = A$$

$$(+Y) \qquad X - \cancel{Y} + \cancel{Y} + Z = A + Y$$

$$X + Z = A + Y$$

$$(-Z) \qquad X + \cancel{Z} - \cancel{Z} = A + Y - Z$$

$$X = A + Y - Z$$

Here are some examples using numbers instead of letters. Each is to be solved for X:

$$15 + X = 7 - 9$$

$$(-15) \qquad \cancel{15} - \cancel{15} + X = 7 - 9 - 15$$

$$X = 7 - 9 - 15$$

$$X = -17$$

$$0.50 = -0.72 + X$$

$$(+0.72) \qquad 0.50 + 0.72 = -\cancel{0.72} + X + \cancel{0.72}$$

$$0.50 + 0.72 = X$$

$$1.22 = X$$

The second technique of rearranging equations involves multiplying or dividing both sides of an equation by the same term. To solve the following equation for X, divide both sides by Y:

$$XY = Z$$

$$(\div Y) \qquad \frac{X\cancel{Y}}{\cancel{Y}} = \frac{Z}{Y}$$

$$X = \frac{Z}{Y}$$

The next equation can be solved for X by multiplying both sides by Z. The entire quantity $(Y + A)$ on the right-hand side must be multiplied by Z:

$$\frac{X}{Z} = (Y + A)$$

$$(\times Z) \qquad \frac{X}{\cancel{Z}} \times \cancel{Z} = Z(Y + A)$$

$$X = Z(Y + A) = ZY + ZA$$

A common equation form that is used in chemistry is

$$\frac{X}{Y} = \frac{W}{Z}$$

To solve for X, multiply both sides by Y:

$$\frac{X}{\cancel{Y}} \times \cancel{Y} = \frac{W}{Z} \times Y$$

$$X = \frac{WY}{Z}$$

To solve the equation for Y, first multiply each side by Y as was done above, then multiply each side by Z/W:

$$\frac{X}{Y} = \frac{W}{Z}$$

$$(\times Y) \qquad X = \frac{WY}{Z}$$

$$\left(\times \frac{Z}{W}\right) \qquad \frac{Z}{W} \times X = \frac{\cancel{W}Y}{\cancel{Z}} \times \frac{\cancel{Z}}{\cancel{W}}$$

$$\frac{ZX}{W} = Y$$

or written the other way around with Y on the left-hand side

$$Y = \frac{ZX}{W}$$

The same kind of operations could be used to solve the equation for W or Z:

$$W = \frac{XZ}{Y} \qquad Z = \frac{WY}{X}$$

Here are some examples of equations that are solved for X:

$$\frac{X}{3} = 12 \qquad\qquad \frac{3}{X} = 12$$

$$\cancel{3} \times \frac{X}{\cancel{3}} = 3 \times 12 \qquad\qquad \cancel{X} \times \frac{3}{\cancel{X}} = 12 \times X$$

$$X = 36 \qquad\qquad \frac{3}{12} = X$$

$$\frac{X}{5} = \frac{2}{3} \qquad\qquad \frac{7}{6} = \frac{4}{X}$$

$$\frac{X}{\cancel{5}} \times \cancel{5} = \frac{2}{3} \times 5 \qquad\qquad X \times \frac{7}{6} = \frac{4}{\cancel{X}} \times \cancel{X}$$

$$X = \frac{10}{3} = 3\frac{1}{3} \qquad\qquad X = \frac{24}{7} = 3\frac{3}{7}$$

Some equations need both kinds of operations to rearrange them, addition and subtraction as well as multiplication and division. Suppose we wished to solve the following equation for X:

$$\frac{X}{Y} + Z = A$$

First subtract Z from both sides,

$$(-Z) \qquad \frac{X}{Y} + \cancel{Z} - \cancel{Z} = A - Z$$

$$\frac{X}{Y} = A - Z$$

Then multiply both sides by Y:

$$(\times Y) \qquad \frac{X}{Y} \times Y = (A - Z) \times Y$$

$$X = Y(A - Z)$$

If the equation was to be solved for Y, you could proceed as before, starting with

$$X = Y(A - Z)$$

Divide both sides by the quantity $(A - Z)$:

$$\div (A - Z) \qquad \frac{X}{(A - Z)} = \frac{Y(A - Z)}{(A - Z)}$$

$$\frac{X}{(A - Z)} = Y$$

To solve the original equation for Z, simply subtract the quantity X/Y from both sides:

$$\left(-\frac{X}{Y} \right) \qquad \frac{X}{Y} - \frac{X}{Y} + Z = A - \frac{X}{Y}$$

$$Z = A - \frac{X}{Y}$$

Carrying out algebraic operations on equations really is not a difficult task. If you carefully think through each step, you will not have any trouble.

D. Practice Problems: Fractions

Perform the following problems involving fractions. Simplify the answers (when possible).

a. $\dfrac{1}{2} + \dfrac{3}{2} =$

b. $\dfrac{5}{6} + \dfrac{1}{3} =$

c. $\dfrac{8}{10} - \dfrac{3}{20} =$

d. $\dfrac{3}{7} \times \dfrac{2}{6} =$

e. $\dfrac{1}{8} \times \dfrac{3}{4} =$

g. $\dfrac{15/24}{1/4} =$

i. $\dfrac{7}{12} \div \dfrac{1}{3} =$

f. $\dfrac{2}{3} \div \dfrac{1}{2} =$

h. $\dfrac{5}{8} - \dfrac{1}{3} =$

j. $\dfrac{8}{10} \div 2 =$

Answers:

a. 4/2 or 2

e. 3/32

h. 7/24

b. 7/6 or $1\dfrac{1}{6}$

f. 4/3 or $1\dfrac{1}{3}$

i. 7/4 or $1\dfrac{3}{4}$

c. 13/20

g. 60/24 or $2\dfrac{1}{2}$

j. 4/10 or 2/5

d. 6/42 or 1/7

E. Practice Problems: Scientific Notation

Express the following numbers in scientific notation.

a. 650
b. 0.044

c. 7.541
d. 0.000099

e. 271.3
f. 100000

Answers: (a) 6.5×10^2, (b) 4.4×10^{-2}, (c) 7.541, (d) 9.9×10^{-5}, (e) 2.713×10^2, (f) 1.0×10^5 or 10^5

F. Practice Problems: Algebraic Operations

Solve the following equations for X.

a. $3X = Y$

d. $\dfrac{W}{X} = Z + A$

g. $X(Y + Z) = A$

b. $XY = Z + W$

e. $\dfrac{B}{Y} = \dfrac{X}{A}$

h. $\dfrac{Y}{XZ} = A - B$

c. $\dfrac{X}{Y} = A$

f. $X - Y = Z - W$

Answers:

a. $X = Y/3$

d. $X = W/(Z + A)$

g. $X = \dfrac{A}{(Y + Z)}$

b. $X = (Z + W)/Y$

e. $X = BA/Y$

c. $X = YA$

f. $X = Z - W + Y$

h. $X = \dfrac{Y}{Z(A - B)}$

The Cell

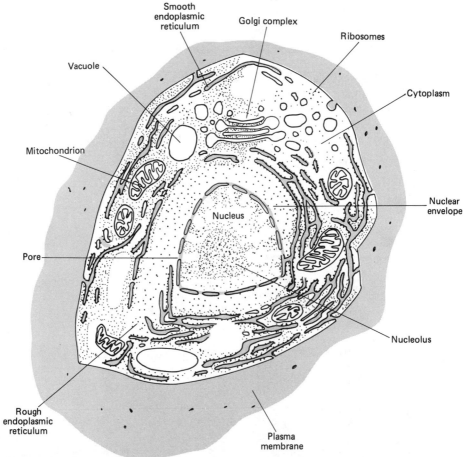

The composition of a typical eukaryotic cell. (From: Wolfe, *Biology, The Foundations*. Wadsworth Publishing Company, 1977, p. 47.)

Answers to Selected Questions

Answers to Questions
Chapter 2

1. **a.** Mass is a measure of the amount of matter in an object, while weight is a measure of the gravitational attraction of the object to another body, in our case the Earth.
   b. Heat is a form of energy. The temperature of an object is determined by the amount of heat it contains.
   c. Density is the mass of an object divided by its volume. Specific gravity is the density of an object divided by the density of a reference, usually water.
   d. Mass is a measure of the amount of matter in an object, while density is the amount of mass per unit volume.

4. (a) kilometers; (b) meters; (c) centimeters or millimeters.

5. It is more convenient. You avoid writing zeroes to place the decimal.

7. The sample with a temperature of 30°C has a larger amount of heat energy contained in it.

8. Using carbon tetrachloride as the reference liquid, the specific gravity of methyl alcohol becomes 0.792 g/ml ÷ 1.59 g/ml, which equals 0.498.

11. **a.** 0.073   **c.** 290000000

12. **a.** $\frac{23}{8}$ or $3\frac{5}{8}$

   c. $\frac{5}{12}$

   d. $\frac{14}{10}$ or 1.4

13. **a.** 3

   c. a/b

   e. $y + 24$

   g. $\left(\dfrac{a + b}{c}\right) - d$

14. **a.** 7.5 cm

   c. 61 cm

   e. 1590 g

   g. 37.8 l

15. **a.** 22°C

   c. 14°F

   d. 326°K

16. **a.** 1.5×10^3 mg

   c. 1.9×10^5 μm

   e. 1.3×10^{-8} cm

   g. 9.080×10^5 g

17. 4.50 g/ml

19. aluminum

21. 53 min

24. $113\frac{1}{2}$ weeks

25. 5.0×10^{-5} l

27. **a.** 1000 cg

   c. 25 lb

   e. $20 \dfrac{mg}{min}$

29. lead

31. 31.5 test tubes calculated, so use 32 test tubes.

Answers to Questions
Chapter 3

2. The stronger the attractive forces, the greater the chance it will be a solid.

4. The particles (atoms or molecules) that make up a gas are far apart and can be squeezed closer together, but in a liquid they are already touching.

5. **a.** sodium + water → sodium hydroxide + hydrogen

   b. sucrose → carbon + water

7. 34.3 g of chlorine

8. In a compound the elements are combined in a definite weight ratio, in a mixture they are not. The mixture would be heterogeneous, and could be separated by physical means. Neither is true for the compound. The identities of the elements are lost when in a compound, not so in a mixture.

10. **a.** Si is silicon; SI, sulfur and iodine.

   c. A molecule containing one sulfur and two oxygen atoms is SO_2; with three oxygen atoms, it's SO_3.

12. 2.6 Å

13. (a) NaF, (c) Al_2O_3, (e) C_2H_6O.

15. They are salts, solid, composed of ions.

16. (a), (c), (e), and (f) are mixtures.

17. (a), (d), and (e) are physical.

Answers to Questions
Chapter 4

2. They are all relative to carbon-12, so they are relative to each other.
4. One mole of anything contains Avogadro's number of the things.
6. (a) 6.02×10^{23}, (b) 6.02×10^{23}, (c) 35.5.
7. 1 mole of H_2 + 1 mole of $I_2 \rightarrow$ 2 moles of HI
   1 molecule of H_2 + 1 molecule of $I_2 \rightarrow$ 2 molecules of HI
   2.0 g of H_2 + 254 g of $I_2 \rightarrow$ 256 g of HI
8. Whether a compound exists as molecules or as aggregates of ions, it will have a formula. Formula weight can be used for both kinds of compounds, whereas molecular weight specifically refers to compounds that exist as molecules.
10. **a.** 1.25 moles **c.** 0.166 mole **e.** 10.2 moles
11. **a.** 18.3 g **c.** 0.352 g **f.** 1530 g
12. 52.2% C, 13.0% H, 34.8% 0
14. **a.** $\dfrac{2 \text{ mole B}}{3 \text{ mole F}_2}$; $\dfrac{2 \text{ mole B}}{2 \text{ mole BF}_3}$; $\dfrac{3 \text{ mole F}_2}{2 \text{ mole BF}_3}$; plus the inverted form of each

   b. $\dfrac{2 \text{ mole Cu}}{1 \text{ mole O}_2}$; $\dfrac{2 \text{ mole Cu}}{2 \text{ mole CuO}}$; $\dfrac{1 \text{ mole O}_2}{2 \text{ mole CuO}}$; plus the inverted form of each
15. **a.** $CH_4 + 3Cl_2 \rightarrow CCl_3H + 3HCl$
    b. $C_3H_8 + 5O_2 \rightarrow 3CO_2 + 4H_2O$
16. 4.00 moles HCl
18. 80 g NaOH
19. 3.6×10^{23} C atoms
20. 1.46 g CO_2

Answers to Questions
Chapter 5

1. Energy
3. The heat energy flows out of the water into the air and table and is eventually lost to the environment.
6. The average KE of the water molecules is directly related to the temperature of the water.
8. The shorter the wavelength the more energetic it is and the more likely that it can do cellular damage.
10. (a) and (d) are kinetic energy; (b), (c), (e), (f) are potential energy
11. 1000
13. (a) and (d) are endothermic; (b) and (c) are exothermic.
16. Exothermic
18. No

19. Photosynthesis converts CO_2, H_2O, and energy from the sun into glucose, which is stored in the plant. Cellular respiration combines the glucose with oxygen forming CO_2 and H_2O, releasing the energy to the body.

20. $Q = (1000\ g)(1.00\ cal/g\ °C)(55°C) = 55000\ cal = 55\ kcal$

22. $Q = (80\ cal/g)(500\ g) = 40000\ cal = 40\ kcal$

24. $T = (0.65\ °C/g)(12\ g) = 7.8°C$; final temperature $= 25.0° - 7.8° = 17.2°C$

25. Benzene with a smaller SH would have a greater temperature change ($12.2°C$) than alcohol ($8.6°C$).

27. $2800\ Cal \times 1\ g/9\ Cal = 311\ g$ or $0.685\ lb$

29. 1 mole $CaCO_3 = 100\ g$; $42.6 \dfrac{kcal}{mole} \times \dfrac{1\ mole}{100\ g} = 4.26 \times 10^{-1} \dfrac{kcal}{g}$

$4.26 \times 10^4\ kcal \times \dfrac{1\ g}{4.26 \times 10^{-1}\ kcal} = 1 \times 10^5\ g\ CaCO_3$

Answers to Questions
Chapter 6

2. Electron $(1-)$, proton $(1+)$, neutron (0). The proton and neutron have about the same mass, and they are about 1840 times heavier than the electron.

3. The atomic number equals the number of protons in the nucleus and the number of electrons about it. The mass number equals the number of protons plus the number of neutrons in the nucleus. It also equals the approximate atomic weight of the atom in amu.

4. boron   5   11   5   5   6
   lithium   3   7   3   3   4
   magnesium   12   24   12   12   12

5. Isotopes are atoms with the same atomic number but different mass numbers. They are atoms of the same element that do not weigh the same since they contain a different number of neutrons. (a) and (c) are isotopes.

9. The energy lost would equal the difference in energy of the two levels and it would be radiated from the atom as light energy.

10. $18e^-$ $n=3$
    $8e^-$ $n=2$
    $2e^-$ $n=1$ lowest energy level

11. **b.** Al

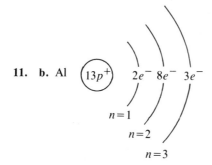

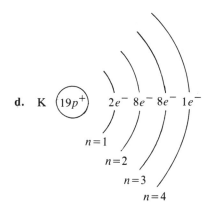

d. K $\left(19p^+\right)$ $2e^-$ $8e^-$ $8e^-$ $1e^-$

$n=1$
$n=2$
$n=3$
$n=4$

12. a. Too many electrons in $n=1$ level, should be $\left(3p^+\right)$ $2e^-$ $1e^-$

$n=1$
$n=2$

14. H·, :C̈l·, :N̈e:, :S̈·, :Ö·, ·N̈·, Na·

16. The group number equals the number of valence electrons in the elements in that group. Representative elements.

17. a. any group I metal
   b. any group VII element
   c. any metal in the center of the Periodic Table in the 4th, 5th, or 6th period
   d. any element in group VIII
   e. B, Al, Ga, In, or Tl
   f. H or He

20. A periodic property is one that reoccurs at regular intervals. In the Periodic Table they are properties that reoccur at regular intervals of atomic numbers.

22. F—smallest, Rb—largest

24. The ionization energies of metals are lower than the ionization energies of non-metals. The rare gases have the highest ionization energies of the elements in each period.

25. †An orbital is a region of space where the probability of finding an electron is high. No more than two electrons can occupy an orbital at the same time.

26. †An s orbital is spherical; a p orbital is dumbbell shaped. The $2s$ orbital is larger than the $1s$ orbital; both are spherical.

27. †There are three p orbitals in the second energy level, p_x, p_y, and p_z. They each point in a different direction along the x, y, and z axes.

28. †**a.** Li $1s^2\ 2s^1$
   b. C $1s^2\ 2s^2\ 2p^2$
   c. F $1s^2\ 2s^2\ 2p^5$

Answers to Questions
Chapter 7

1. the nucleus

3.
alpha	$\alpha, {}^4_2\text{He}$	2+	4 amu	low
beta	$\beta, {}^0_{-1}\text{e}$	1−	0 amu	moderate
gamma	γ	0	0 amu	very high

5. Radiation that will cause matter to break apart into charged fragments called ions, and uncharged fragments called free radicals.

7. Lymphatic tissue, intestinal tract, bone marrow, gonads, and developing embryo.

9. (a) ${}^{14}_6\text{C}$, (b) ${}^{238}_{92}\text{U}$

10. **a.** ${}^{238}_{92}\text{U} \rightarrow {}^{234}_{90}\text{Th} + {}^4_2\text{He}$

    c. ${}^{32}_{15}\text{P} \rightarrow {}^{32}_{16}\text{S} + {}^0_{-1}\text{e}$

    e. ${}^{226}_{88}\text{Ra} \rightarrow {}^{222}_{86}\text{Rn} + {}^4_2\text{He} + \gamma$

12. **a.** ${}^{238}_{92}\text{U} + {}^1_0 n \rightarrow {}^{239}_{92}\text{U} + \gamma$

13. After 6 half-lives, approximately 1.6×10^3 atoms

15. B—long radioactive half-life and long biological half-life

    C—moderately long radioactive and biological half-lives

17. **a.** rem

    b. roentgen

    c. curie

    d. LD_{50}—30 day

21. Intensity at the 12-ft distance is $\frac{1}{4}$ that at 6 ft.

23. (1) keep away from the source when it is radiating

    (2) use proper shielding

    (3) minimum exposure time

25. Iodine is naturally gathered by the thyroid, so I-131 naturally migrates there.

Answers to Questions
Chapter 8

1. **a.** Li^+ **c.** Ba^{2+}

   b. O^{2-}

2. Ionic: $BaCl_2$, MgF_2, NaBr, $CaCO_3$

   Covalent: CS_2, H_2S

4. The ionic bond is the force of attraction between ions of opposite charge in a crystal.

5. Oxidation (a), (b)

   Reduction (c), (d)

6. **a.** $CaBr_2$ **e.** SrO

   c. K_2O

7. **a.** Li_3PO_4 **e.** $NaC_2H_3O_2$
  c. $Al_2(SO_4)_3$

8. **a.** CaI_2 **g.** Al_2S_3
  c. AgCN   **i.** KNO_3
  e. $MgSO_4$ **k.** $CaHPO_4$

9. **a.** calcium chloride   **g.** copper (I) iodide
  c. iron (III) oxide   **i.** silver carbonate
  e. potassium bicarbonate or   **k.** potassium cyanide
     potassium hydrogen carbonate

10. **a.** Fe_2S_3, iron (III) sulfide   **e.** $Cu(NO_3)_2$, copper (II) nitrate
  c. HgO, mercury (II) oxide

14. $H-Br$, $H-\overset{\displaystyle H}{\underset{\displaystyle H}{P}}-H$, $H-\overset{\displaystyle H}{\underset{\displaystyle H}{B}}$, $H-\overset{\displaystyle H}{\underset{\displaystyle H}{C}}-H$

15. (a) $:\!\ddot{F}-\overset{\displaystyle \ddot{N}}{\underset{\displaystyle :\!\ddot{F}\!:}{}}-\ddot{F}\!:$,   (b) $:\!\ddot{C}l-\ddot{B}r\!:$,   (c) $\dot{\ddot{S}}=C=\dot{\ddot{S}}$,   (d) $:\!\ddot{C}l-\overset{\displaystyle \ddot{P}}{\underset{\displaystyle :\!\ddot{C}l\!:}{}}-\ddot{C}l\!:$

17. increasing electronegativity $\longrightarrow$
  a. P, S, Cl
  b. K, Mg, C, F
  c. Ca, Al, P

19. Most electronegative elements—upper right hand corner (Cl, N, O, F).

20. The most polar covalent bond will have the greatest difference in electronegativity between the two bonded atoms. F—F, difference = 0; H—Br, difference is 0.7; $-\overset{|}{\underset{|}{C}}-$Cl, difference is 0.5; $-\overset{|}{N}-$H, difference is 0.9. Most polar bond is $-\overset{|}{N}-$H.

22. (a) $H-\overset{\displaystyle H}{\underset{\displaystyle H}{C}}-\overset{\displaystyle H}{\underset{\displaystyle H}{C}}-\overset{\displaystyle H}{\underset{\displaystyle H}{C}}-H$ (b) $\overset{\displaystyle H}{\diagdown}C=C\overset{\diagup \displaystyle H}{\diagdown}$ with H's   (c) $H-C\equiv C-H$

24. (a) $\left[\,\overset{\displaystyle \ddot{O}=C-\ddot{O}\!:}{\underset{\displaystyle :\!\ddot{O}\!:}{|}}\,\right]^{2-}$,   (b) $H-\overset{\displaystyle \ddot{P}}{\underset{\displaystyle H}{|}}-H$,   (c) $[:\!\ddot{O}-\ddot{O}\!:]^{2-}$,   (d) $:\!C\equiv O\!:$

27. **a.** $H_2C=O$, trigonal planar   **d.** trigonal planar
  b. pyramid   **e.** pyramid
  c. tetrahedral

28. **a.** sulfur tetrafluoride   **g.** phosphorus triiodide
  c. diphosphorus pentoxide   **i.** arsenic pentachloride
  e. nitrogen trichloride

29. **a.** BBr_3
  c. HCl

30. **a.** ionic   **g.** covalent
  c. covalent   **i.** covalent
  e. covalent

Answers to Questions
Chapter 9

2. The electrons that move throughout the metal are very mobile and can flow through the metal as it conducts electricity.

3. **a.** Alloys are homogeneous mixtures of two or more metals.
   b. Alloys can be designed to have properties that are more desirable in a given application than pure metals.

7. (a) Fe^{2+}, (b) Mg^{2+}, (c) Co^{2+}

9. Na^+

11. Ca^{2+}

13. Mg^{2+}

15. Blood plasma must be able to hold the iron as it is absorbed by the body. If the capacity for iron is low in plasma, it is not possible to get the needed iron into the body.

17. vitamin B_{12} and its derivatives

Answers to Questions
Chapter 10

2. Each collision of a molecule with the container wall produces a small force. The sum of all these forces over a given area at any instant is the pressure exerted by the gas. $P = F/A$.

3. The average kinetic energy of a collection of molecules increases as the temperature increases.

6. $V_2 = 12\,1 \left(\dfrac{350 \text{ torr}}{760 \text{ torr}} \right) = 5.5\,1$

8. $27°C = 300°K$; $-25°C = 248°K$; $V_2 = 5.0\,1 \left(\dfrac{248°K}{300°K} \right) = 4.1\,1$

10. $D = \dfrac{44 \text{ g}}{22.4\,1} = 1.96$ g/l at STP

12. **a.** $n = \dfrac{PV}{RT} = \dfrac{(2.00 \text{ atm})\,(100\,1)}{(0.0821\,1 \text{ atm/mole °K})\,(373°K)} = 6.53$ moles

14. Helium: $A \rightarrow B$
    Hydrogen: $B \rightarrow A$

16.

	London	Dipole–dipole	H Bonding
a. Br_2	✔		
c. H_2O	✔	✔	✔
e. CO_2	✔		
g. C_5H_{12}	✔		
i. H_2S	✔	✔	

19. Vapor pressure increases as temperature increases. Normal boiling point is temperature at which the vapor pressure of a liquid equals 760 torr.

21. A solid passing directly into the gas phase.

Answers to Questions
Chapter 11

1. (a) C, (c) D, (d) DR, (f) SR, (h) DR, (j) C, (l) C.

2. (a) C = 4+, (b) Br = 0, (d) Cl = 5+, (f) S = 6+, (h) S = 4+.

3. Reactions (a), (c), and (e) are not redox reactions since no element undergoes a change in oxidation number. In (b), Cu is oxidized, N is reduced. In (d), Zn is oxidized and O is reduced.

4.

	Element oxidized	Element reduced
a.	Na	O
c.	Fe	O
d.	O	Cl

5. (a) oxidized ($-2H$), (b) reduced ($-O$).

7. The minimum amount of energy that must accompany a molecular collision if reactants are to be converted to products.

9. Reduces the activation energy.

11. (a) increase collisions, (b) no effect, (c) increase collisions.

13. 2 million at 30°C, 4 million at 40°C.

15. The rates are equal.

Answers to Questions
Chapter 12

1. A solution is a mixture that can vary in composition. A compound has a fixed composition.

3. These are not exact concentration terms; they only have a relative meaning.

5. **a.** hydrated ethyl alcohol molecules
   b. $K^+_{(aq)}$ and $NO^-_{3(aq)}$
   c. $H^+_{(aq)}$ and $Cl^-_{(aq)}$

6. Only water passes through the membrane in osmosis; water and dissolved solutes can pass through a dialyzing membrane.

7. (a) 1.5% (w/v); (b) 5.0% (w/v).

8. $V_c = (250 \text{ ml}) (1.5M)/(5.0 M) = 75$ ml

9. (a) 286 ml

10. The solution has a higher boiling point and osmotic pressure. Pure water has a higher freezing point and vapor pressure.

13. **a.** 2% NaCl has the higher osmotic pressure.
    b. Net flow of water into 2% NaCl.

14. (a) $1M$ NaCl—2 molar in solute particles; (b) and (c) both contain 1 mole of solute particles—same osmotic pressure.

15. (a) $0.125M$; (b) $0.49M$.

16. (a) 37.3 g KCl; (b) 67.3 g NaCl.

17. (a) 3.0 g KCl; (b) 2.3 g NaCl.

18. (a) 4.8% (w/w).

20. $M_d = (2.3 \text{ l}) (0.50M)/(3.1 \text{ l}) = 0.37M$

Answers to Questions Chapter 13

1. Acid: H^+ Base: OH^-

2. Strong acids
   hydrochloric acid—$HCl_{(aq)}$
   nitric acid—$HNO_{3(aq)}$
   sulfuric acid—$H_2SO_{4(aq)}$

   Strong bases
   sodium hydroxide—NaOH
   potassium hydroxide—KOH
   calcium hydroxide—$Ca(OH)_2$
   magnesium hydroxide—$Mg(OH)_2$

3. The hydronium ion is the combination of one or more water molecules with a hydrogen ion. H_3O^+, $H^+_{(aq)}$

4. Acids: sour taste; turn litmus red; react with certain metals to produce hydrogen; react with metal oxides, carbonates, and bicarbonates; neutralize bases. Bases: bitter taste; turn litmus blue; neutralize acids.

5. A strong acid dissociates completely in solution; a weak acid is only partially dissociated.

8. Wash acid from skin with lots of water.

9. $H_2SO_{4(aq)} \rightarrow H^+_{(aq)} + HSO^-_{4(aq)}$
   $HSO^-_{4(aq)} \rightleftharpoons H^+_{(aq)} + SO^{2-}_{4(aq)}$

10. Concentrated H_2SO_4 has a high affinity for water and can remove water from matter.

11. **a.** $2HCl + Ca(OH)_2 \rightarrow CaCl_2 + 2H_2O$
    b. $HNO_3 + KOH \rightarrow KNO_3 + H_2O$
    c. $H_2SO_4 + 2NaOH \rightarrow Na_2SO_4 + 2H_2O$

12. **a.** $2HCl + K_2O \rightarrow 2KCl + H_2O$
    b. $HNO_3 + NaHCO_3 \rightarrow NaNO_3 + H_2CO_3$
    $\rightarrow CO_2 + H_2O$
    c. $2HCl + Mg \rightarrow MgCl_2 + H_2$

13.

	Acid	Base
KCl	HCl	KOH
$Ca(NO_3)_2$	HNO_3	$Ca(OH)_2$
$Ni(C_2H_3O_2)_2$	$HC_2H_3O_2$	$Ni(OH)_2$

14. **a.** F: $2HNO_{3(aq)} + Ba(OH)_{2(aq)} \rightarrow Ba(NO_3)_{2(aq)} + 2H_2O$
  I: $2H^+_{(aq)} + 2NO^-_{3(aq)} + Ba^{2+}_{(aq)} + 2OH^-_{(aq)} \rightarrow Ba^{2+}_{(aq)} + 2NO^-_{3(aq)} + 2H_2O$
  NI: $2H^+_{(aq)} + 2OH^-_{(aq)} \rightarrow 2H_2O$

15.

	$[H^+_{(aq)}]$	$[OH^-_{(aq)}]$
a.	10^{-3}	10^{-11}
b.	10^{-5}	10^{-9}
c.	10^{-10}	10^{-4}

16. **a.** 3 **e.** 10
  b. 9 **f.** 3

17. **a.** $10^{-5}M$ **e.** $10^{-8}M$
  b. $10^{-10}M$ **f.** $10^{-12}M$

18. **a.** 10^{-3} to $10^{-4}M$ **b.** 10^{-6} to $10^{-7}M$

19. **a.** pH 3 to 4 **b.** pH 7 to 8

20. A solution that minimizes the change in pH when small amounts of acid or base are added to it. Buffers are prepared with (a) a weak acid plus the salt of the weak acid, (b) a weak base plus the salt of the weak base.

21. Carbonate buffer (H_2CO_3/HCO_3^-), phosphate buffer ($H_2PO_4^-/HPO_4^{2-}$), and blood proteins. The carbonate buffer is most important.

22. 7.35–7.45. Acidosis: lower pH than 7.35; alkalosis: higher pH than 7.45.

24. Metabolic acidosis: starvation, ingestion of acids, diabetes, kidney failure, dehydration. Metabolic alkalosis: excessive loss of gastric acid, overdose of antacids, kidney failure.

27. 0.600 l (0.75N) = 0.45 GEW HCl

30. **a.** 0.67N **c.** 2.0N **e.** 0.029N
  b. 1.06N **d.** 1.7N **f.** 1.5N

32. V_b = (500 ml) (1.5N)/3.0N = 250 ml

33. Number of equivalents of acid neutralized equals the number of equivalents of base added, or vice versa.

34. Visual indication of equivalence point.

35. Na = (12.0 ml) (0.0010N)/350 ml = 0.000034N

37. Na = (1000 ml) (0.60N)/0.75N = 800 ml

Answers to Questions
Chapter 14

2. a, c, d

5. **a.** double bond **g.** carbonyl, hydroxyl, triple bond, double
  c. ester bond
  e. three ester groups

6. **a.** ketone   **g.** alcohol, amine, carboxylic acid
    c. amine   **i.** alkene, alcohol
    e. alcohol, carboxylic acid

Answers to Questions
Chapter 15

1. Each carbon atom has four single covalent bonds.
2. alkanes, paraffins
4. C_nH_{2n+2}, $n = 14$; $C_{14}H_{30}$
6. (a), (b), and (d) are identical (1-bromobutane)
    (c) and (e) are identical (2-bromobutane)
9. **a.** 3-methylhexane
    c. 2,2,4,4-tetramethylpentane
    e. 1,6-dichloro-4-ethyl-4-methylheptane
    g. 1-bromo-2-methylpropane
    i. 1,1,4-trichlorocyclohexane
    k. trichloromethane

10. **a.** $CH_3\ \underset{\underset{\textstyle CH_3}{|}}{CH}\ CH_3$

    e. $CH_3\ CH_2-\underset{\underset{\textstyle CH_3}{|}}{\overset{\overset{\textstyle CH_3}{|}}{C}}-CH_2-\underset{\underset{\underset{\textstyle CH_3}{|}}{\overset{\textstyle CH_2}{|}}}{CH}-CH_2\ CH_3$

    c. (cyclopentane ring) Br, Br, Br

    g. $CH_3\ CH_2\ \underset{\underset{\underset{\textstyle CH_3}{|}}{\overset{\textstyle CHCH_3}{|}}}{CH}\ CH_2\ CH_2\ CH_3$

11. $CH_3CH_2CH_2CH_2CH_2CH_3$ hexane

    $CH_3\ \underset{\underset{\textstyle CH_3}{|}}{CH}\ CH_2\ CH_2\ CH_3$ 2-methylpentane

    $CH_3\ CH_2\ \underset{\underset{\textstyle CH_3}{|}}{CH}\ CH_2\ CH_3$ 3-methylpentane

    $CH_3\ \underset{\underset{\textstyle CH_3}{|}}{CH}-\underset{\underset{\textstyle CH_3}{|}}{CH}\ CH_3$ 2,3-dimethylbutane

    $CH_3-\underset{\underset{\textstyle CH_3}{|}}{\overset{\overset{\textstyle CH_3}{|}}{C}}-CH_2\ CH_3$ 2,2-dimethylbutane

13. $C_3H_8 + 5O_2 \rightarrow 3CO_2 + 4H_2O$
    $C_5H_{12} + 8O_2 \rightarrow 5CO_2 + 6H_2O$
    $C_4H_8 + 6O_2 \rightarrow 4CO_2 + 4H_2O$

16. $CH_4 + Cl_2 \xrightarrow{uv} H_3CCl + HCl$
$H_3CCl + Cl_2 \xrightarrow{uv} H_2CCl_2 + HCl$
$H_2CCl_2 + Cl_2 \xrightarrow{uv} HCCl_3 + HCl$
$HCCl_3 + Cl_2 \xrightarrow{uv} CCl_4 + HCl$

17. (b) and (c) are isomers
(d) and (e) are isomers
(g) and (h) are isomers
(a), (f), and (i) do not have isomers

18. cyclopropane, flammable

19.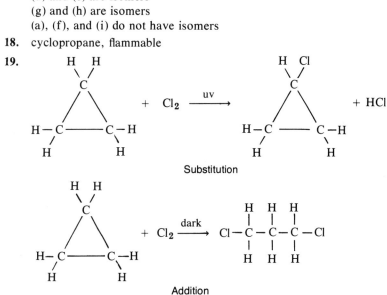

Substitution

Addition

Answers to Questions
Chapter 16

2. alkene: C_7H_{14} C_nH_{2n} $(n=7)$
alkyne: C_5H_8 C_nH_{2n-2} $(n=5)$

3. a. 2-butene
c. 1,4-pentadiene
e. cyclobutene
f. 4-methylcyclopentene
h. 1,3-dichloro-2-hexene

6. a. CH_3CH_3
b. $CH_3-CH-CH_3$
         |
         Br

7. a. polyethylene: $H_2C = CH_2$
b. polyvinylchloride: $H_2C = CHCl$

11. **a.** *meta*
     b. *para*
     c. *ortho*

12. **a.** + HBr

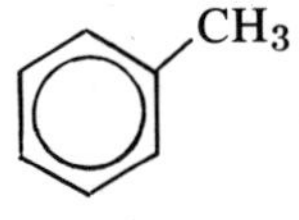

13. **a.** chlorobenzene
     c. methylbenzene (toluene)
     e. 1-bromo-4-methylbenzene (*p*-bromotoluene)
     g. 2,4,6-trinitrotoluene
     i. hydroxybenzene (phenol)

14. **a.** 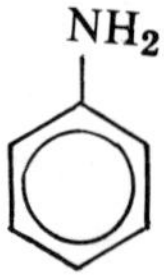   **e.**

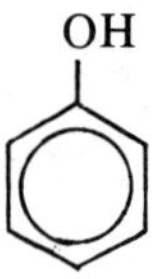

     c.

Answers to Questions
Chapter 17

1. (a) alcohol, (b) ether, (d) phenol, (f) alcohol.
3. **a.** 1-pentanol   **e.** 3-chloro-1-propanol
    c. 2-methyl-2-pentanol   **g.** 2,2-dimethyl-1-propanol
4. (a) secondary, (c) tertiary, (f) secondary.
5. **b.**

$$CH_3-\underset{\underset{CH_3}{|}}{\overset{\overset{OH}{|}}{C}}-CH_2CH_3$$

6. **a.** $CH_3-\underset{\overset{|}{OH}}{CH}-CH_3$

7. **a.** $\longrightarrow CH_3CH_2\underset{\overset{|}{H}}{C}=O \xrightarrow{\text{[O]}} CH_3CH_2\underset{\overset{\diagdown}{OH}}{C}=O$

c. 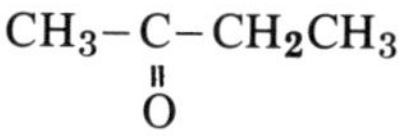 $CH_3-C-CH_2CH_3$
      $\underset{O}{\overset{\|}{}}$

8. a. $CH_3-CH=CH_2$

   c. 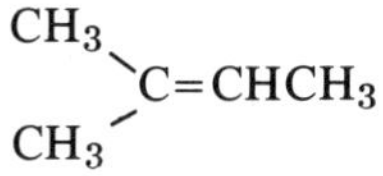 $\underset{CH_3}{\overset{CH_3}{}}\diagdown C=CHCH_3$

   e. 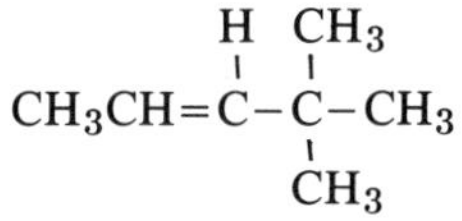 $CH_3CH=\overset{\overset{H}{|}}{C}-\overset{\overset{CH_3}{|}}{\underset{\underset{CH_3}{|}}{C}}-CH_3$

9. a. $CH_3CH_2-O-CH_2CH_3$

13. Methanol is oxidized to formaldehyde and formic acid, both toxic substances.

14.

Most soluble	Least soluble
a. CH_3CH_2OH	$CH_3CH_2CH_3$
b. $HOCH_2CH_2OH$	$CH_3(CH_2)_4CH_2OH$
c. CH_3CH_2OH	$CH_3(CH_2)_6CH_2OH$

15. a. 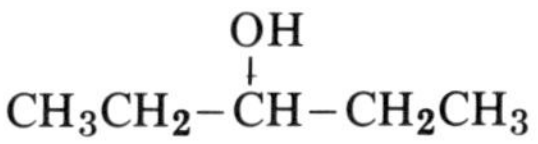 $CH_3CH_2-\overset{\overset{OH}{|}}{CH}-CH_2CH_3$

   c. 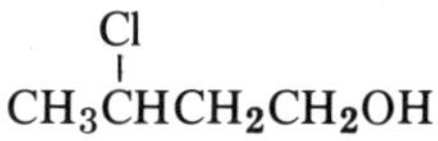 $CH_3\overset{\overset{Cl}{|}}{CH}CH_2CH_2OH$

   e. $CH_3-O-CH_2CH_3$

   g. 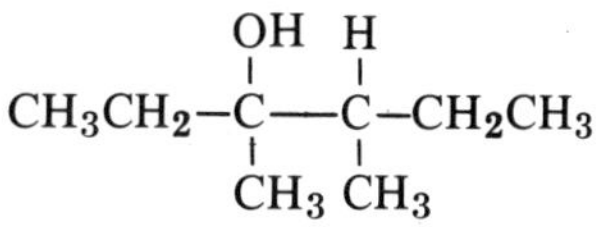 $CH_3CH_2-\overset{\overset{OH}{|}}{\underset{\underset{CH_3}{|}}{C}}-\overset{\overset{H}{|}}{\underset{\underset{CH_3}{|}}{C}}-CH_2CH_3$

   i. 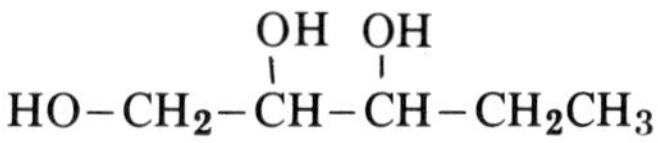 $HO-CH_2-\overset{\overset{OH}{|}}{CH}-\overset{\overset{OH}{|}}{CH}-CH_2CH_3$

   k. 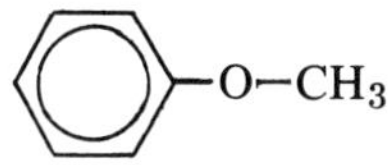 $\langle\bigcirc\rangle-O-CH_3$

Answers to Questions in Chapter 18

2. | Common name | IUPAC name |
|---|---|
| a. formaldehyde | methanal |
| c. acetaldehyde | ethanal |
| d. diethyl ketone | 3-pentanone |
| f. methyl n-propyl ketone | 2-pentanone |

3. a. 3-methyl butanal
   b. 4-methyl-2-hexanone
   c. 2-Bromo-3-pentanone

4. Least soluble:

$$\underset{\text{(phenyl)}}{\bigcirc}-\overset{\overset{\displaystyle O}{\|}}{C}\diagdown_{CH_3} \quad ; \quad CH_3CH_2-\overset{\overset{\displaystyle O}{\|}}{C}-CH_2CH_3 \; ;$$

most soluble: $\quad CH_3-\overset{\overset{\displaystyle O}{\|}}{C}-CH_3$

6. a. $\longrightarrow CH_3COOH$

c. $\longrightarrow H-\overset{\overset{\displaystyle OH}{|}}{\underset{\underset{\displaystyle OCH_3}{|}}{C}}-H$

e. $\longrightarrow CH_3COO^{\ominus}NH_4{}^+ \; + \; 2Ag\downarrow \; + \; 3NH_3 \; + \; H_2O$

f. $\longrightarrow CH_3CH_2-\overset{\overset{\displaystyle OH}{|}}{\underset{\underset{\displaystyle H}{|}}{C}}-CH_3$

8. the ease of aldehyde oxidation

9. a $\longrightarrow CH_3CH_2CHO \; + \; H_2O$

b $\longrightarrow CH_3-\overset{\overset{\displaystyle O}{\|}}{C}-CH_3 \; + \; H_2O$

10.

Common name	**IUPAC name**
a. acetic acid	ethanoic acid
c. ethyl propionate	ethyl propanoate
e. n-butyl formate	n-butyl methanoate

11.

Common name	**IUPAC name**
a. potassium valerate	potassium pentanoate

12. a. $\rightarrow CH_3COO^{\ominus}K^+ + H_2O$
c. $\rightarrow 2\ HCOO^{\ominus}Na^+ + CO_2 + H_2O$

13.

Acid	**Alcohol**
b. acetic acid	n-propyl alcohol
d. benzoic acid	phenol

14. The ester, acetyl salicylic acid, can hydrolyze producing acetic acid which has the vinegar odor.

16. $\longrightarrow CH_3CH_2OH \; + \; CH_3\overset{\overset{\displaystyle O}{\|}}{C}-O^{\ominus}Na^+$

17.   **b** $\quad CH_3CH_2\underset{\underset{\displaystyle CH_3}{|}}{CH}-COOH$ **c** $\quad Cl-\overset{\overset{\displaystyle Cl}{|}}{\underset{\underset{\displaystyle H}{|}}{C}}-CH_2-\overset{\overset{\displaystyle O}{\|}}{C}-H$

g.  $HC{-}O{-}CH_2CH_2CH_2CH_3$ (with O double bonded to C)

j. 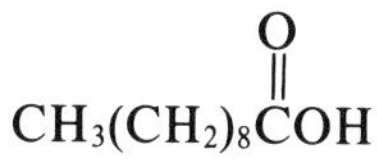 $CH_3(CH_2)_8COH$ (with O double bonded to C)

i.  $CH_3{-}C{-}CH_2CH_2CH_2CH_3$ (with O double bonded to C)

Answers to Questions in Chapter 19

1. **a.** 2°, **b.** 1°, **c.** 3°, **d.** 2°.

2. **b.** *n*-propylamine

    c. aniline

    e. 4-methylaniline

    h. N,N-dimethylaniline

    i. methyl-*n*-propylamine

3. **a.** 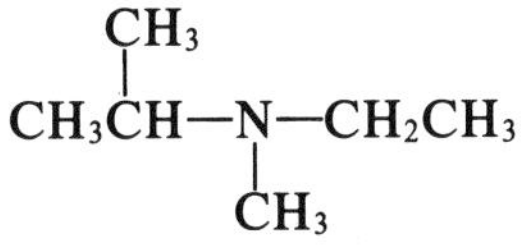 $CH_3CH{-}N{-}CH_2CH_3$ (with CH_3 above CH and CH_3 below N)

    c. 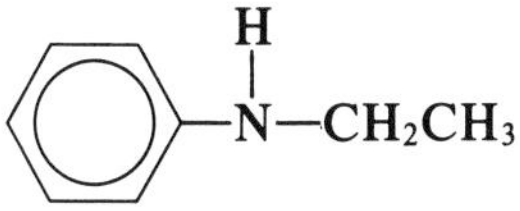 (benzene ring)$-N{-}CH_2CH_3$ (with H above N)

    d. 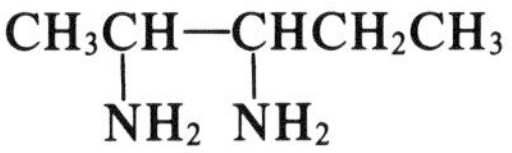 $CH_3CH{-}CHCH_2CH_3$ (with NH_2 below each of the two central carbons)

    g. 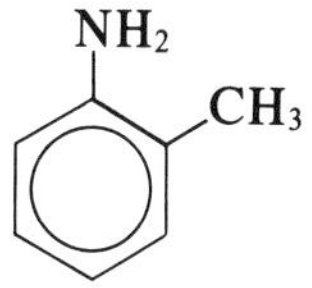 (benzene ring with NH_2 at top and CH_3 on side)

    j. 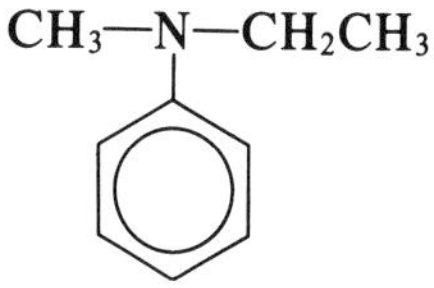 $CH_3{-}N{-}CH_2CH_3$ (with benzene ring below N)

4. **a.** 3-aminohexane, **b.** 2-(dimethylamino) butane.

5. **b.** 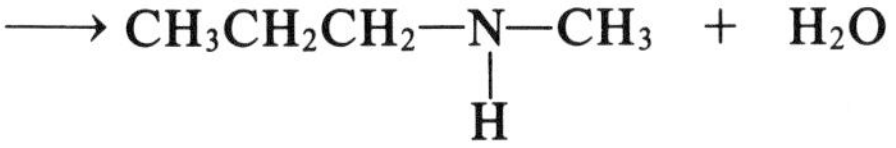 $\longrightarrow CH_3CH_2CH_2{-}N{-}CH_3 + H_2O$ (with H below N)

    d. 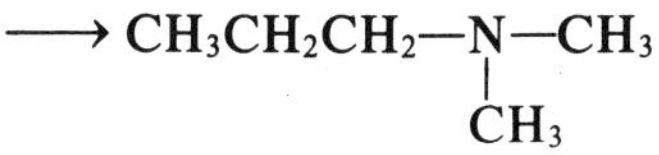 $\longrightarrow CH_3CH_2CH_2{-}N{-}CH_3$ (with CH_3 below N)

7. **a.** $\rightarrow CH_3CH_2NH_3^+ \; Cl^-$ ethylammonium chloride

9. **a.** acetamide

10. **a.** 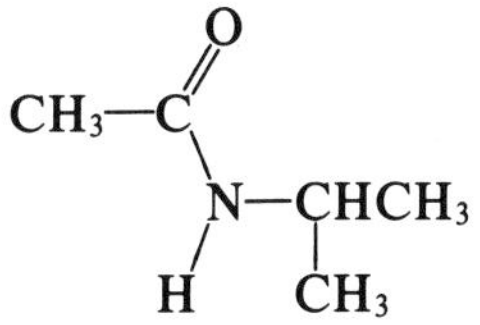 $CH_3{-}C$ (with O double bonded to C) $N{-}CHCH_3$ (with H below N and CH_3 below CH)

    d. 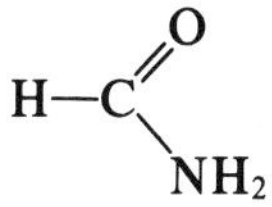 $H{-}C$ (with O double bonded to C and NH_2)

12. No. It has no $-N-H$ bonds.

13. $CH_3COOH + SOCl_2 \longrightarrow CH_3\overset{\overset{\displaystyle O}{\|}}{C}-Cl + SO_2 + HCl$

14. **a.** $CH_3CH_2\overset{\overset{\displaystyle O}{\diagup}}{\underset{\diagdown Cl}{C}} + 2NH_3 \longrightarrow CH_3CH_2\overset{\overset{\displaystyle O}{\diagup}}{\underset{\diagdown NH_2}{C}} + NH_4Cl$

b.

$+ 2\,CH_3NH_2 \longrightarrow$

$+ CH_3NH_3^+Cl^-$

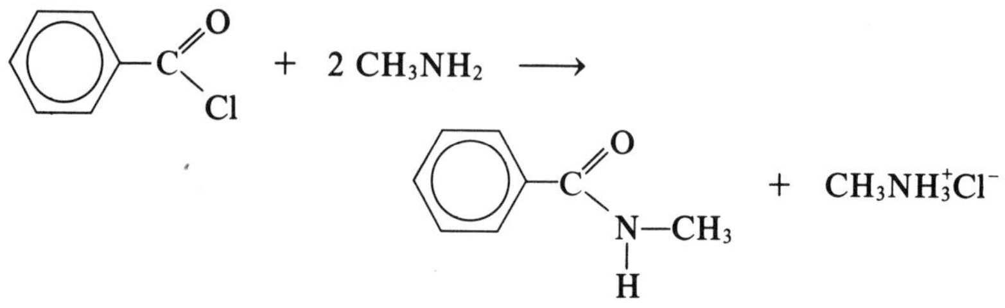

15. **a.** $\rightarrow CH_3COOH + NH_4Cl$
    b. $\rightarrow CH_3COO^{\ominus}Na^+ + NH_3$

16. Both proteins and nylons have repeating amide units in their polymeric structures.

19. **a.** pyrrolidine, **c.** pyridine and pyrrole, **e.** indole.

21. Sedative (sleep aid).

23. Heroin is the diacetate ester of morphine.

Answers to Questions
Chapter 20

1. Solubility. Lipids are soluble in nonpolar solvents but not in polar solvents like water.

3. A long-chain monocarboxylic acid.

4. **a.** $C_nH_{2n+1}COOH$
   b. $R-COOH$, where R has one or more carbon–carbon double bond.

7. Hydrolysis and oxidation cause rancidity. Addition of hygroscopic agents and antioxidants can reduce rancidity.

8. **a.**

$$H_2C-O-\overset{\overset{\displaystyle O}{\|}}{C}-C_{17}H_{33}$$
$$H-\overset{|}{C}-O-\overset{\overset{\displaystyle O}{\|}}{C}-C_{17}H_{35}$$
$$H_2\overset{|}{C}-O-\overset{\overset{\displaystyle O}{\|}}{C}-C_{11}H_{23}$$

b.

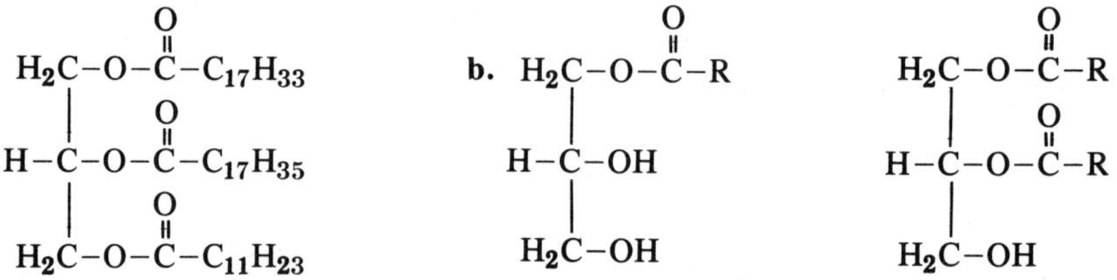

$$H_2C-O-\overset{\overset{\displaystyle O}{\|}}{C}-R$$
$$H-\overset{|}{C}-OH$$
$$H_2\overset{|}{C}-OH$$
monoglyceride

$$H_2C-O-\overset{\overset{\displaystyle O}{\|}}{C}-R$$
$$H-\overset{|}{C}-O-\overset{\overset{\displaystyle O}{\|}}{C}-R$$
$$H_2\overset{|}{C}-OH$$
diglyceride

10.

$$
\begin{array}{c}
\overset{\displaystyle O}{\overset{\displaystyle \|}{}} \\
H_2C-O-C-R \\
\end{array}
$$

$$
\begin{array}{ll}
H_2C-O-\overset{\overset{\textstyle O}{\|}}{C}-R & H_2C-OH \\
| & | \\
H-C-O-\overset{\overset{\textstyle O}{\|}}{C}-R + 3\,NaOH \xrightarrow{\;\Delta\;} H-C-OH + 3\,R-\overset{\overset{\textstyle O}{\|}}{C}-O^{\ominus},\ Na^{\oplus} \\
| & | \qquad\qquad\qquad \text{Soap} \\
H_2C-O-\overset{\overset{\textstyle O}{\|}}{C}-R & H_2C-OH \\
& \text{Glycerol}
\end{array}
$$

13. triglyceride $+\ 3H_2O \xrightarrow{\text{enzymes}}$ glycerol $+$ 3 fatty acids

A triglyceride must be hydrolyzed (in the small intestine) since the body cannot absorb triglycerides but it can absorb glycerol and the fatty acids.

14. $R-\overset{\overset{\textstyle O}{\|}}{C}-O-R'$ Waxes are esters made from long-chain fatty acids and long-chain primary alcohols.

15. A prostaglandin is a large unsaturated fatty acid that contains a ring. They usually contain about 20 carbon atoms. They are used in birth control drugs and in drugs to reduce blood pressure and inflammation.

17. Lecithins and the cephalins. They differ with respect to the nitrogen-containing compound bonded to the oxygen of the phosphate group.

19. Spingolipids

21. **a.** Cholesterol—in cell membranes; used in the synthesis of other body steroids
  b. Bile acids and bile salts—emulsification of fats
  c. Steroids of the adrenal cortex—metabolism and electrolyte balance
  d. Sex hormones—male and female characteristics
  e. Vitamin D—required for absorption of calcium ion from foods and the prevention of rickets

22. **a.** Those vitamins that are soluble in foods that contain fats and oils. Vitamins D, E, A, and K.
  b. Only vitamin D

Answers to Questions
Chapter 21

1. The carbonyl group and the hydroxyl group ($\overset{\displaystyle}{\diagdown}C = O$, $-OH$).

3. (a) Formaldehyde ($H_2C=O$) would be the simplest organic compound that would fit the old definition, but it does not have hydroxyl groups.

4. (b) Aldose, ketose.

6. Simplest aldose: glyceraldehyde (structure is in the text). Simplest ketose: dihydroxyacetone (structure is in the text).

7. $6CO_2 + 6H_2O \xrightarrow[\text{light}]{\text{chlorophyll}} C_6H_{12}O_6 + 6O_2$

10. a.

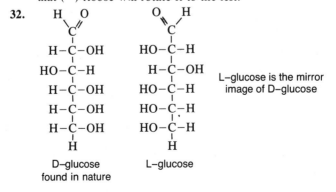

Fischer	Haworth

   H–C–OH
   H–C–OH
   H–C–OH
   H–C——O
   H–C–OH
   H

   b. Deoxyribose does not have an —OH group on carbon 2; it has two hydrogens.

12. Glucose, galactose, fructose. (Structures are in the text.)

14. Glucose can be oxidized to gluconic acid and in doing so will reduce other species, such as Cu^{2+} to Cu^+ in basic solution.

15. The Benedict test can detect the presence of reducing sugars in body fluids. The equation describing the Benedict test is in the text.

18. Maltose reacts with water in the presence of an acid or maltase to form two glucose molecules. Sucrose reacts with water in the presence of an acid or sucrase to form glucose and fructose. Lactose reacts with water in the presence of an acid or lactase to form glucose and galactose.

20. Monosaccharides are very soluble in water; the disaccharides are soluble in water; and the polysaccharides are generally not soluble.

22. Glycogen is a storage form of glucose.

24. Hydrolysis of the oxygen bridges joining the monosaccharides.

28. Plane-polarized light vibrates only in one plane, and a dextrorotatory glucose solution would rotate the plane to the right.

29. A chiral molecule is *not* superimposable on its mirror image.

30. The chiral molecules are (a) and (c). They have a carbon atom with four different groups bonded to it.

31. (+)-Ribose will rotate plane-polarized light to the right the same number of degrees that (−)-ribose will rotate it to the left.

32.

   H   O
    \ //
     C
     |
   H–C–OH
     |
   HO–C–H
     |
   H–C–OH
     |
   H–C–OH
     |
   H–C–OH
     |
     H

   D–glucose
   found in nature

   O   H
    \\ /
     C
     |
   HO–C–H
     |
   H–C–OH
     |
   HO–C–H
     |
   HO–C–H
     |
   HO–C–H
     |
     H

   L–glucose

   L–glucose is the mirror
   image of D–glucose

Answers to Questions
Chapter 22

1. $-\overset{|}{\underset{|}{C}}-\overset{|}{\underset{\underset{NH_2}{|}}{C}}-COOH$ A carboxylic acid with an amino group attached to the carbon atom next to the carboxyl group.

α-Carbon

3. 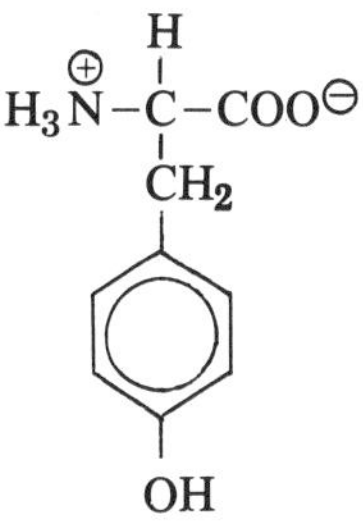 $H_3\overset{\oplus}{N}-\overset{\overset{H}{|}}{\underset{\underset{}{|}}{C}}-COO^{\ominus}$
$\overset{}{\underset{CH_2}{}}$

OH

5. **a.** serine or threonine
   b. aspartic or glutamic acid

8. aspartic and glutamic acid

10. Solutions of strong acids and bases.

12. Primary

15. Peptide—Usually contains less than 50 amino acids combined. Polypeptide—Contains more than 50, but no definite upper limit. Protein—Contains more than 50 amino acid units, but usually the number is very large.

16. Hydrocarbons, —OH, sulfhydryl.

18. 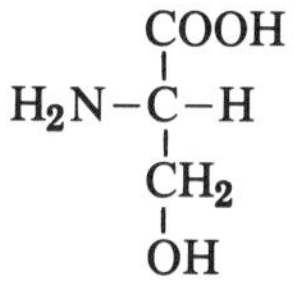 $COOH$
   $H_2N-\overset{|}{\underset{|}{C}}-H$
      $\overset{}{\underset{CH_2}{|}}$
      $\overset{}{\underset{OH}{|}}$

L-Form of serine

20. Colloids. Most of them are between 1 and 100 nm in particle diameter.

22. $Gly-Asp-Gly-Ala-COO^{\ominus}$
   $\overset{}{\underset{CH_2}{|}}$
   $\overset{}{\underset{COO^{\ominus}}{|}}$

The carboxyl groups can bond with the lead ions forming an insoluble substance:

$$-\overset{\overset{O}{\|}}{C}-O^{\ominus} \;+\; Pb^{+2} \;\longrightarrow\; \left(-O-\overset{\overset{O}{\|}}{C}-O^{\ominus}\right)_2 Pb \downarrow$$

23. By the destruction of secondary, tertiary, and quaternary structures of the bacteria's protein.

25. ninhydrin + amino acid → blue color
biuret solution + protein → pink to violet color

27. For the Cys−Cys part

$$\text{H}_2\text{N}-\overset{\overset{\displaystyle \text{H}}{|}}{\underset{\underset{\displaystyle \text{CH}_2}{|}}{\text{C}}}\cdots\cdots\cdots\cdots\cdots\cdots\cdots\cdots\cdots\overset{\overset{\displaystyle \text{O}}{\|}}{\text{C}}-\overset{\overset{\displaystyle \text{H}}{|}}{\text{N}}-\overset{\overset{\displaystyle \text{H}}{|}}{\underset{\underset{\displaystyle \text{CH}_2}{|}}{\text{C}}}\cdots\cdots\cdots\cdots$$

S——S

The rest of the molecule appears as other peptides

29. Hydrogen bonding between $-\overset{}{\underset{\underset{\displaystyle \text{H}}{|}}{\text{N}}}-\overset{\alpha+}{\text{H}}\cdots\cdots\cdots\overset{\alpha-}{\text{O}}=\overset{|}{\underset{|}{\text{C}}}$

30. a. Rolling the hair arranges many of the —S—S— in different places than normal.
b. Mild oxidation breaks many of the —S—S— bonds. (c) Mild reduction reforms *new* —S—S— bonds, which changes the original shape of the hair to the curled shape.

33. Insulin is a relatively small protein. If taken by mouth, it would be digested.

Answers to Questions
Chapter 23

1. Nucleoside—either ribose or deoxyribose and one of the bases T, U, C, G, or U joined to carbon 1′ of the sugar. Nucleotide—a nucleoside with a phosphate group.

2. β-ribose—RNA; β-deoxyribose—DNA. Structures of both are in the text.

3. DNA—A, G, C, T; pyrimidine bases: C,U, T; RNA—A, G, C, U; purine bases: A, G.

5. a. G + ribose, (c) A + deoxyribose.

6. a. Cytidine-5′-monophosphate, ·(c) deoxyadenosine-5′-diphosphate.

7. a.

TMP

8. See figure in text. The diester link joins the sugars together forming the DNA and RNA molecules.

9. DNA: C to G, A to T; RNA: C to G, A to U.

11. In text. Hydrogen bonding between base pairs.

13. In text. You should include unwinding of DNA, base pairing of incoming neucleotide triphosphates with bases on DNA strands, elimination of pyrophosphate, and separation into two DNA molecules.

14. dATP, dCTP, dGTP, and dTTP.

16. Transfer RNA—carrier of amino acids; ribosomal RNA—with protein forms ribosomes; messenger RNA—carries code for protein synthesis from DNA to ribosomes. Directs protein synthesis at ribosomes.

18. A three-base sequence on a mRNA molecule. The anticodon on a tRNA molecule will match only specific codons on mRNA signaling for the amino acid it carries.

20. AUG; on the leading end of mRNA.

21. **a.** cysteine, **c.** glycine.

23. **a.** Loss of $-A-$, **b.** $-gly-lys-try-val-$; **c.** $-gly-asp-gly-tyr-$.

25. Sickle-cell anemia is the result of an alteration in the DNA, the heredity messenger of the cell. A person with the disease has the mutation in both strands of DNA in each DNA molecule; a person with the trait has only one strand affected in each double-helix molecule.

Answers to Questions
Chapter 24

1. Yeast to cause bread to rise or to aid the fermentation of grains to alcohol. Fermentation of cabbage to produce sauerkraut.

3. Both enzymes and ordinary chemical catalysts increase the reaction rate by reducing the activation energy, but enzymes are much more specific. Both form an activated complex.

4. **a.** lyase   **c.** kinase
   b. isomerase

6. A carrier for removal or addition of CO_2.

8. Lock and key and induced-fit theories

10. **a.** group
    b. group

12. When the enzyme is saturated with substrate.

14. Irreversible inhibition.

16. Liver

19. The hypothalamus is stimulated by the nervous system, and as a response secretes releasing agents to generate hormones that will enter the blood.

20. **a.** niacin   **d.** B_{12}
    b. niacin   **e.** folic acid
    c. B_1 **f.** C

Answers to Questions in Chapter 25

1. Catabolism, anabolism.
2. ATP stores chemical energy and transfers energy from those reactions that produce energy to those that require energy.
3. **a.** 7.3 kcal/mole
   b. 10.0 kcal/mole
   c. 7.3 kcal/mole
6. Yes. Energy released in the hydrolysis of creatine phosphate exceeds that released by ATP.
8. **a.** $NAD^{\oplus} + 2e^- + 2H^+ \rightarrow NADH + H^+$
   b. $FAD + 2e^- + 2H^+ \rightarrow FADH_2$
   c. $2H^+ + e^- + \frac{1}{2}O_2 \rightarrow H_2O + energy$
10. $FADH_2$ can enter the ETS at coenzyme Q bypassing the $FMN-FMNH_2$ system.
11. Oxidations are brought about by the removal of $2H^+ + 2e^-$ from molecules.
13. 5 pyruvic acid → 5 acetyl CoA.
14. *Cindy Ann is our kitten. She sure finds mice offensive.*
15. 12ATP

17.
$$CH_3\overset{\overset{O}{\|}}{C}-COOH + CoASH \xrightarrow{enzyme} CH_3\overset{\overset{O}{\|}}{C}-SCoA + CO_2$$

$$NAD^{\oplus} \qquad NADH + H^+$$

$$3ATP \qquad ETS$$

18. 38ATP
21. 30ATP (24 in cycle/ETS and 6 converting pyruvic acid to acetyl CoA).
24. Formation of glucose-6-phosphate.
28. Lactic acid.
30. Under anaerobic conditions, ATP is produced via the EM pathway with the production of lactic acid. As the lactic acid level rises—acidosis sets in—the body gulps for air but insufficiently fast to restore blood pH—eventual collapse will result.
32.

aerobic glycolysis	8 ATP
2(pyruvic acid) → 2(acetyl CoA)	6 ATP
2(acetyl CoA) → $4CO_2 + 4H_2O$	24 ATP
	38 ATP

36. 6 moles of acetyl-CoA are formed in 5 turns around the cycle.

Activation of fatty acid	−2 ATP
5 turns around FA cycle	25 ATP
	23 ATP

38. Carbohydrate metabolism—the EM pathway.

Glossary

The number(s) in parentheses following each term indicates the chapter in which the term is introduced.

α-**Amino acid**   (22) A carboxylic acid that has an amino group ($—NH_2$) bonded to the carbon atom adjacent to the carboxyl group.

α-**Helix**   (22) A right-handed coil structure that represents the predominate secondary structure of proteins.

α-**Radiation**   (7) A form of nuclear radiation. A stream of helium nuclei. Symbolized α or $_2^4$He.

Absolute temperature scale   (2, 10) The temperature scale that begins at the absolute zero of temperature. Therefore, there are no negative temperatures on the absolute scale. Temperatures are expressed in degrees Kelvin (°K), and are related to Celsius temperatures as: °K = °C + 273.

Acid   (13) A substance that produces hydronium ions when dissolved in water (Arrhenius).

Acid chloride   (19) An organic compound that contains the $—\overset{\displaystyle O}{\overset{\displaystyle \|}{C}}—Cl$ functional group.

Acidosis   (13) The condition brought about by the lowering of the blood pH to a value less than 7.35.

Activated complex   (11) A high energy species in a reaction that is neither reactant nor product, but intermediate between them.

Activation energy   (11) The minimum energy that must be attained by reactants if they are to become products in a chemical change.

Active site   (24) That place on an enzyme at which its catalytic activity takes place.

Acyl-Coenzyme A (25) The species formed when coenzyme A combines with a fatty acid acyl group. $R—\overset{\overset{\textstyle O}{\|}}{C}—SCoA$

Addition reaction (15) A reaction in which atoms or groups of atoms are added to a molecule.

Aerobic (25) In the presence of oxygen.

Alcohol (14, 17) An organic compound containing the hydroxyl ($—\overset{|}{\underset{|}{C}}—OH$) functional group.

Aldehyde (14, 18) An organic compound containing the aldehyde functional group, $—\overset{\overset{\textstyle O}{\|}}{C}—H$.

Aldose (21) A monosaccharide that, in its open chain form, contains an aldehyde functional group (such as glucose).

Aliphatic (14) Any organic compound or group that contains straight or branched chains of carbon atoms, or nonaromatic rings of carbon atoms.

Alkaline (13) Refers to a solution that is basic or a compound that can form a basic solution (pH > 7).

Alkali metal (6) Any metal in Group I on the periodic table.

Alkaloids (19) Complex heterocyclic amines isolated from plants.

Alkalosis (13) The condition brought about by the elevation of the blood pH above 7.45.

Alkane (14, 15) A hydrocarbon that contains only single carbon–carbon bonds of general formula C_nH_{2n+2}.

Alkene (14, 16) A hydrocarbon that contains at least one carbon–carbon double bond. The general formula for alkenes with one double bond is C_nH_{2n}.

Alkyl group (15) The organic group that results when a hydrogen atom is removed from an alkane; symbolized R—.

Alkyne (14, 16) A hydrocarbon that contains at least one carbon–carbon triple bond. The general formula for an alkyne with one triple bond is C_nH_{2n-2}.

Allosteric enzyme (24) An enzyme that has a specific end product inhibitor site called an allosteric site.

Alloy (9) A mixture of elements, mostly metals, that has the characteristic properties of a metal.

Amalgam (9) An alloy of mercury.

Amide (14, 19) An organic compound containing the $—\overset{\overset{\textstyle O}{\|}}{C}—NH_2$ functional group.

Amide bond (19, 22) The carbon-nitrogen bond in the $—\overset{\overset{\textstyle O}{\|}}{C}—\overset{\overset{\textstyle H}{|}}{N}—$ group. It is also called the peptide bond or peptide linkage.

Amine (14, 19) An organic compound containing the amine functional group; $—NH_2$, $—NHR$, $—NR_2$. *See* primary, secondary, and tertiary amine.

Amorphous solid (10) One that does not have a highly organized arrangement of particles, such as rubber, glass, and most plastics.

Amphetamines (19) Nonalkaloid drugs synthesized to mimic the action of epinephrine and norepinephrine.

Amphoteric (22) A term used to describe a substance that can act as an acid or base.

Anabolism (25) That type of metabolism that involves the synthesis of larger molecules from smaller molecules.

Anaerobic (25) Not in the presence of oxygen.

Anion (3) An ion that bears a negative charge.

Anticodon (23) The three base code region in a transfer RNA that is paired with the corresponding codon of a messenger RNA in protein synthesis.

Apoenzyme (24) The protein portion of a conjugated enzyme.

Aqueous solution (12) One in which water is the solvent.

Aromatic (14, 16) A class of organic compounds that includes benzene, derivatives of benzene, and those with more than one benzene structure in the molecule.

Artificial radioactivity (7) That displayed by isotopes made radioactive in a nuclear reactor or accelerator.

Atmosphere (10) A unit of pressure equivalent to 760 mmHg.

Atom (3) The smallest particle of an element that can exist and still have the chemical properties of that element.

Atomic mass unit (4) A relative unit of mass used to compare masses of atoms, ions, and molecules. The mass of the C-12 isotope is assigned a value of exactly 12 amu. One amu equals 1.6×10^{-24} g.

Atomic number (6) The number of protons in the nucleus of an atom and also the number of electrons about the nucleus.

Atomic weight (4) The relative weight (mass) of an atom compared to that of C-12 (12.000 amu).

Avogadro's law (10) Equal volumes of gas at the same temperature and pressure contain the same number of molecules (or moles) of gas.

Avogadro's number (4) 6.02×10^{23}. The number of particles (atoms, molecules, or ions) in one mole of the species.

β-Radiation (7) A kind of nuclear radiation. A stream of high-speed electrons. Symbolized β or $_{1}^{0}e$.

Background radiation (7) The low level ionizing radiation that is always present in the environment from both natural and man-made sources.

Balanced equation (4) One that has the same number of atoms of each element on both sides of the equation.

Barbiturates (19) Drugs that are derivatives of barbituric acid which depress the central nervous system.

Barometer (10) A device used to measure the pressure of gases.

Base (13) A substance that produces hydroxide ions when dissolved in water (Arrhenius).

Benedict's reagent (18, 21) A basic solution of copper (II) sulfate used to detect the presence of an aldehyde.

Beta-oxidation (25) The oxidation of fatty acids in the fatty acid cycle to produce acetyl coenzyme A.

Bile salts (20) Salts of the steroid bile acids.

Biological half-life (7) Concerning radioisotopes in the body—it is the time required for the body to eliminate one-half of a quantity of radioactive species by biological means.

Biological work (5) Work that occurs within a living organism; transport, synthesis, and mechanical work.

Bond angle   (8) The angle between adjacent bonds.

Bond energy   (8) The energy released as a bond is formed, also the energy needed to break the bond.

Bond length   (8) The distance between the nuclei of two bonded atoms.

Boyle's law   (10) The volume of a fixed sample of any gas varies inversely with pressure as long as the temperature of the gas does not change. $P_1V_1 = P_2V_2$.

Buffer   (13) A solution that resists changes in pH as small amounts of acid or base are added to it. Buffers are prepared from weak acids and their salts or weak bases and their salts.

Calcification   (22) The deposition of calcium-containing minerals in a network of collagen fibers to form bone and teeth.

Calorie   (5) A unit of energy equal to the amount of heat needed to raise the temperature of one gram of water one degree Celsius.

Carbohydrate   (21) Polyhydroxyaldehydes and polyhydroxyketones of simple formula CH_2O, or substances that yield these on hydrolysis.

Carbonyl group   (14, 18) The $-\overset{\overset{\displaystyle O}{\|}}{C}-$ functional group.

Carboxylic acid   (14, 18) An organic acid that contains the carboxyl functional group, $-\overset{\overset{\displaystyle O}{\|}}{C}-OH$.

Catabolism   (25) The breaking down of larger molecules into smaller molecules in metabolism. These are usually energy-producing changes.

Catalyst   (11) A substance that increases the rate of a chemical reaction without being consumed in the process.

Cation   (3) An ion that bears a positive charge.

Cellular respiration   (5) In living organisms, the conversion of glucose into carbon dioxide, water, and energy.

Cellulose   (21) A polymer of glucose found in plants in which the glucose units are joined by β-1,4 linkages.

Celsius temperature scale   (2) The temperature scale used in the metric system.

Cephalins   (20) The phosphatidyl ethanolamine and the phosphatidyl serine phospholipids.

Charles' law   (10) The volume of a fixed quantity of gas varies directly with its absolute temperature (°K), as long as the pressure on the gas remains constant. $V_1/T_1 = V_2/T_2$.

Chemical change   (3) One that is accompanied by a change in the chemical composition of matter; chemical reactions.

Chemical equation   (3) A way of expressing a chemical change using formulas and symbols of the reactants and products.

Chemical equilibrium   (11) The condition reached in a reversible reaction in which the rates of the forward and reverse reactions are equal.

Chemical properties   (3) Those properties of matter determined in chemical changes.

Chiral center   (21) In organic compounds, a carbon atom in an optical isomer that has four nonidentical groups bonded to it.

Cholesterol   (20) The most abundant steroid in the human body.

Chromosome (23) The DNA–protein complex in cells that controls the genetic character of an organism.

Cis-trans isomers (16) The geometrical isomers that arise based on the arrangement of like atoms or groups on the same side (cis-) or opposite sides (trans-) of a carbon–carbon double bond. 2-Butene displays cis-trans isomerism.

Coagulation (22) The irreversible denaturation of a protein.

Codon (23) Any one of the codes formed by a specific sequence of three nitrogen bases in mRNA that calls for a specific tRNA bound amino acid in protein synthesis.

Cofactor (24) The nonprotein part of a conjugated protein.

Collagen (22) The protein that is the principal component of connective tissue. It is also the protein network in bone and teeth.

Colligative property (12) A physical property of a solution that depends only on the concentrations of solute particles and not on their identity. Boiling point elevation, osmotic pressure, freezing point lowering, and vapor pressure lowering are colligative properties.

Colloid (12) A dispersion of particles that range from 1–100 nm in diameter in a dispersing medium.

Combination reaction (11) One in which two or more elements and/or compounds combine to produce a new compound. Also called a synthesis reaction.

Combined gas law (10) A combination of Boyle's and Charles' law. $P_1V_1/T_1 = P_2V_2/T_2$.

Combustion (15) The rapid reaction of oxygen with another substance to produce heat, light, and the oxides of the elements in the consumed substance.

Competitive inhibition (24) The inhibition of an enzyme by molecules that are similar to the substrate normally acted on by the enzyme.

Complete protein (22) One that contains all the essential amino acids.

Compound (3) A pure substance composed of two or more elements combined in a definite weight ratio.

Compound lipid (20) A subclass of lipids that yield an alcohol, fatty acids, and other complex compounds upon hydrolysis. The phospholipids and sphingolipids.

Condensation (10) A physical change in which a gas is converted to a liquid or solid. The reverse of vaporization and sublimation.

Conjugated protein (22) A protein that yields amino acids plus a nonprotein unit upon hydrolysis.

Coordinate covalent bond (8) A covalent bond in which both of the shared electrons are provided by only one of the bonded atoms.

Covalent bond (8) The bond holding two atoms together that arises from the sharing of one or more pairs of electrons between the atoms.

Covalent compound (8) One in which the atoms are joined together by covalent bonds.

Covalently bonded solid (10) One composed of atoms covalently bonded to each other to form a hard, rigid network of atoms, as in diamond and quartz.

Crenation (12) The collapse of red blood cells.

Curie (7) A unit of radioactivity equal to 37 billion disintegrations per second.

Cycloalkane (15) A hydrocarbon in which three or more carbon atoms are joined in a ring. All carbon–carbon bonds are single bonds.

Dalton's law (10) The total pressure exerted by a mixture of gases is the sum of the pressures exerted by each gas (the partial pressures) if present alone under the same conditions.

Decomposition reaction (11) One in which a compound is decomposed to form simpler substances.

Dehydration (17) The removal of water.

Denaturation (22) The destruction of the quaternary, tertiary, or secondary structure of a protein.

Denatured alcohol (17) Ethyl alcohol to which a small amount of toxic material is added to make it unsuitable as a beverage.

Density (2) The mass of an object divided by its volume.

Detergents (20) Compounds with cleansing action like soap but unlike soap do not form insoluble precipitates (scum) with metal ions normally found in hard water.

Dextrans (21) Linear polymers of glucose units joined together by α-1,6 linkages.

Dextrins (21) Fragments composed of several connected glucose units formed in the early stages of starch hydrolysis.

Dialysis (12) The separation of dissolved ions and smaller molecules from larger colloidal particles in a solution by use of a semipermeable membrane. The kidney machine uses dialysis to remove wastes from blood.

Diene (16) An unsaturated organic compound containing two carbon–carbon double bonds.

Diol (17) An alcohol that contains two hydroxyl functional groups.

Dipole–dipole force (10) The electrostatic force of attraction between neighboring polar molecules (dipoles).

Disaccharide (21) A carbohydrate consisting of two monosaccharide units joined together. Sucrose, lactose, and maltose are disaccharides.

Distillation (12) A method of purifying liquids or separating liquid mixtures based on differences in volatility.

DNA (23) Deoxyribonucleic acid. It contains the genetic information of an organism.

DNA replication (23) The manner of synthesis of DNA in an organism.

Double bond (8,16) A covalent bond that shares two pairs of electrons between two bonded atoms.

Double helix (23) The secondary structure of DNA.

Double replacement reaction (11) One in which two compounds react to produce two other compounds by the exchange of ions, atoms, or groups of atoms.

Dynamic equilibrium (10) The condition in which two opposing changes occur at the same time and at the same rate. See also chemical equilibrium.

Edema (12) The abnormal retention of water in the body.

Effective collision (11) In terms of chemical reactions, a collision between reacting species that can result in the formation of products.

Effective half-life (7) In terms of radioisotopes in the body, it is the time required for the amount of radioactivity in a patient to decrease by one-half through both normal radioactive decay and biological elimination of the radioisotope.

Electrolyte (12) Any substance that when dissolved in water will produce a solution that conducts an electric current. Acids, bases, and soluble salts are electrolytes.

Electromagnetic spectrum (5) A display of electromagnetic radiation from that of very short wavelengths (gamma rays and cosmic rays) to that of very long wavelengths (radio waves).

Electron (6) A subatomic particle of unit negative charge. They are located around the nucleus in atoms. Symbolized e^-.

Electron dot symbol (6) A symbol of an element that is surrounded by dots (·) which represent the valence electrons: ·$\overset{..}{\underset{.}{C}}$·, Na·

Electronegativity (8) A measure of the tendency of an atom to attract the electrons it shares in a covalent bond to itself.

Electronic configuration (6) The pattern of electron distribution in the energy levels of an atom.

Electron-sea model (9) A theory of bonding in metals in which the metal loses one or more valence electrons and exists as a positive ion imbedded in a highly mobile sea of valence electrons.

Electron transport system (25) *See* ETS.

Electrophoresis (22) A method of separating compounds based on differences in migration rates in an electric field.

Electrostatic attraction (10) The force of attraction that naturally exists between species that bear opposite electrical charges (+ and −).

Element (3) Those pure substances that cannot be decomposed to simpler substances in chemical changes.

Embden-Meyerhof pathway (25) A series of chemical reactions in the cytoplasm of cells that converts glucose to pyruvic acid (in the presence of oxygen) or to lactic acid (in the absence of oxygen).

Emission spectrum (6) The light of specific wavelengths that is emitted from excited atoms.

Endocrine system (24) The system of glands in the body that produce hormones.

Endothermic change (5) A physical or chemical change that absorbs heat from the surroundings as it takes place.

End-product inhibition (24) The inhibition of an enzyme by the end product of a series of reactions which then reduces the rate of formation of the end product.

Energy (5) The ability to do work.

Enzyme (24) A biological catalyst.

Enzyme-substrate complex (24) The combination of an enzyme and its substrate.

Equilibrium vapor pressure (10) The pressure exerted by the vapor of a substance as it exists in equilibrium with the solid or liquid phase.

Equivalence point (13) That point in a titration in which the number of equivalents of acid neutralized exactly equals the number of equivalents of base added. The visual determination of the equivalence point by use of an indicator is called the end-point.

Ester (14, 18) An organic compound that contains the $-\overset{\displaystyle O}{\overset{\|}{C}}-OR$ functional group.

Ether (14, 17) An organic compound that contains the $-\overset{|}{\underset{|}{C}}-O-\overset{|}{\underset{|}{C}}-$ functional group.

ETS (25) The electron transport system. A complex series of reactions that convert ADP to ATP and consumes oxygen to produce water in cellular respiration.

Evaporation (10) The escape of molecules from the liquid phase into the gas phase. The opposite of condensation.

Exothermic change (5) A physical or chemical change that releases heat to the surroundings as it proceeds.

Extracellular water (12) The water located between the cells in tissue. Also called interstitial water.

Fahrenheit temperature scale   (2) The temperature scale used in English-speaking countries. The Fahrenheit and Celsius temperature scales are related through the equation: $°F = 1.8(°C) + 32$.

Fat   (20) A triglyceride lipid that is solid at room temperature.

Fat-soluble vitamins   (20) Those vitamins soluble in triglycerides, vitamins D, E, A, and K.

Fat solvents   (20) Nonpolar organic solvents that dissolve lipids such as benzene, diethyl ether, and carbon tetrachloride.

Fatty acid   (20) Carboxylic acids with large aliphatic groups usually with an even number of carbon atoms.

Fehling's reagent   (18) A basic solution of copper (II) sulfate that is used to determine the presence of aldehydes.

Fermentation   (25) The conversion of glucose to ethyl alcohol and carbon dioxide by the enzymes in yeast.

Film badge   (7) A plastic badge containing layers of photographic film worn by those who work around ionizing radiation. Once developed and analyzed, the amount of ionizing radiation an individual has been exposed to can be determined.

Fischer structure   (21) Structures that can be used to represent monosaccharides.

Formula   (3) A symbolic way of representing the elemental composition of compounds and polyatomic ions.

Formula unit   (3, 8) The smallest part of an ionic compound that represents the simplest ratio of anions to cations. A formula unit of sodium chloride is one Na^+ and one Cl^-.

Formula weight   (4) The sum of the atomic weights of all the atoms in a formula. Means the same as molecular weight for molecular compounds.

Forward reaction   (11) In an equation that describes an equilibrium, the reaction that proceeds from left to right as the equation is written.

Functional group   (14) In organic compounds, those characteristic atoms or groups of atoms that are responsible for most of the chemical properties of the compound.

γ-Radiation   (7) Gamma radiation is a highly penetrating form of nuclear radiation, similar to x-rays.

Gas   (3) One of the three physical states of matter. Gases are composed of widely separated particles. They have no definite shape and completely fill the container in which they are stored.

Geiger counter   (7) An electrical device used to detect and measure ionizing radiation.

Gene   (23) That specific portion of a chromosome that codes for a specific protein.

Genetic code   (23) The heredity message stored in the sequential arrangement of nitrogenous bases in DNA.

Genetic disease   (23) A disease that is transferable to offspring due to errors in the structure of the DNA of one or both parents.

Geometric isomers   (16) Compounds that differ only in the spatial arrangements of atoms or groups of atoms. Cis- and trans-2-butene are geometrical isomers.

Glucosuria   (21) The condition characterized by the presence of glucose in the urine.

Glycogen   (21) The storage form of glucose in animals, a polymer of glucose joined through α-1,4 and α-1,6 ether linkages.

Glycogenesis   (21) The conversion of glucose to glycogen in the body.

Glycogenolysis   (21) The conversion of glycogen to glucose in the body.

Glycolysis   (25) The conversion of glucose via the Embden-Meyerhof pathway to produce either pyruvic acid or lactic acid.

Gram atomic weight   (4) An amount of an element equal to its atomic weight in grams. One GAW is one mole of an element, and it contains Avogadro's number (6.02×10^{23}) of atoms.

Gram equivalent weight   (13) In terms of acids and bases, the number of grams of an acid that can provide one mole of hydrogen ion in a reaction; or the number of grams of a base that can provide one mole of hydroxide ion in a reaction.

Gram formula weight   (4) An amount of a compound equal to its formula weight in grams. One GFW of any compound is one mole of that compound and it contains Avogadro's number of molecules or formula units.

Group   (6) The columns of elements on the periodic table. Also called families. Elements in the same group have similar chemical properties.

Half-life   (7) As it concerns radioactive isotopes, it is the time required for the radioactivity of a sample to decrease by half.

Hallucinogen   (19) A drug that affects the perception of mood, time, space, the senses, and self. Also called a psychedelic drug.

Haloenzyme   (24) An active enzyme unit formed by the combination of an apoenzyme and the proper cofactor(s).

Halogen   (6) The elements in Group VII of the periodic table, F, Cl, Br, I.

Haworth structures   (21) Structural formulas that represent the cyclic forms of carbohydrates.

Heat   (5) A form of energy related to the average kinetic energy of the particles that make up a sample of matter. Heat is that form of energy that flows from higher to lower temperature regions.

Heat of fusion   (5) The amount of heat required to convert one gram of a solid at its melting point to one gram of liquid at the same temperature.

Heat of vaporization   (5) The amount of heat required to convert one gram of a liquid at its boiling point to one gram of vapor at the same temperature.

Hemodialysis   (12) The removal of metabolic wastes from the blood by dialysis.

Hemolysis   (12) The rupturing of red blood cells.

Heterocyclic compound   (14) A cyclic compound that contains atoms of at least two different elements in the ring. One of the three broad classes of organic compounds.

Heterogeneous mixture   (3) One in which visible boundaries indicate the presence of more than one substance in the sample. Examples are sand, granite, and blood.

Hexose   (21) A six-carbon monosaccharide, such as glucose, galactose, and fructose.

Holding site   (24) The region on the surface of an enzyme that grips the substrate molecule, holding it in the active site.

Homogenous mixture   (3) One in which two or more different substances are so intimately mixed that visible boundaries separating them are not evident. Solutions are homogeneous mixtures.

Hormone   (24) Substances produced in the endocrine system that serve as chemical messengers to stimulate a response in the body.

Hydration   (12) The interaction of a dissolved molecule or ion with the solvent water. (17) The addition of water to an alkene.

Hydrocarbon   (14, 15) An organic compound made up of only carbon and hydrogen.

Hydrocolloids   (21) Polysaccharides that form gelatinous or near-solid colloids in water.

Hydrogenation   (16) The addition of hydrogen to an unsaturated compound.

Hydrogen bond   (10) The force of attraction between a hydrogen atom bonded to N, O, or F and a second atom of N, O, or F. Hydrogen bonds are responsible for the secondary structure of DNA and the proteins.

Hydrolase   (24) A type of enzyme that catalyzes a hydrolysis reaction.

Hydronium ion   (13) The proton–water complex that represents the condition of the hydrogen ion in water. It is commonly symbolized H_3O^+ or $H^+_{(aq)}$.

Hyperglycemia   (21) The condition characterized by a consistent blood glucose level greater than about 130 mg/100 ml.

Hypertonic solution   (12) One with an osmotic pressure greater than that of some reference solution.

Hypoglycemia   (21) The condition characterized by a consistent blood glucose level less than about 45 mg/100 ml.

Hypotonic solution   (12) One with an osmotic pressure less than that of some reference solution.

Ideal gas law   (10) $PV = nRT$

Imine   (18) An organic compound with the $\overset{\diagdown}{\underset{\diagup}{C}}$=NH functional group.

Indicator   (13) A substance that undergoes a color change in response to changes in hydronium ion concentration.

Induced fit theory   (24) A theory of enzyme action which holds that a substrate may induce an enzyme to change shape slightly so the active site fits the substrate properly.

Interferon   (23) A body protein that inhibits viral infection.

Intermolecular force   (10) A force between molecules.

Intracellular water   (12) Water within cells.

Ion   (3) Any chemical species that bears an electrical charge.

Ion exchange   (12) The exchange of ions which are associated with a stationary resin for those in solution. Ion exchange may be used to purify water by exchanging hydronium and hydroxide ions (which react to form water) for metal ions and anions in water.

Ionic bond   (8) The force of attraction between ions of opposite charge in a solid.

Ionic compound   (8) One composed of ions, such as NaCl, KNO_3, and MgF_2.

Ionic equation   (13) For reactions that take place in solution, an ionic equation represents each compound that dissociates in solution as ions. NaCl in water would be represented as $Na^+_{(aq)}$ plus $Cl^-_{(aq)}$. *See also* net-ionic equation.

Ionization energy   (6) The energy required to completely remove an electron from a gaseous atom. $M_{(g)} + energy \rightarrow M^+_{(g)} + e^-$.

Ionizing radiation   (7) Radiation that is capable of breaking molecules apart into ions or free radicals. Nuclear radiation, x-rays, and cosmic rays are types of ionizing radiation.

Ion-product equation   (13) For water, it is the product of the molar concentrations of the hydronium ion and the hydroxide ion in an aqueous solution. $K_w = [H^+_{(aq)}][OH^-] = 1 \times 10^{-14}$. K_w is called the ion-product constant.

Irreversible inhibition   (24) The inhibition of an enzyme that permanently destroys the catalytic property of the enzyme.

Isoelectric point (22) The pH at which an amino acid exists as a zwitterion. Also called the isoelectric pH.

Isomerase (24) A type of enzyme that catalyzes the intramolecular rearrangement of a substrate compound.

Isomers (15) Compounds with the same molecular formulas but different structures. Ethyl alcohol and dimethyl ether are isomers; CH_3CH_2OH, CH_3OCH_3.

Isotonic solution (12) A solution with the same osmotic pressure as some reference solution.

Isotopes (6) Atoms of the same element (same atomic number) that have different mass numbers, such as Cl–35 and Cl–37.

IUPAC The International Union of Pure and Applied Chemistry. One function of this body is to standardize the language of chemistry throughout the world.

Kelvin temperature scale (2, 10) *See* absolute temperature scale.

Ketoanemia (25) The condition characterized by abnormally high levels of ketone bodies in the blood.

Ketone (14, 18) An organic compound of general formula $R—\overset{\displaystyle O}{\overset{\displaystyle \|}{C}}—R'$.

Ketone bodies (25) Acetone, acetoacetic acid, and β-hydroxybutyric acid, compounds that can accumulate in the body as a result of reduced carbohydrate metabolism. *See* ketosis.

Ketonuria (25) The condition characterized by increased levels of ketone bodies in the urine.

Ketose (21) A monosaccharide that, in its open chain form, contains a ketone structure.

Ketosis (25) A condition brought on by reduced carbohydrate metabolism resulting in the abnormal buildup of ketone bodies.

Kilocalorie (5) One thousand calories. The nutritional calorie (the Calorie with a capital C) is the kilocalorie.

Kinetic energy (5) Energy possessed by a body by virtue of its motion. $KE = \frac{1}{2}m(v)^2$.

Kinetic molecular theory (10) A set of postulates that serves as a model of the gaseous state of matter.

Krebs cycle (25) A cycle of ten chemical reactions that converts acetyl coenzyme A to CO_2 and H_2O while producing the electrons and protons needed by the electron transport system to produce ATP.

Law of conservation of energy (5) Energy is neither created nor destroyed in a chemical reaction.

Law of conservation of mass (5) There is neither a gain nor loss in the total mass of matter in a chemical or physical change.

Law of definite composition (3) All samples of a given pure compound are composed of the same elements in the same, definite weight ratio.

LD_{50}–30 day (7) As applied to ionizing radiation, it is the amount of radiation required by each member of a group to kill 50% of them within 30 days.

LeChatelier's principle (13) If a stress is applied to a system in equilibrium, the system will change to minimize the effect of that stress.

Lewis structure (8) A structural formula of a molecule or polyatomic ion that shows the distribution of all valence electrons about and between the atoms in the species.

Ligase (24) A type of enzyme that catalyzes the joining of two molecules.

Lipase (20) A type of enzyme that catalyzes the hydrolysis of triglycerides.

Lipids (20) A class of biological compounds grouped together by the fact that they are insoluble in water but soluble in the nonpolar fat solvents.

Lipoproteins (20) A combination of a lipid and a protein forming a species that is used to transport lipids through the lymph system and bloodstream.

Liquid (3) The state of matter in which molecules are free to move around each other while being close together in random arrangements. Liquids have definite volume but assume the shape of the container in which they are kept.

Liter (2) The metric unit of volume equal to the volume of 1 kg of water at 4°C.

Lock and Key theory (24) A theory of enzyme action that holds that an enzyme and substrate must fit together in a specific way if the catalytic function of the enzyme is to be realized. *See also* Induced fit theory.

London force (10) An intermolecular force that arises from the formation of instantaneous dipoles in neighboring molecules.

Lyase (24) A type of enzyme that catalyzes the addition or removal of groups to or from a substrate without the use of hydrolysis or oxidation-reduction reactions.

Markovnikov's rule (16) The addition of a compound of general formula HX to a carbon–carbon multiple bond will take place with the hydrogen joining to the carbon atom with the greater number of hydrogens already bonded to it, as X joins to the other carbon atom.

Mass (2) A measure of the quantity of matter contained in an object.

Mass number (6) The sum of the number of neutrons plus protons in the nucleus of an atom.

Matter (3) Anything that occupies space and has mass.

Melting point (10) The temperature at which a solid substance is in equilibrium with its liquid phase.

Mercaptan (14, 17) An organic compound with the —SH functional group. Also called a thiol.

Metabolism (25) The complex series of reactions that transform absorbed food into cellular compounds and energy.

Metals (3) Those elements located to the left of the stairstep line on the periodic table. Metals have shiny surfaces, are good conductors of electricity and heat, and nearly all are malleable and ductile. Metals tend to lose electrons in chemical reactions with nonmetals.

Metal ion activator (24) The specific metal ion required by an enzyme if it is to act as a biological catalyst.

Metallic bond (9) The bonding in metals. *See* electron-sea model.

Meter (2) The fundamental unit of length in the metric system.

Milligram % (12) A weight/volume percent equal to the number of milligrams of a particular substance in 100 ml of solution.

Mixture (3) A sample of matter composed of two or more pure substances combined in a weight ratio that can vary. Each substance retains its own identity and can be separated by physical means.

mmHg (10) Millimeters of mercury. A unit of pressure expressed as the height of the column of mercury in a barometer.

Molar volume (10) The volume occupied by one mole of any gas at STP, 22.4 l.

Molarity (12) A solution concentration term which is equal to the number of moles of solute in one liter of solution. Symbolized M.

Mole (4) Avogadro's number of particles (atoms, molecules, ions, formula units, electrons, etc.). *See also* gram atomic weight and gram formula weight.

Molecular solid (10) One composed of individual molecules held together by intermolecular forces in a crystal.

Molecular weight (4) The sum of the atomic weights of all the atoms in a molecule.

Molecule (3) The smallest unit of a molecular species that retains the chemical properties of that species.

Monomer (16) The small molecules which react with each other and join together to form a polymer. Glucose is the monomer of starch.

Monosaccharide (21) A carbohydrate that cannot be broken down by acid hydrolysis into simpler carbohydrates.

mRNA (23) Messenger ribonucleic acid. The nucleic acid that carries the genetic message from DNA to the ribosomes where it directs protein synthesis.

Mutation (23) Any alteration in the base sequence in DNA.

Narcotics (19) Powerful pain-killing drugs that are also very addictive.

Natural radioactivity (7) That displayed by radioisotopes that are found in nature.

Net-ionic equation (13) A chemical equation that includes only those species (atoms, molecules, or ions) that are directly involved in the reaction. Spectator ions are not included.

Neutralization (13) The reaction of an acid with a base to produce a salt and water.

Neutron (6) The subatomic particle with a mass nearly the same as that of the proton that has no electrical charge. Found in the nucleus of atoms. Symbolized n.

Nitrogen balance (25) The state in which the nitrogen intake of a body equals the nitrogen excreted.

Noncompetitive inhibition (24) The reversible inhibition of an enzyme by a substance that combines with the enzyme at a point other than the active site.

Nonelectrolyte (12) A substance that when dissolved in water forms a solution that does not conduct an electric current. Also, any substance that itself does not conduct electricity.

Nonmetals (3) Those elements located to the right of the stairstep line on the periodic table. They may be solids, liquids, or gases and they tend to gain electrons in chemical reactions with metals. Compounds composed of only nonmetals most often exist as molecules: CO_2, H_2O, NH_3, PCl_3.

Nonpolar covalent bond (8) A covalent bond in which the bonding electrons are shared equally, as in covalent bonds between atoms of the same elements, H:H.

Normal boiling point (10) The temperature at which the equilibrium vapor pressure of a liquid equals 760 torr.

Normality (13) A method of expressing the concentration of a solution which is equal to the number of gram equivalent weights (equivalents) of solute in one liter of solution. Symbolized N.

Nucleic acids (23) RNA and DNA. Compounds that carry the genetic code of an organism (DNA and RNA) or participate in protein synthesis in a direct way (tRNA, rRNA). Polymers of nucleotides.

Nucleoside (23) Compounds that yield a pentose sugar plus a heterocyclic nitrogen base upon hydrolysis. When combined with phosphoric acid, they form nucleotides.

Nucleotide (23) Compounds that yield a pentose sugar, a heterocyclic nitrogen base, and phosphoric acid upon hydrolysis. The repeating units in the nucleic acids.

Nucleus (6) The small, positively charged, massive portion of an atom composed of protons and neutrons. Also, the principal membrane-bound component in a eukaryotic cell that contains the chromosomes.

Nutritional calorie (5) *See* kilocalorie.

Octet rule (8) In a chemical reaction, an atom will lose, gain, or share the necessary number of electrons so that the highest occupied energy level ends up with 8 electrons (an octet of electrons).

Oil (20) A triglyceride in which the fatty acid groups are unsaturated. Oils are liquids at room temperature.

Olefin (16) An alkene.

Oligosaccharide (21) Carbohydrates composed of from two to ten joined monosaccharide units.

Opium (19) The milky sap obtained from the oriental poppy which is a rich source of narcotic drugs.

Optical isomers (21) Compounds with a chiral center of the same formula but different structure. They can rotate plane polarized light passing through their solutions to the left (levorotatory isomer) or to the right (dextrorotatory isomer).

Optimum temperature (24) The temperature at which an enzyme is most active.

Optimum pH (24) The pH at which an enzyme is most active.

Orbit (6) In the Bohr model of the hydrogen atom, the path traveled by the electron about the nucleus.

Osmosis (12) The net movement of water through a semipermeable membrane from a more dilute solution into a more concentrated solution.

Osmotic pressure (12) A colligative property of solutions equal to the pressure that must be applied to the more concentrated solution to just halt the net flow of water during osmosis.

Oxidation (8, 11) The loss of one or more electrons, the loss of hydrogen atoms, or the gain of oxygen atoms by a species. The algebraic increase in oxidation number of an atom.

Oxidation number (11) An arbitrarily assigned number that indicates the number of electrons an atom has either completely or partially lost or gained as it became an ion or incorporated into a compound.

Oxidation-reduction reaction (11) A reaction in which the oxidation numbers of at least two atoms change; one change indicating oxidation, the other reduction. Also called a redox reaction.

Oxidative deamination (25) The removal of the α-amino group ($-NH_2$) from an amino acid as the α-carbon is oxidized to a carbonyl group ($\diagdown C{=}O$).

Oxidative phosphorylation (25) The synthesis of ATP from ADP + Pi in the electron transport system (ETS) during aerobic respiration.

Oxidizing agent (11) The species that is reduced in a redox reaction. The species that is reduced can be regarded as bringing about the oxidation of another substance.

Oxidoreductase (24) A type of enzyme that catalyzes an oxidation-reduction reaction.

G15
GLOSSARY

Oxygen debt (25) Describes the condition of oxygen need in cells following anaerobic glycolysis to oxidize accumulated lactic acid to pyruvic acid.

Paraffin (15) An alkane.

Partial pressure (10) The pressure exerted by one specific gas in a mixture of gases. *See* Dalton's law.

Parts per million (ppm) (12) The number of grams of one component in a mixture per million grams of that mixture. Also can be used in terms of volume.

Pentose (21) A five-carbon monosaccharide, such as ribose and deoxyribose.

Peptide bond (22) The amide bond formed as amino acids join together in a dehydration reaction to form proteins (polypeptides).
$$-\overset{\overset{\displaystyle O}{\|}}{C}-\overset{\overset{\displaystyle H}{|}}{N}-$$

Percent composition (4) A method of expressing the composition of a mixture or compound equal to the percent of the weight (or volume) of a sample of matter that is contributed by one particular substance. *See* weight/volume, weight/weight, and volume/volume percent.

Period (6) The horizontal rows of elements on the periodic table, Figure 6.5.

Periodic law (6) The periodic properties of the elements are a function of their atomic numbers.

Periodic properties (6) Those properties of elements that follow repeating trends at regular intervals as the elements are scanned in order of increasing atomic number. Ionization energy, metallic character, and atomic size all vary in a periodic way.

Periodic table (3, 6) A display of the elements in order of increasing atomic number which allows similarities and relationships among the elements to be seen more easily. *See* Figure 6.5.

pH (13) A method of expressing the hydronium ion concentration of a solution equal to the negative logarithm of the molar hydronium ion concentration. $pH = -\log [H^+_{(aq)}]$.

pH scale (13) The range of pH values from 0 to 14 that span from very acidic to very basic solutions. As pH values get smaller than 7, the solutions become increasingly acidic; as they increase to values greater than 7, solutions become increasingly basic. A pH of 7 is that of a neutral solution.

Phenols (17) Aromatic organic compounds which have a hydroxyl group (—OH) bonded to a carbon atom in a benzene ring.

Phospholipid (20) A class of lipids that yield glycerol, fatty acids, phosphoric acid, and a nitrogen-containing alcohol on hydrolysis.

Photosynthesis (5) The conversion of CO_2 and H_2O to glucose in the presence of light and chlorophyll in green plants.

Physical change (3) A change in matter that takes place without a change in chemical composition. Melting, bending, evaporation, and condensation are physical changes.

Physical properties (3) Those properties of a substance that can be determined without changing the chemical composition of the substance. Hardness, color, viscosity, physical state, and melting temperature are all physical properties.

Plane-polarized light (21) Electromagnetic radiation that is vibrating in only one plane.

pOH (13) A way of expressing the hydroxide ion concentration of a solution equal to the negative logarithm of the molar hydroxide concentration. The sum of the pH and pOH of a given solution equals 14. Not as widely used as pH.

Polar covalent bond (8) A covalent bond in which the electrons are shared unequally by the bonded atoms. Covalent bonds between atoms of different electronegativity are polar.

Polarimeter (21) A device used to measure the degree and direction of rotation of plane-polarized light passing through a solution of an optically active compound.

Polyatomic ion (3) A species composed of two or more atoms covalently bonded together that carries an electrical charge. Most polyatomic ions are anions. Examples are NO_3^-, OH^-, $C_2H_3O_2^-$, and NH_4^+.

Polymer (16) A large molecule composed of many repeating units such as polyvinyl-chloride, polystyrene, and starch.

Polymerization (16) The synthesis of polymers from monomers, the units that combine to form polymers.

Polypeptide (22) A polymer made up of many amino acid units joined together by peptide bonds.

Polysaccharide (21) A polymeric carbohydrate composed of a large number of monosaccharide units joined together. Starch and cellulose are polysaccharides composed of repeating glucose units.

Polyunsaturated (20) A term describing organic compounds that have several carbon–carbon multiple bonds in the molecule.

Potential energy (5) The energy possessed by an object by virtue of its position, condition, or composition.

Pressure (10) Force per unit area. The pressure exerted by a gas is equal to the force of the molecular collisions on a given surface each instant divided by the area of the surface receiving the force.

Primary alcohol (17) An alcohol which has only one organic group joined to the carbon atom that bears the —OH group, R—CH_2OH. Symbolized 1°-alcohol. Methanol is classed as a 1°-alcohol.

Primary amine (19) An amine that has only one organic group bonded to the nitrogen atom, RNH_2. Symbolized 1°-amine.

Primary structure (22, 23) The sequence of amino acids in a polypeptide (protein), or the sequence of nucleotides in a nucleic acid.

Product (3) The substances produced in a chemical change.

Proenzyme (24) An enzyme that is secreted by cells in an inactive form that is activated elsewhere in the body.

Property (3) Any characteristic of matter that can be used to determine its identity.

Prosthetic group (22) The nonprotein part of a conjugated protein. In terms of enzymes, a prosthetic group is a small molecule, tightly bound to the active site which is required for enzyme activity. On the other hand, coenzymes are also frequently required for enzyme action, but are more weakly bound.

Proteins (22) Polypeptides. The large, nitrogen-containing organic compounds found in the cells of all living things.

Proton (6) The subatomic particle of unit positive charge found in the nucleus of all atoms. Symbolized p^+. The atomic number of an atom equals the number of protons in the nucleus of that atom.

Quaternary amine (19) A cationic amine in which four organic groups are bonded to the nitrogen atom, R_4N^+. Solutions of certain quaternary amines are used as disinfectants.

Quaternary structure (22) The highest level of structure in proteins in which two or more specific protein units (each displaying tertiary structure) assemble to form a larger protein unit which has properties different from those of the individual protein units. The quaternary structure of hemoglobin is an assemblage of four protein units.

R (10) The ideal gas law constant: 0.0821 liter-atm/mole-°K.

Racemic mixture (21) A mixture containing equal amounts of the (+) and (−) optical isomers of an optically active compound.

Rad (7) Radiation absorbed dose. An amount of absorbed radiation that produces 100 ergs of energy per gram of tissue. A unit for measuring quantities of absorbed ionizing radiation.

Radioactive decay (7) The decrease in the amount of radiation emitted by a sample of radioactive material as time passes. It also describes a radioactive event. *See* radioactivity.

Radioactivity (7) The spontaneous decay of a nucleus with the emission of particles and/or radiation as it transforms to an isotope of another element.

Reactants (3) Those substances that are consumed in a chemical reaction.

Reaction rate (11) The speed with which a reaction occurs expressed in terms of the change in concentration of a reactant or product per unit of time.

Recombinant DNA (23) The incorporation of a portion of DNA from one organism into the DNA of another.

Redox reaction (11) See oxidation-reduction reaction.

Reducing agent (11) The substance that is oxidized in a redox reaction.

Reducing sugar (21) A carbohydrate that can reduce copper (II) to copper (I) in Fehling's or Benedict's reagent or reduce Ag (I) to silver metal in Tollen's reagent because it contains an oxidizable aldehyde group.

Reduction (8, 11) The gain of one or more electrons by a species or the algebraic decrease in the oxidation number of an element in a reaction. Also, the gain of hydrogen atoms or loss of an oxygen atom by a compound.

Releasing factors (24) Specific hormones that are released by the hypothalamus in response to nerve impulses that in turn cause the release of other hormones elsewhere in the body.

Rem (7) A unit used to express quantities of absorbed ionizing radiation which takes into account how the different types of radiation affect tissue.

Resonance (8) The condition in which the electrons in a multiple bond of a Lewis structure can be placed between two or more identical atoms to generate two or more equivalent bonding patterns called resonance structures. This suggests that, in the molecule, the pair of electrons can be shared by more than two atoms simultaneously, adding stability to the species.

Resonance hybrid (8) An approximate Lewis structure which attempts to show the distribution of valence electrons in species that display resonance. The hybrid would be an average of the resonance forms that could be drawn for the species.

Respiratory chain (25) The electron transport system.

Reverse reaction (11) In an equation describing a reversible reaction ($\rightleftarrows$), the equation that is read from right to left ($\leftarrow$) describes the reverse reaction.

Reversible reaction (11) As the equation is written, one that can proceed from right to left or from left to right, or in both directions simultaneously. *See* dynamic equilibrium.

Ribosome (23) The site of protein synthesis in cells.

Roentgen (7) A unit used to express quantities of x-ray or gamma ionizing radiation.

rRNA (23) Ribosomal ribonucleic acid. They are part of the ribosomes in cells.

Salt (13) An ionic compound composed of one or more positively charged metal ions (or polyatomic cations) and one or more negatively charged nonmetal ions (or polyatomic anions, except OH^-). The ionic compound formed in an acid–base neutralization reaction.

Saponification (18, 20) The hydrolysis of a triglyceride in an aqueous solution of strong base to produce soap and glycerol.

Saturated hydrocarbon (15) A hydrocarbon that contains only single bonds—an alkane.

Saturated solution (12) A solution in contact with undissolved solute that contains the maximum amount of dissolved solute possible at that temperature.

Secondary alcohol (17) An alcohol that has two organic groups bonded to the hydroxyl bearing carbon, R_2HCOH. Abbreviated 2°-alcohol.

Secondary amine (19) An amine that has two organic groups bonded to the nitrogen atom, R_2NH. Abbreviated 2°-amine.

Secondary structure (22, 23 The second level of structure of proteins and nucleic acids, brought about by hydrogen bonding interactions. Two secondary structures adopted by proteins are the α-helix and the pleated sheet. The secondary structure of DNA is the double helix.

Semipermeable membrane (12) A membrane that allows only certain species to pass through while barring others, usually larger species. Used in osmosis and dialysis.

Shared pair (8) The pair of electrons shared by two atoms in a covalent bond.

Sickle cell anemia (22) A genetic disease characterized by sickle shaped red blood cells.

Simple lipid (20) One that yields only alcohols and fatty acids upon hydrolysis. Waxes and triglycerides are two categories of simple lipids.

Simple protein (22) A protein that yields only amino acids upon hydrolysis.

Simple sugar (21) A monosaccharide.

Single bond (8) A covalent bond in which only one pair of electrons is shared between two atoms.

Single replacement reaction (11) A redox reaction in which one element (A) replaces another in a compound (BX) to produce a different compound (AX) and the replaced element (B). Usually takes place in solution. $A + BX \rightarrow AX + B$.

Soap (20) A soluble salt of a fatty acid that displays a cleansing action. Household soaps are usually the sodium or potassium salts of fatty acids.

Solid (3) A physical state of matter which has an ordered arrangement of particles packed closely together. Solids have definite shapes and are not easily deformed.

Solute (12) The component of a solution that is dissolved in the solvent.

Solution (3, 12) A homogeneous mixture of two or more substances. May be solid, liquid, or gas. *See* solute and solvent.

Solvation (12) The interaction of a dissolved molecule or ion with the solvent. If the solvent is water, it is called hydration.

Solvent (12) The component of a solution in which the solute is dissolved.

Specific gravity (2) The ratio of the density of a liquid to the density of some reference liquid, usually water at 4°C.

Specific heat (5) The amount of heat required to raise the temperature of one gram of a substance one degree Celsius. The specific heat of water is 1.00 cal/g°C.

Spectator ion (13) Those ions in solution not directly involved in a chemical reaction. Spectator ions are omitted from net-ionic equations.

Sphingolipid (20) A lipid that yields a fatty acid, a nitrogen containing alcohol, sphingosine, and sometimes phosphoric acid upon hydrolysis.

Sphingosine (20) A complex unsaturated nitrogen-containing alcohol found in the sphingomyelins.

Standard pressure (10) For gases, 760 torr (1 atm).

Standard temperature (10) For gases, 0°C (273°K).

Starch (21) A polysaccharide found in plants made up of branched chains of glucose molecules joined through α-1,4 and α-1,6 linkages.

Steroid (20) A class of lipids that all contain a characteristic fused four ring system called the steroid nucleus.

STP (10) Standard temperature and pressure for gases; 0°C, 760 torr.

Strong acid (13) An acid that completely or nearly completely dissociates into ions in solution. The common strong acids are HCl, HNO_3, and H_2SO_4.

Subatomic particles (6) The fundamental particles that compose atoms. The electron, proton, and neutron.

Sublimation (10) The conversion of a solid directly to a gas at constant temperature without passing through the liquid phase.

Substance (3) A homogeneous sample of matter. There are two kinds of pure substances, elements and compounds.

Substitution reaction (15) A reaction in which an atom, ion, or group of atoms substitutes for (replaces) an atom, ion, or group in a compound.

Substrate (24) In terms of enzyme action, the species acted on by an enzyme.

Substrate level phosphorylation (25) The transfer of a phosphate group from one compound to another.

Sulfhydryl group (17) The —SH functional group.

Surface tension (10) The force that draws the surface of a liquid together and causes the surface to act like a thin elastic "skin."

Surfactant (10) A substance that reduces the surface tension of a liquid.

Suspension (12) A dispersion of particles that are larger than about 100 nm in diameter. The particles settle out in time, and can also be removed by simple filtration through filter paper. Suspensions, if transparent, display the Tyndall effect.

Symbol (3) A one- or two-letter abbreviation that stands for one atom or one mole of an element.

Systematic name (8) A name of a compound or ion that results when a set of accepted, systematic rules are followed when deriving the name. IUPAC names are ystematic names.

Target cell (24) The specific cell acted on by a specific hormone.

Temperature (2) A measure of the hotness of an object which may be expressed in degrees Kelvin, degrees Celsius, or degrees Fahrenheit.

Terpene (20) A class of lipids that can be considered as being derived from isoprene, CH_2=CCH_3CH=CH_2.

Tertiary alcohol (17) An alcohol which has three organic groups bonded to the carbon atom that bears the hydroxyl group, R_3COH. Symbolized 3°-alcohol.

Tertiary amine (19) An amine that has three organic groups bonded to nitrogen, R_3N. Symbolized 3°-amine.

Tertiary structure (22) The third level of structure in proteins brought on by the folding of the α-helical secondary structure into a particular three-dimensional shape.

Thiol (17) An organic compound that contains the —SH functional group. Also called a mercaptan.

Titration (13) A volumetric procedure that can be used to determine the concentration of either an acid or a base in an unknown solution. *See* equivalence point and indicator.

Tollen's reagent (18) A solution of silver nitrate in aqueous ammonia used to detect the presence of an aldehyde group in organic compounds (as in the reducing sugars).

Torr (10) A unit of pressure equal to the millimeter of mercury. 1 torr = 1 mmHg = 1/760 atm.

Transamination (25) The reaction between an α-amino acid and an α-keto acid to form a different α-amino acid and a different α-keto acid. The α-amino carbon on the α-amino acid is oxidized to an α-keto carbon and the α-keto carbon on the keto acid is reduced to an amine.

Transcription (23) The process by which a gene code on DNA is copied to give the complementary code on RNA.

Transferase (24) A type of enzyme that catalyzes the transfer of a group of atoms from one compound to another.

Translation (23) The process by which a protein is synthesized at the ribosomes under the direction of mRNA.

Transmutation (7) A nuclear process in which one element is converted into another element.

Triene (16) An unsaturated organic compound containing three carbon–carbon double bonds.

Triglyceride (20) A simple lipid that yields glycerol and fatty acids upon hydrolysis. Triglycerides are triesters of glycerol and three fatty acids. *See* fat and oil.

Triol (17) An alcohol that contains three hydroxyl functional groups. Glycerol is a triol.

Triose (21) A three-carbon monosaccharide.

Triple bond (8) A covalent bond in which three pairs of electrons are shared between two atoms.

Turnover number (24) The number of substrate molecules that one enzyme unit can act on (turnover) per unit of time.

Tyndall effect (12) The visible beam of light seen passing through a transparent colloidal dispersion or a suspension. The light beam is not visible as it passes through a solution.

Unsaturated compound (16) Organic compounds that have one or more carbon–carbon multiple bonds.

Unsaturated solution (12) A solution that contains a smaller amount of solute than would be present in a saturated solution.

Unshared pair (8) A pair of electrons in the valence shell of an atom that is not involved in covalent bonding.

Urea cycle (25) A series of reactions in the liver that converts ammonia into urea. Also called the ornithine cycle.

Uremia (25) A condition characterized by the inability of the kidneys to release urea into the urine.

Valence electrons (6) The electrons in the outermost energy level of an atom. The valence electrons are those directly involved in the chemistry of an atom.

Vapor pressure (10) *See* equilibrium vapor pressure.

Virus (23) A biological unit made up of either DNA or RNA surrounded by a protein coat. A virus can enter a cell and reproduce by taking control of the synthesis processes.

Viscosity (10) A measure of the resistance to flow by a liquid.

Vitamers (20) Similar compounds that display the same vitamin function.

Vitamin (20, 24) Substances in foods that are essential in trace amounts for good health.

Volume/volume % (12) A method of expressing solution concentrations which is equal to the number of milliliters of solute per 100 ml of solution. Frequently used for solutions in which both solute and solvent are liquids.

Wavelength (5) The distance between adjacent crests on a wave.

Wax (20) As ester of a long-chain carboxylic acid and a long-chain alcohol. A type of simple lipid.

Weak acid (3) An acid that dissociates only to a small extent when dissolved in water. Acetic acid is a weak acid.

Weight (2) The force of attraction between two objects by virtue of their mass. On the earth, the weight of an object is the force of attraction of that object for the earth. The weight of an object decreases as its distance from the center of the earth increases, though its mass remains constant.

Weight/volume % (12) A method of expressing solution concentrations which is equal to the number of grams of solute per 100 ml of solution.

Weight/weight % (12) A method of expressing solution concentrations which is equal to the number of grams of solute per 100 g of solution.

Work (5) That which results in a change in position or arrangement of a sample of matter.

Zwitterion (22) A dipolar ion of an amino acid. The form of the amino acid at its isoelectric point, $\overset{\oplus}{H_3N}-\overset{\overset{\displaystyle H}{|}}{\underset{\underset{\displaystyle R}{|}}{C}}-COO^{\ominus}$.

Index

Absolute temperature scale 24, 222
Absolute zero 222
Acetal 470
Acetaldehyde 472
Acetaldehyde syndrome 3
Acetaminophen 516
Acetanilide 515
Acetic acid 484
Acetoacetic acid 775
Acetone 474, 775
Acetylcholine 505
Acetyl coenzyme A
    Krebs cycle, and 746
    metabolism intermediate, as 773-774
    from pyruvic acid 747
    structure of 709
Acetylsalicylic ácid 491
Acid
    amino 610, 618
    carboxylic 475-485
    common 308-311
    defined 306
    diprotic 309
    fatty 534, 539
    monoprotic 309
    properties of 307
    strong and weak 311-312
    triprotic 310
Acid chloride 510
Acidic solution 320
Acidosis
    defined 332
    metabolic 333
    respiratory 333
Activated complex 260-261
Activation energy 259
Active site, enzyme 715
Acyl coenzyme A 769
Addition reactions
    of alkenes 399-402
    of alkynes 406-407
    to carbonyl group 469-471
    of cycloalkanes 385
    defined 385

Adenine 651
Adenosine 738
Adenosine triphosphate 737-740
Adipose tissue 565, 768
Aerobic glycolysis 758
Alanine 612
Alcohols
    chemical properties 431-439
    common. list of 439-441
    dehydration of 431-434
    denatured 440
    hydrogen bonding in 426
    grain 422, 440
    naming 422-425
    oxidation of 435-437, 466-467
    physical properties 425-427
    preparation 427-431
    structure 420
    subclasses 421
    wood 439
Aldehydes
    chemical properties 467-471
    functional group 455-456
    naming 456-460
    physical properties 462-465
    preparation of 466-467
    reduction of 429
Aldose 572
Algebraic operations 794-798
Alkali metals 120
Alkaline earths 120
Alkaloid 520
Alkalosis
    defined 332
    metabolic 333
    respiratory 333
Alkanes
    defined 362
    chemical properties 378-382
    cracking of 398
    halogenated 381

naming 367-373
physical properties 376
table of 363
writing formula for 365-366, 373-374
Alkenes
    chemical properties 399-403
    defined 390
    naming 391-395
    physical properties 399
    sources 398-399
Alkylation (of benzene) 410
Alkyl group 367-368
Alkynes
    chemical properties 405-407
    defined 403
    naming 403-405
    table of 405
Allosteric enzyme 724
Allosteric site 724
Alloy 203-204
Alpha radiation 136
Amalgams 203-204
Amide linkage 507
Amides
    functional group 506-507
    naming 507-508
    physical properties 509-510
    preparation 510-511
Amines
    chemical properties 504
    functional groups 499
    naming 499
    physical properties 502
    preparation 502
    subclasses of 499
    as weak bases 504
Amine salts
    naming 504-505
    preparation 504-505
Amino acid
    abbreviations for 612-616
    electrophoresis of

621-622
    essential 617
    metabolism of 777-779
    optical activity of 618-619
    pool 776-777
    structure 610-611
    20 common 612-616
    zwitterion form 619-620
Amino group 500
Ammonia, aqueous 314
Amniocentesis 688
Amorphous solids 244
AMP, cyclic 730
Amphetamine 520-521
Amphoteric behavior 620
Amylase 594
Amylopectin 590
Amylose 590
Anabolism 736
Anaerobic glycolysis 758, 763-764
Anemia, pernicious 709-710
Angstrom 18
Anhenius, Svante 304-306
Anion 53
Anomeric carbon 575
Antibiotics 690
Anticodon 673
Antimetabolites 690
Antioxidants 446, 547
Apoenzyme 704
Arginine 616
Aromatic class of compounds 354
Aromatic hydrocarbons
    chemical properties 409-410
    defined 408-409
    naming 411-413
    polycyclic 415
A-site 677
Asparagine 615
Aspartic acid 615
Aspirin 491
Atherosclerosis 558
Atmosphere
    composition of 230

Atmosphere *(cont.)*:
  pressure unit 217
Atomic mass unit 63
Atomic number 108-109
Atomic size 124-125
Atomic structure
  Bohr model 113-116
  modern model 127-130
Atomic weight
  of common elements 63
  defined 62
Atom
  central, in molecule 192
  composition of 107-110
  defined 48-49
  electronic structure of 113-116, 127-130
  relative weight of 62-63
  size of 49, 124-125
ATP
  energy source, as 739-740
  production in ETS 744-745
  production in Krebs-ETS 754
  structure 738
  yield in fatty acid metabolism 772
  yield in glycolysis 764-766
Avogadro's Law 226-227
Avogadro's Number 67

B₁, vitamin 707
B₂, vitamin 711-712
B₃, vitamin 712-713
B₆, vitamin 708
B₁₂, vitamin 709-711
Bacquerel, Henri 134-135
Balance 9-10
Balanced equations
  calculations based on 75-79
  meaning of 74-75
  mole ratios from 75
Ball and stick models 354
Barbituates 522-523
Barbituric acid 522
Barometer 216-217
Base
  common 313-314
  defined 306
  properties 312-313
Base pairing
  in DNA 662-663
  in RNA 671
  in transcription 676
Basic solution 320
Benedict's test 467-468, 582
Benzedrine 520
Benzene

bonding in 408
naming derivatives of 411-413
properties of 408
reactions of 409-410
structure of 409
β-hydroxybutyric acid 775
Beta-oxidation. *See* fatty-acid cycle
Beta radiation 137-138
BHA-BHT 446, 547
Bile salts 558-559
Biological half-life 143
Biotin 706
Biuret test 644.
Blood buffers 331-334
Blood gases 232-234
Bohr atom
  description of 113-116
  electronic transitions in 115
  energy levels in 113-116
Boiling point, normal 242
Bond, covalent
  coordinate covalent 185-186
  defined 173
  double 177
  energy 173-174, 177
  length 193
  nonpolar 180-181
  polar 180-181
  in polyatomic ions 186-187
  single 175
  triple 177
Bond, ionic 163
Bond angle
  defined 193
  tetrahedral 193, 348
  trigonal 193, 349
  linear 193
Bond energy 173-174, 177
Boyle's Law 218-221
Buffer
  blood 331-334
  carbonate 331
  composition 329
  defined 329
  operation 329-330
  phosphate 331
  proteins as 332, 620-621
Buret 23, 338

C, vitamin 705-706
Caffeine 519
Calcification 641
Calcium
  in body fluids 274
  as nutrient 208-209
Calcium hydroxide 313

Calorie
  as energy unit 92
  nutritional 101
Carbohydrate
  hydrolases 697
Carbohydrates
  classes of 571-572
  definition of 571
  digestion of, summary 755
  metabolism of 754-767
Carbolic acid 442
Carbon
  chemical properties of 345-348
  compounds, nature of 346
  electronic structure of 117
Carbon monoxide, toxicity of 745
Carbonyl group
  adddition to 469-471
  description 454
  hydrogen bonding to 463
  polarity of 462
  reduction of 462
Carboxylic acids
  chemical properties 481-483
  functional group 475
  naming 476-479
  physical properties 480-481
  preparation 480
  reduction of 429
  tables of 477-479
  as weak acids 481-482
Caries, dental 641
Catabolism 736
Catalyst 262-264
Cation 53
Cell membrane 554-556
Cellulose 592-593
Cephaline 552-553
Charles' Law 221-224
Chemical change
  definition 40
  types of 249-252
Chemical property 39
Chemistry
  areas of 2
  definition 1
Chiral 598-601
Chitin 593
Chlorination
  of alkanes 379-380
  of water 301
Chlorophyll 206
Cholesterol 558
Choline 505

Chromatography
  description of 625-626
  types of 626-628
Chromosomes 665-666
Cis-trans isomerism 396-397
Citric acid 485
Citric acid cycle. *See* Krebs cycle
Clinistix 583-584
Cobalt as nutrient 210
Codeine 524
Codons, genetic 676
Coefficient 52, 71
Coenzyme 704
Coenzyme A
  pantothenic acid in 709-711
  structure of 709
Cofactor, enzyme 704
Collagen 640-642
Colligative properties
  boiling point elevation 290-291
  defined 290
  freezing point lowering 291-292
  osmotic pressure 292-296
  vapor pressure lowering 290
Collision, molecular
  effective 259-260
  nature of 215
  pressure from 216
Colloid
  defined 296
  dispersion 296
  types 297
Colorimeter 98-99
Combination reactions 249
Combined Gas Law 225-226
Combustion 378, 399
Compound
  covalent 173
  definition of 49
  ionic 53, 160
  molecular models of 354
  percent composition of 69
Condensation 241
Condensed formula 352-353
Configuration, D and L 601
Conjugated protein 610
Conservation of Energy, Law of 84
Conservation of Mass, Law of 41
Conversion factor 15
Coordinate covalent bond 185-186
Copper, as nutrient 210
Cortisone 560

Covalent bond. *See* Bond, covalent
Covalent compounds
  comparison with ionic 196-197
  defined 173
  naming inorganic 184-185
  predicting formulas for 175-176
Covalently bonded solids 244-245
Cracking alkanes 398
Crenation 295
Cresols 444
Crick, Francis 661
Curie 146, 148
Cyanide ion, toxicity of 745
Cyanocobalamin (B_{12}) 709-711
Cyanohydrin 469
Cyclamate 588
Cyclichemiacetal 574-575
Cyclichemiketal 580
Cycloalkanes 382-385
Cycloalkenes
  examples of 395
  naming 394-395
Cysteine 614
Cytochrome 207, 744-745
Cytosine 651

Dalton's Law 229-232
Datril® 516
D-configuration 601
DDT, in humans 381
Decarboxylases 702
Decimals 787
Decomposition reactions 250
Definite Composition, Law of 50
Dehydration
  of alcohols 431-434
  of the body 272-274, 309
Dehydrogenases 698
Dehydrases 701
Demerol® 524-525
Denaturation of protein 643-644
Denatured alcohol 440
Denominator 786
Density
  of common substances 31
  definition 30
  of gases at STP 227
Deoxyribonucleic acid (DNA)
  base pairing in 662-663
  composition of 652
  mutations in 683-686

recombinant 689
replication 666-671
role in transcription 674-676
secondary structure of 661-664
Deoxyribose 574-578, 650
Detergent 549-550
Deuterium 111
Dextrans 593
Dextrins 591
Dextrorotatory 598
Diabetes
  and blood sugar levels 595-596
  ketosis in 775-776
  role of insulin in 628
Diagnosis
  use of enzymes in 726-728
  use of radioisotopes in 151-153
Dialysis 298
Dienes 390-394
Diethylstilbestrol 445
Digestion and absorption of
  carbohydrates 594-595, 755
  lipids 565, 767
  proteins 644-645
Dilution of solutions 288-289
Dipole 235-236
Dipole-dipole forces 235-236, 281
Disaccharides
  definition of 585
  lactose 586-587
  maltose 585-586
  sucrose 587-589
Disease
  genetic 686-688
  vitamin deficiency 705
Dispersion, colloidal 296
Dispersion force 236-237
Distillation 299
  description of 299
  fractional 375
Disulfide bond 636-637
DNA. *See* deoxyribonucleic acid
Dose, drug
  calculation of 28-29
Dose equivalent 148
Dosimeter 144, 146
Double bond 177
Double helix 662-664
Double replacement reactions 250-251

Edema 274
Electrolyte 276

Electromagnetic spectrum 86
Electron 107
Electron-dot symbols 122-123
Electronegativity
  defined 181
  difference and bond polarity 182-183
  periodic trends 182
  values 182
Electronic configuration of atoms
  defined 116
  of several elements 117-122
Electron pairs
  shared 175
  unshared 175
Electron-sea model 202-203
Electrons, valence 118
Electron transport system (ETS)
  ATP production in 744-745
  diagrammed 744
  description 744-745
  effect of poisons on 745
  NADH, $FADH_2$ oxidation in 744
Electrophoresis 621-622
Electrostatic force 234
Elements
  composition of body 45
  definition 42
  diatomic 52
  list of 42-45
  symbols of 46
Embden-Meyerhof pathway
  ATP yield from 764-766
  description 757-758
  diagrammed 760
  step by step description 759-763
Emission spectrum 112-113
Enatiomers 600
Endocrine system
  description 728-730
  location of glands 728
Endothermic 97-98
End-product inhibition 724
Energy
  activation 259
  cellular 737-741
  definition 83
  forms of 85-88
  kinetic 88
  potential 88
Enzyme
  active site 715
  as catalysts 714-718
  classes of 697-703

composition of functioning unit 704-705
  use in diagnosis 726-728
  factors influencing 719-722
  holding site 716
  Induced Fit Theory 718
  inhibition 722-726
  Lock and Key Theory 716
  specificity 718
Enzyme activity, effect of
  cofactor concentration 722
  enzyme concentration 721-722
  pH 720
  substrate concentration 720-721
  temperature 719
Epinephrine 730, 732
Equation, chemical
  balancing 70-74
  calculations based on 75-79
  coefficients in 71
  formula 315
  ionic 316
  meaning of 74-75
  net-ionic 316
  writing 40
Equations, nuclear 136-139
Equilibrium
  chemical 266-268
  defined 266
  dynamic 241
  saturated solution 277-278
  vapor pressure 241-242
Equivalence point 338
Equivalent weight 334-335
Esters
  chemical properties 489
  as flavoring agents 489
  functional group 486
  inorganic 437-438
  naming 486-487
  physical properties 488-489
  preparation of 437-438, 483
  of salicylic acid 491-492
Estrogens 560-561, 733
Ethers
  as anesthetics 449-450
  cyclic 448
  naming 447
  preparation of 433-434
  properties of 448-449
Ethyl alcohol
  denaturing 440

Ethyl alcohol *(cont.)*:
    description of 440
    preparation of 430
    proof of 430
Ethylene glycol 441
ETS. *See* Electron
    Transport System
Eugenol 445
Evaporation 240-241
Exothermic 96-97
Exponential notation 790-794
Extracellular water 273-274

Factor-label method 12-17
FAD-FADH$_2$ 711-712
Family, chemical 120
Fat 544
Fat solvent 532
Fatty acids
    activation of 769
    chemical properties 538-
    539
    essential 537
    metabolism of 769-773
    physical properties
    535-536
    salts of 539, 546-547
    saturated 535
    unsaturated 536
Fatty-acid cycle
    diagrammed 770
    step by step
    description 770-772
Fehling's test 467-468, 582
Fermentation 430, 763-764
Ferritin 205
Film badge 144, 146
Fischer, Emil 601
Flouride, in water 284
Fluoride treatment, dental
    642
FMN 711
Folic acid 707
Force
    dipole-dipole 235-236,
    281
    dispersion 236-237
    electrostatic 234
    intermolecular 234-239
    of molecular collisions
    216
    Van der Waals 238
Formaldehyde 472
Formalin 472
Formic acid 484
Formula
    condensed 352
    Lewis 175-180, 187-190
    meaning 50
    molecular 351
    rules for writing 50

    structural 351
Formula equation 315
Formula unit 53, 163
Formula weight 64
Forward reaction 264-265
Fractional distillation 375
Fractions, in calculations 787-
    789
Friedel-Crafts reaction 410
Fructose 579-580
Fuel value of foods 102
Functional group
    definition of 355
    organic, table of 356-357
Fusion, heat of 94-95

Galactose 579
Galactosemia 579
Gamma radiation 138-139
Gas laws
    Avogadro's 226-227
    Boyle's 218-221
    Charles' 221-224
    Combined 225-226
    Dalton's 229-232
    Ideal 227-229
Gas tension 231
Gases
    blood 232-234
    density at STP 227
    description of 38
    KMT of 214-216
    laws describing 218-232
    molar volume of 226-227
    pressure of 216-218
Geiger-Muller counter
    144-145
Genes 666
Genetic code 666
Genetic disease 686-688
Genetic engineering 688
Genetics 665
Glucose
    description 580-582
    and diabetes 595-596
    metabolism of 755-767
    tests for 582-584
Glucose-6-phosphate 458, 756
Glucose tolerance test 595-
    596
Glucosuria 595
Glutamic acid 615
Glutamine 615
Glycerol
    description 441
    metabolism of 769
    in triglycerides 541
Glycine 612
Glycogen
    breakdown 757, 768
    description of 592

    formation 592, 756
Glycogenesis 592, 756
Glycogenolysis 592, 757
Glycolipids 557
Glycoside bond 586
Grain alcohol 422, 440
Gram 9
Gram atomic weight 65
Gram formula weight 65
Greek prefixes, numerical 184
Group, chemical 120
Group number and valence
    electrons 121
Guanine 651

Half-life
    biological 143-144
    effective 143-144
    of protein 776
    radioactive 143-144
Hallucinogen 525
Halogenation
    of alkenes 400
    of alkynes 406
    of benzene 410
    defined 400
Halogens 120
Halothane 450
Haworth structures 575
Heart attack, diagnosis of 727
Heat energy
    definition 24, 85
    measurement of 92-94
Heat of fusion 94-95
Heat of vaporization 95-96
Helix
    alpha 634-635
    double 662
    triple 640
Heme 205, 518, 631
Hemiacetal 469
Hemiketal 469
Hemodialysis 298
Hemoglobin
    description of 631-632,
    640
    electromicrograph of 608
Hemolysis 294
Hemosiderin 205
Heparin 593-594
Heredity 665
Heroin 524
Heterocyclic compound
    definition of 355
    nitrogen containing 516-
    520
    oxygen containing 448
Heterogeneous mixtures 56
Hexachlorophene 443
Hexoses 578-582
Hexyl resorcinol 443

Histidine 616
Holoenzyme 704
Homogeneous 56
Homologous series 363
Hormones
    of adrenal cortex 560
    of anterior pituitary 729
    function 728
    releasing factors 729
    sex 560-562
    table of 731-733
Hydrases 701
Hydration
    of alkenes 427-429
    of ions 279
Hydrocarbons
    aliphatic 354
    aromatic 408-414
    defined 354
    models of 354
    saturated 362
    sources of 374-377
    unsaturated 389
Hydrochloric acid 308-309
Hydrocolloids 593
Hydrogenation
    of alkenes 401
    of alkynes 406
    defined 401
    of fats and oils 401, 546
Hydrogen bonding
    in alcohols 426
    in amides 509
    in carboxylic acids 480
    description of 237-238
    in nucleic acids 661-664
    in proteins 634-639
    in water 238
Hydrolases 697-698
Hydrolysis
    of amides 511-513
    of carbohydrates 585,
    591
    definition of 273
    of esters 489-490
    of nucleic acids 649-650
    of proteins 513
Hydronium ion 306
Hyperglycemia 595
Hypertonic solution 295
Hypoglycemia 595
Hypothalamus 729

Ideal Gas Law 227-229
Ideal Gas Law Constant 228
Imidazol 518
Imine 471, 503
Incomplete combustion 378
Indicators 326, 338
Indole 517-518
Induced-Fit Theory 718

Infrared light 86
Inhibition, enzyme
    competitive 722-723
    described 722
    end-product 724-725
    irreversible 725
    noncompetitive 724
Inhibitor, enzyme 722
Inorganic phosphate 738
Insulin
    amino acid sequence of
    629
    from different sources
    630
    as hormone 732
    structure of 629-630
Intercellular water 273
Interferon 692-693
Intermolecular forces
    definition 234
    dipole-dipole 235-236
    hydrogen bond 237-238
    London 236-237
Intracellular water 273-274
Intrinsic factor 711
Invert sugar 588
Ion
    definition 53
    electron-dot symbols for
    167-168
    polyatomic 53, 166
    simple 53, 165
    size 164
    spectator 316
    table of 54, 166
Ion exchange 300-301
Ion-product constant 320
Ion-product equation 320
Ionic bond 163
Ionic compounds
    comparison with
    covalent 196-197
    defined 53, 160
    formation 160-163
    naming 169-172
    predicting formulas of
    166-169
    solutions of 279-280
Ionic equation 316
Ionization energy 124-125
Ionization of water 319-320
Ionizing radiation
    affect on water 141
    alpha 136
    beta 137-138
    biological effect 140
    defined 140
    diagnosis, use in 151-153
    gamma 138-139
    natural background 149
    safety 148-150

therapy, use in 150-151
threshold exposure level
142
x-rays 138-139
Iron
    in cytochromes 744-745
    as nutrient 209
Isoelectric pH 620
Isoleucine 613, 617
Isomerases 702
Isomerism
    cis-trans 396-397
    geometric 396-397
    optical 595-602, 618
    structural 364-366
Isopropyl alcohol 440-441
Isotonic solution 294
Isotopes
    carbon-12 61-63, 111
    contribution to atomic
    weight 111
    definition 111
    metastable 138
    natural abundance of
    111
    parent-daughter 136
    radioactive 135-139
    symbolized 110
IUPAC 367

Kekule, August 408
Kelvin temperature 24, 222
Keratin, structure 635
Ketal 470
Ketoanemia 776
Ketone
    chemical properties 467-
    471
    functional group 455-
    456
    naming 460-461
    physical properties 462-
    465
    preparation of 465-467
    reduction of 429-430
Ketone bodies 775
Ketose 572
Ketosis 775
Kidney machine 298
Kilogram 9
Kinases 700
Kinetic energy
    average kinetic energy
    88-90
    definition 88
    distribution 90, 215-216,
    241, 262
Kinetic Molecular Theory
    214-215
Kornberg, Arthur 671
Krebs cycle

diagrammed 748-749
equation summarizing
752
with ETS, ATP produc-
tion in 754
memory device for recal-
ling 752-753
metabolism, in 746
step by step description
of 748-752
Krebs, Hans 753
K_w 320

Lactic acid 485
Lactose 586-587
Lattice, crystal 54, 162
L-configuration 601
LD_{50}-30 day 148
Le Chatelier's Principle 329
Lecithin 552-553
Length
    metric-English conver-
    sions 18
    metric, units of 17-18
Leucine 613, 617
Levorotatory 598
Lewis, G. N. 175
Lewis structures
    defined 175-176
    deriving 178-180,
    187-190
    predicting resonance
    with 190
Ligases 702
Lipase 565, 697
Lipid
    compound 551-557
    defined 532
    digestion of 767
    fraction 532
    metabolism of 767-773
    serum levels 768
    simple 540-551
    steroids 557-562
    subclasses of
    533-534
Lipoprotein 768
Liquids 38, 239-243
Liter 20
Litmus 307, 312
Lock and Key Theory 716
Logarithms 322-323
London forces 236-237
LSD 525
Lyases 701
Lysine 616, 617

Magnesium
    in body fluids 274
    as nutrient 208-209
Magnesium hydroxide 313

Maltose 585-586
Manganese, as nutrient 209-
210
Mannitol 589
Markovnikov's Rule 401-402,
428-429
Mass
    definition 8
    measurement of 9
    metric-English conver-
    sions 9
    units of 12
Mass number 109-110
Matter 37
Melting point 244
Membrane
    cell 554-556
    dialyzing 298
    osmotic 292
    semipermeable 292
Menstrual cycle, hormone
    control of 560-561
Meperidine 524-525
Mercury
    amalgams 203-204
    toxicity of 204
Mescaline 525
Metabolism
    amino acid 777-779
    carbohydrates 754-767
    defined 736
    ETS in 741-746
    lipid 767-773
    protein 776-779
Metal-ion activation 704
Metallic bond 202-203
Metallic solids 245
Metalloids 48
Metals
    of biological importance
    207-211
    bonding in 202-203
    described 205
Metals, biologically important
    calcium 208-209, 274
    cobalt 210
    copper 210-211
    iron 209
    magnesium 208-209, 274
    manganese 209
    molybdenum 211
    potassium 115, 207-208,
    274
    sodium 115, 207-208, 274
    zinc 211
Metastable isotope 138
Meter 17
Methionine 614, 617
Methyl alcohol
    description of 439-440
    preparation of 430

Methyl alcohol (cont.):
  toxic nature 439
Methyl salicylate 491-492
Metric prefixes 8
Metric system 7
Microwaves 86-87
Micelle 549
Milliequivalent 336
Milligram percent 283
Miscible 426
Mitochondria 741
Mixture
  definition 55
  gaseous 229-232
  heterogeneous 56
  homogeneous 56
  physical separation of 56
mmHg 216
Molar volume 226-227
Molarity 284-287
Molecular models 354
Molecular shapes
  described 193
  predicting 192-196
Molecular solids 244
Molecular weight 64
Mole
  calculation of 66-67
  defined 66
Molecule 50
Molybdenum, as nutrient 211
Monomer 402
Monosaccharides
  alpha and beta forms 576
  definition of 571
  Fischer structures 576
  Haworth structures 575
  hexoses 578-582
  open-chain structure 574
  pentoses 574-578
  trioses 573-574
Morphine 522-523
Morton, William 449
Mucopolysaccharides 594
Mutarotation 578
Mutations 683-686

$NAD^+ - NADH + H^+$
  EM pathway involvement in 761, 763
  in fatty acid cycle 771
  fermentation, involvement in 764
  in formation of acetyl CoA 747
  Krebs cycle, involvement in 750-752
  oxidation in ETS 744
  in redox reactions 743-744

Naphthalene 414
Narcotics 522
Neothyl 450
Net-ionic equation 316
Neutral solution 319
Neutralization 307, 314-317
Neutron 108
Niacin 712-714
Nicotinamide 518, 712
Nicotinic acid 518, 712
Ninhydrin test 644
Nitration (of benzene) 410
Nitric acid 310
Nitrile, reduction of 503
Nitrogen balance 777
Nitroglycerin 437-438
Nobel, Alfred 438
Nonbonding electrons 175
Nonelectrolyte 276
Nonmetal 47
Norepinephrine 730, 732
Nuclear equations
  balancing 136
  writing 136-139
Nuclear radiation 135-138
Nucleases 698
Nucleic acids
  base pairing in 662-663
  composition 649-652, 658
  formation of 657-660
  inhibition of synthesis 690-692
  nitrogen bases in 651-652
  sugars in 650-651
  See also DNA and RNA
Nucleoprotein 649
Nucleosides
  abbreviations of 653
  composition 652
  formation of 653
  naming 652
Nucleotides
  abbreviation of 653, 657
  composition 654
  formation of 655-656
  naming 655
  nucleic acids from 657-660
Nucleus (of atom) 108
Numerator 786
Nylon 514-515

Octet rule
  described 158-159
  exceptions to 191-192
Oil 544
Olefins 390

Oligosaccharide 572
Opium 522
Optical isomerism 595-602
Optimum pH 720
Optimum temperature 719
Orbit, electron 113
Orbital
  defined 127
  p-orbital 127-128
  s-orbital 127-128
Organic chemistry, defined 345
Orlon 404
Ornithine cycle. See urea cycle
Ortho, meta, para 411-412
Osmosis 292
Osmotic membrane 292
Osmotic pressure 293
Oxalic acid 485
Oxidation
  of alcohols 435-437, 466-467
  of aldehydes 467-468
  defined 251, 255, 257
  in ETS 744
  Beta- 769-772
Oxidation number
  changes in reactions 255
  determination of 252-255
  meaning of 251-252
Oxidation-reduction reactions 251-257
Oxidative deamination 778
Oxidative phosphorylation 741-746
Oxidizing agent 251
Oxidoreductase 698
Oxygen debt 763

Pantothenic acid 708-709
Paraffin 361
Paraformaldehyde 472
Paraldehyde 472
Parkinson's Disease 602
Partial pressure 229-232
Parts per billion 284
Parts per million 284
Pauling, Linus 706
Pellagra 712-713
Penicillin 690-691
Pentases 574-578
Pepsin 644
Peptide 622-624
Peptide bond 622
Percent
  milligram 283
  volume/volume 283
  weight/volume 282-283
  weight/weight 284

Percent composition
  compounds 69-70
  solutions 282-284
Period 120
Periodic Law 119
Periodic properties
  atomic size 124-125
  defined 124
  ionization energy 124-125
Periodic table
  derivation 119
  displayed 120, 121, inside front cover
Petroleum 375
pH
  of blood 325, 329
  of common solutions 325
  defined 322
  determination of 323-324
  isoelectric 620
  meter 327
  paper 326
  scale 324-325
Phenacetin 516
Phencyclidine 526
Phenobarbital 523
Phenols
  as germicides 443-444
  naming 442
Phenyl group 499
Phenylalanine 613, 617
Phenylketonuria (PKU) 687
Phosphates, cellular 742-743
Phosphatides 552-554
Phosphatidic acids 551
$3', 5'$-Phosphodiester linkage 657-658
Phospholipids 551-556
Phosphoric acid 310
Phosphorglation
  defined 740-741
  oxidative 741-746
  substrate level 741
Photosynthesis 101, 569-570
Physical change 40
Physical property 39
P_i 738
Picric acid 443-444
Plasmalogens 553-554
Plastics 402-404
Pleated sheet structure 634, 636
pOH 327-328
Polarimeter 598
Polarized light 597-598
Polyamide 515
Polycyclic aromatics 414-415
Polyethylene 403-404
Polymerase 658

Polymerization
  of alkenes 402-403
  of amino acids 622-624
  forming nylon 515
Polymers, table of 404
Polyribosomes 683
Polysaccharides
  cellulose 592-593
  definition of 590
  glycogen 592
  miscellaneous 593-594
  starch 590-592
Polystyrene 404
Polyvinylchloride 403-404
Potassium
  in body fluids 274
  detection of 115
  as nutrient 207-208
Potassium hydroxide 313
Potential energy 90
P$_i$ 740, 756
Pressure
  definition 216
  of gaseous mixtures 229-232
  measurement of 216
  osmotic 292-295
  partial 229
  standard 224
  units of 216-217
Primary
  alcohol 421
  amine 499
Primary structure
  of nucleic acids 657-660
  of proteins 632-634
Problem solving 12-17
Product, reaction 40
Proenzyme 705
Progesterone 560-561
Proline 614
Property
  chemical 39
  definition 39
  physical 39
Prostaglandins 538
Prosthetic group 610, 704
Proteases 698
Proteins
  classification of 610
  complete 617
  composition of 607-609
  description of 606
  denaturization of 643-644
  functions of 609-610
  H-bonding in 634-639
  hydrophobic attractions in 638
  ionic attractions in 637

primary structure of 632-634
  properties of 642-644
  quaternary structure 639
  secondary structure 634-635
  tertiary structure 635
  tests for 644
  turnover rate 776
Protein synthesis
  regulation of 683
  summarized 685
  translation 673-683
Proton 107
Psilocybin 526
P-site 677
Ptyalin 594
Purine 519
Purine bases 651
Pyridine 518
Pyridoxine 708
Pyrimidine 519
Pyrimidine bases 651
Pyrophosphate 658, 740, 756, 769
Pyrrole 517
Pyrrolidine 517

Quaternary
  amines 505
  structure of proteins 639-640
Quantum number 114
Quinine 519
Quinoline 519

R 228
Racemic mixture 598
Rad 147-148
Radiation, nuclear 135-138
Radiation, safety 148-150
Radiation sickness 142
Radiation, units
  curie 146, 148
  dose equivalent-rem 148
  LD$_{50}$-30 day 148
  rad 147-148
  roentgen 146-147
Radioactive half-life 143
Radioactivity
  artificial 139-140
  natural 139-140
Radioisotope 135
Rare gases 121
Rate of reaction 258-264
Reactant, chemical 40
Reaction rates
  in chemical equilibrium 267

defined 258
  factors affecting 261-264
Reactions, chemical
  how they occur 258-260
  rate of 258, 261-264
  types of 249-257
Reactions, reversible 264-266
Recombinant DNA 689
Redox reactions 251-257
Reducing agent 251
Reducing sugar 582
Reduction
  of aldehydes 468
  of amides 502
  defined 165, 251, 255-257
  of imides 503
  of ketones 468
  of nitriles 503
  of nitrobenzene 503
Releasing factors 729
Rem 148
Renal threshold 595
Replication, DNA 666-671
Representative elements 121
Resonance
  in Benzene 408
  defined 190
  predicting 190
Respiratory chain. See Electron Transport System
Reverse reaction 265
Riboflavin 711-712
Ribonucleic acids (RNA)
  base pairing in 671
  composition 652, 671
  formation in cells 673-676
  messenger 671-673
  ribosomal 671-673
  transfer 671-673
Ribose 574-578, 650
Ribosomes 676
RNA, messenger
  codons 676
  description 671-673
  formation in cells 673-676
  role in protein synthesis 677-683
RNA, Ribosomal description of 671-673
RNA, Transfer
  anticodon 673-674
  description of 671-673
  role in protein synthesis 677-683
Roentgen 146-147

Saccharin 588-589

Safety radiation 148-150
Salicylic acid 491
Salol 492
Salts
  amine 504-506
  of carboxylic acids 482
  formation of 317-318
Saponification 313, 490, 547-548
Saran 404
Saturated
  hydrocarbon 361
  solution 277-278
Sanger, Frederick 628
Scientific notation 790-794
Scintillation camera 144-147
Scurvy 705
Secondary
  alcohol 421
  amine 499
Secondary structure
  of DNA 661-664
  of proteins 634-635
Serine 613
Shapes of molecules
  predicting 195-196
  types 193
Sickle-cell anemia 632, 687
Sickle-cell trait 632, 687
Simple protein 610
Single bond 175
Single replacement reactions 250
Site
  A- 677
  allosteric 724
  P- 677
Skeleton structure 187
Soaps
  cleansing action of 549-550
  preparation 547-548
  scum formation 548
Sodium
  in body fluids 274
  detection of 115
  as nutrient 207-208
Sodium hydroxide 313
Solids
  amorphous 244
  covalently bonded 244-245
  described 38
  metallic 245
  molecular 244
Solute 276
Solution
  acidic 320
  aqueous 276
  basic 320

Solution *(cont.)*:
    characteristics of 276-277
    concentrated 281
    dilute 281
    dilution of 288-289
    hypertonic 295
    hypotonic 294-295
    isotonic 294
    neutral 319
    preparing 286
    properties of 289-296
    saturated 277
    types 275
    unsaturated 278
Solution concentration
    molarity 284-287
    normality 336-338
    volume/volume % 283
    weight/volume % 282-283
    weight/weight % 284
Solvation 279
Solvent 276
Sorbitol 589
Space-filling model 354
Specific gravity 32
Specific heat
    definition 92
    values of 93
Specificity, enzyme 718
Spectator ion 316
Spectrum
    electromagnetic 86
    visible emission 112-113
Sphingolipids 556-557
Sphingomyelins 556
Sphingosine 556
Standard pressure 224
Standard temperature 224
Starch 590-592
Stereoisomers 597
Steroid nucleus 558
Steroids 557-562
STP 224-225
Streptomycin 691
Subatomic particles
    definition 107
    electron 107
    neutron 108
    proton 107
Sublimation 244
Substance, pure 41
Substitution reactions
    of alkanes 379-381
    of aromatic hydrocarbons 409-410
    defined 379
Substituent 367-368
Substrate 696, 720-721
Substrate-level phosphorylation 741

Sucrose 587-589
Sulfonation, of benzene 410
Sulfanilamide 691, 723
Sulfuric acid 309-310
Surface tension 240
Surfactant 240
Suspensions 297-298
Sweetness of substances 589
Symbols
    electron-dot 122-123, 161
    of elements 46
    of isotopes 111
    nuclear 136-138
Synthesis reactions 249

Target cells 729
Tay-Sachs Disease 557
Teflon 404
Temperature
    conversions 25, 222
    definition 24
    and Kinetic energy 90, 215-216, 262
    standard 224-225
Temperature scales
    Absolute 24, 222
    Celsius 24
    comparison of 26
    Fahrenheit 24
Tension, gas 231
Terpenes 563
Tertiary
    alcohol 421
    amine 499
Tertiary structure
    of proteins 635-639
Testosterone 560, 733
Tetracycline 691
Tetrahedral bond angle 193
Tetrahedral structure 193, 348
Tetrahydrocannabinol 445
Therapy, radiation use in 150-151
Thermometer 25
Thiamine 707
Thiols 441-442
Thionyl chloride 510
Threonine 613, 617
Thymine 651
Thymol 443
Tincture 440
Titration 338-340
Tollen's test 467
Tooth, structure of 642
Torr 217
Transacylases 700
Transaminases 700
Transamination 777-778
Transcription 673-676
Transferases 700

Transition elements 122
Transition state complex
    with enzymes 717
    described 260-261
Translation 676-683
Transmutation 140
Triglycerides
    chemical properties 546-548
    fats and oils 544
    hydrogenation 541
    hydrolysis 546
    mixed 542
    naming 542-543
    physical properties 545
    saponification of 547-548
    simple 542
Trioses 573-574
Triple bond 177
Tritium 111
Tryptophan 614, 617
Turnover number, enzyme 715
Tylenol® 516
Tyndall effect 296
Tyrosine 614

Ultraviolet light 86-87, 643
Unsaturated
    compounds 390
    soluion 277
Uracil 651
Urea 344, 514
Urea cycle
    diagrammed 781
    step by step description 780-783
Urushiol 444

Valence electrons 118, 130
Valine 612, 617
Van der Waals Forces 238
Vanillin 444
Vaporization, heat of 95-96
Vapor pressure, equilibrium 241-242
Vinegar 311
Vinethene 450
Vinyl group 404
Viral infection 692
Virus
    description of 689-690
    mode of operation 690
Viscosity 239
Vitamer 562
Vitamins
    A 564
    biotin 706
    C 705-706
    cyanocobalamin (B_{12})

    709-711
    D 562-563
    E 564
    as enzyme cofactors 705-714
    fat soluble 562-565
    folic acid 707
    K 565
    Niacin (B_3) 712-714
    pantothenic acid 708-709
    pyridoxine, pyridox-amine, pyridoxal (B_6 family) 708
    riboflavin (B_2) 711-712
    thiamine (B_1) 707
    water soluble 705-714
Volatility 242
Volume
    devices for measuring of gases 218-227
    metric-English conversions 21
    metric units of 22
    molar 226-227
Volume/Volume % 283

Water
    average adult intake 2
    deionized 300
    dipole 278
    distilled 300
    distribution in body 272-273
    extracellular 273-274
    hydrogen bonding in 23
    intercellular 273
    intracellular 273
    ionization of 319-320
    properties of 271
    purification of 299-301
    structure of 278
Watson, James D. 661
Wavelength 85-86
Waxes 540
Weight 8
Weight/volume % 282-283
Weight/weight % 284
Wilson's disease 210
Wohler, Fredrick 344
Wood alcohol 422, 430, 439
Work
    biological 84-85
    definition 83

Xanthoproteic test 644
Xylitol 589
Xylocaine 515

Zinc, as nutrient 211
Zwitterion 619-620
Zymogen 705

P

P
P
F